Bat Biology and Conservation

Bat

Biology and Conservation

Edited by

Thomas H. Kunz

and

Paul A. Racey

SMITHSONIAN INSTITUTION PRESS
Washington and London

Copy editor: Susan A. Kreml
Production editor: Deborah L. Sanders
Designer: Janice Wheeler

Library of Congress Cataloging-in-Publication Data

Bat biology and conservation / edited by Thomas H. Kunz and
Paul A. Racey.
   p. cm.
"This volume derives from four symposia convened at the Tenth
International Bat Research Conference, which was held at Boston
University on August 6–11, 1995."—Pref.
Includes bibliographical references and indexes.
ISBN 1-56098-825-8 (hardcover : alk. paper)
1. Bats—Congresses. 2. Wildlife conservation—Congresses.
I. Kunz, Thomas H. II. Racey, P. A. III. International Bat Re-
search Conference (10th : 1995 : Boston University)
QL737.C5B367 1998
599.4—dc21   98-22944

British Library Cataloguing-in-Publication Data available

Manufactured in the United States of America
05 04 03 02 01 00 99 98 5 4 3 2 1

♾ The paper used in this publication meets the minimum requirements of the American National Standard for Information Sciences—Permanence of Paper for Printed Library Materials ANSI
Z39.48-1984.

Contents

List of Contributors   ix
Preface   xiii

Part One: Phylogeny and Evolution   1

Introduction by Nancy B. Simmons and Suzanne J. Hand

1. **A Reappraisal of Interfamilial Relationships
   of Bats**   3
   Nancy B. Simmons

2. **In the Minotaur's Labyrinth: Phylogeny
   of the Bat Family Hipposideridae**   27
   Wieslaw Bogdanowicz and Robert D. Owen

3. **Phylogeny of Neotropical Short-Tailed Fruit Bats,
   Carollia spp.: Phylogenetic Analysis of Restriction Site
   Variation in mtDNA**   43
   Burton K. Lim and Mark D. Engstrom

4. **Phylogenetic Accuracy, Stability, and Congruence:
   Relationships within and among the New World Bat Genera
   Artibeus, Dermanura, and *Koopmania***   59
   Ronald A. Van Den Bussche, Jeremy L. Hudgeons, and Robert J. Baker

5. **A Southern Origin for the Hipposideridae (Microchiroptera)?
   Evidence from the Australian Fossil Record**   72
   Suzanne J. Hand and John A. W. Kirsch

v

Part Two: Functional Morphology   91

Introduction by Sharon M. Swartz and Ulla M. Norberg

6. Morphological Adaptations for Flight in Bats   93
   Ulla M. Norberg

7. Skin and Bones: Functional, Architectural, and Mechanical
   Differentiation in the Bat Wing   109
   Sharon M. Swartz

8. Chiropteran Muscle Biology: A Perspective from Molecules
   to Function   127
   John W. Hermanson

9. Form, Function, and Evolution in Skulls and Teeth of Bats   140
   Patricia W. Freeman

10. Chiropteran Hindlimb Morphology and the Origin
    of Blood Feeding in Bats   157
    William A. Schutt, Jr.

11. Interspecific and Intraspecific Variation in Echolocation
    Call Frequency and Morphology of Horseshoe Bats,
    Rhinolophus and *Hipposideros*   169
    Charles M. Francis and Jörg Habersetzer

Part Three: Echolocation   181

Introduction by Hans-Ulrich Schnitzler and Cynthia F. Moss

12. How Echolocating Bats Search and Find Food   183
    Hans-Ulrich Schnitzler and Elisabeth K. V. Kalko

13. How Echolocating Bats Approach and Acquire Food   197
    Elisabeth K. V. Kalko and Hans-Ulrich Schnitzler

14. Computational Strategies in the Auditory Cortex
    of the Big Brown Bat, *Eptesicus fuscus*   205
    Steven P. Dear

15. Sensorimotor Integration in Bat Sonar   220
    Doreen E. Valentine and Cynthia F. Moss

16. Adaptation of the Auditory Periphery of Bats
    for Echolocation   231
    Marianne Vater

Part Four: Conservation Biology   247

Introduction by Elizabeth D. Pierson and Paul A. Racey

17. Ecology of European Bats in Relation to Their Conservation   249
    Paul A. Racey

18. Impacts of Ignorance and Human and Elephant Populations
    on the Conservation of Bats in African Woodlands   261
    M. Brock Fenton and I. L. Rautenbach

19. Conservation Biology of Australian Bats: Are Recent Advances
    Solving Our Problems?   271
    Gregory C. Richards and Leslie S. Hall

20. Brazilian Bats and Conservation Biology: A First Survey   282
    Jader Marinho-Filho and Ivan Sazima

21. **The Middle American Bat Fauna: Conservation in the Neotropical–Nearctic Border**   295
Héctor T. Arita and Jorge Ortega

22. **Tall Trees, Deep Holes, and Scarred Landscapes: Conservation Biology of North American Bats**   309
Elizabeth D. Pierson

23. **Conservation of Bats on Remote Indo-Pacific Islands**   326
William E. Rainey

24. **Geographic Patterns, Ecological Gradients, and the Maintenance of Tropical Fruit Bat Diversity: The Philippine Model**   342
Ruth C. B. Utzurrum

Taxonomic Index   355
Subject Index   363

Contributors

Héctor T. Arita is a research professor and head of the Department of Natural Resources at the Instituto de Ecologia, Universidad Nacional Autonoma de Mexico, Mexico City. His research on bats focuses on the local and regional processes that determine the composition and structure of bat communities.

Robert J. Baker is Horn Professor of Biology and Museum Science at Texas Tech University, Lubbock, and curator of mammals and vital tissues and director of the Natural Science Research Lab at the museum. His research interests include systematics of phyllostomid bats, factors that affect chromosomal evolution, and genotoxicology of Chernobyl.

Wieslaw Bogdanowicz is director of research at the Museum and Institute of Zoology, Polish Academy of Sciences, Warsaw. His research on bats, conducted in Europe, North America, and Asia, largely focuses on phylogeny, systematics, evolution, and functional morphology.

Steven P. Dear is an assistant professor in the Department of Neuroscience and Anatomy at Penn State College of Medicine and Department of Acoustics at Penn State University, State College, Pennsylvania. His research on bats focuses on understanding the neural basis of echolocation behavior.

Mark D. Engstrom is curator of mammals in the Centre for Biodiversity and Conservation Biology at the Royal Ontario Museum and associate professor of zoology at the University of Toronto. His research focuses on systematics, speciation, and phylogenetic relationships of New World mammals, primarily rodents and bats.

M. Brock Fenton is professor of biology at York University, in North York, Ontario. His research on bats focuses on behavior and ecology and is based on field studies in North, Central, and South America and in Africa and Australia.

Charles M. Francis is a senior scientist at Bird Studies Canada/Long Point Bird Observatory, and is a research associate with the Wildlife Conservation Society and the Royal Ontario Museum, Ottawa. His research on bats focuses on distribution and biogeography, taxonomy, echolocation behavior, community ecology, and conservation.

Patricia W. Freeman is curator of zoology and associate professor of the University of Nebraska State Museum and School of Biological Sciences at the University of Nebraska, Lincoln. She is especially interested in form and function in a wide range of living mammals, including how the diversity of crania, jaws, and teeth within the order Chiroptera is correlated with dietary habits.

Jörg Habersetzer is a research scientist at the Senckenberg Research Institute in Frankfurt, Germany. His early research and publications largely focused on the echolocation and foraging behavior of horseshoe bats in India. Jörg currently uses microradiography to investigate the morphology of fossil bats from Messel, Germany.

Leslie S. Hall is an associate professor in the Department of Veterinary Pathobiology and a member of the Centre for Conservation Biology at the University of Queensland, Brisbane, Australia. His main interest in bats is in their ecology, morphology, and conservation.

Suzanne J. Hand is a senior research scientist in the School of Biological Science at the University of New South Wales, Sydney. For the past 15 years Suzanne has been investigating the ancient history of Australia's unique bat fauna.

John W. Hermanson is an associate professor of anatomy at the College of Veterinary Medicine and director of the graduate program in zoology at Cornell University, Ithaca, New York. His research focuses on muscle biology, including flight musculature in bats, locomotor muscles in horses, and paleontology of horses and rodents.

Jeremy L. Hudgeons graduated from Texas Tech University, Lubbock, in 1996 with a bachelor's degree in biology. He was a Goldwater Scholar and a Howard Hughes Fellow during his tenure at Texas Tech.

Elisabeth K. V. Kalko is assistant professor in the Department of Animal Physiology at the University of Tubingen in Tubingen, Germany. Her field and laboratory research on bats focuses on sensory ecology, foraging and echolocation behavior, and community ecology. She is a research affiliate of the Smithsonian Tropical Research Institute in Panama and a research associate with the National Museum of Natural History, Washington, D.C.

John A. W. Kirsch is a professor of zoology and director of the University of Wisconsin Zoological Museum, Madison. He is known mostly for his extensive research on marsupials; his interest and enthusiasm for bats was provoked largely by an impatience with marsupials for failing to evolve flight.

Thomas H. Kunz is a professor of biology and director of the Center for Ecology and Conservation Biology at Boston University. His research on bats focuses on reproductive biology, physiological ecology, behavioral ecology, and conservation biology.

Burton K. Lim is assistant curator in the Centre for Biodi-

versity and Conservation Biology at the Royal Ontario Museum in Toronto. His research interests are in the areas of morphological and molecular systematics of bats, incorporating phylogenetic applications and investigating methodological approaches.

Jader Marinho-Filho is a professor in the Departmento de Zoologia at the Universidade de Brasilia, Brazil. His bat research concentrates on frugivory, seed dispersal, biogeography, and conservation biology.

Cynthia F. Moss is an associate professor of psychology and directs the Auditory Neuroethology Laboratory at the University of Maryland, College Park. Echolocating bats serve as model systems for her research, including studies of auditory perception, adaptive motor control, and the neural mechanisms supporting sensorimotor integration.

Ulla M. Norberg is an associate professor of zoology at the University of Gothenburg, Sweden. Her research focuses on the ecological and evolutionary morphology and the biomechanics and energetics of flight in bats and birds.

Jorge Ortega is a doctoral candidate at the Instituto de Ecologia, Universidad Nacional Autonoma de Mexico, Mexico City, where his research focuses on social behavior and mate guarding in polygynous species of fruit-eating bats. His research also addresses questions on biogeography and ecomorphology.

Robert D. Owen is an associate professor of biology at Texas Tech University, Lubbock. His research on bats has included phylogenetic evaluations of the subfamily Stenodermatinae and the families Rhinolophidae and Hipposideridae. His current research focuses on the systematics and biogeography of Neotropical bats.

Elizabeth D. Pierson has conducted research on the ecology and conservation biology of bats for the past 20 years, primarily in the western United States and on oceanic islands of the western Pacific. She has worked with state and federal agencies to incorporate conservation-related bat research into governmental wildlife and habitat management practices. She is based in Berkeley, California.

Paul A. Racey is Regius Professor of Natural History at the University of Aberdeen, Aberdeen, Scotland. His early research on bats was in reproductive biology and endocrinology, but his focus has since switched to mostly ecology and conservation biology.

William E. Rainey is research associate in the Department of Integrative Ecology at the University of California, Berkeley. He has conducted research for the past 30 years on both terrestrial and marine tropical island vertebrates, combining ecological research with consultation on science-based conservation policy and initiatives.

I. L. (Naas) Rautenbach is director of the Transvaal Museum of Natural History in Pretoria, South Africa, and for the past 25 years has served as the curator of mammals. As curator he focused on building the museum collections and conducted research on mammalian diversity.

Gregory C. Richards has conducted research on bats in Australia for nearly three decades, focusing primarily on ecology, conservation biology, and taxonomy. He is currently a fauna consultant primarily assisting the mining industry with bat conservation problems, and is working toward a doctorate in forest bat ecology.

Ivan Sazima is a professor in the Departamento de Zoologia at the Universidade Estadual de Campinas, Brazil. He is one of the leading researchers on vertebrate ecology and behavior in this country. His research and publications focus mainly on life histories and feeding behavior of Brazilian vertebrates.

Hans-Ulrich Schnitzler is a professor of animal physiology at the Zoological Institute of the University of Tubingen in Tubingen, Germany. His research and publications have emphasized the neuronal basis of auditory-initiated behavior in mammals, especially echolocating bats, dolphins, and rats. His field research focuses on comparative studies on sensory ecology of bats. He is a research affiliate of the Smithsonian Tropical Research Institute in Panama.

William A. Schutt, Jr., is an assistant professor of biology at Southampton College, near New York City. He is also a Coleman Research Fellow at the American Museum of Natural History (Department of Mammalogy). Bill's research on bats focuses on functional morphology (especially on the hindlimb) and evolution.

Nancy B. Simmons is an associate curator in the Department of Mammalogy at the American Museum of Natural History, New York, and holds adjunct faculty positions at City University of New York and Columbia University. Her research focuses on the evolutionary biology of bats and the origins of bat diversity.

Sharon M. Swartz is an associate professor of biology and medicine in the Department of Ecology and Evolutionary Biology at Brown University, Providence, Rhode Island. Her research focuses on the evolution of the mammalian musculoskeletal system, particularly the evolution and diversification of bat flight.

Ruth C. B. Utzurrum is an assistant professor and cofounder of the Center for Tropical Conservation Studies at Silliman University in the Philippines. She is a field associate of the Field Museum of Natural History in Chicago and is a member of the IUCN/SSC Chiroptera Specialist Group. Ruth's research and publications have largely focused on the processes underlying community structure, spatiotemporal patterns of abundance and diversity, feeding ecology of fruit bats, and conservation of tropical vertebrates.

Doreen E. Valentine earned a doctorate in neuroscience from Harvard University in 1995. Her work on spatial perception and orientation in bats was conducted in the Bioacoustics Research Laboratory at Harvard.

Ronald A. Van Den Bussche is an assistant professor of zoology at Oklahoma State University, Stillwater. His research on bats largely focuses on evolutionary relationships based on DNA sequence data.

Marianne Vater is professor of zoology at the University of Potsdam, Germany. Her field of research is auditory neurobiology, with special interest in comparative aspects of cochlear and auditory brainstem processing in echolocating bats.

Preface

During the past three decades, there have been enormous advances in the study of bats, manifested by a proliferation of symposia, published research articles, reviews, and monographs on topics including physiology, ecology, social behavior, echolocation, biogeography, conservation, morphology, genetics, paleontology, and evolution. These contributions have added measurably to a literature that sometimes seems overwhelming, even to the most ardent consumers of scientific information. Notwithstanding, knowledge of bats continues to advance at an unprecedented rate, and it is for this reason that this book was conceived. This volume derives from four symposia convened at the Tenth International Bat Research Conference, which was held at Boston University on August 6–11, 1995. The editors invited the symposium organizers on the basis of their expertise and experience in the study of bats; in collaboration with the editors, the organizers then invited each of the contributors. Each author or group of authors was asked to present new findings or to summarize research topics that had not been treated in other recent works.

As is true in most scientific disciplines, observations and experiments for the study of bats have improved with important advances in the design and application of new instrumentation, development of new observational techniques, use of advanced analytical methods, and use of sophisticated experimental protocols. Research on bats that was once based for the most part on dusty museum specimens has advanced to include collections which today are rightly preserved for such uses as histology, postcranial anatomy, chromosomal morphology, and DNA libraries. In turn, these and more traditional characters have been increasingly used to test phylogenetic hypotheses. Morphological studies that were once based solely on traditionally preserved specimens have advanced to include functional analysis of both captive and free-ranging animals—enhanced by sophisticated instrumentation, three-dimensional imaging, and computer simulations. Studies of echolocation, once largely the purview of laboratory biologists, also have been extended to include research on free-ranging animals, facilitated in part by the development and use of portable recording instruments and multiflash photography. Highly sophisticated studies of echolocation that were once considered unthinkable now employ dynamic models and supercomputers to record and simulate the production, detection, and analysis of sounds produced by echolocating bats.

As we approach the twenty-first century, the loss of natural habitats and the associated species continues unabated, largely as a consequence of human population growth and the related effects of deforestation and pollution. These losses highlight the need to protect the earth's fragile ecosystems by focusing new efforts on inventorying and monitoring bat faunas at both regional and global scales. Increased efforts are needed to identify and preserve what remains of critical habitats, not only in tropical regions where the diversity of bats is greatest, on oceanic islands and archipelagoes where endemism is highest, but also in temperate regions where some of the largest maternity and hibernating colonies of bats have been reported. The need for conservation-based research, and for efforts that focus on the relationships between bats and humans, will become increasingly important as the world's human population continues to spiral out of control.

This volume would not have been possible without the enthusiasm and dedication of the organizers and conveners of the four symposia: Nancy Simmons and Sue Hand, "Phylogeny and Evolution" (Part One); Sharon Swartz and Ulla Norberg, "Functional Morphology" (Part Two); Hans-Ulrich Schnitzler and Cindy Moss, "Echolocation in Bats" (Part Three); and Elizabeth Pierson and Paul Racey, "Conservation Biology" (Part Four). These individuals not only were instrumental in organizing the symposia but also as-

sumed important responsibilities in the editorial process and prepared an introduction to the chapters in their part of the volume. Each manuscript was reviewed by at least two anonymous reviewers, and to these individuals we are most grateful. The following individuals reviewed one or more manuscripts: Hans Baagøe, Mark Brigham, Keith Condon, Marian Dagosto, Dave Dalton, Ginny Dalton, Ken Dial, Peggy Eby, Brock Fenton, Tim Flannery, Ted Fleming, Jiri Gaisler, Bill Henson, John Kirsch, Karl Koopman, Tom Kunz, Mitch Masters, Juan Carlos Morales, Cindy Moss, Ulla Norberg, Bruce Patterson, Colin Pennycuick, Elizabeth Pierson, Paul Racey, Bill Rainey, Bernard Sige, Nancy Simmons, Don Thomas, Ron Van Den Bussche, Blaire Van-Valkenburgh, Jeff Wenstrup, John Wible, Gary Wiles, Don Wong, Dave Worthington, and John Zook. Alison Lavender of Boston University assisted us in the final preparation of manuscripts.

We thank Peter Cannell, who has been enthusiastic about this project even from the time it was little more than an idea, for his patience and forbearance. We are grateful to the staff at the Smithsonian Institution Press for their assistance during the editorial and production phases. Finally, we are indebted to our wives, Margaret and Cilla, for their continued support, for accepting our unbridled passion for bats, and for tolerating our physical and sometimes mental absences.

Bat Biology and Conservation

Part One
Phylogeny and Evolution

In recent years it has become clear that phylogenetic hypotheses provide a critical framework for interpreting patterns of evolution in diverse biological systems. As knowledge of bat morphology, genetics, ecology, and behavior grows, so does the need for well-substantiated phylogenetic hypotheses. One effect of increased interest in the evolutionary history of bats has been broad-scale reevaluation of traditional hypotheses of relationship. The monophyly of many higher-level taxa has been questioned recently, including Chiroptera (>900 extant species), Microchiroptera (>700 species), Vespertilionidae (>300 species), Vespertilioninae (>250 species), *Hipposideros* (>60 species), and *Artibeus* (>15 species). As demonstrated in Chapters 1 through 5, such controversies almost always lead to marked improvement in our understanding of relationships and evolutionary patterns, even when subsequent analyses support traditional groupings.

Recent advances in understanding bat relationships and evolution have come about largely through improvements in phylogenetic methods and increased attention to problematic taxa. Discovery of new fossils and development of new data sets have also contributed to rapidly improving resolution of relationships at several taxonomic levels. Each of the chapters in Part One, "Phylogeny and Evolution," addresses a distinct phylogenetic question: interrelationships of subfamilies and families of bats (Simmons), relationships of genera and species of hipposiderid bats (Bogdanowicz and Owen; Hand and Kirsch), relationships within *Artibeus* sensu lato (Van Den Bussche et al.), and relationships of populations and species of *Carollia* (Lim

1

and Engstrom). Although these contributions focus on different systematic problems at different taxonomic levels, several themes emerge.

The 1990s have seen an increase in the types of data available for phylogenetic analysis. Molecular techniques have dramatically improved, and information from both mitochondrial and nuclear genes is frequently used in systematic studies. At the same time there has been a resurgence of interest in morphology as systematists have attempted to reconcile alternative phylogenetic hypotheses. Most workers now agree that there is no single "best" data set for answering phylogenetic questions. Molecular data, although highly informative, are subject to most of the same problems that complicate interpretation of morphological data (e.g., homoplasy), and molecular systematists must also contend with problems concerning alignment (determination of homologous sites for comparison) and a low "signal-to-noise ratio" in some sequence data sets.

Choice of data set always depends on the systematic problem under consideration. As demonstrated in the chapters of Part One, quickly evolving genes (e.g., cytochrome *b*) and anatomical systems (e.g., dental morphology) appear to be well suited for studying relationships among species of bats, while more slowly evolving genes (e.g., 12S rDNA) and morphological systems (e.g., postcranial osteology) may be better suited for addressing questions at higher taxonomic levels. However, evolution has not necessarily occurred at the same rate in all chiropteran lineages, and types of data that are highly informative in some parts of the phylogenetic tree may not be informative in other parts. Many workers now combine several types of data in their analyses to improve both resolution and stability of the resulting phylogenetic hypotheses. Some contributions in this first part of the volume combine different types of molecular data; others include morphological characters from different anatomical systems, and still other studies combine both morphological and molecular data. Maximizing information has become a high priority in phylogenetic studies at all taxonomic levels.

Closely related to data sampling is the problem of taxonomic sampling. Systematists increasingly recognize that dense taxonomic sampling is necessary to avoid analytical biases (e.g., long-branch attraction). Dense sampling is particularly important in studies of higher-level relationships in which the use of exemplar taxa may significantly underestimate the amount of taxonomic polymorphism within diverse groups. Inclusion of fossil taxa in analyses of living forms (and vice versa) provides reciprocal illumination of relationships and may also contribute important information on divergence times and historical biogeography. Modern phylogenetic software packages can accommodate both missing data and cases of taxonomic polymorphism. This analytical flexibility has facilitated studies on the effects of taxonomic sampling (including the importance of fossils), data sampling, outgroup choice, and other factors on the outcome of phylogenetic analyses.

Recent methodological advances have also included development of methods for evaluating the relative amount of support for various groupings in phylogenetic trees. These techniques, which include bootstrap analysis and Bremer support or decay analyses, are being used increasingly in studies of chiropteran phylogeny. By distinguishing relationships that are well supported from those which remain problematic, workers can make informed decisions about changes in classification and identify areas requiring additional study. Subsequent evaluation of character transformations and biogeographic and ecological hypotheses (all of which either benefit from or require a phylogenetic context) can thus take into account both the strengths and weaknesses of the phylogenetic hypotheses.

Although the contributions to this part focus on resolution of long-standing problems in bat systematics, they also help to illustrate the value of phylogenetic hypotheses for understanding diverse biological patterns. Several chapters utilize phylogenies to evaluate biogeographic hypotheses, reaching new and exciting conclusions concerning the time and place of major evolutionary radiations of bats (see Hand and Kirsch; Simmons). Other chapters provide insight into the nature of molecular change associated with speciation events in bats (see Lim and Engstrom; Van Den Bussche et al.), and one discusses the evolution of karyotypes and specific morphological characters in a phylogenetic context (see Bogdanowicz and Owen). Taken together, these contributions help to demonstrate that phylogenies are not just tools for classification but rather hypotheses that provide a critical framework for comparative studies in other fields of biology. This seems particularly appropriate in the context of this volume, which highlights recent advances in diverse areas of chiropteran biology.

NANCY B. SIMMONS AND SUZANNE J. HAND

1
A Reappraisal of Interfamilial Relationships of Bats

NANCY B. SIMMONS

Higher-level relationships of bats remain poorly understood despite recent advances in our understanding of relationships at lower taxonomic levels. Although most bat species can be assigned easily to one of several monophyletic families, there is little agreement concerning relationships among the latter taxa. This is not just a problem of resolution: Previous hypotheses of interfamilial relationships are largely incongruent. Lack of a well-substantiated phylogeny has hampered systematic studies at lower taxonomic levels, hindered attempts to understand patterns of morphological, karyotypic, and molecular change, and complicated interpretation of biogeographic and ecological data. In this context, the purpose of the current study is to provide a working hypothesis of interfamilial relationships of bats based upon a "total-evidence" analysis of available morphological and molecular data.

Background

Most workers now agree that Chiroptera (Megachiroptera + Microchiroptera), Megachiroptera (=Pteropodidae), and Microchiroptera are each monophyletic (for a review, see Simmons 1994). Within Microchiroptera, most families

can be diagnosed by unique apomorphies (Table 1.1) and/ or combinations of plesiomorphic and apomorphic features that strongly suggest monophyly (Miller 1897, 1907; Hill 1974; Van Valen 1979; Koopman 1984, 1994; Griffiths and Smith 1991; Corbet and Hill 1992; Griffiths et al. 1992; Griffiths 1994). The principal exception is Vespertilionidae, a diverse group that comprises between five and eight subfamilies in different classifications: Kerivoulinae, Miniopterinae, Murininae, Tomopeatinae, Vespertilioninae, Nyctophilinae (included within Vespertilioninae by Koopman [1994] and Volleth and Heller [1994]), Antrozoinae (named as a subfamily by Miller [1897] but subsequently included either within Nyctophilinae [e.g., Hill and Smith 1984] or Vespertilioninae [e.g., Koopman 1993, 1994]), and Myotinae (usually included as a tribe within Vespertilioninae [Koopman 1994] but here raised to subfamily rank following the suggestion of Volleth and Heller [1994]; see footnote *a* to Table 1.2).

Systematists have long recognized that Vespertilionidae as so defined is diagnosed by few if any apomorphies, and some data have been cited as evidence that the family is not monophyletic. For example, *Tomopeas ravus* (the only member of the subfamily Tomopeatinae) shares immunological

Table 1.1

Apomorphies Diagnosing Families within Microchiroptera

Emballonuridae
   Postorbital processes unusually long and thin
   M. sternohyoideus attached to posterior larynx by connective tissue
Craseonycteridae
   Nasal branches of premaxilla fused to each other posteriorly, forming rectangular plate that overlies nasals and maxillae
   Uropatagium well-developed but tail and calcar absent
Rhinopomatidae
   Slit-like valvular nostrils that can be closed
   Long tail protrudes from short uropatagium
Nycteridae
   Saucer-shaped depression on rostrum supporting extensive chamber with several partitions; chamber opens dorsally via slit with dermal foliations
   Fibula absent
   T-shaped terminal caudal vertebra
Megadermatidae
   Premaxillae reduced to tiny, cartilaginous rods (representing nasal process) or absent
   Ears very large, fused along inner margins for at least one-third of their length; tragus very large, bifid
Rhinolophidae
(=Rhinolophinae sensu Koopman 1993, 1994)
   Complex noseleaf with horseshoe, sella, connecting process, and triangular or subtriangular lancet
Hipposideridae
(=Hipposiderinae sensu Koopman 1993, 1994)
   Complex noseleaf with horseshoe and posterior foliations; lancet rounded, elliptical, or tridentate in outline; sella apparently absent
   Seventh cervical, first thoracic, and second thoracic vertebrae fused
   Pubic spine fused to anterior ilium to form preacetabular foramen

Noctilionidae
   Muzzle pointed with projecting dermal pad
   Distinct cheek-pouches
   Foot with very long, bony calcar supported by distally expanded, flattened calcaneum
Mormoopidae
   Lower lip with plate-like outgrowths
Phyllostomidae
   Oviductal folds restricted to extramural oviduct
   Friction lock in digits of feet
Mystacinidae
   Claws of toes with basal talon
Molossidae (sensu Koopman 1993, 1994)
   Greater trochanter of femur with hooklike process
   Fringe of "spoon-shaped" bristles present on toes I and V
   Uropatagium freely movable on long tail
Natalidae
   Subcutaneous "natalid" organ on forehead of adult males
Furipteridae
   Thumb reduced, almost entirely enclosed in propatagium
   Thoracic nipples located near xiphoid process of sternum
Thyropteridae
   Large, pedicellate suction discs on base of thumb and ankle
   Soft tissues of digit III and IV of hind foot fused to form single thick digit
Myzopodidae
   Tragus fused to pinna; external auditory meatus partially obstructed by mushroom-shaped process
   Large, non-pedicellate suction pads on thumb and ankle
Vespertilionidae (sensu Koopman 1993, 1994)
   No unique features

Sources: Miller 1907; Hill 1974; Van Valen 1979; Hood and Smith 1982, 1983; Gopalakrishna and Chari 1983; Hill and Smith 1984; Koopman 1984, 1994, personal communication; Griffiths and Smith 1991; Corbet and Hill 1992; Griffiths 1994, unpublished data; Simmons and Quinn 1994; Simmons, personal observation.

Note: Here "apomorphies" refers to characteristics not found in any other microchiropteran taxon.

---

affinities and several derived morphological and molecular traits with molossids (Miller 1907; Davis 1970; Barkley 1984; Pierson 1986; Sudman et al. 1994). Miniopterines, which differ from other vespertilionids in several aspects of fetal development and adult morphology, show immunological affinity with molossids (Mein and Tupinier 1977; Gopalakrishna and Chari 1983; Pierson 1986). Kerivoulines share derived morphological traits with Natalidae, Thyropteridae, Furipteridae, and Myzopodidae (Sigé 1974; Van Valen 1979). These patterns have led various workers to suggest removal of one or more subfamilies from Vespertilionidae (Davis 1970; Sigé 1974; Mein and Tupinier 1977; Van Valen 1979; Gopalakrishna and Chari 1983; Barkley 1984; Sudman et al. 1994). Each of the eight vespertilionid subfamilies just listed appears to be monophyletic, with the exception of Vespertilioninae, which may be paraphyletic. In terms of morphology, Vespertilioninae is "perhaps best characterized by the absence of the special modifications that distinguish the other groups" (Miller 1907, p. 197). Volleth and Heller (1994) described several derived karyotype features that diagnose Vespertilioninae (including Nyctophilinae but excluding Myotinae), but their study did not include any members of Antrozoinae or Lasiurini.

Previous Phylogenetic Hypotheses

Smith (1976, 1980) reviewed the history of chiropteran classification from the seventeenth century through the 1970s. Although many early classification schemes were intended to portray "natural" relationships among taxa, Smith (1976) provided one of the first explicit phylogenetic hypotheses (Figure 1.1, left). Smith's (1976, p. 56) cladogram was intended to represent the then "generally accepted view" of bat phylogeny. This hypothesis was presumably based principally upon consideration of skin and skeletal characters as discussed by Miller (1907). Smith's (1976) tree indicated monophyly of several higher-level taxa including four extant superfamilies: Emballonuroidea (Emballonuridae + Rhinopomatidae + Craseonycteridae), Rhinolophoidea (Nycteridae + Megadermatidae + Rhinolophidae + Hipposideridae), Noctilionoidea (=Phyllostomoidea; Noctilionidae + Mormoopidae + Phyllostomidae), and Vespertilionoidea (Mystacinidae + Molossidae + Myzopodidae + Thyropteridae + Furipteridae + Natalidae + Vespertilionidae). Two clades subsequently named as infraorders by Koopman (1985) also appeared in Smith's tree: Yinochiroptera (Emballonuroidea + Rhinolophoidea) and Yangochiroptera (Noctilionoidea + Vespertilionoidea). Monophyly of Vespertilionidae was apparently assumed.

Van Valen (1979) challenged Smith's (1976) phylogeny with a new hypothesis based upon a cladistic analysis of diverse morphological traits (Figure 1.1, right). Although details of the data set and analysis were never published, Van Valen (1979) apparently considered myology and traits of fetal membranes in addition to osteological features. The resulting phylogeny (see Figure 1.1) supported monophyly of three of the four superfamilies (Rhinolophoidea, Noctilionoidea, and Vespertilionoidea). Emballonuroidea, Yinochiroptera, and Yangochiroptera did not appear as monophyletic groups in the context of this hypothesis. Kerivoulinae was removed from Vespertilionidae and placed in a separate clade together with Myzopodidae, Thyropteridae, Furipteridae, and Natalidae.

Novacek (1980a) and Luckett (1980a) published cladograms of bat relationships based on analyses of single organ systems (auditory region morphology and fetal membranes, respectively). In part because the tree derived from auditory characters differed significantly from all previous hypotheses of higher-level relationships, Novacek (1980a) warned against using that cladogram as a basis for a new phylogenetic reconstruction or classification. Luckett (1980a) was unable to resolve many relationships but provided limited support for monophyly of Noctilionoidea. A close relationship between Megadermatidae and a clade containing Vespertilionidae and Thyropteridae was also suggested (Luckett 1980a). Monophyly of Vespertilionidae was apparently assumed by both Novacek (1980a) and Luckett (1980a).

Pierson (1986) proposed a phylogeny of bats based on an analysis of transferrin immunological distance data (Figure 1.2, left). This tree differed significantly from previous hypotheses in grouping Tomopeatinae and Miniopterinae with Molossidae rather than Vespertilioninae, placing Rhi-

Table 1.2

Apomorphies Diagnosing Subfamilies of Vespertilionidae

Antrozoinae
  Muzzle surmounted by low, horseshoe-shaped dermal ridge
Kerivoulinae
  Ears funnel-shaped
  Tendon locking mechanism on digits of feet with plicae but no tubercles
  Ceratohyal and epihyal both absent, replaced by fibrous strand connecting stylohyal to basihyal; m. ceratohyoideus inserts on stylohyal
Miniopterinae
  Second phalanx of digit III of wing greatly elongated
  Tendon locking mechanism absent from digits of feet
  Unique pattern of development of chorioallantoic placenta
Myotinae, new rank[a]
  State I of chromosome 12 (faint GTG-positive band present near telomere)
Murininae
  Nostrils elongated to form tubes
  Anterior upper premolar very large, roughly equivalent to canine in cross section
  State II of chromosome 13 (minute GTG-negative arm present)
Tomopeatinae
  Seventh cervical vertebra fused with first thoracic vertebra
Vespertilioninae (including Nyctophilinae)
  State II of chromosome 1/2 (GTG-negative and RBG-positive region close to the centromere in arm 1)
  State II of chromosome 7 (minute, GTG-negative, early replicating short arm absent)

Sources: Gopalakrishna and Chari 1983; Koopman 1984, 1994, personal communication; Corbet and Hill 1992; Simmons and Quinn 1994; Volleth and Heller 1994; T. Griffiths, unpublished data; W. Schutt, unpublished data.

Note: Here "apomorphies" refers to characteristics not found in any other vespertilionid taxon.

[a]Myotinae as defined here comprises *Myotis* + *Lasionycteris*. This taxon was named by Tate (1942) as a tribe within Vespertilioninae. Myotini originally comprised *Myotis* + *Lasionycteris* + *Plecotus* + *Corynorhinus* + *Idonycteris* + *Euderma*, although Tate (1942) noted that the latter four genera might alternatively be referred to their own tribe, Plecotini. This suggestion was followed by subsequent authors, and Myotini has been restricted to *Myotis* + *Lasionycteris* in recent classifications (e.g., Hill and Harrison 1987; Koopman 1994). Volleth and Heller (1994) described karyological data which indicate that Myotini represents a lineage distinct from Vespertilioninae, and therefore they suggested that Myotini be raised to the rank of subfamily. Despite their suggestion, Volleth and Heller (1994) did not use the name Myotinae in their publication, so formal usage of the name at the subfamily rank dates from the current study.

nopomatidae within Rhinolophoidea, and associating Mystacinidae with Noctilionoidea. In the context of Pierson's (1986) phylogeny, Yinochiroptera is monophyletic but Yangochiroptera and the four superfamilies are not.

Baker et al. (1991) analyzed variation in rDNA restriction sites in a study designed to test bat monophyly. Although their data set could not resolve the interrelationships of

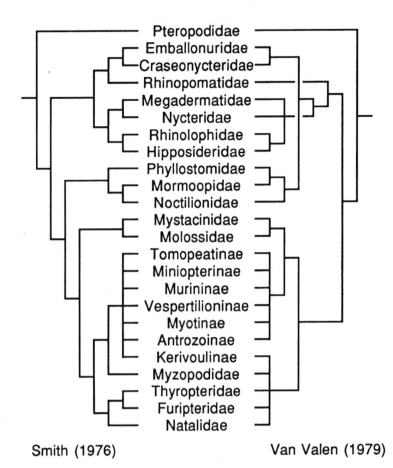

Smith (1976) Van Valen (1979)

Figure 1.1. Phylogenetic hypotheses presented by Smith (1976) and by Van Valen (1979). Smith's tree represents the then "generally accepted view" of bat relationships (Smith 1976: 56); Van Valen's tree is based on a cladistic analysis of diverse morphological data.

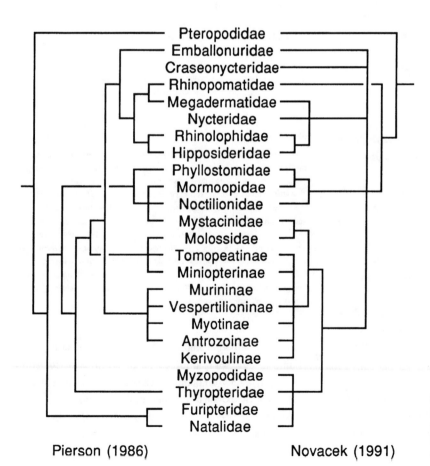

Pierson (1986) Novacek (1991)

Figure 1.2. Phylogenetic hypotheses presented by Pierson (1986) and by Novacek (1991). Pierson's tree is based on immunological distance data; Novacek's tree is based on diverse morphological data.

bats, support was found for several groups including Yinochiroptera, Noctilionoidea, and a clade containing Mormoopidae and Noctilionidae (Baker et al. 1991).

Novacek (1991) used a phylogeny of bats (Figure 1.2, right) as a framework for discussion of evolution of cochlear features. This tree was based on consideration of the morphological characters discussed in Koopman (1984), but no explicit character analysis was presented. Novacek's (1991) tree indicated paraphyly of Yinochiroptera and Emballonuroidea, left unresolved the monophyly of Yangochiroptera, and suggested monophyly of the remaining three superfamilies. Vespertilionid monophyly was assumed.

Griffiths and colleagues analyzed features of the hyoid region and presented two alternative phylogenies for yinochiropteran bats (Griffiths and Smith 1991; Griffiths et al. 1992). Neither tree supported monophyly of Emballonuroidea or Rhinolophoidea; the only clade appearing in both trees was Rhinolophidae + Hipposideridae (Griffiths et al. 1992).

In summary, higher-level phylogenies of bats proposed in the last two decades are largely incongruent. Most workers agree that Rhinolophidae and Hipposideridae are sister-taxa and that Phyllostomidae, Mormoopidae, and Noctilionidae are closely related, but there is little agreement beyond these points.

Methods

Taxonomic Sampling and Data

Based on Tables 1.1 and 1.2, and the considerations just discussed, 24 chiropteran taxa were chosen for phylogenetic analysis (see Appendix 1.2). Monophyly of Vespertilioninae (including Nyctophilinae) was assumed for practical reasons; the implications of this assumption are discussed later.

Only one outgroup is necessary to root a phylogenetic tree (Nixon and Carpenter 1993), but at least two outgroups are usually included in cladistic analyses to establish character polarities and permit testing of ingroup monophyly (Maddison et al. 1984). Megachiroptera (=Pteropodidae) is the appropriate proximal outgroup for addressing interfamilial relationships among microchiropterans if one accepts the monophyly of Chiroptera, Megachiroptera, and Microchiroptera. Testing microchiropteran monophyly, however, requires the inclusion of additional outgroups. The identity of the sister-group of bats is still the subject of considerable debate, but a close relationship between Chiroptera and Dermoptera (colugos) is strongly supported by morphological data (Simmons 1993a, 1994, 1995; Simmons and Quinn 1994), 12S rDNA sequences (sampled from more

than 150 species representing 20 orders; Vrana 1994), and combined morphological data and cytochrome oxidase II gene sequences (Novacek 1994). Accordingly, Dermoptera was included as an outgroup in the current study. Scandentia (tree shrews) was included as an additional outgroup to facilitate testing of the effect of outgroup choice on the outcome of the analysis.

Discrete character data available for addressing higher-level relationships of bats consist principally of morphological features and information on rDNA restriction site variation. Appendix 1.1 lists the characters employed in the current study and provides references for each character; the data matrix is given in Appendix 1.2. Although some amino acid and nucleotide sequence data exist for bats, sampling has been restricted to fewer than half the extant families and no published data set includes representatives of more than 7 of the 24 taxa included in the current study (e.g., Pettigrew et al. 1989; Adkins and Honeycutt 1993; Stanhope et al. 1993). While it is clear that sequence data hold potential for resolving higher-level relationships of bats, the data available were judged too incomplete to warrant inclusion in this study.

Phylogenetic Methods

In recent years there has been much discussion of the relative merits of "total-evidence" (character congruence) versus taxonomic congruence methods for resolving systematic problems (Kluge 1989; Barrett et al. 1991, 1993; Bull et al. 1993; de Queiroz 1993; Eernisse and Kluge 1993; Kluge and Woolf 1993; Nelson 1993; Chippindale and Wiens 1994; Hulsenbeck et al. 1994; de Queiroz et al. 1995). A total-evidence approach was adopted in this study because most of the relevant data consist of discrete morphological characters that can be included in a single analysis (see discussion in Simmons 1993a). With few exceptions, previous phylogenetic hypotheses were based upon overlapping data sets, limiting the value of taxonomic congruence methods.

Available restriction site data cannot resolve interfamilial relationships in the absence of other data (Baker et al. 1991). The restriction site data can be easily included in a total-evidence analysis together with discrete morphological characters, however, which is the approach taken here. The only significant data that cannot be included in such a total-evidence analysis are Pierson's (1986) immunological distances. While some workers might argue that discrete character data (such as morphological and restriction site data) can be transformed into distances and combined with immunological or DNA–DNA hybridization data, any benefits gained from such an approach are greatly outweighed by the loss of homology information, amplification of problems associated with missing data, and the probability that such a data set would not satisfy the underlying assumptions of the statistical models used in distance analyses (Swofford and Olsen 1990; Cracraft and Helm-Bychowski 1991; Miyamoto and Cracraft 1991).

In this study, 192 discrete characters were scored for phylogenetic analysis and the resulting data matrix was analyzed using PAUP (Phylogeny Analysis Using Parsimony) version 3.1.1 (Swofford 1993). All transformations were unordered. A heuristic search with a random-addition sequence and 1,000 repetitions was used to find most-parsimonious trees. Near-most-parsimonious trees (1–8 steps longer) were identified in subsequent heuristic searches using the same parameters, and a decay analysis was performed following the methods of Bremer (1988). Decay values for strongly supported clades were obtained using constrained heuristic analyses to identify the shortest trees that did not include a particular clade. A bootstrap analysis using heuristic methods (random-addition sequence, 10 repetitions for each of 1,000 bootstrap replicates) was also used to evaluate the relative support for various groupings.

MacClade version 3.0 (Maddison and Maddison 1992) was used for data entry, comparison of alternative topologies, and some graphics.

Results and Discussion

Phylogenetic Relationships

Heuristic analysis of the complete data set (Appendix 1.2) resulted in discovery of two equally most-parsimonious trees (607 steps each; consistency index, CI = 0.410; retention index, RI = 0.593). A strict consensus of these trees is shown in Figure 1.3 together with results of the decay and bootstrap analyses.

Monophyly of many (but not all) higher-level taxa was supported. Microchiropteran monophyly was confirmed, with this clade appearing in the strict-consensus tree and 100% of the bootstrap replicates. A minimum of 13 additional steps was required to collapse this clade (i.e., make it nonmonophyletic) in the decay analysis.

Yinochiroptera appeared paraphyletic in the strict-consensus tree, with Emballonuridae placed outside a clade

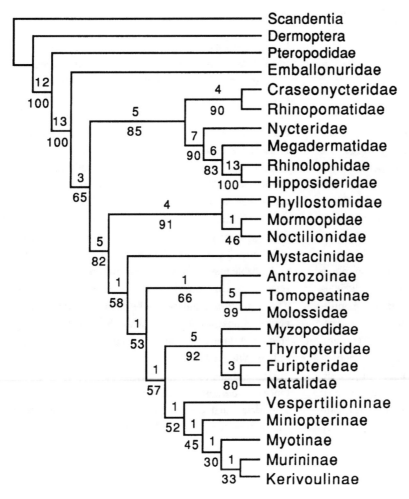

Figure 1.3. Strict consensus of two equally most-parsimonious trees (607 steps each; consistency index = 0.410; retention index = 0.593) found in a heuristic analysis of the data set given in Appendix 1.2. Numbers below internal branches represent the percentage of bootstrap replicates in which the clade appeared; numbers above the branches are decay values (the minimum number of additional steps required to collapse the clade).

containing all other microchiropteran taxa. A minimum of three additional steps was required to collapse the latter clade in the decay analysis (bootstrap value, BV = 65%). Placement of Emballonuridae within a monophyletic Yinochiroptera was supported in only 15% of the bootstrap replicates. Emballonuroidea appeared paraphyletic in the strict-consensus tree, and only 4% of the bootstrap replicates supported monophyly of this group. In contrast, monophyly of a clade composed of Rhinopomatidae + Craseonycteridae was very strongly supported. This clade appeared in the strict-consensus tree and 90% of the bootstrap replicates, and a minimum of four additional steps was required to collapse the clade in the decay analysis. Rhinopomatidae + Craseonycteridae grouped with Rhinolophoidea (rather than with Emballonuridae) in the strict-consensus tree and 85% of the bootstrap replicates, and a minimum of five additional steps was required to collapse this clade.

Rhinolophoid monophyly was also strongly supported in the current study. This grouping appeared in the strict-consensus tree and 90% of the bootstrap replicates, and a minimum of 7 additional steps was required to collapse the clade in the decay analysis. Relationships within Rhinolophoidea were well resolved with high bootstrap and decay values. A clade containing Rhinolophidae + Hipposideridae was very strongly supported (bootstrap value, BV = 100%; 13 additional steps to collapse the clade) with Megadermatidae as its sister-group (BV = 83%; minimum 6 additional steps to collapse the clade). Little support was found for a Nycteridae + Rhinolophidae + Hipposideridae clade (BV = 9%) or a Nycteridae + Megadermatidae clade (BV = 5%).

Yangochiropteran monophyly was strongly supported in the strict-consensus tree and the bootstrap and decay analyses (bootstrap value, BV = 82%; minimum five additional steps to collapse the clade). Within this group, noctilionoid monophyly was strongly supported (BV = 91%; minimum four additional steps to collapse the clade), and a sister-group relationship between Noctilionidae and Mormoopidae received weak support (BV = 46%; minimum one additional step to collapse the clade). In contrast, little support was found in the bootstrap analysis for an alternative hypothesis grouping Mormoopidae + Phyllostomidae (BV = 35%) or Phyllostomidae + Noctilionidae (BV = 13%).

Monophyly of Vespertilionoidea sensu Koopman (1994) was indicated in the strict-consensus tree, but this grouping received only weak or moderate support in bootstrap and decay analyses (minimum one additional step to collapse the clade; bootstrap value, BV = 58%). Mystacinidae appeared as the basal branch within the vespertilionoid clade in the strict-consensus tree. Alternative hypotheses that received limited support in the bootstrap analysis included placement

of Mystacinidae as the sister-group of Noctilionoidea (BV = 35%) and placement of Mystacinidae as the sister-group of Molossidae + Tomopeatinae + Antrozoinae (BV = 24%).

Very strong support was found for a clade containing Myzopodidae + Thyropteridae + Furipteridae + Natalidae (minimum five steps to collapse the clade; bootstrap value, BV = 92%). Within this group, the Neotropical taxa (Thyropteridae + Furipteridae + Natalidae) formed a clade in one of the two most-parsimonious trees and 48% of the bootstrap replicates. Alternatively, Thyropteridae and Myzopodidae grouped together in the other most-parsimonious trees and 47% of the bootstrap replicates. A close relationship between Furipteridae and Natalidae was indicated in both the strict-consensus tree and the bootstrap and decay analyses (BV = 80%; minimum three additional steps to collapse the clade).

Very strong support was found for a sister-group relationship between Molossidae and the vespertilionid subfamily Tomopeatinae. These taxa grouped together in the strict-consensus tree and 99% of the bootstrap replicates; a minimum of five additional steps were required to collapse the clade in the decay analysis. This level of support was among the highest found for any clade in the current study. Support was also found for a close relationship between this clade (Tomopeatinae + Molossidae) and another of the vespertilionid subfamilies, Antrozoinae. This grouping appeared in both the strict-consensus and bootstrap trees (bootstrap value, BV = 66%; minimum one additional step to collapse the clade). No support was found for monophyly of Vespertilionidae as traditionally defined (i.e., including all seven subfamilies; BV < 1%), and little support was found for a clade comprising the non-tomopeatine vespertilionid subfamilies (BV = 27%). The results clearly indicated that at least one subfamily (Tomopeatinae) and probably a second (Antrozoinae) have closer affinities with molossids than they do with other "vespertilionid" taxa.

The remaining subfamilies (Vespertilioninae + Miniopterinae + Myotinae + Murininae + Kerivoulinae) formed a clade in the strict-consensus tree, but support for this grouping was relatively weak (bootstrap value, BV = 52%; minimum one additional step to collapse the clade). Within this group, Murininae + Kerivoulinae formed a clade in the strict-consensus tree and received limited support in the bootstrap and decay analyses (BV = 33%; minimum one additional step to collapse the clade). Myotinae appeared as the sister-group of the latter clade (BV = 30%; minimum one additional step to collapse the clade). Miniopterinae was placed as the sister-group of the Myotinae + Murininae + Kerivoulinae clade (BV = 45%; minimum one additional step to collapse the clade).

A variety of relationships not portrayed in the strict-consensus tree also require discussion. Other than those previously mentioned, the only groupings that appeared in 10% or more of the bootstrap replicates were a clade containing Miniopterinae + Kerivoulinae (bootstrap value, BV = 16%) and a larger clade comprising Miniopterinae + Myotinae + Kerivoulinae (BV = 14%). Very little support was found for a Myotinae + Vespertilioninae clade (BV = 6%), despite the fact that these taxa are generally considered to be closely related. Similarly, there was essentially no support for an Antrozoinae + Vespertilioninae clade (BV = 1%), an Antrozoinae + Vespertilioninae + Myotinae clade (BV < 1%), or an Antrozoinae + Myotinae clade (BV < 1%). These findings indicate that "Vespertilioninae" sensu Koopman (i.e., including myotines and antrozoines) is not a monophyletic group.

This observation aside, Vespertilioninae is still a problematic taxon even when restricted (as in this study) to Pipistrellini + Eptesicini + "Nycticeiini" + Plecotini + Lasiurini + Vespertilionini (including Nyctophilini). Vespertilioninae as thus defined was assumed to be monophyletic at the outset of the current study despite the lack of conclusive evidence supporting this hypothesis. Vespertilioninae was polymorphic for almost 10% of the characters included in the data set and may eventually prove to be polymorphic for many more when gaps in the data set are filled. This relatively high degree of taxonomic polymorphism may be partially responsible for the low bootstrap and decay values associated with the vespertilionid section of the tree. If Vespertilioninae is not monophyletic, various members of the subfamily may prove to be more closely related to other taxa (e.g., molossids or other vespertilionid subfamilies) than to each other. Resolution of these issues requires detailed taxonomic and character sampling beyond the scope of this study.

Effect of Outgroup Choice

As described in *Methods,* Dermoptera and Scandentia were used as consecutive outgroups to root trees and polarize characters. Although chiropteran monophyly is now well established, many workers disagree concerning the relationships of dermopterans and bats. To test the effect of inclusion of Dermoptera in the current study, an heuristic search (random-addition sequence, 100 repetitions) was conducted with Dermoptera excluded and Scandentia used to root the tree. This analysis resulted in two equally parsimonious trees (588 steps; CI = 0.421, RI = 0.581) that proved to be identical in topology to those found in the earlier analysis of the complete data set. A similar result was obtained when Scandentia was excluded and Dermoptera was used to root the tree (two trees, 582 steps each; CI = 0.413, RI = 0.580), and when both Scandentia and Der-

moptera were removed and Pteropodidae was used to root the tree (two trees, 543 steps each; CI = 0.426, RI = 0.588). These results suggest that outgroup selection is not likely to be a significant source of bias in this study.

Is Microchiroptera Monophyletic?

Although most workers agree that Microchiroptera is monophyletic, the current study was designed to provide a test of this hypothesis. Recent DNA–DNA hybridization studies by Kirsh and his colleagues (Hutcheon and Kirsch 1995; Kirsch 1995; Pettigrew and Kirsch 1995) have suggested that Rhinolophoidea and Pteropodidae may be sister-taxa, implying microchiropteran paraphyly. As described, this hypothesis was refuted in this study. Rhinolophoidea did not group with Pteropodidae in any of the 1,000 bootstrap replicates. A constrained heuristic analysis (random-addition sequence, 1,000 repetitions) indicated that the shortest trees containing a Rhinolophoidea + Pteropodidae clade are 23 steps longer than the most-parsimonious trees. These results suggest that Kirsh (1995) is correct in his hypothesis that this is a case of adenine–thymine (AT) bias in which "molecular data [are] positively misleading."

Comparisons with Previous Hypotheses of Interfamilial Relationships

The current data set provides an opportunity to compare previous hypotheses of higher-level relationships in terms of the number of additional steps required by each topology. Trees consistent with Pierson's (1986) phylogeny are a minimum of 62 steps longer than the most-parsimonious trees discussed previously. Van Valen's (1979) phylogeny requires a minimum of 56 additional steps and Smith's (1976) a minimum of 45 steps; Novacek's (1991) phylogeny requires 34 additional steps. All previous phylogenetic hypotheses are thus substantially longer (6%–10%) than the most-parsimonious trees found here.

The phylogenies of Smith (1976), Van Valen (1979), and Novacek (1991) were based on subsets of the morphological data considered in the current study. In contrast, Pierson's (1986) tree represents an independent data set based entirely on immunological distance data. Comparisons of Pierson's tree (Figure 1.2) with the strict-consensus tree generated in the current study (Figure 1.3) reveal several interesting points of congruence. Both studies found support for a close relationship between Rhinopomatidae and Rhinolophoidea, a sister-group relationship between Rhinolophidae and Hipposideridae, a link between *Tomopeas* and molossids, and a sister group relationship between Furipteridae and Natalidae. Points of incongruence between the two studies may stem from the inability of distance methods to distinguish between shared, derived similarity (synapomorphy) and primitive similarity (symplesiomorphy). For example, the immunological similarities between Mystacinidae and noctilionoids perceived by Pierson (1986) may reflect retention of primitive yangochiropteran immunological traits rather than shared, derived features that indicate a sister-group relationship.

Conclusions

Higher-Level Classification of Bats

The results of this study provide support for many higher-level groupings proposed by previous workers. Two of the four superfamilies appear to be monophyletic (Rhinolophoidea, Noctilionoidea), as does one of the two infraorders (Yangochiroptera). Nonmonophyly of one superfamily (Emballonuroidea) and infraorder (Yinochiroptera) is a result of perceived relationships of one family, Emballonuridae. Rather than grouping with Craseonycteridae and Rhinopomatidae, Emballonuridae appears to represent the most basal branch within Microchiroptera. Strong support was found for a clade containing Rhinopomatidae + Craseonycteridae, and this group appears to be the sister-taxon of Rhinolophoidea. In this context, I recommend that Emballonuroidea be restricted to Emballonuridae, and that the name Rhinopomatoidea (= Rhinopomina Bonaparte, 1838) be employed for the clade comprising Rhinopomatidae + Craseonycteridae (Table 1.3). Yinochiroptera should be restricted to Rhinopomatoidea + Rhinolophoidea, leaving Emballonuroidea as infraorder incertae sedis for the time being. Reexamination of a number of Eocene bats often considered to represent basal microchiropterans (e.g., *Palaeochiropteryx, Archaeonycteris, Icaronycteris, Hassianycteris*) should be completed before proposing any additional infraordinal groupings; this work will be presented elsewhere (Simmons and Geisler 1998).

Within Rhinolophoidea, monophyly of a clade comprising Rhinolophidae + Hipposideridae was very strongly supported. Although many researchers continue to recognize these as separate families (Corbet and Hill 1992; Bogdanowicz and Owen, Chapter 2, and Hand and Kirsch, Chapter 5, this volume), I prefer to follow Koopman (1984, 1993, 1994) and recognize the close relationship between these taxa by placing them as subfamilies (Rhinolophinae and Hipposiderinae) within a single family (Rhinolophidae). This classification facilitates reference to both groups as well as to the larger clade that they comprise.

Relationships within Yangochiroptera were well resolved in the current study, but many clades received only limited support in the bootstrap and decay analyses. Strong support was found for monophyly of only three groups: (1) Noctilionoidea; (2) a clade containing Myzopodidae + Thyropteridae + Furipteridae + Natalidae; and (3) a clade containing Molossinae + Tomopeatinae. Monophyly of Vespertilionoidea sensu Koopman (1994) was not strongly supported because of uncertainty about the relationships of Mystacinidae. Given this situation, I recommend that Mystacinidae be referred to Yangochiroptera incertae sedis pending further study.

Monophyly of a clade containing Myzopodidae + Thyropteridae + Furipteridae + Natalidae was strongly supported in the current study. A possible close relationship among these taxa has been recognized by many workers (Thomas 1904; Miller 1907; Van Valen 1979; Koopman 1984) but has not been reflected in most recent classifications (Hill and Smith 1984; Koopman 1984, 1993, 1994). To remedy this situation and provide a name for this distinctive group, I recommend that the name Nataloidea (= Natalinia Gray, 1866) be applied to the clade comprising Myzopodidae + Thyropteridae + Furipteridae + Natalidae.

The current study strongly supported monophyly of a clade containing Tompeatinae + Molossidae. This result confirms the findings of Barkley (1984) and Sudman et al. (1994) and supports their decision to move Tomopeatinae from Vespertilionidae to Molossidae. This change compels recognition of two subfamilies in Molossidae: Molossinae (= Molossidae of Figure 1.1 and of Koopman 1984, 1993, 1994) and Tomopeatinae. Further subdivisions should be recognized at the tribal level (e.g., Tadarini, Molossini, Cheiromelini).

Another putative vespertilionid subfamily, Antrozoinae, also appears to be more closely related to molossids than to other vespertilionids. To recognize this relationship while at the same time avoiding "overlumping" taxa, I propose that Antrozoinae be raised to family rank and that the name Molossoidea be applied to the clade comprising Antrozoidae + Molossidae. It is possible that Mystacinidae may eventually be shown to be more closely related to this clade than to other lineages; if so, it could be transferred to Molossoidea to reflect that relationship.

These changes effectively restrict Vespertilionidae (and Vespertilionoidea) to Vespertilioninae + Minopterinae + Myotinae + Murininae + Kerivoulinae. Reconsideration of the rank status of the subfamilial groups within Vespertilionidae must await future resolution of relationships among these groups. Of primary importance is the issue of monophyly of Vespertilioninae, which has yet to be adequately demonstrated. It is anticipated that future studies will recognize more than one family within Vespertilionoidea as defined earlier, which remains a large and com-

Table 1.3

Higher-Level Classification of Recent Bats

Koopman (1994)	This study
ORDER CHIROPTERA	ORDER CHIROPTERA
Suborder Megachiroptera	**Suborder Megachiroptera**
Family Pteropodidae	Family Pteropodidae
Suborder Microchiroptera	**Suborder Microchiroptera**
Infraorder Yinochiroptera	Infraorder incertae sedis
SUPERFAMILY EMBALLONUROIDEA	SUPERFAMILY EMBALLONUROIDEA
Family Emballonuridae	Family Emballonuridae
Family Craseonycteridae	Infraorder Yinochiroptera
Family Rhinopomatidae	SUPERFAMILY RHINOPOMATOIDEA
SUPERFAMILY RHINOLOPHOIDEA	Family Craseonycteridae
Family Nycteridae	Family Rhinopomatidae
Family Megadermatidae	SUPERFAMILY RHINOLOPHOIDEA
Family Rhinolophidae	Family Nycteridae
Subfamily Rhinolophinae	Family Megadermatidae
Subfamily Hipposiderinae	Family Rhinolophidae
Infraorder Yangochiroptera	Subfamily Rhinolophinae
SUPERFAMILY NOCTILIONOIDEA	Subfamily Hipposiderinae
Family Mormoopidae	Infraorder Yangochiroptera
Family Noctilionidae	SUPERFAMILY INCERTAE SEDIS
Family Phyllostomidae	Family Mystacinidae
SUPERFAMILY VESPERTILIONOIDEA	SUPERFAMILY NOCTILIONOIDEA
Family Mystacinidae	Family Noctilionidae
Family Molossidae	Family Mormoopidae
Family Vespertilionidae	Family Phyllostomidae
Subfamily Tomopeatinae	SUPERFAMILY MOLOSSOIDEA, NEW RANK
Subfamily Kerivoulinae	Family Antrozoidae, new rank
Subfamily Vespertilioninae	Family Molossidae
Subfamily Miniopterinae	Subfamily Tomopeatinae
Subfamily Murininae	Subfamily Molossinae
Family Myzopodidae	SUPERFAMILY VESPERTILIONOIDEA
Family Thyropteridae	Family Vespertilionidae
Family Furipteridae	Subfamily Vespertilioninae
Family Natalidae	Subfamily Miniopterinae
	Subfamily Myotinae, new rank
	Subfamily Murininae
	Subfamily Kerivoulinae
	SUPERFAMILY NATALOIDEA, NEW RANK
	Family Myzopodidae
	Family Thyropteridae
	Family Furipteridae
	Family Natalidae

plex taxon (≥225 species) despite the taxonomic changes proposed here.

Biogeographic Implications

The phylogenetic hypothesis presented in Figure 1.3 has several biogeographic implications. The origin of Chiroptera and the two suborders seems to be rooted in the Old World, as evidenced by the exclusively Old World distributions of Pteropodidae, Yinochiroptera, and both outgroups. Emballonuridae has a pantropical distribution, but morphological, allozyme, and immunological data indicate that the New World taxa form a clade derived from within an Old World radiation (Robbins and Sarich 1988; Griffiths and Smith 1991).

The fossil record is largely consistent with an Old World origin of microchiropteran bats, with many of the oldest and most primitive forms occurring in the Old World (e.g., *Palaeochiropteryx, Archaeonycteris,* and *Hassianycteris* from the Early Eocene of Europe; Habersetzer and Storch 1987; Novacek 1987; Simmons and Geisler 1998). One apparent exception is *Icaronycteris index* from the ?Late Paleocene/Early Eocene of North America. However, this form seems to be closely related to a lineage of European Eocene bats (i.e., Archaeonycteridae; Habersetzer and Storch 1987), suggesting that it may represent an early Tertiary derivative of an Old World clade.

Yangochiroptera has both Old World and New World members. Noctilionoidea is exclusively Neotropical in distribution, as are three of the four families that constitute Nataloidea (Thyropteridae, Furipteridae, and Natalidae). The oldest putative nataloid *(Hornovits tsuwape)* is from the Early Eocene of North America (Beard et al. 1992), which is consistent with a New World origin for this clade. In contrast, Myzopodidae has a fossil representative in east Africa but is today endemic to Madagascar (Hill and Smith 1984). This distribution raises interesting questions concerning how and when the myzopodid lineage might have dispersed to Africa from the New World.

The origin of molossids, which have a pantropical/subtropical distribution, has long been the subject of debate. The earliest molossid fossils are from the Late Eocene of North America, with slightly younger taxa appearing in the Late Eocene of Europe (Hand 1990). Three different areas of origin have been proposed for molossids: the Neotropics, the Ethiopian region, and the Indo-Australian region (Freeman 1981). Most recent workers have favored an Old World origin (e.g., Legendre 1984; Hand 1990), but the topology of the tree generated in the current study suggests an alternative hypothesis. Both Antrozoidae and Tomopeatinae are endemic to the New World, suggesting that molossids may have originated in the Neotropics rather than Africa or Asia.

The origins and patterns of dispersal of various lineages of Vespertilionidae are too complex to address in the context of this study. Each subfamily has representatives on several continents, and a much finer grained phylogeny (e.g., at the genus or species level) is needed before much can be said about the biogeographic history of these taxa. Consideration of fossils is also important because some groups have fossil members in regions where they are no longer found.

One interesting possibility raised by the current study is that the basal radiation of Yangochiroptera occurred in the New World. This hypothesis is supported by optimization of geographic distribution on the cladogram shown in Figure 1.3. If the yangochiropteran lineage originated in the Neotropics, then several invasions of Old World areas by New World taxa are implied: (1) dispersal of the ancestral lineage of Mystacinidae from South America to New Zealand (a hypothesis previously suggested by Pierson et al. 1986); (2) dispersal of the ancestral lineage of Myzopodidae to Africa; (3) at least one dispersal of molossids to the Old World; and (4) one and perhaps several dispersal events involving different vespertilionid lineages. Numerous "reverse" dispersal events (i.e., from the Old World back to the New World) are also likely. These and other biogeographic hypotheses require more extensive and detailed testing in future studies.

Directions for Future Research

Future research on higher-level relationships of bats should include several components. Many of the characters used in the current study have not been adequately sampled in all relevant taxa (see Appendix 1.2). In particular, more work is needed on myology, wrist and ankle morphology, anatomy of viscera and the urogenital tract, morphology and development of fetal membranes, and variation in rDNA restriction sites. Many new characters will undoubtedly be discovered as work in these areas progresses. Nucleotide sequence data also hold great promise, and future "total-evidence" analyses should include sequence data as well as the character set employed in this study.

Taxonomic sampling in future studies should be designed to address problematic issues such as the relationships of Mystacinidae, monophyly of Vespertilioninae, and interrelationships of vespertilionid subfamilies and tribes. Inclusion of fossil taxa (e.g., *Icaronycteris, Palaeochiropteryx*) in future studies may be revealing because these forms preserve combinations of character states not seen in living taxa (Novacek 1987; Simmons and Geisler 1998). The distributions of fossil bat taxa may also contribute significant information for analyses of the biogeographic history of various lineages.

In summary, this study does not represent the final word on higher-level relationships of bats. Instead, the phylogenetic and biogeographic hypotheses presented here should be interpreted as working hypotheses that should be tested with additional data in future studies.

Appendix 1.1.
Descriptions of the 192 Characters

1. Ear pinnae: (0) not funnel-shaped; (1) more or less funnel-shaped. (Corbet and Hill 1992; Koopman 1994; Simmons, personal observation)

2. Tragus: (0) absent; (1) present. (Corbet and Hill 1992; Koopman 1994; Simmons, personal observation)

3. Narial structures: (0) none; (1) dermal ridge dorsal to nostrils; (2) noseleaf; (3) dermal foliations with central slit. (Hill 1974; Corbet and Hill 1992; Simmons, personal observation)

4. M. occipitofrontalis: (0) inserts into connective tissue and skin over nasal region; (1) inserts onto nasal cartilage via common tendon with contralateral muscle. (Winge 1941; H. Whidden, unpublished data)

5. Nasopalatine duct: (0) present; (1) absent. (Loo and Kanagasuntheram 1972; Bhatnagar and Wible 1994; Wible and Bhatnagar 1997)

6. Paraseptal cartilage (=vomeronasal cartilage): (0) C-, J-, or U-shaped; (1) bar-shaped; (2) absent. (Loo and Kanagasuntheram 1972; Bhatnagar and Wible 1994; Wible and Bhatnagar 1997)

7. Vomeronasal epithelial tube: (0) well-developed, with neuroepithelium; (1) rudimentary, without neuroepithelium; (2) absent. (Loo and Kanagasuntheram 1972; Bhatnagar and Wible 1994; Wible and Bhatnagar 1997)

8. Accessory olfactory bulb: (0) present; (1) absent. (Loo and Kanagasuntheram 1972; Bhatnagar and Wible 1994; Wible and Bhatnagar 1997)

9. Articulation of premaxilla with maxilla: (0) via sutures; (1) by fusion; (2) via ligaments—premaxilla freely movable. (Hill 1974; Koopman 1994; Simmons, personal observation)

10. Nasal branches of premaxillae: (0) well-developed; (1) reduced or absent. (Hill 1974; Koopman 1994; Simmons, personal observation)

11. Development of palatal branches of premaxillae: (0) well-developed; (1) reduced or absent. (Hill 1974; Koopman 1994; Simmons, personal observation)

12. Palatal branches of premaxillae: (0) not fused with one another across midline; (1) fused at midline. This character cannot be scored in taxa that lack palatal branches. (Hill 1974; Simmons, personal observation)

13. Emargination in anterior palate, between medial incisors: (0) absent (i.e., medial incisors directly adjacent to one another); (1) present. (Koopman 1994; Simmons, personal observation)

14. Extent of anterior palatal emargination: (0) shallow, extends posteriorly no farther than anterior edge of canines; (1) deep, extends posteriorly at least as far as posterior edge of canines. This character cannot be scored in taxa that lack an anterior palatal emargination. (Koopman 1994; Simmons, personal observation)

15. Hard palate: (0) extends posteriorly into interorbital region; (1) terminates either at or anterior to level of zygomatic roots. (Simmons, personal observation)

16. Number of upper incisors in each side of jaw: (0) two; (1) one; (2) none. (Koopman 1994; Simmons, personal observation)

17. Number of lower incisors in each side of jaw: (0) three; (1) two; (2) one; (3) none. (Koopman 1994; Simmons, personal observation)

18. Number of upper premolars in each side of jaw: (0) three; (1) two; (2) one. (Koopman 1994; Simmons, personal observation)

19. Number of lower premolars in each side of jaw: (0) three; (1) two. (Koopman 1994; Simmons, personal observation)

20. Structure of lower first and second molars: (0) postcristid connects hypoconid with hypoconulid (nyctalodonty); (1) postcristid bypasses hypoconulid to connect with entoconid (myotodonty); (2) cusps and cristae not distinguishable (teeth modified for fruit- and/or nectar-feeding). (Menu and Sigé 1971; Novacek 1987; Simmons, personal observation)

21. Elongate angular process on lower jaw: (0) present; (1) absent. (Novacek 1987; Simmons, personal observation)

22. Projection of angular process: (0) at or below level of occlusal plane of toothrow, well below coronoid process; (1) above level of occlusal plane of toothrow, at same level as the coronoid process. This character cannot be scored in taxa that lack an angular process. (Simmons, personal observation)

23. Postorbital process: (0) present; (1) absent. (Koopman 1994; Simmons, personal observation)

24. Attachment of periotic to basisphenoid: (0) sutured; (1) loosely attached via ligaments. (Novacek 1991; Simmons, personal observation)

25. Basal turn of cochlea: (0) not enlarged; (1) greatly enlarged. (Novacek 1980a, 1980b, 1987, 1991; Simmons, personal observation)

26. Type of cochlea: (0) cryptocochlear; (1) phanerocochlear. (Novacek 1980b, 1991; Simmons, personal observation)

27. Lateral process of tympanic bone: (0) absent or weak; (1) well-developed—forms external auditory meatus. (Novacek 1980a, 1980b; Simmons, personal observation)

28. Tympanic annulus: (0) inclined; (1) semivertical in orientation. (Novacek 1980a, 1980b; Simmons, personal observation)

29. Epitympanic recess: (0) shallow and broad; (1) deep, often constricted in area. (Novacek 1980a, 1980b; Simmons, personal observation)

30. Fossa for m. stapedius: (0) indistinct; (1) shallow and broad; (2) deep, constricted in area, often a crescent-shaped fissure. (Novacek 1980a, 1980b; Simmons, personal observation)

31. Fenestra rotundum: (0) small or of moderate size; (1) enlarged. (Novacek 1980a; Simmons, personal observation)

32. Aquaeductus cochleae: (0) large; (1) small or absent. (Novacek 1980a; Simmons, personal observation)

33. M. tensor tympani muscle: (0) spindle-shaped, inserts via single tendon onto tubercular processus muscularis of malleus; (1) two-headed, inserts via two tendons onto processus muscularis and accessory process; (2) broad sheet of fibers, inserts into crestlike processus muscularis; (3) absent. (Doran 1878; Wassif 1950; Wassif and Madkour 1963; Novacek 1980a; Zeller 1986)

34. Orbicular apophysis: (0) absent or small; (1) large. (Doran 1878; Novacek 1987; Simmons, personal observation)

35. Laryngeal echolocation: (0) absent; (1) present. (Novick 1962; Simmons 1980; Fenton and Bell 1981; Weid and Helversen 1987; Surlykke et al. 1993; Göpfert and Wasserthal 1995; E. Kalko, unpublished data)

36. Number of submaxillary glands: (0) one pair; (1) two pairs. (Robin 1881; Le Gros Clark 1926)

37. Division of right lung: (0) into four lobes; (1) into three lobes; (2) into two lobes; (3) undivided. (Robin 1881)

38. Division of left lung: (0) into two lobes; (1) undivided. (Robin 1881)

39. Tracheal rings: (0) approximately equal in diameter throughout length of trachea; (1) one ring enlarged to form tracheal expansion just posterior to larynx; (2) two to eight rings enlarged to form tracheal expansion just posterior to larynx; (3) nine or more rings enlarged to form a tracheal expansion that is separated from larynx by four or five rings of normal diameter. (Griffiths 1982, 1983, 1994, and unpublished data; Griffiths and Smith 1991; Griffiths et al. 1992)

40. Midline hyoid strap musculature (m. sternohyoideus, m. geniohyoideus, and m. hyoglossus): (0) all strap muscles directly attached to basihyal via fleshy fibers; (1) all indirectly attached to basihyal via basihyal tendon—a "free-floating" strap muscle condition; (2) no basihyal tendon and no connection between strap muscles and basihyal; (3) m. geniohyoideus and m. hyoglossus attached to basihyal via basihyal tendon, but m. sternohyoideus not connected to basihyal. (Griffiths 1982, 1983, 1994, and unpublished data; Griffiths and Smith 1991; Griffiths et al. 1992)

41. Deep division of m. mylohyoideus: (0) absent; (1) present—runs dorsal to midline strap musculature and inserts on basihyal. (Griffiths 1982, 1983, 1994, and unpublished data; Griffiths and Smith 1991; Griffiths et al. 1992)

42. Insertion of m. mylohyoideus: (0) inserts on basihyal and basihyal raphe (runs ventral to midline strap musculature); (1) inserts on basihyal, basihyal raphe, and thyrohyal. (Griffiths 1982, 1983, 1994, and unpublished data; Griffiths and Smith 1991; Griffiths et al. 1992)

43. Form of m. mylohyoideus: (0) aponeurotic anteriorly; (1) fleshy for entire width from mandibular symphysis to at least basihyal region. (Griffiths 1982, 1983, 1994, and unpublished data; Griffiths and Smith 1991; Griffiths et al. 1992)

44. M. mandibulohyoideus (=medial part of anterior digastric): (0) absent; (1) present. (Griffiths 1982, 1983, 1994, and unpublished data; Griffiths and Smith 1991; Griffiths et al. 1992)

45. Development of m. mandibulohyoideus: (0) well-developed; (1) reduced to small muscle with tendon; (2) reduced to tendon only. This character cannot be scored in taxa that lack m. mandibulohyoideus. (Griffiths 1982, 1983, 1994, and unpublished data; Griffiths and Smith 1991; Griffiths et al. 1992)

46. Superficial slip of m. stylohyoideus: (0) present—passes superficial to digastric muscles; (1) absent. (Griffiths 1982, 1983, 1994, and unpublished data; Griffiths and Smith 1991; Griffiths et al. 1992)

47. Deep slip of m. stylohyoideus: (0) present—passes deep to digastic muscles; (1) absent. (Griffiths 1982, 1983, 1994, and unpublished data; Griffiths and Smith 1991; Griffiths et al. 1992)

48. Origin of m. geniohyoideus: (0) from flat posterior surface of mandible lateral to symphysis; (1) from pronglike process that extends posteroventrally from symphysis region. (Griffiths 1982, 1983, 1994, and unpublished data; Griffiths and Smith 1991; Griffiths et al. 1992)

49. Attachment of m. geniohyoideus at origin (mandible): (0) by long tendon; (1) by very short tendon; (2) partly (medial muscle fibers only) by tendon, and partly (lateral fibers only) directly from the bone of the mandible; (3) entirely by fleshy fibers from the bone. (Griffiths 1982, 1983, 1994, and unpublished data; Griffiths and Smith 1991; Griffiths et al. 1992)

50. Origin of m. genioglossus: (0) immediately lateral to mandibular symphysis; (1) extended laterally onto medial surface of mandible, occupying anterior one-quarter to one-third of medial mandibular surface. (Griffiths 1982, 1983, 1994, and unpublished data; Griffiths and Smith 1991; Griffiths et al. 1992)

51. M. genioglossus and m. geniohyoideus: (0) not fused; (1) ventralmost fibers of m. genioglossus fused to fibers from caudal portion of m. geniohyoideus. (Griffiths 1982, 1983, 1994, and unpublished data; Griffiths and Smith 1991; Griffiths et al. 1992)

52. Insertion of m. genioglossus: (0) all fibers insert into posterior tongue; (1) ventralmost fibers insert onto basihyal, the other fibers into the posterior tongue. (Griffiths 1982, 1983, 1994, and unpublished data; Griffiths and Smith 1991; Griffiths et al. 1992)

53. Origin of m. hyoglossus: (0) from entire lateral basihyal and thyrohyal in broad, unbroken sheet; (1) from lateral basihyal and lateral thyrohyal in two sheets separated by a space; (2) from antimere in part, and from lateral basihyal and thyrohyal; (3) from lateral basihyal (no thyrohyal origin). (Griffiths 1982, 1983, 1994, and unpublished data; Griffiths and Smith 1991; Griffiths et al. 1992)

54. Form of m. styloglossus: (0) one belly; (1) two bellies separated by lateral part of m. hyoglossus. (Griffiths 1982, 1983, 1994, and unpublished data; Griffiths and Smith 1991; Griffiths et al. 1992)

55. Origin of m. styloglossus: (0) from expanded tip of stylohyal and/or adjacent surface of skull; (1) from ventral surface of midpoint of stylohyal. (Griffiths 1982, 1983, 1994, and unpublished data; Griffiths and Smith 1991; Griffiths et al. 1992)

56. Insertion of m. ceratohyoideus onto ceratohyal: (0) present; (1) absent. (Griffiths 1982, 1983, 1994, and unpublished data; Griffiths and Smith 1991; Griffiths et al. 1992)

57. Insertion of m. ceratohyoideus onto epihyal: (0) present;

(1) absent. (Griffiths 1982, 1983, 1994, and unpublished data; Griffiths and Smith 1991; Griffiths et al. 1992)

58. Insertion of m. ceratohyoideus onto stylohyal: (0) present; (1) absent. (Griffiths 1982, 1983, 1994, and unpublished data; Griffiths and Smith 1991; Griffiths et al. 1992)

59. Insertion of m. thyrohyoideus: (0) onto thyrohyal; (1) onto thyrohyal and basihyal (muscle enlarged). (Griffiths 1982, 1983, 1994, and unpublished data; Griffiths and Smith 1991; Griffiths et al. 1992)

60. Origin of m. sternohyoideus: (0) from entire anterodorsal surface of manubrium and dorsal surface of first costal cartilage; (1) from entire anterodorsal surface of manubrium and medial tip of clavicle; (2) from entire anterodorsal surface of manubrium—no costal or clavicular origin; (3) restricted to medialmost surface of manubrium in vicinity of keel. (Griffiths 1982, 1983, 1994, and unpublished data; Griffiths and Smith 1991; Griffiths et al. 1992)

61. Form of m. sternohyoideus: (0) relatively broad; (1) reduced to a narrow strip of muscle. (Griffiths 1982, 1983, 1994, and unpublished data; Griffiths and Smith 1991; Griffiths et al. 1992)

62. Origin of m. sternothyroideus: (0) entirely from lateral manubrium; (1) entirely from the medial tip of clavicle; (2) from both the medial tip of clavicle and the lateral manubrium. (Griffiths 1982, 1983, 1994, and unpublished data; Griffiths and Smith 1991; Griffiths et al. 1992)

63. M. omohyoideus: (0) originates from scapula; (1) originates from clavicle; (2) absent. (Griffiths 1982, 1983, 1994, and unpublished data; Griffiths and Smith 1991; Griffiths et al. 1992)

64. Form of basihyal: (0) transverse, unadorned bar or plate; (1) curved bar with apex directed anteriorly. (Griffiths 1982, 1983, 1994, and unpublished data; Griffiths and Smith 1991; Griffiths et al. 1992)

65. Shape of curved basihyal: (0) V-shaped; (1) U-shaped. This character cannot be scored in taxa that lack a curved basihyal. (Griffiths 1982, 1983, 1994, and unpublished data; Griffiths and Smith 1991; Griffiths et al. 1992)

66. Entoglossal process of basihyal: (0) absent; (1) present. (Griffiths 1982, 1983, 1994, and unpublished data; Griffiths and Smith 1991; Griffiths et al. 1992)

67. Size of entoglossal process of basihyal: (0) small; (1) very large, resulting in T-shaped basihyal. This character cannot be scored in taxa that lack an entoglossal process. (Griffiths 1982, 1983, 1994, and unpublished data; Griffiths and Smith 1991; Griffiths et al. 1992)

68. Ceratohyal: (0) approximately equal in length to epihyal; (1) reduced to half the length of epihyal, (2) reduced to tiny element or completely absent. (Griffiths 1982, 1983, 1994, and unpublished data; Griffiths and Smith 1991; Griffiths et al. 1992)

69. Epihyal: (0) approximately equal in length to ceratohyal; (1) reduced to half the length of ceratohyal; (2) reduced to very tiny element or completely absent. (Griffiths 1982,

1983, 1994, and unpublished data; Griffiths and Smith 1991; Griffiths et al. 1992)

70. Form of stylohyal: (0) gently curved bar with no enlargement or other modification to the lateral edge or cranial tip; (1) gently curved bar with cranial tip slightly expanded; (2) gently curved bar with bifurcated tip; (3) gently curved bar with large lateral expansion or "foot"; (4) gently curved bar with very large axe-shaped enlargement at cranial tip; (5) entire lateral half swollen. (Griffiths 1982, 1983, 1994, and unpublished data; Griffiths and Smith 1991; Griffiths et al. 1992; Simmons, personal observation)

71. Posteriorly directed, ventral accessory processes on centra of cervical vertebrae: (0) none present; (1) present on second and third cervical vertebrae; (2) present on second through fifth or through sixth cervical vertebrae. (Novacek 1987; Simmons, personal observation)

72. Seventh cervical vertebra and first thoracic vertebra: (0) not fused; (1) fused. (A. Peffley, in manuscript; Simmons, personal observation)

73. First and second thoracic vertebrae: (0) not fused; (1) fused. (A. Peffley, in manuscript; Simmons, personal observation)

74. Fusion of anterior ribs to vertebrae: (0) no ribs fused to vertebrae; (1) first rib fused to vertebrae; (2) at least first five ribs fused to vertebrae. (A. Peffley, in manuscript; Simmons, personal observation)

75. Width of first rib: (0) similar to width of other ribs; (1) first rib at least twice the width of other ribs. (J. Geisler, unpublished data; Simmons, personal observation)

76. Articulation of first rib: (0) rib articulates with first costal cartilage, rib not fused to manubrium; (1) costal cartilage absent or ossified, rib fused to manubrium. (J. Geisler, unpublished data; A. Peffley, in manuscript; Simmons, personal observation)

77. Point at which second rib articulates with sternum: (0) at manubrium–mesosternum joint; (1) at manubrium—no contact between rib (or costal cartilage) and mesosternum. (J. Geisler, unpublished data; Simmons, personal observation)

78. Means by which second rib articulates with sternum: (0) via costal cartilage; (1) by fusion (costal cartilage absent or ossified). (J. Geisler, unpublished data; Simmons, personal observation)

79. Number of costal cartilages posterior to second rib that articulate with mesosternum: (0) five or more; (1) four; (2) three. (J. Geisler, unpublished data; Simmons, personal observation)

80. Anterior laminae on ribs: (0) absent; (1) present. (J.Geisler, unpublished data; Simmons, personal observation)

81. Width of anterior laminae on ribs: (0) less than width of main body of rib; (1) equal to or wider than main body of rib. This character cannot be scored for taxa that lack anterior lamellae on the ribs. (J. Geisler, unpublished data; Simmons, personal observation)

82. Posterior laminae on ribs: (0) absent; (1) present. (J. Geisler, unpublished data; Simmons, personal observation)

83. Width of posterior laminae on ribs: (0) less than width of main body of rib; (1) equal to or wider than main body of rib. This character cannot be scored for taxa that lack posterior lamellae on the ribs. (J. Geisler, unpublished data; Simmons, personal observation)

84. Lateral processes of manubrium: (0) extend laterally only to clavicular joint; (1) extend laterally well beyond clavicular joint. (J. Geisler, unpublished data; Simmons, personal observation)

85. Anterior face of manubrium: (0) small; (1) broad, triangular, and defined by elevated ridges. (J. Geisler, unpublished data; Simmons, personal observation)

86. Ventral process of manubrium: (0) absent; (1) present, and distal tip blunt or rounded; (2) present, and distal tip laterally compressed. (J. Geisler, unpublished data; Simmons, personal observation)

87. Angle between axis of ventral process and body of manubrium: (0) acute; (1) approximately 90°; (2) obtuse; (3) ventral process bilobed with one acute and one obtuse process. This character cannot be scored in taxa that lack a ventral process on the manubrium. (J. Geisler, unpublished data; Simmons, personal observation)

88. Length of manubrium posterior to lateral processes: (0) greater than 2.5 times transverse width; (1) less than 2 times transverse width. (J. Geisler, unpublished data; Simmons, personal observation)

89. Width of mesosternum: (0) average width less than half the distance between clavicles at claviculosternal joint; (1) average width greater than three-quarters the distance between clavicles. (J. Geisler, unpublished data; Simmons, personal observation)

90. Xiphisternum: (0) without keel; (1) with prominent longitudinal keel. (J. Geisler, unpublished data; Simmons, personal observation)

91. Posterior xiphisternum: (0) with wide lateral flare; (1) not laterally flared. (J. Geisler, unpublished data; Simmons, personal observation)

92. Acromion process: (0) without medial shelf; (1) with shelf that projects medially over supraspinous fossa. (J. Geisler, unpublished data; Simmons, personal observation)

93. Triangular anteromedial projection on tip of acromion process: (0) absent; (1) present. (J. Geisler, unpublished data; Simmons, personal observation)

94. Triangular posterolateral projection on tip of acromion process: (0) absent; (1) present. (J. Geisler, unpublished data; Simmons, personal observation)

95. Dorsal articular facet of scapula (for trochiter of humerus): (0) absent; (1) present. (J. Geisler, unpublished data; Simmons, personal observation)

96. Form of dorsal articular facet: (0) small groove on anteromedial rim of glenoid fossa; (1) oblique and oval, on anteromedial rim of glenoid fossa; (2) large and flat, clearly separated from glenoid fossa. This character cannot be scored in taxa that lack a dorsal articular facet. (J. Geisler, unpublished data; Simmons, personal observation)

97. Width of infraspinous fossa of scapula: (0) narrow—length equal to or greater than 2 times width; (1) wide—length less than or equal to 1.5 times width. (J. Geisler, unpublished data; Simmons, personal observation)

98. Number of facets on infraspinous fossa: (0) one; (1) two; (2) three. (J. Geisler, unpublished data; Simmons, personal observation)

99. Relative widths of infraspinous facets: (0) intermediate infraspinous facet narrower than posterolateral facet; (1) facets approximately equal in width; (2) intermediate facet wider than posterolateral facet. This character cannot be scored in taxa that have only one or two infraspinous facets. (J. Geisler, unpublished data; Simmons, personal observation)

100. Lateral or posterolateral facet of infraspinous fossa: (0) restricted—does not extend into infraglenoid region anteriorly or wrap around intermediate facet at posterior (caudal) angle of scapula; (1) extends into infraglenoid region and wraps around caudal end of intermediate facet. This character cannot be scored in taxa that have only one infraspinous facet. (J. Geisler, unpublished data; Simmons, personal observation)

101. Axillary border of scapula: (0) with thick lip; (1) with thick lip that has blade-like lateral edge; (2) with no thick lip—axillary border flat or slightly upturned. (J. Geisler, unpublished data; Simmons, personal observation)

102. Pit on scapula for attachment of clavicular ligament: (0) absent; (1) present anterior and medial to glenoid fossa. (J. Geisler, unpublished data; Simmons, personal observation)

103. Medial superior border of scapula: (0) without ventral projection; (1) with triangular ventral projection. (J. Geisler, unpublished data; Simmons, personal observation)

104. Shape of coracoid process: (0) relatively short and stout; (1) long and thin. (J. Geisler, unpublished data; Simmons, personal observation)

105. Direction of curve of coracoid process: (0) ventrolateral curve; (1) ventral curve; (2) ventromedial curve. (J. Geisler, unpublished data; Simmons, personal observation)

106. Tip of coracoid process: (0) not flared, approximately same width as shaft; (1) distinctly flared; (2) bifurcated. (J. Geisler, unpublished data; Simmons, personal observation)

107. Suprascapular process: (0) present; (1) absent. (J. Geisler, unpublished data; Simmons, personal observation)

108. Articulation of clavicle: (0) clavicle articulates with or lies in contact with acromion process; (1) clavicle is suspended by ligaments between acromion and coracoid processes; (2) clavicle articulates with or lies in contact with coracoid process. (Strickler 1978; J. Geisler, unpublished data; Simmons, personal observation)

109. Origin of m. subclavius: (0) from first costal cartilage only; (1) from first costal cartilage and lateral process of manubrium. (Leche 1886; Strickler 1978)

110. Origin of anterior division of m. pectoralis profundus: (0) from first costal cartilage only; (1) from first costal carti-

lage and clavicle; (2) from clavicle only. (Leche 1886; Strickler 1978)

111. M. dorsi patagialis: (0) absent; (1) present. (Leche 1886; Le Gros Clark 1924, 1926; Strickler 1978)

112. M. humeropatagialis: (0) absent; (1) present. (Leche 1886; Vaughan 1959, 1970; Strickler 1978)

113. M. occipitopollicalis: (0) absent; (1) present. (Leche 1886; Le Gros Clark 1924, 1926; Vaughan 1970)

114. Tendinous attachments of m. occipitopollicalis to the anterior division of m. pectoralis profundus: (0) absent; (1) present. This character cannot be scored in taxa that lack m. occipitopollicalis. (Strickler 1978)

115. Tendinous attachments of m. occipitopollicalis to the posterior division of m. pectoralis profundus: (0) absent; (1) present. This character cannot be scored in taxa that lack m. occipitopollicalis. (Strickler 1978)

116. Form of m. occipitopollicalis: (0) without muscle belly between cranial muscle belly and band of elastic tissue; (1) with small intermediate muscle belly. This character cannot be scored in taxa that lack m. occipitopollicalis. (Strickler 1978)

117. M. occipitopollicalis insertional complex: (0) includes muscle fibers distal to band of elastic tissue; (1) entirely tendinous distal to band of elastic tissue. This character cannot be scored in taxa that lack m. occipitopollicalis. (Strickler 1978)

118. Insertion of m. occipitopollicalis: (0) into pollex and metacarpal of digit II; (1) into pollex only. This character cannot be scored in taxa that lack m. occipitopollicalis. (Strickler 1978)

119. Differentiation of m. spinotrapezius from trapezius complex: (0) not differentiated; (1) clearly differentiated. (Leche 1886; Le Gros Clark 1924, 1926; Strickler 1978)

120. Differentiation of m. clavotrapezius from trapezius complex: (0) not differentiated; (1) clearly differentiated. (Leche 1886; Le Gros Clark 1924, 1926; Strickler 1978)

121. Origin of m. levator scapulae: (0) from the atlas; (1) from three to five of any of the second through the sixth cervical vertebrae; (2) from the fourth and fifth cervical vertebrae only. (Leche 1886; Le Gros Clark 1924, 1926; Strickler 1978)

122. M. omocervicalis: (0) absent; (1) present. (Leche 1886; Strickler 1978)

123. Origin of m. omocervicalis: (0) from ventral arch of the second cervical vertebra; (1) from transverse processes of the second cervical vertebra; (2) from transverse processes of the third and fourth cervical vertebra. This character cannot be scored in taxa that lack m. omocervicalis. (Leche 1886; Strickler 1978)

124. Insertion of m. omocervicalis: (0) into acromion process of scapula; (1) into clavicle. This character cannot be scored in taxa that lack m. omocervicalis. (Leche 1886; Strickler 1978)

125. Origin of m. serratus anterior: (0) from six or more ribs; (1) from four or five ribs; (2) from two ribs. (Davis 1938; Le Gros Clark 1924, 1926; Strickler 1978)

126. Insertion of m. latissimus dorsi: (0) into ventral ridge of humerus; (1) into ventral ridge and distal pectoral crest (muscle divided). (Leche 1886; Le Gros Clark 1924, 1926; Strickler 1978; Pierson 1986)

127. Insertion of m. teres major: (0) into ventral ridge; (1) into pectoral crest. (Leche 1886; Le Gros Clark 1924, 1926; Strickler 1978)

128. Origins of m. acromiodeltoideus: (0) from acromion process plus a quarter or less of the transverse scapular ligament; (1) from acromion plus more than half of transverse scapular ligament. (Leche 1886; Le Gros Clark 1924, 1926; Davis 1938; Strickler 1978)

129. Origin of m. spinodeltoideus: (0) from vertebral border of scapula plus transverse scapular ligament; (1) from vertebral border only; (2) muscle absent. (Leche 1886; Le Gros Clark 1924, 1926; Davis 1938; Strickler 1978)

130. Extent of trochiter: (0) does not extend to level of head of humerus; (1) extends just to level of head of humerus; (2) extends well beyond level of head of humerus. (Walton and Walton 1970; Hill 1974; Simmons, personal observation)

131. Shape of head of humerus: (0) round in outline; (1) oval or elliptical. (Hill 1974; Simmons, personal observation)

132. Displacement of distal articular surfaces from line of humerus shaft: (0) not displaced; (1) displaced. (Hill 1974; Simmons, personal observation)

133. Epitrochlea: (0) broad; (1) relatively narrow. (Hill 1974; Simmons, personal observation)

134. Sesamoid element dorsal to magnum–trapezium articulation: (0) absent; (1) present. (Cypher 1996)

135. Sesamoid element ventral to unciform–magnum articulation: (0) absent; (1) present. (Cypher 1996)

136. Sesamoid element dorsal to lunar–radius articulation: (0) absent; (1) present. (Cypher 1996)

137. Sesamoid element dorsal to unciform–magnum articulation: (0) absent; (1) present. (Cypher 1996)

138. Sesamoid element dorsal to trapezium–metacarpal I articulation: (0) absent; (1) present. (Cypher 1996)

139. First (proximal) phalanx of wing digit II: (0) ossified; (1) unossified or absent. (Walton and Walton 1970; Hill 1974; Strickler 1978)

140. Second phalanx of wing digit II: (0) ossified; (1) unossified or absent. (Walton and Walton 1970; Hill 1974; Strickler 1978)

141. Third (ungual) phalanx of wing digit II: (0) ossified; (1) unossified or absent. (Walton and Walton 1970; Hill 1974; Strickler 1978)

142. Third (ungual) phalanx of wing digit III: (0) completely ossified; (1) ossified only at the base; (2) unossified or absent. (Walton and Walton 1970; Hill 1974; Strickler 1978)

143. Folding of wings: (0) all phalanges in digits III–V flex anteriorly toward the underside of the wing; (1) proximal phalanx of digits III and IV fold posteriorly, and distal phalanges fold anteriorly; (2) distal phalanges of digits III and IV fold anteriorly, and proximal phalanges not folded. This character cannot be scored in taxa that lack wings. (Walton and Walton 1970; Hill 1974)

144. Vertebral fusion in posterior thoracic and lumbar series: (0) none; (1) at least three vertebrae fused. (A. Peffley, in manuscript; Simmons, personal observation)

145. Extent of sacrum (defined as all vertebrae that articulate with the pelvis or are fused with those that articulate with the pelvis): (0) sacrum terminates anterior to acetabulum; (1) sacrum extends posteriorly to at least the midpoint of the acetabulum. (A. Peffley, in manuscript; Simmons, personal observation)

146. Sacral laminae (thin plates of bone that extend laterally from the sacral vertebrae posterior to the iliosacral joint): (0) absent or small; (1) broad. (A. Peffley, in manuscript; Simmons, personal observation)

147. Iliosacral articulation: (0) not expanded dorsoventrally; (1) expanded dorsoventrally, and ilium has large dorsomedial flange. (Simmons, personal observation)

148. Form of ischium: (0) large dorsal tuberosity projects from posterior horizontal ramus; (1) no dorsal tuberosity. (Simmons, personal observation)

149. Pubic spine: (0) absent; (1) straight; (2) tip bent sharply dorsally. (Simmons, personal observation)

150. Articulation between pubes in male: (0) covers broad area, and symphysis long in anteroposterior dimension; (1) restricted to small area and consists of an ossified interpubic ligament or short symphysis. (Leche 1886; Simmons, personal observation)

151. Obturator foramen: (0) foramen normal, and rim well-defined; (1) foramen partially filled in with thin, bony sheet along posteroventral rim. (Simmons, personal observation)

152. Form of m. psoas minor: (0) tendinous for approximately half of length; (1) thick and fleshy along entire length. (MacAlister 1872; Leche 1886; Le Gros Clark 1924, 1926; Davis 1938; Vaughan 1959, 1970; Simmons, personal observation)

153. Form of m. gluteus superficialis: (0) single muscle mass; (1) differentiated into m. gluteus maximus and m. tensor fascia femoris. (Humphry 1869; MacAlister 1872; Leche 1886; Le Gros Clark 1924, 1926; Davis 1938; Vaughan 1959, 1970; Simmons, personal observation)

154. M. piriformis: (0) present; (1) absent. (Humphry 1869; MacAlister 1872; Leche 1886; Le Gros Clark 1924, 1926; Davis 1938; Vaughan 1959, 1970; Simmons, personal observation)

155. Shaft of femur: (0) straight; (1) bent such that distal shaft is directed dorsally. (Simmons, personal observation)

156. Form of fibula: (0) complete and well-developed; (1) thin and threadlike; (2) entirely absent. (Walton and Walton 1968, 1970; Simmons, personal observation)

157. Calcar: (0) absent; (1) present. (Walton and Walton 1970; Simmons, personal observation)

158. Number of phalanges on each of digits II–V of foot: (0) three; (1) two. (Walton and Walton 1970; Simmons, personal observation)

159. Digital tendon locking mechanism: (0) absent; (1) consists of tubercles on flexor tendon and plicae on adjacent tendon sheath; (2) consists of plicae but no tubercles. (Simmons and Quinn 1994; W. Schutt, unpublished data)

160. Baculum: (0) absent; (1) present. (Matthews 1941; Wassif and Madkour 1963; Brown et al. 1971; Agrawal and Sinha 1973; Luckett 1980a, 1980b; Smith and Madkour 1980; Hill and Smith 1984; Hill and Harrison 1987; Madkour 1989)

161. Form of baculum: (0) saddle-shaped or slipper-shaped; (1) elongated, with long central shaft. (Matthews 1941; Wassif and Madkour 1963; Brown et al. 1971; Agrawal and Sinha 1973; Luckett 1980a, 1980b; Smith and Madkour 1980; Hill and Smith 1984; Hill and Harrison 1987; Madkour 1989)

162. Pubic nipples in females: (0) absent; (1) one pair present. (Simmons 1993b, and personal observation)

163. Vulval opening: (0) oriented transversely; (1) oriented anteroposteriorly. (Matthews 1941; Loo and Kanagasuntheram 1972; Luckett 1980a; Pierson et al. 1986; Simmons, personal observation)

164. Form of clitoris: (0) small, not elongated anteroposteriorly (0); clitoris elongated (1). (Leche 1886; Matthews 1941; Luckett 1980a; Pierson et al. 1986; Simmons, personal observation)

165. External uterine fusion: (0) fusion minimal—uterine horns more than 70% length of common uterine body; (1) fusion more extensive—uterine horns less than 50% length of common uterine body. (Leche 1886; Le Gros Clark 1924, 1926; Matthews 1941; Hood and Smith 1982, 1983; Madkour 1989; Simmons, personal observation)

166. Internal uterine fusion: (0) fusion absent—two cervical openings into vagina; (1) fusion present—common uterine lumen. (Leche 1886; Matthews 1941; Hood and Smith 1982, 1983; Simmons, personal observation)

167. Common uterine lumen: (0) lumen short in comparison to length of cornual lumina; (1) lumen large, and cornual lumina either join immediately within common uterine body or are reduced to tubular intramural uterine cornua. This character cannot be scored in taxa that lack a common uterine lumen. (Leche 1886; Matthews 1941; Hood and Smith 1982, 1983; Simmons, personal observation)

168. Uterotubal junction: (0) with oviductal papillae or complex folds; (1) simple—no papillae or complex folds. (Hood and Smith 1982, 1983)

169. Location of blastocyst stage: (0) attained in uterus; (1) attained in oviduct. (Luckett 1980a)

170. Implantation: (0) superficial; (1) secondarily interstitial. (Luckett 1980a)

171. Orientation of embryonic disc: (0) disc oriented toward uterotubal junction; (1) disc consistently antimesometrial. (Luckett 1980a, 1980b, 1980c, 1993)

172. Yolk sac development: (0) mesoderm spreads and exocoelom expands to cover embryonic half of yolk sac only; yolk sac endoderm does not become vascular or exhibit hypertrophy; (1) mesoderm spreads and exocoelom expands to cover embryonic half of yolk sac only; embryonic half

becomes vascular and exhibits endoderm hypertrophy; (2) mesoderm spreads over entire yolk sac, but exocoelom expands to cover only embryonic half of yolk sac; hypertrophy occurs in vascular yolk sac at embryonic pole; (3) mesoderm spreads over entire surface of yolk sac; exocoelom expands to separate yolk sac completely from chorion; the free, vascular yolk sac subsequently collapses and endodermal cells hypertrophy. (Luckett 1980a, 1980c, 1993)

173. Allantoic vesicle: (0) large—occupies at least half the circumference of chorion during limb-bud stage; (1) small—occupies less than half the circumference of chorion; (2) does not form—allantois remains tubular and vestigial throughout gestation. (Luckett 1980a, 1980b, 1980c, 1993)

174. Primordial amniotic cavity: (0) does not form at any stage in development—amniogenesis is by folding; (1) does form but is transitory, lost in later development; (2) persists as definitive amniotic cavity. (Luckett 1980a, 1980b, 1980c, 1993)

175. Definitive chorioallantoic placenta: (0) endotheliochorial; (1) hemochorial. (Luckett 1980a, 1980b, 1980c, 1993; Bhiwgade et al. 1992)

176. Angle between dorsal horns of spinal cord: (0) 70°–80°; (1) 35°–50°; (2) 0°–25°. (Johnson and Kirsch 1993)

177. Relative size of inferior colliculus: (0) significantly smaller than superior colliculus; (1) larger than superior colliculus. (Johnson and Kirsch 1993)

178. Fusion of left central lobe of liver: (0) fused with left lateral lobe; (1) separate from other lobes or partially fused with right central lobe. (Robin 1881; Davis 1938)

179. Location of gallbladder: (0) in right lateral fissure of liver; (1) in umbilical fissure. (Robin 1881; Davis 1938)

180. Caecum: (0) present; (1) absent. (Robin 1881; Luckett 1980a, 1980b)

181. rDNA restriction site 20: (0) present; (1) present in some individuals, absent from others (polymorphic). (Agrawal and Sinha 1973)

182. rDNA restriction site 28: (0) absent; (1) present. (Baker et al. 1991)

183. rDNA restriction site 29: (0) absent; (1) present. (Baker et al. 1991)

184. rDNA restriction site 37: (0) absent; (1) present. (Baker et al. 1991)

185. rDNA restriction site 38: (0) absent; (1) present. (Baker et al. 1991)

186. rDNA restriction site 40: (0) absent; (1) present. (Baker et al. 1991)

187. rDNA restriction site 43: (0) absent; (1) present. (Baker et al. 1991)

188. rDNA restriction site 44: (0) absent; (1) present. (Baker et al. 1991)

189. rDNA restriction site 45: (0) absent; (1) present. (Baker et al. 1991)

190. rDNA restriction site 46: (0) absent; (1) present. (Baker et al. 1991)

191. rDNA restriction site 47: (0) absent; (1) present; (2) polymorphic. (Baker et al. 1991)

192. rDNA restriction site 50: (0) absent; (1) present. (Baker et al. 1991)

Appendix 1.2.
Data Matrix

The numbers following each taxonomic name refer to character states described in Appendix 1.1. Cases of taxonomic polymorphism are indicated by the presence of more than one character state within curly brackets (e.g., {01}). A question mark indicates that no data were available for a particular character; a dash indicates that the character could not be scored (i.e., was not applicable) in that taxon for reasons described in Appendix 1.1.

Scandentia
```
00000  00000  000-0  00000  00000  00000  0?300  00000  0000-  00000  00000  00000
0000-  0-000  00000  00000  -0-00  0?000  00000  -00--  00000  00000  000-?  ---00
20--0  00020  000??  ???00  00-00  10100  00{01}00  00000  -{01}?10  10???  00000
10001  ?????  ?????  ??
```

Dermoptera
```
000?0  00000  000-1  0011?  1-000  00100  00000  ???00  0000-  11000  01000  10101
0120-  0-005  00000  00000  -0-01  12100  00000  -00--  00000  00011  010-?  ---00
0100?  00000  001??  ???00  00-00  00001  00000  00010  -0000  0-??0  00111  10001
00000  01000  00
```

Pteropodidae
```
00000  121{012}0  000-0  {012}{123}{01}02  1-000  00001  10000  {01}000{013}  01011
11000  01100  00102  0000-  0-000  {01}0000  000{01}{01}  0100{01}  2{12}{01}00
{01}0000  -01-0  00000  0{01}001  {01}{01}100  00000  01212  000{01}0  00000  00000
02001  00011  000{01}0  01011  00000  0-010  13020  00100  00{01}00  00100  {02}0
```

Emballonuridae

```
01000  {01}{12}120  10{01}1{01}  {01}{01}110  00001  10002  01{02}11  120{012}0
0110-  01000  00{01}01  0{01}{01}01  0110-  0-10{12}  20000  0000{01}  01{01}00
2{12}101  {01}000{01}  00210  {01}{01}000  {01}{01}202  01110  10{01}1{01}  01012
000{01}1  11010  00011  12101  00011  0???0  11011  00000  0-0?0  13?10  21100
01000  01000  10
```

Rhinopomatidae

```
01100  01120  100-1  11210  00111  10002  11111  03130  01010  01020  00001  00103
1210-  11202  20000  10001  11110  11100  00000  -01-0  20000  01001  00111  01111
01012  11001  000??  ??00  12001  00011  1???0  10011  01000  10?00  ?3010  ??001
00000  00000  00
```

Craseonycteridae

```
011??  ???20  11101  1121?  00111  1?0??  ????1  ??30  0100-  01020  00001  10103
1010-  11201  21010  10011  01010  2110?  0000?  ?01-0  2?000  0?2??  ?????  ?????
?????  ????2  000??  ??01  12211  00011  0???0  1002?  ?100?  ?????  ?????  ?????
?????  ?????  ??
```

Nycteridae

```
01301  22121  000-1  00210  {01}0011  {01}1110  00211  13110  0010-  01010  00201
0{01}102  0{01}10-  100{01}0  2{01}001  01011  11100  23011  00001  001-0  21000
01212  01101  01111  01111  11011  00011  00011  12000  10011  0??1  31011  10000
0-???  ?????  ??000  11100  01100  01
```

Megadermatidae

```
012?0  0112{01}  10101  21{12}10  00011  01101  01111  0200{01}  00010  01030  00201
0{01}102  0010-  10000  21001  11011  11111  10101  00001  {12}0221  21000  01212
11101  01111  11112  11111  100??  ??01  120{01}1  00021  01001  11011  11000
10?00  ?2110  ??001  10001  10000  00
```

Rhinolophidae

```
00200  01121  000-1  11100  00111  {01}0110  01011  {01}31{0123}0  00010  10000
00201  00102  00{12}0-  10103  21021  11110  -1111  10101  10001  20221  21000
01211  10101  01111  10--2  10112  100??  ??11  12000  01020  1??01  11011  11000
10000  ?3010  21010  00100  01000  00
```

Hipposideridae

```
00200  01121  000-{01}  11{12}10  00111  {01}0110  01{01}11  131{0123}0  00010
10000  0020{01}  00102  00{012}0-  10{12}03  21121  11110  -1111  {12}0100  00001
{02}0221  21000  00211  ?0???  ??11  ?0--2  ??02  10000  10011  120{01}0  01020
11001  11111  {01}1000  10???  ?????  21010  01000  01000  00
```

Phyllostomidae

```
0120{01}  {01}{01}{01}10  010-{01}  {01}{123}{012}{01}{012}  00111  {01}0{01}02
10011  1000{12}  10{01}0-  110{013}0  00{12}0{01}  {01}0{01}0{12}  01{02}0-
{01}0{01}01  20000  000{01}{01}  0100{01}  {12}{01}{01}0{01}  00001  {12}1{12}00
1{01}000  1{01}20{12}  {01}{01}1{01}1  0{01}011  01112  011{01}2  00011  00101
10001  00011  0101{01}  {01}{01}000  -0111  1111{01}  11221  21100  00000  00101  00
```

Noctilionidae

```
01001  12110  110-0  02211  00111  00112  10?11  10001  1010-  11030  00101  00101
0100-  10201  20000  00001  11000  22100  00100  -1200  10000  00102  11101  00011
01112  00111  10001  11101  12201  00011  0???0  11010  -0111  10110  11220  21100
00011  00011  00
```

Mormoopidae

```
{01}100{01}  {01}{012}{01}10  010-0  01100  00111  {01}0110  10011  ??21  1000-
11000  00111  00101  01{02}0-  10{01}01  2{01}000  00001  11100  2{12}101  00101
1{01}210  11000  10202  11111  01011  01112  01102  1{01}011  01101  10011  00011
0???0  1100{01}  ?0111  111??  ?????  ?????  00011  00010  00
```

Mystacinidae

```
01001  22110  010-0  12111  00111  10010  11011  ???01  10111  11030  00111  10103
0020-  10101  20000  00000  -0-00  12101  00101  211-0  00000  ?01??  ?????  ?????
?????  ????2  111??  ???01  10101  00111  0???0  01001  ?0010  100??  ?????  ?????
?????  ?????  ??
```

Antrozoinae

```
011??  ???10  10110  1{01}211  00111  10002  01011  ???0?  ?????  ?????  ?????
?????  ?????  ????2  20000  00010  -1000  12100  00101  211-0  00100  001??  ?????
?????  ?????  ????2  011??  ???01  11001  10011  0???0  11011  00000  10???  ?????
?????  ?????  ?????  ??
```

Tomopeatinae

```
010??  ???10  10110  11211  00111  10002  01?11  ???03  10011  11000  00211  10002
00211  0-201  21000  00000  -1000  12000  00111  211-0  00112  011??  ?????  ?????
?????  ????2  111??  ???01  11001  10111  0???0  0101?  ?000?  ?????  ?????  ?????
?????  ?????  ??
```

Molossidae

```
01011  {01}{12}110  {01}{01}{01}1{01}  1{012}{12}1{01}  00111  {01}0002  01{01}11
10003  10{01}??  11000  00211  10002  00211  0-201  21000  0000{01}  0{01}00{01}
{12}2000  0{01}111  2{01}2{01}0  001{01}2  10102  01111  01010  01001  00102  11110
01001  11101  10111  00110  0101{01}  00000  10000  03111  21100  00000  01000  21
```

Myzopodidae

```
110??  ???10  01100  00001  01111  10001  01011  ???01  01012  01131  11211  00112
0020-  0-004  20000  00001  01100  22100  10101  20210  1?100  011??  ?????  ?????
?????  ????2  000??  ???11  10001  10011  0???0  1110?  ?0000  10???  ?????  ?????
?????  ?????  ??
```

Thyropteridae

```
11011  01110  01100  00001  01111  10001  01?11  ???01  1000-  11131  11211
00{01}?2  0?20-  11021  21100  00011  11100  21111  01101  10200  00000  10101
10101  11011  01012  00002  000??  ???11  10001  10011  1???0  1110?  ?0001  101?1
021?1  ?????  00000  00000  00
```

Furipteridae

```
110?1  12110  1010{01}  00100  01111  10101  01011  ???01  10011  11111  11211
00113  0000-  11021  21000  00011  11100  23111  11101  111-0  20011  001??  ?????
?????  ?????  ????2  000??  ???11  11011  00011  0???0  1100?  ?0001  10???  ?????
?????  00000  10000  00
```

Natalidae

```
110?0  12110  01100  00000  00111  10000  01011  ???01  10011  11111  11211  00013
0020-  10001  21000  00001  11001  20111  01001  111-0  20111  01102  10101  11010
01211  00102  00000  01011  11011  00011  0???0  11001  00000  10???  ?????  ?????
00000  10000  20
```

Vespertilioninae

```
01{02}10  12110  10110  {01}0{12}{01}{01}  00111  1000{12}  01011  1{01}101  10012
11010  00111  10112  01210  0-201  20000  000{012}1  01{01}00  {12}21{01}{01}  00101
211-0  0010{012}  {012}0102  10110  10110  00--1  00102  011??  ???01  11001  10011
0???0  1101{01}  {01}0000  10000  02111  ?1100  ?????  ?????  ??
```

Minopterinae

```
01010  {01}{02}{01}10  10110  00100  00111  10002  01011  10101  10112  11010  00111
10012  01010  0-201  20000  00011  11100  23111  01101  211-0  00102  101??  ?????
?????  ?????  ????2  111??  ???01  11201  10011  0???0  11000  -0000  10???  ????1
??100  ?????  ?????  ??
```

Myotinae

```
01010  12110  10110  00{01}{01}{01}  00111  10001  01011  ???01  10112  11010  00111
10012  01210  0-201  20000  000{01}1  01100  21111  00101  21210  00101  0010{01}
101{01}{01}  {01}0010  01001  00102  0110?  0??01  11001  10011  00110  11011  00000
10000  02111  21???  00000  00000  00
```

Murininae
```
010??  ???10  10110  00110  00111  10002  01?11  ???01  10112  11010  00211  10011
0120-  10201  20000  00021  01100  21111  00001  20220  20100  011??  ?????  ?????
?????  ????2  011??  ???01  11001  10011  0???0  11011  00000  10???  ?????  ?????
?????  ?????  ??
```

Kerivoulinae
```
110??  ???10  10110  00001  00111  00002  01?11  1??01  1010-  11000  00111  11012
01210  0-221  20000  00021  01100  21111  01101  21210  00100  011??  ?????  ?????
?????  ????2  011??  ???01  11001  10011  0???0  11021  00000  10???  ?????  ??100
?????  ?????  ??
```

Acknowledgments

Special thanks go to K. Bhatnagar, J. Cypher, J. Geisler, T. Griffiths, K. Koopman, A. Peffley, W. Schutt, and J. Wible for sharing unpublished data. J. Geisler, T. Griffiths, J. Kirsch, K. Koopman, M. McKenna, M. Novacek, A. Peffley, R. Voss, and J. Wible read earlier drafts of this manuscript, and I thank them for their comments. Thanks also go to M. Carleton and L. Gordon (U.S. National Museum), M. Hafner (Louisiana State Museum of Zoology), P. Jenkins (British Museum of Natural History), and B. Patterson (Field Museum of Natural History) for permission to study specimens in their care. This study was supported by National Science Foundation grant DEB-9106868.

Literature Cited

Adkins, R. M., and R. L. Honeycutt. 1993. A molecular examination of archontan and chiropteran monophyly. *In* Primates and Their Relatives in Phylogenetic Perspective, R. D. E. MacPhee, ed., pp. 227–249. Advances in Primatology Series. Plenum Press, New York.

Agrawal, V. C., and Y. P. Sinha. 1973. Studies on the bacula of some Oriental bats. Anatomischer Anzeiger 133:180–192.

Baker, R. J., R. L. Honeycutt, and R. A. Van Den Bussche. 1991. Examination of monophyly of bats: Restriction map of the ribosomal DNA cistron. Bulletin of the American Museum of Natural History 206:42–53.

Barkley, L. J. 1984. Evolutionary relationships and natural history of *Tomopeas ravus* (Mammalia: Chiroptera). Master's thesis, Louisiana State University, Baton Rouge.

Barrett, M., M. J. Donoghue, and E. Sober. 1991. Against consensus. Systematic Zoology 40:486–493.

Barrett, M., M. J. Donoghue, and E. Sober. 1993. Crusade? A reply to Nelson. Systematic Biology 42:216–217.

Beard, K. C., B. Sigé, and L. Krishtalka. 1992. A primitive vespertilionoid bat from the early Eocene of central Wyoming. Comptes Rendus de l'Académie des Sciences, Série III, Sciences de la Vie 314:735–741.

Bhatnagar, K. P., and J. R. Wible. 1994. Observations on the vomeronasal organ of the colugo *Cynocephalus* (Mammalia, Dermoptera). Acta Anatomica 151:43–48.

Bhiwgade, D. A., A. B. Singh, A. P. Manekar, and S. N. Menon. 1992. Ultrastructural development of chorioallantoic placenta in the Indian miniopterus bat, *Miniopterus schreibersii fuliginosus* (Hodgson). Acta Anatomica 145:248–264.

Bremer, K. 1988. The limits of amino acid sequence data in angiosperm phylogenetic reconstructions. Evolution 42:795–803.

Brown, R. E., H. H. Genoways, and J. K. Jones. 1971. Bacula of some Neotropical bats. Mammalia 35:456–464.

Bull, J. J., J. P. Hulsenbeck, C. W. Cunningham, D. L. Swofford, and P. J. Waddell. 1993. Partitioning and combining data in phylogenetic analysis. Systematic Biology 42:384–397.

Chippindale, P. T., and J. J. Wiens. 1994. Weighting, partitioning, and combining characters in phylogenetic analysis. Systematic Biology 43:278–287.

Corbet, G. B., and J. E. Hill. 1992. The Mammals of the Indomalayan Region: A Systematic Review. Oxford University Press, Oxford.

Cracraft, J., and K. Helm-Bychowski. 1991. Parsimony and phylogenetic inference using DNA sequences: Some methodological considerations. *In* Phylogenetic Analysis of DNA Sequences, M. M. Miyamoto and J. Cracraft, eds., pp. 184–220. Oxford University Press, New York.

Cypher, J. L. 1996. Phylogenetic analysis of the order Chiroptera using carpal morphology. Master's thesis, Eastern Michigan University, Ypsilanti.

Davis, D. D. 1938. Notes on the anatomy of the treeshrew *Dendrogale*. Zoology Series of the Field Museum of Natural History 20: 383–404.

Davis, W. B. 1970. *Tomopeas ravus* Miller (Chiroptera). Journal of Mammalogy 51:244–247.

de Queiroz, A. 1993. For consensus (sometimes). Systematic Biology 42:368–372.

de Queiroz, A., M. J. Donoghue, and J. Kim. 1995. Separate versus combined analysis of phylogenetic evidence. Annual Review of Ecology and Systematics 26:657–681.

Doran, A. H. G. 1878. Morphology of the mammalian ossicula auditus. Transactions of the Linnean Society of London, 2nd Series 1:371–497.

Eernisse, D. J., and A. G. Kluge. 1993. Taxonomic congruence versus total evidence, and amniote phylogeny inferred from fossils, molecules, and morphology. Molecular Biology and Evolution 10:1170–1195.

Fenton, M. B., and G. P. Bell. 1981. Recognition of insectivorous bats by their echolocation calls. Journal of Mammalogy 62:233–243.

Freeman, P. W. 1981. A multivariate study of the family Molossidae (Mammalia, Chiroptera): Morphology, ecology, evolution. Fieldiana Zoology 7:1–173.

Göpfert, M. C., and L. T. Wasserthal. 1995. Notes on the echolocation calls, food and roosting behavior of the Old World sucker-footed bat, *Myzopoda aurita* (Chiroptera, Myzopodidae). Zeitschrift für Saeugetierkunde 60:1–8.

Gopalakrishna, A., and G. C. Chari. 1983. A review of the taxonomic position of *Miniopterus* based on embryological characters. Current Science (Bangalore) 52:1176–1180.

Griffiths, T. A. 1982. Systematics of the New World nectar-feeding bats (Mammalia, Phyllostomidae), based on the morphology of the hyoid and lingual regions. American Museum Novitates 2742:1–45.

Griffiths, T. A. 1983. Comparative laryngeal anatomy of the big brown bat, *Eptesicus fuscus*, and the mustached bat, *Pteronotus parnellii*. Mammalia 47:377–394.

Griffiths, T. A. 1994. Phylogenetic systematics of slit-faced bats (Chiroptera, Nycteridae), based on hyoid and other morphology. American Museum Novitates 3090:1–17.

Griffiths, T. A., and A. L. Smith. 1991. Systematics of emballonuroid bats (Chiroptera: Emballonuridae and Rhinopomatidae), based on hyoid morphology. Bulletin of the American Museum of Natural History 206:62–83.

Griffiths, T. A., A. Truckenbrod, and P. J. Sponholtz. 1992. Systematics of megadermatid bats (Chiroptera, Megadermatidae), based on hyoid morphology. American Museum Novitates 3041:1–21.

Habersetzer, J., and G. Storch. 1987. Klassifikation und funktionelle Flügelmorphologie paläogener Fledermäuse (Mammalia, Chiroptera). Courier Forschunginstitut Senckenberg 91:11–150.

Hand, S. J. 1990. First Tertiary molossid (Microchiroptera: Molossidae) from Australia: Its phylogenetic and biogeographic implications. Memoirs of the Queensland Museum 28:175–192.

Hill, J. E. 1974. A new family, genus, and species of bat (Mammalia: Chiroptera) from Thailand. Bulletin of the British Museum (Natural History) Zoology 27:301–336.

Hill, J. E., and D. L. Harrison. 1987. The baculum in the Vespertilioninae (Chiroptera: Vespertilionidae) with a systematic review, a synopsis of *Pipistrellus* and *Eptesicus*, and the descriptions of a new genus and subgenus. Bulletin of the British Museum (Natural History) Zoology 52:225–305.

Hill, J. E., and J. D. Smith. 1984. Bats: A Natural History. University of Texas Press, Austin.

Hood, C. S., and J. D. Smith. 1982. Cladistical analysis of female reproductive histomorphology in phyllostomoid bats. Systematic Zoology 31:241–251.

Hood, C. S., and J. D. Smith. 1983. Histomorphology of the female reproductive tract in phyllostomoid bats. Occasional Papers of the Museum, Texas Tech University 86:1–38.

Hulsenbeck, J. P., D. L. Swofford, C. W. Cunningham, J. J. Bull, and P. J. Waddell. 1994. Is character weighting a panacea for the problem of data heterogeneity in phylogenetic analysis? Systematic Biology 43:288–291.

Humphry, G. M. 1869. The myology of the limbs of *Pteropus*. Journal of Anatomy and Physiology 3:294–334.

Hutcheon, J., and J. A. W. Kirsch. 1995. Interfamilial relationships within the Microchiroptera: A preliminary study using DNA hybridization. Bat Research News 36:73–74.

Johnson, J. I., and J. A. W. Kirsch. 1993. Phylogeny through brain traits: Interordinal relationships among mammals including primates and Chiroptera. *In* Primates and Their Relatives in Phylogenetic Perspective, R. D. E. MacPhee, ed., pp. 293–331. Advances in Primatology Series. Plenum Press, New York.

Kirsch, J. A. W. 1995. Bats are monophyletic; megabats are monophyletic; but are microbats also? Bat Research News 36:78.

Kluge, A. G. 1989. A concern for evidence and a phylogenetic hypothesis of relationships among *Epicrates* (Boidae, Serpentes). Systematic Zoology 38:7–25.

Kluge, A. G., and A. J. Wolf. 1993. Cladistics: What's in a word? Cladistics 9:183–195.

Koopman, K. F. 1984. Bats. *In* Orders and Families of Recent Mammals of the World, S. Anderson and J. K. Jones, eds., pp. 145–186. Wiley, New York.

Koopman, K. F. 1985. A synopsis of the families of bats, part VII. Bat Research News 25:25–29 (dated 1984 but issued in 1985).

Koopman, K. F. 1993. Order Chiroptera. *In* Mammal Species of the World, 2nd Ed., D. E. Wilson and D. M. Reeder, eds., pp. 137–241. Smithsonian Institution Press, Washington, D.C.

Koopman, K. F. 1994. Chiroptera: Systematics. Handbook of Zoology, Vol. 8, Part 60: Mammalia. Walter de Gruyter, Berlin.

Leche, W. 1886. Uber die säugthiergattung *Galeopithecus*: Eine morphologische untersuchung. Kungliga Svenska Vetenskaps-Akademiens Handlingar 21:1–92.

Legendre, S. 1984. Essai de biogéographie phylogénique des molossidés (Chiroptera). Myotis Mitteilungsblatt für Fiedermauskundler 21–22:30–36.

Le Gros Clark, W. E. 1924. The myology of the tree-shrew (*Tupaia minor*). Proceedings of the Zoological Society of London 1924: 461–497.

Le Gros Clark, W. E. 1926. On the anatomy of the pen-tailed tree-shrew (*Ptilocercus lowii*). Proceedings of the Zoological Society of London 1926:1179–1309.

Loo, S. K., and R. Kanagasuntheram. 1972. The vomeronasal organ in the tree shrew and slow loris. Journal of Anatomy 112: 165–172.

Luckett, W. P. 1980a. The use of fetal membrane characters in assessing chiropteran phylogeny. *In* Proceedings of the Fifth International Bat Research Conference, D. E. Wilson and A. L. Gardner, eds., pp. 245–266. Texas Tech University Press, Lubbock.

Luckett, W. P. 1980b. The suggested evolutionary relationships and classification of tree shrews. *In* Comparative Biology and Evolutionary Relationships of Tree Shrews, W. P. Luckett, ed., pp. 3–31. Plenum Press, New York.

Luckett, W. P. 1980c. The use of reproductive and developmental features in assessing tupaiid affinities. *In* Comparative Biology and Evolutionary Relationships of Tree Shrews, W. P. Luckett, ed., pp. 245–266. Plenum Press, New York.

Luckett, W. P. 1993. Developmental evidence from the fetal membranes for assessing archontan relationship. *In* Primates and Their Relatives in Phylogenetic Perspective, R. D. E. MacPhee, ed., pp. 149–186. Advances in Primatology Series. Plenum Press, New York.

MacAlister, A. 1872. The myology of the Cheiroptera. Philosophical Transactions of the Royal Society of London 162:125–171.

Maddison, W. P., and D. R. Maddison. 1992. MacClade, version 3.0. (Computer program distributed on floppy disk.) Sinauer Associates, Sunderland, Mass.

Maddison, W. P., M. J. Donoghue, and D. R. Maddison. 1984. Outgroup analysis and parsimony. Systematic Zoology 33:83–103.

Madkour, G. 1989. Uro-genitalia of Microchiroptera from Egypt. Zoologischer Anzeiger 222:337–352.

Matthews, L. H. 1941. Notes on the genitalia and reproduction of some African bats. Proceedings of the Zoological Society of London Series B 111:289–346.

Mein, P., and Y. Tupinier. 1977. Formule dentaire et position systématique du Minioptère (Mammalia, Chiroptera). Mammalia 41:207–211.

Menu, H., and B. Sigé. 1971. Nyctalodontie et myotodontie, importants caractères de grades évolutifs chez les chiroptères entomophages. Comptes Rendus de l'Académie des Sciences, Paris, Series III, Life Sciences 272:1735–1738.

Miller, G. S. 1897. Revision of the North American bats of the family Vespertilionidae. North American Fauna 13:5–135.

Miller, G. S. 1907. The families and genera of bats. U.S. National Museum Bulletin 57:1–282.

Miyamoto, M. M., and J. Cracraft. 1991. Phylogenetic inference, DNA sequence analysis, and the future of molecular systematics. *In* Phylogenetic Analysis of DNA Sequences, M. M. Miyamoto and J. Cracraft, eds., pp. 3–17. Oxford University Press, New York.

Nelson, G. 1993. Why crusade against consensus? A reply to Barrett, Donoghue, and Sober. Systematic Biology 42:215–216.

Nixon, K. C., and J. M. Carpenter. 1993. On outgroups. Cladistics 9:413–426.

Novacek, M. J. 1980a. Phylogenetic analysis of the chiropteran auditory region. *In* Proceedings of the Fifth International Bat Research Conference, D. E. Wilson and A. L. Gardner, eds., pp. 317–330. Texas Tech University Press, Lubbock.

Novacek, M. J. 1980b. Cranioskeletal features in tupaiids and selected Eutheria as phylogenetic evidence. *In* Comparative Biology and Evolutionary Relationships of Tree Shrews, W. P. Luckett, ed., pp. 35–93. Plenum Press, New York.

Novacek, M. J. 1987. Auditory features and affinities of the Eocene bats *Icaronycteris* and *Palaeochiropteryx* (Microchiroptera, *incertae sedis*). American Museum Novitates 2877:1–18.

Novacek, M. J. 1991. Aspects of morphology of the cochlea in microchiropteran bats: An investigation of character transformation. Bulletin of the American Museum of Natural History 206:84–100.

Novacek, M. J. 1994. Morphological and molecular inroads to phylogeny. *In* Interpreting the Hierarchy of Nature: From Systematic Patterns to Evolutionary Process Theories, L. Grande and O. Rieppel, eds., pp. 85–131. Academic Press, New York.

Novick, A. 1962. Orientation in Neotropical bats. I. Natalidae and Emballonuridae. Journal of Mammalogy 43:449–455.

Pettigrew, J. D., and J. A. Kirsch. 1995. Flying primates revisited: DNA hybridization with fractionated, GC-enriched DNA. South African Journal of Science 91:477–482.

Pettigrew, J. D., B. G. M. Jamieson, S. K. Robson, L. S. Hall, K. I. McAnally, and H. M. Cooper. 1989. Phylogenetic relations between microbats, megabats, and primates (Mammalia: Chiroptera and Primates). Philosophical Transactions of the Royal Society of London B 325:489–559.

Pierson, E. D. 1986. Molecular systematics of the Microchiroptera: Higher taxon relationships and biogeography. Ph.D. dissertation, University of California, Berkeley.

Pierson, E. D., V. M. Sarich, J. M. Lowenstein, M. J. Daniel, and W. E. Rainey. 1986. A molecular link between the bats of New Zealand and South America. Nature (London) 6083:60–63.

Robbins, L. W., and V. M. Sarich. 1988. Evolutionary relationships in the family Emballonuridae (Chiroptera). Journal of Mammalogy 69:1–13.

Robin, H. A. 1881. Recherches anatomiques sur les mammifères des chiroptères. Annales des Sciences Naturelles, 6th Series 12:1–180.

Sigé, B. 1974. Donées nouvelles sur le genre Stehlinia (Vespertilionoidea, Chiroptera) du Paléogène d'Europe. Palaeovertebrata 6:253–272.

Simmons, J. A. 1980. Phylogenetic adaptations and the evolution of echolocation in bats. *In* Proceedings of the Fifth International Bat Research Conference, D. E. Wilson and A. L. Gardner, eds., pp. 267–278. Texas Tech University Press, Lubbock.

Simmons, J. A. 1993a. The importance of methods: Archontan phylogeny and cladistic analysis of morphological data. *In* Primates and Their Relatives in Phylogenetic Perspective, R. D. E. MacPhee, ed., pp. 1–61. Plenum Press, New York.

Simmons, J. A. 1993b. Morphology, function, and phylogenetic significance of pubic nipples in bats (Mammalia: Chiroptera). American Museum Novitates 3077:1–37.

Simmons, J. A. 1994. The case for chiropteran monophyly. American Museum Novitates 3103:1–54.

Simmons, J. A. 1995. Bat relationships and the origin of flight. Symposia of the Zoological Society of London 67:27–43.

Simmons, N. B., and J. G. Geisler. 1998. Phylogenetic relationships of *Icaronycteris, Archaeonycteris, Hassianycteris,* and *Palaeochiropteryx* to extant bat lineages, with comments on the evolution of echolocation and foraging strategies in Microchiroptera. Bulletin of the American Museum of Natural History 235:1–182.

Simmons, N. B., and T. H. Quinn. 1994. Evolution of the digital tendon locking mechanism in bats and dermopterans: A phylogenetic perspective. Journal of Mammalian Evolution 2:231–254.

Smith, J. D. 1976. Chiropteran evolution. *In* Biology of Bats of the New World Family Phyllostomatidae, Part I, R. J. Baker, J. K. Jones, and D. C. Carter, eds., pp. 49–69. Special Publications, No. 10, The Museum, Texas Tech University.

Smith, J. D. 1980. Chiropteran phylogenetics: Introduction. *In* Proceedings of the Fifth International Bat Research Conference,

D. E. Wilson and A. L. Gardner, eds., pp. 233–244. Texas Tech University Press, Lubbock.

Smith, J. D., and G. Madkour. 1980. Penial morphology and the question of chiropteran phylogeny. *In* Proceedings of the Fifth International Bat Research Conference, D. E. Wilson and A. L. Gardner, eds., pp. 347–365. Texas Tech University Press, Lubbock.

Stanhope, M. J., W. J. Bailey, J. Czelusniak, M. Goodman, J.-S. Si, J. Nickerson, J. G. Sgouros, G. A. M. Singer, and T. K. Kleinschimidt. 1993. A molecular view of primate supraordinal relationships from the analysis of both nucleotide and amino acid sequences. *In* Primates and Their Relatives in Phylogenetic Perspective, R. D. E. MacPhee, ed., pp. 251–292. Advances in Primatology Series. Plenum Press, New York.

Strickler, T. L. 1978. Functional osteology and myology of the shoulder in the Chiroptera. Contributions to Vertebrate Evolution 4:1–198.

Sudman, P. D., L. J. Barkley, and M. S. Hafner. 1994. Familial affinity of *Tomopeas ravus* (Chiroptera) based on protein electrophoretic and cytochrome B sequence data. Journal of Mammalogy 75:365–377.

Surlykke, A., L. A. Miller, B. Møhl, B. B. Andersen, J. Christensen-Dalsgaard, and M. B. Jørgensen. 1993. Echolocation in two very small bats from Thailand: *Craseonycteris thonglongyai* and *Myotis siligorensis*. Behavioral Ecology and Sociobiology 33:1–12.

Swofford, D. L. 1993. Phylogenetic Analysis Using Parsimony (PAUP), version 3.1.1. (Computer program distributed on floppy disk.) Smithsonian Institution, Washington, D.C.

Swofford, D. L., and G. J. Olsen. 1990. Phylogeny reconstruction. *In* Molecular Systematics, D. M. Hillis and C. Moritz, eds., pp. 411–501. Sinauer Associates, Sunderland, Mass.

Tate, G. H. H. 1942. Results of the Archbold Expeditions, No. 47. Review of the vespertilionine bats, with special attention to genera and species of the Archbold collections. Bulletin of the American Museum of Natural History 80:221–297.

Thomas, O. 1904. On the osteology and systematic position of the rare Malagasy bat, *Myzopoda aurita*. Proceedings of the Zoological Society of London 2:2–6.

Van Valen, L. 1979. The evolution of bats. Evolutionary Theory 4:104–121.

Vaughan, T. A. 1959. Functional morphology of three bats: *Eumops, Myotis, Macrotus*. University of Kansas, Publications of the Museum of Natural History 12:1–153.

Vaughan, T. A. 1970. The muscular system. *In* Biology of Bats, Vol. I, W. A. Wimsatt, ed., pp. 140–194. Academic Press, New York.

Volleth, M., and K.-G. Heller. 1994. Phylogenetic relationships of vespertilionid genera (Mammalia: Chiroptera) as revealed by karyological analysis. Zeitschrift für Zoologische Systematik und Evolutionsforschung 32:11–34.

Vrana, P. B. 1994. Molecular approaches to mammalian phylogeny. Ph.D. dissertation, Columbia University Graduate School of Arts and Sciences, New York.

Walton, D. W., and G. M. Walton. 1968. Comparative osteology of the pelvic and pectoral girdles of the Phyllostomatidae (Chiroptera; Mammalia). Journal of the Graduate Research Center, Southern Methodist University 37:1–35.

Walton, D. W., and G. M. Walton. 1970. Post-cranial osteology of bats. *In* About Bats: A Chiropteran Biology Symposium, B. H. Slaughter and D. W. Walton, eds., pp. 93–126. Fondren Science Series, Vol. 11. Southern Methodist University Press, Dallas, Tex.

Wassif, K. 1950. The tensor tympani muscle of bats. Annals and Magazine of Natural History 3:811–812.

Wassif, K., and G. Madkour. 1963. Studies on the osteology of the rat-tailed bats of the genus *Rhinopoma* found in Egypt. Bulletin of the Zoological Society of Egypt 18:56–80.

Weid, R., and O. V. Helversen. 1987. Ortungsrufe Europäischer Fledermäuse beim Jagdflug im Freiland. Myotis 25:5–27.

Wible, J. R., and K. P. Bhatnagar. 1997 [article date: 1996]. Chiropteran vomeronasal complex and the interfamilial relationships of bats. Journal of Mammalian Evolution 3:285–314.

Winge, H. 1941. The Interrelationships of Mammalian Genera, Vol. 1. Monotremata, Marsupialia, Insectivora, Chiroptera, Edentata. C. A. Reitzels, Copenhagen.

Zeller, U. 1986. Ontogeny and cranial morphology in the tympanic region of the Tupaiidae, with special reference to *Ptilocercus*. Folia Primatologica 47:61–80.

2

In the Minotaur's Labyrinth
Phylogeny of the Bat Family Hipposideridae

WIESLAW BOGDANOWICZ AND ROBERT D. OWEN

The family Hipposideridae is composed of nine Recent genera with about 65 species, which are widespread throughout warm areas of the Old World from western Africa east to the New Hebrides and extend only marginally into the Palaearctic (Corbet and Hill 1991, 1992; Koopman 1994). The genus *Hipposideros* has about 50 species; the other genera either are monotypic *(Anthops, Cloeotis, Paracoelops, Rhinonycteris)* or have 2 species *(Asellia, Aselliscus, Coelops, Triaenops)*. Hipposiderid fossils are known from the middle Eocene of Europe (Sigé and Legendre 1983; Sigé 1991), the early Oligocene of Arabo-Africa (Sigé et al. 1994), the late Oligocene of Australia (Archer et al. 1994), and probably the Miocene of Asia (K. F. Koopman, in litt.).

During the past 150 years hipposiderids have attracted the attention of numerous taxonomists (summarized by Hill 1963). The most prominent studies in this century were those by Tate (1941) and Hill (1963) of the genus *Hipposideros* and by Hill (1982) of the genera *Rhinonycteris, Cloeotis,* and *Triaenops.* Their work resulted in recognition of 11 (Tate 1941) or 6 (Hill 1963) supraspecific groups within the genus *Hipposideros.* Most of the more recent researchers (e.g., Jenkins and Hill 1981; Kock and Bhat 1994; Koopman 1994) either have accepted Hill's point of view or have made only minor changes to his 1963 classification. However, the question arises as to what extent Hill's carefully arranged, but nevertheless intuitive, species groups reflect phylogenetic history.

More importantly, none of the previous studies evaluated phylogenetic affinities within the entire family. The aim of our chapter is to fill this gap, although the lack of well-preserved materials for some taxa makes our analysis incomplete. Nevertheless, it is a first step toward a comprehensive revision of phylogenetic relationships among hipposiderids. We also evaluated different hypotheses concerning the geographic center of origin for the family and the monophyly of the genus *Hipposideros.* These assessments were made through phylogenetic analyses of metrical and discrete-state characters of the cranium, dentition, and external morphology.

Materials and Methods

Species, Specimens, and Measurements

Our study was based on 57 species and 702 adult specimens (skins and skulls), each of which had no or few missing char-

Table 2.1

Common-Part-Removed Analysis

Species	n (sexed + unsexed)	R^2	Species	n (sexed + unsexed)	R^2
Anthops ornatus	3 + 1	0.936	*Hipposideros inexpectatus*	1 + 0	0.952
Asellia patrizii	3 + 0	0.977	*Hipposideros jonesi*	17 + 0	0.986
Asellia tridens	10 + 0	0.970	*Hipposideros lankadiva*	8 + 0	0.964
Asselliscus stoliczkanus	9 + 0	0.984	*Hipposideros larvatus*	29 + 0	0.985
Asselliscus tricuspidatus	10 + 0	0.983	*Hipposideros lekaguli*	3 + 0	0.991
Cloeotis percivali	11 + 0	0.970	*Hipposideros lylei*	12 + 0	0.985
Coelops frithi	18 + 0	0.854	*Hipposideros macrobullatus*	4 + 0	0.988
Coelops robinsoni	2 + 0	0.796	*Hipposideros maggietaylorae*	13 + 0	0.976
Hipposideros abae	15 + 0	0.986	*Hipposideros marisae*	9 + 0	0.989
Hipposideros armiger	26 + 0	0.978	*Hipposideros magalotis*	11 + 0	0.978
Hipposideros ater	11 + 0	0.987	*Hipposideros muscinus*	11 + 0	0.983
Hipposideros beatus	16 + 0	0.992	*Hipposideros obscurus*	12 + 0	0.987
Hipposideros bicolor	8 + 0	0.984	*Hipposideros papua*	8 + 0	0.985
Hipposideros caffer	59 + 0	0.989	*Hipposideros pomona*	11 + 0	0.977
Hipposideros calcaratus	11 + 0	0.985	*Hipposideros pratti*	14 + 0	0.976
Hipposideros camerunensis	3 + 0	0.979	*Hipposideros pygmaeus*	10 + 0	0.992
Hipposideros cervinus	30 + 0	0.993	*Hipposideros ridleyi*	5 + 0	0.992
Hipposideros cineraceus	10 + 0	0.988	*Hipposideros ruber*	45 + 0	0.986
Hipposideros commersoni	11 + 0	0.969	*Hipposideros sabanus*	3 + 0	0.978
Hipposideros corynophyllus	1 + 0	0.969	*Hipposideros semoni*	7 + 0	0.979
Hipposideros curtus	8 + 0	0.987	*Hipposideros speoris*	11 + 0	0.984
Hipposideros cyclops	12 + 0	0.978	*Hipposideros stenotis*	4 + 0	0.984
Hipposideros diadema	27 + 0	0.977	*Hipposideros terasensis*	25 + 0	0.979
Hipposideros dinops	12 + 0	0.969	*Hipposideros turpis*	11 + 0	0.983
Hipposideros dyacorum	10 + 0	0.983	*Hipposideros wollastoni*	3 + 0	0.991
Hipposideros fuliginosus	10 + 0	0.992	*Rhinonycteris aurantius*	10 + 0	0.987
Hipposideros fulvus	10 + 0	0.986	*Triaenops furculus*	5 + 4	0.989
Hipposideros galeritus	12 + 0	0.994	*Triaenops persicus*	25 + 0	0.977
Hipposideros halophyllus	2 + 0	0.982			

Notes: This table lists ingroup taxa (Hipposideridae), number of specimens *(n)*, and coefficient-of-determination (R^2) values adjusted for degrees of freedom from linear regression of each ingroup taxon on a multiple-member outgroup (*Rhinolophus celebensis*, n = 12; *R. clivosus*, n = 10; *R. hipposideros*, n = 12; *R. malayanus*, n = 8; and *R. sinicus*, n = 9). When multiplied by 100, R^2 indicates percentage of original variance explained by the outgroup.

acters (Table 2.1). Adults were recognized by fused epiphyses in wing bones. *Hipposideros schistaceus* was found to be morphologically very similar to or even indistinguishable from *H. lankadiva* (P. J. J. Bates, personal communication; our observations); consequently, it was not treated as a distinct species. Space limitation precludes a list of specimens examined, but this list is available on request from the first author via Internet at wieslawb@robal.miiz.waw.pl.

A total of 45 cranial, dental, and external characters were measured with a Sylvac electronic caliper directly connected to a PC-compatible laptop computer for automatic data capture. The width and height of the foramen magnum (not included in Table 2.2) were used to calculate the foramen magnum area according to the formula given by Radinsky (1967). Measurements were taken to the nearest 0.01 mm for cranial and dental characters and to the nearest 0.1 mm for external characters. The length of the hindfoot

included the claws. Other measurements followed Freeman (1981), Rautenbach (1986), and Bogdanowicz (1992). Forearm lengths given in *Results* are from Koopman (1994).

Transformations

All continuous values were transformed to their natural logarithms, and the value of each character was calculated as the average of the means for males and females. Where possible, samples from single or neighboring populations were used for each species. For two endemic species without evident sexual dimorphism *(Anthops ornatus* and *Triaenops furculus)*, unsexed specimens also were used.

The common-part-removed transformation (Wood 1983) was applied to remove the portion of variance accounted for by regression onto selected rhinolophid taxa. Because Rhinolophidae is the sister-taxon to the Hipposideridae

Table 2.2

Size-Out Analysis

Character	PC1	PC2	PC3	PC4
Greatest skull length	0.997	−0.023	0.006	−0.026
Condylocanine length	0.997	−0.015	0.022	−0.020
Least interorbital breadth	0.822	0.300	0.256	−0.004
Zygomatic breadth	0.984	0.059	−0.074	−0.053
Mastoid breadth	0.976	0.116	0.115	−0.029
Breadth of braincase	0.976	0.094	0.117	−0.020
Breadth of nasal swellings	0.964	−0.076	−0.126	0.162
Height of braincase (excluding bullae)	0.975	−0.027	0.100	−0.091
Length of maxillary toothrow	0.991	−0.029	−0.051	−0.030
Length of upper molariform row	0.990	−0.023	−0.080	−0.020
Width across upper canines	0.976	−0.043	−0.132	0.054
Height of upper canine	0.974	0.078	−0.027	−0.041
Width across upper third molars	0.981	−0.029	−0.106	0.020
Length of upper third molar	0.947	−0.150	−0.115	0.017
Width of upper third molar	0.970	−0.080	−0.102	−0.008
Supraorbital length	0.938	−0.140	−0.026	0.153
Palatal length	0.926	0.136	−0.101	0.008
Area of foramen magnum	0.922	0.115	0.209	−0.187
Bullar length	0.955	0.010	0.164	0.176
Bullar width	0.898	0.200	0.141	0.246
Greatest length of mandible	0.995	−0.016	−0.059	−0.013
Length of mandibular toothrow	0.992	−0.015	−0.053	−0.024
Postdental length	0.991	0.015	−0.060	−0.000
Height of mandibular ramus	0.968	0.019	−0.140	−0.047
Height of lower canine	0.968	0.119	−0.096	−0.006
Coronoid–angular distance	0.977	0.021	−0.143	−0.027
Length of moment arm of temporal	0.989	0.038	−0.066	−0.015
Length of moment arm of masetter	0.951	−0.077	−0.239	−0.037
Forearm length	0.983	0.002	−0.014	−0.015
Third digit, metacarpal length	0.972	−0.027	−0.021	−0.010
Third digit, first phalanx length	0.901	0.340	0.011	−0.041
Third digit, second phalanx length	0.909	−0.248	0.088	−0.178
Fourth digit, metacarpal length	0.987	−0.036	0.038	−0.030
Fourth digit, first phalanx length	0.959	0.065	−0.022	−0.117
Fourth digit, second phalanx length	0.919	−0.241	0.139	−0.091
Fifth digit, metacarpal length	0.972	−0.124	0.143	−0.087
Fifth digit, first phalanx length	0.952	0.143	−0.070	−0.095
Fifth digit, second phalanx length	0.922	−0.192	0.128	−0.156
Head and body length	0.976	0.042	0.027	−0.012
Tail length	0.522	0.753	−0.152	0.126
Ear length	0.828	0.064	0.375	0.231
Tibia length	0.978	0.004	0.078	−0.018
Hindfoot length	0.970	−0.127	0.040	−0.030
Greatest breadth of anterior noseleaf	0.856	−0.366	0.015	0.272
Greatest breadth of horseshoe (including secondary leaflets)	0.901	−0.288	−0.081	0.247
All characters, variance explained	(89.9%)	(3.1%)	(1.5%)	(1.1%)

Note: Data are character loadings for the first four principal components (PC)—and, in parentheses at the end of the table, the percentages of variance they explain—from principal-components analysis of the correlation matrix. Analysis was based on 57 taxa of the Hipposideridae and the average of five species of *Rhinolophus (R. celebensis, R. clivosus, R. hipposideros, R. malayanus,* and *R. sinicus).*

(e.g., Novacek 1991), this regression function provided an estimate of ancestral hipposiderid morphology. In this study, the vector of character values for each hipposiderid species was regressed separately onto the means of those for *Rhinolophus celebensis, R. clivosus, R. hipposideros, R. malayanus,* and *R. sinicus.* For each ingroup taxon, the vector of residuals was retained. These vectors were combined and used in further calculations in the form of a transformed data matrix.

Another method used was a "size-out" procedure. After averaging the $\log_e$-transformed values, a correlation matrix was calculated and a principal component analysis was performed on the matrix. From this, the matrix of projections of each species of each component was calculated. The first principal component, which primarily reflected size relationships, was then deleted from the matrix, and the remaining principal component scores were taken to be character-state values for a newly created suite of characters. The foregoing calculations were made using the multiple linear regression analysis and factor analysis procedures from the SPSS for Windows package (Narušis 1993).

Phylogenetic Analyses

MAXIMUM-LIKELIHOOD METHOD.   The common-part-removed and size-out data matrices were subjected to the CONTML procedure of PHYLIP (Felsenstein 1993), which estimates phylogenies by the restricted maximum-likelihood method based on the Brownian motion model. This algorithm, used primarily for genetic distance data, makes four assumptions concerning the data: (1) the lineages evolve independently; (2) after lineages separate, their genetic (and morphometric) evolution proceeds independently; (3) drift, rather than selection, is the cause of evolutionary change; and (4) each character drifts independently. These assumptions, although never met absolutely in our study (or probably in any phylogenetic study), have been discussed more thoroughly by Felsenstein (1981) and by Bogdanowicz and Owen (1992). In each case, we used global optimization in search for the best tree, which resulted in about 15,000 tree topologies being compared each time. For both analyses, *Rhinolophus celebensis, R. clivosus, R. hipposideros, R. malayanus,* and *R. sinicus* were included to provide a root for the tree (Novacek 1991). In the common-part-removed data, *Rhinolophus* spp. were represented by a vector of zeros (Bogdanowicz and Owen 1992). Because results are dependent on the order in which the species are encountered in the data set, each analysis was repeated 50 times with different orderings of the input taxa. To compare trees and define areas of congruence, we used the majority-rule consensus procedure. The majority-rule consensus tree consists of all groups that occur more than 50% of the time (CONSENSE program of PHYLIP; Felsenstein 1993).

PARSIMONY ANALYSIS.   The analysis was based on a set of as many as 30 possibly discrete-state cranial, dental, and external characters (Appendix 2.1), scored from each adult specimen. The ingroup included 57 species and eight of the nine extant genera of the Hipposideridae. Multiple outgroup taxa were used to polarize the character states, enhancing the prospect of correctly identifying autapomorphic features in the outgroup. The characters for analysis were selected after their extensive evaluation from a large series of specimens. The outgroup was composed of the sister-family Rhinolophidae, represented by *Rhinolophus celebensis, R. coelophyllus, R. hipposideros, R. luctus, R. malayanus,* and *R. sinicus.* Megadermatidae *(Cardioderma cor, Megaderma lyra,* and *M. spasma)* and Nycteridae *(Nycteris hispida, N. grandis,* and *N. thebaica)* were used as further outgroup taxa. The Rhinopomatidae *(Rhinopoma microphyllum)* completed the outgroup (see Novacek 1991). A hypothetical ancestor of the Hipposideridae was designated, and all character states were polarized by the outgroup comparison method of Maddison et al. (1984).

Cladograms were constructed using the branch-and-bound algorithm (Hendy and Penny 1982), with the option for reconsidering an order of species that is included in PHYLIP version 3.5 under UNIX (Felsenstein 1993). As many as 31,286 most-parsimonious trees were obtained during a single run for 10,000,000 possible trees. To compare trees and define areas of congruence, we again used the majority-rule consensus procedure. In a final consensus tree, however, we also included groups that occur less than 50% of the time, working downward in their frequency of occurrence, so long as they continue to resolve the tree and do not contradict more frequent groups (CONSENSE program of PHYLIP; Felsenstein 1993). In this respect, the method is similar to the Nelson consensus method (Nelson 1979) although not identical (Felsenstein 1993). Tree lengths, consistency, and retention indices were calculated by Hennig86 version 1.5, using the *mhennig** and *bb** options (Farris 1988). These parameters may have been slightly underestimated because of an overflow of the available tree space in the software used.

Morphological Dispersion

Morphological dispersion of the fauna may be determined by calculating each taxon's average phenetic distance from every other taxon in the fauna, summing the averages, and computing the faunal average (Findley 1976; Freeman 1981; Bogdanowicz 1992). In our studies, the average taxonomic distances (NTSYS-pc package; Rohlf 1993) between every pair of species in a fauna were computed based on a matrix of standardized residuals. The Kruskall–Wallis nonparametric test was used to evaluate differences among average faunal values. Because no nonparametric multiple-range test exists for unequal sample sizes, the Mann–Whitney *U*-test was conducted on all pairwise combinations of four faunas (nonparametric tests procedure of the SPSS package; Narušis 1993). Geographic affiliations of hipposiderid groups were based on Wallace's zoogeographic divisions (Lincoln et al. 1982).

Results

Common-Part-Removed Analysis

Multiple linear regressions of character vectors of each species on that of the outgroup (Rhinolophidae) revealed

that the portion of the vector variance accounted for by the outgroup ranged from 79.6% *(Coelops robinsoni)* to 99.4% *(Hipposideros galeritus)* (see Table 2.1). Thus, the maximum-likelihood analyses were performed on residuals vectors representing from 20.4% to 0.6% of the original variance in the data from each species.

The majority consensus tree, computed from 50 original cladograms that were produced by changing the order of input taxa, indicates the existence of six relatively stable groups within the family (Figure 2.1): two of these are monotypic, two are characterized by 3–4 species, and two comprise as many as 24–25 taxa. The monotypic groups are composed of the Philippine species *Hipposideros pygmaeus* and the Australasian *Aselliscus tricuspidatus*. Within the multispecies clusters, the first one contains four small taxa (forearm length, <50 mm) from the genus *Hipposideros*, which are limited in their present distribution to New

Figure 2.1. Majority consensus cladogram from 50 maximum-likelihood analyses of common-part-removed continuous-state data. Stars mark the taxonomic groupings that appear in 95% or more of the trees. Letters that follow taxon names denote species-group membership according to Hill (1963) and Koopman (1994): A, *armiger*; B, *bicolor*; C, *cyclops*; D, *diadema*; M, *megalotis*; P, *pratti*; S, *speoris*.

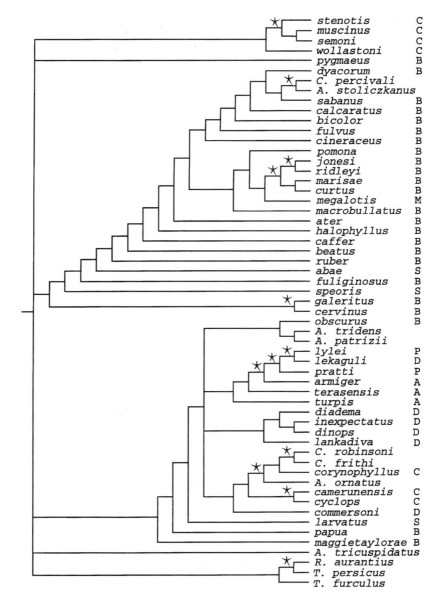

Guinea or northern Australia. The next two assemblages are formed by both *Hipposideros* and non-*Hipposideros* species from different regions of the Old World. Interestingly, all *Hipposideros* occurring in the first of these assemblages are characterized by small or medium forearm lengths, which frequently are less than 50 mm and almost never exceed 66 mm (*H. fuliginosus,* 56–64 mm; *H. abae,* 55–66 mm). The second assemblage is dominated by large and medium bats such as *H. pratti* (forearm length, 81–89 mm), *H. inexpectatus* (100–101 mm), *H. dinops* (93–97 mm), and *H. commersoni* (77–115 mm). The only exception is the small *H. obscurus* (40–52 mm) from the Philippines. The last polytypic cluster is formed by the Australian species *Rhinonycteris aurantius,* the Malagasy *Triaenops furculus,* and the primarily African *T. persicus.* All three taxa have similar noseleaf structure and greatly expanded zygoma.

Size-Free Analysis

The first principal component of $\log_e$-transformed morphometric data explains 89.8% of the variation, and all characters have high positive loadings on this component (see Table 2.2). The second through fourth components explain 3.1%, 1.5%, and 1.1%, respectively; each of the remaining components account for less than 1.0% of the variance. After removal of the first-component projections, the maximum-likelihood analysis was conducted on vectors from the remaining 44 components.

The majority consensus tree based on 50 size-free cladograms indicates very few groups that maintain stable structure under reordering of taxa in the data set (Figure 2.2). A few groups contain pairs of species morphologically similar to each other (e.g., *H. camerunensis* and *H. cyclops;*

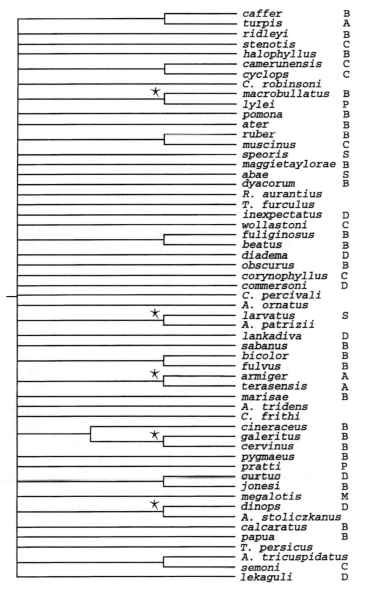

Figure 2.2. Majority consensus cladogram from 50 maximum-likelihood analyses of size-free continuous-state data. Stars mark the taxonomic groupings that appear in 95% or more of the trees. Letters that follow taxon names denote species-group membership according to Hill (1963) and Koopman (1994): A, *armiger;* B, *bicolor;* C, *cyclops;* D, *diadema;* M, *megalotis;* P, *pratti;* S, *speoris.*

and *H. galeritus* and *H. cervinus*). The majority, however, show no clear connections, even in the case of clusters with the 95% repeatability of branches (e.g., *H. macrobullatus* and *H. lylei*; and *H. dinops* and *A. stoliczkanus*).

Parsimony Analysis of Discrete-State Characters

The branch-and-bound algorithm gave 31,286 most-parsimonious trees with length of 99, consistency index of 0.32, and retention index of 0.69. In general, a low consistency index indicates that the data matrix does not "fit" the tree well (i.e., contains much homoplasy). A fairly good retention index value suggests, however, that many of the characters used are only partly homoplasious and that their transformation series show some synapomorphy for the particular tree topology. This determination is reflected in the relative stability of several clades present in the majority consensus tree, albeit this stability was exhibited mainly at the top of the tree. The affinities of the basal clades are less likely to be consistent, and it seems that at least four species could be treated as basal taxa: *Hipposideros bicolor*, *H. fulvus*, *H. semoni*, and *H. muscinus* (Figure 2.3). These hipposiderids differ from their hypothetical ancestor in that the center of the posterior border of hard palate is anterior to the posterior curvature of the palate bone (character 9) or is spiculated (character 10) (but see *H. muscinus*), and that metacarpal IV is longer than metacarpal V (character 29) (Appendix 2.2). In fact, the latter character is a synapomorphic feature of all the taxa studied, although it has been reversed in the two *Coelops* species and *H. corynophyllus*.

Above the four basal taxa, a trichotomy is formed from

Figure 2.3. Nelson-like consensus cladogram from parsimony analysis of discrete-state data. Numbers at the forks indicate the percentage of times that the group consisting of the species which are to the right of that fork occurred among the 31,286 trees. Letters that follow taxon names denote species-group membership according to Hill (1963) and Koopman (1994): A, *armiger*; B, *bicolor*; C, *cyclops*; D, *diadema*; M, *megalotis*; P, *pratti*; S, *speoris*.

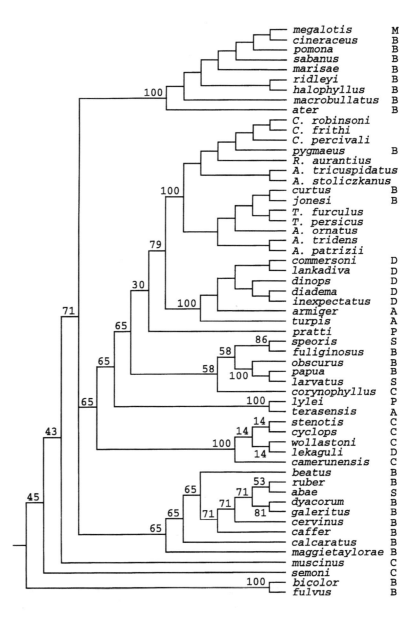

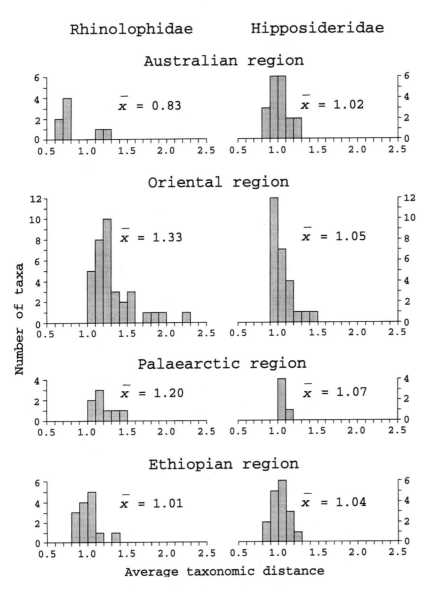

Figure 2.4. Distribution of average taxonomic distances among rhinolophid faunas (Bogdanowicz 1992) and hipposiderid faunas (this study) in four zoogeographic regions. $\bar{x}$ = overall mean taxonomic distance.

three polytypic lineages, one of which (upper lineage in Figure 2.3) can be found in all original trees. This clade comprises nine fairly small Asian and African species (forearm length, ≤48 mm), including large-eared *H. megalotis* from Kenya and Ethiopia. These species are closely related chiefly because they have one or both of the following features: a cranium that is relatively broad across the mastoids (character 4) and a foramen ovale of medium size (character 11). Both characters are partly homoplasious, and the latter especially shows strong parallelism with the second lineage (largest, central lineage), grouping all the remaining clades that are common to all competing trees. The other characteristic feature of this lineage is the presence of both *Hipposideros* and non-*Hipposideros* taxa and sister-group relationships of large hipposiderid bats (e.g., *H. commersoni, H. lankadiva, H. dinops,* and *H. inexpectatus*) and genera other than *Hipposideros*. Within the third lineage,

relationships are not stable and several different tree topologies may exist (see Figure 2.3).

Morphological Dispersion and Origin of the Hipposideridae

For hipposiderids from the Ethiopian, Palaearctic, Oriental, and Australian regions, the average taxonomic distance values range from 1.02 to 1.07 (Figure 2.4) and the differences among the four faunas are not statistically significant (Kruskall–Wallis H-test, $\alpha = 0.05$). Pairwise comparisons also indicate that none of the studied faunas is more dispersed than another, although a nearly significant difference was observed between the Palaearctic and Australian faunas (Mann–Whitney U-test, $p = 0.051$), with the Palaearctic fauna being more dispersed morphologically.

In contrast, the distribution of average values is different in different faunas. The Ethiopian and Australian regions are characterized by values more or less symmetrically distributed (skewness, 0.25 and 0.97, respectively), whereas those distributions in the Oriental and Palaearctic regions are skewed to the right (skewness, 1.52 and 1.88, respectively). These results suggest that the Oriental and Palaearctic faunas are composed of a majority of species that morphologically are close to their nearest neighbor. The most distinctive species is *Coelops robinsoni* (average taxonomic distance, 1.41) from the Oriental region, which together with *C. frithi* can easily be identified by the presence of a rudimentary tail and unusually short ears.

Discussion

Phylogenetic Relationships among Hipposiderids and Monophyly of the Genus *Hipposideros*

For the two metrical data sets used in this study, the transformation to remove the common part resulted in significantly higher stability of clades than did the transformation to remove size. The relationships suggested by the common-part-removed cladogram were more or less in good agreement with Hill's (1963) arrangements of *Hipposideros* taxa, grouping the majority of bats from the *bicolor* group into one clade and those from *diadema*, *pratti*, and *armiger* into the other (see Figure 2.1). This was not the case with the size-free majority consensus tree (see Figure 2.2), where even the clades with the highest repeatability frequently contained taxa without clear phylogenetic connections.

Interestingly, evolutionary affinities within the sister-family Rhinolophidae were also better explained by the common-part-removed cladogram than by the size-free cladogram (Bogdanowicz and Owen 1992). Wood (1983) reached similar conclusions, although about phenetic relationships, in his studies of storks (Ciconiidae) and cranes (Gruidae). In combination, these studies suggested that the use of the common-part-removed method may be applicable to the phylogenetic classifications of at least those vertebrates, such as bats and birds, that exhibit determinate growth.

The discrete-state consensus cladogram (see Figure 2.3) did not corroborate current systematic arrangements, and several taxa traditionally thought to be close systematically, such as those from the *bicolor* group, occurred in different clades; this may have resulted from a lack of sufficient material. A low character-to-taxon ratio (30:57) and the absence of some data (see Appendix 2.2) reduce the support for clades that include relatively incomplete taxa. On the other hand, the obtained consistency index values of original most-parsimonious trees, given the large number of

taxa in the analysis, are only slightly less than the expected value (0.32 versus 0.34). Missing data, however, may also mask the presence of homoplasy and give higher consistency index values for matrices, with many cells scored as "?" (Sanderson and Donoghue 1989).

Despite these possible limitations, the consensus cladograms derived from the common-part-removed and discrete-state matrices have several features in common. First, both confirm close phylogenetic relationships between members of the *diadema* and *armiger* groups (Hill 1963). Second, the three species from the *speoris* group are much like certain species from the *bicolor* group. In our opinion, the taxonomic status of both groups needs to be redefined and revised (see also Kock and Bhat 1994). Third, large-eared *H. (Syndesmotis) megalotis* of the *megalotis* group, which is believed to be a single living descendant of the middle Miocene *Syndesmotis* lineage (Legendre 1982), constitutes a clade together with some bats of the *bicolor* group (see Figures 2.1 and 2.3) belonging to the subgenus *Hipposideros*. Such a phylogenetic position of *H. megalotis* contradicts both its subgeneric and its supraspecific group validity (see Hill 1963; Legendre 1982). Fourth, primarily African *Asellia* species and Southeast Asian *Coelops* species, as well as *Anthops ornatus* from the Solomon Islands, consistently occur within the clades composed of the *Hipposideros* taxa. In our opinion, this suggests that the genus *Hipposideros* does not comprise all the descendants of an ancestor and most probably should be treated as a paraphyletic group.

On the basis of fossil evidence, Legendre (1982) noted that species of the extinct subgenus *Hipposideros (Pseudorhinolophus)* are morphologically similar to some large Recent taxa, such as *H. armiger*, *H. diadema*, and *H. commersoni*, and could be ancestral to *Asellia* (see also Sigé 1968). Both consensus cladograms (see Figures 2.1 and 2.3) may support this hypothesis.

The status of the other non-*Hipposideros* genera seems to be more complicated. *Rhinonycteris aurantius* is thought to be closely related to *Brachipposideros nooraleebus* from the middle Miocene deposits from Riversleigh, Australia (Sigé et al. 1982; Hand 1993; Archer et al. 1994). The common-part-removed and discrete-state cladograms do not contradict this interpretation, although the position of *Rhinonycteris* in the second of these cladograms may indicate its closer relationships with large rather than small *Hipposideros* species. In the light of the microcomplement fixation transferrin data, however, *Rhinonycteris* is outside both the *Hipposideros* and *Rhinolophus* genera (Pierson 1986). Its distance from these genera (103 and 110 units, respectively) was substantially greater than the distance between them (74 and 84 units). *Aselliscus* was much closer (44 versus 81 units) to *Rhinolophus* than to *Hipposideros,* and about the same distance from *Hipposideros* as was *Rhinolophus* (81 and 84 units, respectively). Evidently, the results of morphological and immunological data may not be comparable at all levels in the phylogeny. Optimum resolution at different evolutionary levels by albumin, electrophoretic, chromosomal, and morphological characters was shown by Arnold et al. (1982) in a study on phyllostomoid bats.

Morphological Dispersion, Cladograms, and Center of Origin

Findley (1976) reasoned that older bat faunas are phenotypically more diverse. Our analysis indicated that, unlike rhinolophids (Bogdanowicz and Owen 1992), hipposiderids have no significant differences in morphological dispersion among the Ethiopian, Palaearctic, Oriental, and Australian regions (see Figure 2.4). Our results suggest (albeit weakly) that the Palaearctic fauna might be older than the Australian fauna (Mann–Whitney U-test, $p = 0.051$), and this hypothesis, while considering the age of fossil taxa, agrees with palaeontological findings.

The maximum-likelihood cladograms do not give any constructive suggestion about the center of origin for the family, because at least six clades might be ancestral (see Figures 2.1 and 2.2). On the other hand, the most basal hipposiderids on the consensus cladogram derived from the discrete-state data (Figure 2.3) are recently known from Afghanistan, Pakistan, India, and Sri Lanka *(H. fulvus),* and from Thailand, Malaysia, Indonesia, and the Philippines *(H. bicolor).* The next two basal species *(H. semoni* and *H. muscinus)* are limited in their present distribution to New Guinea or northeastern Queensland in Australia (Corbet and Hill 1992; Koopman 1994). However, it is generally assumed that Australia was colonized by bats migrating from Asia rather than from South America (Hamilton-Smith 1975; Flannery 1989; Hand et al. 1994), although their appearance in Australia predates the final breakup of Gondwana (Hand et al. 1994). The arrival of hipposiderids into New Guinea most probably occurred during the Miocene *(Hipposideros* spp.) and Pliocene *(Aselliscus tricuspidatus)* (Flannery 1990).

To date it has been generally accepted that the family originated somewhere in the Old World tropics, probably in Africa or Asia (Koopman 1970; Sigé 1991), and only recently Hand et al. (1994) suggested, although indirectly, that it may have evolved in the Southern Hemisphere. The Tertiary karstic fissure-fills of western Europe show that from the late Eocene to at least the middle Oligocene, hipposiderids were the most diverse and numerous group of the cave-dwelling bats and that their distribution in the past was

wider than it is today (Hand 1984; Sevilla 1990). Their oldest remains are known from the late middle Eocene of Europe (e.g., Sigé and Legendre 1983; Sigé 1991). Based on fossil evidence, hipposiderids were present in Arabo-Africa and Australia by the early and late Oligocene (Archer et al. 1994; Sigé et al. 1994), respectively. In contrast, their known fossil remains in the Oriental region are very few and relatively young, dating back only to the Neogene (Hand 1984, p. 883; K. F. Koopman, in litt.). In our opinion, however, the lack of older Oriental material should provisionally be treated as an artifactual product of the general unavailability of well-examined fossil material from this region, because we find support from the neontological data to indicate that the Hipposideridae, like their sister-family Rhinolophidae (Bogdanowicz and Owen 1992), most probably originated in Asia, not in Africa (but see Sigé 1991). This hypothesis of origin would also be consistent with the paleoclimatic evidence of tropical rainforest development in Southeast Asia during the Tertiary (Heaney 1991), as it is generally agreed that the family Hipposideridae would have developed and radiated in such conditions (B. Sigé, in litt.).

Karyology and Phylogenetic Relationships

As far as we are aware, karyotypes of 22 hipposiderid species have been described (Table 2.3). Six diploid chromosome numbers have been encountered for these species, with only two within the genus *Hipposideros* ($2n = 32, 52$).

It is very important for phylogenetic considerations to determine the mode and direction of karyotypic change in the family studied. First, at the level of nondifferentially stained karyotypes, the genus *Hipposideros* shows considerable karyotypic conservatism, and for all but one species, $2n = 32$ and the number of autosomal arms (FN) = 60. Rautenbach et al. (1993) suggested that the chromosomal complement of *H. commersoni* ($2n = 52$, FN = probably 60) may have evolved from the most common hipposiderid state by 10 centric fissions, producing change in diploid number but not in fundamental number. On the basis of data on the G-banded chromosomes, Sreepada et al. (1993) proposed that the ancestral lineage of *Hipposideros* derived from a rhinolophoid ancestor whose karyotype was similar to that of *R. luctus* ($2n = 32$, FN = 60; Naidu and Gururaj 1984; Harada et al. 1985b; Hood et al. 1988). However, direct comparison of the banding pattern in chromosomes of *R. luctus* and *Hipposideros* spp. is not available. Homology of these karyotypes is thus not clear, and the arm combination in metacentric autosomes may differ.

Second, karyological data suggest that the autosomes for the ancestral karyotype in the sister-family Rhinolophidae contained all acrocentric elements with $2n = 62$ and FN =

Table 2.3

Synopsis of Karyotypes of Hipposiderids

Species	2n	FN	X	Y	Reference
Asellia tridens	50	62	SM	A	Baker et al. 1974
Aselliscus stoliczkanus	30	56	SM	A	Harada et al. 1985a
Cloeotis percivali	40	?	?	?	Rautenbach et al. 1993
Coelops frithi	30	56	ST	ST	Andō et al. 1980
Hipposideros armiger	32	60	SM	ST	Hood et al. 1988; Qumsiyeh et al. 1988
Hipposideros ater	32	60	SM	A	Ray-Chaudhuri et al. 1971; Sreepada et al. 1993
Hipposideros bicolor	32	60?	?	?	Ray-Chaudhuri and Pathak 1966
Hipposideros caffer	32	60	ST	A	Peterson and Nagorsen 1975
	32	60	SM	A	Rautenbach et al. 1993
	32	60	?	?	Dulić and Mutere 1974
Hipposideros cervinus	32	60	SM	A	Harada and Kobayashi 1980
Hipposideros cineraceus	32	60	SM	A	Sreepada et al. 1993
Hipposideros commersoni	52	60?	?	?	Rautenbach et al. 1993
Hipposideros diadema	32	60	SM	A	Harada and Kobayashi 1980
Hipposideros fulvus	32	60	SM	A	Ray-Chaudhuri et al. 1971; Harada et al. 1985a; Hood et al. 1988
	32	60	M	A	Sreepada et al. 1993
	32	60	M	ST	Handa and Kaur 1980
Hipposideros hypophyllus	32	60	M	A	Sreepada et al. 1993
Hipposideros lankadiva	32	60	M	A	Sreepada et al. 1993
	32	60	M	ST	Handa and Kaur 1980
Hipposideros larvatus	32	60	SM	ST	Harada et al. 1982; Hood et al. 1988
Hipposideros lekaguli	32	60	SM	ST	Harada et al. 1982; Hood et al. 1988
Hipposideros pratti	32	60	M	ST	Zhang 1985
Hipposideros speoris	32	60	ST	A	Sreepada et al. 1993
	32	60?	SM	A	Dulić 1984 (Figure 3)
	32	60	SM	ST	Handa and Kaur 1980
Hipposideros terasensis	32	60	ST	A	Andō et al. 1980
Hipposideros turpis	32	60	ST	A	Andō et al. 1980
Triaenops persicus	36	60	M	ST	Dulić and Mutere 1977

Notes: 2n, diploid number of chromosomes; FN, total number of autosomal arms; X, Y, sex chromosomes (states: A, acrocentric; M, metacentric; SM, submetacentric; ST, subtelocentric). The karyotypes of *Hipposideros cervinus, H. hypophyllus,* and *H. terasensis* were originally described as those of *H. galeritus labuanensis, H. pomona,* and *H. armiger terasensis,* respectively (reviewed by Jenkins and Hill 1981; Yoshiyuki 1991; Kock and Bhat 1994).

60 (summarized by Zima et al. 1992; see also Bogdanowicz and Owen 1992). This interpretation also is supported by phylogenetic analyses of morphological characters (Bogdanowicz and Owen 1992). An acrocentric composition of the primitive karyotype, and the trend toward low diploid chromosome numbers, have also been suggested for the families Phyllostomidae and Vespertilionidae (Baker and Bickham 1980). Third, a newly described fossil genus *Vaylatsia,* although a member of the Hipposideridae, probably represents the stem group of *Rhinolophus* (Sigé 1990).

In light of these considerations, it seems that the autosomes of the common ancestor of Rhinolophidae and Hipposideridae consisted of all acrocentrics rather than metacentrics and that the trend was toward low, rather than high, diploid numbers. A comparison between the G-banded chromosomes of *R. acuminatus* and *H. armiger* also indicated that some non-Robertsonian processes must have occurred during the evolution from the ancestral karyotype to the karyotypes we find today (Qumsiyeh et al. 1988). This assessment does not contradict phylogenetic relationships suggested by the common-part-removed cladogram and seems to be in partial agreement with the positions of mainly non-*Hipposideros* species shown in the discrete-state cladogram. Their karyotypes have probably undergone extensive Robertsonian and non-Robertsonian changes, which resulted in the variable number of chromosomes (2n = 30–50) and autosomal arms (FN = 56–62) (see Table 2.3).

Conclusions

Several standard and novel analyses of a morphological data set, supplemented with karyotypic information, were used in search of a robust hypothesis for the phylogeny and the center of origin of the bat family Hipposideridae. The results suggest that phylogenetic affinities among Recent species are not expressed accurately by current systematic arrangements based on Hill's (1963) supraspecific groupings and that the genus *Hipposideros* might be a paraphyletic taxon. Morphological dispersion analysis failed to reveal any significant differences in morphological diversification among hipposiderid faunas from four zoogeographic regions, and no center of origin could be inferred from the analyses of phenetic data. From phylogenetic evidence, however, it appears that the family most probably originated somewhere in the Oriental region. Our analyses also suggest that the common ancestor of the sister-families Rhinolophidae and Hipposideridae had all acrocentric rather than metacentric autosomes, and that these families independently followed a pattern of Robertsonian fusions, with the result that low diploid numbers are the derived condition in both lineages.

The phylogenetic relationships suggested by this study (Figures 2.1 and 2.3) are supported by metric and nonmetric morphological data and should be considered tentative, as working hypotheses. Future study must adopt the total-evidence approach, combining morphological, genetic, and biochemical character sets and determining the most-parsimonious outcome of the pooled matrix. We still have a long way to go in the field of hipposiderid phylogeny, but, like Theseus in the Minotaur's labyrinth, we have few landmarks to guide us.

Appendix 2.1.
The 30 Characters in the Parsimony Analysis of the Hipposideridae

1. Hornlike crest in middle of dorsal part of premaxillae: (0) absent; (1) present.
2. Location of greatest neurocranial breadth, dorsal view: (0) anterior cranium or middle of cranium; (1) posterior cranium.
3. Position of braincase, excluding sagittal crest: (0) evidently higher than rostrum; (1) almost as high as or lower than rostrum.
4. Distance between mastoids: (0) less than or equal to zygomatic breadth; (1) greater than zygomatic breadth.
5. Zygoma, lateral view: (0) not expanded at all to moderately expanded; (1) expanded into a wide plate.
6. Anterior end of sagittal crest: (0) extends to interorbital constriction; (1) extends past interorbital constriction.
7. Lambdoidal crest: (0) absent or weak; (1) extremely well-developed.
8. Perforations behind nasal swellings: (0) zero to five foramina present; (1) more than five foramina present.
9. Location of central portion of posterior border of hard pal-

ate, ventral view: (0) behind or at the level of posterior curvature of palate bone; (1) anterior to posterior curvature of palate bone.
10. Posterior nasal spine (i.e., spicule at the center of posterior edge of hard palate): (0) absent or inconspicuous; (1) well-developed.
11. Foramen ovale: (0) small; (1) medium—about half the size of glenoid fossa; (2) almost as large as glenoid fossa. The linear character transformation is hypothesized to be $0 > 1 > 2$.
12. Type of cochlea: (0) phanaerocochlear; (1) cryptocochlear. This character and its states correspond to those distinguished by Novacek (1991), although some of our findings differ; we found both states present in more species, and a state different from that reported by Novacek (1991) for several species.
13. Least basioccipital width: (0) less than or equal to the least width of the sphenoidal bridge; (1) greater than the least width of the sphenoidal bridge.
14. Foramen magnum: (0) elliptical; (1) oval.
15. Shape of hamular process of the pterygoids, lateral view: (0) strongly notched; (1) weakly or not at all indented.
16. Size of hamular process of the pterygoids: (0) long, practically reaching glenoid fossa; (1) short.
17. Anterior edge of the ascending mandibular ramus: (0) posterior to or at the middle of last upper molar; (1) anterior to the middle of last upper molar.
18. Position of mental foramen, lateral view: (0) anterior to or in the middle of first premolar; (1) posterior to the middle of first premolar.
19. Bone connection between angular and condyloid processes, lateral view: (0) strongly notched; (1) shallow.
20. Posterior cusp on upper canines (usually one-quarter to one-half the height of canine): (0) absent; (1) present.
21. Heel of second upper molar: (0) well-developed; (1) inconspicuous.
22. Shape of third upper molar, buccal view: (0) pentagonal; (1) triangular.
23. Anterior segment (parastyle–paracone–mesostyle triangle) of third upper molar, occlusal view: (0) almost as large as that of the second upper molar; (1) much smaller than that of the second upper molar.
24. Lobes on the lower inner incisors: (0) three full lobes; (1) two full lobes plus a third, rudimentary lobe; (2) only two lobes. The linear character transformation is hypothesized to be $0 > 1 > 2$.
25. Diastema between first lower incisors: (0) absent; (1) present.
26. Diastema between the last lower incisors and lower canines: (0) absent; (1) present.
27. Horizontal ribs on outer parts of ears: (0) present; (1) absent.
28. Third metacarpal: (0) shorter than fifth metacarpal; (1) longer than fifth metacarpal.
29. Fourth metacarpal: (0) shorter than fifth metacarpal; (1) longer than fifth metacarpal.
30. Tail: (0) well-developed; (1) rudimentary.

Appendix 2.2.
Distribution of States for the 30 Characters within Taxa of the Family Hipposideridae

Characters and their states are as defined in Appendix 2.1. For all characters, state 0 is plesiomorphic, and states 1 and 2 are apomorphic. In the data matrix shown here, the letter "B" indicates that both states 0 and 1 are present (this was coded as "?" in Hennig86), and a question mark means that the state is unknown.

Taxon	1	2	3	4	5	6	7	8	9	10	11	12	13	14	15	16	17	18	19	20	21	22	23	24	25	26	27	28	29	30
Ancestor	0	0	0	0	0	0	0	0	0	0	0	0	0	0	0	0	0	0	0	0	0	0	0	0	0	0	0	0	0	0
Anthops ornatus	0	0	0	0	0	0	0	1	0	1	0	0	0	0	0	1	0	0	0	1	0	1	0	0	0	0	0	0	1	1
Asellia patrizii	0	0	1	0	0	1	0	1	1	0	0	0	0	0	0	0	0	0	1	0	1	1	0	0	0	0	0	1	1	0
Asellia tridens	0	0	1	0	0	1	0	1	1	0	0	B	0	0	0	0	0	0	0	1	1	1	1	0	0	0	0	1	1	0
Aselliscus stoliczkanus	0	0	0	0	0	0	0	0	0	1	0	0	1	0	1	0	0	1	1	0	0	0	0	0	0	1	0	1	1	0
Aselliscus tricuspidatus	0	0	0	0	0	0	0	0	0	1	0	0	B	0	0	0	0	1	1	0	0	0	0	0	0	B	0	1	1	0
Cloeotis percivali	1	0	0	0	0	0	0	1	0	1	2	0	1	0	0	1	0	1	1	1	1	0	0	0	0	0	1	1	1	0
Coelops frithi	0	1	0	0	0	0	0	1	1	B	2	0	0	1	0	0	0	B	0	1	0	0	0	0	0	1	1	0	0	1
Coelops robinsoni	0	1	0	0	0	0	0	1	1	1	2	0	0	B	0	0	0	1	0	1	0	0	0	0	0	1	1	0	0	1
Hipposideros abae	0	0	0	0	0	0	0	1	0	0	0	1	0	0	0	0	0	0	1	0	0	0	0	0	0	0	0	1	1	0
Hipposideros armiger	0	0	0	0	0	1	0	1	0	0	B	0	0	0	0	0	0	0	0	1	0	0	0	0	0	0	0	1	1	0
Hipposideros ater	0	0	0	1	0	0	0	0	0	0	0	1	1	0	0	0	0	0	0	0	0	0	0	0	0	0	0	0	1	0
Hipposideros beatus	0	0	0	0	0	0	0	0	0	0	0	0	1	0	0	?	0	0	0	1	0	1	0	0	0	0	0	1	1	0
Hipposideros bicolor	0	0	0	0	0	0	0	1	0	0	0	B	0	0	0	0	0	0	0	0	0	0	0	0	0	0	0	0	1	0
Hipposideros caffer	0	0	1	0	0	0	0	0	0	0	B	1	0	0	0	0	0	0	1	0	0	0	0	0	0	B	0	1	1	0
Hipposideros calcaratus	0	0	0	0	1	0	0	0	0	0	0	1	0	0	0	0	0	0	1	0	0	0	0	0	0	0	0	0	1	0
Hipposideros camerunensis	0	0	0	0	0	0	0	0	0	1	0	0	0	0	0	0	0	0	0	0	0	0	0	0	B	0	0	1	1	0
Hipposideros cervinus	0	0	0	0	0	0	0	0	0	1	0	0	1	0	0	0	0	0	1	0	0	0	0	0	0	0	0	1	1	0
Hipposideros cineraceus	0	0	0	1	0	0	0	0	1	0	0	0	0	0	0	0	0	0	0	0	0	0	0	0	0	0	0	0	1	0
Hipposideros commersoni	0	0	1	0	0	1	1	0	1	0	0	B	0	0	0	B	0	0	0	0	1	0	0	B	0	0	0	1	1	0
Hipposideros corynophyllus	0	0	0	0	0	0	0	0	0	0	0	0	0	0	?	0	1	0	0	0	0	1	0	0	0	0	0	1	0	0
Hipposideros curtus	0	0	1	0	0	0	0	1	1	1	0	B	1	0	0	0	0	0	1	1	0	0	0	0	0	0	0	0	1	0
Hipposideros cyclops	0	0	0	0	0	0	0	1	0	0	0	1	0	0	0	0	0	0	0	0	0	0	0	0	1	0	0	1	1	0
Hipposideros diadema	0	0	1	0	0	0	1	0	0	0	0	0	0	0	0	0	0	0	0	1	0	1	0	0	0	0	0	1	1	0
Hipposideros dinops	0	0	1	0	0	1	1	0	1	0	0	0	0	0	1	0	B	0	0	0	0	1	0	0	0	0	0	1	1	0
Hipposideros dyacorum	0	0	0	0	0	0	0	1	0	0	0	1	0	0	0	0	0	0	0	0	0	0	0	0	0	0	0	1	1	0
Hipposideros fuliginosus	0	0	0	0	0	0	0	1	0	0	B	1	0	0	0	0	0	0	1	0	1	0	0	0	0	0	0	1	1	0
Hipposideros fulvus	0	0	0	0	0	0	0	1	0	0	0	0	0	0	0	0	0	0	0	0	0	0	0	0	0	0	0	0	1	0
Hipposideros galeritus	0	0	0	0	0	0	0	1	0	0	B	?	0	0	0	0	0	0	0	0	0	0	0	0	0	0	0	1	1	0
Hipposideros halophyllus	0	0	1	0	0	0	0	1	0	2	0	1	0	?	0	0	0	0	0	0	0	0	0	0	0	0	0	1	1	0
Hipposideros inexpectatus	0	0	1	0	0	1	1	0	0	1	0	0	0	0	1	0	0	0	0	0	1	0	1	0	0	0	0	1	1	0
Hipposideros jonesi	0	0	1	0	0	0	1	1	1	1	0	1	0	0	0	0	0	1	1	0	0	0	0	0	0	0	0	1	1	0
Hipposideros lankadiva	0	0	1	0	0	1	1	0	0	0	0	0	0	0	0	0	0	0	0	0	0	1	0	0	0	0	0	1	1	0
Hipposideros larvatus	0	0	0	0	0	0	0	1	1	0	0	0	0	0	0	0	1	0	0	0	0	1	0	1	0	0	0	1	1	0
Hipposideros lekaguli	0	0	0	0	0	0	0	1	0	0	0	0	0	0	0	0	0	0	0	0	0	0	0	0	0	0	0	1	1	0
Hipposideros lylei	0	0	0	0	0	1	0	0	0	0	0	0	0	0	0	0	0	0	0	0	0	0	0	0	0	0	0	1	1	0
Hipposideros macrobullatus	0	0	0	1	0	0	0	0	0	0	1	0	1	1	0	0	0	0	0	0	0	0	0	0	0	0	0	0	1	0
Hipposideros maggietaylorae	0	0	0	0	0	0	0	1	0	0	0	1	0	0	0	0	0	0	1	0	0	0	0	0	0	0	0	0	1	0
Hipposideros marisae	0	0	1	0	0	0	0	1	1	1	0	1	0	?	0	0	0	0	0	0	0	0	0	0	0	0	0	0	1	0
Hipposideros megalotis	0	0	1	0	0	0	0	1	0	1	0	0	0	0	0	0	0	1	0	0	0	0	0	0	B	0	0	0	1	0
Hipposideros muscinus	?	0	0	0	0	0	0	0	0	0	0	0	0	0	0	0	0	0	0	0	0	0	0	0	0	0	0	0	1	0

Taxon (continued)	Character																													
	1	2	3	4	5	6	7	8	9	10	11	12	13	14	15	16	17	18	19	20	21	22	23	24	25	26	27	28	29	30
Hipposideros obscurus	0	0	0	0	0	0	0	0	1	0	B	0	0	0	0	1	0	0	0	0	1	0	0	0	0	0	0	1	1	0
Hipposideros papua	0	0	0	0	0	0	0	1	1	0	0	0	0	0	0	B	0	0	1	0	1	0	0	0	B	0	0	1	1	0
Hipposideros pomona	0	0	0	1	0	0	0	0	1	0	1	0	0	0	?	0	0	0	0	0	0	0	0	0	0	0	0	0	1	0
Hipposideros pratti	0	0	0	0	0	0	0	0	0	0	0	0	1	0	0	0	0	0	0	0	0	1	0	0	0	0	0	1	1	0
Hipposideros pygmaeus	0	0	0	1	0	0	0	1	0	1	2	0	0	0	0	0	0	0	0	1	0	0	0	0	0	0	0	1	1	0
Hipposideros ridleyi	0	0	0	1	0	1	0	0	1	0	1	0	1	0	0	0	0	0	0	0	0	0	0	0	0	0	0	0	1	0
Hipposideros ruber	0	0	0	0	0	0	0	0	1	0	0	0	1	0	0	0	0	0	0	1	0	0	0	0	0	0	0	1	1	0
Hipposideros sabanus	0	0	0	0	0	0	0	1	1	0	1	0	1	0	?	0	?	0	B	0	0	0	1	0	0	0	0	0	1	0
Hipposideros semoni	0	0	0	0	0	0	0	0	0	1	0	0	0	0	?	0	0	0	0	0	0	0	0	0	0	0	0	0	1	0
Hipposideros speoris	0	0	0	0	0	0	0	0	1	0	0	0	1	0	0	0	1	0	0	1	0	0	0	0	0	0	0	1	1	0
Hipposideros stenotis	0	0	0	0	0	0	0	0	?	1	0	0	0	0	?	0	?	0	0	0	0	0	0	0	0	1	0	1	1	0
Hipposideros terasensis	0	0	0	0	0	0	1	0	0	0	0	0	0	0	0	0	0	0	0	0	B	0	0	0	0	0	0	1	1	0
Hipposideros turpis	0	0	0	0	0	0	0	0	1	0	0	0	0	0	?	0	0	0	0	0	0	0	1	0	0	0	0	1	1	0
Hipposideros wollastoni	0	0	0	0	0	0	0	0	0	1	0	0	0	0	0	0	0	0	0	0	0	0	0	0	0	0	0	1	1	0
Rhinonycteris aurantius	1	0	0	0	1	0	0	0	0	1	1	0	0	0	0	0	0	0	0	1	0	0	0	0	0	0	0	1	1	0
Triaenops furculus	1	0	1	0	1	B	0	1	0	1	0	0	0	0	0	0	0	0	0	1	0	0	0	2	0	0	0	1	1	0
Triaenops persicus	1	0	1	0	1	B	0	1	0	1	0	0	0	0	0	0	0	0	0	1	0	0	0	2	0	0	0	1	1	0

Acknowledgments

We acknowledge S. J. Hand, K. F. Koopman, J. Juste-B., S. Kasper, T. Ge, and L. L. Gordon for critically reading earlier drafts of this paper. We are also grateful to J. Zima, M. Volleth, and K.-G. Heller for their detailed comments about the karyology section. J. Felsenstein kindly gave us the information necessary to modify his programs in the PHYLIP package. We also thank S. Kasper for running many of the likelihood analyses. Special thanks are extended to the following institutions and curators who made material in their care available to us: American Museum of Natural History, New York (N. B. Simmons and K. F. Koopman); Harrison Zoological Museum, Sevenoaks, Great Britain (D. L. Harrison and P. J. J. Bates); Instituut voor Taxonomische Zoölogie, Amsterdam (P. J. H. van Bree); Natural History Museum of Los Angeles County, Los Angeles, California (S. B. George); Muséum National d'Histoire Naturelle, Paris (M. Tranier); National Museum of Natural History, Smithsonian Institution, Washington, D.C. (M. D. Carleton and C. O. Handley, Jr.); Nationaal Natuurhistorisch Museum, Leiden, the Netherlands (C. Smeenk); Natural History Museum, London (P. D. Jenkins); Royal Ontario Museum, Toronto (J. L. Eger); and Senckenberg Museum, Frankfurt am Main, Germany (G. Storch).

This research was supported by fellowships to the first author from the Fulbright Foundation and the Royal Society of London–Polish Academy of Sciences exchange program. Travel grants to New York and London were kindly provided by the American Museum of Natural History and the Mammal Research Institute of the Polish Academy of Sciences.

Literature Cited

Andō, K., T. Tagawa, and T. A. Uchida. 1980. Karyotypes of Taiwanese and Japanese bats belonging to the families Rhinolophidae and Hipposideridae. Cytologia (Tokyo) 45:423–432.

Archer, M., S. J. Hand, and H. Godthelp. 1994. Riversleigh. The Story of Animals in Ancient Rainforests of Inland Australia. Reed Books, Sydney.

Arnold, M. L., R. L. Honeycutt, R. J. Baker, V. M. Sarich, and J. K. Jones, Jr. 1982. Resolving a phylogeny with multiple data sets: A systematic study of phyllostomoid bats. Occasional Papers of the Museum, Texas Tech University 77:1–15.

Baker, R. J., and J. W. Bickham. 1980. Karyotypic evolution in bats: Evidence of extensive and conservative chromosomal evolution in closely related taxa. Systematic Zoology 29:239–253.

Baker, R. J., B. L. Davis, R. G. Jordan, and A. Binous. 1974. Karyotypic and morphometric studies of Tunisian mammals: Bats. Mammalia 38:695–710.

Bogdanowicz, W. 1992. Phenetic relationships among bats of the family Rhinolophidae. Acta Theriologica 37:213–240.

Bogdanowicz, W., and R. D. Owen. 1992. Phylogenetic analyses of the bat family Rhinolophidae. Zeitschrift für Zoologische Systematik und Evolutionsforschung 30:142–160.

Corbet, G. B., and J. E. Hill. 1991. A World List of Mammalian Species, 3rd Ed. Natural History Museum Publications and Oxford University Press, London.

Corbet, G. B., and J. E. Hill. 1992. The Mammals of the Indomalayan Region. Oxford University Press, Oxford.

Dulić, B. 1984. Chromosomes of four Indian species of Microchiroptera. *In* Proceedings of the Sixth International Bat Research Conference, E. E. Okon and A. E. Caxton-Martins, eds., pp. 25–37. University of Ife Press, Ile-Ife.

Dulić, B., and F. A. Mutere. 1974. The chromosomes of two bats from East Africa: *Rhinolophus clivosus* Cretzschmar, 1828 and *Hipposideros caffer* (Sundevall, 1846). Periodicum Biologorum 76:31–34.

Dulić, B., and F. A. Mutere. 1977. Chromosomes of some East African bats. Säugetierkundliche Mitteilungen 25:231–233.

Farris, J. S. 1988. Hennig86, version 1.5. Computer program and documentation. Port Jefferson Station, N.Y.

Felsenstein, J. 1981. Evolutionary trees from gene frequencies and quantitative characters: Finding maximum likelihood estimates. Evolution 35:1229–1242.

Felsenstein, J. 1993. PHYLIP (Phylogeny Inference Package), version 3.5p. Software and documentation. University of Washington, Seattle.

Findley, J. S. 1976. The structure of bat communities. American Naturalist 110:129–139.

Flannery, T. F. 1989. Origins of the Australo-Pacific land mammal fauna. Australian Zoological Reviews 1:15–24.

Flannery, T. F. 1990. Mammals of New Guinea. Robert Brown, Carina, Queensland, Australia.

Freeman, P. W. 1981. A multivariate study of the family Molossidae (Mammalia, Chiroptera): Morphology, ecology, evolution. Fieldiana Zoology 1316:1–173.

Hamilton-Smith, E. 1975. Gondwanaland and the Chiroptera. Journal of the Australian Mammal Society 1:382–383.

Hand, S. 1984. Bat beginnings and biogeography: A southern perspective. *In* Vertebrate Zoogeography and Evolution in Australia: Animals in Space and Time, M. Archer and G. Clayton, eds., pp. 853–904. Hesperian Press, Carlisle, Australia.

Hand, S. J. 1993. First skull of a species of *Hipposideros (Brachipposideros)* (Microchiroptera: Hipposideridae), from Australian Miocene sediments. Memoirs of the Queensland Museum 33:179–182.

Hand, S., M. Novacek, H. Godthelp, and M. Archer. 1994. First Eocene bat from Australia. Journal of Vertebrate Paleontology 14:375–381.

Handa, S. M., and S. Kaur. 1980. Chromosome studies on three species of *Hipposideros* (Hipposideridae: Chiroptera). Caryologia 33:537–549.

Harada, M., and T. Kobayashi. 1980. Studies on the small mammal fauna of Sabah, East Malaysia. II. Karyological analysis of some Sabahan mammals (Primates, Rodentia, Chiroptera). Contributions from the Biological Laboratory, Kyoto University 26:83–95.

Harada, M., M. Minezawa, S. Takada, S. Yenbutra, P. Nunpakdee, and S. Ohtani. 1982. Karyological analysis of 12 species of bats from Thailand. Caryologia 35:269–278.

Harada, M., S. Yenbutra, K. Tsuchiya, and S. Takada. 1985a. Karyotypes of seven species of bats from Thailand (Chiroptera, Mammalia). Experientia (Basel) 41:1610–1611.

Harada, M., S. Yenbutra, T. H. Yoshida, and S. Takada. 1985b. Cytogenetical study of *Rhinolophus* bats (Chiroptera, Mammalia) from Thailand. Proceedings of the Japan Academy 61B: 455–458.

Heaney, L. R. 1991. A synopsis of climatic and vegetational change in Southeast Asia. Climatic Change 19:53–61.

Hendy, M. D., and D. Penny. 1982. Branch and bound algorithms to determine minimal evolutionary trees. Mathematical Biosciences 59:277–290.

Hill, J. E. 1963. A revision of the genus *Hipposideros*. Bulletin of the British Museum (Natural History) Zoology 11:1–129.

Hill, J. E. 1982. A review of the leaf-nosed bats *Rhinonycteris*, *Cloeotis*, and *Triaenops* (Chiroptera: Hipposideridae). Bonner Zoologische Beiträge 33:165–186.

Hood, C. S., D. A. Schlitter, J. I. Georgudaki, S. Yenbutra, and R. J. Baker. 1988. Chromosomal studies of bats (Mammalia: Chiroptera) from Thailand. Annals of Carnegie Museum 57:99–109.

Jenkins, P. D., and J. E. Hill. 1981. The status of *Hipposideros galeritus* Cantor, 1846 and *Hipposideros cervinus* (Gould, 1854) (Chiroptera: Hipposideridae). Bulletin of the British Museum (Natural History) Zoology 41:279–294.

Kock, D., and H. R. Bhat. 1994. *Hipposideros hypophyllus* n. sp. of the *H. bicolor* group from peninsular India (Mammalia: Chiroptera: Hipposideridae). Senckenbergiana Biologica 73:25–31.

Koopman, K. F. 1970. Zoogeography of bats. *In* About Bats: A Chiropteran Biology Symposium, B. H. Slaughter and D. W. Walton, eds., pp. 29–50. Southern Methodist University Press, Dallas.

Koopman, K. F. 1994. Chiroptera: Systematics. *In* Handbook of Zoology, Vol. 8. Mammalia, Part 60. de Gruyter, Berlin.

Legendre, S. 1982. Hipposideridae (Mammalia: Chiroptera) from the Mediterranean Middle and Late Neogene and evolution of the genera *Hipposideros* and *Asellia*. Journal of Vertebrate Paleontology 2:386–399.

Lincoln, R. J., G. A. Boxshall, and P. F. Clark. 1982. A Dictionary of Ecology, Evolution, and Systematics. Cambridge University Press, Cambridge.

Maddison, W. P., M. J. Donoghue, and D. R. Maddison. 1984. Outgroup analysis and parsimony. Systematic Zoology 33:83–103.

Naidu, K. N., and M. E. Gururaj. 1984. Karyotype of *Rhinolophus luctus*. Current Science (Bangalore) 53:825–826.

Narušis, M. J. 1993. SPSS for Windows, release 6.0. Software and documentation. SPSS, Inc., Chicago.

Nelson, G. 1979. Cladistic analysis and synthesis: Principles and definitions, with a historical note on Adanson's *Familles des Plantes* (1763–1764). Systematic Zoology 28:1–21.

Novacek, M. J. 1991. Aspects of the morphology of the cochlea in microchiropteran bats: An investigation of character transformation. Bulletin of the American Museum of Natural History 206:84–100.

Peterson, R. L., and D. W. Nagorsen. 1975. Chromosomes of fifteen species of bats (Chiroptera) from Kenya and Rhodesia. Life Sciences Occasional Papers, Royal Ontario Museum 27:1–14.

Pierson, E. D. 1986. Molecular systematics of the Microchiroptera: Higher taxon relationships and biogeography. Ph.D. dissertation, University of California, Berkeley.

Qumsiyeh, M. B., R. D. Owen, and R. K. Chesser. 1988. Differential rates of genic and chromosomal evolution in bats of the family Rhinolophidae. Genome 30:326–335.

Radinsky, L. 1967. Relative brain size: A new measure. Science 155:836–838.

Rautenbach, I. L. 1986. Karyotypical variation in southern African Rhinolophidae (Chiroptera) and non-geographic morphometric variation in *Rhinolophus denti* Thomas, 1904. Cimbebasia 8A:130–139.

Rautenbach, I. L., G. N. Bronner, and D. A. Schlitter. 1993. Karyotypic data and attendant systematic implications for the bats of southern Africa. Koedoe 36:87–104.

Ray-Chaudhuri, S. P., and S. Pathak. 1966. Studies on the chromosomes of bats: List of worked out Indian species of Chiroptera. Mammalian Chromosome Newsletter 22:206.

Ray-Chaudhuri, S. P., S. Pathak, and T. Sharma. 1971. Karyotypes of five Indian species of Microchiroptera. Caryologia 24:239–245.

Rohlf, F. J. 1993. NTSYS-pc: Numerical Taxonomy and Multivariate Analysis System, Version 1.80. Exeter Software, Setauket, N.Y.

Sanderson, M. J., and M. J. Donoghue. 1989. Patterns of variation in levels of homoplasy. Evolution 43:1781–1795.

Sevilla, P. 1990. Rhinolophoidea (Chiroptera, Mammalia) from the upper Oligocene of Carrascosa del Campo (central Spain). Geobios (Lyon) 23:173–188.

Sigé, B. 1968. Les chiroptères du Miocène inférieur de Bouzigues. I. Etude systématique. Palaeovertebrata 1:65–133.

Sigé, B. 1990. Nouveaux chiroptères de l'Oligocène moyen des phosphorites du Quercy, France. Comptes Rendus de l'Académie des Sciences, Paris, Série II 310:1131–1137.

Sigé, B. 1991. Rhinolophoidea et Vespertilionoidea (Chiroptera) du Chambi (Eocène inférieur de Tunisie). Aspects biostratigraphique, biogéographique et paléoécologique de l'origine des chiroptères modernes. Neues Jahrbuch für Geologie und Paläontologie Abhandlungen 182:355–376.

Sigé, B., and S. Legendre. 1983. L'histoire des peuplements de chiroptères du basin méditerranéen: L'apport comparé des remplissages karstiques et des dépôts fluvio-lacustres. Mémoires de Biospéologie 10:209–225.

Sigé, B., S. J. Hand, and M. Archer. 1982. An Australian Miocene *Brachipposideros* (Mammalia, Chiroptera) related to Miocene representatives from France. Palaeovertebrata 12:149–171.

Sigé, B., H. Thomas, S. Sen, E. Gheerbrant, J. Roger, and Z. Al-Sulaimani. 1994. Les chiroptères de Taqah (Oligocène inférieur, Sultanat d'Oman). Premier inventaire systématique. Münchner Geowissenschaftliche Abhandlungen 26A:35–48.

Sreepada, K. S., K. N. Naidu, and M. E. Gururaj. 1993. Trends of karyotypic evolution in the genus *Hipposideros* (Chiroptera: Mammalia). Cytobios 75:49–57.

Tate, G. H. H. 1941. Results of the Archbold expeditions. No. 35. A review of the genus *Hipposideros* with special reference to Indo-Australian species. Bulletin of the American Museum of Natural History 78:353–371.

Wood, D. S. 1983. Character transformations in phenetic studies using continuous morphometric variables. Systematic Zoology 32:125–131.

Yoshiyuki, M. 1991. Taxonomic status of *Hipposideros terasensis* Kishida, 1924 from Taiwan (Chiroptera, Hipposideridae). Journal of the Mammalogical Society of Japan 16:27–35.

Zhang, W. 1985. A study on karyotypes of the bats *Tadarida teniotis insignis* Blyth and *Hipposideros pratti* Thomas. Acta Theriologica Sinica 5:189–193.

Zima, J., M. Volleth, I. Horáček, J. Cerveny, A. Cervená, K. Prucha, and M. Macholán. 1992. Comparative karyology of rhinolophid bats (Chiroptera: Rhinolophidae). *In* Prague Studies in Mammalogy, I. Horáček and V. Vohralík, eds., pp. 229–236. Charles University Press, Prague.

3
Phylogeny of Neotropical Short-Tailed Fruit Bats, *Carollia* spp.
Phylogenetic Analysis of Restriction Site Variation in mtDNA

BURTON K. LIM AND MARK D. ENGSTROM

Short-tailed fruit bats *(Carollia)* are among the most abundant species of mammals found in Neotropical rainforests. Although ecologically and behaviorally they include some of the better-known bats from this area (Fleming 1988), there is no corroborated phylogeny hypothesizing relationships among species within the genus. *Carollia* belongs to the New World family Phyllostomidae, which is characterized by a fleshy noseleaf structure presumed to function in echolocation. *Carollia* is differentiated from other genera of phyllostomid bats by several characteristics: an incomplete zygomatic arch, dentition with reduced size and cusp pattern, the distinctive tribanding pattern of the fur, and the presence of a short tail.

Although it has not been demonstrated in a phylogenetic context, *Carollia* has traditionally been included with the smaller *Rhinophylla* in the subfamily Carolliinae (Jones and Carter 1976), which more recently has been assigned to the level of subtribe within the subfamily Phyllostominae (Baker et al. 1989). *Rhinophylla* also possesses an incomplete zygomatic arch and an even greater reduction in size of dentition and cusp pattern but lacks a tail and the tribanding pattern of the fur. On the basis of a cladistic analysis of female reproductive histomorphology, stenodermatines are

considered the sister-taxon to carolliines, as suggested by a uniquely modified ovarian ligament (Hood and Smith 1982). Within the subfamily Phyllostominae, glossophagines, stenodermatines, and carolliines also share an uniquely derived fundic oviductal entry into the uterus (Hood and Smith 1982).

On a generalized scale of decreasing abundance and restricted distribution (Pine 1972; Hall 1981; Koopman 1982), the four species of *Carollia* currently recognized are *C. perspicillata* from the Atlantic versant in central Mexico, through Central America, and into South America to southeastern Brazil and extreme northeastern Argentina (Barquez et al. 1993); *C. brevicauda* from the Atlantic versant in central Mexico, through Central America, and into South America to eastern Brazil and Bolivia; *C. castanea* from the Atlantic versant in eastern Honduras, through Central America, and into South America to Venezuela, central Brazil, and Bolivia; and *C. subrufa* from the Pacific versant in Colima, Mexico, south through Guatemala to the Atlantic coast in northwestern Honduras, throughout El Salvador, and into western Nicaragua and northwestern Costa Rica.

The species of *Carollia* can usually be defined on the basis of cranial morphology, dentition, and banding pattern of the dorsal hairs (Pine 1972). However, there is considerable overlap and intraspecific variation even within Middle America (Owen et al. 1984) where most of the primary research has been done. An analysis of cranial morphometrics was able to distinguish most species of *Carollia* in canonical space (McLellan 1984). *Carollia castanea* was the most distinctive, and *C. brevicauda* and *C. perspicillata* were the most difficult to differentiate. Morphological variation within and among species is complex as the result of sexual and geographic variation.

Chromosome morphology and diploid number are very similar among the species. With one exception, *Carollia* share an X-autosome translocation (females, $2n = 20$; males, $2n = 21$; Baker 1979). Unlike populations found in Costa Rica (Patton and Gardner 1971) and Colombia (Baker and Bleier 1971), karyotypes of *C. castanea* from Peru lack this translocation and both sexes have the same diploid number of 22. This latter karyotype has been hypothesized to be the ancestral condition for the genus (Patton and Gardner 1971); however, polarity has not been rigorously determined and these data provide little resolution of relationships among species within the genus. *Carollia castanea* from Peru also lack heterochromatin in comparison to *C. brevicauda* and *C. perspicillata,* but other geographic samples of *C. castanea* were not available for examination (Stock 1975).

Although *Carollia* has been one of the more thoroughly studied genera of Neotropical bats, the systematic limits are still poorly known. The species are readily distinguishable in Central America, but identification keys based on Central American specimens fail in South America where there is an overall size reduction and the banding pattern of the dorsal fur is not so distinctive as in Central America. As emphasized by the recent discovery of an undescribed species in Peru (Pacheco et al. 1995), the taxonomy, distribution, and systematics of *Carollia* need revision. The impetus for this study was our difficulty in assigning specimens of *Carollia* to species during extensive fieldwork in Guyana in addition to the lack of a species-level phylogeny from previous morphological and karyological studies.

The primary objectives of this study are to (1) examine systematic relationships and taxonomic affinities between Central and South American populations of individual species (i.e., for broadly distributed species, do morphologically distinctive Central and South American populations represent single monophyletic lineages?); (2) propose an hypothesis of evolutionary relationships within *Carollia;* and (3) discuss the biogeographic implications of this phylogeny. To supplement previous studies that incorporated morphological and chromosomal data in a comparative approach, our analysis relies on cladistic analysis of molecular data.

Materials

In our analysis, we used 145 individuals (Appendix 3.1) from 21 general localities in Central and South America including Mexico, Guatemala, El Salvador, Belize, Panama, Ecuador, and Guyana (Figure 3.1). Tissue samples and voucher specimens have been deposited at the Royal Ontario Museum (ROM) in Toronto, Angelo State University Natural History Collection (ASNHC) in Texas, Florida State Museum (FSM) at the University of Florida in Gainesville, and Museo Ecuatoriano de Ciencias Naturales (MECN) in Quito, Ecuador. If available, at least 5 individuals of each species from each of several localities were sampled to assess variation within and among populations. In addition to the four recognized species of *Carollia, Rhinophylla pumilio* and *R. fischerae* (carolliines), *Artibeus lituratus* (stenodermatine), and *Glossophaga soricina* (glossophagine) within the subfamily Phyllostominae (sensu Baker et al. 1989) were used as increasingly distant outgroup taxa.

Methods

Mitochondrial DNA (mtDNA) is considered suitable for phylogenetic analysis because it has a high rate of nucleotide base-pair substitutions but a low rate of sequence rearrangements (Dowling et al. 1990). In molecular systematics, restriction site analysis offers a balance between resolution and applicability. Difficulties such as sequence alignment and character weighting are usually associated with direct sequencing of rapidly evolving DNA (Avise 1994). These methodological problems can be avoided by using indirect methods of surveying DNA sequences such as partial endonuclease digest mapping (PEDM).

The PEDM technique was recently modified from the traditional double-digest restriction fragment length polymorphism (RFLP) method; it has the advantage of increased resolution from end-labeling for postulating homologous cut sites (Morales et al. 1993). End-labeled partial digestions from the PEDM technique permit direct mapping of homologous cut sites at a higher resolution than is possible using only multiple complete digestion fragments as in RFLP. Traditional RFLP data have a disadvantage in that fragments are not independent because a gain of a cut site results in two smaller fragments from the original larger fragment. This problem is avoided in PEDM by using mapped cut sites, rather than presence or absence of fragments, as characters. By using PEDM, a broader segment of the mtDNA genome can also be assayed for larger sample sizes at reduced time and expense relative to mtDNA sequencing. This advantage facilitates the study of widely distributed taxa and rapid assessment of polymorphism

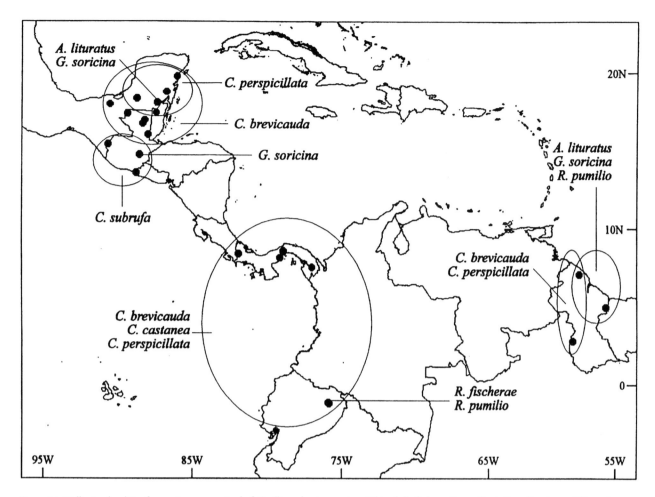

Figure 3.1. Collecting localities for specimens examined of *Carollia* and outgroup taxa *(Rhinophylla fischerae, R. pumilio, Artibeus lituratus,* and *Glossophaga soricina)* in Mexico, Belize, Guatemala, El Salvador, Panama, Ecuador, and Guyana.

levels within and between populations. PEDM has been used successfully in phylogenetic studies on deer mice (*Peromyscus;* van Coeverden de Groot 1995), rhinoceros (Rhinocerotidae; Morales and Melnick 1994), tree bats (*Lasiurus;* Morales and Bickham 1995), and elephants (*Loxodonta;* Cherfas 1989).

We used the partial endonuclease digest mapping procedure as formalized by Morales et al. (1993) and as refined by van Coeverden de Groot (1995). Briefly, it involved incubation with STE (sodium chloride, Tris-HCl, EDTA) and proteinase K and a phenol and chloroform extraction procedure. The isolated genomic mtDNA was amplified for a primer-specific gene region by polymerase chain reaction (PCR). The resultant product was completely digested with restriction endonucleases, and fragments were separated by electrophoresis on 1.5% agarose TBE (Tris-borate + EDTA) gels. A serial dilution test identified the enzyme concentration required to obtain the full complement of partial fragments. Unique individuals were reamplified using a biotinylated primer and partially digested with the appropriate

concentration. Resultant fragments of various lengths were separated by gel electrophoresis and transferred by Southern blotting onto a nylon membrane. After washing the membrane with appropriate reagents, nonradioactive streptavidin and a phosphatase substrate were applied, which caused the biotin-labeled fragments to chemiluminesce; these were then detected on autoradiography film. The restriction endonuclease cut sites were mapped using an internal biotinylated DNA size standard for the partial digests and cross-referenced to the corresponding complete digest fragments. The presence or absence of each cut site was recorded as a character state for cladistic analysis.

A 2,400-base-pair (bp) region of mtDNA that included the protein coding genes ND3, ND4L, and ND4 (ND3–4) was amplified by primers 772 and 773 (LGL Ecological Genetics). During PCR, the reactants were heated at 94°C for 2 min and amplified for 34 cycles by denaturing at 94°C for 1 min, annealing at 52°C for 2 min, and extending at 72°C for 3 min; at the end of the 34 cycles was a 7-min soak at 75°C. Development and composition of the primers were as described by Cronin et al. (1993); the sequence for 772 is 5'-TAA(C/T)TAGTACAG(C/T)TGACTTCCAA-3' and for 773 is 5'-TTTTGGTTCCTAAGACCAA(C/T)GGAT-3'. This gene region has proven useful in phylogenetic analyses of deer mice (*Peromyscus;* van Coeverden de Groot 1995), lizards (*Sceloporus;* Arévalo et al. 1994), and salmon (*Oncorhynchus;* Cronin et al. 1993). Twelve restriction endonucleases with unique recognition sequences of 4 bp (*RsaI, BstUI, HaeIII, MboI,* and *TaqI*), 5 bp (*BstNI, DdeI, HinfI, Sau96I,* and *ScrFI*), and 6 bp (*EcoRI* and *HindIII*) were used to digest the ND3–4 gene region under incubation.

To hypothesize phylogenetic relationships, restriction site variation was analyzed by an heuristic search under Dollo parsimony using PAUP version 3.1.1 (Swofford 1993) on an Apple Macintosh Quadra 650 computer. The following search options were used to ensure a thorough analysis: random-addition sequence, 20 trees held at each step, tree bisection and reconnection (TBR) swapping algorithm, zero-length branches collapsed into polytomies, multiple most-parsimonious tree saving during branch swapping, and steepest descent for each round of swapping. To test for tree stability, values of branch support were calculated for the major lineages within *Carollia.* A converse-constraints approach, appropriate for large data sets and time-consuming analyses (Bremer 1994), includes calculating the shortest tree without a specific clade and subtracting the length of the most-parsimonious tree.

Dollo parsimony has been suggested as a suitable model of evolution for mtDNA because there may be asymmetry in the probability of gaining and losing a restriction site; that is, losses may be more prevalent than gains (DeBry and Slade 1985). Wagner parsimony assumes an equal chance of forward and reverse changes, which may not be the best estimator of the likelihood of mtDNA restriction site variation. In contrast, Dollo parsimony permits a unique character gain but the only type of homoplasy allowed is a reversal to the ancestral condition. Although Dollo permits only convergent losses, which may be too restrictive, it is a more reasonable weighting scheme given the greater likelihood that shared losses rather than shared gains will be homoplasious.

The consistency index (CI) and retention index (RI) measure the extent of homoplasy, but they have different interpretations and applications (Farris 1989b). The low CI in our analysis was not unexpected because CI typically decreases as the number of taxa increases (Archie 1989). In contrast, the RI was high, indicating the fraction of observed synapomorphy compared to the maximum possible amount of synapomorphy (Farris 1989a). For analyses with large numbers of taxa, the RI is a better comparative indicator because it measures the relative amount of homoplasy as opposed

to the absolute amount. For example, for the derived state of a binary character that appears on three independent branches of a 75-terminal-taxa cladogram, CI = 0.33 and RI = 0.97. In contrast, for the derived state of a binary character that appears on three independent branches of a 15-terminal-taxa cladogram, CI = 0.33 but RI = 0.86. As indicated by the RI but not the CI, two extra steps on a cladogram with many taxa is not as problematic phylogenetically as two extra steps on a cladogram with few taxa.

Results

For the 145 individuals used in this analysis, there were 142 restriction cut sites (Appendix 3.2) from the following 12 enzymes (number of sites in parentheses): *BstNI* (8), *HaeIII* (18), *HinfI* (19), *MboI* (15), *RsaI* (16), *Sau96I* (12), *ScrFI* (11), *TaqI* (12), *BstUI* (5), *HindIII* (2), *DdeI* (24), and *EcoRI* (0). A total of 75 unique haplotypes with 123 restriction sites was phylogenetically informative and used in a Dollo parsimony analysis (PAUP; Swofford 1993). The analysis produced 769 equally parsimonious trees of length 393, CI of 0.313, and RI of 0.960 after 65 hr of execution. Although the CI was low, as expected for data sets with many taxa, an RI of 0.96 represents a good fit of character data to the most-parsimonious tree.

The 50%-majority-rules consensus tree (Figure 3.2) places *C. subrufa* and *C. perspicillata* as monophyletic sister-taxa. This clade forms a monophyletic trichotomy with two populations of *C. brevicauda,* one representing central Panama south into South America and the second representing northern Central America south to western Panama. *Carollia castanea* is a monophyletic basal lineage of the genus with samples from Panama and Ecuador forming discrete clades. Within the outgroups, *G. soricina, A. lituratus,* and *Rhinophylla* each form monophyletic clades with *R. pumilio* and *R. fischerae* as sister-species. These taxa were rooted as a monophyletic outgroup.

Some equally parsimonious trees differed from the majority-rules consensus tree topology primarily because of the placement of one sample (cbgy7) from Guyana, which was originally assigned to *C. brevicauda* on the basis of size of skull and body. In 43% of the equally parsimonious trees, this sample clustered within *C. perspicillata* in a clade with three haplotypes from Guyana and one haplotype from Panama (Figure 3.3). In many of the equally parsimonious trees, however, all haplotypes of *C. brevicauda* appeared as a monophyletic clade, including sample cbgy7, which grouped with another Guyana haplotype (Figure 3.4).

Branch support values (Bremer 1994) are 0 for the monophyly of *C. brevicauda, C. perspicillata,* and the clade for *C. perspicillata* and *C. subrufa* (Figure 3.5). Lack of branch

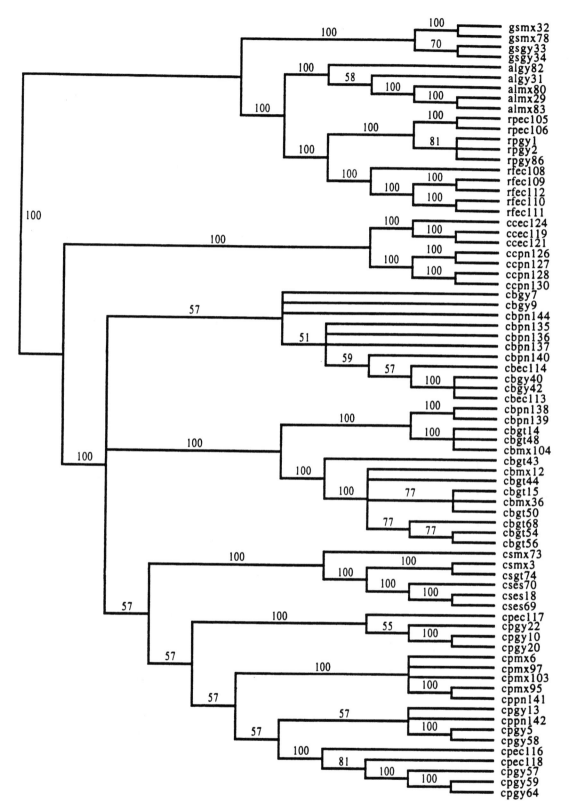

Figure 3.2. The 50%-majority-rules consensus cladogram for 769 equally parsimonious trees (393 steps each; consistency index = 0.313; retention index = 0.960) from a Dollo parsimony analysis of *Carollia* and related outgroup taxa. The numbers above the branches indicate the percentage of equally parsimonious trees that support that particular clade. In the terminal-taxa codes, the first two letters are for species: al, *Artibeus lituratus*; cb, *C. brevicauda*; cc, *C. castanea*; cp, *C. perspicillata*; cs, *Carollia subrufa*; gs, *Glossophaga soricina*; rf, *Rhinophylla fischerae*; rp, *R. pumilio*. The second two letters are for countries: ec, Ecuador; es, El Salvador; gt, Guatemala; gy, Guyana; mx, Mexico; pn, Panama. The trailing numerals identify individual specimens.

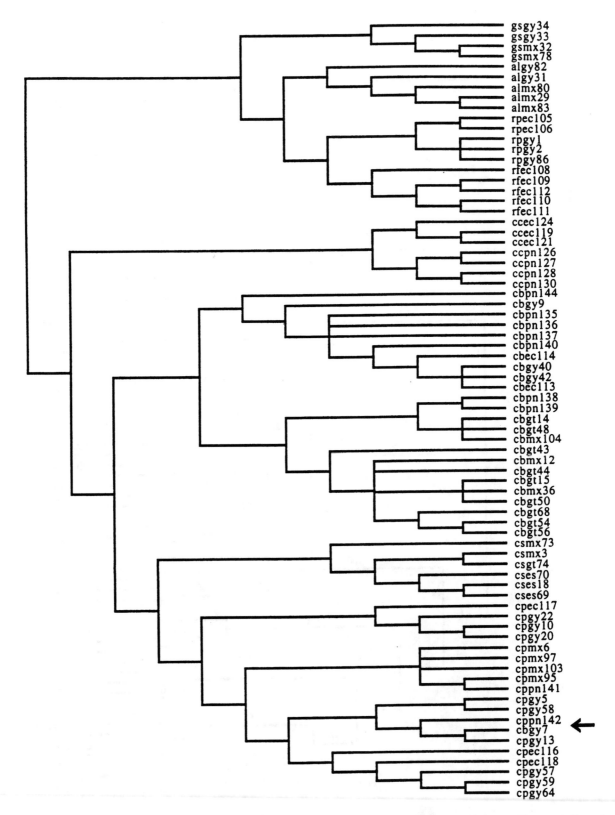

gsgy34
gsgy33
gsmx32
gsmx78
algy82
algy31
almx80
almx29
almx83
rpec105
rpec106
rpgy1
rpgy2
rpgy86
rfec108
rfec109
rfec112
rfec110
rfec111
ccec124
ccec119
ccec121
ccpn126
ccpn127
ccpn128
ccpn130
cbpn144
cbgy9
cbpn135
cbpn136
cbpn137
cbpn140
cbec114
cbgy40
cbgy42
cbec113
cbpn138
cbpn139
cbgt14
cbgt48
cbmx104
cbgt43
cbmx12
cbgt44
cbgt15
cbmx36
cbgt50
cbgt68
cbgt54
cbgt56
csmx73
csmx3
csgt74
cses70
cses18
cses69
cpec117
cpgy22
cpgy10
cpgy20
cpmx6
cpmx97
cpmx103
cpmx95
cppn141
cpgy5
cpgy58
cppn142
cbgy7
cpgy13
cpec116
cpec118
cpgy57
cpgy59
cpgy64

Figure 3.3. One of the equally parsimonious trees that had a different topology from the 50%-majority-rules consensus cladogram. The arrow indicates the problematic sample of *Carollia brevicauda* from Guyana that can appear within the *C. perspicillata* clade. Terminal-taxa codes are as defined for Figure 3.2.

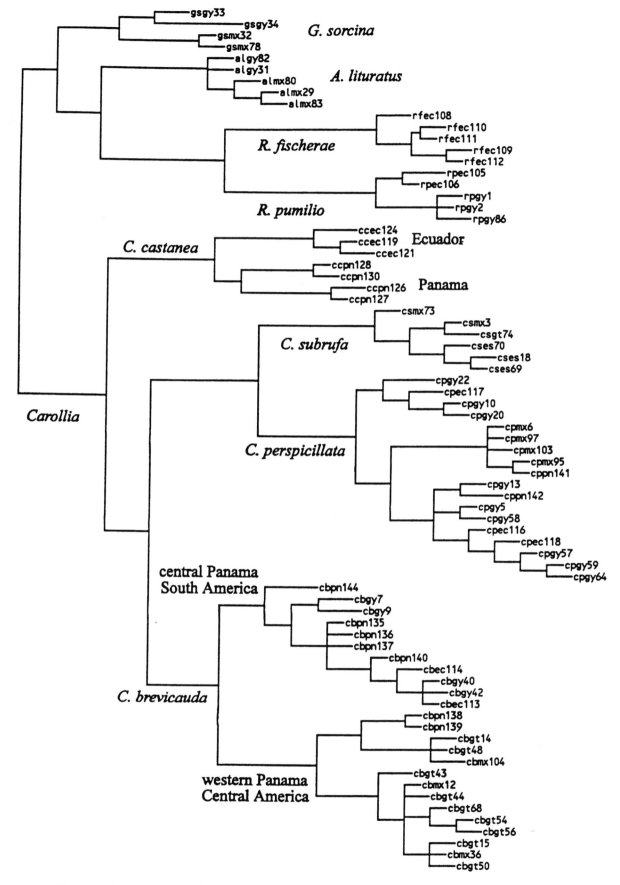

Figure 3.4. Cladogram of an equally parsimonious tree consistent with the 50%-majority-rules consensus cladogram. This shows all four species of *Carollia* as monophyletic and indicates geographic structuring within *C. brevicauda* and *C. castanea*. Terminal-taxa codes are as defined for Figure 3.2.

support for these clades results from the inconsistent placement of the one problematic sample of *C. brevicauda* (cbgy7). However, the monophyly of *C. castanea* and *C. subrufa* is well supported with values of 16 and 14, respectively. Branch support values for the other major lineages were 8 for the genus *Carollia* and 2 for the clade comprising *C. brevicauda*, *C. perspicillata*, and *C. subrufa*. Branch lengths for the major lineages within *Carollia* are also shown in Figure 3.5. Synapomorphies defining major clades range from 6 (*C. perspicillata*; *C. perspicillata* and *C. subrufa*) to 16 (*C. subrufa*). There is a proportionately higher occurrence of unique gains over convergent losses at more basal nodes, and losses are more common than gains at more terminal nodes. Unique gains are forced down the tree, resulting in convergent losses accounting for the homoplasy.

Discussion

Our estimate of phylogenetic relationships within *Carollia* is best described in Figure 3.4. Although one sample from Guyana (cbgy7) caused the monophyly of *C. brevicauda* to collapse in a 50%-majority-rules consensus tree (see Figure 3.2), many most equally parsimonious trees still retained a monophyletic clade for this species (Figure 3.4). Morphologically, this sample falls within *C. brevicauda*, but its inclusion within that species is weakly supported from the molecular analysis as evidenced by branch support values of zero for the independence of lineages of *C. brevicauda* and *C. perspicillata* (see Figure 3.5). Sample cbgy7, from northwestern Guyana, might represent a morphologically cryptic lineage, specifically distinct from either *C. brevicauda* or *C. perspicillata*. Alternatively, an ancestral haplotype originally present as a polymorphism in the (*brevicauda* [*perspicillata*, *subrufa*]) clade, might have become fixed in this population while being lost in other taxa because of random mtDNA lineage sorting and extinction. Whatever the exact history of that haplotype, this result suggests that systematic relationships within *Carollia* are more complex than predicted by their conservative morphology. We noted, however, that relationships among all other samples of *C. brevicauda* and *C. perspicillata* included in the study are stable, forming separate monophyletic groups.

Without explicitly hypothesizing phylogenetic relationships, Pine (1972) suggested that *C. subrufa* and *C. brevicauda* were more similar to each other, on the basis of size, than either was to *C. perspicillata* or to *C. castanea*. Two quantitative morphometric studies (McLellan 1984; Owen et al. 1984) also showed decreasing phenetic similarity among *C. subrufa*, *C. brevicauda*, *C. perspicillata*, and *C. castanea*. Our cladistic analysis of molecular data recognized *C. perspicillata* and *C. subrufa* as sister-species and *C. brevicauda* and *C.*

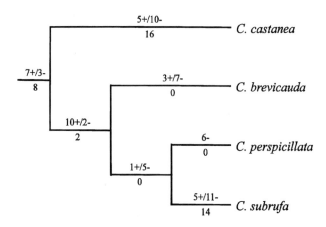

Figure 3.5. Summary cladogram of the major lineages within *Carollia*. The numbers of restriction site gains (+) and losses (-) supporting the species-level nodes are indicated above the branches, and the branch support values of Bremer (1994) are given below the branches.

castanea as successively distantly related taxa, in contrast to relationships predicted from overall similarity.

Morphometric analyses have also demonstrated, to varying degrees, a decreasing size cline from north to south for all species of *Carollia* (McLellan 1984), which has confused identifications in northern South America. In our study, all recognized taxa were included within monophyletic clades (see Figure 3.4) in most reconstructions. Restriction site mapping confirmed the species boundaries and affinities of the common and wide-ranging sympatric species *C. brevicauda* and *C. perspicillata*. These appear to be the only two species of *Carollia* in Guyana, although *C. castanea* has been reported from Amazonian Brazil (Uieda 1980) and Venezuela (Linares 1987). Using length of forearm and tibia as representative of size, both *C. brevicauda* and *C. perspicillata* from Guyana are smaller than other respective conspecific populations (Table 3.1). This shift in relative size, and the indistinctive banding pattern of the fur, account for the

Table 3.1

Lengths of Forearms and Tibiae of Two *Carollia* spp., by Geographic Subgroup

Specimen group	n	Length of forearm (mm)	Length of tibia (mm)
C. brevicauda			
Guyana localities	6	38.0 (37–39)	16.3 (16–17)
Other localities	35	40.0 (38–42)	17.4 (17–19)
C. perspicillata			
Guyana localities	9	40.9 (39–44)	18.8 (17–20)
Other localities	16	43.7 (42–46)	21.0 (19–23)

Notes: n = number of bats sampled. Length data given are mean values, with ranges in parentheses.

taxonomic uncertainty in identifications and distributions of these species within the Guianas (Husson 1978; Eisenberg 1989; Koopman 1993).

Within *C. brevicauda,* a size cline was observed from a cranial canonical variation analysis (McLellan 1984). A larger northern group included samples from Costa Rica north to Mexico, and a smaller southern group included samples from eastern Panama and Ecuador. This pattern was congruent with geographic structuring evident from our molecular analysis. Individuals from western Panama into Middle America form a monophyletic group, separate from individuals from Central Panama into South America (Figure 3.4).

Morphologically, *C. castanea* is distinctive, and it is the only chromosomally polymorphic species in the genus. The systematic import of this polymorphism is difficult to determine without further geographic sampling. For example, the distributional extent of the $2n = 22$ karyotype with the hypothesized primitive XY system, currently known only from Peru (Patton and Gardner 1971), is unknown. The karyotype of our samples of *C. castanea* from the Amazonian lowlands of eastern Ecuador is $2n = 20/21$ with the X-autosome translocation (Lim and Engstrom, unpublished data), which is the same as that from Colombia and Costa Rica (Baker and Bleier 1971; Patton and Gardner 1971).

Interpretation of the $2n = 22$ karyotype as ancestral is not inconsistent with our data, which place the *C. castanea* lineage as basal within *Carollia.* When mapped onto the phylogeny, the presence of both primitive and derived states in this species then suggest that it is paraphyletic with some populations sharing a common ancestor with other species of *Carollia* after divergence from the $2n = 22$ form. By this interpretation, the Peruvian group represents a distinct species. It could also be hypothesized, however, that the $2n = 22$ system was uniquely derived within this population of *C. castanea* (the polarity of the X-autosome translocation has not been rigorously established), or less parsimoniously, that it represents a reversal to the ancestral condition. Our analysis did not include samples of the $2n = 22$ Peruvian form. A more thorough geographic sampling of both karyotypes and mtDNA genotypes is needed to determine which hypothesis is most consistent with phylogenetic relationships and thus the most appropriate taxonomic rank for populations currently included within *C. castanea.*

The oldest known phyllostomid fossil bat is *Notonycteris magdalenensis* (Savage 1951) from the Miocene of Colombia. There are no other pre-Pleistocene phyllostomid fossils (Koopman 1976), which has led to considerable conjecture about the origin of this family. Based on a study of allozyme variation, Straney et al. (1979) estimated that the Phyllostomidae evolved in the early Oligocene. Using species rich-

ness as an indicator of time of divergence, Koopman (1976) suggested that the family was the first bat lineage to appear in South America. Because South America was essentially an insular continent from the early Tertiary to the Quaternary (Simpson 1950), it is assumed that diversification occurred primarily in isolation. However, island-hopping (Simpson 1950) or waif dispersal (Hershkovitz 1969) has been proposed as a possible mode of mammalian interchange between North and South American fauna, although the exact timing and scenarios of dispersal are disputed. For bats, the only mammals that possess powered flight, bodies of water may not necessarily pose so great a barrier as for a terrestrial animal.

Consideration of the geological history of the Americas (Simpson 1950; Hershkovitz 1969; Savage 1974), the chiropteran fossil record (Savage 1951), interpretation of present phyllostomid diversity and distribution (Koopman 1976), karyology (Baker 1979), and our data suggest a possible biogeographic scenario for the genus. If *Carollia* originated in South America along with other phyllostomid higher taxa before the closure of the Panamanian Isthmus, the Panamanian Marine Portal (Savage 1974) would have been available as a dispersal route to protypical Central America. *Carollia* are not especially vagile, having a mean home range of about 1 km² (Fleming 1988). Therefore, occasional dispersal to the volcanic islands that were the forerunners of Central America might have been possible while still providing a barrier sufficient to restrict the amount of gene flow between the founders and ancestral South American populations following any colonization events.

Carollia brevicauda and *C. castanea* appear as successively basal taxa in the genus, and both display primary geographic structuring between Central and South American populations consistent with this dispersal hypothesis. On the basis of the lack of similar patterns of geographic variation in *C. perspicillata,* the geographic range of its ancestor (contemporary with *C. brevicauda* and *C. castanea*) probably was not expanded into Central America until the establishment of a permanent land connection. *Carollia subrufa* is distributed entirely in xeric environments in Middle America and is the probable sister-taxon to *C. perspicillata.* It likely subsequently speciated in situ in Central America in association with the uplift of the Continental Divide and increased aridity on the Pacific versant.

The foregoing hypothesis contrasts with the pattern of speciation suggested by Pine (1972). On the basis of their similar size, largely allopatric distributions, and different habitat requirements, Pine (1972) proposed that *C. subrufa* and *C. brevicauda* were allopatrically derived sister-taxa. Although it is occasionally sympatric with *C. brevicauda* and *C. perspicillata,* *C. subrufa* usually is parapatrically or allopatrically distributed relative to all other species of *Carollia*. On the basis of distribution, therefore, a similar biogeographic origin could be hypothesized between *C. subrufa* and any of the other recognized species of *Carollia*, including *C. perspicillata*, the most likely candidate based on our analysis.

Conclusions

Species boundaries within *Carollia* were clarified with restriction site analysis, which indicated the following points.

1. *Carollia brevicauda* and *C. perspicillata* are the only species present in Guyana and most probably throughout the Guianas and eastern Brazil. Although both species are relatively small in size compared to other populations in other parts of the species ranges, molecular data substantiated that each taxon represents a monophyletic lineage despite geographic variation in size that has previously confused identifications, particularly in South America.

2. Our hypothesis of phylogenetic relationships suggests that *C. perspicillata* and *C. subrufa* are sister-taxa with *C. brevicauda* and *C. castanea* as successively distant basal lineages.

3. These data and patterns of geographic variation within species are consistent with a biogeographic scenario proposing the origin of *Carollia* in South America with dispersal of *C. castanea* and *C. brevicauda* to Central America before the establishment of a fixed link between the continents. The immediate ancestor of *C. perspicillata* expanded its range into Central America after completion of the Panamanian Isthmus, wherein *C. subrufa* speciated in xeric habitats along the Pacific versant of Middle America.

Combining the cladistic analysis of female reproductive histomorphology (Hood and Smith 1982) at higher taxonomic levels within phyllostomid bats and our molecular study of species-level relationships within *Carollia*, Figure 3.6 can be used as a working hypothesis for the phylogeny of *Carollia* and related outgroup taxa. An experimental design for future analyses should concentrate on wider taxonomic and geographic representation. To properly test the sister-group relationship between *Carollia* and *Rhinophylla*, samples of *R. alethina* should be included as well as additional outgroups and relatively conservative characters. Wider geographic coverage of *Carollia* will include samples of the chromosomal race of *C. castanea* and the recently discovered new species from Peru.

Clarification of phylogenetic relationships within *Carollia* will facilitate the comparison of ecological and behav-

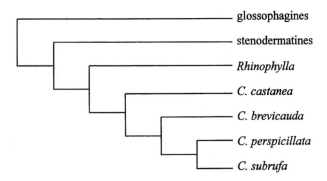

Figure 3.6. Hypothesized phylogeny for *Carollia* and related outgroup taxa, based on Hood and Smith (1982) and Baker et al. (1989) for basal lineages and molecular data from this study for more terminal relationships.

ioral studies in an evolutionary context. For example, diet might result opportunistically from current resource availability or be constrained by evolutionary modifications in physiology. These competing hypotheses could be used to examine the coevolution of *Carollia* and its primary food source, *Piper*. In addition, measures of biodiversity such as taxonomic dispersion (Williams et al. 1993) that rely on phylogenetic divergence, taxonomic representation, and species richness to develop conservation priorities require hypotheses of phylogenetic relationships. Because it is one of the most prevalent seed dispersers of the colonizing plant genus *Piper* (Fleming 1988), *Carollia* can have a profound impact on regeneration and sustainable use of rainforests. It is important in this context to recognize species boundaries and evolutionary relationships within *Carollia* and to evaluate each species as an unique contributor to its environment.

Appendix 3.1.
The 145 Specimens Examined in This Study

After the locality information, the sample size and museum collection are given in parentheses. The museums are Royal Ontario Museum (ROM) in Toronto, Angelo State University Natural History Collection (ASNHC) in Texas, Florida State Museum (FSM) at the University of Florida in Gainesville, and Museo Ecuatoriano de Ciencias Naturales (MECN) in Quito, Ecuador.

Carollia brevicauda (total, 50)

Belize

Orange Walk; Las Milpas; 17°45′ N, 89°00′ W (1 FSM)

Ecuador

Napo; Parque Nacional Yasuni, Estación Cientifica Onkone Gare, 38 km S Pompeya Sur; 00°39′ S, 76°27′ W (3 ROM)

Guatemala

El Petén; Biotopo Cerro Cahuí, El Remate; 120 m; 17°00' N, 89°44' W (5 ROM)

El Petén; Campo Los Guacamayos, Biotopo Laguna del Tigre, 40 km NE Naranjo; 17°36' N, 90°49' W (5 ROM)

El Petén; 1.5 km S and 7 km W Poptún; 515 m; 16°18' N, 89°20' W (5 ROM)

El Petén; Tikal; 210 m; 17°12' N, 89°37' W (5 ROM)

Guyana

Barima-Waini; Kwabanna; 07°34' N, 59°09' W (4 ROM)

Upper Takutu–Upper Essequibo; Kuma River, 9 km S and 8 km E Lethem, Kanuku Mountains; 03°16' N, 59°43' W (1 ROM)

Mexico

Campeche; 7 km W Escárcega; 18°36' N, 90°14' W (2 ASNHC)

Quintana Roo; Kohunlich; 18°25' N, 88°48' W (4 ASNHC)

Quintana Roo; Laguna Noh-Bec, 2 km W Noh-Bec; 19°08' N, 88°11' W (5 ROM)

Quintana Roo; Tulum; 20°10' N, 87°29' W (1 ROM)

Quintana Roo; Tabasco, 5 km N Jonuta; 18°08' N, 92°07' W (2 ROM)

Panama

Chiriquí; Ojo de Agua, 2 km N Santa Clara; 1500 m; 08°52' N, 82°45' W (2 ROM)

Darién; Parque Nacional Darién, Estación Pirre; 100 m; 08°06' N, 77°43' W (1 ROM)

Panamá; Parque Nacional Altos de Campana; 850 m; 08°41' N, 79°56' W (4 ROM)

Carollia castanea (total, 15)

Ecuador

Napo; Parque Nacional Yasuni, Estación Cientifica Onkone Gare, 38 km S Pompeya Sur; 00°39' S, 76°27' W (3 ROM; 3 MECN)

Panama

Canal Zone; Parque Nacional Soberanía; 09°08' N, 79°44' W (2 ROM)

Chiriquí; Ojo de Agua, 2 km N Santa Clara; 1500 m; 08°52' N, 82°45' W (1 ROM)

Chiriquí; Santa Clara; 1300 m; 08°50' N, 82°45' W (2 ROM)

Darién; Parque Nacional Darién, Estación Pirre; 100 m; 08°06' N, 77°43' W (2 ROM)

Panamá; Parque Nacional Altos de Campana; 850 m; 08°41' N, 79°56' W (2 ROM)

Carollia perspicillata (total, 37)

Belize

Orange Walk; Las Milpas; 17°45' N, 89°00' W (2 FSM)

Ecuador

Napo; Parque Nacional Yasuni, Estación Cientifica Onkone Gare, 38 km S Pompeya Sur; 00°39' S, 76°27' W (2 ROM)

Napo; Parque Nacional Yasuni, 42 km S and 1 km E Pompeya Sur; 00°41' S, 76°25' W (1 ROM)

Guyana

Barima-Waini; Kwabanna; 07°34' N, 59°09' W (6 ROM)

Upper Takutu–Upper Essequibo; Kuma River, 9 km S and 8 km E Lethem, Kanuku Mountains; 03°16' N, 59°43' W (5 ROM)

Mexico

Campeche; 7 km W Escárcega; 18°36' N, 90°14' W (1 ASNHC; 1 ROM)

Quintana Roo; Kohunlich; 18°25' N, 88°48' W (5 ASNHC)

Quintana Roo; Laguna Noh-Bec, 2 km W Noh-Bec; 19°08' N, 88°11' W (5 ROM)

Quintana Roo; Tulum; 20°10' N, 87°29' W (1 ROM)

Panama

Canal Zone; Gamboa; 09°06' N, 79°42' W (2 ROM)

Chiriquí; Santa Clara; 1300 m; 08°50' N, 82°45' W (3 ROM)

Darién; Parque Nacional Darién, Estación Pirre; 100 m; 08°06' N, 77°43' W (2 ROM)

Panamá; Parque Nacional Altos de Campana; 850 m; 08°41' N, 79°56' W (1 ROM)

Carollia subrufa (total, 13)

El Salvador

Ahuachapán; El Refugio, El Imposible; 240 m; 13°48' N, 90°00' W (5 ROM)

Guatemala

El Progreso; Rio Uyús, 5 km E San Cristobal Acasaguastlán; 240 m; 14°57' N, 89°50' W (5 ROM)

Mexico

Chiapas; 18 km S Frontera Comalapa; 720 m; 15°31' N, 92°07' W (3 ROM)

Rhinophylla fischerae (total, 5)

Ecuador

Napo; Parque Nacional Yasuni, Estación Cientifica Onkone Gare, 38 km S Pompeya Sur; 00°39' S, 76°27' W (2 ROM)

Napo; Parque Nacional Yasuni, 42 km S and 1 km E Pompeya Sur; 00°41' S, 76°25' W (2 MECN; 1 ROM)

Rhinophylla pumilio (total, 9)

Ecuador

Napo; Parque Nacional Yasuni, Estación Cientifica
 Onkone Gare, 38 km S Pompeya Sur; 00°39′ S, 76°27′ W
 (3 ROM)

Guyana

Barima-Waini; Kwabanna; 07°34′ N, 59°09′ W (4 ROM)
East Berbice-Corentyne; Mapenna Creek, 6 km from Corentyne
 River; 05°23′ N, 57°22′ W (2 ROM)

Artibeus lituratus (total, 9)

Guyana

Barima-Waini; Santa Cruz; 07°40′ N, 59°14′ W (1 ROM)
East Berbice-Corentyne; Mapenna Creek, 6 km from Corentyne
 River; 05°23′ N, 57°22′ W (3 ROM)

Mexico

Quintana Roo; Laguna Noh-Bec, 2 km W Noh-Bec; 19°08′ N,
 88°11′ W (5 ROM)

Glossophaga soricina (total, 7)

Guyana

Barima-Waini; Kwabanna; 07°34′ N, 59°09′ W (1 ROM)
East Berbice-Corentyne; Orealla Creek mouth; 05°20′ N,
 57°20′ W (1 ROM)

Mexico

Quintana Roo; Laguna Noh-Bec, 2 km W Noh-Bec; 19°08′ N,
 88°11′ W (3 ROM)

Guatemala

El Progreso; Rio Uyús, 5 km E San Cristobal Acasaguastlán;
 240 m; 14°57′ N, 89°50′ W (2 ROM)

Appendix 3.2.
Data Matrix for 145 Samples of *Carollia* and Related Outgroup Taxa

The numbers denote the presence (1) or absence (0) of 142 cut sites from 12 restriction endonucleases for the 2,400-bp mtDNA sequence including the protein-coding genes ND3, ND4L, and ND4. Data are not shown for the restriction endonuclease *Eco*RI, because it produced no cut sites. Asterisks identify the 75 unique haplotypes used in the Dollo parsimony analysis. A sample without an asterisk coded the same as the sample immediately preceding it. There were 123 phylogenetically informative sites. Nineteen characters are autapomorphies or nonvariable sites, here identified by the order in which they are listed under their enzyme: *Bst*NI site 7; *Hae*I sites 3 and 17; *Hin*fI sites 1, 5, 7, and 17; *Mbo*I sites 1, 8, 9, and 15; *Rsa*I sites 10, 13, and 16; *Sau*96I site 7; *Scr*FI site 9; *Bst*UI site 5; and *Dde*I sites 3 and 23. Under each enzyme, cut sites are listed in order of relative closeness to the labeled primer 772 (leftmost being the closest, rightmost the farthest).

In the specimen codes, the first two letters denote species: al, *Artibeus lituratus*; cb, *Carollia brevicauda*; cc, *Carollia castanea*; cp, *Carollia perspicillata*; cs, *Carollia subrufa*; gs, *Glossophaga soricina*; rf, *Rhinophylla fischerae*; rp, *Rhinophylla pumilio*. The second two letters are for country: bl, Belize; ec, Ecuador; es, El Salvador; gt, Guatemala; gy, Guyana; mx, Mexico; pn, Panama. The trailing numerals in the specimen codes identify individual specimens.

Specimen	BstNI	HaeI	HinfI	MboI	RsaI	Sau96I	ScrFI	TaqI	BstUI	Hind III	DdeI
*gsmx32	01000001	000001001000010000	100000010111110100	010001000110001	0011000100011000	000001000000	0000000001	000000010110	00000	01	100100000000010010000000
gsmx77	01000001	000001001000010000	100000010111110100	010001000110001	0011000100011000	000001000000	0000000001	000000010110	00000	01	100100000000010010000000
gsgt88	01000001	000001001000010000	100000010111110100	010001000110001	0011000100011000	000001000000	0000000001	000000010110	00000	01	100100000000010010000000
*gsgy33	01000001	000001000000110000	100000010010000100	010001000110001	0011000100010000	001000000000	0000000001	100000010110	00000	01	100100010000000010000000
*gsgy34	00000000	000001000010110000	100000010010000100	010000000010001	0011000100010000	000100101000	0000000000	100000010110	00000	01	100100010000000010000000
*gsmx78	01000001	000001000010000	100000010111110100	010001000110001	0011000100010000	000001000000	0000000001	000000010110	00000	01	100100000000010010000000
gsgt89	01000001	000000001000010000	100000010111110100	010001000110001	0011000100010000	000001000000	0000000001	000000010110	00000	01	100100000000010010000000
*almx29	00000000	100010000010110000	100000000010000100	010001000100001	0010000000110000	000000000000	0000000000	000010011110	00000	01	100100000000100000110100
algy30	00000000	100010000010110000	100000000010000100	010001000100001	0010000000110000	000000000000	0000000000	000010011110	00000	01	100100000000100000110100
almx79	00000000	100010000010110000	100000000010000100	010001000100001	0011000000110000	000000000000	0000000000	000010011110	00000	01	100100000000100000110100
algy81	00000000	100010000010110000	100000000010000100	010001000100001	0011000000110000	000000000000	0000000000	000010011110	00000	01	100100000000100000110100
*algy31	00000000	100010000110110000	100000000010000100	010001000100001	0011000000110000	000000000000	0000000000	000010011110	00000	01	100100000000100000110100
*almx80	00000000	100010000010110000	100000000010000100	010001000100001	0011000000110000	000000000000	0000000000	000010011110	00000	01	100100000000100000110100
*algy82	00000000	100010000010110000	100000000010000100	010001000100011	0011000000110000	000000000000	0000000000	000010011110	00000	01	100100000000100000110100
almx84	00000000	100010000010110000	100000000010000100	010001000100011	0011000000110000	000000000000	0000000000	000010011110	00000	01	100100000000100000110100
*almx83	00000000	100010000010100000	100000000010000100	010001000100001	0011000000110000	000000000000	0000000000	000010011110	00000	01	100100000000100000110100
*rpgy1	10000000	000000010000001000	100000010100011100	010000100110001	1001000000000010	000000000000	1000000000	000000001100	11000	00	110000100011010001000000
rpgy18	10000000	000000010000001000	100000010100011100	010000100110001	1001000000000010	000000000000	1000000000	000000001100	11000	00	110000100011010001000000
rpgy85	10000000	000000010000001000	100000010100011100	010000100110001	1001000000000010	000000000000	1000000000	000000001100	11000	00	110000100011010001000000

Specimen (cont'd)	BstNI	HaeI	HinfI	MboI	RsaI	Sau96I	ScrFI	TaqI	BstUI	Hind III	DdeI
*rpgy2	10000000	000000010000001000	1000000010100011100	010000100110001	1001000001000010	00000000000	10000000000	000000001100	11000	00	11000010001100000100000
rpgy27	10000000	000000010000001000	1000000010100011100	010000100110001	1001000001000010	00000000000	10000000000	000000001100	11000	00	11000010001100000100000
*rpgy86	10000000	000000000000001000	1000000010100011100	010000100110001	1001000000000010	00000000000	10000000000	000000001100	11000	00	11000010001011010001000000
*rpec105	10000010	000000010000001000	1000000010100011100	010000100110001	1000000000000010	00000000000	10000011100	000000001100	11000	00	01000010001100000100000
*rpec106	10000000	001000010000001001	1100000000100000100	010000100110001	1000000000000010	00000000000	10000000000	000000001100	11000	00	01000010001100000100000
rpec107	10000000	001000010000001001	1100000000100000100	010000100110001	1000000000000010	00000000000	10000000000	000000001100	11000	00	01000010001100000100000
*rfec108	00000000	000010010000110000	1001000001000000110	011010100110001	0000000000010000	00010000000	00000000000	101011001110	01010	01	00100100111001010001001
*rfec109	00001000	000010010000110000	1001000001000100110	011010100110001	0000000000010000	00010000000	00000010000	101011001110	01000	01	00000100101100101001001
*rfec110	00001000	000010010000110000	1001001000100000110	011010100110001	0000000000010000	00000000000	00000010000	101011001110	01010	01	00000000101100101001001
*rfec111	00001000	000010010000110000	1001000000100000110	011010101110001	0000000000010000	00000000000	00000010000	101011001110	01010	01	00000100101100101001001
*rfec112	00001000	000010010000110000	1001000000100100110	011010100110001	0000000000010000	00000000000	00000010000	101011001110	01010	01	00000100101100101001001
*ccec119	00000001	100111000010100000	1110000001000000100	010000000110001	0011010000010100	101000010001	00000000000	010000000010	00100	11	00011000001000010000000
ccec120	00000001	100111000010100000	1110000001000000100	010000000110001	0011010000010100	101000010001	00000000000	010000000010	00100	11	00011000001000010000000
ccec122	00000001	100111000010100000	1110000001000000100	010000000110001	0011010000010100	101000010001	00000000000	010000000010	00100	11	00011000001000010000000
ccec123	00000001	100111000010100000	1110000001000000100	010000000110001	0011010000010100	101000010001	00000000000	010000000010	00100	11	00011000001000010000000
*ccec121	00000001	100111000010100000	1110000001000000100	010000000100001	0011010000010100	101000010101	00000000000	010000000010	00100	11	00011000001000010000000
*ccec124	00000001	100111000110100000	1110000001000000100	110000011110001	0011010000010100	101000010001	00000000000	010000000010	00100	11	00011000001000010000000
*ccpn126	00000001	100110100010100000	1110000000111000100	010000000110001	0011010000010100	100100010001	00000000000	000000101111	00000	11	00011000000000010000000
ccpn129	00000001	100110100010100000	1110000000111000100	010000000110001	0011010000010100	100100010001	00000000000	000000101111	00000	11	00011000000000010000000
ccpn133	00000001	100110100010100000	1110000000111000100	010000000110001	0011010000010100	100100010001	00000000000	000000101111	00000	11	00011000000000010000000
ccpn134	00000001	100110100010100000	1110000000111000100	010000000110001	0011010000010100	100100010001	00000000000	000000101111	00000	11	00011000000000010000000
*ccpn127	00000001	100110100010100000	1110000000111000100	010000000110001	0011010000010100	000100010000	00000000000	000000101111	00000	11	00011000000000010000000
*ccpn128	00000001	000010000010100010	1110100001010000100	010000000100001	0011010000010100	000000010000	00000000000	010000000010	00000	11	00011000001000010000000
ccpn131	00000001	000010000010100010	1110100001010000100	010000000100001	0011010000010100	000000010000	00000000000	010000000010	00000	11	00011000001000010000000
ccpn132	00000001	000010000010100010	1110100001010000100	010000000100001	0011010000010100	000000010000	00000000000	010000000010	00000	11	00011000001000010000000
*ccpn130	00000001	000010000010100110	1110100001010000100	010000000100001	0011010000010100	000000010000	00000000000	010000000010	00000	11	00011000001000010000000
*cbgy7	01000000	100100000000100000	1100000101110000100	000001000100001	0011100000010000	010000000100	01000000000	000000001100	00000	11	10010000000000100000000
*cbgy9	01000000	100100000000100000	1100000000111000100	000000000100111	0011100000010000	010100000100	01000000000	000000001100	00000	11	10010000000000100000000
*cbmx12	01000100	100100000011100100	1100000000111000100	010101000001011	0011110000010000	010000000111	01100001000	000100000100	00000	01	10010000010100100000000
cbmx37	01000100	100100000011100100	1100000000111000100	010101000001011	0011110000010000	010000000111	01100001000	000100000100	00000	01	10010000010100100000000
cbmx38	01000100	100100000011100100	1100000000111000100	010101000001011	0011110000010000	010000000111	01100001000	000100000100	00000	01	10010000010100100000000
cbmx39	01000100	100100000011100100	1100000000111000100	010101000001011	0011110000010000	010000000111	01100001000	000100000100	00000	01	10010000010100100000000
cbgt53	01000100	100100000011100100	1100000000111000100	010101000001011	0011110000010000	010000000111	01100001000	000100000100	00000	01	10010000010100100000000
cbb190	01000100	100100000011100100	1100000000111000100	010101000001011	0011110000010000	010000000111	01100001000	000100000100	00000	01	10010000010100100000000
cbmx98	01000100	100100000011100100	1100000000111000100	010101000001011	0011110000010000	010000000111	01100001000	000100000100	00000	01	10010000010100100000000
cbmx99	01000100	100100000011100100	1100000000111000100	010101000001011	0011110000010000	010000000111	01100001000	000100000100	00000	01	10010000010100100000000
cbmx100	01000100	100100000011100100	1100000000111000100	010101000001011	0011110000010000	010000000111	01100001000	000100000100	00000	01	10010000010100100000000
*cbgt14	01000100	000100000011100100	1000000000111000100	010101000011111	0011110000010000	100000001000	01100001000	000100000100	00000	01	10010000010000100000000
cbgt46	01000100	000100000011100100	1000000000111000100	010101000011111	0011110000010000	100000001000	01100001000	000100000100	00000	01	10010000010000100000000
cbgt55	01000100	000100000011100100	1000000000111000100	010101000011111	0011110000010000	100000001000	01100001000	000100000100	00000	01	10010000010000100000000
*cbgt15	01000100	000100000011100100	1000000000111000100	010101000001011	0011110000010000	010000000111	01100001000	000100000100	00000	01	10010000010100100000000
cbmx8	01000100	000100000011100100	1000000000111000100	010101000001011	0011110000010000	010000000111	01100001000	000100000100	00000	01	10010000010100100000000
cbmx11	01000100	000100000011100100	1000000000111000100	010101000001011	0011110000010000	010000000111	01100001000	000100000100	00000	01	10010000010100100000000
cbgt16	01000100	000100000011100100	1000000000111000100	010101000001011	0011110000010000	010000000111	01100001000	000100000100	00000	01	10010000010100100000000
cbgt17	01000100	000100000011100100	1000000000111000100	010101000001011	0011110000010000	010000000111	01100001000	000100000100	00000	01	10010000010100100000000
cbmx35	01000100	000100000011100100	1000000000111000100	010101000001011	0011110000010000	010000000111	01100001000	000100000100	00000	01	10010000010100100000000
cbgt45	01000100	000100000011100100	1000000000111000100	010101000001011	0011110000010000	010000000111	01100001000	000100000100	00000	01	10010000010100100000000
cbgt47	01000100	000100000011100100	1000000000111000100	010101000001011	0011110000010000	010000000111	01100001000	000100000100	00000	01	10010000010100100000000
cbgt51	01000100	000100000011100100	1000000000111000100	010001000001011	0011110000010000	010000000111	01100001000	000100000100	00000	01	10010000010100100000000
cbgt52	01000100	000100000011100100	1000000000111000100	010101000001011	0011110000010000	010000000111	01100001000	000100000100	00000	01	10010000010100100000000
cbgt67	01000100	000100000011100100	1000000000111000100	010101000001011	0011110000010000	010000000111	01100001000	000100000100	00000	01	10010000010100100000000
cbmx102	01000100	000100000011100100	1000000000111000100	010101000001011	0011110000010000	010000000111	01100001000	000100000100	00000	01	10010000010100100000000
*cbmx36	01000100	000100000011100100	1100000000111000100	010101000001011	0011110000010000	010000000111	01100001000	000100000100	00000	01	10010000010100100000000
*cbgy40	01000000	100100000000100000	1100000000111000100	000001000100001	0011110000010000	010000000100	01000000000	000000001100	00000	11	10010000000000100000000
cbgy41	01000000	100100000000100000	1100000000111000100	000001000100001	0011110000010000	010000000100	01000000000	000000001100	00000	11	10010000000000100000000
*cbgy42	01000000	100100000000100000	1100000000100000100	000001010100001	0011110000010000	010000000100	01000000000	000000001100	00000	11	10010000000000100000000
*cbgt43	01000100	100100000011100100	1000000000111000100	010101000001011	0011110000010000	010000000111	01100001000	000100000100	00000	11	10010000010100100000000
*cbgt44	01000100	100100000011100100	1100000000111000100	010100000001011	0011110000010000	010000000111	01100001000	000100000100	00000	01	10010000010100100000000
cbgt49	01000100	100100000011100100	1100000000111000100	010100000001011	0011110000010000	010000000111	01100001000	000100000100	00000	01	10010000010100100000000
cbmx101	01000100	100100000011100100	1100000000111000100	010100000001011	0011110000010000	010000000111	01100001000	000100000100	00000	01	10010000010100100000000
*cbgt48	01000100	100100000011100100	1000000000111000100	010101000011111	0011110000010000	100000001000	01100001000	000100000100	00000	01	10010000010000100000000
*cbgt50	01000100	000100000011100100	1100000000111000100	010101000001011	0011110000010000	010000000111	01100001000	000100000100	00000	01	10010000010000100000000
*cbgt54	01000100	100100000011100100	1000000000111000100	010101000001011	0011110000010000	010000000111	01100001000	000100000100	00000	01	10010000010100100000000
*cbgt56	01000100	000100000011100100	1000000000111000100	010101000001011	0011110000010000	010000000111	01100001000	000100000100	00000	01	10010000010100100000000
*cbgt68	01000100	100100000011100100	1100000000111000100	010101000001011	0011110000010000	010000000111	01100001000	000100000100	00000	01	10010000010100100000000
*cbmx104	01000100	100100000011100100	1000000000111000100	010001000001011	0011110000010000	100000001000	01100001000	000100000100	00000	01	10010000010000100000000
*cbec113	01000000	100100000000100000	1100000000111000100	000001000100001	0011110000010000	010000000100	01000000000	000000001100	00000	11	10010000000000100000000
cbec115	01000000	100100000000100000	1100000000111000100	000001000100001	0011110000010000	010000000100	01000000000	000000001100	00000	11	10010000000000100000000

Specimen (cont'd)	BstNI	HaeI	HinfI	MboI	RsaI	Sau96I	ScrFI	TaqI	BstUI	Hind III	DdeI
*cbec114	01000000	100100000000100000	110000000111000100	000001000100001	0011110000010000	010000000100	01000000000	000000001100	00000	11	100100001000000100000000
*cbpn135	01000000	100100000000100000	110000000111000100	010001000100011	0011110000010000	010000000100	01000000000	000000001100	00000	11	100100001001000100000000
*cbpn136	01000000	100100000000100000	110000000111000000	010001000100011	0011110000010000	010000000100	01000000000	000000001100	00000	11	100100001001000100000000
*cbpn137	01000000	100100000000100000	100000000111000100	010001000100011	0011110000010000	010000000100	01000000000	000000001100	00000	11	100100001001000100000000
*cbpn138	01000000	100100000011100100	110000000101100100	010101000011111	0011111000010000	100000000100	01100001000	000000001100	00000	01	100100000001000100000000
*cbpn139	01000100	100100000011100100	110010001011000100	010101000011111	0011111000010000	100000000100	01100001000	000000001100	00000	01	100100000001000100000000
*cbpn140	01000000	100100000000100000	110000000111000100	000001000100001	0011110000010000	010000000100	01000000000	000000001100	00000	11	100100000001000100000000
*cbpn144	01000000	100111000000100000	110000000111000100	000001000100011	0011110000010000	010000000100	01000000000	000000001100	00000	11	100100001001000100000000
*csmx3	00110101	110001000011110000	110000010110000100	010001000111001	0011110010010000	100000000010	00011101011	000000000100	00000	01	110100000100000000000000
csgt4	00110101	110001000011110000	110000010110000100	010001000111001	0011110010010000	100000000010	00011101011	000000000100	00000	01	110100000100000000000000
csgt23	00110101	110001000011110000	110000010110000100	010001000111001	0011110010010000	100000000010	00011101011	000000000100	00000	01	110100000100000000000000
csmx24	00110101	110001000011110000	110000010110000100	010001000111001	0011110010010000	100000000010	00011101011	000000000100	00000	01	110100000100000000000000
csgt75	00110101	110001000011110000	110000010110000100	010001000111001	0011110010010000	100000000010	00011101011	000000000100	00000	01	110100000100000000000000
csgt76	00110101	110001000011110000	110000010110000100	010001000111001	0011110010010000	100000000010	00011101011	000000000100	00000	01	110100000100000000000000
*cses18	00110101	110001000011110000	110000010110000101	010001000111001	0011110010010000	100000000010	00011001001	000000001100	00000	01	110100000100000000000000
*cses69	00110101	110001000011110000	110000010110000101	010001000111001	0011110010010000	100000000010	00011001001	000000001100	00000	01	110100000100000000000000
cses72	00110101	110001000011110000	110000010110000101	010001000111001	0011110010010000	100000000010	00011001001	000000001100	00000	01	110100000100000000000000
*cses70	00110101	110001000011110000	110000010110000101	010001000111001	0011110010010000	100000000010	00011001001	000000001100	00000	01	110100000100000000000000
cses71	00110101	110001000011110000	110000010110000101	010001000111001	0011110010010000	100000000010	00011001001	000000001100	00000	01	110100000100000000000000
*csmx73	00010101	110001000011110000	110000010110000100	010001000111001	0011110010010000	100000000010	00001001001	000000000100	00000	01	110100010100000000000000
*csgt74	00110101	110001000011110000	110000010110000100	010001000111001	0011110010010000	100000000010	00011101011	000000000110	00000	01	110100000100000000000000
*cpgy5	01000001	100110000001110000	100000010111000100	000001000100011	0011110000010000	110000000110	01000000001	000000001100	00000	11	100100000000000100000000
*cpmx6	01000100	100110000001110000	110000010110000100	000001000110011	0011110000010000	110000000010	01100001000	000000001100	00000	11	100000010000000100000000
cpmx19	01000100	100110000001110000	110000010110000100	000001000110011	0011110000010000	110000000010	01100001000	000000001100	00000	11	100000010000000100000000
cpmx21	01000100	100110000001110000	110000010110000100	000001000110011	0011110000010000	110000000010	01100001000	000000001100	00000	11	100000010000000100000000
cpmx60	01000100	100110000001110000	110000010110000100	000001000110011	0011110000010000	110000000010	01100001000	000000001100	00000	11	100000010000000100000000
cpmx61	01000100	100110000001110000	110000010110000100	000001000110011	0011110000010000	110000000010	01100001000	000000001100	00000	11	100000010000000100000000
cpmx62	01000100	100110000001110000	110000010110000100	000001000110011	0011110000010000	110000000010	01100001000	000000001100	00000	11	100000010000000100000000
cpmx63	01000100	100110000001110000	110000010110000100	000001000110011	0011110000010000	110000000010	01100001000	000000001100	00000	11	100000010000000100000000
cpb191	01000100	100110000001110000	110000010110000100	000001000110011	0011110000010000	110000000010	01100001000	000000001100	00000	11	100000010000000100000000
cpb192	01000100	100110000001110000	110000010110000100	000001000110011	0011110000010000	110000000010	01100001000	000000001100	00000	11	100000010000000100000000
cpmx93	01000100	100110000001110000	110000010110000100	000001000110011	0011110000010000	110000000010	01100001000	000000001100	00000	11	100000010000000100000000
cpmx94	01000100	100110000001110000	110000010110000100	000001000110011	0011110000010000	110000000010	01100001000	000000001100	00000	11	100000010000000100000000
cpmx96	01000100	100110000001110000	110000010110000100	000001000110011	0011110000010000	110000000010	01100001000	000000001100	00000	11	100000010000000100000000
cppn143	01000100	100110000001110000	110000010110000100	000001000110011	0011110000010000	110000000010	01100001000	000000001100	00000	11	100000010000000100000000
cppn145	01000100	100110000001110000	110000010110000100	000001000110011	0011110000010000	110000000010	01100001000	000000001100	00000	11	100000010000000100000000
cppn146	01000100	100110000001110000	110000010110000100	000001000110011	0011110000010000	110000000010	01100001000	000000001100	00000	11	100000010000000100000000
cppn147	01000100	100110000001110000	110000010110000100	000001000110011	0011110000010000	110000000010	01100001000	000000001100	00000	11	100000010000000100000000
cppn148	01000100	100110000001110000	110000010110000100	000001000110011	0011110000010000	110000000010	01100001000	000000001100	00000	11	100000010000000100000000
cppn149	01000100	100110000001110000	110000010110000100	000001000110011	0011110000010000	110000000010	01100001000	000000001100	00000	11	100000010000000100000000
*cpgy10	01000001	100111000001110000	110000010101000100	000001000100011	0011110000010000	110000000110	01000000001	000000001100	00000	11	100100000000000100000000
*cpgy13	01000001	100111000001110000	110000010101000100	000001000100001	0011110000010000	110000000110	01000000001	000000001100	00000	11	100100000000000100000000
*cpgy20	01000001	100111000001110000	110000010101000100	000001000100011	0011110000010000	110000000010	01000000001	000000001100	00000	11	100100000000000100000000
*cpgy22	01000101	110001000001100000	110000010101000100	000000000100011	0011110000010000	110000000110	01000001001	000000001100	00001	11	100100000000000100000000
*cpgy57	01000001	100110000001110000	110000010101000100	000001000100011	0011110000010000	110000000010	01000000001	000000001100	00000	11	100100010000000100000000
*cpgy58	01000001	100110000001110000	100000010111000100	000001000100011	0011110000010000	110000000010	01000000001	000000001100	00000	11	100100000000000100000000
*cpgy59	01000001	100110000001110000	110000010101000100	000001000100011	0011110000010000	110000000010	01000000001	000000001100	00000	11	100100000000000100000000
*cpgy64	01000001	100110000001110000	110000010101000100	000001000100000	0011110000010000	110000000010	01000000001	000000001100	00000	11	100100000000000100000000
cpgy65	01000001	100110000001110000	110000010101000100	000001000100011	0011110000010000	110000000010	01000000001	000000001100	00000	11	100100000000000100000000
cpgy66	01000001	100110000001110000	110000010101000100	000001000100011	0011110000010000	110000000010	01000000001	000000001100	00000	11	100100000000000100000000
*cpmx95	01000100	100110000001110000	110000010110000100	000001000110011	0011110000010000	110000000010	01100001000	000000001100	00000	11	100000010000000100000000
*cpmx97	01000100	100110000001110000	110000010110000100	000001000110011	0011110000010001	110000000010	01100001000	000000001100	00000	11	100000010000000100000000
*cpmx103	01000100	100110000001110000	110000010110000100	000001000110011	0011110000010000	011000000010	01100001000	000000001100	00000	11	100000010000000100000000
*cpec116	01000001	100110000001110000	110000010101000100	000001000100011	0011110000010000	110000000110	01000000001	000000001100	00000	11	100100010000000100000000
*cpec117	01000101	100110000001110000	110000010101000100	000001000100011	0011110000010000	110000000110	01000001001	000000001100	00000	11	100100010000000100000000
*cpec118	01000001	100110000001110000	110000010101000100	000001000100011	0011110000010000	110000000110	01000000001	000000001100	00000	11	100100010000000100000000
*cppn141	01000100	100110000001110000	110000010110000100	000001000110011	0011110000010000	110000000010	01100001000	000000001100	00000	11	100000010000000100000000
*cppn142	01000001	100110000001110000	110000010101000100	010001000100001	0011110000010000	100000000110	01000000001	000000001100	00000	11	100100000000000100000000

Acknowledgments

We thank Brock Fenton for comments throughout the course of the project and Peter de Groot for instruction on the partial endonuclease digest mapping technique. Fiona Reid, Robert Dowler, Susan Woodward, Charles Robertson, Eamon O'Toole, Francisco Sornoza, Peter de Groot, Yolanda Hortelano, Jim Cathey, and Jenna Dunlop helped to obtain samples in the field. We appreciate the cooperation of the national authorities who provided permits and approved our collecting and research work within their countries. Many people with Youth Challenge International provided logistic support and field assistance in Guyana. Bob Murphy, Amy

Lathrop, and Allan Baker kindly made available computer time to use PAUP, and Cary Gilmore assisted with lab supplies. Nancy Simmons, Juan Carlos Morales, Laurence Packer, Ron Pearlman, and an anonymous reviewer provided useful comments during revision of the manuscript. Fieldwork and laboratory analysis were generously supported by the ROM Foundation of the Royal Ontario Museum through the Department of Museum Volunteers and the ROM Reproductions Association. This is Contribution No. 35 from the Centre for Biodiversity and Conservation Biology at the Royal Ontario Museum.

Literature Cited

Archie, J. W. 1989. Homoplasy excess ratios: New indices for measuring levels of homoplasy in phylogenetic systematics and a critique of the consistency index. Systematic Zoology 38:253–269.

Arévalo, E., S. K. Davis, and J. W. Sites, Jr. 1994. Mitochondrial DNA sequence divergence and phylogenetic relationships among eight chromosome races of the *Sceloporus grammicus* complex (Phrynosomatidae) in Central Mexico. Systematic Biology 43:387–418.

Avise, J. C. 1994. Molecular Markers, Natural History, and Evolution. Chapman and Hall, New York.

Baker, R. J. 1979. Karyology. *In* Biology of Bats of the New World Family Phyllostomatidae, Part III, R. J. Baker, J. K. Jones, Jr., and D. C. Carter, eds., pp. 107–156. No. 10, Special Publications, The Museum, Texas Tech University, Lubbock.

Baker, R. J., and W. J. Bleier. 1971. Karyotypes of bats of the subfamily Carolliinae (Mammalia; Phyllostomatidae) and their evolutionary implications. Experientia (Basel) 27:220–222.

Baker, R. J., C. S. Hood, and R. L. Honeycutt. 1989. Phylogenetic relationships and classification of the higher categories of the New World bat family Phyllostomidae. Systematic Zoology 38:228–238.

Barquez, R. M., N. P. Giannini, and M. A. Mares. 1993. Guide to the Bats of Argentina: Guia de los Murcielagos de Argentina. Oklahoma Museum of Natural History, Norman.

Bremer, K. 1994. Branch support and tree stability. Cladistics 10:295–304.

Cherfas, J. 1989. Science gives ivory a sense of identity. Science 246:1120–1121.

Cronin, M. A., W. J. Spearman, R. L. Wilmot, J. C. Patton, and J. W. Bickham. 1993. Mitochondrial DNA variation in chinook salmon *(Oncorhynchus tshawytscha)* and chum salmon *(O. keta)* detected by restriction enzyme analysis of polymerase chain reaction (PCR) products. Canadian Journal of Fisheries and Aquatic Sciences 50:708–715.

DeBry, R. W., and N. A. Slade. 1985. Cladistic analysis of restriction endonuclease cleavage maps within a maximum-likelihood framework. Systematic Zoology 34:21–34.

Dowling, T. E., C. Moritz, and J. D. Palmer. 1990. Nucleic acids. II. Restriction site analysis. *In* Molecular Systematics, D. M. Hillis and C. Moritz, eds., pp. 250–317. Sinauer Associates, Sunderland, Mass.

Eisenberg, J. F. 1989. Mammals of the Neotropics. Vol. 1., The Northern Neotropics: Panama, Colombia, Venezuela, Guyana, Suriname, French Guiana. University of Chicago Press, Chicago.

Farris, J. S. 1989a. The retention index and the rescaled consistency index. Cladistics 5:417–419.

Farris, J. S. 1989b. The retention index and homoplasy excess. Systematic Zoology 38:406–407.

Fleming, T. H. 1988. The Short-Tailed Fruit Bat: A Study in Plant–Animal Interactions. University of Chicago Press, Chicago.

Hall, E. R. 1981. The Mammals of North America, Vol. 1, 2nd Ed. Wiley, New York.

Hershkovitz, P. 1969. The evolution of mammals on southern continents. Quarterly Review of Biology 44:1–70.

Hood, C. S., and J. D. Smith. 1982. Cladistical analysis of female reproductive histomorphology in phyllostomatoid bats. Systematic Zoology 31:241–251.

Husson, A. M. 1978. The mammals of Suriname. Zoölogische Monographieën van het Rijksmuseum van Natuurlijke Historie 2:1–569.

Jones, J. K., Jr., and D. C. Carter. 1976. Annotated checklist, with keys to subfamilies and genera. *In* Biology of Bats of the New World Family Phyllostomatidae, Part I, R. J. Baker, J. K. Jones, Jr., and D. C. Carter, eds., pp. 7–38. No. 10, Special Publications, The Museum, Texas Tech University, Lubbock.

Koopman, K. F. 1976. Zoogeography. *In* Biology of Bats of the New World Family Phyllostomatidae, Part I, R. J. Baker, J. K. Jones, Jr., and D. C. Carter, eds., pp. 39–47. No. 10, Special Publications, The Museum, Texas Tech University, Lubbock.

Koopman, K. F. 1982. Biogeography of the bats of South America. *In* Mammalian Biology in South America, M. A. Mares and H. H. Genoways, eds., pp. 273–302. No. 6, Special Publication Series, Pymatuning Laboratory of Ecology, University of Pittsburgh, Pittsburgh.

Koopman, K. F. 1993. Order Chiroptera. *In* Mammal Species of the World: A Taxonomic and Geographic Reference, 2nd Ed., D. E. Wilson and D. M. Reeder, eds., pp. 137–241. Smithsonian Institution Press, Washington, D.C.

Linares, O. J. 1987. Murcielagos de Venezuela. Cuadernos Lagoven, Venezuela.

McLellan, L. J. 1984. A morphometric analysis of *Carollia* (Chiroptera, Phyllostomidae). American Museum Novitates 2791:1–35.

Morales, J. C., and J. W. Bickham. 1995. Molecular systematics of the genus *Lasiurus* (Chiroptera: Vespertilionidae) based on restriction-site maps of the mitochondrial ribosomal genes. Journal of Mammalogy 76:730–749.

Morales, J. C., and D. J. Melnick. 1994. Molecular systematics of living rhinoceros. Molecular Phylogenetics and Evolution 3:128–134.

Morales, J. C., J. C. Patton, and J. W. Bickham. 1993. Partial endonuclease digestion mapping of restriction sites using PCR-amplified DNA. PCR Methods and Applications 2:228–233.

Owen, J. G., D. J. Schmidly, and W. B. Davis. 1984. A morphometric analysis of three species of *Carollia* (Chiroptera, Glossophaginae) from Middle America. Mammalia 48:85–93.

Pacheco, V., H. de Macedo, E. Vivar, C. Ascorra, R. Arana-Cardó, and S. Solari. 1995. Lista anotada de los mamíferos Peruanos. Occasional Papers in Conservation Biology 2:1–35.

Patton, J. L., and A. L. Gardner. 1971. Parallel evolution of multiple sex-chromosome systems in the phyllostomatid bats, *Carollia* and *Choeroniscus*. Experientia (Basel) 27:105–106.

Pine, R. H. 1972. The bats of the genus *Carollia*. Technical Monograph, Texas Agricultural Experiment Station, Texas A&M University 8:1–125.

Savage, D. E. 1951. A Miocene phyllostomatid bat from Colombia, South America. University of California Publications, Bulletin of the Department of Geological Sciences 28:357–366.

Savage, J. M. 1974. The isthmian link and the evolution of Neotropical mammals. Natural History Museum of Los Angeles County Contributions in Science 260:1–51.

Simpson, G. G. 1950. History of the fauna of Latin America. American Scientist 38:361–389.

Stock, A. D. 1975. Chromosome banding pattern, homology, and its phylogenetic implications in bat genera *Carollia* and *Choeroniscus*. Cytogenetics and Cell Genetics 14:34–41.

Straney, D. O., M. H. Smith, I. F. Greenbaum, and R. J. Baker. 1979. Biochemical genetics. *In* Biology of Bats of the New World Family Phyllostomatidae, Part III, R. J. Baker, J. K. Jones, Jr., and D. C. Carter, eds., pp. 157–176. No. 10, Special Publications, The Museum, Texas Tech University, Lubbock.

Swofford, D. L. 1993. PAUP: Phylogenetic Analysis Using Parsimony, Version 3.1. Computer program distributed by the Illinois Natural History Survey, Champaign.

Uieda, W. 1980. Ocorrência de *Carollia castanea* na Amazônia Brasileira (Chiroptera, Phyllostomidae). Acta Amazonica 10: 936–938.

van Coeverden de Groot, P. J. 1995. Phylogenetic systematics and speciation in highland deer mice of the *Peromyscus mexicanus* species group. Master's thesis, University of Toronto, Toronto.

Williams, P. H., R. I. Vane-Wright, and C. J. Humphries. 1993. Measuring biodiversity for choosing conservation areas. *In* Hymenoptera and Biodiversity, J. LaSalle and I. D. Gauld, eds., pp. 309–328. C.A.B. International, Wallingford, U.K.

4
Phylogenetic Accuracy, Stability, and Congruence

Relationships within and among the New World Bat Genera *Artibeus*, *Dermanura*, and *Koopmania*

RONALD A. VAN DEN BUSSCHE, JEREMY L. HUDGEONS, AND ROBERT J. BAKER

Fig-eating bats (*Artibeus*, sensu Andersen 1908) of the New World leaf-nosed family Phyllostomidae represent a complex that is both a common and important component of mammalian biodiversity of the New World tropics. These bats therefore are important in several issues such as the nature of speciation, biogeography, and habitat loss. Although it has long been recognized that the approximately 20 species can be placed in two groups on the basis of morphology, karyology, and allozymes, these taxa have been divided into three genera (*Artibeus*, *Dermanura*, and *Koopmania*; Owen 1987, 1991). Moreover, it has been proposed that two of these genera are more closely related to other Stenodermatini genera than to each other (Owen 1987). Therefore, because these taxa are important to the mammalian fauna of Central and South America, it is critical to produce a robust phylogenetic tree accurately reflecting the evolutionary history of the New World leaf-nosed bats *Artibeus*, *Dermanura*, and *Koopmania*.

It is generally agreed that an accurate phylogeny of life could provide meaningful order to major biological questions related to evolution, medicine, ecology, and biodiversity (Gouy and Li 1989; Durfy and Willard 1990; Baker 1994; Hillis et al. 1994). Although trees that estimate relatedness are commonly generated, how to set confidence limits on the branches of such trees has been difficult to resolve (Felsenstein 1985; Li et al. 1987; Saitou and Nei 1987; Bremer 1988; Nei 1991; Donoghue et al. 1992; Rzhetsky and Nei 1992; Hillis and Bull 1993; Hillis et al. 1992, 1994; Felsenstein and Kishino 1993; Zharkikh and Li 1993). We compared two methods, bootstrap (Felsenstein 1985) and Bremer support (Bremer 1988) analyses, to examine the extent to which each of the two methods retrieves the phylogenetic relationships estimated by the other.

Historical Perspective of the Taxonomy of *Artibeus*

On the basis of the number of molar teeth, Gervais (1855) proposed dividing the species of *Artibeus* into two genera (*Artibeus* and *Dermanura*). However, Andersen (1908) clearly illustrated that this character, because it has a high degree of variability, provided an artificial classification that did not reflect the phylogenetic relationships among stenodermatine bats and thus relegated *Dermanura* as a synonym of *Artibeus*. It has long been recognized that the approximately 20 species of *Artibeus* (sensu Andersen 1908) represent two groups based on morphological, karyotypic, and allozymic characters (Baker 1973; Straney et al. 1979). However, on the basis of analysis of morphological characters from essentially all extant stenodermatine taxa, if the 20 species represented a monophyletic genus at least 15 additional stenodermatine genera would have to be included (Owen 1987; Figure 17). Therefore, Owen (1987) suggested that *Artibeus* be limited to the larger species and resurrected *Dermanura* (Gervais 1855) for the smaller taxa. Owen (1991) further restricted the latter genus by recognizing *D. concolor* as a monotypic genus, *Koopmania*.

Enchisthenes (Andersen 1906) is a monotypic genus that appears more similar to *Artibeus* than to *Uroderma* with the exception of the position and relative size of the upper third molar and the relative size of the lower third molar, which are more similar to the condition found in *Uroderma* (Andersen 1906, 1908; Miller 1907). Additionally, based on a few dental characters and a tragus with a pointed projection on the inner margin and near the tip, *Enchisthenes* appears more closely allied with *Artibeus* and deserves generic status (Andersen 1906, 1908; Miller 1907). Determining the sister-taxon to *Enchisthenes* has been problematic for nearly all systematic studies designed to elucidate its phylogenetic placement with other stenodermatine bats. Molecular evidence has demonstrated that *Enchisthenes hartii* is not a member of *Dermanura* and is an outlier to the *Artibeus-Dermanura* complex (Van Den Bussche et al. 1993). Thus, we follow Van Den Bussche et al. (1993) and recognize *Enchisthenes* as a distinct genus, as originally proposed by Andersen (1906), on the basis of morphological characters.

Van Den Bussche et al. (1993) examined the phylogenetic distribution of an *Eco*RI-defined nuclear satellite DNA repeat and 402 base pairs (bp) of DNA sequence variation from the mitochondrial cytochrome *b* gene to test the validity of the genera *Artibeus*, *Dermanura*, *Koopmania*, and *Enchisthenes*. Both the nuclear heterochromatic DNA and the mitochondrial sequence data were best interpreted as indicating that *Artibeus*, *Koopmania*, and *Dermanura* shared a recent common ancestor after diverging from other Stenodermatine genera. Thus, the generic distinctions within this clade rest on the magnitude of differences among sublineages, not on the argument of paraphyly as suggested by Owen (1987). Most current data, including characters as diverse as karyology (Baker 1973), morphology (Smith 1976), and both nuclear and mitochondrial DNA markers (Van Den Bussche et al. 1993), have been interpreted as supporting a close evolutionary relationship of *Artibeus*, *Dermanura*, and *Koopmania*. However, the validity of these three genera is neither clear nor widely accepted (Koopman 1993).

Within *Artibeus* and *Dermanura*, sister-group relation-

ships and alpha taxonomy are also in a state of flux. *Artibeus* (sensu Owen 1987) contains approximately 10 species (*amplus*, *fimbriatus*, *fraterculus*, *hirsutus*, *inopinatus*, *intermedius*, *jamaicensis*, *lituratus*, *planirostris*, *obscurus*) while there are 8 recognized species of *Dermanura* (*anderseni*, *azteca*, *cinerea*, *glauca*, *gnoma*, *phaeotis*, *tolteca*, *watsoni*). Within both genera numerous questions exist concerning the validity of several named taxa as well as sister-group relationships. For example, several taxa have been recently described (*A. amplus* and *D. gnoma*) or rediscovered (*A. fimbriatus*), and the validity of several others is currently being evaluated (Handley 1987; Koopman 1993). In the original description of the genus *Artibeus*, Gray (1838) included 7 species. Of these, *A. lobatus* was never subsequently cited and may be a synonym for *A. jamaicensis*; also, the validity of his *A. fuliginosus* is questionable and has been synonymized with *A. obscurus* (Handley 1989). Other questions at the alpha taxonomic level concern (1) whether *A. planirostris* is a subspecies of the wide-ranging *A. jamaicensis*, as proposed by Hershkovitz (1949) and Kraft (1982); (2) whether *A. intermedius* (Davis, 1984) is a valid species or rather a subspecies of *A. lituratus*, as maintained by Koopman (1993); and (3) the validity of *A. fimbriatus* (Gray 1838), a taxon that has been ignored for 150 years or so but which Handley (1989) and Patterson et al. (1992) argued is the most distinctive species of *Artibeus*. With the exception of Owen (1987) and Patterson et al. (1992), there have been few attempts to define phylogenetic relationships either among taxa or within either genus.

The taxonomic objective of this study is to provide resolution to the phylogenetic relationships within and among the genera *Artibeus*, *Dermanura*, and *Koopmania*. These relationships are addressed by examining the phylogenetic distribution and restriction fragment pattern of an *Eco*RI-defined satellite DNA repeat and DNA sequence variation from the entire mitochondrial cytochrome *b* gene for essentially all extant species of *Artibeus* and *Dermanura* as well as *Koopmania*. The satellite DNA has been shown to be restricted to *Artibeus*, *Dermanura*, and *Koopmania* (Van Den Bussche et al. 1993), whereas DNA sequence variation from the cytochrome *b* gene has provided resolution of the phylogenetic relationships of several other phyllostomid taxa (Van Den Bussche and Baker 1993; Van Den Bussche et al. 1993; Baker et al. 1994).

Materials and Methods

DNA Isolation and Phylogenetic Distribution of Satellite DNA

DNA was extracted from frozen heart, liver, kidney, or muscle tissue following standard protocols (Bingham et al.

1981; Strauss 1987; Longmire et al. 1991). To evaluate the phylogenetic distribution of the satellite DNA repeat, approximately 5 μg of genomic DNA from all taxa under study was digested with the restriction endonuclease *Eco*RI, separated by electrophoresis on a 0.8% agarose gel, and transferred to a nylon membrane (Southern 1975). The approximately 900-bp *Eco*RI-defined satellite monomer from *A. lituratus* was radioactively labeled and hybridized to the transferred DNA following the protocol described by Van Den Bussche et al. (1993).

Cytochrome *b* and Phylogenetic Analyses

The entire 1,140-bp cytochrome *b* gene was amplified via polymerase chain reaction (PCR) (Saiki et al. 1986, 1988). Conditions for amplification and sequencing were described by Baker et al. (1994). Cytochrome *b* DNA sequences were entered into MacClade (Maddison and Maddison 1992) and aligned by hand. Quantitative comparisons for all pairwise comparisons among taxa were made for the cytochrome *b* gene. These comparisons included transition and transversion substitutions for the first-, second-, and third-codon positions, as well as percent sequence divergence. Values of sequence divergence were corrected for multiple substitutions using the two-parameter model of Kimura (1980). To test whether nucleotide substitution rates of the cytochrome *b* gene differed within Stenodermatini bats or within *Artibeus* or *Dermanura* relative to the outgroups *(Uroderma* and *Chiroderma)*, we employed the relative rates test (Sarich and Wilson 1967) and compared these values to a binomial distribution (Mindell and Honeycutt 1990; Allard and Honeycutt 1992).

Amino acid residues and DNA sequences were analyzed as discrete characters. MacClade (Maddison and Maddison 1992) was utilized to translate DNA sequences into amino acid residues and then to create a step matrix for a proto-parsimony analysis using PAUP version 3.0s (Swofford 1991). Phylogenetic relationships also were evaluated by coding DNA sequences as discrete, unordered characters, employing various search algorithms (heuristic, branch-and-bound, exhaustive), and applying several character-state-transformation weighting matrices. The various weighting schemes used in the phylogenetic analyses were as follows: all substitutions; first- and second-codon positions only; transversion parsimony analysis (weighing transversions over transitions by 1:0, 2:1, 5:1, and 15:1); and, finally, a weighted parsimony in which the weight applied to each of the 12 nucleotide substitutions was inversely proportional to the frequency of occurrence of each substitution in the data set utilizing the "All Changes" option in MacClade (Maddison and Maddison 1992). The frequency of these changes was based on the frequency of all 12 substitutions among the single most-parsimonious tree as well as all other trees within two steps of the most-parsimonious topology.

To evaluate which data sets under the various weighting schemes contained significant levels of phylogenetic structure, the distribution of 10,000 randomly drawn trees was examined with the g_1 statistic (Hillis 1991; Huelsenbeck 1991). Resulting g_1 values were compared to critical values (Hillis and Huelsenbeck 1992). All data sets that contained significantly greater levels of phylogenetic signal than noise were used in a phylogenetic analysis. To ascertain the relationships among *Artibeus*, *Koopmania*, *Dermanura*, and *Enchisthenes*, all taxa were included in a heuristic search with character-state changes polarized by designating *Uroderma* and *Chiroderma* as outgroups. These taxa were chosen as outgroups on the basis of a previous study (Van Den Bussche et al. 1993). The confidence or accuracy of each clade was evaluated by heuristic bootstrap analysis with 1,000 iterations. Finally, MacClade was used to evaluate alternative topologies of *Artibeus*, *Koopmania*, *Dermanura*, and *Enchisthenes*.

Relationships among species within the *Artibeus* and *Dermanura* clades were evaluated by employing all character-state transformation matrices just listed with the exception of amino acid residues and only first- and second-codon positions. Polarity of character-state changes was established using *Enchisthenes* and a member of the sister-clade as outgroups. Branch-and-bound searches were performed to identify the most-parsimonious topology(ies), and the confidence or accuracy of each clade was evaluated by bootstrap analysis with a branch-and-bound option and 1,000 iterations. To evaluate further the robustness of clades on the most-parsimonious trees, a Bremer support analysis was performed (Bremer 1988; Donoghue et al. 1992; Kallersjo et al. 1992; Eernisse and Kluge 1993; Lundrigan and Tucker 1994).

Specimens Examined

Tissues from the following specimens were obtained from the frozen tissue collections at The Museum, Texas Tech University (TK); Museum of Southwestern Biology, University of New Mexico (NK); Museum of Vertebrate Zoology, Berkeley, California (MVZ); The Field Museum, Chicago; and the American Museum of Natural History (AMNH):

Artibeus fimbriatus (TK 18991)—Brazil, São Paulo, Estaçao Biológica de Boraceia
Artibeus fraterculus (TK 16631, MVZ 168913)—Peru, Department of Lambayeque, about 12 km NNW Olmos

Artibeus hirsutus (NK 11128)—Mexico, Sonora, about 8 km NW San Carlos

Artibeus inopinatus (TK 40184)—Honduras, Valle, about 13.7 km SSW San Lorenzo

Artibeus intermedius (TK 31924)—Costa Rica, Guanacaste, Finca La Pacifica

Artibeus jamaicensis (TK 17073)—Suriname, Nickerie, Kayserberg air field strip, 3°06′ N, 56°29′ W

Artibeus jamaicensis (TK 18788, AMNH 267202)—French Guiana, Paracou, near Sinnamary

Artibeus lituratus (TK 25029)—Trinidad, St. George, SIMLA Research Center

Artibeus obscurus (TK 17308)—Suriname, Para, Zanderij, 5°27′ N, 55°12′ W

Artibeus obscurus (TK 18787, AMNH 267210)—French Guiana, Paracou, near Sinnamary

Artibeus planirostris (TK 16633, MVZ 170016)

Chiroderma villosum (TK 25052)—Trinidad, St. George, SIMLA Research Center, 6.4 km N Arimas

Dermanura anderseni (NK 14319)—Bolivia, Pando, Rio Madre de Dios

Dermanura azteca (TK 4723)—Mexico, Sinaloa, 3.2 km NE Rosario on road to Matatan

Dermanura cinerea (TK 18790, AMNH 267197)—French Guiana, Paracou, near Sinnamary

Dermanura glauca (TK 16636, MVZ 173952)—Peru, Department of Cuzco, 2 km NE Amaybamba

Dermanura gnoma (TK 18789, AMNH 267200)—French Guiana, Paracou, near Sinnamary

Dermanura phaeotis (TK 5411)—Nicaragua, Managua, about 1.2 km N Masachapa

Dermanura tolteca (TK 22579)—Panama, Darién, 6 km SW Cana

Dermanura watsoni (TK 7877)—Nicaragua, Zelaya, 3 km NW Rama

Enchisthenes hartii (TK 22690)—Peru, Huanúco, 11 km N, 6 km E Tingo Maria

Koopmania concolor (TK 10378)—Suriname, Commewijne, Nieuwe Grand Plantation, 5°53′ N, 54°54′ W

Koopmania concolor (TK 11240)—Suriname, Brokopondo, 1 km N Rudi, Kappelvliequeld 300 m, 3°44′ N, 56°08′ W

Uroderma bilobatum (TK 25256)—Trinidad, Mayaro, about 6.4 km N Arima

Results

The 900-bp *Eco*RI fragment from *A. lituratus* (Van Den Bussche et al. 1993) was hybridized to *Eco*RI-digested genomic DNA from representatives of all 19 species of *Artibeus*, *Dermanura*, and *Koopmania* examined in this study. As previously shown by Van Den Bussche et al. (Figure 2; 1993), this *Eco*RI-defined satellite DNA repeat occurs in all members of the genera *Artibeus*, *Dermanura*, and *Koopmania*. Moreover, all members of the genus *Dermanura* are characterized by the presence of 500-bp and 400-bp *Eco*RI fragments (data not shown).

Of the 1,140 bp of the cytochrome *b* gene, 738 bp (64.7%) were identical among all taxa. Of the 402 variable sites, 97 were autapomorphous, leaving 305 positions as potentially informative. Percent sequence divergence for all pairwise comparisons, corrected for multiple substitutions (Kimura 1980), are presented above the diagonal in Table 4.1, and the transition-to-transversion ratio for each pairwise comparison is entered below the diagonal. Stenodermatine taxa demonstrated nucleotide divergence values ranging from 19.2% for the comparison of *Uroderma* and *Enchisthenes* to 1.2% between two individuals of *A. jamaicensis* (Table 4.1). The relative rates test indicated that pairwise comparisons of ingroup taxa relative to *Uroderma* and to *Chiroderma* were not significant at the 5% level, based on the binomial distribution (these results may be obtained from the senior author). Phylogenetic analyses were performed to evaluate the relationships among and within *Artibeus*, *Dermanura*, and *Koopmania*. All weighting schemes contained significant phylogenetic structure when compared to critical values based on the g_1 statistic.

The topology presented in Figure 4.1 is the result of a heuristic search utilizing all variable positions and applying equal weight to all substitutions. A single most-parsimonious tree of 1,251 steps and a highly significant g_1 statistic of -0.581 ($p < 0.01$) were generated (Hillis and Huelsenbeck 1992). Numbers above each lineage in Figure 4.1 are the assigned branch lengths from PAUP, and numbers below each lineage are the percentage of 1,000 bootstrap iterations in which each clade was detected. All other weighting schemes resulted in similar topologies concerning the relationships among genera. The most-parsimonious trees from the analyses with differential weights assigned to the various types of character-state changes differed according to the level of resolution provided to terminal taxa.

Figure 4.2A–C presents simplified versions of the most-parsimonious tree for the relationships of *Artibeus*, *Dermanura*, *Koopmania*, and *Enchisthenes* based on cytochrome *b*, as well as less-parsimonious relationships for these taxa, assuming that these four genera shared a most recent common ancestor after diverging from all other stenodermatines. The single most-parsimonious tree based on this study is 1,251 steps (Figure 4.2A); the topology in Figure 4.2B requires only 2 additional steps and probably does not

Table 4.1

Sequence Divergence and Transition-to-Transversion Ratios for Pairwise Comparisons

	Uroderma	Chiroderma	Enchisthenes	A. obscurus	A. obscurus	A. jamaicensis	A. jamaicensis	A. planirostris	A. lituratus	A. intermedius	A. inopinatus	A. hirsutus	A. fraterculus	A. fimbriatus	Koopmania	Koopmania	D. gnoma	D. glauca	D. anderseni	D. cinerea	D. tolteca	D. phaeotis	D. watsoni	D. azteca
Uroderma	—	14.96	19.18	16.73	16.01	17.55	17.99	17.09	16.81	16.54	17.04	16.63	16.56	15.83	17.34	18.09	18.15	17.29	18.56	17.90	17.09	16.99	17.87	18.98
Chiroderma	2.6	—	17.16	16.11	15.66	16.10	15.98	16.55	16.04	15.20	15.00	14.64	15.80	14.19	15.55	15.34	16.26	16.07	15.34	15.56	15.31	15.43	14.65	15.96
Enchisthenes	3.0	3.6	—	16.19	16.87	17.64	17.87	15.58	17.24	16.26	16.23	17.26	16.24	15.89	16.50	15.80	16.84	16.54	15.46	15.80	15.35	15.67	17.70	18.14
A. obscurus	2.6	2.7	2.3	—	3.26	6.30	6.68	6.81	8.40	7.89	9.22	9.83	9.82	10.11	10.93	10.71	14.08	13.66	12.69	12.93	12.91	13.26	14.21	14.47
A. obscurus	2.6	2.6	2.4	2.6	—	5.50	6.07	6.30	7.47	7.06	8.77	8.99	9.07	8.95	10.87	10.14	14.06	13.40	13.05	12.54	12.01	12.45	13.49	14.18
A. jamaicensis	2.6	2.8	2.8	4.2	7.6	—	1.24	3.91	6.87	6.67	10.49	9.59	9.26	9.04	12.01	11.48	13.96	13.95	13.28	13.71	13.28	12.44	15.00	14.38
A. jamaicensis	2.7	2.8	2.9	4.5	8.4	6.0	—	4.48	6.78	6.57	11.11	9.59	9.46	9.54	12.22	11.90	14.29	14.17	13.71	14.47	13.71	12.86	14.89	14.27
A. planirostris	2.6	2.9	2.4	2.8	4.2	2.6	3.1	—	6.79	5.98	9.44	8.64	6.45	9.12	10.95	10.42	12.78	13.26	11.75	12.07	11.79	11.57	13.57	13.39
A. lituratus	2.2	2.4	2.4	3.7	5.2	5.2	5.1	4.2	—	2.52	9.59	9.00	8.07	9.86	12.03	12.55	14.56	13.41	12.98	12.77	13.11	12.12	13.94	13.97
A. intermedius	2.3	2.6	2.6	3.9	5.9	6.2	6.1	7.1	2.5	—	8.74	8.05	7.15	9.11	11.03	11.44	12.97	13.15	11.34	11.75	11.75	11.85	12.69	12.94
A. inopinatus	2.3	2.4	2.7	3.6	5.2	6.9	7.3	9.0	5.3	8.3	—	6.85	7.73	8.74	10.84	10.94	13.29	13.49	12.70	10.92	11.77	12.18	13.01	13.58
A. hirsutus	2.5	2.5	2.9	3.9	4.6	5.3	5.3	8.2	4.3	6.2	8.3	—	6.35	8.64	12.31	11.77	12.84	12.32	11.63	11.34	11.46	11.97	13.01	13.15
A. fraterculus	2.2	2.5	2.6	4.2	5.9	6.5	6.7	13.0	4.7	7.6	10.9	12.8	—	8.82	10.71	10.91	13.39	11.87	11.00	10.60	11.35	11.13	11.84	12.71
A. fimbriatus	2.3	2.4	2.7	5.2	7.6	8.6	9.1	11.1	7.7	15.2	8.3	8.2	12.4	—	10.48	10.58	12.73	11.96	11.11	11.11	11.23	10.29	11.62	11.86
Koopmania	2.5	2.6	2.5	3.0	4.1	4.9	5.0	5.0	4.4	5.8	4.9	5.7	6.0	7.5	—	2.41	13.28	12.76	10.52	11.35	11.80	11.15	12.91	12.65
Koopmania	2.4	2.5	2.5	3.1	4.0	5.0	5.2	5.1	4.9	6.4	5.3	5.8	6.6	8.3	8.0	—	13.37	13.71	11.34	12.07	11.90	11.78	13.33	12.74
D. gnoma	2.7	2.5	2.5	5.0	5.5	5.2	5.3	4.7	4.4	5.3	5.5	5.9	5.9	5.9	5.8	6.2	—	10.62	9.67	10.18	10.59	11.15	12.60	12.61
D. glauca	2.8	2.6	2.6	4.6	5.5	5.5	5.5	6.6	5.2	6.5	5.9	5.4	6.2	6.8	5.2	6.0	5.5	—	9.43	10.25	10.92	9.46	12.90	13.34
D. anderseni	2.5	3.0	2.4	4.2	6.1	5.8	6.0	6.6	5.0	6.4	7.2	7.6	7.8	7.3	5.5	6.4	6.8	11.5	—	7.13	9.35	8.45	10.47	10.37
D. cinerea	2.7	2.5	2.5	3.9	5.1	6.0	6.4	6.8	5.0	6.6	6.1	6.4	6.4	7.3	5.9	6.8	7.2	9.8	11.8	—	8.37	8.06	9.26	10.18
D. tolteca	2.6	2.6	2.4	4.3	4.9	5.8	6.0	5.1	4.6	6.6	5.8	5.6	5.9	6.3	4.8	5.2	7.5	6.1	7.3	5.4	—	5.41	9.99	10.92
D. phaeotis	2.5	2.6	2.5	3.8	4.8	5.1	5.3	5.3	5.0	7.2	6.4	6.3	6.3	6.2	5.4	5.4	5.4	6.7	7.2	5.6	5.6	—	9.17	9.78
D. watsoni	2.3	2.4	3.1	4.5	5.9	7.0	6.9	7.2	5.8	7.7	7.9	7.9	7.8	8.3	7.3	8.1	7.1	9.0	6.5	6.0	5.9	5.9	—	4.84
D. azteca	2.3	2.3	2.7	4.0	5.3	5.6	5.6	5.9	4.9	6.4	6.7	6.5	6.7	6.7	5.2	5.6	6.6	7.6	8.1	6.6	6.1	5.9	6.6	—

Notes: Data above the diagonal are percent sequence divergence for all pairwise comparisons corrected for multiple substitution by the two-parameter model of Kimura (1980), with the transition-to-transversion ratio set at 5:1. Below the diagonal are the values for number of transitions divided by number of transversions for all pairwise comparisons. *A.*, *Artibeus*; *D.*, *Dermanura*. Two individuals of *A. obscurus*, *A. jamaicensis*, and *Koopmania* were examined to evaluate intraspecific variation based on cytochrome *b* DNA sequence analysis.

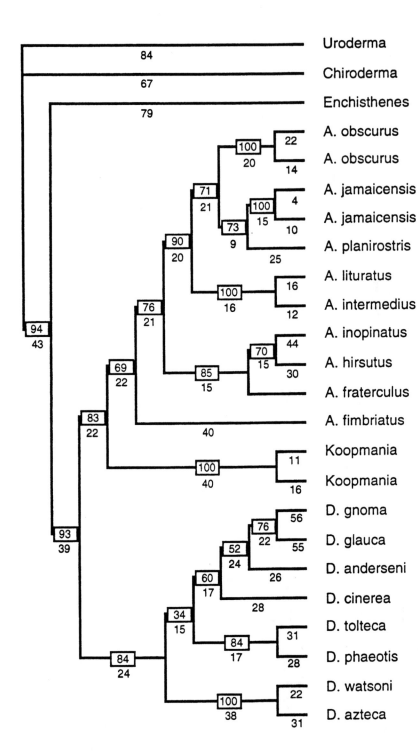

Figure 4.1. Topology of the single most-parsimonious tree of 1,251 mutational events depicting the phylogenetic relationships among *Artibeus*, *Dermanura*, and *Koopmania*. For this analysis, equal weights were assigned to all characters. Numbers below each lineage are the assigned branch lengths from PAUP (Phylogenetic Analysis Using Parsimony); numbers in boxes on the internal lineages are the percentage of 1,000 bootstrap iterations in which the clade was detected. The g_1 statistic for tree length distributions of 10,000 randomly drawn trees from these data was highly significant (−0.581). A., *Artibeus*; D., *Dermanura*.

represent a significantly greater number of ad hoc assumptions. Moreover, the topologies presented in Figure 4.2A and 4.2B are both consistent with the phylogenetic distribution of the *Eco*RI-defined satellite DNA (see Figure 2 of Van Den Bussche et al. 1993).

The topology in Figure 4.2C, proposing a close relationship of *Enchisthenes* with *Dermanura* and *Koopmania*, required 1,270 steps or 19 more steps than the most-parsimonious tree from the cytochrome *b* data. Additionally, this

topology is not consistent with the phylogenetic relationships for these taxa based on the distribution of the *Eco*RI satellite DNA. The topology in Figure 4.2C requires the evolution of the *Eco*RI-defined satellite repeat in the common ancestor of *Artibeus*, *Dermanura*, *Koopmania*, and *Enchisthenes*, followed by the loss of hundreds of copies of this repetitive family in the *Enchisthenes* lineage, and, finally, the evolution of the additional *Eco*RI site, producing the 500-bp and 400-bp bands in the common ancestor of *Dermanura*

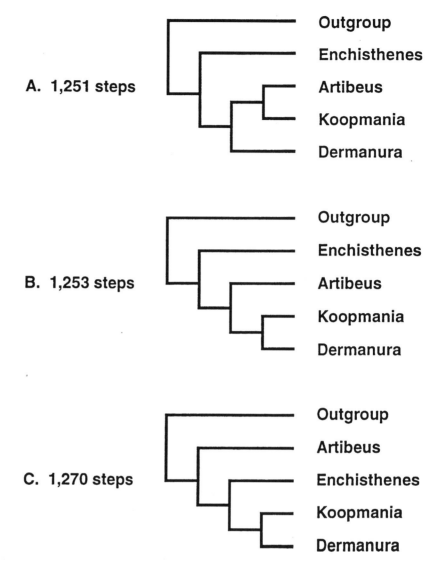

Figure 4.2. Topologies of three phylogenetic hypotheses for the relationships among *Artibeus, Dermanura, Koopmania,* and *Enchisthenes.*

after diverging from the most recent common ancestor of *Dermanura* and *Enchisthenes.*

Because *Koopmania* did not group within either the *Artibeus* or the *Dermanura* clades (see Figure 4.1), *Koopmania* was considered as an ingroup member of both genera in analyzing the relationships among species of *Artibeus* and *Dermanura.* To polarize character-state changes for the *Artibeus* clade, *D. cinerea* and *Enchisthenes* were designated as outgroups. The topology in Figure 4.3 is the result of a branch-and-bound analysis of the unweighted data, which resulted in a single most-parsimonious tree of 673 steps (consistency index, CI = 0.571). The g_1 statistic from 10,000 randomly drawn trees was highly significant (-0.819; $p <$ 0.01). Numbers in boxes along the internal lineages reflect the percentage of 1,000 bootstrap iterations with the branch-and-bound option in which each clade was detected (number in the upper portion of each box) and the number of additional steps from the most-parsimonious tree re-

quired for a particular clade to collapse (Bremer support analysis) (Bremer 1988; Donoghue et al. 1992). All character-state transformation matrices resulted in similar topologies to that shown in Figure 4.3, with the exception of less resolution among closely related taxa.

Figure 4.4 is the single most-parsimonious tree of 697 steps (CI = 0.540) depicting the phylogenetic relationships among the eight species of *Dermanura.* This topology results from a branch-and-bound search with the character-state changes polarized relative to *A. planirostris* and *Enchisthenes.* The g_1 statistic of 10,000 randomly drawn trees was highly significant (-1.256; $p < 0.01$). As with Figure 4.4, the accuracy (Hillis et al. 1994) or confidence (Felsenstein and Kishino 1993) of each clade was examined using both bootstrap and Bremer support analyses. Numbers in the upper portion of the boxes along the internal branches reflect the percentage of 1,000 bootstrap iterations with the branch-and-bound option in which each clade was detected, and

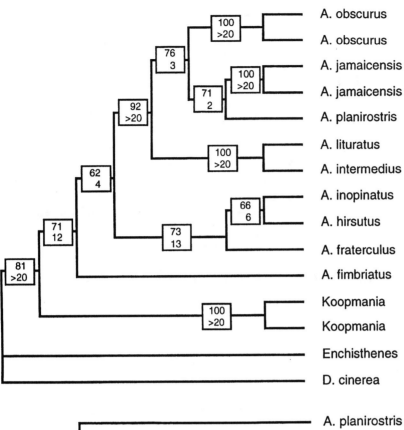

Figure 4.3. Topology of the single most-parsimonious tree of 673 steps depicting the phylogenetic relationships within *Artibeus*. This depiction resulted from a branch-and-bound search of the cytochrome *b* data. The g_1 statistic from 10,000 randomly drawn trees was highly significant (-0.819). Within each box on the internal lineages, the top number indicates the percentage of 1,000 bootstrap iterations with the branch-and-bound option in which the clade was detected; the bottom number is the cladistic-stability value determined by a Bremer support analysis (Bremer 1988). A., *Artibeus*; D., *Dermanura*.

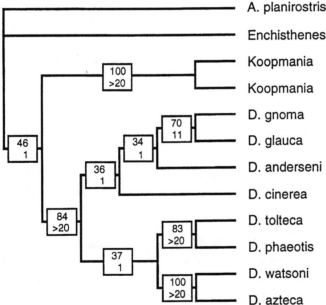

Figure 4.4. Topology of the single most-parsimonious tree of 697 steps depicting the phylogenetic relationships within *Dermanura*. This depiction resulted from a branch-and-bound search of the cytochrome *b* data and had a significant phylogenetic signal-to-noise ratio ($g_1 = -1.256$). Within each box on the internal lineages, the top number indicates the percentage of 1,000 bootstrap iterations with the branch-and-bound option in which the clade was detected; the bottom number is the cladistic-stability value determined by a Bremer support analysis (Bremer 1988). A., *Artibeus*; D., *Dermanura*.

the numbers in the lower portion of the boxes are from the Bremer support analysis (Bremer 1988)

Discussion

Rate Heterogeneity

Understanding the rate of nucleotide substitutions among independent lineages is important for assessing phyloge-netic analyses. Differences in rates of substitutions among independent lineages, if large enough, can alter the resultant phylogenetic relationships (Swofford and Olsen 1990). Using the relative rates test and binomial distribution, none of the rate differences among lineages were significant at the $p = 0.05$ level, and only three comparisons had a *p*-value between 0.1 and 0.05. These three cases occurred when the relative rates of nucleotide substitution along the *Enchis-thenes* lineage and one of the lineages of *A. obscurus* were compared relative to *Uroderma* ($p = 0.0873$); *Enchisthenes* and *A. fimbriatus* compared relative to *Chiroderma* ($p = 0.0746$); and *A. fimbriatus* and *D. azteca* relative to *Chiroderma* ($p = 0.0938$). These results indicate that there is minimal heterogeneity in the rate of nucleotide substitutions among the various lineages.

Patterns of Nucleotide Substitution

It has been repeatedly demonstrated that the vertebrate mitochondrial genome is characterized by a strong bias toward transition substitutions (Brown et al. 1982; Moritz et al. 1987). The ratios of transitions to transversions for the taxa examined herein show the expected trend for a typical cytochrome *b* gene: these ratios are high for closely related taxa and decrease for more distantly related taxa (see Table 4.1). Transition-to-transversion ratios range from a high of 15.2 for the comparison of *A. fimbriatus* with *A. intermedius* to a low of 2.2 for the comparison of *Uroderma* with either *A. lituratus* or *A. fraterculus*. These data are interpreted to indicate that problems of homoplastic changes associated with multiple substitutions should be minimal in the phylogenetic analysis.

As seen in other studies utilizing the cytochrome *b* gene, percent sequence divergence was partitioned in a hierarchical fashion. Percent sequence divergence within species was 1.2% for *A. jamaicensis*, 2.4% for *Koopmania concolor*, and 3.3% for *A. obscurus* whereas the average percent sequence divergence within genera was 7.8% among species of *Artibeus* and 10.4% among *Dermanura* species. Finally, average sequence divergence among all stenodermatines examined was 12.2%. These levels of percent sequence divergence are similar to levels reported for other studies on phyllostomid bats based on the cytochrome *b* gene (Van Den Bussche and Baker 1993; Van Den Bussche et al. 1993; Baker et al. 1994).

Reliability of Clades

Much debate has occurred over setting confidence levels on the clades in a phylogenetic tree (Felsenstein 1985; Penny and Hendy 1985, 1986; Felsenstein and Kishino 1993; Hillis and Bull 1993). Felsenstein (1985) proposed a bootstrap method for placing confidence limits on the different clades, but Hillis and Bull (1993) suggested that bootstrap values are biased estimates of the degree of confidence that can be placed on a particular clade. Hillis and Bull (1993) contended that if one makes certain assumptions concerning the tempo and mode of evolution of the characters under study, bootstrap values of 70% are probably conservative estimates of a particular clade being correct. An alternative approach for examining the stability of various clades is the Bremer support analysis (Bremer 1988; Donoghue et al.

1992). This method examines successively larger trees than the most-parsimonious tree, one step at a time, and determines where various clades collapse or decay.

Bootstrap and Bremer support analyses were performed to evaluate the degree to which the two methods provide compatible results (see Figures 4.3 and 4.4). Both methods provide identical results under conditions of bootstrap values of 80% or more. For example, in Figures 4.3 and 4.4, all clades supported in more than 80% of the 1,000 bootstrap iterations were present in all trees up to 20 steps longer than the most-parsimonious tree. However, for clades supported in fewer than 80% of the bootstrap iterations, there was no correlation of results from these two approaches. For example, of the clades supported in 70%–79% of the bootstrap iterations, two clades collapse with the addition of 2 and 3 steps, although 12 and 13 steps were required to collapse two other clades with similar bootstrap values (Figure 4.3). Finally, for those clades supported in fewer than 70% of the bootstrap iterations, most collapsed with the addition of only a few steps, the exception being one clade in Figure 4.3 that was supported in 66% of the bootstrap iterations and required 6 additional steps over the most-parsimonious topology before decaying.

Beyond this study's obvious implications of evaluating the evolutionary relationships among a complex group of phyllostomid bats, the trees produced may also hold broader implications for the field of systematics concerning the identification of robust clades in a phylogenetic tree. More specifically, these data suggest that rather than placing confidence on clades supported by bootstrap values of 70% or greater (Hillis and Bull 1993), a more conservative estimate would be to set this limit at bootstrap values of 80% or more. The lack of correlation between the results of the bootstrap analysis and the Bremer support analysis for clades detected in fewer than 80% of the bootstrap iterations may reflect the actual composition as well as the location of the clades on the tree. These data illustrate that such analyses allow identification of the most phylogenetically robust clades through bootstrap values and the number of additional steps required to collapse a particular clade. Data from other taxa now should be used to test where bootstrap methods (Felsenstein 1985) and Bremer support methods (Bremer 1988) provide congruent results.

Taxonomic Implications at the Generic Level

Two independent, nonlinked data sets (cytochrome *b* sequence variation and the phylogenetic distribution of the *Eco*RI-defined satellite DNA) are concordant in supporting the monophyly of *Artibeus*, *Koopmania*, and *Dermanura* relative to all other taxa. That these relationships were supported by all character-weighting schemes including amino acid residues and first- and second-codon positions, as well as applying differential weights to the various nucleotide substitutions, lends considerable support for the robustness of the phylogenetic hypothesis depicted in Figure 4.1.

One interesting relationship predicted from the cytochrome *b* data is the closer phylogenetic affinity of *Koopmania* with *Artibeus*. When alternative phylogenetic hypotheses for the placement of *Koopmania* were examined (see Figure 4.2), the differences in tree lengths between sister-group relationships of *Artibeus* and *Koopmania* or *Koopmania* and *Dermanura* are insignificant (1,251 steps versus 1,253 steps). Moreover, the affinity of *Koopmania* with *Artibeus* rather than *Dermanura* is clearly depicted in the analyses of the smaller data sets (see Figures 4.3 and 4.4). When the relationships among *Dermanura* are evaluated with *Koopmania* considered an ingroup taxon, the clade uniting these taxa is nonsignificant.

When the relationships among *Artibeus* are examined and *Koopmania* is considered an ingroup, the clade uniting *Koopmania* with *Artibeus* is supported in 81% of 1,000 bootstrap iterations. Moreover, trees more than 20 steps longer than the most parsimonious must be examined before this clade collapses. Finally, the topologies (Figure 4.2A,B) are consistent with the relationships based on the *Eco*RI-defined satellite DNA (Figure 2 of Van Den Bussche et al. 1993). We interpret these data as documenting a close phylogenetic relationship of *K. concolor* with *Artibeus* and question the validity of *Koopmania* as a distinct genus. Until additional data are presented to the contrary, we suggest *Koopmania concolor* (Owen 1991) be recognized as *Artibeus concolor*.

The only phylogenetic hypothesis for the relationships among *Artibeus*, *Koopmania*, *Dermanura*, and *Enchisthenes* that seems incompatible with the cytochrome *b* and satellite DNA is that *Enchisthenes* is closely related to *Dermanura* (Figure 4.2C). This tree requires 19 additional steps in the cytochrome *b* data and the loss of hundreds of copies of the satellite DNA after *Enchisthenes* diverged from its most recent common ancestor with *Artibeus*, *Koopmania*, and *Dermanura*. We interpret these data as supporting recognition of *Enchisthenes* as a lineage distinct from *Artibeus*, *Dermanura*, and *Koopmania*, as argued by Andersen (1908) and supported by Van Den Bussche et al. (1993).

Taxonomic Implications within *Artibeus* and *Dermanura*

A high degree of confidence can be placed on most sister-group relationships depicted in Figure 4.3, judging from bootstrap and Bremer support analyses. The sister-group relationship of *A. inopinatus* and *A. hirsutus* was detected in only 66% of the 1,000 bootstrap iterations; however, six

additional steps are required for this clade to collapse. The clade uniting *A. jamaicensis* and *A. planirostris* and the clade uniting *A. obscurus* and *A. jamaicensis* with *A. planirostris* were detected in 71% and 72% of the 1,000 bootstrap iterations, respectively, and these clades collapse with the addition of two and three steps respectively, from the most-parsimonious tree.

The clade uniting *A. inopinatus*, *A. hirsutus*, and *A. fraterculus* as well as the clade uniting all species of *Artibeus* are only marginally supported by bootstrap analysis, yet these clades do not collapse until trees 12 and 13 steps longer than the most-parsimonious tree are examined. Additional support for the phylogenetic relationships depicted in Figure 4.4 comes from the observation that this topology passes the test of character congruence (Kluge 1989), having identical nodes in common with a tree constructed using DNA sequence data from subunits 6 and 8 of the mitochondrial adenosine triphosphatase (ATPase) genes (Patterson et al. 1992). Although the mitochondrial genome is inherited as a single linked marker, that different genes sequenced from independent individuals support the same phylogenetic tree should be viewed as supporting these relationships based on multiple maternal markers.

Concerning alpha taxonomic questions, we examined two individuals of *A. obscurus*, one individual from a site near the type locality of *A. obscurus* and a second from the traditional range of *A. fuliginosus*. The sister-group relationship for these taxa is strongly supported by cytochrome *b* data (see Figure 4.3). Moreover, the percent sequence divergence for these taxa is less than that seen between other closely related species (see Table 4.1). These data are interpreted as supporting Handley's (1989) synonymy of *A. fuliginosus* with *A. obscurus*.

Only two additional steps collapse the sister-group relationship of *A. planirostris* and *A. jamaicensis*. Additional studies, covering a broader geographic area and more individuals, are needed before these taxa are synonymized. Although this study was not designed to specifically address alpha taxonomic questions, the cytochrome *b* data strongly support a sister-group relationship of *A. intermedius* and *A. lituratus*, although the degree of sequence divergence between these taxa is as great as that seen among other species of *Artibeus*. As with *A. planirostris*, further studies covering a broader geographic area and more individuals are necessary before these taxa are synonymized.

Cytochrome *b* data are compatible with the conclusions of Handley (1989) and Patterson et al. (1992) documenting the uniqueness of *A. fimbriatus* and thereby supporting the specific status of this taxon. Finally, cytochrome *b* data support a close relationship of *A. fraterculus*, *A. hirsutus*, and *A. inopinatus* (see Figure 4.3). These taxa occur on the Pacific slope of the Andean and Middle American cordilleras, and their close relationship has also been hypothesized from analysis of morphological characters by Patten (1970) and of genetic characters by Patterson et al. (1992).

Cytochrome *b* sequence data support the monophyly of *Dermanura;* this clade was supported in 84% of the 1,000 bootstrap iterations, coupled with the observation that topologies more than 20 steps longer than the most parsimonious are required to collapse this clade (see Figure 4.4). The only other strongly supported clades predict sister-group relationships for *D. gnoma* and *D. glauca,* for *D. tolteca* and *D. phaeotis,* and for *D. watsoni* and *D. azteca.* Koop and Baker (1983) examined the phylogenetic relationships among several of these species and detected no fixed allozymic differences. Allozymic and cytochrome *b* data are interpreted as indicating that these taxa may have evolved via a rapid radiation, or burst, from their common ancestor, as has been suggested for other bats (Baker et al. 1991). If such a scenario is true, the short time that many of these lineages shared a common ancestor may make it difficult to find synapomorphies documenting sister-group relationships within *Dermanura.*

Because the phylogenetic history of extant organisms may never be known, our next best situation is to have strongly supported phylogenies based on several independent data sets and characters. Such robust phylogenies stand as our best approximations of the true evolutionary history of taxa (Penny and Hendy 1986; Allard and Miyamoto 1992; Miyamoto et al. 1994). Implicit in this assumption is that we have confidence in the data as well as in the methods used to reconstruct evolutionary history.

For several reasons we have a high degree of confidence in the phylogenetic relationships of these taxa based on satellite DNA and cytochrome *b* sequence data. First, irrespective of the methods used to weight various characters and character-state transformations, nearly identical topologies were obtained by the two methods for the phylogenetic relationships. Second, statistical indicators for the phylogenetic information content of the data (g_1 statistic) as well as the confidence in the particular branching patterns (bootstrap and Bremer support values) identified several strongly supported clades. Finally, for estimation of the phylogenetic relationships among *Artibeus, Dermanura,* and *Koopmania* as well as relationships within *Artibeus,* multiple independent data sets support identical phylogenetic relationships.

Conclusions

We have examined the phylogenetic distribution of an *Eco*RI-defined nuclear satellite DNA repeat and DNA se-

quence variation from the mitochondrial cytochrome *b* gene (1,140 bp) for fig-eating bats of the New World genus *Artibeus* to better understand the systematic relationships within this diverse group and to estimate the level of resolution supporting each clade. To determine which clades are robust and should receive the highest confidence, we compared bootstrap and Bremer support analyses to examine the extent to which each method corroborates the phylogenetic relationships estimated by the other. All clades supported by at least 80% of the bootstrap iterations required more than 20 steps before they decayed. Bootstrap clades identified by 70%–79% collapsed within 2–13 additional steps.

Taxonomic conclusions from this study are as follows: (1) *Artibeus, Dermanura,* and *Koopmania* as recognized by Owen (1987, 1991) form a monophyletic group to the exclusion of other Stenodermatini genera; (2) both *Artibeus* and *Dermanura* are monophyletic within this larger group; (3) DNA sequence data of the cytochrome *b* gene are interpreted as failing to support recognition of the genus *Koopmania,* and we concluded that this taxon should be recognized as *Artibeus concolor;* and (4) DNA sequence variation of the mitochondrial cytochrome *b* gene provides resolution to many alpha taxonomic-level relationships within this controversial yet important component of the South American fauna.

Acknowledgments

Robert D. Bradley, James C. Cathey, Burhan M. Gharaibeh, Alec Knight, Bruce D. Patterson, John A. Peppers, Travis Perry, Cheryl A. Schmidt, and Amanda J. Wright critically reviewed earlier drafts of this manuscript. Rodney L. Honeycutt and David M. Hillis provided helpful discussions concerning data analysis, and R. L. Honeycutt provided a computer program for testing significance of rate heterogeneity. Specimens for this study were provided from frozen tissue collections at The Museum, Texas Tech University; Museum of Southwestern Biology, University of New Mexico (Terry L. Yates); Museum of Vertebrate Zoology, Berkeley, California (Jim Patton); The Field Museum, Chicago (Bruce D. Patterson); and the American Museum of Natural History (Nancy B. Simmons). This study was supported in part by a National Science Foundation grant to R. J. Baker. Jeremy L. Hudgeons was supported by a grant from the Howard Hughes Medical Institute through the Undergraduate Biological Science Education Program. These sequences generated for this study have been deposited in GenBank under accession numbers U66498–U66519.

Literature Cited

Allard, M. W., and R. L. Honeycutt. 1992. Nucleotide sequence variation in the mitochondrial 12S rRNA gene and the phylogeny of African mole-rats (Rodentia: Bathyergidae). Molecular Biology and Evolution 9:27–40.

Allard, M. W., and M. M. Miyamoto. 1992. Testing phylogenetic approaches with empirical data, as illustrated with the parsimony method. Molecular Biology and Evolution 9:778–786.

Andersen, K. 1906. Brief diagnosis of a new genus and ten new forms of stenodermatous bats. Annual Magazine of Natural History, Series 7, 18:419–423.

Andersen, K. 1908. A monograph of the Chiroptera genera *Uroderma, Enchisthenes,* and *Artibeus.* Proceedings of the Zoological Society of London 1908:204–319.

Baker, R. J. 1973. Comparative cytogenetics of the New World leaf-nosed bats (Phyllostomatidae). Periodicum Biologorum 75:37–45.

Baker, R. J. 1994. Some thoughts on conservation, biodiversity, museums, molecular characters, systematics, and basic research. Journal of Mammalogy 75:277–287.

Baker, R. J., R. L. Honeycutt, and R. A. Van Den Bussche. 1991. Examination of monophyly of bats: Restriction map of the ribosomal DNA cistron. *In* Contributions to Mammalogy in Honor of Karl F. Koopman, T. A. Griffiths and D. Klingener, eds., pp. 42–53. Bulletin of the American Museum of Natural History 206:1–432.

Baker, R. J., V. A. Taddei, J. L. Hudgeons, and R. A. Van Den Bussche. 1994. Systematic relationships within *Chiroderma* (Chiroptera: Phyllostomidae) based on cytochrome b sequence variation. Journal of Mammalogy 75:321–327.

Bingham, P., R. Levis, and G. M. Rubin. 1981. Cloning of DNA sequences from the *white* locus of *D. melanogaster* by a novel and general method. Cell 25:693–704.

Bremer, K. 1988. The limits of amino acid sequence data in angiosperm phylogenetic reconstruction. Evolution 42:795–803.

Brown, W. M., E. M. Prager, A. Wang, and A. C. Wilson. 1982. Mitochondrial DNA sequences of primates: Tempo and mode of evolution. Journal of Molecular Evolution 18:225–239.

Davis, W. B. 1984. Review of the large fruit-eating bats of the *Artibeus* "lituratus" complex (Chiroptera: Phyllostomidae) in middle America. Occasional Papers of the Museum, Texas Tech University 93:1–16.

Donoghue, M. J., R. G. Olmstead, J. F. Smith, and J. D. Palmer. 1992. Phylogenetic relationships of dipscales based on *rbc*L sequences. Annals of the Missouri Botanical Garden 79:333–345.

Durfy, S. J., and H. F. Willard. 1990. Concerted evolution of primate alpha satellite DNA. Evidence for an ancestral sequence shared by gorilla and human X chromosome alpha satellite. Journal of Molecular Evolution 216:555–565.

Eernisse, D. J., and A. G. Kluge. 1993. Taxonomic congruence versus total evidence, and the phylogeny of amniotes inferred from fossils, molecules, and morphology. Molecular Biology and Evolution 10:1170–1195.

Felsenstein, J. 1985. Confidence limits on phylogenies: An approach using the bootstrap. Evolution 39:783–791.

Felsenstein, J., and H. Kishino. 1993. Is there something wrong with the bootstrap analysis? A reply to Hillis and Bull. Systematic Biology 42:193–200.

Gervais, M. P. 1855. Cheiroptères sud-américans. *In* Animaux Nouveaux ou Rares de l'Amérique du Sud . . . Mammifères, F. Casttlenau, ed., pp. 25–28. Chez P. Bertrand, Paris. (The publication date is listed as 1856 by some authors.)

Gouy, M., and W.-H. Li. 1989. Molecular phylogeny of the kingdoms Animalia, Plantae, and Fungi. Molecular Biology and Evolution 6:109–122.

Gray, J. E. 1838. A revision of the genera of bats (Vespertilionidae), and the description of some new genera and species. Magazine of Zoology and Botany 2:483–505.

Handley, C. O., Jr. 1987. New species of mammals from northern South America: Fruit-eating bats, genus *Artibeus* Leach. *In* Studies in Neotropical Mammalogy: Essays in Honor of Philip Hershkovitz, B. D. Patterson and R. M. Timm, eds., pp. 163–172. Fieldiana Zoology New Series No. 39.

Handley, C. O., Jr. 1989. The *Artibeus* of Gray 1838. Advances in Neotropical Mammalogy 1989:443–468.

Hershkovitz, P. 1949. Mammals of northern Colombia. Preliminary report no. 5: Bats (Chiroptera). Proceedings of the U.S. National Museum 99:429–454.

Hillis, D. M. 1991. Discriminating between phylogenetic signal and random noise in DNA sequences. *In* Phylogenetic Analysis of DNA Sequences, M. M. Miyamoto and J. Cracraft, eds., pp. 278–294. Oxford University Press, New York.

Hillis, D. M., and J. J. Bull. 1993. An empirical test of bootstrapping as a method for assessing confidence in phylogenetic analysis. Systematic Biology 42:182–192.

Hillis, D. M., and J. P. Huelsenbeck. 1992. Signal, noise, and reliability in molecular phylogenetic analysis. Journal of Heredity 83:189–195.

Hillis, D. M., J. P. Huelsenbeck, and C. W. Cunningham. 1994. Application and accuracy of molecular phylogenies. Science 264:671–677.

Hillis, D. M., J. J. Bull, M. E. White, M. R. Badgett, and I. J. Molineux. 1992. Experimental phylogenetics: Generation of a known phylogeny. Science 255:589–592.

Huelsenbeck, J. P. 1991. Tree-length distribution skewness: An indicator of phylogenetic information. Systematic Zoology 40:257–270.

Kallersjo, M., J. S. Farris, A. G. Kluge, and C. Bult. 1992. Skewness and permutation. Cladistics 8:275–287.

Kimura, M. 1980. A simple method for estimating rate of base substitutions through comparative studies of nucleotide sequences. Journal of Molecular Evolution 16:111–120.

Kluge, A. G. 1989. A concern for evidence and a phylogenetic hypothesis of relationships among *Epicrates* (Boidae, Serpentes). Systematic Zoology 38:7–25.

Koop, B. F., and R. J. Baker. 1983. Electrophoretic studies of relationships of six species of *Artibeus* (Chiroptera: Phyllostomidae). Occasional Papers of the Museum, Texas Tech University 83:1–12.

Koopman, K. F. 1993. Order Chiroptera. *In* Mammal Species of the World: A Taxonomic and Geographic Reference, D. E. Wilson and D. M. Reeder, eds., pp. 137–242. Smithsonian Institution Press, Washington, D.C.

Kraft, R. 1982. Notes on the type specimens of *Artibeus jamaicensis planirostris* (Spix, 1823). Spixiana 5:311–316.

Li, W.-H., K. H. Wolfe, J. Sourdis, and P. M. Sharp. 1987. Reconstruction of phylogenetic trees and estimation of divergence times under nonconstant rates of evolution. Cold Spring Harbor Symposia on Quantitative Biology 52:847–856.

Longmire, J. L., R. E. Ambrose, N. C. Brown, T. J. Cade, T. L. Maechtle, W. S. Seeger, F. P. Ward, and C. M. White. 1991. Use of sex-linked minisatellite fragments to investigate genetic differentiation and migration of North American populations of the peregrine falcon *(Falco peregrinus)*. *In* DNA Fingerprinting: Approaches and Applications, T. Burke, G. Dolf, A. Jeffreys, and R. Wolff, eds., pp. 217–229. Birkhauser Verlag, Basel.

Lundrigan, B. L., and P. K. Tucker. 1994. Tracing paternal ancestry in mice, using the Y-linked, sex-determining locus, *Sry*. Molecular Biology and Evolution 11:483–492.

Maddison, W. P., and D. R. Maddison. 1992. MacClade: Analysis of phylogeny and character evolution. Version 3.0. Sinauer Associates, Sunderland, Mass.

Miller, G. S. 1907. The families and genera of bats. Bulletin of the U.S. National Museum 57:1–282.

Mindell, D. P., and R. L. Honeycutt. 1990. Ribosomal RNA in vertebrates: Evolution and phylogenetic applications. Annual Review of Ecology and Systematics 21:541–566.

Miyamoto, M. M., M. C. Allard, R. M. Adkins, L. L. Janecek, and R. L. Honeycutt. 1994. A congruence test of reliability using linked mitochondrial DNA sequences. Systematic Biology 43: 236–249.

Moritz, C., T. E. Dowling, and W. M. Brown. 1987. Evolution of animal mitochondrial DNA: Relevance for population biology and systematics. Annual Review of Ecology and Systematics 18: 269–292.

Nei, M. 1991. Relative efficiencies of different tree-making methods for molecular data. *In* Phylogenetic Analysis of DNA Sequences, M. M. Miyamoto and J. Cracraft, eds., pp. 90–128. Oxford University Press, Oxford.

Owen, R. D. 1987. Phylogenetic analyses of the bat subfamily Stenodermatinae (Mammalia: Chiroptera). Special Publications, The Museum, Texas Tech University 26:1–65.

Owen, R. D. 1991. The systematic status of *Dermanura concolor* (Peter, 1865) (Chiroptera: Phyllostomidae), with a description of a new genus. *In* Contributions to Mammalogy in Honor of Karl F. Koopman, T. A. Griffiths and D. Klingener, eds., pp. 18–25. Bulletin of the American Museum of Natural History 206:1–432.

Patten, D. R. 1970. A review of the large species of *Artibeus* (Chiroptera: Phyllostomatidae) from western South America. Ph.D. dissertation, Texas A&M University, College Station.

Patterson, B. D., V. Pacheco, and M. V. Ashley. 1992. On the origins of the western slope region of endemism: Systematics of fig-eating bats, genus *Artibeus*. Memorias del Museo de Historia Natural, U. N. M. S. M. (Lima) 21:189–205.

Penny, D., and M. Hendy. 1985. Testing methods of evolutionary tree construction. Cladistics 1:266–278.

Penny, D., and M. Hendy. 1986. Estimating the reliability of evolutionary trees. Molecular Biology and Evolution 3:403–417.

Rzhetsky, A., and M. Nei. 1992. A simple method for estimating and testing minimum-evolution trees. Molecular Biology and Evolution 9:945–967.

Saiki, R. K., T. L. Bugawan, G. T. Horn, K. B. Mullis, and H. A. Erlich. 1986. Analysis of enzymatically amplified beta-globin HLA-DQ alpha DNA with allele-specific oligonucleotide probes. Nature (London) 324:163–166.

Saiki, R. K., D. H. Gelfand, S. Stoffel, S. J. Scharf, R. Higuchi, G. T. Horn, K. B. Mullis, and H. A. Erlich. 1988. Primer-directed enzymatic amplification of DNA with thermostable DNA polymerase. Science 239:487–491.

Saitou, N., and M. Nei. 1987. The neighbor-joining method: A new method for reconstructing phylogenetic trees. Molecular Biology and Evolution 4:406–425.

Sarich, V. M., and A. C. Wilson. 1967. Immunological time scale for hominid evolution. Science 158:1200–1202.

Smith, J. D. 1976. Chiropteran evolution. Special Publications, The Museum, Texas Tech University 10:46–49.

Southern, E. M. 1975. Detection of specific sequences among DNA fragments separated by gel electrophoresis. Journal of Molecular Biology 98:503–517.

Straney, D. O., M. H. Smith, I. F. Greenbaum, and R. J. Baker. 1979. Biochemical genetics. *In* Biology of Bats of the New World Family Phyllostomatidae, Part III, R. J. Baker, J. K. Jones, Jr., and D. C. Carter, eds., pp. 157–176. No. 16, Special Publications, The Museum, Texas Tech University, Lubbock.

Strauss, W. M. 1987. Preparation of genomic DNA from mammalian tissue. *In* Current Protocols in Molecular Biology, F. M. Ausubel, R. Brent, R. E. Kingston, D. D. Moore, J. A. Smith, J. G. Seidman, and K. Struhl, eds., pp. 2.2.2–2.2.3. Greene and Wiley-Interscience, New York.

Swofford, D. L. 1991. PAUP: Phylogenetic Analysis Using Parsimony. User's manual. Illinois Natural History Survey, Champaign.

Swofford, D. L., and G. J. Olsen. 1990. Phylogeny reconstruction. *In* Molecular Systematics, D. M. Hillis and C. Moritz, eds., pp. 411–501. Sinauer Associates, Sunderland, Mass.

Van Den Bussche, R. A., and R. J. Baker. 1993. Molecular phylogenetics of the New World bat genus *Phyllostomus* based on cytochrome *b* DNA sequence variation. Journal of Mammalogy 74:793–802.

Van Den Bussche, R. A., R. J. Baker, H. A. Wichman, and M. J. Hamilton. 1993. Molecular phylogenetics of Stenodermatini bat genera: Congruence of data from nuclear and mitochondrial DNA. Molecular Biology and Evolution 10:944–959.

Zharkikh, A., and W.-H. Li. 1993. Inconsistency of the maximum-parsimony method: The case of five taxa with a molecular clock. Systematic Biology 42:113–125.

5

A Southern Origin for the Hipposideridae (Microchiroptera)?

Evidence from the Australian Fossil Record

SUZANNE J. HAND AND JOHN A. W. KIRSCH

The Old World tropical and subtropical bat family Hipposideridae Flower and Lydekker, 1891 contains 65 living species. The genus *Hipposideros* has 53 species, and eight other genera have 1 or 2 species: *Rhinonycteris, Coelops, Paracoelops, Triaenops, Cloeotis, Anthops, Asellia,* and *Aselliscus* (Koopman 1994). The Hipposideridae is often regarded to be a subfamily of the Rhinolophidae Gray, 1825 (e.g., Koopman 1994). Pierson (1986), however, found that although species of *Rhinolophus* and *Hipposideros* are morphologically similar, they are immunologically as distinct as other taxa placed in separate families and are recognized as such here (also following Miller 1907; Corbet and Hill 1992). In the most recent systematic summary of the group, Koopman (1994) recognized two tribes: Coelopsini, for *Coelops* and *Paracoelops,* and Hipposiderini, containing two subtribes, Hipposiderina *(Hipposideros, Anthops, Aselliscus, Asellia)* and Rhinonycterina *(Rhinonycteris, Cloeotis, Triaenops).*

Within the speciose genus *Hipposideros,* most authors (Gray 1866; Peters 1871; Tate 1941a; Hill 1963) have recognized a number of species groups or even subgenera (e.g., *Syndesmotis* Peters, 1871). Recent authors (e.g., Koopman 1994) generally accept Hill's (1963) seven species groups

(i.e., the *H. megalotis, bicolor, cyclops, speoris, pratti, armiger,* and *diadema* species groups), with some also recognizing the subgenus *Syndesmotis* for *H. megalotis* (Legendre 1982). Hill (1963) grouped *Hipposideros* species into three primary divisions (Appendix 5.1), interpreted to represent three distinct evolutionary trends within the genus.

Hipposiderids have a long and relatively good fossil record extending back to the middle Eocene of Europe, early Oligocene of Arabia, late Oligocene of Australia, early Miocene of Africa, and Pleistocene of Asia. As predominantly cave-dwelling bats, they have commonly been preserved in Cenozoic karstic sediments. Their geographic range once extended further north, as the rich European hipposiderid record testifies, with as many as five hipposiderid taxa in three genera represented in many early–middle Tertiary European sites; for example, in the late Eocene to late Oligocene Quercy phosphorite localities (Remy et al. 1987). Tertiary hipposiderids (2–65 million years old) are referred to the genera *Palaeophyllophora* (late Eocene–late Oligocene taxa), *Asellia* (middle Miocene–Recent), *Vaylatsia* (late Eocene–?early Miocene), *Hipposideros* (early Miocene–Recent), and the subgenera *Hipposideros (Pseudorhinolophus)* (middle Eocene–middle Miocene), *H. (Brachipposideros)* (early Oligocene–middle Miocene), and *H. (Syndesmotis)* (middle Miocene–Recent).

In the Oligo-Miocene and Pliocene sequences of freshwater and karstic limestones on Riversleigh Station, northwestern Queensland, Australia, hipposiderids are particularly diverse and well represented, constituting 12 of 35 bat species identified to date, with as many as eight taxa occurring syntopically in deposits such as Upper Site (Archer et al. 1994). Many taxa are known from complete or nearly complete skulls as well as disarticulated but complete postcranial material. Several hipposiderid genera or subgenera are represented (e.g., *Brachipposideros, Hipposideros, Rhinonycteris*). New hipposiderid genera recently described from Riversleigh include *Miophyllorhina, Xenorhinos,* and *Riversleigha* (Hand 1998, n.d.); a number of other distinctive but currently less well represented taxa await description. Although some extinct European taxa also occur at Riversleigh (e.g., *Brachipposideros*), others are conspicuously absent (e.g., *Pseudorhinolophus, Palaeophyllophora, Asellia, Vaylatsia*). The Recent Australian hipposiderid fauna is composed of 5 species of *Hipposideros* and the endemic *Rhinonycteris aurantius.*

In an effort to interpret the history and radiation of hipposiderids in the Australian region, we conducted a study of the phylogenetic relationships of the Riversleigh taxa among themselves and to other hipposiderid genera and species groups, living and extinct. Most previous systematic studies of extant hipposiderids have been based on craniodental and postcranial morphology as well as external morphology (especially of the noseleaf and ear). Recent taxonomic studies, which have generally focused on individual species groups, have additionally used electrophoretic data, morphometrics, and bacular morphology (Hill et al. 1986; Zubaid and Davison 1987; Kitchener et al. 1992; Flannery and Colgan 1993; Kitchener and Maryanto 1993). Karyological, immunomolecular, and rDNA restriction-site data are also available for some species (Ando et al. 1980; Pierson 1986; Baker et al. 1991). Our analyses were restricted to craniodental characters but the results are, in broad terms, in agreement with those of other studies.

Materials and Methods

Forty extinct and extant hipposiderid species and 3 extant rhinolophids were scored for 59 discrete characters (36 cranial, 20 dental, and 3 skeletal), 16 of which were multistate (Appendix 5.2). Of 65 living hipposiderid species, 30 were chosen as representative of previously recognized genera and species groups, although hindsight suggests that these taxa may not represent all lineages. Also, while characters used were chosen for their intraspecific invariability, not all subspecies for all taxa could be screened. Of the 7 extinct

hipposiderid species included in the study, 5 were from Australian Oligo-Miocene deposits (*Hipposideros bernardsigei, Brachipposideros nooraleebus, Rhinonycteris tedfordi, Riversleigha williamsi, Xenorhinos halli*), 1 was from European Eocene deposits (*Palaeophyllophora quercyi*), and 1 from the European Miocene (*Pseudorhinolophus bouziguensis*). These species were represented by complete or nearly complete craniodental and skeletal material, with missing data restricted to between 1 and 7 characters each.

Polarity of character states was determined by outgroup comparison, and a hypothetical ancestor was used to root the trees. Following Miller (1907) and Simmons (see Chapter 1, this volume), megadermatids (*Lavia frons, Megaderma spasma,* the European Eocene *Necromantis adichaster*) and nycterids (*Nycteris javanica*) were interpreted as sistergroups of a hipposiderid-rhinolophid clade. When the polarity of character states in rhinolophoids (hipposiderids, rhinolophids, megadermatids, and nycterids) was unclear, emballonurids (*Taphozous georgianus, Mosia nigrescens*), a rhinopomatid (*Rhinopoma hardwickei*), vespertilionids (*Pipistrellus tasmaniensis, Kerivoula picta*), and pteropodids (*Rousettus amplexicaudatus, Pteropus scapulatus*) also were examined.

The data set was subjected to several PAUP 3.1.1 analyses (Swofford 1993), using heuristic searches (simple stepwise addition) of all 59 characters, as well as three subsets of the characters: skeletal + cranial (characters 1–39), skeletal + dental (characters 1–3 + 40–59), and cranial + dental (characters 4–59). Characters were treated as both unordered and ordered, with seven exceptions (characters 4, 25, 31, 35, 45, 49, and 50, in which transformation series could not be specified). Nonminimum tree-length experiments were conducted on the best-resolved trees (e.g., all characters, unordered and ordered), and bootstrapping (heuristic search, random addition, 100 replicates) was attempted. Also analyzed (all characters, ordered and unordered) was the subset of hipposiderid taxa common to our study and the phylogenetic study by Bogdanowicz and Owen (see Chapter 2, this volume). In the parsimony analysis by Bogdanowicz and Owen, up to 30 discrete-state characters in 57 extant hipposiderid species (but no fossil taxa) were examined. Figure 5.1 shows strict-consensus trees generated by PAUP analysis of (a) all taxa, all characters, unordered; (b) all taxa, all characters, ordered (with the exception of the seven characters just noted); and (c) taxa common to both studies, all characters, unordered.

Given the low tree consistencies and variable topological results of our PAUP analyses (Table 5.1 and Figure 5.1), we also conducted a parallel series of analyses with Hennig86 (version 1.5; Farris 1988), using that program's successive-weighting option in the hope of improving and stabilizing resolution among the species. Because of the high degree of homoplasy and large number of taxa, exhaustive searching for the most-parsimonious trees was not practical. Instead, we used the following strategy: A heuristic search using *mhennig** was carried out, followed by invocation of *xsteps w* (weighting based on the initial character consistencies) and another round of searching with *mhennig**; this sequence (*xsteps w* followed by *mhennig**) was repeated until the tree consistency did not increase further. At this point *bb** (more thorough branch-swapping) was applied to the resulting tree(s) and a strict consensus on the outcome was calculated, if necessary; the character-weightings for calculating the *bb** trees, as well as the usual tree statistics, were recorded. In some cases the number of equally parsimonious *bb** trees (more than about 2,200) exceeded available memory. Table 5.2 presents the results of all Hennig86 analyses, and Figure 5.2 shows the strict-consensus topologies for the runs using (a) all unordered characters and all taxa and (b) all characters (again unordered) with the reduced suite of taxa.

Results

PAUP Analyses

Table 5.1 summarizes the statistics for the PAUP analyses. Trees based on all 59 characters (unordered and ordered) were better resolved (87.5% and 82.5%, respectively) than trees based on subsets of the data (20%–32.5%), with the exception of the ordered and unordered cranial + dental subset (82.5% and 70%, respectively). Tree consistencies were low (maximum consistency index, CI, was 0.25, where available memory enabled all trees to be found), but analyses using the full character set resulted in few equally parsimonious trees (four for unordered characters and eight for ordered), with strict and 50% majority-rule consensus trees being the same in those cases. Ordering of characters did not always improve percent resolution or other tree statistics (Table 5.1), probably because only nine multistate characters could be confidently ordered. Near-most-parsimonious trees to best-resolved trees (all characters, unordered and ordered, all taxa; Figure 5.1a,b) indicated poor support for many clades, all clades being divided by three additional steps. Clades best supported in these analyses were a unified Hipposideridae; a group composed of *muscinus*-group taxa; a *Hipposideros-Asellia-Pseudorhinolophus-Palaeophyllophora* clade; and the following clades of *Hipposideros* species: *calcaratus-maggietaylorae, fulvus-ruber, lylei-pratti-turpis,* and *lankadiva-larvatus.* Not surprisingly these same clades gained the highest bootstrap values (>70%), although bootstrapping might have been severely limited by available memory (maximum of 25,000 trees).

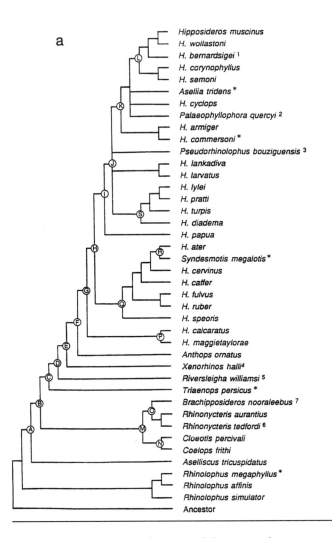

Figure 5.1. Cladograms resulting from PAUP (Phylogenetic Analysis Using Parsimony) conducted on all 59 characters. (**a**) Strict consensus for all 40 taxa plus a hypothetical ancestor; all characters unordered. (**b**) Strict consensus for all 40 taxa plus a hypothetical ancestor; some characters ordered. (**c**) Strict consensus for the 30 taxa common to both this study and that of Bogdanowicz and Owen (Chapter 2, this volume),

Topologies and degrees of resolution of the strict-consensus trees resulting from the eight 41-taxon PAUP analyses varied greatly. However, the following results appeared regularly if not consistently: Hipposideridae monophyletic, with Rhinolophidae as its plesiomorphic sister-group; *Aselliscus tricuspidatus* basal among hipposiderids; *Hipposideros* paraphyletic with species of *Asellia, Palaeophyllophora,* and *Pseudorhinolophus* nesting within it, and these three species commonly associated with Division 2 taxa; Division 1 taxa as the most plesiomorphic division of *Hipposideros;* and, almost always basal to a *Hipposideros*-dominated clade, species of *Xenorhinos, Riversleigha, Triaenops, Rhinonycteris, Brachipposideros, Cloeotis,* and *Coelops.* The positions of the following taxa were most variable among trees: *Anthops*

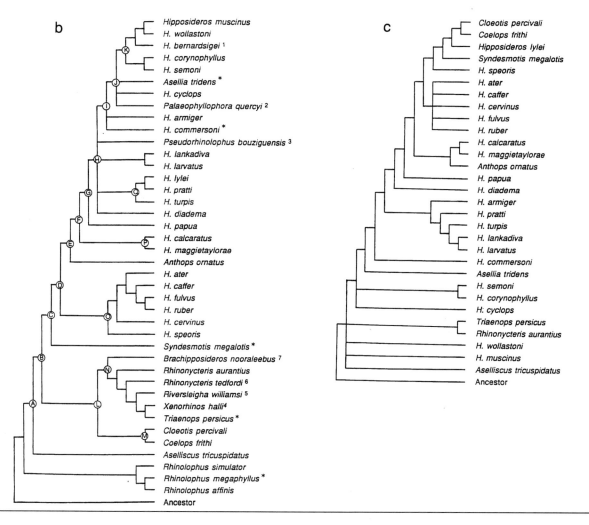

plus a hypothetical ancestor; all characters unordered. Superscript numbers 1–7 indicate the extinct taxa; taxon ages and authors are as follows: [1]*Hipposideros bernardsigei*, early Miocene (Hand 1997a); [2]*Palaeophyllophora quercyi*, late Eocene (Revilliod 1917); [3]*Pseudorhinolophus bouziguensis*, early Miocene (Sigé 1968); [4]*Xenorhinos halli*, early Miocene (Hand 1998); [5]*Riversleigha williamsi*, early Miocene (Hand n.d.);

[6]*Rhinonycteris tedfordi*, early Miocene (Hand 1997b); [7]*Brachipposideros nooraleebus*, early Miocene (Sigé et al. 1982). In (a) and (b), an asterisk indicates that the taxon includes Tertiary representatives; circled letters A–S refer to clades diagnosed in Appendix 5.4.

ornatus, *Syndesmotis megalotis*, *Hipposideros papua*, *H. speoris*, and the *H. calcaratus–H. maggietaylorae* clade.

Only in one PAUP tree was Hipposideridae not monophyletic: in the tree based on unordered cranial + dental characters, *Cloeotis percivali* formed the plesiomorphic sister-species of a multitomy of *Rhinolophus* spp. Hipposiderid paraphyly was also suggested in the parallel run of the Hennig86 analyses (unordered cranial + dental characters), where *H. semoni* grouped with *Rhinolophus* spp. (see following). These results reflect the removal from analyses of the skeletal characters that so cleanly separate hipposiderids and rhinolophids, as well as the unordering of some multistate characters. Apparent apomorphies uniting *Cloeotis percivali* with rhinolophids involve reduction of the

palate anteriorly and posteriorly and, perhaps concomitantly, a change in P[4] shape (i.e., derived states of characters 11, 12, 13, 45, and 46).

The apparent relationships of Riversleigh fossil taxa varied significantly among the 41-taxon trees (compare Figure 5.1a and 5.1b). Most PAUP trees (and Hennig86 trees; see following) suggested that four of the five Riversleigh fossil taxa (*Xenorhinos*, *Riversleigha*, *Brachipposideros*, and *Rhinonycteris tedfordi*) formed part of a near-basal clade that also included *Triaenops persicus* and *Rhinonycteris aurantius*; within this group, *Brachipposideros nooraleebus* tended to be most plesiomorphic and *Xenorhinos* and *Triaenops* distal sister-species (Figure 5.1b). Two unambiguous apomorphies grouped Riversleigh taxa and *Triaenops persicus* (Figure 5.1b):

Table 5.1

Results of PAUP Analyses

Analysis	No. of trees	Tree length	CI	RI	RC	% resolution
41 taxa; characters unordered						
All characters	4	364	0.25	0.54	0.13	87.5
Skeletal + cranial	2,795	240	0.25	0.55	0.14	32.5
Skeletal + dental	>13,700	101	0.32	0.66	0.21	30.0
Cranial + dental	142	361	0.24	0.53	0.12	70.0
41 taxa; some characters ordered						
All characters	8	380	0.23	0.55	0.13	82.5
Skeletal + cranial	2,480	257	0.23	0.55	0.13	22.5
Skeletal + dental	>13,700	103	0.31	0.67	0.21	20.0
Cranial + dental	8	377	0.23	0.54	0.12	82.5
31 taxa; all characters						
Unordered	24	289	0.30	0.48	0.14	83.3
Some ordered	24	297	0.29	0.51	0.15	53.3

Notes: Analyses were done for all 41 taxa and for the 31 taxa common to this study and that of Bogdanowicz and Owen (chapter 2, this volume). Tree length, consistency index (CI), retention index (RI), rescaled consistency index (RC), and percent resolution refer to the strict-consensus trees resulting in each case. (Cases that had more than 13,600 trees exceeded memory limitations and are therefore not a complete sample.) Percent resolution on the strict consensus of these trees was calculated as the ratio of internal nodes recovered divided by the maximum possible ($n - 1$, where n is the number of taxa), times 100.

Table 5.2

Results of Hennig86 Analyses

Analysis	No. of xsteps w iterations	No. of mhennig* trees	No. of bb* trees	Tree length	CI	RI	RC	Percentage of characters for which: CW = 0	CW = 10	% resolution
41 taxa; characters unordered										
All characters	3	4	128	248	0.50	0.78	0.39	51	8.5	57.5
Skeletal + cranial	4	1	24	181	0.55	0.81	0.45	54	10.3	55.0
Skeletal + dental	3	9	>2,274	145	0.74	0.93	0.69	43.5	30.4	30.0
Cranial + dental	4	3	429	227	0.44	0.74	0.33	53.6	3.6	67.5
41 taxa; some characters ordered										
All characters	2	4	276	249	0.49	0.79	0.39	50.8	8.5	50.0
Skeletal + cranial	2	10	>2,273	195	0.59	0.83	0.49	51.3	10.3	35.0
Skeletal + dental	5	12	>2,271	148	0.74	0.94	0.70	34.8	30.4	32.5
Cranial + dental	3	1	6	243	0.41	0.74	0.30	51.8	3.6	82.5
31 taxa; all characters										
Unordered	4	5	20	277	0.58	0.74	0.43	40.7	18.6	73.3
Some ordered	3	3	3	270	0.59	0.76	0.45	44.1	18.6	80.0

Notes: Analyses were done for all 41 taxa and for the 31 taxa common to this study and that of Bogdanowicz and Owen (chapter 2, this volume). Tree length, consistency index (CI), retention index (RI), rescaled consistency index (RC), and percent resolution refer to the bb* trees resulting in each case. (Cases that had more than 2,200 trees exceeded memory limitations and are therefore not a complete sample.) Character-weightings (CW) of 0 and 10 are those used in calculating the bb* trees. Percent resolution on the strict consensus of these trees was calculated as the ratio of internal nodes recovered divided by the maximum possible ($n - 1$, where n is the number of taxa), times 100.

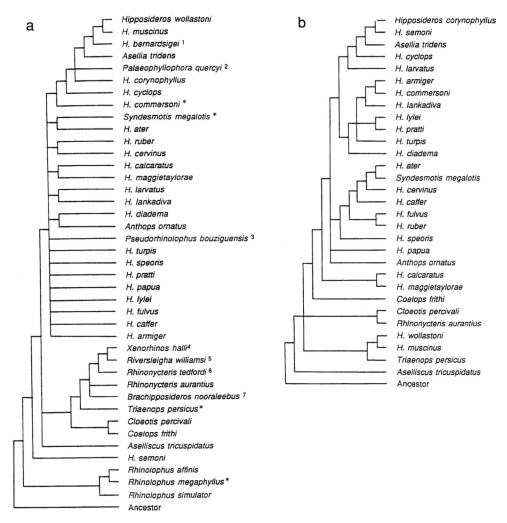

Figure 5.2. Cladograms resulting from Hennig86 analyses conducted on all 59 characters, unordered. Superscripting and other conventions are as defined for Figure 5.1. (a) Strict consensus for all 40 taxa plus a hypothetical ancestor. (b) Strict consensus for the 30 taxa common to both this study and that of Bogdanowicz and Owen (Chapter 2, this volume), plus a hypothetical ancestor.

crested, thickened premaxillae, and the posterior position (dorsal to M³) of the infraorbital foramen (derived states of characters 10 and 17, respectively). However, in the best-resolved tree (all characters, unordered; Figure 5.1a) the Riversleigh fossil taxa were basal but not monophyletic: species of *Brachipposideros* and *Rhinonycteris* formed a clade, but *Xenorhinos, Riversleigha,* and *Triaenops* formed increasingly distant sister-taxa to a *Hipposideros*-dominated group. In both trees (Figure 5.1a,b), *Coelops* and *Cloeotis* together formed the sister-group of a clade containing at least *Rhinonycteris* and *Brachipposideros*.

Consensus trees resulting from PAUP analysis of all characters (unordered and ordered) of taxa common to our study and that of Bogdanowicz and Owen (Figure 5.1c) differed strikingly from the latter's Figure 2.3. In our 31-taxon trees, *Coelops, Cloeotis, Aselliscus, Rhinonycteris,* and *Triaenops* were basal taxa rather than distal, Division 1 taxa

were relatively distal, and *H. maggietaylorae* and *H. calcaratus* were sister-species. In our study and that of Bogdanowicz and Owen, *Hipposideros* was not monophyletic, Division 3 taxa tended to be distal, at least some Division 2 taxa were relatively basal, and *Rhinonycteris* grouped with *Cloeotis*. Overall, our 31-taxon PAUP trees were more similar to our 41-taxon PAUP trees than to Bogdanowicz and Owen's Figure 2.3. However, there were also striking differences between our 41- and 31-taxon trees; for example, in the latter, *Triaenops* grouped with some Division 2 taxa, *Coelops* with *Hipposideros* species (synapomorphically sharing a narrow rostrum, short sphenorbital fissure, and small P4, i.e., derived states of 5, 27, 44), and *lankadiva* and *larvatus* were separated by other *Hipposideros* species. Tree statistics improved with removal of the seven fossil taxa and three rhinolophids (CI = 0.30), although percent resolution declined (see Table 5.1).

Hennig86 Analyses

Results of the Hennig86 analyses are summarized in Table 5.2. As anticipated, use of the Hennig86 successive-weighting option greatly increased the consistencies of the trees, raising them to 0.74 in some cases, albeit only where the number of trees exceeded memory. However, this increase had a cost in information: Between 35% and 54% of the characters used in each run received zero weight, while no more than 30% were given the maximum weight of 10. The three skeletal characters (1–3; synapomorphic in their derived states for Hipposideridae) were always among the latter, and their high weight is clearly the reason that *Rhinolophus* species were consistently the sister-group to all other taxa.

A run of unordered characters omitting the skeletal ones produced an association of *H. semoni* with *Rhinolophus;* that with some ordered features placed *H. semoni* as the sister-group to all other species including *Rhinolophus*. Percent resolution varied among the strict-consensus trees. This index was highest for the trees of cranial + skeletal data on 41 taxa and for those of 31 taxa (species common to our analysis and that of Bogdanowicz and Owen), in the latter case probably less because of the fewer missing data than because inclusion of fossil and other living taxa in the 41-taxon trees generated additional homoplasy. Although 11 characters were "perfect" for the 31-taxon trees, derived states of 3 of these (8, 19, and 33) were autapomorphous in that context and 1 (character 56, uniting the absent *Rhinolophus* spp. in its derived state) was uninformative. Otherwise, the 41-taxon trees based on all 59 characters or the unordered skeletal + cranial data had the highest percentages of resolution; the two trees based on the many fewer skeletal + dental characters (23) were the least resolved. We take it that, for our data at least, more data give more detailed estimates of phylogeny; for this reason our conclusions from the Hennig86 runs rest mainly on the all-characters trees, using the ones based on unordered features for 41 or 31 taxa (Figure 5.2a,b, the strict-consensus topologies for these runs).

Topologically, the 41-taxon trees of course differed greatly, but there are some commonalities among them. First, rhinolophids and hipposiderids were almost always cleanly separated. Second, in most cases, *Hipposideros semoni* was the first branch among Hipposideridae (*H. semoni* is similarly near the base in Bogdanowicz and Owen's study). Third, *Aselliscus tricuspidatus* was one of three major clades among the remaining species in the all-characters topologies or part of a larger multitomy in analyses with fewer characters. Fourth, *Coelops frithi* and *Cloeotis percivali* were paired in seven of eight cases (the exception being the tree

based on unordered skeletal + dental features, where the 2 species were independently part of a basal multitomy). Fifth, in the all-characters trees, *Coelops, Cloeotis,* and *Triaenops persicus* were affiliated with a group of Riversleigh fossil species (*Riversleigha williamsi, Xenorhinos halli, Brachipposideros nooraleebus,* and *Rhinonycteris tedfordi*) that also included *Rhinonycteris aurantius;* the subgrouping of *Triaenops* and the Riversleigh taxa was supported by a strict synapomorphy in character 10 (the derived state, a very thick and crested premaxilla). Sixth, and as implied, the paraphyletic nature of *Hipposideros* and lack of support for its putative species-groups were echoed from the PAUP analyses throughout the Hennig86 runs. Nevertheless, in Figure 5.2a there is a large, albeit unresolved, group of all *Hipposideros* species excepting *H. semoni* that includes a terminal clade (of which *H. cyclops* is the first branch) consisting of the remaining Division 2 species of Hill (1963). *Asellia tridens* and *Pseudorhinolophus bouziguensis* are included in the larger assemblage of *Hipposideros* spp. The tree based on all characters with some ordered differed little from Figure 5.2a and is therefore not shown.

The consensus for the 31-taxon analysis (Figure 5.2b), based on all (unordered) characters and lacking species not represented in Bogdanowicz and Owen's study, is quite different from our corresponding tree of all 41 taxa (Figure 5.2a) and from Figure 2.3 in Bogdanowicz and Owen's chapter with respect to common taxa. *Aselliscus* is part of a basal, unresolved group (Bogdanowicz and Owen placed it in a distal clade; again, compare with our PAUP analyses), but *Hipposideros semoni* is three nodes from the base of the hipposiderid multitomy (it is the second branch in Figure 2.3 of Bogdanowicz and Owen). *Coelops* and *Cloeotis* are, however, again paired (but distally) in both studies, while *Triaenops* is affiliated with *Rhinonycteris aurantius* (in our analyses only; *Rhinonycteris* is several steps removed from *Triaenops* in Bogdanowicz and Owen's Figure 2.3). At the same time, both our trees and those of Bogdanowicz and Owen once more indicate the paraphyly of *Hipposideros* and call into question the reality of putative sister-groups within the genus. However, two uniquely derived character states (of characters 58 and 59; respectively, the conspicuous elongation of M1–2 postmetacristae and lateral compression of the M1 trigonid) are synapomorphic in our tree for *H. maggietaylorae* with *H. calcaratus*. Another strict synapomorphy (the shared, derived state of character 5, a narrow rostrum) unites the group of mostly Division 1 species of which *H. papua* is the first branch and *Coelops* and *Cloeotis* a terminal pair; *Anthops* is also a member of this clade. Only two Division 3 species (*H. lylei* and *H. speoris*) are included in this group; the rest of the Division 3 taxa are singly (*H. diadema*) or in combination sister-groups to it, and all Division 2 taxa are more basal in the tree, with *Asellia* separating these from the clade of combined Division 1 and 3 taxa. Thus, the apparent direction of change is generally reversed in the 31-taxon tree, which lacks fossils, from that in the 41-taxon consensus: *Coelops* and *Cloeotis* are terminal, not basal, while the Division 2 species are basal rather than terminal. (However, the 31-taxon tree based on ordered characters did place Division 2 species with the distal clade of mostly Division 3 taxa.)

Three European fossil taxa surveyed for information but not included in the final PAUP or Hennig86 analyses were *Brachipposideros collongensis* Depéret, 1892 (early Miocene), *Pseudorhinolophus schlosseri* (Revilliod 1917) (late Eocene–early Oligocene), and *Vaylatsia* cf. *prisca* (Revilliod 1920; but see Sigé 1978, 1990) (late Eocene–early Oligocene). Ultimately these species were not included because characters could not be confidently scored on the basis of the material available for examination, not because of problems with missing data. However, some preliminary PAUP results suggested that (1) *B. collongensis* is a plesiomorphic taxon not necessarily closely related to Australia's *B. nooraleebus;* (2) based on craniodental morphology alone, *Vaylatsia* cf. *prisca* is the plesiomorphic sister-group of extant rhinolophids (see *Discussion*); and (3) *P. schlosseri* is more closely related to *H. cyclops* than to *P. bouziguensis*, suggesting that *Pseudorhinolophus* is possibly para- or polyphyletic (although the identity of the "*P. schlosseri*" cranial material we examined is in doubt).

Discussion

Phylogenetic Relationships

Hypotheses of paraphyly of the genus *Hipposideros* have been advanced previously by palaeontologists Hugueney (1965), Sigé (1968), and Legendre (1982). In 1968, B. Sigé erected two *Hipposideros* subgenera for fossil taxa: *Pseudorhinolophus*, once regarded as a genus, was given subgeneric status in recognition of its possible ancestry to some Division 3 *Hipposideros* taxa (e.g., *armiger, diadema,* and *commersoni*), while the subgenus *H. (Brachipposideros)* was erected for fossil species showing affinities with small Recent species of *Hipposideros* such as *H. caffer*. Legendre (1982) noted a possible relationship between species of *Pseudorhinolophus* and *Asellia* and between *Brachipposideros* and *Syndesmotis* species (see also Sigé et al. 1982). Some, although not all, of these theories are supported by our study. Among the probable descendants of a *Hipposideros* clade are species of *Asellia*, *Palaeophyllophora,* and *Pseudorhinolophus*. However, no special relationship was found between Australian *Brachipposideros* and *Syndesmotis* or members of the *bicolor* group,

including *H. caffer*. Cranial material of European *Brachipposideros* could not be included in these analyses, and the close relationship of European and Australian *Brachipposideros* remains untested. Certainly Australian *Brachipposideros* groups tightly with endemic *Rhinonycteris*. A special relationship between *Asellia*, Division 2 taxa, and the highly autapomorphic *Palaeophyllophora* is suggested for the first time here, although the latter's similarities to *H. cyclops, H. commersoni*, and *Asellia* have been discussed by, among others, Revilliod (1917).

Among Division 1 taxa, we found no clear division between Hill's (1963) *bicolor* and *megalotis* groups or *bicolor* and *galeritus* subgroups, with, for example, *fulvus* and *ruber* occurring as sister-species. Hill (1963, p. 16) stressed the unity of the *bicolor* group (*galeritus* plus *bicolor* subgroups), concluding that "no definitive line of separation can be found between the subgroups, and they are linked by species exhibiting their respective characteristics in differing combinations." In fact, Hill (1963, Figure 3) divided these taxa into subgroups according to their degree of specialization rather than shared derived characters, so that sister-taxa, as now understood in the light of phylogenetic systematics, might be expected to belong to different subgroups. For example, species such as *ater* and *fulvus*, which Hill regarded to be among the simplest members of the *bicolor* group, were placed in the *bicolor* subgroup, while *galeritus* (including *cervinus*) and *caffer* (including *ruber*), which he regarded to be among the most specialized, *although not necessarily closely related to each other,* were placed in the *galeritus* subgroup. Relationships among Division 3 taxa were also poorly resolved by our study. However, it is possible that Division 3 species groups might also be paraphyletic, being traditionally grouped according to degree of specialization rather than relationship. For example, *H. lankadiva* and *H. larvatus* consistently appear in our analyses as sister-taxa, but Hill (1963) considered them to be the least modified members of the *diadema* and *speoris* groups, respectively.

The phylogenetic relationships of Division 2 taxa were recently examined by Hand (1997a), who described Riversleigh's early Miocene *H. bernardsigei* as the first fossil member of the New Guinea–centered *muscinus* group (i.e., *muscinus, wollastoni, corynophyllus, edwardshilli, semoni,* and *stenotis*). Hill (1963) proposed that the least specialized members of the *cyclops* group (*cyclops* and *camerunensis*) show affinities with the *bicolor* group and that their origins lie remotely with that species group; in contrast, our analyses suggested that their orgins might lie with the Division 3 taxa *Asellia* and *Palaeophyllophora* (see Figures 5.1 and 5.2). The possibility that at least one member—*H. semoni*—might be basal among hipposiderids (as suggested here) needs further examination. *Hipposideros semoni* and its Australian sister-species *H. stenotis* are usually interpreted to be the most derived Division 2 taxa (Hill 1963; Flannery and Colgan 1993; Hand 1997a), being distinguished among hipposiderids by their extremely enlarged petrosals, broad rostrum, deep frontal depression, and very wide sphenorbital fissure, as well as large M3 and m3, very reduced P3 and p3, and very tall c1 and p4. Many of these features almost certainly relate to the transmission and reception of their acoustically highly specialized ultrasonic calls, and probably also to diet, and are generally assumed to convergently resemble those characterizing most but not all rhinolophids.

Species of *Rhinonycteris*, *Brachipposideros*, *Xenorhinos*, *Riversleigha*, *Triaenops*, *Coelops*, and *Cloeotis* appear to represent early but specialized offshoots of the hipposiderid radiation. Hill (1982) placed *Rhinonycteris*, *Triaenops*, and *Cloeotis* in a small group characterized principally by features of the noseleaf. Gray (1866) also recognized the distinctiveness of *Rhinonycteris* and separated it from other rhinolophids and hipposiderids (the Rhinolophina) as the sole member of the Rhinonycterina. As pointed out by Hill (1982), Gray's definition of this group would also include the subsequently described *Triaenops* and *Cloeotis;* the three taxa comprise Koopman's (1994) Rhinonycterina. Pierson's (1986) preliminary transferrin immunological distance data on the relationship of *Rhinonycteris* to *Rhinolophus* and *Hipposideros* suggested it is far removed from both. Pierson (1986) also concluded that *Aselliscus,* according to our study one of the most plesiomorphic hipposiderids, grouped more closely with *Rhinolophus* than with *Hipposideros*.

Perhaps one of the most surprising results of our analyses is the apparent association between the geographically widely separated *Coelops* and *Cloeotis*. Most authors have recognized the highly autapomorphic nature of *Coelops*; Koopman (1994), for example, placed species of *Coelops* and *Paracoelops* (the latter is known only from the badly damaged type specimen) in a separate tribe, the Coelopsini. Tate (1941b) referred *Coelops* to a separate subfamily, the Coelopsinae, on the basis of characters of the tail, pinna, noseleaf, metacarpals, and four craniodental features. Two of the latter features (i.e., greatly extended canine-bearing portion of the maxilla; U-shaped symphysial portion of combined mandibles) are autapomorphies shared by no other hipposiderid taxa, whereas two (C1 with anterior and posterior accessory cusps; much fenestrated basicranium) are shared with *Cloeotis.*

Timing and Place of Radiation

Fossil taxa provide a minimum age estimate for hipposiderid radiation. Species of *Pseudorhinolophus* and *Palaeophyllophora* first occur in the fossil record in the middle and late Eocene of Europe. They are interpreted here to be relatively derived hipposiderids, a conclusion also reached by Revilliod (1917; but see Sigé 1978 for discussion regarding the dentition of *Palaeophyllophora*). The divergence between the hipposiderid and rhinolophid lineages was until recently thought to have occurred before the middle Eocene. However, there is now doubt that Paleogene species referred to the genus *Rhinolophus* (e.g., *R. priscus* Revilliod, 1920) are true rhinolophids. Sigé (1990) erected the genus *Vaylatsia* in the Hipposideridae for taxa with dental morphology very close to that of Neogene (and extant) *Rhinolophus* species but with the distal end of the humerus exhibiting morphology characteristic of hipposiderids. Analysis of craniodental characters of a crushed, partial skull of *Vaylatsia* cf. *prisca* from the early Oligocene Quercy deposit of Mas de Got indicates that this group is the plesiomorphic sister-group of extant rhinolophids. Joint study (with B. Sigé) of a more complete skull and postcranial material of *V.* cf. *prisca* may help resolve the phylogenetic position of *Vaylatsia*. Certainly, the European fossil record provides clear evidence that hipposiderids had diverged from megadermatids (*Necromantis* species), a sister-group at least once removed, before the middle Eocene (Revilliod 1922; Marandat et al. 1993).

Because hipposiderids occur suddenly in the middle Eocene European fossil record, the early radiation of the group is thought to have occurred outside Europe, possibly in Asia (Sigé 1977, 1991; Legendre 1982; Sigé et al. 1982; Sigé and Legendre 1983). The Tertiary record for hipposiderids outside Europe supports this hypothesis. Two Asian species have been described: *Brachipposideros khengkao* and *Hipposideros felix* from early Miocene deposits at Li Mae Long, Thailand (Mein and Ginsburg 1997). Other Tertiary bats from Asia (plus the Indian subcontinent) include pteropodids, emballonurids, vespertilionoids, and molossids (Rich et al. 1983; Qiu et al. 1985; Legendre et al. 1988; Ducrocq et al. 1993), but no Tertiary hipposiderids have yet been described.

The African record is better. Butler (1969, 1984) reported three hipposiderid taxa from the Miocene of east Africa, at least one of which is described as being very similar to Europe's *Pseudorhinolophus bouziguensis*. *Syndesmotis vetus* is recorded from the middle Miocene deposit of Beni Mellal in Morocco (Lavocat 1961; Sigé 1976; Legendre 1982). From a Pliocene deposit in Madagascar, Sabatier and Legendre (1985) reported the presence of *Triaenops persicus*, *H. commersoni*, and a smaller, unidentified hipposiderid not now living on the island. Sigé et al. (1994) described an early Oligocene bat fauna from Taqah, in the Sultanate of Oman, Arabian Peninsula (part of a common Arabo-African area at the time). Of eight bat taxa recorded from Taqah, two are hipposiderids: a species of *Brachipposideros* and an indeterminate form perhaps more similar to *Pseudorhinolophus*.

In North Africa, the earliest representatives of two modern bat clades make their appearance: a rhinolophoid (nycteridid) and vespertilionoid (phillisid) from the late early Eocene deposit of Chambi in Tunisia (Sigé 1991). Because modern bats did not occur in Europe until the middle Eocene, migration of such bats from Africa to Europe is suggested (Sigé 1991). However, early Oligocene bats from Taqah and from the Fayum (Sigé 1985) exhibit a high degree of endemism compared with European bats of similar age (Sigé et al. 1994) and are quite derived with respect to their European contemporaries, which appear to have evolved gradually in situ. Sigé et al. (1994) concluded that modern bats first entered Europe from Africa in the middle Eocene but that the Arabo-African and western European bat faunas were subsequently separated for a long period of the Eocene and Oligocene. During the late Oligocene and throughout the Neogene there was an influx of modern types into Europe, including species of *Brachipposideros* and *Asellia* among hipposiderids, as well as nycteridids, pteropodids, emballonurids, and megadermatids.

The Australian Tertiary record includes a minimum of 12 hipposiderid taxa from a sequence of freshwater and karstic limestones at Riversleigh in northwestern Queensland. Riversleigh's bat-bearing sediments span the period from late Oligocene to middle Miocene and also include Pliocene and Quaternary cave deposits (Archer et al. 1994). *Brachipposideros, Hipposideros, Rhinonycteris, Myophyllorhina, Xenorhinos,* and *Riversleigha* are represented in these deposits, most by more than one species (Sigé et al. 1982; Hand 1997a, 1997b, 1998, n.d.). As yet there is no evidence of the European and Arabo-African (i.e., western) taxa *Pseudorhinolophus, Palaeophyllophora, Asellia,* or *Vaylatsia* in the Australian record, but species of *Brachipposideros* and *Hipposideros* seem to occur in all three areas. Other hipposiderid genera represented at Riversleigh *(Rhinonycteris, Xenorhinos,* and *Riversleigha)* are known only from Australia but appear to be closely related to African and Asian taxa *(Triaenops, Cloeotis,* and *Coelops).*

Area cladograms for extinct and extant members of the family Hipposideridae, in which taxon names on a phylogeny are replaced by the geographic regions in which the taxa occur (Platnick and Nelson 1978; Nelson and Rosen 1981), reveal no clear biogeographic pattern of hipposiderid radiation. In many cases, immediate sister-taxa are from the same region (e.g., *calcaratus-maggietaylorae, lylei-pratti, muscinus* clade), but in others they are geographically widely separated (e.g., *Coelops-Cloeotis*). Within the *Hipposideros-Asellia-Palaeophyllophora-Pseudorhinolophus* clade, Ethiopian or Palearctic taxa are among the most plesiomorphic members, while in other clades these are Australian or Asian taxa.

An alternative method for visualizing historical bio-geography is to fit parsimoniously a multistate, unordered character representing geographic occurrence onto the species cladogram, using, for example, MacClade (Maddison and Maddison 1992). We did this, coding the distributions of species following Hill (1963) and Koopman (1994) and mapping the resulting scores onto the consensus cladograms of Figures 5.1a,b, and 5.2a, with *Rhinolophus* spp. and the hypothetical ancestor omitted; the four zoogeographic regions (Australian, Ethiopian, Palearctic, and Indo-Malayan) are those defined by Koopman (1970). Figure 5.3 shows the result using one PAUP-derived tree (Figure 5.1b; based on all characters with some ordered). At the base of the figure, Australian taxa are much the commonest *(Aselliscus, Brachipposideros, Rhinonycteris, Riversleigha, Xenorhinos),* so that, despite subsequent character-state changes, the common ancestor of hipposiderids is parsimoniously interpreted as having been Australian. Only one group, the *H. muscinus* clade, appears to be a "back-migrant," being separated from other Australian taxa by several non-Australian lineages. Figure 5.1a and the much less resolved Hennig86 equivalent, Figure 5.2a, imply similar conclusions.

Thus, the MacClade mappings indicate an ultimate Australian origin for Hipposideridae, with subsequent dispersal or vicariant events leading to the occupation of (especially) Indo-Malaya by several lineages. Of course, modern distributions of taxa might not wholly reflect past distributions, and the hipposiderid fossil record is biased geographically and temporally. Moreover, character fittings are influenced by the choice (or omission) of both ingroup and outgroup species, as well as by the uncertainties of parsimony analyses themselves (in all our trees, however, Australian taxa are among the first to branch from the base). Nevertheless, the relationships shown in Figures 5.1 through 5.3 do suggest that the primary radiation for hipposiderids occurred south of Europe, probably in the Australian region during or before the middle Eocene. During the Tertiary, hipposiderids appear to have been widely distributed throughout lower latitudes of the Old World, with regional endemism apparent at various times in different areas but with little evidence of profound, long-lasting barriers to dispersal. Palaeogeographic reconstructions published since the mid-1980s (Audley-Charles 1987) have suggested that island stepping-stones were available in the region for good dispersers for the last 100 million years, with the palaeobotanical record showing interchange of plants between, for example, Australia and Asia since the Cretaceous (Truswell et al. 1987).

A Southern Hemisphere origin for the world's extant bat radiation was argued by Pierson (1986) on the basis of distributions of endemic bat families. Hershkovitz (1972) also suggested a Gondwanan origin for bats, and Sigé (1991) proposed that modern bat groups evolved from isolated

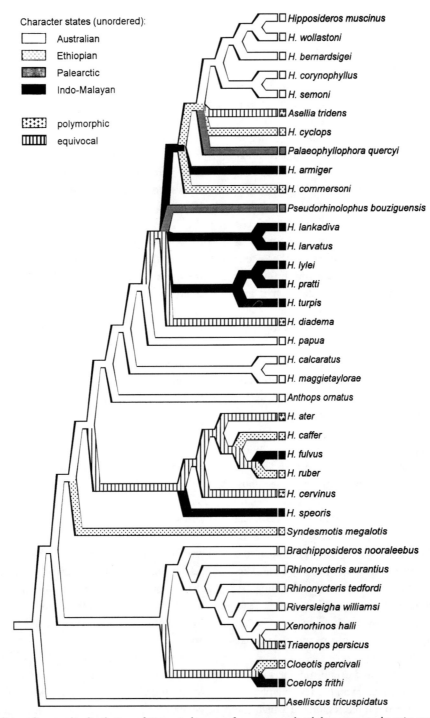

Figure 5.3. MacClade fitting of geographic distributions of 37 terminal taxa as a four-state unordered character onto the strict-consensus PAUP cladogram of Figure 5.1b, with *Rhinolophus* spp. and the hypothetical ancestor omitted. Character states (geographic regions) are depicted visually on this cladogram. Five taxa occurring in two areas were polymorphic: two Palearctic-Ethiopian *(Asellia tridens, Triaenops persicus)* and three Indo-Malayan–Australian species *(Hipposideros ater, H. cervinus, H. diadema).* Tree length = 17 steps; consistency index = 0.47; retention index = 0.44; resolution = 21%.

immigrant archaic groups in the Southern Hemisphere in the early Eocene. Africa was believed by Schlosser (1911; see Sigé 1985, 1991 for discussion) to be the source for America's phyllostomoids (=noctilionoids). Ducrocq et al. (1993) suggested that megachiropterans diverged from primitive bats within the Paleocene–early Eocene interval in southern Asia. Archaic bats (i.e., archaeonycteridids) were widely distributed by the early Eocene, being now known from the fossil records of the Northern and Southern hemispheres. The oldest representatives of modern bat groups are the Tunisian vespertilionoid and nycteridid from the late early Eocene discussed earlier (Sigé 1991), a natalid from the late Wasatchian of central Wyoming (Beard et al. 1992), and a pteropodid from the late Eocene of Thailand (Ducrocq et al. 1993). It appears that modern bat clades were established early in the Paleogene and overlapped widely with more primitive bats both in time and in space.

Archaeonycteridids occurred in the Australian region in the early Eocene (Hand et al. 1994), before Australia separated from Antarctica and South America approximately 35–45 million years ago (Veevers 1984). As yet, it is unclear whether they first entered Australia from South America or southeast Asia, but their presence puts archaic bats in the right place at the right time to provide support for a Southern Hemisphere origin for some bat clades, including perhaps rhinolophoids. The evolutionary relationships of Australia's 55-million-year-old *Australonycteris clarkae* are currently under study by B. Sigé and S. Hand, and possible relationships to modern clades are being assessed. A gap in the Australian fossil mammal record between 55 and approximately 25 million years ago (Archer et al. 1994) prevents documentation of key events in the evolution of bats in the Australian region, but the later (middle Tertiary) Australian record shows a diverse hipposiderid fauna indicative of earlier radiation and probable divergence from other hipposiderids by the middle to late Eocene.

Of potential sister-group taxa to hipposiderids, rhinolophids and nycteridids are absent from the Australian fossil record but Tertiary megadermatids are relatively common. The latter includes species of *Megaderma* and *Macroderma* (Hand 1985, 1995, 1996) that have been interpreted as relatively derived taxa by Hand (1985) but as plesiomorphic by Griffiths et al. (1992). Molossids, vespertilionids, and the ancestors of New Zealand's mystacinids are known from Australian Oligo-Miocene deposits and emballonurids from Pliocene sediments, but megachiropterans have yet to be discovered in the pre-Pleistocene record of Australia (Hand 1990; Archer et al. 1994; Hand et al. 1998). Bogdanowicz (1992) and Bogdanowicz and Owen (1992) argued for a Southeast Asian origin for rhinolophids and proposed that *Rhinolophus megaphyllus* from the Malay Peninsula, New Guinea, Australia, New Ireland, and New Britain is the most basal member.

In the Australian region, hipposiderid endemism is greater in the islands of New Guinea, the Bismark Archipelago, Solomon Islands, and North Moluccas than in Australia, the island nature and vertically stratified habitats of the New Guinea region probably fostering speciation and endemism in the group. The New Guinea region has one endemic monotypic genus and nine endemic species (*Anthops,* and *Aselliscus tricuspidatus, Hipposideros papua, demissus, muscinus, wollastoni, edwardshilli, corynophyllus, maggietaylorae,* and *calcaratus;* Flannery 1990), while Australia has one of each *(Rhinonycteris aurantius* and *H. stenotis); H. semoni* is shared by Australia and New Guinea. *Aselliscus tricuspidatus,* found today in the Moluccas and islands east to the New Hebrides, is evidently one of the most plesiomorphic hipposiderids.

In Australia, *Rhinonycteris aurantius* and five *Hipposideros* species (representing the *bicolor, cyclops,* and *diadema* groups) inhabit warm, humid caves in tropical-subtropical latitudes. Only two of the four extant Australian hipposiderid lineages (*Rhinonycteris* and the *H. muscinus* group) are as yet represented in the Oligo-Miocene Riversleigh deposits, and this presumably reflects a longer history for these groups in northern Australia; both are endemics of the Australian region. Species of the *Rhinonycteris-Brachipposideros* clade are common in Riversleigh's late Oligocene to middle Miocene sediments and also occur in the Pliocene Rackham's Roost deposit; *R. aurantius* is found in Riversleigh's Quaternary cave deposits. A marked decline in *Rhinonycteris* diversity occurred in the Riversleigh area sometime after the middle Pliocene (a drop from at least four species in the Rackham's Roost deposit to a single living species), and this perhaps coincides with colonization of Australia by the Asian-centered *bicolor* and *diadema* groups. *Hipposideros ater* is a common cave species in the area today, and a possible member of the *bicolor* group occurs in Riversleigh's Pliocene Rackham's Roost deposit. No Tertiary or Quaternary representatives of the *diadema* group are known from Riversleigh, although *H. diadema* occurs in ?Pleistocene cave sediments at Chillagoe, northeastern Queensland. Today, two subspecies of *H. diadema* inhabit northeastern Queensland and the Northern Territory, as part of the species' current broad Indo-Australian distribution. In the Riversleigh region, the *H. muscinus* group is represented by *H. bernardsigei* in a number of early to middle Miocene sites and in the modern fauna by *H. stenotis,* but the group is not represented in Pliocene or Pleistocene sediments. A shared Australian–New Guinean early Miocene distribution for the group is likely (Flannery 1990; Hand 1997a). The age of Riversleigh's *H. bernardsigei* (20 million years old) provides a minimum date for the radiation of the *muscinus* group and its separation from African relatives *H. cyclops* and *Asellia*.

Although species of the *Rhinonycteris* group (sensu lato) are the most common bats in Riversleigh's Oligo-Miocene and Pliocene deposits, there is no evidence that they ever inhabited New Guinean landmasses. This is particularly curious because they are contemporaries of Riversleigh's *H. bernardsigei* (see earlier discussion), they persisted in the Riversleigh region throughout the Neogene, and close relatives are spread as widely as the Palearctic and Ethiopian regions (e.g., species of *Triaenops* and *Brachipposideros*). There appears to have been little physical impediment for the group to reach New Guinea, although estimates for the date of first emergence for New Guinea vary from the Cretaceous (Lloyd 1992) to 30 million years ago (Flannery 1990), and the extent and dates of subsequent episodes of submergence (and presumably extinction) are also unclear. Ecological rather than geographic barriers appear to have been at least partly responsible, with a lack of preferred habitat (including microclimate) and competition for available resources perhaps being factors.

Conclusions

In summary, virtually all our analyses, whether based on PAUP or Hennig86, (1) demonstrate the unity of the Hipposideridae, (2) indicate some surprising associations of geographically separated taxa (e.g., of *Coelops* and *Cloeotis*, and of *Triaenops* with these, *Rhinonycteris*, and Riversleigh taxa), (3) generally support the supposed relationships of many fossils but call into question others, (4) suggest that the genus *Hipposideros* is paraphyletic, and (5) raise serious doubts about the composition of previously proposed divisions or their subsumed species-groups within *Hipposideros*.

Notwithstanding the last point, there is often a clear separation of Hill's Division 2 species from the other two divisions of *Hipposideros*. *Aselliscus* is usually basal to other Hipposideridae, as is *H. semoni* in some trees, perhaps because of the possibly convergent resemblance of the latter to *Rhinolophus* in characters relating to their shared and highly specialized echolocation systems. Certainly, further examination of *H. semoni* (and the closely related Australian *H. stenotis*) should receive high priority in any future study of hipposiderid phylogeny.

Similarly, the association of *Brachipposideros nooraleebus* with other Riversleigh fossils must be tested further, particularly in view of its importance for temporal correlation with hipposiderids elsewhere. Finally, our comparison of trees with and without fossils as part of the suite of taxa has demonstrated once more the significant effect that inclusion of extinct forms may have not only on specific relationships but also on the inferred directionality of anatomical evolution (Kirsch and Archer 1982).

Craniodental data alone, as used here, will probably not be sufficient to interpret the phylogenetic history of the family Hipposideridae. The addition of data from postcranial, noseleaf, and pinna morphology as well as molecular studies may help resolve phylogenetic relationships in the Hipposideridae, especially within the genus *Hipposideros*, while information from the rapidly growing fossil bat record should help fill in gaps in the larger picture of bat evolution and radiation in the Southern Hemisphere.

Appendix 5.1.
Hipposideros Species Groupings Used in This Study

This categorization of *Hipposideros* spp. is from Hill (1963). See also Koopman (1994) and Flannery (1995).

Division 1	*H. coronatus*	*H. fuliginosus*
megalotis group	*H. ridleyi*	*H. caffer*
H. megalotis	*H. jonesi*	*H. lamottei*
bicolor group	*H. dyacorum*	*H. ruber*
bicolor subgroup	*H. sabanus*	*H. beatus*
H. bicolor	*H. doriae*	*H. coxi*
H. pomona	*H. obscurus*	*H. papua*
H. macrobullatus	*H. marisae*	
H. ater	*galeritus* subgroup	**Division 2**
H. fulvus	*H. pygmaeus*	*cyclops* group
H. halophyllus	*H. galeritus*	*H. cyclops*
H. cineraceus	*H. cervinus*	*H. camerunensis*
H. nequam	*H. crumeniferus*	*H. muscinus*
H. calcaratus	*H. breviceps*	*H. wollastoni*
H. maggietaylorae	*H. curtus*	*H. corynophyllus*

H. edwardshilli	*armiger* group	*H. lekaguli*
H. semoni	*H. armiger*	*H. lankadiva*
H. stenotis	*H. turpis*	*H. schistaceus*
	speoris group	*H. diadema*
Division 3	*H. abae*	*H. dinops*
pratti group	*H. larvatus*	*H. demissus*
H. pratti	*H. speoris*	*H. inexpectatus*
H. lylei	*diadema* group	*H. commersoni*

Appendix 5.2.
The 59 Characters in the Phylogenetic Analysis of Rhinolophoid Taxa

The states are denoted as follows: (0) interpreted plesiomorphic condition; (1–4) apomorphic states. Characters 1–3 are from Miller (1907), characters 9 and 10 are from Hill (1963), and all others are from personal observation.

1. Pelvis: (0) without postacetabular foramen; (1) with postacetabular foramen.
2. Number of phalanges in each of toes 2–5: (0) three; (1) two.
3. Pectoral girdle: (0) last cervical and first dorsal vertebrae fused plus ribs and sternum; (1) further modified such that last cervical and first two dorsal vertebrae fused plus ribs and sternum.
4. Interorbital region: (0) only moderately constricted (more than or equal to half rostral width); (1) markedly constricted (less than half rostral width); (2) very markedly constricted (much less than half rostral width); (3) barely constricted (greater than or equal to rostral width).
5. Rostral width: (0) wide (more than half the mastoid width); (1) narrow (less than or equal to half the maximum braincase width).
6. Rostrum height: (0) about as high as braincase; (1) much lower than braincase.
7. Rostral inflation: (0) moderate; (1) small; (2) large.
8. Central nasals: (0) not especially inflated; (1) conspicuously inflated.
9. Anterior palatal foramina: (0) closed laterally by maxillae; (1) encircled by premaxillae.
10. Premaxillae: (0) not especially thickened; (1) very thick with distinct crest on dorsal surface at line of contact.
11. Shape of anteroventral palate: (0) V-shape; (1) shape of blunt V; (2) U-shape.
12. Posterior extent of palate midline in relation to M3: (0) posterior to or at the level of M3; (1) anterior to M3.
13. Posterior extent of palate midline in relation to posterolateral incisura: (0) posterior to or at the level of posterolateral palatal incisura; (1) anterior to posterolateral incisura.
14. Frontal depression: (0) deep or conspicuous; (1) shallow or absent.
15. Supraorbital ridges: (0) present; (1) absent.
16. Supraorbital crest: (0) weakly developed; (1) strongly developed.
17. Position of infraorbital foramen: (0) dorsal to M1–2; (1) dorsal to M2; (2) dorsal to M3.
18. Shape of infraorbital foramen: (0) round; (1) subrounded; (2) elongate (oval).
19. Lacrimal foramen size: (0) small; (1) slightly enlarged; (2) enlarged.
20. Anteroventral rim of orbital floor: (0) rounded; (1) raised and sharp.
21. Zygomatic width: (0) less than or equal to maximum braincase width; (1) greater than maximum braincase width.
22. Height of zygomatic process: (0) low; (1) moderate; (2) very tall.
23. Height of sagittal crest: (0) low; (1) tall.
24. Form of sagittal crest: (0) incomplete; (1) complete.
25. Sphenorbital bridge: (0) constricted (width posterior to pterygoid processes is three-quarters the anterior width); (1) wide (lateral margins posterior to the pterygoid processes are parallel); (2) greatly constricted (width posterior to pterygoid processes is half the anterior width); (3) very greatly constricted (width posterior to pterygoid processes is less than half the anterior width); (4) not constricted.
26. Sphenorbital fissure: (0) fissure and optic foramen small, paired; (1) fissure elongated.
27. Posterior extent of sphenorbital fissure: (0) level with glenoid fossa; (1) anterior to glenoid fossa.
28. Optic foramen: (0) completely separated from sphenorbital fissure by bar of bone; (1) incompletely separated from sphenorbital fissure; (2) confluent with sphenorbital fissure.
29. Mesopterygoid roof: (0) without groove; (1) with groove opening into basisphenoid depression; (2) with deep, broad groove terminating in excavation anterior to basisphenoid (no depression).
30. Protrusion of vomer posterior to palate: (0) inconspicuous; (1) conspicuous.
31. Position of hamular processes: (0) middle third of sphenoidal bridge; (1) posterior third of sphenoidal bridge; (2) middle to posterior third of sphenoidal bridge; (3) anterior third of sphenoidal bridge.
32. Glenoid fossa shape: (0) round; (1) wide (conspicuously wider than long).
33. Basisphenoid depression: (0) present; (1) absent.
34. Basisphenoid shape: (0) long and broad; (1) diamond-shaped; (2) extremely short.
35. Petrosal expansion, in terms of width of cavity for periotic:

(0) cavity width is six times the interperiotic distance (basioccipital width); (1) cavity width is one to three times the interperiotic distance; (2) cavity width is four times the interperiotic distance; (3) cavity width is five times the interperiotic distance; (4) cavity width is eight times the interperiotic distance.

36. Foramen ovale position: (0) level with anterior to mid glenoid fossa; (1) level with mid to posterior glenoid fossa.

37. Foramen ovale size: (0) less than half the glenoid area; (1) half to two-thirds the glenoid area; (2) more than two-thirds the glenoid area.

38. Accessory foramina to foramen ovale (subovale?): (0) absent; (1) present.

39. Paroccipital processes: (0) slender; (1) broad and laterally expanded.

40. C1 posterior accessory cusp: (0) absent or very low; (1) tall (conspicuous but less than half the height of main cusp); (2) very tall (equal to or greater than half the height of main cusp).

41. C1 anterior accessory cusp: (0) absent or low; (1) tall.

42. P2 reduction: (0) P2 present but reduced; (1) P2 absent.

43. P2 extrusion: (0) P2 in toothrow (not extruded); (1) P2 extruded from toothrow but still partially separates C1 and P4; (2) P2 greatly extruded.

44. P4 size: (0) approximately equal to size of M1; (1) conspicuously smaller than M1.

45. P4 shape: (0) about as wide as long; (1) wider than long; (2) longer than wide.

46. P4 anterior margin: (0) narrow; (1) wide.

47. P4 anterolingual cingular cusp: (0) present; (1) absent.

48. P4–M2 posterior cingula: (0) narrow; (1) thick.

49. M1 structure: (0) protofossa open, with or without dihedral crest; (1) protofossa closed, without dihedral crest; (2) protofossa closed, with weak dihedral crest; (3) protofossa closed, with strong dihedral crest; (4) protofossa closed by joining posterior cingulum rather than base of metacone.

50. M3 w-pattern: (0) little reduced (premetacrista half to three-quarters the paracrista length); (1) reduced (premetacrista as small as one-third the preparacrista length); (2) very reduced (premetacrista less than one-third the preparacrista length); (3) complete.

51. M3 reduction (in relation to size of P4): (0) little reduced; (1) moderately reduced in length and/or width; (2) extremely reduced.

52. Ascending ramus height: (0) short (less than twice the m3 height); (1) tall (more than twice the m3 height).

53. Lower incisor size: (0) i1–2 approximately equal in size; (1) i2 conspicuously larger than i1.

54. i1 structure: (0) trilobed; (1) bilobed.

55. P4 reduction: (0) reduced; (1) very reduced (much less than half p4 height).

56. p4: (0) absent; (1) present.

57. m3 talonid width: (0) complete; (1) narrow.

58. Elongation of M1–2 postmetacristae: (0) absent; (1) conspicuous.

59. Lateral compression of M1 trigonid: (0) absent; (1) conspicuous.

Appendix 5.3.
Data Matrix for PAUP and Hennig86 Analyses

The 59 characters and their states are as defined in Appendix 5.2.

Taxon	1–5	6–10	11–15	16–20	21–25	26–30	31–35	36–40	41–45	46–50	51–55	56–59
Ancestor	00000	00000	00000	00000	00000	000?0	00000	00000	00000	00000	00000	0000
Anthops ornatus	11111	10000	01110	01201	01000	11000	01102	10001	10110	00021	11000	0100
Asellia tridens	11100	00000	20010	11201	11110	10010	01002	00000	01?11	11002	11100	0000
Aselliscus tricuspidatus	11100	10000	00010	01001	01100	10110	00001	00000	00000	00000	00000	0000
Brachipposideros nooraleebus	1?110	10001	00010	12001	?200?	1?000	?0001	00001	00101	00000	00000	0000
Cloeotis percivali	11111	10000	21111	?0000	02004	10010	00001	01000	10101	11020	00000	0000
Coelops frithi	11111	11000	10010	10001	00000	11010	00001	02011	10112	00040	00000	0000
Hipposideros armiger	11110	01000	20011	?1201	11111	11011	21001	00010	00210	00022	11100	0100
Hipposideros ater	11101	11000	10010	01201	01003	11111	10011	01000	00000	00002	11000	0100
Hipposideros bernardsigei	11100	100??	10110	01201	11000	102?0	00123	00000	00201	00021	11??0	0100
Hipposideros caffer	11101	10000	00010	01201	12012	11110	20011	00101	00213	10002	11000	0100
Hipposideros calcaratus	11131	11000	00011	?0201	01003	01011	11001	00101	00010	00002	11000	0111
Hipposideros cervinus	11101	10000	00010	01201	01003	11110	00011	20001	00210	00001	10000	0100
Hipposideros commersoni	11120	01010	20010	11201	12111	11010	01001	00000	00211	00122	21000	0100
Hipposideros corynophyllus	11110	02000	01000	11101	11100	10210	01012	00000	00101	00102	11000	0100
Hipposideros cyclops	11110	00010	21010	02101	11010	11010	21002	10000	00201	11012	11000	0100
Hipposideros diadema	11110	00010	01010	11201	01101	11000	01001	00100	00110	01012	11110	0100

Taxon *(continued)*	Characters 1–5	6–10	11–15	16–20	21–25	26–30	31–35	36–40	41–45	46–50	51–55	56–59
Hipposideros fulvus	11101	10000	?0010	00201	01100	100?1	10011	00110	00110	00002	10100	0100
Hipposideros lankadiva	11110	00010	?0010	10201	11114	11011	01001	00110	00110	01022	211?1	0100
Hipposideros larvatus	11110	00010	01010	00101	11100	11?11	21001	00010	00210	00002	21001	0100
Hipposideros lylei	11111	10010	20000	10101	01101	11?11	00001	???00	00101	10002	11101	0100
Hipposideros maggietaylorae	11131	11000	?0011	?0201	11010	11010	01001	00001	00112	10002	21100	0111
Hipposideros muscinus	11100	02000	11010	02201	11002	10211	00003	00000	00101	00021	00000	0000
Hipposideros papua	11111	00000	00010	01201	11000	11010	01001	00101	00110	00002	11000	0100
Hipposideros pratti	11110	00010	20000	10201	01100	11?11	00001	00110	00110	00002	11100	0100
Hipposideros ruber	11101	10000	00010	00201	01003	11010	10011	00101	00113	00001	10100	0100
Hipposideros semoni	11120	02000	11000	10010	10103	10210	21024	00000	00201	11120	01000	0000
Hipposideros speoris	11101	10010	00010	11201	0100?	11?10	?0001	10?00	00110	00002	21000	0100
Hipposideros turpis	11110	00000	20010	10201	0?101	11110	01???	001?0	00103	00002	21100	0100
Hipposideros wollastoni	11110	02000	11011	?2101	11012	10211	20000	00000	00101	00031	10000	0100
Palaeophyllophora quercyi	???11	000??	20011	?1201	10110	10010	01002	101?0	00101	00012	21000	0111
Pseudorhinolophus bouziguensis	11110	00000	00010	10101	1111?	11010	11001	00100	00210	00022	11000	0100
Rhinolophus affinis	00001	12100	21100	12010	00003	10020	11124	00000	00001	10000	00000	1000
Rhinolophus megaphyllus	00001	10100	21100	02010	00003	10220	11104	00100	00001	10040	00000	1000
Rhinolophus simulator	00000	10100	21110	02000	00000	10?20	11124	10000	00001	10040	000?1	1000
Rhinonycteris aurantius	11110	10001	00011	?2101	02102	10000	00001	00001	00201	10000	00000	0000
Rhinonycteris tedfordi	1?110	100??	11100	12101	02110	10000	00001	00001	00100	00000	0??00	0000
Riversleigha williamsi	1?110	000??	00010	11201	02110	10000	01001	00001	00110	00020	01100	0000
Syndesmotis megalotis	11101	11100	20?11	?1101	0000?	10011	00002	21001	01?01	00002	11100	0100
Triaenops persicus	11110	01001	00110	02001	02010	10110	21002	00001	00103	00021	00110	0000
Xenorhinos halli	1?100	110??	11110	03111	02004	11001	31001	00001	00010	00020	01?00	0000

Appendix 5.4.
Diagnostic Apomorphies of the Major Clades

The major clades (A–S) were diagnosed by PAUP analysis of all taxa and all characters (ordered and unordered). The few synapomorphies included here are unambiguous (i.e., they occurred at the same position of the tree in both ACCTRAN and DELTRAN optimizations). Characters and their states are as defined in Appendix 5.2. Numbers in parentheses specify character states 2–4, state 1 being otherwise assumed, except in the case of reversal (denoted by "r"). Clade A is the Hipposideridae.

Characters Unordered

A. Pelvis with acetabular foramen; toes with two phalanges; last cervical and first two dorsal vertebrae fused, plus ribs and sternum; frontal depression shallow or absent; anteroventral rim of orbital floor raised and sharp; petrosals only moderately expanded. Characters 1, 2, 3, 14, 20, 35

B. Interorbital region markedly constricted; zygomatic process very tall; C1 posterior accessory cusp very tall; P2 extruded from toothrow but still partially separating C1 and P4. Characters 4, 22(2), 40(2), 43

C. Glenoid fossa conspicuously wider than long. Character 32

D. Median position (dorsal to M2) of infraorbital foramen; infraorbital foramen elongate; mesopterygoid roof without groove; P4 small; ascending ramus tall. Characters 17(1)r, 18(2), 29(0)r, 44, 52

E. Posterior extent of sphenorbital fissure anterior to glenoid fossa. Character 27

F. Narrow rostrum; zygomatic process moderately tall; M3 reduced; m3 talonid narrow. Characters 5, 22(1)r, 51, 57

G. Mesopterygoid roof with groove opening into basisphenoid depression; M1 protofossa open; M3 very reduced such that premetacrista is less than one-third the preparacrista length. Characters 29, 49(0)r, 50(2)

H. C1 posterior accessory cusp tall. Character 40(1)r

I. Rostrum tall. Character 6(0)r

J. Rostrum wide; supraorbital crests strong; sagittal crest tall; C1 posterior accessory cusp low or absent. Characters 5(0)r, 16, 23, 40(0)r

K. Posterior extent of sphenorbital fissure level with glenoid fossa; periotic cavity width four times the interperiotic width; P4 approximately equal in size to M1. Characters 27(0)r, 35(2), 44(0)r

L. Anteroventral palate the shape of a blunt V; sagittal crest incomplete; optic foramen confluent with sphenorbital fissure. Characters 11(1)r, 24(0)r, 28(2)

M. Supraorbital crest development strong; P4 wider than long. Characters 16, 45

N. Rostrum narrow; anterior position of infraorbital foramen (i.e., dorsal to M1–2); C1 anterior accessory cusp tall. Characters 5, 17(0)r, 41

O. Thickened, crested premaxillae; mesopterygoid roof without groove. Characters 10, 29(0)r

P. Interorbital region barely constricted; little rostral inflation; supraorbital ridges absent; anterior position of infraorbital foramen (i.e., dorsal to M1–2); elongation of M1–2 post-metacristae; lateral compression of M1 trigonid. Characters 4(3), 7, 15, 17(0)r, 58, 59

Q. Interorbital constriction moderate; glenoid fossa round. Characters 4(0)r, 32(0)r

R. Little rostral inflation; vomer protruding conspicuously posterior to palate; foramen ovale half to two-thirds the glenoid area; P4 and M1 approximately equal in size. Characters 7, 30, 37, 44(0)r

S. Sphenorbital bridge wide; i2 conspicuously larger than i1. Characters 25, 53

Characters Ordered (Except for the Seven Discussed in the Text)

A. Pelvis with acetabular foramen; toes with two phalanges; last cervical and first two dorsal vertebrae fused, plus ribs and sternum; frontal depression shallow or absent; anteroventral rim of orbital floor raised and sharp; zygomatic process moderately tall; petrosals only moderately expanded; C1 posterior accessory cusp tall. Characters 1, 2, 3, 14, 20, 22, 35, 40

B. C1 posterior accessory cusp very tall; P2 extruded from toothrow but still partially separating C1 and P4. Characters 40(2), 43

C. Infraorbital foramen subrounded; foramen ovale level with mid to posterior glenoid fossa; M3 very reduced in overall shape (premetacrista length less than one-third the preparacrista length); M3 reduced in length and width; ascending ramus tall; m3 talonid narrow. Characters 18, 36, 50(2), 51, 52, 57

D. Infraorbital foramen elongate; posterior extent of sphenorbital fissure anterior to glenoid fossa; P4 smaller than M1. Characters 18(2), 27, 44

E. Interorbital region markedly constricted; glenoid fossa wide. Characters 4, 32

F. Foramen ovale less than half the area of glenoid fossa. Character 36(0)r

G. Rostrum tall; C1 posterior accessory cusp tall. Characters 6(0)r, 40(1)r

H. Rostrum wide; supraorbital crest strong; sagittal crest tall; C1 posterior accessory cusp low or absent. Characters 5(0)r, 16, 23, 40(0)r

I. Little rostral inflation; anteroventral palate U-shaped; accessory foramina to foramen ovale absent. Characters 7, 11(2), 38(0)r

J. Posterior extent of sphenorbital fissure level with glenoid fossa; periotic cavity width four times the interperiotic width; P4 approximately equal in size to M1. Characters 27(0)r, 35(2), 44(0)r

K. Anteroventral palate shape a blunt V; sagittal crest incomplete; optic foramen confluent with sphenorbital fissure. Characters 11(1)r, 24(0)r, 28(2)

L. Interorbital region markedly constricted; supraorbital crest development strong. Characters 4, 16

M. Anteroventral palate shape a blunt V; anterior position of infraorbital foramen (i.e., dorsal to M1–2); foramen ovale half to two-thirds the area of glenoid fossa; C1 anterior accessory cusp tall. Characters 11, 17(0)r, 37, 41

N. Thickened, crested premaxillae; posterior position of infrabital foramen (i.e., dorsal to M3); mesopterygoid roof without groove. Characters 10, 17(2), 29(0)r

O. C1 posterior accessory cusp tall. Character 40(1)r

P. Interorbital region barely constricted; little rostral inflation; supraorbital ridges absent; anterior position of infraorbital foramen (i.e., dorsal to M1–2); elongation of M1–2 post-metacristae; lateral compression of M1 trigonid. Characters 4(3), 7, 15, 17(0)r, 58, 59

Acknowledgments

This study would not have been possible without the help and encouragement given by M. Archer, H. Godthelp, and B. Sigé. We also thank the following people who kindly provided access to comparative specimens in their institutions: B. Engesser, H. Felten, T. Flannery, W. Fuchs, L. Gibson, J. E. Hill, M. Hugueney, P. Jenkins, D. Kitchener, K. Koopman, P. Mein, R. Rachl, B. Sigé, N. B. Simmons, G. Storch, M. Sutermeister, and S. Van Dyck. The Riversleigh project has been supported by the Australian Research Council, the Department of the Environment, Sport and Territories, National Estate Programme Grants (Queensland), Queensland National Parks and Wildlife Service, the Australian Geographic Society, the Linnean Society of New South Wales, ICI Australia, the Queensland Museum, and the University of New South Wales. An American Museum of Natural History (AMNH) Study Grant made it possible for S.H. to examine many specimens included in this analysis; J.A.W.K.'s work in Madison was supported by private donors, and collaborative efforts in Sydney were facilitated by an Australian Research Council Co-operative Research Centre (CRC) grant to Jack Pettigrew. Finally, we thank B. Sigé, K. Koopman, and J. Hutcheon for many helpful comments on earlier versions of this manuscript.

Literature Cited

Ando, K., T. Tagawa, and T. A. Uchida. 1980. Karyotypes of Taiwanese and Japanese bats belonging to the families Rhinolophidae and Hipposideridae. Cytologia (Tokyo) 45:423–432.

Archer, M., S. J. Hand, and H. Godthelp. 1994. Riversleigh: The Story of Animals in the Ancient Rainforests of Inland Australia. Reed Books, Sydney.

Audley-Charles, M. G. 1987. Dispersal of Gondwanaland: Relevance to the evolution of angiosperms. In Biogeographical Evolution of the Malay Archipelago, T. C. Whitmore, ed., pp. 5–25. Clarendon Press, Oxford.

Baker, R. J., R. L. Honeycutt, and R. A. Van den Bussche. 1991.

Examination of monophyly of bats: Restriction map of the ribosomal DNA cistron. Bulletin of the American Museum of Natural History 206:42–53.

Beard, K. C., B. Sigé, and L. Krishtalka. 1992. A primitive vesper-tilionoid bat from the early Eocene of central Wyoming. Comptes Rendus de l'Académie des Sciences, Série II, Sciences de la Terre et des Planètes 314:735–741.

Bogdanowicz, W. 1992. Phenetic relationships among bats of the family Rhinolophidae. Acta Theriologia 37:213–240.

Bogdanowicz, W., and R. D. Owen. 1992. Phylogenetic analyses of the bat family Rhinolophidae. Zeitschrift für Zoologische Systematik und Evolutionsforschung 30:142–160.

Butler, P. M. 1969. Insectivores and bats from the Miocene of East Africa: New material. *In* Fossil Vertebrates of Africa, Vol. 1, L. S. B. Leakey, ed., pp. 1–37. Academic Press, New York.

Butler, P. M. 1984. Macroscelidea, Insectivora, and Chiroptera from the Miocene of East Africa. Palaeovertebrata 14:117–200.

Corbet, G. B., and J. E. Hill. 1992. The Mammals of the Indo-malayan Region: A Systematic Review. Oxford University Press, Oxford.

Ducrocq, S., J.-J. Jaeger, and B. Sigé. 1993. Un mégachiroptère dans l'Eocène supérieur de Thaïlande: Incidence dans la discussion phylogénique du groupe. Neues Jahrbuch für Geologie und Paläontologie Monats Hefte 1993:561–575.

Farris, J. S. 1988. Hennig86, version 1.5. Computer program and documentation. Port Jefferson Station, N.Y.

Flannery, T. F. 1990. Mammals of New Guinea. Robert Brown and Associates, Carina, Queensland.

Flannery, T. F. 1995. Mammals of the South-West Pacific and Moluccan Islands. Australian Museum and Reed Books, Sydney.

Flannery, T. F., and D. J. Colgan. 1993. A new species and two new subspecies of *Hipposideros* (Chiroptera) from western Papua New Guinea. Records of the Australian Museum 45:43–57.

Gray, J. E. 1866. A revision of the genera of Rhinolophidae, or horseshoe bats. Proceedings of the Zoological Society of London 1866:81–83.

Griffiths, T. A., A. Truckenbrod, and P. J. Sponholtz. 1992. Systematics of megadermatid bats (Chiroptera, Megadermatidae), based on hyoid morphology. American Museum Novitates 3041:1–21.

Hand, S. J. 1985. New Miocene megadermatids (Megadermatidae, Chiroptera) from Australia with comments on megadermatid phylogenetics. Australian Mammalogy 8:5–43.

Hand, S. J. 1990. Australia's first Tertiary molossid (Microchiroptera: Molossidae): Its phylogenetic and biogeographic implications. Memoirs of the Queensland Museum 28:175–192.

Hand, S. J. 1995. First record of the genus *Megaderma* Geoffroy 1810 (Microchiroptera: *Megaderma*) from Australia. Palaeovertebrata 24:47–66.

Hand, S. J. 1996. New Miocene and Pliocene megadermatids (Microchiroptera) from Australia, with broader comments on megadermatid evolution. Geobios (Lyon) 29:365–377.

Hand, S. J. 1997a. *Hipposideros bernardsigei,* a new hipposiderid (Microchiroptera) from the Miocene of Australia and a reconsideration of the monophyly of related species groups. Münchner Geowissenschaftliche Abhandlungen A 34:73–92.

Hand, S. J. 1997b. New Miocene leaf-nosed bats (Microchiroptera, Hipposideridae) from Riversleigh, northwestern Queensland. Memoirs of the Queensland Museum 41:335–349.

Hand, S. J. 1998. *Xenorhinos,* a new genus of Old World leaf-nosed bats (Microchiroptera: Hipposideridae) from the Australian Miocene. Journal of Vertebrate Paleontology 18:430–439.

Hand, S. J. n.d. *Riversleigha williamsi,* a large new Miocene hipposiderid from Riversleigh, Queensland. Alcheringa (in press).

Hand, S. J., P. Murray, D. Megirian, M. Archer, and H. Godthelp. 1998. Mystacinid bats (Microchiroptera) from the Australian Tertiary. Journal of Paleontology 72:538–545.

Hand, S. J., M. Novacek, H. Godthelp, and M. Archer. 1994. First Eocene bat from Australia. Journal of Vertebrate Paleontology 14:375–381.

Hershkovitz, P. 1972. The recent mammals of the neotropical region: A zoogeographic and ecological review. *In* Evolution, Mammals, and Southern Continents, A. Keast, F. C. Erk, and B. Glass, eds., pp. 311–432. State University of New York Press, Albany.

Hill, J. E. 1963. A revision of the genus *Hipposideros.* Bulletin of the British Museum (Natural History), Zoology 11:1–129.

Hill, J. E. 1982. A review of the leaf-nosed bats *Rhinonycteris, Cloeotis,* and *Triaenops* (Chiroptera: Hipposideridae). Bonner Zoologishe Beiträge 33:165–186.

Hill, J. E., A. Zubaid, and G. W. H. Davison. 1986. The taxonomy of leaf-nosed bats of the *Hipposideros bicolor* group (Chiroptera: Hipposideridae) from southeastern Asia. Mammalia 50:535–540.

Hugueney, M. 1965. Les chiroptères du Stampien supérieur de Coderet-Branssat. Documents du Laboratoire Géologique de la Faculté des Sciences de Lyon 9:97–127.

Kirsch, J. A. W., and M. Archer. 1982. Polythetic cladistics, or, when parsimony's not enough: The relationships of carnivorous marsupials. *In* Carnivorous Marsupials, M. Archer, ed., pp. 595–619. Royal Zoological Society of New South Wales, Sydney.

Kitchener, D. J., and I. Maryanto. 1993. Taxonomic reappraisal of the *Hipposideros larvatus* species complex (Chiroptera: Hipposideridae) in the Greater and Lesser Sunda Islands, Indonesia. Records of the Western Australian Museum 16:119–173.

Kitchener, D. J., R. A. How, N. K. Cooper, and A. Suyanto. 1992. *Hipposideros diadema* (Chiroptera, Hipposideridae) in the Lesser Sunda Islands, Indonesia: Taxonomy and geographic morphological variation. Records of the Western Australian Museum 16:1–60.

Koopman, K. F. 1970. Zoogeography of bats. *In* About Bats, B. H. Slaughter and D. W. Walton, eds., pp. 29–50. Southern Methodist University Press, Dallas, Tex.

Koopman, K. F. 1994. Chiroptera: Systematics. *In* Handbook of Zoology, Vol. 8, Mammalia, Part 60. Walter de Gruyter, Berlin.

Lavocat, R. 1961. Le gisement de vertébrés miocènes de Beni Mellal (Maroc.). Etude systématique de la faune des mammifères et conclusions générales. Notes et Mémoires du Service Géologique de Maroc 155:1–122.

Legendre, S. 1982. Hipposideridae (Mammalia: Chiroptera) from the Mediterranean Middle and Late Neogene and evolution of the genera *Hipposideros* and *Asellia.* Journal of Vertebrate Paleontology 2:386–399.

Legendre, S., T. H. V. Rich, P. V. Rich, G. J. Knox, P. Punyaprasiddhi, D. M. Trumpy, J. Wahlert, and P. Napawongse Newman. 1988. Miocene vertebrate fossils from Thai Shell Exploration and Production Company's Nong Hen I (A) exploration well, Phitsanulok Basin, Thailand. Journal of Vertebrate Paleontology 8:278–289.

Lloyd, A. R. 1992. The geology, biostratigraphy, and hydrocarbon potential of the Papuan Basin, Papua New Guinea. Alan R. Lloyd and Associates, Duncraig, Australia.

Maddison, W. P., and D. R. Maddison. 1992. MacClade Analysis of Phylogeny and Character Evolution, version 3. Sinauer Associates, Sunderland, Mass.

Marandat, B., J.-Y. Crochet, M. Godinot, J.-L. Hartenberger, S. Legendre, J. A. Remy, B. Sigé, J. Sudre, and M. Vianey-Liaud. 1993. Une nouvelle faune à mammifères d'âge éocène moyen (Lutétien supérieur) dans les phosphorites du Quercy. Geobios (Lyon) 26(5): 617–623.

Mein, P., and L. Ginsburg. 1997. Les mammifères du gisement miocène inférieur de Li Mae Long, Thailand: Systématique, biostratigraphie et paléoenvironnement. Geodiversitas 19:783–844.

Miller, G. S. 1907. The families and genera of bats. Bulletin of the U.S. National Museum 57:1–282.

Nelson, G., and D. E. Rosen. 1981. Systematics and Biogeography: Cladistics and Vicariance. Columbia University Press, New York.

Peters, W. 1871. Über die Gattungen und Arten der Hufeisennasen. Rhinolophi, pp. 301–332. Monatsberichte K. Preuss, Akademie der Wissenschaft.

Pierson, E. D. 1986. Molecular systematics of the Microchiroptera: Higher taxon relationships and biogeography. Ph.D. dissertation, University of California, Berkeley.

Platnick, N. I., and G. Nelson. 1978. A method of analysis for historical biogeography. Systematic Zoology 27:1–16.

Qiu, Z., D. Han, G. Qi, and Y. Lin. 1985. A preliminary report on a micromammalian assemblage from the hominoid locality of Lufeng, Yunnan. Acta Anthropologia Sinica 4:14–32.

Remy, J. A., J.-Y. Crochet, B. Sigé, J. Sudre, L. de Bonis, M. Vianey-Liaud, M. Godinot, J.-L. Hartenberger, B. Lange-Badre, and B. Comte. 1987. Biochronologie des phosphorites du Quercy: Mise à jour des listes fauniques et nouveaux gisements de mammifères fossiles. Münchner Geowissenschaftliche Abhandlungen A 10:169–188.

Revilliod, P. 1917. Contribution à l'étude des chiroptères des terrains tertiaires. 1. Mémoires de la Société Paléontologique Suisse 43:3–57.

Revilliod, P. 1920. Contribution à l'etude des chiroptères des terrains tertiares. 2. Mémoires de la Société Paléontologique Suisse 44:63–129.

Revilliod, P. 1922. Contribution a l'étude des chiroptères des terrains tertiaires. 3. Mémoires de la Société Paléontologique Suisse 45:133–195.

Rich, T. H. V., Y.-P. Zhang, and S. J. Hand. 1983. Insectivores and a bat from the early Oligocene Caijiachong Formation of Yunnan, China. Australian Mammalogy 6:61–75.

Sabatier, M., and S. Legendre. 1985. Une faune à rongeurs et chiroptères plio-pléistocènes de Madagascar. Congrès National des Sociétés Savantes, Sciences 4:21–28.

Schlosser, M. 1911. Beiträge zür Kenntnis der oligozänen Landsäugetiere aus dem Fayum. Beiträge zür Paläontologie und Geologie von Oesterreich-Ungarn Orient 24:51–167.

Sigé, B. 1968. Les chiroptères du Miocene inférieur de Bouzigues. I. Etude systématique. Palaeovertebrata 1:65–133.

Sigé, B. 1976. Les Megadermatidae (Chiroptera, Mammalia) miocènes de Beni Mellal, Maroc. Géologie Méditerranéenne 3:71–86.

Sigé, B. 1977. Les insectivores et chiroptères du Paléogène moyen d'Europe dans l'histoire des faunes de mammifères sur ce continent. Jurij A. Orlov Memorial. Journal of the Palaeontological Society of India 20:178–190.

Sigé, B. 1978. La poche à phosphate de Ste-Néboule (Lot) et sa faune de vertébrés du Ludien supérieur. 8. Insectivores et chiroptères. Palaeovertebrata 8:243–268.

Sigé, B. 1985. Les chiroptères oligocènes du Fayum, Egypte. Geologica et Palaeontologica 19:161–189.

Sigé, B. 1990. Nouveaux chiroptères de l'Oligocène moyen des phosphorites du Quercy, France. Compte Rendus de l'Académie des Sciences, Série II, Sciences de la Terre et des Planètes 310: 1131–1137.

Sigé, B. 1991. Rhinolophoidea et Vespertilionoidea (Chiroptera) du Chambi (Eocène inférieur de Tunisie): Aspects biostratigraphiques, biogéographiques, et paléoécologiques de l'origine des chiroptères modernes. Neues Jahrbuch für Geologie und Paläontologie Abhandlungen 182:355–376.

Sigé, B., and S. Legendre. 1983. L'histoire des peuplements de chiroptères du bassin Méditerranéen: L'apport comparé des remplissages karstiques et des dépôts fluvio-lacustres. Mémoirs Biospéologie 10:209–225.

Sigé, B., S. J. Hand, and M. Archer. 1982. An Australian Miocene Brachipposideros (Mammalia, Chiroptera) related to Miocene representatives from France. Palaeovertebrata 12:149–171.

Sigé, B., H. Thomas, S. Sen, E. Gheerbrandt, J. Roger, and Z. Al-Sulaimani. 1994. Les chiroptères de Taquah (Oligocène inférieur, Sultanat d'Oman): Premier inventaire systématique. Münchner Geowissenschaftliche Abhandlungen A 26:35–48.

Swofford, D. L. 1993. PAUP: Phylogenetic Analysis Using Parsimony, version 3.1. Computer program distributed by the Illinois Natural History Survey, Champaign.

Tate, G. H. H. 1941a. A review of the genus Hipposideros with special reference to Indo-Australian species. Bulletin of the American Museum of Natural History 78:353–393.

Tate, G. H. H. 1941b. Results of the Archbold Expeditions, No. 36: Remarks on some Old World leaf-nosed bats. American Museum Novitates 1140:1–11.

Truswell, E. M., A. P. Kershaw, and I. R. Sluiter. 1987. The Australian–Southeast Asian connection: Evidence from the palaeobotanical record. In Biogeographical Evolution of the Malay Archipelago, T. C. Whitmore, ed., pp. 32–49. Clarendon Press, Oxford, U.K.

Veevers, J. J., ed. 1984. Phanerozoic Earth History of Australia. Clarendon Press, Oxford, U.K.

Zubaid, A., and G. W. H. Davison. 1987. A comparative study of the baculum in peninsular Malaysian hipposiderines. Mammalia 51:139–144.

Part Two
Functional Morphology

The ancient science of morphology, the study of animal structure, has experienced a vibrant rebirth during the past few decades. New approaches to two- and three-dimensional imaging, modern experimental methods for the study of the function of structures of biological interest, and advances in computational analysis techniques have had an enormous impact on the kinds of data morphologists collect and on the ways in which morphological information can be analyzed. Of equal or greater importance, morphology is no longer a primarily descriptive science. Performance capabilities, functional interpretations, ecological consequences, and evolutionary significance of structure are now primary concerns.

Functional analysis has progressed far beyond armchair hypothesizing about the roles and capabilities of particular structural designs to detailed empirical analyses of living animals and sophisticated computer modeling that integrate methods and approaches from physiology and engineering. The true biological context of structures and the behaviors that they permit and facilitate has become a central concern as scholars work to link more detailed information about organismal performance in natural environments to anatomical information. Structures are no longer viewed in isolation; functional, genetic, and developmental integration among anatomical regions and among distinct tissues has become critical. More detailed and careful attention is being paid to diversity within the group of interest and its relatives, and thus the historical phylogenetic context of structural design has come to the forefront.

The chapters in Part Two, "Functional Morphology," represent the changing face of morphology. They broadly sample the structure of bats, ranging from the details of cochlear structure and size (Francis and Habersetzer), to the morphology of dentition (Freeman), to diverse aspects of the locomotor system (Norberg; Hermanson; Schutt; Swartz). The contributors have employed a wide range of sophisticated techniques for data collection: digital acquisition and processing of ultrasonic communication and ultrahigh-resolution radiography (Francis and Habersetzer), finite-element modeling and photoelastic stress analysis (Freeman), electromyography and histochemistry (Hermanson), high-speed cinematography (Norberg), and in vivo strain analysis and mechanical testing (Swartz). In many cases, these data have been analyzed with special attention to the ecology of the species under study (Francis and Habersetzer; Norberg; Schutt), the historical biology of bats (Freeman; Hermanson; Schutt), the mechanics and physiology of changing body size (Freeman; Norberg; Swartz), and to a variety of principles from physics and engineering including materials science and aerodynamics (Francis and Habersetzer; Hermanson; Freeman; Norberg; Swartz).

The new approaches that these contributions exemplify illustrate the great impact morphological sciences can have on our understanding of the biology of bats. Norberg summarizes evidence demonstrating the role of flight mechanics and energetics in imposing constraints on bat behavior; these behavioral constraints, spelled out in the morphology of the wings, in turn constrain all aspects of a species' ecology. Hermanson demonstrates that the primary muscular "motor" of the flight apparatus of bats appears to be subject to rather rigid constraints and has varied little throughout the diversification of bats and their flight modes. Schutt integrates morphological, phylogenetic, ecological, and behavioral data to develop an alternative hypothesis for the evolution of blood feeding in bats. Free-

man explicates the complex patterns of structural similarity and divergence among bats of greatly differing dietary preference, providing both functional and historical contexts for the evolution of the teeth and skulls of megachiropterans and microchiropterans. Francis and Habersetzer show that the structure of the inner ear apparatus is strongly but imperfectly related to the frequency of a species' echolocation calls, and they suggest explanations for both the overall pattern and the outlying exceptions. Swartz demonstrates that the bat wing can be partitioned into several functional units, each of which possesses distinctive structural design in terms of skeletal function, bone composition, and wing membrane architecture.

The work presented here also points clearly to future directions by which the study of morphology can change our understanding of the biology of bats. Our ever-improving knowledge of the structural design of key aspects of important functional complexes, in combination with the rapidly advancing understanding of chiropteran phylogeny and ecology, suggest that the time is ripe for more in-depth historical and ecomorphological analyses of bat structural diversity. These bodies of information, once integrated, may allow us to ask and answer key questions: How have the unique morphological features of bats, found in virtually all anatomical regions, diversified during the phylogenesis of the Chiroptera? How are the ecology and the morphology of various taxa interrelated? Are certain aspects of bat design evolutionarily conservative, and if so, can we offer robust hypotheses to explain this conservatism? If particular aspects of bat structure are highly plastic, what are the mechanisms underlying this plasticity and what is its ecophysiological and evolutionary significance? In this way, we not only will advance knowledge of chiropteran biology but can provide insights of broad significance in organismal biology.

SHARON M. SWARTZ AND ULLA M. NORBERG

6
Morphological Adaptations for Flight in Bats

ULLA M. NORBERG

Because flight is energetically very expensive, bats require highly advanced morphological adaptations in their wing apparatus for efficient flight. Aerodynamic and inertial properties determine the structure and function of the muscular and skeletal systems, of tendons and the flight membrane. Wings must be resistant to bending but light in weight to keep inertial forces low. Muscles and tendons must be arranged to transmit a great deal of power to the wings. Bats have several morphological arrangements in the wings that increase resistance to aerodynamic forces and increase aerodynamic performance (Norberg 1969, 1972a; Altenbach and Hermansson 1987). Variation in wing shape is correlated with different flight modes, kinematics, and flight speeds (see, for example, Norberg 1981, 1987, 1990a, 1994; Norberg and Rayner 1987; Norberg and Fenton 1988; Thollesson and Norberg 1991).

Muscle System

Muscle Types

Different flight types and kinematics put different demands on the size, arrangement, and biochemistry of the flight muscles. Locomotory movements in vertebrates are produced by striated muscle, which shows diverse structure, physiology, and biochemistry. The structure and function of muscle fiber types in bats and birds have been reviewed by Norberg (1990b), and bat muscle fibers are described in detail by J. Hermanson in Chapter 8 (this volume) and therefore are only briefly treated here.

Flapping flight involves mainly isotonic contractions; that is, the fibers exert a constant force while shortening (McMahon 1984). In gliding flight the wings are held down in an outstretched position, which involves mainly isometric contraction; that is, the muscle fibers develop force (tension) without appreciable change in length.

There are two main types of striated muscle fibers, tonic fibers and twitch (phasic) fibers. Tonic fibers are slow contracting and do not occur in bats; twitch fibers are either slow- or fast contracting. The smallest bats that have been investigated have mainly fast oxidative (FO) or fast oxidative glycolytic (FOG) fibers in their pectoral muscles. Both fiber types are fast contracting, fatigue resistant, adapted to fast, repetitive movements such as wing flapping, and can recover fairly quickly so that they are suitable for sustained flight. The rather small molossids have one type of FO fibers.

Rhinolophids have FOG as well as slow-twitch (SO) fibers. Slow fibers are fatigue resistant and are responsible for maintaining posture and for carrying out slow, repetitive movements; these are therefore suitable for sustained flight and gliding. The intermediate FOG fibers (I fibers), which also are useful for maintaining posture, are common in many soaring birds and indicate a certain degree of flexibility in contractile capacities (Johnston 1985). FOG fibers have moderate oxidative capacity, typical of slow fibers (high efficiency and low rate of energy consumption). Although rhinolophids have slow fibers, they do not glide. But some *Pteropus* species glide, and the one *Pteropus* that has been investigated has FOG fibers (George 1965). The connection between fiber type, wing shape, and foraging mode is discussed in Norberg (1990b).

Main Flight Muscles

The flight muscles in bats have been described in detail in a number of species (Vaughan 1959, 1966, 1970a; Norberg 1970, 1972b; Kovtun 1970, 1981, 1984; Strickler 1978; Altenbach 1979; Hermanson 1979). Electromyographic analyses of the flight muscles during flight in bats (Altenbach 1979; Hermanson and Altenbach 1981, 1983, 1985) support the results on muscle action from anatomical studies.

In contrast to the wing movements of birds, the wing movements of bats are powered and controlled by several muscles. In bats, birds, and pterosaurs, the pectoral muscle is the main adductor (depressor) and usually a pronator (nose-down rotator) of the wing. The direction of pull of the main flight muscles in bats is schematically shown in Figure 6.1. The pectoralis, serratus anterior, subscapularis, clavodeltoideus, and latissimus dorsi are the main downstroke muscles. The pectoralis and the serratus anterior are the main adductors of the wings, whereas the clavodeltoideus mainly extends the humerus pronated by the latissimus dorsi. The upstroke is controlled mainly by the spinodeltoideus and the acromiodeltoideus, and by the trapezius group and the long and lateral head of triceps brachii. The trapezius muscles move the scapula toward the vertebral column, thus acting as indirect wing elevators. Several muscles are bifunctional, controlling humeral orientation and stabilizing the shoulder joint. This complex wing muscular system in bats, unlike that in birds, reflects the use of the forelimbs in climbing and terrestrial locomotion in protobats as well as extant bats.

Muscle Arrangements

The effectiveness of muscles can be increased by, for example, force lever systems and skeletal arrangements providing rigidity. Insertions of muscles on ridges and tuberosities of the humerus augment rotational forces and provide larger attaching surfaces; attachments on tubercles, extending from the main axis of the bone, increase the mechanical outcome.

Figure 6.2 shows two skeletomuscular force lever systems in bats. The first system (Figure 6.2a) involves musculus extensor carpi radialis longus, which pulls the second (leading-edge) digit forward. This muscle passes along the anterior side of the forearm, and its tendon passes anterior to the carpal bones at a distance from the fulcrum of the second metacarpal. The trapezium of the carpus projects dorsally and prevents the tendon from sliding posteriorly when the muscle contracts. A similar force lever system occurs in the fifth digit (Figure 6.2b). Contraction of the musculus abductor digiti quinti pulls the fifth metacarpal ventrally, maintaining the chordwise camber of the wing during the downstroke for efficient lift production.

In the examples shown in Figure 6.2, F is taken to be the applied force of the force lever system in the direction of the muscle tendon at the origin, and l is the force lever arm (perpendicular distance from the fulcrum to the tendon). The moment of action then is $l \times F$ and hence increases proportionally to l. Such muscular arrangements, which are also present at other locations in the wing, increase the efficacy of the muscles, reducing the demands for heavy muscle bellies and contributing to the low weight of the wings. Several skeletal arrangements that are discussed later also contribute to low weight.

Contraction Rate and Wingbeat Frequency

Wingbeat frequency is determined by the aerodynamic and inertial properties of the wings and body, and the properties of the muscles must match this frequency (Pennycuick and Rezende 1984). Maximum and minimum wingbeat frequencies can be used to predict the maximum size of a flying animal (Pennycuick 1975). In a diagram of wingbeat frequency versus body mass, the maximum and minimum wingbeat frequency lines converge as body mass increases until a point at which flapping at the maximum frequency is the only way to achieve flight.

In geometrically similar bats the maximum wingbeat frequency, f_{max}, is inversely proportional to the time taken for the stroke, which in turn is proportional to the wing length. The frequency should thus vary with the negative one-third power of the body mass, M: $f_{max} \propto M^{-0.33}$ (Hill 1950). The lower limit of the wingbeat frequency is associated with the need to provide sufficient airflow over the wings to supply lift and thrust in slow flight and hovering, and should vary inversely with the square root of any representative length; otherwise stated, $f_{min} \propto M^{-0.67}$ (Pennycuick 1975).

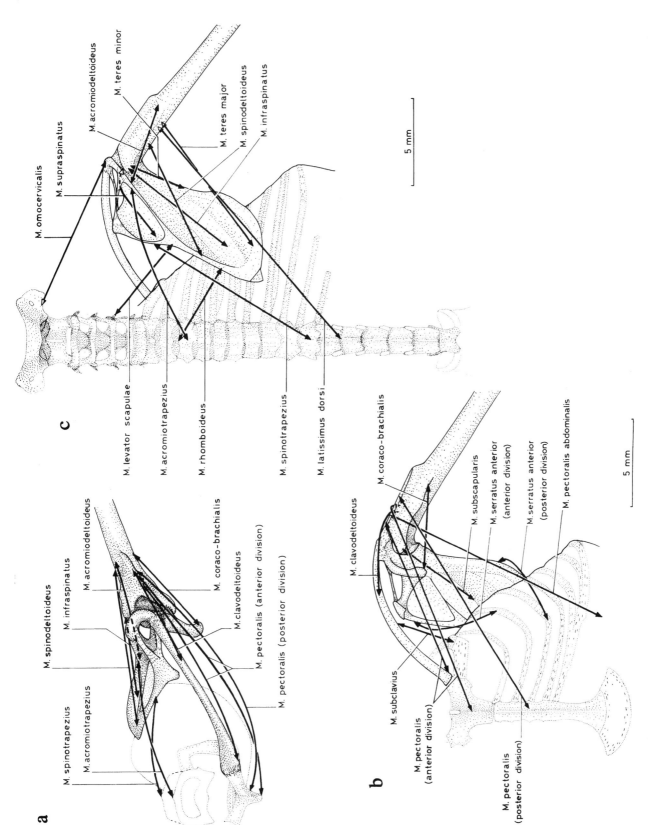

Figure 6.1. Direction of pull (shown by arrows) of the flight muscles in bats. (a) Anterior view of the left shoulder region, showing the direction of pull of muscles moving the arm. (b) Ventral view of the left shoulder region, showing the direction of pull of downstroke muscles acting on the clavicle, the scapula, and the humerus. (c) Dorsal view of the right shoulder region, showing the direction of pull of upstroke muscles acting on the scapula and the humerus. (From Norberg 1970.)

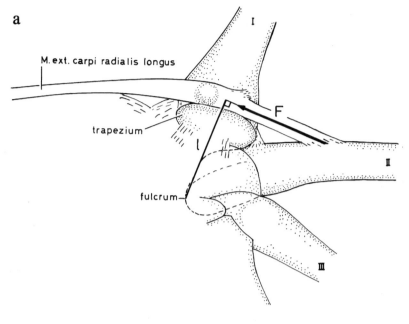

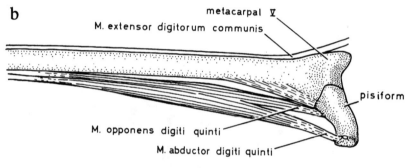

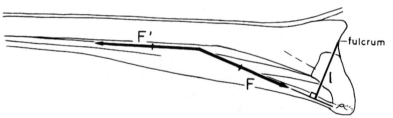

Figure 6.2. Two skeletomuscular force lever systems in bats, where F is the applied force ($=F'$), l is the force lever arm on which F acts, and the moment of action is $F \times l$. (a) The wrist, where an extensor muscle pulls the second metacarpal forward. (b) The metacarpal of the fifth digit, where contraction of an abductor muscle pulls the metacarpal ventrally. (From Norberg 1970.)

The largest bats have a mass of about 1.5 kg, which is used as the mass at which the lines converge in Figure 6.3. However, it is difficult to say whether this is near the maximum possible bat mass; it may be unreliable to predict an absolute upper limit for flight, because muscle efficiency varies among different-sized bats and the structure and function of muscles and wings may differ among species.

Wingbeat frequency and flight speed were measured for 19 and 21 species of bats, respectively, belonging to nine families (Norberg, unpublished data). Wingbeat frequency is plotted against body mass in Figure 6.3, which shows that $f_w \propto M^{-0.27}$. Rayner (1988) compiled data for various birds and the resulting correlations vary among different groups; for all birds except hummingbirds the frequency varies ac-

cording to the equation $f_w = 3.98M^{-0.27}$. Thus, there is a striking similarity between bats and birds in this respect. Bats have a lower y-intercept (lower frequency for their size), which may depend on their generally lower wing loadings. The observed frequencies indicate that the maximum size for bats would be approximately 2 kg if the frequency alone determines this size.

Skeleton System and Wing Membrane

Trunk

Bats, birds, and pterosaurs have (or had) short, streamlined, and stiff trunks. The vertebrae are fused in different re-

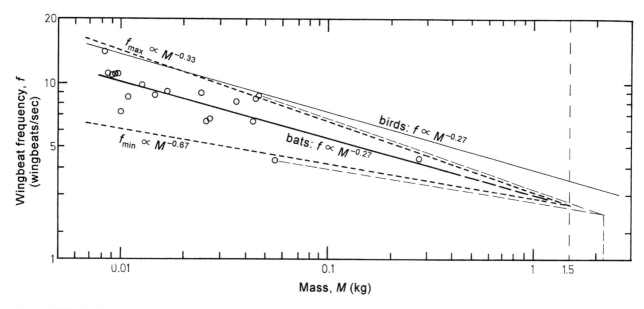

Figure 6.3. Wingbeat frequency in bats. Maximum and minimum frequencies converge as mass increases until a point at which maximum frequency is the only way to achieve flight. Observed wingbeat frequencies lie within the predicted maximum and minimum frequency lines that converge at a body mass of 2–3 kg.

gions; in bats fusion is common between adjacent trunk vertebrae (Vaughan 1970b; Strickler 1978), and when not fused, vertebrae are shaped to limit or prevent motion. The ribs are usually flattened and provide an origin area for flight muscles.

Shoulder Girdle and Sternum

The shoulder girdle in bats is composed of a well-developed and highly movable scapula and a clavicle that is attached to the sternum. Unlike that of birds and pterosaurs, the scapula in bats has three facets (see Figure 6.1c), which provide large attaching surfaces for several flight muscles. The coracoid is reduced to the acromion process on the scapula.

The size of the sternum has often been associated with the size of the pectoralis muscle and thus with the ability to fly. The lack of a well-developed sternal ridge in pterosaurs, which provides a large surface for the origins of the pectoral muscles, has led to the suggestion that pterosaurs could not produce a downstroke as powerful as that of birds. Although some bats have a well-developed sternal ridge, others have scarcely any and still are excellent fliers. In bats, the anteriormost part of the manubrium of the sternum usually has a ventrally projecting tubercle from which a median ligamentous sheet passes ventrally, and together with the sternum this sheet forms a surface to which the pectoralis muscle can attach (Norberg 1970) (Figure 6.4). The sternal ridge together with this ligamentous

sheet are analogous to the sternal ridge in birds, forming an area of attachment for the pectoralis muscle. Because ligamentous sheets do not fossilize, we do not know if pterosaurs had a similar arrangement.

In some microchiropteran species, the scapula and humerus interlock as the humerus is abducted during the upstroke by the scapular muscles (Vaughan 1959; Altenbach and Hermanson 1987). The lock prevents the wing from elevating above the horizontal when the humerus is fully protracted, although the arm can be raised when retracted. Albatrosses have a similar lock on the humerus (Pennycuick 1982). A locking arrangement reduces the cost of flight by relieving wing-depressing muscles from much of the work of starting the downstroke.

Wing Skeletal and Membrane Arrangements

The force of the airstream subjects the wings to great forces during flight. Special arrangements reduce the demands for powerful muscles and large cross-sectional areas of the wing bones, reducing the wing mass and thus their inertial loads and the total mass of the bat.

The leading edge of the wing is subjected to especially great strains during flight, and the wing skeleton forms a relatively stable system in the arm- and handwings (Figure 6.5) (Norberg 1985). This arrangement provides good support for the anterior part of the wings and provides a curved wing profile when the wing is outstretched. A triangular unit in the armwing is formed by the humerus, the

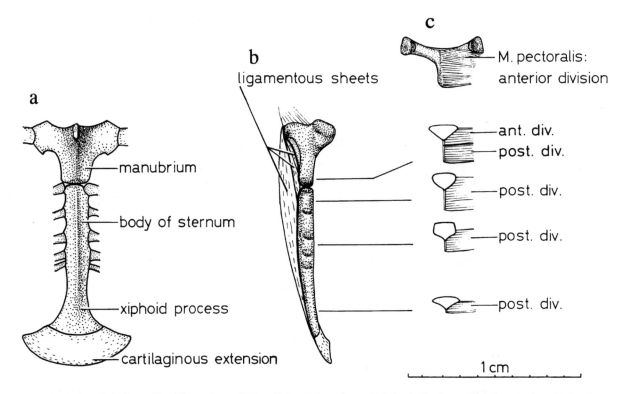

Figure 6.4. Sternum in the long-eared bat, *Plecotus auritus*. (a) Ventral view. (b) Lateral view with the median ligamentous sheets in place. (c) Anterior view (top) and cross sections with musculus pectoralis schematically in place. (From Norberg 1970.)

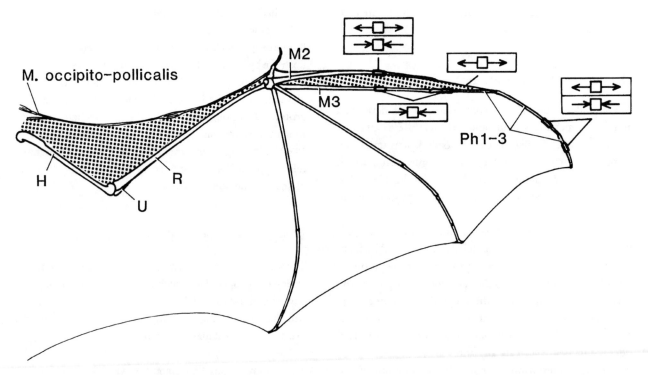

Figure 6.5. General arrangement of the bat wing. The stippled areas show the stay systems formed by skeleton and tendons. A special arrangement of the second and third digits makes the leading edge of the handwing rigid in the membrane plane. Arrows indicate compression or tension forces in the plane of the wing. H, humerus; R, radius; U, ulna; M, metacarpal; Ph, phalanx. (From Norberg 1970, 1985; copyright 1985 by the President and Fellows of Harvard College. Reprinted by permission of Harvard University Press.)

radius (and the very rudimentary ulna), and the occipitopollicalis muscle. The humerus forms an angle less than 180° with the forearm. Because the elbow joint is more elevated than the shoulder and wrist when the wing is outstretched laterally, and because the muscle along the leading edge of the armwing tightens the membrane anterior to the arm, the chordwise profile of the armwing becomes convex, which promotes lift production. The muscle together with the propatagium prevent the angle between the humerus and the ulna and radius from opening excessively.

The aerodynamic forces cause the wing membrane to bulge during flight, and the tension set up in the membrane pulls at the lines of attachments in the leg, arm, and digits in the plane of the membrane (Norberg 1972a, 1990b). The strain is especially powerful on the leading-edge digits, which stretch out the membrane and lead the wing movements. A special arrangement of the second and third digits form a rigid leading edge of the handwing (Norberg 1969) (see Figure 6.5). Determining characters are the ligamentous connection between the second and third digits and the bending of the third metacarpophalangeal joint, making the joint angle somewhat less than 180° in the membrane plane anterior to the third digit. Because of the convexity, this arrangement constitutes a rigid unit in the membrane plane between the carpals and the joint between the first and second phalanges of the third digit when the second digit extends forward in the downstroke. This rigid unit is especially broad in large and broad-winged bats. There is also a similar rigid unit between the third and fourth digits in megachiropteran bats, which are broad-winged, and this releases the wingtip from large tension forces and acts to keep the joints of the fourth digit steady without involving powerful muscles (Norberg 1972a, 1990b) (Figure 6.6).

The bat digits are further shaped so that their greatest cross-sectional diameters are in these planes where the bending forces are largest, which keeps their mass low while still maintaining rigidity in the important directions (Norberg 1970, 1972a, 1990b). This formation is generally more pronounced in small bats than in larger ones (mainly megachiropterans), in which the membrane curvature is less pronounced and hence tauter, which means larger tension forces at the lines of attachment. In most bats, the tips of the third to fifth digits are cartilaginous and flexible and the tips of the fourth and fifth digits are usually bifurcated to increase the area of attachment of the wing membrane. The flexibility and bifurcation are safety factors preventing rupture at the wing's trailing edge during flight. Flexible wingtips may also function to save energy during flight (see following).

Wing Adaptations Enhancing Flight Performance

Bat wings must have particular properties to produce enough lift at slow flight speeds. The highly movable wings permit the bat to change planform, geometry, and aerodynamic characteristics of the wing to improve the flight performance in a desired manner: At one flight speed a particular wing profile is suitable and at another speed a different profile is better. Bats do not have wing slots, like birds, but they do have other wing characteristics for increasing lift during slow flight (Norberg 1972a).

Wing Camber

Bats have the ability to vary the anteroposterior curvature (camber) of the wing, mainly by lowering the propatagium and dactylopatagium brevis (by lowering the thumb) and the fifth digit. The lift coefficient, and hence lift, increase with the camber of the wing (but along with an increase of the drag coefficient). Megachiropteran bats usually have relatively broader patagia anterior to the arm and third digit than do long-winged and narrow-winged microchiropteran bats. Megachiropterans also have shorter metacarpals in relation to the total digit length than do microchiropterans, which enhances their ability to camber the wing more strongly. The short metacarpals also permit them to flex the handwing more during the upstroke to reduce drag when the third to fifth digits flex (Figure 6.7).

Wing Flaps

Leading-edge flaps in airplane wings are high-lift devices keeping the flow laminar over the wing at higher angles of attack. This design permits higher lift coefficients without flow separation, particularly for thin-section wings with sharp leading edges. In bats, the propatagium and dactylopatagia brevis and minus together may function as a leading-edge flap when lowered by the first and second digits (Figure 6.8). These patagia are especially large in slow-flying bats with broad wings, and the efficiency increases with decreasing thickness of the leading edge, which is very thin in bats.

Turbulence Generators

Another way to delay stalling in slow flight is to make the boundary layer of the wing turbulent. The boundary layer is the layer very near the wing surface, where the air movement is retarded by friction. It can be either laminar, turbulent, or laminar anteriorly and turbulent posteriorly. Every

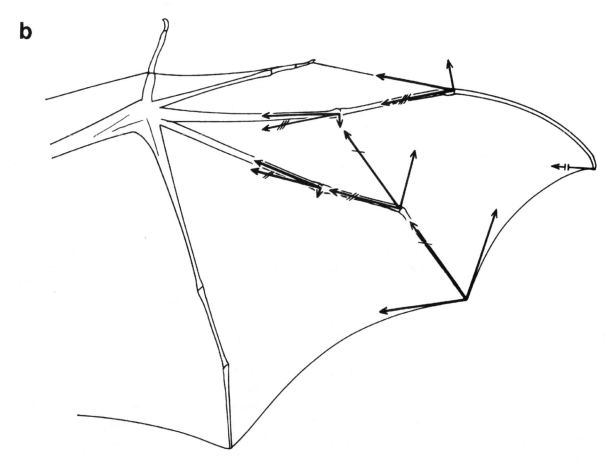

Figure 6.6. Wing adaptations in the megachiropterans. (a) Wings of the megachiropteran *Rousettus aegyptiacus*. The second to fourth digits are angled so that the membrane around the proximal part of these digits is kept taut without need of large muscular forces. Large wrinkles in the dactylopatagium indicate location and direction of the greatest tension forces between the digits. (b) Forces (represented by arrows) that act on the third and fourth digits in a megachiropteran bat. The fourth metacarpophalangeal joint is angled backward and the fourth interphalangeal joint forward. This arrangement tightens the proximal part of the membrane between the third and fourth digits and keeps the digital joints very steady without need of large muscular forces. (Photo by the author; from Norberg 1972a.)

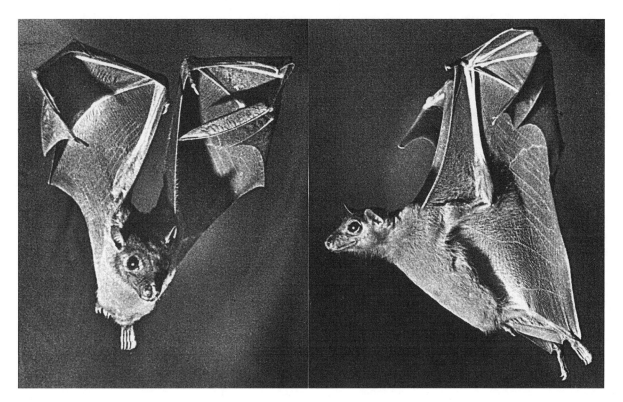

Figure 6.7. The megachiropteran *Rousettus aegyptiacus* in flight. The phalanges of the third to fifth digits are strongly flexed to reduce drag during the later part of the upstroke. (Photo by the author; from Norberg 1972b, 1990b.)

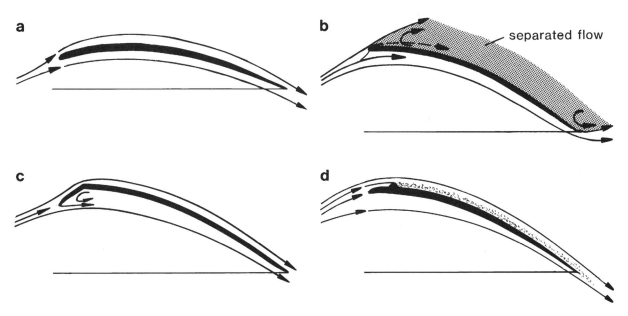

Figure 6.8. Airflow (arrows) over the ventral and dorsal surfaces of the bat wing. (**a**) Laminar flow. (**b**) Airflow separation (stippling) over the upper surface of the wing. (**c**) Leading-edge flap keeping the airflow "attached" to the wing surface. (**d**) Transition from laminar to turbulent flow of the boundary layer behind a protuding structure that is acting as a turbulence generator. (From Norberg 1985; copyright 1985 by the President and Fellows of Harvard College. Reprinted by permission of Harvard University Press.)

pressure increase in the flow direction is unfavorable for keeping the boundary layer laminar, particularly at high Reynolds numbers. When the Reynolds number lies below a critical value, which is different for different profiles, the aerodynamic lift coefficient can be improved by induced turbulence of the boundary layer. With a turbulent boundary layer there is a constant interchange of momentum between the rapid outer layers and the slow inner ones, so that the inner layers receive kinetic energy from the free external flow. Therefore, a turbulent flow can continue against a considerable increase of pressure and can remain attached to the wing surface at higher angles of attack and give more lift. With a laminar boundary layer, flow separation occurs, resulting in a loss of lift (wing stalling) (see Norberg 1972a, 1990b for further details).

In bats, projection of the digits and arm above the dorsal wing surface, roughening the upper wing surface near the leading edge, may act to generate a turbulent boundary layer. The digits project more on the upper surface of the wing than on the lower side in many microchiropteran bats, and the projections can be very sharp. These characteristics may have evolved to act as "turbulence generators" (Pennycuick 1971; Norberg 1972a) (see Figure 6.8), particularly in slow-flying bats. In fast flights the wings do not stall as easily as in slow flight, so for fast fliers a streamlined wing profile would instead be more useful because it decreases drag. In fact, the cross-sectional shape of the forearm in many broad-winged slow-flying bats is almost rounded, whereas it is dorsoventrally flattened in the fast-flying molossid bats (Vaughan and Bateman 1980).

Energy-Saving Elastic Systems

Work must be done to accelerate the wings at the beginning of the downstroke, but at the end of the downstroke in fast flight the kinetic energy of the wing can be transferred to the air and provide lift, the incident air slowing down the flapping movement. Part of the work accelerating the wing at the beginning of the downstroke is therefore recoverable at the bottom.

In hovering and slow flight this transfer of energy is not so easily achieved, because the relative airspeed at the turning points of the wingstroke is very small, and the forward speed component is small or nonexistent. Therefore, the loss of inertial power should be important in hovering and slow flight unless kinetic energy can be removed and stored by some other means.

Insects have a highly elastic, energy-saving mechanism at the wing hinge, based on the highly elastic protein resilin, without which many insects would not be able to produce the power necessary to fly (Weis-Fogh 1960). In birds, the clavicles may store elastic energy in the first part of the downstroke and release it at the top and bottom of the wingstroke where it might be transformed into aerodynamic work (Schaefer 1975; Jenkins et al. 1988; Goslow et al. 1989). Likewise, the primary feathers might increase the efficiency of transfer of the wing's kinetic energy to the air toward the end of the downstroke when the feathers unbend (Pennycuick and Lock 1976). The wing membranes of bats are highly elastic structures containing the protein elastin, and in most bats the tips of the third, fourth, and fifth digits and the tail tip are cartilaginous and flexible. These structures may absorb kinetic energy, store it as elastic energy as they are deformed, and transfer it into aerodynamic work at the top and bottom of the wingstroke (Norberg 1990b). Without any elastic energy-saving mechanism, the inertial power makes up about 56% of the total power in a 10-g bat that is hovering and 30% in slow forward flight, so the savings are important (Norberg et al. 1993).

Wing Design and Ecology

Flight mechanics impose significant constraints on behavior, determining a bat's niche, and the influence of these constraints can be traced in wing adaptation. Most animal-eating bats hawk insects in the air, in open spaces, or within vegetation, and need to be highly agile and maneuverable. Several species are frugivorous and use flight mostly for commuting between the roost and foraging place. Some species are nectarivorous, some are sanguivorous, and still others are carnivorous. Widely differing wing structures are required by bats with different foraging behavior for minimum energetic costs for the required flight mode.

Wing and Body Size

The size of the wing can be described by wing loading, WL, which is the weight, Mg ($=$mass times the acceleration of gravity), divided by the wing area, S: WL $= Mg/S$. Wing loading is thus the ratio of a force to an area (given in newtons per square meter, $N \cdot m^{-2}$) and is equivalent to a pressure. The mean pressure and suction over the wings must match the wing loading. Because aerodynamic pressure varies as speed squared, any characteristic flight speed (V), such as minimum power speed and maximum range speed, is proportional to the square root of the wing loading: $V \propto (Mg/S)^{0.5}$. Thus, slow-flying bats have low wing loadings (large wings), and bats with small wings have to fly faster for their body size. For geometrically similar bats, wing loading is proportional to mass to the one-third power, so larger bats have to fly faster than smaller ones when real speeds are considered. Wing loading in bats ranges from about 4 N·m⁻² in disk-winged bats *(Thyroptera)* to about 40 N·m⁻² in some flying foxes *(Pteropus)*.

Flight Speed

Flight speeds were measured from high-speed ciné films (200 frames/sec) in 21 morphologically diverse bat species, representing nine families, and ranging in body mass from 8 to 270 g. Norberg (1987), using data from the literature, plotted speed against body mass for flight in open air and in caves and for flight indoors and within vegetation. Taking these data together, flight speed is here plotted against wing loading (Figure 6.9) for 27 species of bats flying in open air or in a large cave (2 species) and for 21 species flying indoors or within vegetation. For open-field and cave flights, the flight speed varies with wing loading as follows:

$$V = 2.46(Mg/S)^{0.48};$$

$$r = 0.58.$$

The slope is similar to that predicted for geometric similarity (0.50). It is also similar to those predicted for minimum power and maximum range speeds, but the elevation is much higher (Figure 6.9). The minimum power and maximum range speeds can be calculated from Equations 4 and 5 and the regression for wing loading versus body mass (all bats) in Table 4 in Norberg and Rayner (1987) and are as follows: $V_{mp} = 1.18(Mg/S)^{0.48}$ and $V_{mr} = 1.59(Mg/S)^{0.48}$ (broken lines in Figure 6.9). The measured open-field flight speeds are thus much higher than predicted for any of these optimal flight speeds. Indoor and within-vegetation flight speeds are, however, slower than open-field speeds and vary with the wing loading according to the equation

$$V = 2.12(Mg/S)^{0.26};$$

$$r = 0.47.$$

Wing Shape

The overall shape of the wing can be described by the aspect ratio (AR), which is wingspan *(b)* divided by the mean wing chord *(c)*: AR = *b/c*. The aspect ratio is thus an indication of the narrowness of the wing. The taper of the wing makes calculating the mean wing chord difficult, but because wing area *(S)* equals *bc,* then the equation can be written as AR = $(b \times b)/(b \times c) = b^2/S$. A higher AR value means greater aerodynamic efficiency and lower energy losses in flight, particularly at slow flight speeds.

Variation in wingtip shape can be described by three wingtip indices: the tip length ratio, T_l; the tip area ratio, T_s; and the tip shape index, T_i (Norberg and Rayner 1987). The tip length ratio is the ratio of the handwing length, l_{hw}, to the armwing length, l_{aw}: $T_l = l_{hw}/l_{aw}$. The tip area ratio is the ratio of the handwing area, S_{hw}, to the armwing area, S_{aw}: $T_s = S_{hw}/S_{aw}$. The tip shape index is determined by the relative size of the handwing and armwing and is a measure of wingtip angle and shape, independent of the extent of the handwing, and is given by the equation $T_i = T_s/(T_l - T_s)$. Low indices indicate pointed wingtips, high values rounded tips. Infinity would indicate a rectangular wing, and the tips would be triangular when $T_i = 1$.

Most bats have rounded wings, with $T_i > 1$. The most rounded wingtips are found in nycterids, rhinolophids, hipposiderids, noctilionids, natalids, and some phyllostomids. Species with pointed wingtips ($T_i < 1$) are primarily pteropodids with long wingtips or microchiropterans with

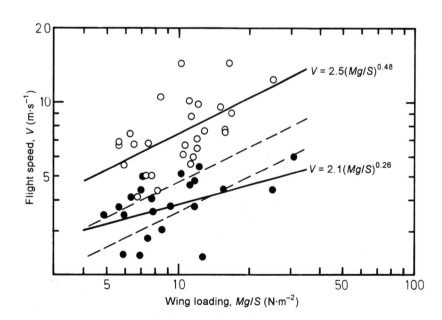

Figure 6.9. Flight speed plotted against wing loading for open-field flights (including two flights in and outside large caves; open circles) and for flights within vegetation or indoors (closed circles). The broken lines are the predicted regressions for maximum range speed (top) and minimum power speed (bottom).

Low flight speed ←——————————————→ **High flight speed**

⬇   ⬇

FOR HIGH AGILITY (HIGH ROLL ACCELERATION)

Increased Q:
large b, large S
large T_i (broad, rounded wingtips)
low AR (broad wings)

Increased Q:
high v
high WL (small wing area)

Increased influence on J_{roll}:
small T_l (short handwings for high
    wing flexibility)

Decreased J_{roll}:
large T_l (long handwings)
small T_i (pointed wingtips)
high AR (narrow wings)

- -

FOR HIGH MANEUVERABILITY (SMALL RADIUS OF TURN)

Decreased WL:
low body mass
large wing area

Decreased WL:
low body mass

- -

EXAMPLES

Plecotus auritus

Pipistrellus pipistrellus
Eptesicus nilssoni

Nyctalus noctula
Vespertilio murinus
Otomops martiensseni

large T_i, small T_l,
low AR, low WL

very large T_i, small T_l,
low AR, average WL

small T_i, large T_l,
high AR, average to high WL

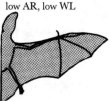

Figure 6.10. Hypothetical selection pressures for high agility and high maneuverability in slow-flying compared with fast-flying insectivorous bats: $a_{roll} = Q/J_{roll} \propto V^2 Sb/J_{roll}$, where Q is the aerodynamic roll moment (torque), V is flight speed, S is wing area, b is wingspan, and J_{roll} is the roll moment of inertia of body and wings combined. T_i is wingtip shape index; T_l, wingtip length index; AR, aspect ratio; WL, wing loading.

narrow wingtips, including *Macroderma gigas*, *Phyllostomus hastatus*, *Mimetillus moloneyi*, *Nyctalus noctula*, and *Tadarida teniotis* (see further Norberg and Rayner 1987).

Maneuverability and Agility

Many bats forage in cluttered environments, and most insectivorous species pursue flying prey. This habit has strongly influenced flight adaptations such as maneuverability and agility. Prey catching, avoidance of obstacles, and landing require different types of maneuvers. Flying bats may detect insects at ranges of a few meters so they must make rapid maneuvers to pursue and catch a prey. Maneuverability refers to the minimum radius of turn the bat can attain without loss of speed or momentum, and

agility refers to the maximum roll acceleration during the initiation of a turn and measures the rapidity with which the flight path can be altered (Norberg and Rayner 1987). The radius of a banked turn is proportional to the square root of the wing loading and inversely proportional to the lift coefficient, so high maneuverability is obtained by bats with low body mass, large wings, and the ability to control camber of the wings to increase the lift coefficient.

To initiate a turn, a net rolling moment must be produced. This can be achieved by differential twisting or flexing of the wings, or by unequal flapping of the two wings, giving asymmetrical aerodynamic roll moments (torques) (Norberg 1976; Thollesson and Norberg 1991). The torque (Q) is proportional to speed squared, wing area, and wingspan. The fastest entry into a turn is achieved at the maximum angular acceleration available to the bat, a_{roll}, which is the aerodynamic torque divided by the total roll moment of inertia of body and wings, $a_{roll} = Q/J_{roll}$ (Andersson and Norberg 1981). To enhance rapid maneuvers, a bat should thus have a large Q or a small J_{roll}.

The morphological correlates with maneuverability and agility are summarized in Figure 6.10. In slow-flying species a large Q is obtained by broad wings (low AR), broad, rounded wingtips (large T_i), a long wingspan, and a large wing area. Short handwings (small T_l) increase wing flexibility and have increased influence on J_{roll}. The majority of bats have more rounded wingtips with $T_i > 1$, and there is considerable consistency within families (see Norberg and Rayner 1987). Families with very rounded wingtips and low wing loadings are the Nycteridae, Rhinolophidae, Hipposideridae, Noctilionidae, Natalidae, and some Phyllostomidae. Hipposiderids and rhinopomatids also have very short handwings.

In fast-flying species a high wing loading increases Q, whereas a thin body, short and narrow wings (high AR), pointed wingtips (small T_i), and long handwings (large T_l) decrease J_{roll} for high roll acceleration (see Thollesson and Norberg 1991). Molossids have high AR and high WL, and the wingtips are slightly longer than average and slightly rounded. The most pointed wingtips occur in *Hipposideros speoris* and *Chrotopterus auritus*.

Ecomorphology of Flight

Certain combinations of aspect ratio and wing loading index that are independent of body mass constrain bats to use only certain kinds of flight behavior and ecology (Norberg 1981; Norberg and Rayner 1987). Therefore, we can predict the predominant flight mode and performance of a bat from its body mass, wingspan, and wing area.

Norberg and Rayner (1987) used a principal components analysis for an investigation of bat wing shapes including 215 species from 16 families to obtain indices for wing loading (PC 2) and aspect ratio (PC 3) to clarify the functional basis of the ecomorphological correlations in bats. In such an analysis the components are not quite independent of body mass, although the method corrects for deviations from geometric similarity within bat groups. A more straightforward method is to plot the aspect ratio against a wing loading index that is made independent of body mass. In geometrically similar bats, wing loading is proportional to body mass raised to one-third, $Mg/S \propto M^{0.33}$. This gives that $(Mg/S)/M^{0.33}$ is equal to a constant, k_1, and so $(M/S)/M^{0.33} = M^{0.67}/S = k_2$. The latter expression can be taken as a nondimensional wing loading index, RWL (relative wing loading), equal to $M^{0.67}/S$.

Figure 6.11 shows AR plotted against RWL for 210 species of bats from 16 families (data from Norberg and Rayner 1987). High wing loading and large aspect ratio are characteristic of bats with fast sustained flight, such as molossids, some emballonurids *(Taphozous)*, and some vespertilionids *(Tylonycteris, Pipistrellus)*. Other emballonurids, hipposiderids, noctilionids, and some vespertilionids *(Lasionycteris, Nyctophilus, Eptesicus)* have low wing loadings and slow, enduring, and inexpensive flight. Most of these bats are insectivorous and hunt insects in the air in open spaces.

Bats with low RWL can fly slowly for their size and can carry heavy prey. A low body mass and low AR (short wings) also enable these bats to fly very slowly within vegetation with high maneuverability. Nycterids, megadermatids, rhinolophids, and many vespertilionids belong to this group. Several are insectivorous gleaners and hoverers or carnivores. Most frugivorous and nectarivorous species (belonging to Pteropodidae and Phyllostomidae) have short wings with high wing loadings and low aspect ratio, characteristic of fast but not enduring flight. Because the tail membrane is usually reduced, their total wing area (which includes the tail membrane) becomes smaller in comparison with bats of other families, and their wing loading becomes larger. For a more detailed description of wing shape and flight performance in bats, see Norberg and Rayner (1987).

Summary

Because flight is energetically very expensive, bats have highly advanced morphological adaptations in their wings for efficient flight. Flight mechanics impose significant constraints on behavior that are responsible for shaping the bat's niche, and the influence of these constraints may be traced in wing adaptation. A brief review is given here of myological and osteological arrangements improving rigidity and flight performance (Norberg 1969, 1972a, 1972b).

Variation in wing shape can be correlated with different flight modes, kinematics, and speed. Flight speeds and wingbeat frequencies were measured from high-speed ciné films (200 frames/sec) in 21 morphologically diverse bat species, representing 9 families, and ranging in body mass from 8 to 270 g. Least-squares regression showed that wingbeat frequency is proportional to body mass, $f_w \propto M^{-0.27}$, and flight speed to wing loading, $V \propto (Mg/S)^{0.48}$. The latter regression included data from the literature of flight speeds of another 27 species.

Maximum and minimum wingbeat frequencies set bounds to the maximum size of animals with aerobic flight (Pennycuick 1975). The observed frequencies indicate that the maximum size for bats would be about 2–3 kg if frequency alone determined this size. The size and shape of

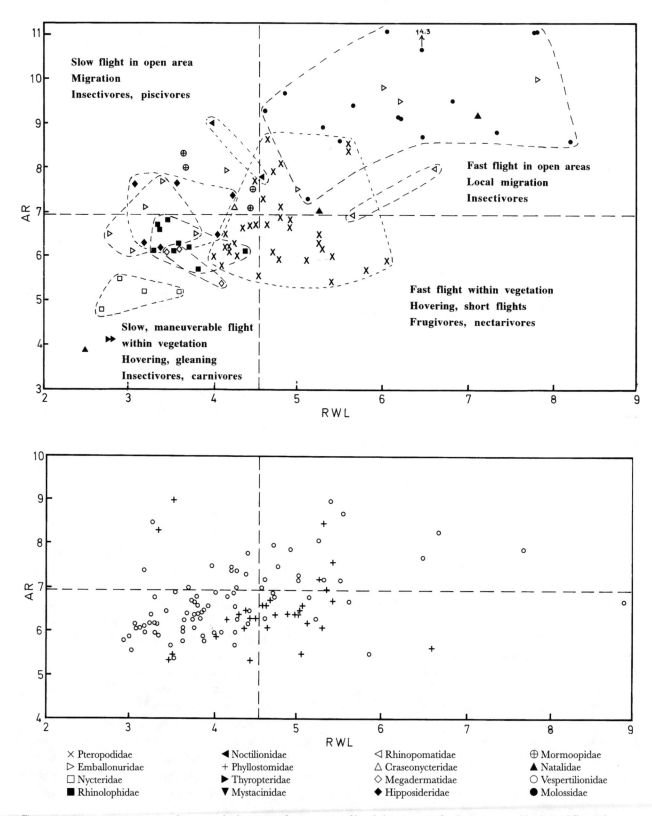

Figure 6.11. Aspect ratio (AR) versus relative wing loading (RWL) for 210 species of bats belonging to 16 families (represented by the 16 different shapes and directionals). The families are separated into two diagrams for the sake of clarity (14 families on the top graph; 2 on the bottom). The horizontal and vertical broken lines are drawn along the mean values for AR and RWL, respectively.

the wings can be quantified by three parameters—wing loading, aspect ratio, and wingtip shape index—and used to predict flight performance, agility, and maneuverability of the different species (Norberg and Rayner 1987; Thollesson and Norberg 1991). A simplified way of grouping bats according to aspect ratio and a wing loading index that is independent of body size is demonstrated.

Acknowledgments

I am indebted to Å. Norberg and S. Swarz for commenting on the manuscript, and to all those who helped me in the field to catch and fly bats: S. Altenbach, A. Brooke, C. Diaz, A. Rodríguez-Durán, and S. Rice. This work was supported by grants from the Swedish Natural Science Research Council.

Literature Cited

Altenbach, J. S. 1979. Locomotor morphology of the vampire, *Desmodus rotundus*. Special Publication No. 6. American Society of Mammalogists, Lawrence, Kans.

Altenbach, J. S., and J. W. Hermanson. 1987. Bat flight muscle function and the scapulohumeral lock. *In* Recent Advances in the Study of Bats, M. B. Fenton, P. Racey, and J. M. V. Rayner, eds., pp. 100–118. Cambridge University Press, Cambridge.

Andersson, M., and R. Å. Norberg. 1981. Evolution of reversed sexual size dimorphism and role partitioning among predatory birds, with a size scaling of flight performance. Biological Journal of the Linnean Society 15:105–130.

George, J. C. 1965. The evolution of the bird and bat pectoral muscles. Pavo 3:131–142.

Goslow, G. E., Jr., K. P. Dial, and F. A. Jenkins, Jr. 1989. The avian shoulder: An experimental approach. American Zoologist 29:287–301.

Hermanson, J. W. 1979. The forelimb morphology of the pallid bat, *Antrozous pallidus*. Master's thesis, Northern Arizona University, Flagstaff.

Hermanson, J. W., and J. S. Altenbach. 1981. Functional anatomy of the primary downstroke muscles in the pallid bat *Antrozous pallidus*. Journal of Mammalogy 62:795–800.

Hermanson, J. W., and J. S. Altenbach. 1983. The functional anatomy of the shoulder of the pallid bat *Antrozous pallidus*. Journal of Mammalogy 64:62–75.

Hermanson, J. W., and J. S. Altenbach. 1985. Functional anatomy of the shoulder and arm of the fruit-eating bat *Artibeus jamaicensis*. Journal of Zoology (London) 205:157–177.

Hill, A. V. 1950. The dimensions of animals and their muscular dynamics. Science Progress (London) 38:209–230.

Jenkins, F. A., Jr., K. P. Dial, and G. E. Goslow, Jr. 1988. A cineradiographic analysis of bird flight: The wishbone in starlings is a spring. Science 241:1495–1498.

Johnston, I. A. 1985. Sustained force development: Specializations and variation among the vertebrates. Journal of Experimental Biology 115:239–251.

Kovtun, M. F. 1970. Morfofunkcional'nyj analiz mysc pleca letucich mysej v svjazi s ich poletom. (Morpho-functional analysis of the shoulder muscles in bats in relation to their flight.) Vestnik Zoologii 1:18–22 (in Russian).

Kovtun, M. F. 1981. Sravnitel 'naja morfologija i évoljucija organov lokomocii rukokrylych. (Comparative morphology and evolution of locomotor organs in bats.) D.Sc. thesis, Academy of Sciences of the Ukrainian SSR, Kiev (in Russian).

Kovtun, M. F. 1984. Strojenie i evolucja organow lokomocji rukokrylych. (Structure and evolution of the locomotor organs in bats.) Nauk Dumka, Kiev (in Russian).

McMahon, T. A. 1984. Muscles, Reflexes, and Locomotion. Princeton University Press, Princeton.

Norberg, U. M. 1969. An arrangement giving a stiff leading edge to the hand wing in bats. Journal of Mammalogy 50:766–770.

Norberg, U. M. 1970. Functional osteology and myology of the wing of *Plecotus auritus* Linnaeus (Chiroptera). Arkiv för Zoologi 22:483–543.

Norberg, U. M. 1972a. Bat wing structures important for aerodynamics and rigidity. Zeitschrift für Morphologie der Tiere 73:45–62.

Norberg, U. M. 1972b. Functional osteology and myology of the wing of the dog-faced bat *Rousettus aegyptiacus* (E. Geoffroy) (Pteropodidae). Zeitschrift für Morphologie der Tiere 73:1–44.

Norberg, U. M. 1976. Some advanced flight manoeuvres of bats. Journal of Experimental Biology 64:489–495.

Norberg, U. M. 1981. Allometry of bat wings and legs and comparison with bird wings. Philosophical Transactions of the Royal Society of London B 292:359–398.

Norberg, U. M. 1985. Flying, gliding, and soaring. *In* Functional Vertebrate Morphology, M. Hildebrand, D. M. Bramble, K. F. Liem, and D. B. Wake, eds., pp. 129–158, 391–393. Harvard University Press, Cambridge.

Norberg, U. M. 1987. Wing form and flight modes in bats. *In* Recent Advances in the Study of Bats, M. B. Fenton, P. Racey, and J. M. V. Rayner, eds., pp. 43–56. Cambridge University Press, Cambridge.

Norberg, U. M. 1990a. Ecological determinants of bat wing shape and echolocation call structure with implications for some fossil bats. *In* European Bat Research 1987, V. Hanák, I. Horácek, and J. Gaisler, eds., pp. 197–211. Charles University Press, Prague.

Norberg, U. M. 1990b. Vertebrate Flight: Mechanics, Physiology, Morphology, Ecology, and Evolution. Zoophysiology Series, Vol. 27. Springer, Berlin.

Norberg, U. M. 1994. Wing design, flight morphology, and habitat use in bats. *In* Ecological Morphology, P. C. Wainwright and S. M. Reilly, eds., pp. 205–239. University of Chicago Press, Chicago.

Norberg, U. M., and M. B. Fenton. 1988. Carnivorous bats? Biological Journal of the Linnean Society 33:383–394.

Norberg, U. M., and J. M. V. Rayner. 1987. Ecological morphology and flight in bats (Mammalia; Chiroptera): Wing adaptations, flight performance, foraging strategy, and echolocation. Philosophical Transactions of the Royal Society of London B 316:335–427.

Norberg, U. M., T. H. Kunz, J. F. Steffensen, Y. Winter, and O. von Helversen. 1993. The cost of hovering and forward flight in a nectar-feeding bat, *Glossophaga soricina,* estimated from aerodynamic theory. Journal of Experimental Biology 182:207–227.

Pennycuick, C. J. 1971. Gliding flight of the dog-faced bat *Rousettus aegyptiacus* observed in a wind tunnel. Journal of Experimental Biology 55:833–845.

Pennycuick, C. J. 1975. Mechanics of flight. *In* Avian Biology, Vol. V, D. S. Farner and J. R. King, eds., pp. 1–75. Academic Press, New York.

Pennycuick, C. J. 1982. The flight of petrels and albatrosses (Procellariiformes) observed in south Georgia and its vicinity. Philosophical Transactions of the Royal Society of London B 300: 75–106.

Pennycuick, C. J., and A. Lock. 1976. Elastic energy storage in primary feather shafts. Journal of Experimental Biology 64: 677–689.

Pennycuick, C. J., and M. A. Rezende. 1984. The specific power output of aerobic muscle, related to the power density of mitochondria. Journal of Experimental Biology 108:377–392.

Rayner, J. M. V. 1988. Form and function in avian flight. *In* Current Ornithology, Vol. 5, R. F. Johnston, ed., pp 1–66. Plenum Press, New York.

Schaefer, G. W. 1975. Dimensional analysis of avian forward flight. *In* Symposium on biodynamics of animal locomotion. Cambridge, U.K., September 1975 (unpublished manuscript).

Strickler, T. L. 1978. Functional Osteology and Myology of the Shoulder in Chiroptera. Contributions to Vertebrate Evolution, Vol. 4. Karger, Basel.

Thollesson, M., and U. M. Norberg. 1991. Moments of inertia of bat wings and body. Journal of Experimental Biology 158: 19–35.

Vaughan, T. A. 1959. Functional morphology of three bats: *Eumops, Myotis, Macrotus.* Publications of the Museum of Natural History, University of Kansas 12:1–153.

Vaughan, T. A. 1966. Morphology and flight characteristics of molossid bats. Journal of Mammalogy 47:259–260.

Vaughan, T. A. 1970a. The muscular system. *In* Biology of Bats, Vol. 1, W. A. Wimsatt, ed., pp. 139–194. Academic Press, London.

Vaughan, T. A. 1970b. The skeletal system. *In* Biology of Bats, Vol. 1., W. A. Wimsatt, ed., pp. 97–138. Academic Press, London.

Vaughan, T. A., and M. M. Bateman. 1980. The molossid wing: Some adaptations for rapid flight. *In* Proceedings of the Fifth International Bat Research Conference, pp. 195–216. Texas Tech Press, Lubbock.

Weis-Fogh, T. 1960. A rubber-like protein in insect cuticle. Journal of Experimental Biology 37:889–907.

7
Skin and Bones
Functional, Architectural, and Mechanical Differentiation in the Bat Wing

SHARON M. SWARTZ

Engineers and, increasingly, biologists recognize that the functional behavior of mechanical systems is dictated by the combination of material properties and structural geometry. What happens to a vertebrate limb when it experiences a load during locomotion, for example, is a consequence of both the size and shape of the support elements, composed of bone, cartilage, and ligament, and also of the particular mechanical properties of the tissues comprising those elements. Biologists, even those sensitive to the role of material properties in determining mechanical function, typically proceed immediately to morphology, bypassing considerations of materials science altogether. This bias probably arises from our empirical experience with patterned variation observable in vertebrates: Tissue properties tend to vary little among even distantly related and functionally divergent groups, whereas anatomical variation is observable at all levels in the taxonomic hierarchy.

Indeed, it seems clear that most evolutionary changes in the design of vertebrates appear to occur through changes in spatial deployment of tissues whose characteristics are generally quite uniform. For mammals and birds, it is widely believed that bones, muscles, ligaments, and other anatomical components are composed of the same mate-

rial, even in diverse taxa. It is often assumed that this conservativeness arises from constraints on the developmental systems of organisms; from this flows the assumption that selection is far more likely to achieve the ends of functional novelty by building new structures from old materials rather than through the modification of basic tissue types.

Here, I argue that the unique demands of powered flight have exerted strong selection on the design of wings. As a consequence, bats have diverged from their relatives not only in aspects of anatomical design such as the elongated bones of the forelimb and enhanced pectoral musculature, but also in a number of mechanically and energetically crucial aspects of both structural architecture and material composition.

Differentiation among Regions of the Bat Wing

General Functions of Regions of the Wing

The premises for our investigations of functional, material, and geometric specializations in bat wings are that they are distinctive in design and mechanical usage from all other mammalian limbs, and also that wings are not single homogeneous entities but rather can be subdivided into several major structural units or regions. I examine (1) differences among these regions, (2) skeletal function in two of three of these regions during flight, (3) patterns of variation in structural geometry of skeletal elements of each region, (4) variation among regions in material properties of the skeleton, (5) structural and material variation among regions in wing membrane skin, and (6) energetic considerations that may underlie the structural design of wings.

I have focused on three primary subdivisions of the wing: the plagiopatagium, dactylopatagium, and uropatagium. Each portion of the wing is somewhat distinct, not only in its anatomy but also, to varying degrees, in functional performance. Anatomically, each of these regions is primarily supported by particular skeletal structures: the humerus, radius, and body wall for the plagiopatagium; the carpals, metacarpals, and phalanges for the dactylopatagium; and the hindlimb and caudal skeleton for the uropatagium.

Functionally, the wing's center of lift is located within the plagiopatagium, approximately 75% from the distal end of the wing, although its precise location varies according to the particular wing shape and distribution of body weight in a species and with the timing of the wingbeat cycle in an individual (Norberg 1976, 1990). The plagiopatagium primarily generates lift and serves to translate lift forces to the humerus and the animal's body, although the absolute value of this lift force may be low relative to that produced by the distal handwing because of its significantly lower relative airspeed (Norberg 1976). With a limited number of skeletal supports and joints, the plagiopatagium is limited to some degree in the way that the skeletal elements can dictate the shape of the wing membrane surface and thus plays a relatively small role in controlling wing camber.

The dactylopatagium is highly maneuverable, because of the large number of relatively independently moving bones (carpals, metacarpals, and phalanges) and the nature of the surface configurations of the joint and the attachments of controlling musculature. Various portions of the dactylopatagium can be placed at a wide range of angles to the primary directions of downstroke movement, either together as a unit or with a considerable degree of independence. This region of the wing generates enormous lift given its relative high airspeed (Norberg 1976) and is believed to play the major role in the generation of thrust and in the control and fine-tuning of flight motions.

The uropatagium, while less important for flight mechanics than the other regions of the wing, nonetheless generates lift as well as contributing significantly to maneuverability (Norberg and Rayner 1987). It fulfills an additional important and unique function in some insectivorous bats by assisting in the capture of flying prey items, acting as a biological catapult to project airborne prey into the mouth (see Kalko and Schnitzler, Chapter 13, this volume).

Skeletal Function

To determine how the wing bones function during flight and to investigate functional differences among bones in distinct regions of the wing, we have conducted a series of investigations on skeletal loading during flight in the gray-headed flying-fox, *Pteropus poliocephalus* (Swartz et al. 1992, 1993). We selected this species for in-depth analysis because (1) its large body size (adult mass, 450–900 g) is associated with bones large enough to successfully attach instruments (Swartz 1991; Biewener 1992), (2) this species is locally abundant and easily trapped along much of the coastline in eastern Australia, and (3) individuals are readily trained to fly on command in a controlled setting (30-m linear flight cage).

To directly assess bone loading, we used in vivo bone strain recording (Swartz et al. 1992, 1993). We surgically implanted miniaturized strain gauges, highly sensitive length transducers, on the surfaces of bones (for technical details, see Perry and Lissner 1962; Dally and Riley 1978; Swartz 1991; Biewener 1992). However, because rosette gauges are constructed from a combination of three single-element gauges, these require significantly larger surfaces for the attachment of bones, three times the implanted wire needed for a single element, and three amplifier/data channels rather than one. These factors limit their applicability for many biological applications, and thus prohibit their use on bones smaller than 6–7 mm in diameter. We attached rosette strain gauges to the humerus and radius, and single-element gauges to the metacarpals and proximal phalanges III and V, placing three to four gauges around the bone circumference.

Because of the small size of the more distal bones, we were unable to obtain data from the more distal phalanges. We obtained recordings from each gauge site from 2 to 4 animals, using a total of 11 animals in the course of the study. We allowed the animals to recover from surgery, and then recorded patterns of strain along the bone with simultaneous video recordings of the patterns of wing movement during a series of flights along the length of a 30-m flight cage. For each flight, the bat was allowed to drop from an investigator's hand; this method produced takeoffs that were kinematically similar to those filmed during natural behavior in the field.

Certain aspects of the strain profiles from each bone are shared among all recordings from all bones and virtually all gauge sites (Figure 7.1). Before a bat begins to fly, recordings from the strain gauge are close to zero. Immediately at

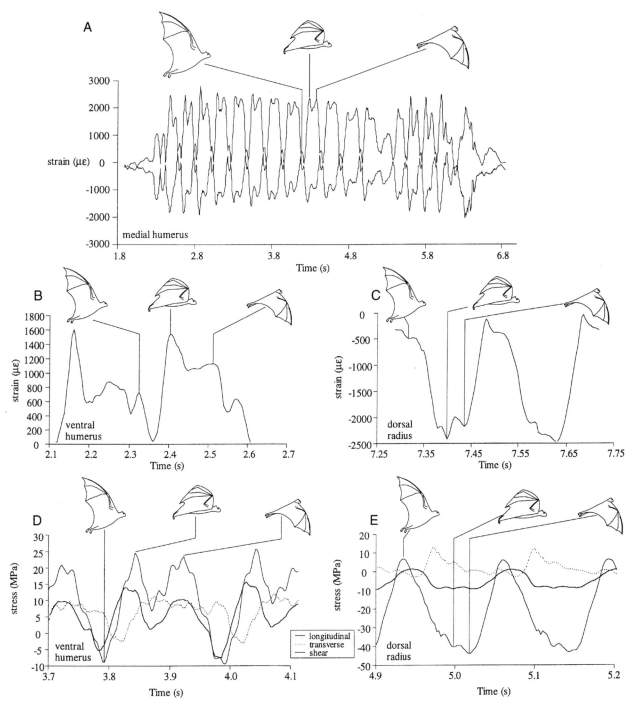

Figure 7.1. Representative principal strain and stress on areas of the humerus and radius during flight, based on videotapes (60 frames/sec) of one bat for a single flight. The schematic bats indicate representative phases of one wingbeat (top of upstroke, mid-downstroke, and bottom of downstroke). Magnitudes of strain are reported in microstrain, $\mu\epsilon$ ([change in length/original length] $\cdot$ 10^{-6}); tensile (elongating) strains are positive and compressive (shortening) strains are negative. Values of stress (in megapascals, MPa) are calculated from raw strain data, with the assumption that material characteristics of bat bone are comparable to those of other mammals and using a transversely orthotropic model for bone stiffness. **(A)** Maximum and minimum principal strains on the medial humerus during a complete flight sequence, from takeoff (small peaks, far left) through a number of regular wingbeat cycles and a postural shift in preparation for landing, followed by five decelerating wingbeats and landing. Note that the maximum and minimum principal strains (oriented at 90° to one another) are similar in magnitude but opposite in sign, which is typical for structures in relatively simple loading regimes (as the bone elongates, it simultaneously broadens). Bone strains typically reach greatest absolute values at mid-downstroke and lessen at the bottom of the downstroke and again at the top of the upstroke. During some phases of the locomotor cycle, strain values approach zero. **(B, C)** Maximum principal strains on the ventral humerus and dorsal radius during two wingbeats. **(D, E)** Maximum stresses (longitudinal, transverse, and shear) in the ventral humerus and dorsal radius during two wingbeats. Peaks in longitudinal stress correspond to those in principal strain; transverse and shear stress peaks may occur at somewhat different points in the wingbeat cycle. (Modified from Swartz et al. 1992.)

takeoff, each gauge site undergoes cyclical changes in the magnitude of strain, with one complete cycle of strain changes corresponding to one wingbeat. The precise shape of the changing strain trace varies among gauge sites, with specific portion of the flight, and, to a lesser extent, among individuals. During one to three wingbeats during takeoff, strain values were always quite low. Although intuition suggests that strains would be greater at takeoff than during level flight, video analysis demonstrated that these low strains during wingbeats correspond to the initial period of acceleration and gain in elevation. Thus, low strain values may result from the passive use of gravitational acceleration to generate airflow during takeoff.

Takeoff is followed by a series of relatively high strain peaks (10–15) corresponding to level flight. As the bat prepares to land, the shape of the strain trace changes slightly and begins to decrease in magnitude, dropping off sharply during the last few wingbeats before landing; at landing, wingbeats abruptly stop. Strain recordings from most gauge sites show two to four distinct peaks during the wingbeat cycle, typically three peaks per wingbeat. These can correspond to mid-downstroke, bottom of the downstroke, and top of the upstroke, but this pattern is not uniform for all bones studied.

BONES OF THE PLAGIOPATAGIUM.   Recordings from all sites on the two bones of the plagiopatagium (the humerus and the radius) indicated that their greatest strains occurred near the middle of the downstroke, when the wing is almost fully extended and nearly parallel to the ground, and during a second large peak at the transition of downstroke to upstroke (see Figure 7.1). The magnitude of maximum strain was well within the range observed during vigorous locomotor activity in other mammals and birds, and both the humerus and the radius showed principal strains oriented between 27° and 35° to the long axes of the bones (Swartz et al. 1992) (Figure 7.2). If the direction of bone loading is perfectly aligned with the anatomical axes of the bones, the orientation of the principal strain is close to 0°; if the bone surfaces experience pure shear from torsional loading, the orientation of the principal strain is 45°. The high angle of principal strain orientation for the humerus and radius indicates an unusually large degree of shear, most likely because of torsion. In addition, although the humerus shows little sign of bending superimposed on the torsional loading, the principal radial strains are positive (elongation) on the ventral surface and negative (compression) on the dorsal surface of the bone, demonstrating a significant bending moment that tends to bend the bone in an upwardly concave fashion at mid-downstroke.

Assuming that the material properties of the bone comprising the shafts of large bat humeri and radii are similar to that reported from compact bone of other mammal limbs, we can use directly measured values of strain to estimate the stresses developed within the bones. Calculation of stress depends on the elastic moduli (stiffnesses) of the bone tissue and on Poisson's ratio, a factor that quantifies for a particular material the amount of deformation that will occur at 90° to the axis of loading relative to the deformation along the loading axis. Here, we have assumed that the modulus of compact cortical bone is 18 GPa and that its value for Poisson's ratio is 0.3 (Currey 1984c) and have estimated the longitudinal, transverse, and shear stresses at the recording sites (Carter 1978). The longitudinal stresses we calculated in this manner are comparable in magnitude to those recorded from other mammals during vigorous activities such as fast cantering, galloping, and jumping. Transverse stresses, however, are significantly larger, particularly at the bottom of the downstroke where they approach 30%–40% of the maximum longitudinal stresses for the radius and 50%–120% of that observed in the humerus. The magnitude of shear stress is also unusually high in comparison to other mammals. We estimated shear stresses of more than 10–16 megapascals (MPa) in the humerus and 8–12 MPa in the radius. In comparison, values no greater than 6 MPa have been recorded during walking and running in humans and dogs (Carter 1978; Carter et al. 1980).

Torsional loading is particularly critical to the design of bones. Bone tissue, like many stiff and relatively brittle materials, is poorly suited to resisting shear. Indeed, the shear strength of bone is only one-third to one-quarter its strength in tension or compression (Nordin and Frankel 1989). This implies that it is selectively advantageous for bones that experience significant torsion to be designed to minimize the shear stress resulting from a given load.

How does this torsional load arise? All bats, and indeed, all flying vertebrates (see also Pennycuick 1967; Norberg 1990; Biewener and Dial 1995) have wings in which the stiff structural axis of the proximal wing skeleton is positioned relatively close to the leading edge of the wing, with a large region of wing surface caudal to the bony axis. The center of pressure is located somewhere within this region of wing membrane skin and is thus displaced relative to the bones. The upward force exerted during the downstroke at this center of pressure thus has a large moment about the bones and will tend to twist the wing as a whole in a pronating direction. At least some of this torsion may not be borne directly by the soft tissues surrounding the wing joints and is, instead, transmitted within the wing bones themselves. Torsional loading of the type that would produce the strains which we recorded is consistent with an earlier theoretical model of bone loading in bird wings

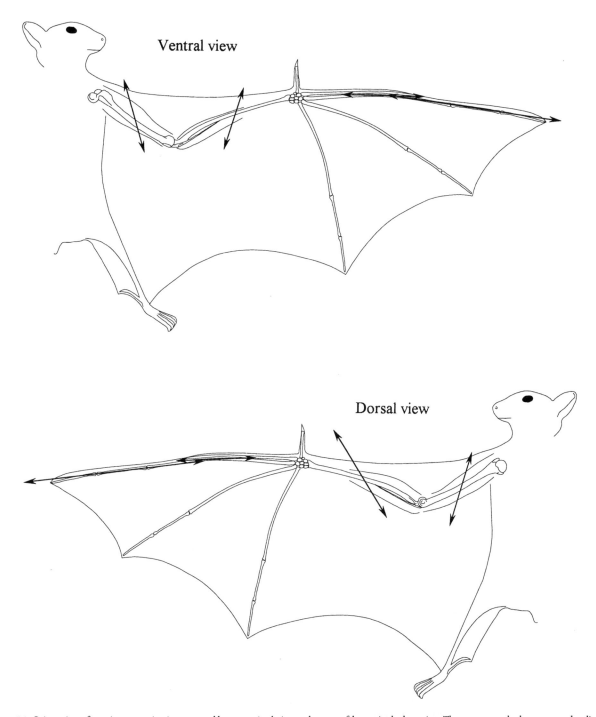

Ventral view

Dorsal view

Figure 7.2. Orientation of maximum strains (represented by arrows) relative to the axes of bones in the bat wing. The vectors on the humerus and radius represent maximum principal strains, and those on the metacarpal and phalanx are estimates of maximum strains.

(Pennycuick 1967), which has also been observed in birds (Biewener and Dial 1995).

BONES OF THE DACTYLOPATAGIUM. The general pattern of change in bone strain during the wingbeat cycle is similar for the proximal and distal wing regions (compare Figure 7.1A with Figure 7.3). However, in the dactylopatagium, the timing of the peak of strain differs significantly from that observed in the forewing. In the bones of the dactylopatagium, the metacarpals and the phalanges, the two largest peaks of strain coincide with the transition between downstroke and upstroke and a point of wingtip directional change during upstroke (Figure 7.4). These peaks coincide with the times of maximum wing acceleration and suggest that inertial forces, those arising from the acceleration and deceleration of wing mass, dictate the

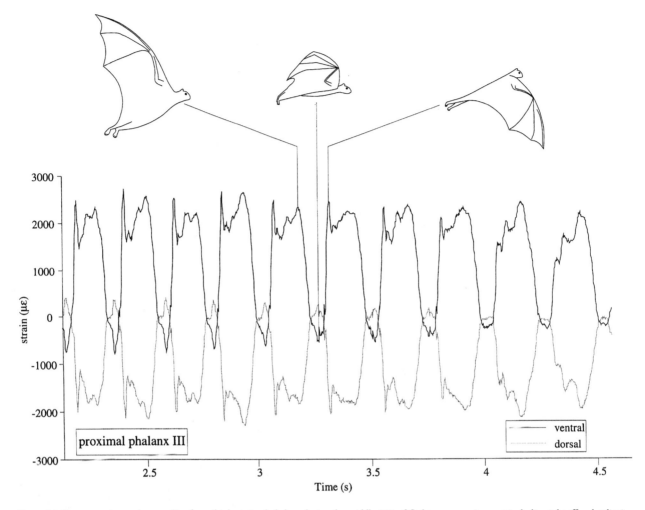

Figure 7.3. Representative strain recording from third proximal phalanx during the middle 65% of flight sequence (i.e., not including takeoff or landing). Although similar to Figure 7.1A, this graph illustrates the ventral and dorsal longitudinal strains rather than the maximum and minimum principal strains at a single anatomical location. Strains on opposing bone surfaces are opposite in sign and similar in magnitude, indicating nearly pure bending. Note that the direction of bending is reversed in the middle of the downstroke (ventral strains become compressive and dorsal strains positive).

loading of these bones (see also Norberg 1976; Norberg et al. 1993). This is consistent with the view that although the plagiopatagium and the dactylopatagium move in a coordinated fashion during each wingbeat, they function in a distinctive manner.

The individual bone loadings were also distinctive in this region of the wing. Strains along the long axes of the bones were three to four times the strains recorded perpendicular to the bone axis, suggesting largely axial bone loading or pure bending (Figure 7.5A). Indeed, the Poisson's ratio of typical mammalian compact cortical bone would produce longitudinal-to-transverse-strain ratios of 3.33 in the absence of off-axis loading, corresponding closely to our data (3.78) (Figure 7.5B). Moreover, the strains on the dorsal and ventral surfaces of bone were large, similar in magnitude, and opposite in sign: tensile on the dorsal surface at the beginning of the upstroke, and compressive at the end of

the upstroke; and compressive and tensile on the ventral surface at the beginning and end of the upstroke, respectively. On the medial and lateral surfaces of the bones, where one would expect to find the neutral axis of a beam loaded in bending, strains approached zero, again matching the expectation for beams loaded in bending with little or no torsion (Figure 7.5A).

The bending we observe in the distal wing does not remain constant in magnitude along a proximodistal axis. Strains in the metacarpus were significantly lower than those recorded from the phalanges, and within each of these bones, proximal strains were lowest, followed by those from the midshaft and distal regions in turn (Figure 7.5C). This strain increase along the axis of the bone reaches extreme values, with some strains exceeding 6,000 $\mu\epsilon$ ($\mu\epsilon$ = [change in length/original length] $\times 10^{-6}$), approaching those typically recognized as yield strains for compact bone

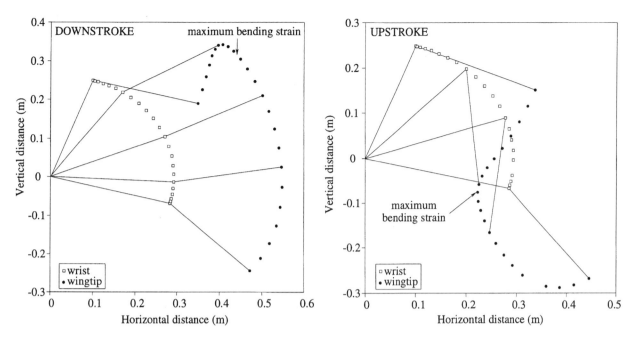

Figure 7.4. Horizontal and vertical motion of the carpus and wingtip of *Pteropus poliocephalus* during a typical wingbeat cycle. With the shoulder as the center of the coordinate system, the points represent the position of the joints in frontal view; line segments represent proximal and distal wing segments at selected times during the wingbeat. Downstroke (left) and upstroke (right) are defined by the motions of the plagiopatagium (shoulder-to-carpus segment). Note that the wingtip moves great distances and thus undergoes large accelerations and decelerations; also, movements of the dactylopatagium lag significantly behind those of the proximal wing. Peak strains (arrows) on the distal wing occur when the direction and speed of the wingtip change.

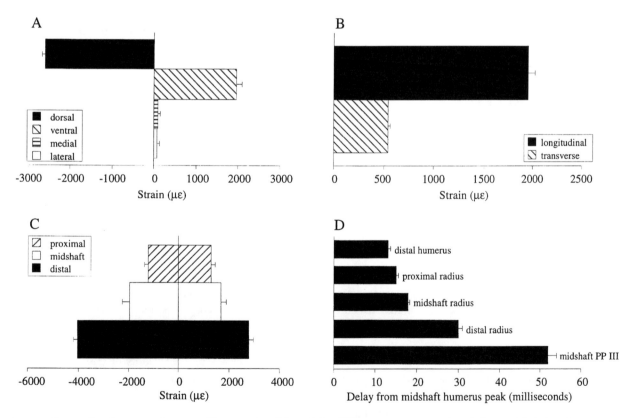

Figure 7.5. Pattern of peak strains on the wings of *Pteropus poliocephalus* in flight. All bars represent means and standard errors. (A) Peak strains on different surfaces of the third metacarpal. In keeping with nearly pure bending, dorsal and ventral strains are large, similar in magnitude, and opposite in direction; medial and lateral strains are close to zero. (B) Peak strains along and perpendicular to the axis of the third metacarpal. The ratio of longitudinal to transverse strains is close to the value predicted from material properties (Poisson's ratio) of compact cortical bone. (C) Peak strains along the length of the third proximal phalanx (PP III). Strains increase significantly toward the distal end of the bone. (D) Difference between the humeral midshaft (most proximal recording site) and more distal wing bones in timing of the peak principal or longitudinal strain. Timing delay increases along the wing's proximodistal axis.

(Currey 1984c; Keaveny and Hayes 1993). This is particularly noteworthy given that we were unable to implant gauges on the distal phalanges, presumably the sites of the most extreme bending, and that our subjects flew at moderate speeds in level, nonturning flight. Greater velocity, ascending flight, and turning maneuvers all have the potential to exert greater aerodynamic forces on the wing and therefore to generate even higher strains within the wing skeleton.

Strains reached their peak values at different times depending on their location along the wing's proximodistal axis (Figure 7.5D). We can display this offset in timing in relation to the timing of the maximum strain developed in the humeral midshaft; the strain peak is reached only a few milliseconds later in the distal portion of the bone, but significantly later in the metacarpals and phalanges, showing progressive delay along the wing bone. This lag suggests that the bending of the wing skeleton can be visualized as a wave passing down the wing from the shoulder to the tip during about one-half the total time of the downstroke.

Skeletal Geometry

The functions of the skeleton of the proximal and distal parts of the wing as just described differ profoundly. Proximal wing bones are loaded primarily in torsion, to a degree unknown elsewhere among vertebrate limb bones; this torsion arises through the displacement of the wing's center of pressure from the bony structural axis of the wing. Distal wing bones also experience a unique degree of dorsoventral bending in coordination with the acceleration and deceleration of the distal wing. This observation leads us to ask: To what degree are these very distinctive mechanical loading regimes reflected in the architecture of the skeleton? Using beam analysis, we have focused on the whole shape of the bone, particularly aspects of cross-sectional geometry, as indicators of the ability of a bone to resist particular kinds of loading.

TORSIONAL RESISTANCE IN THE STRUCTURE OF THE PROXIMAL WING. The surface shear stress, τ, develops in a beamlike structure subjected to a torsional moment, T, such that

$$\tau = Tr/J,$$

where r is the distance from the torsional axis to the beam surface and J is the polar moment of inertia, a measurement of both the cross-sectional area and its distribution about the torsional axis. For a circular cross section,

$$J = [\pi(R^4 - r^4)]/2,$$

where R and r are, respectively, the outer and inner radii of the cross section. For a given cross-sectional area, J is maximized when R is maximized and $R - r$, the wall thickness, is minimized. Hence, to most effectively resist torsional loading, bones should be constructed as very thin-walled, large-diameter cylinders.

Bats have extremely thin-walled humeri and radii, in direct contrast to the bones both of the more distal part of the wing and of the hindlimb (Figure 7.6). In comparison to a diverse group of terrestrial mammals (Currey and Alexander 1985), bats have proximal wing bones thinner than those of any other taxa. Like bats, birds and pterosaurs have thin-walled wing bones, and like bats, other flying vertebrates possess wings whose structural axis is found near the wing's leading edge, displaced from the center of pressure, producing torsion on the wing bone.

The common perception of the wing bones of bats is that they are extremely slender. We have used allometric analysis to demonstrate that, although all bat wing bones are relatively elongated, they may have strikingly larger diameters than those of other mammals of comparable body mass (Swartz 1997). The humerus and radius, in particular, have significantly greater diameters than those of comparably sized mammals based on comparisons with a sample of nonvolant mammals drawn from dermopterans, primates, rodents, marsupials, insectivores, and small carnivores and ungulates (Figure 7.7). Thus, although this increase in diameter tends to increase bone mass in addition to the increase in bone mass associated with bone elongation, in combination with a reduction in wall thickness it will increase J and hence reduce t for a given load. Thus, the geometry of the humeral and radial shafts is unusually well suited to resist large torsional forces.

PROMOTION OF BENDING IN THE BONES OF THE DISTAL WING. If a beam experiences primarily bending forces oriented within a single plane, the cross-sectional geometries best suited to resist that force are those in which the material is distributed as far as possible from the beam's neutral plane of bending, typically halfway between the surfaces undergoing the greatest tension and those with the greatest compression. These geometries minimize both beam surface stresses caused by bending and the deflections of the beam ends under load. The I-beam is a notable engineering exemplar of this idea, with the majority of the beam cross-sectional area concentrated in the two plates furthest from its neutral plane; the remaining material, the central web, serves the function of keeping the flanges apart, thus producing the classic "I" shape. For bones subjected to dorsoventral bending, stresses would be minimized with a structural cross section that maximizes dorsoventral diameter and minimizes mediolateral diameter, with material as far as possible from the neutral axis. This design is well

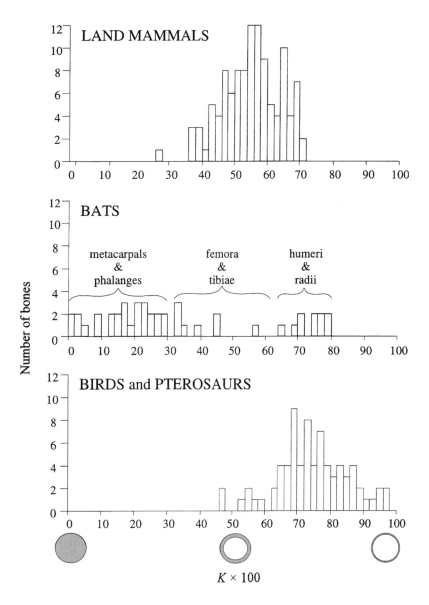

Figure 7.6. Frequency distribution of cortical thickness of limb bones in a variety of vertebrates. Data on terrestrial mammals and on birds and pterosaurs are from Currey and Alexander 1985; bat data were collected from radiographs of this study's subjects. Cortical thickness is expressed as K, and marrow cavity thickness equals KR, where R is the radius of the bone cross section (after Currey and Alexander 1985). When K is 0, the bone cross section is solid, and K approaches 1 as wall thickness approaches 0. Distributions for land mammals and for birds plus pterosaurs show some overlap but are clearly distinct, with flying vertebrates having larger mean K values. Bats are distinct from other mammals in that the walls of their humeri and radii are less thick than those of other mammals, falling well within the bird and pterosaur range. Other wing bones of bats (metacarpals and phalanges) deviate from the typical mammalian pattern in a different way, having much greater cortical thickness. (Modified from Swartz et al. 1992.)

realized by a transversely flattened, dorsoventrally expanded ellipse of large outer dimensions and thin walls.

In the distal portion of the wing, however, the typical cross-sectional geometry does not match this prediction. Outer diameters are not expanded in the fashion seen in the humerus and radius, and as more distally on the wing, the trend for increased diameter relative to nonvolant mammals is completely reversed, with the distalmost elements greatly reduced in diameter (see Figure 7.7). The proximal-to-distal trend is also seen within the architecture of the individual bones; the majority of wing bones of bats, and all distal wing bones, taper in diameter from 25% to 50%–75% of the bone's proximal-to-distal length, with the intensity of tapering generally greater for the distal than the proximal bones (Swartz 1997). Similarly, wall thickness is greatly enhanced in the distal wing bones (see Figure 7.6), with both metacarpals and, particularly, phalanges outside

the range for other mammalian limb bones. This tremendous increase in wall thickness, so extreme that in many bats the absolute wall thickness of distal phalanges is greater than the absolute thickness of the far larger humerus, may reach the extreme of total obliteration of the medullary cavity in the distal elements. Extremes of wall thickness are limited, however, to the bones of the wing; values for the femur and tibia are well within the range for land mammals (Figure 7.6).

The geometry of the distal wing bones does not match the pattern predicted for resistance to bending, suggesting that the structural architecture of these bones is driven by other considerations. Here, the geometry of the skeleton may actually serve to enhance deflections of the distal wing rather than to resist bending forces and to reduce the effect of bending moments. The wingtip, particularly the phalangeal region, deflects markedly during some portions of

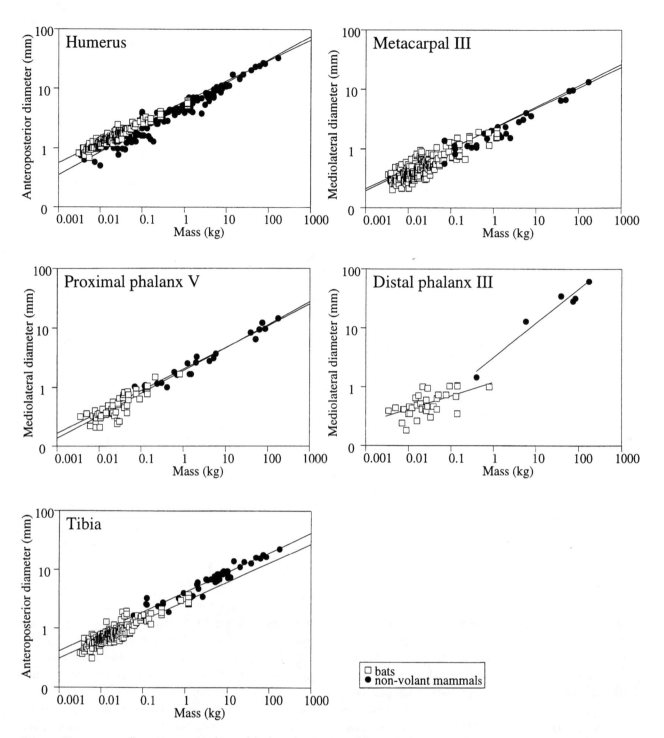

Figure 7.7. Representative allometric regression plots (with log-log reduced major axes) for midshaft diameters of bones in bats and in nonvolant mammals. The pattern for the radius is similar to that for the humerus; all metacarpals and proximal phalanges are similar, and all distal phalanges are similar.

the wingbeat cycle, particularly when bats turn or execute other spatially complex maneuvers. Additional flexibility in the bones of the distal wing, conferred in part by structural geometry and in part by material properties (see following), may enhance a bat's ability to adopt the appropriate three-dimensional configuration of the wing for controlling these movements. The large deflections possible in these bones as

a consequence of great elongation, coupled with reduced diameter and second moment of area, may even contribute to the development of high camber, although the greatest part of the chordwise wing camber certainly arises through flexion at the metacarpophalangeal and interphalangeal joints of the fourth and fifth digits (Norberg 1970, 1972). The deflection of these bones under aerodynamic forces may significantly influence wing curvature, counteracting the camber developed by joint movement.

Material Properties

There is relatively little variation in the mechanical properties of the tissue in the compact cortical bone tissue of the limb skeleton, even among mammals of diverse locomotor mode and over a large range of body sizes (Biewener 1982; Currey 1984a, 1987). However, given the tremendous range of variation in patterns of bone loading and in structural geometry in the wing skeleton of bats, we have asked whether bats may differ from the typical mammalian pattern of relatively uniform mechanical properties of bone.

As a first step in assessing variation in mechanical properties, we have assayed bone mineralization or ash content, looking in depth at variation among bones and throughout postnatal ontogeny in the Mexican free-tailed bat *Tadarida brasiliensis* (Papadimitriou et al. 1996). Previous work has shown that mineralization, readily assessed by careful determination of the ash content of bone specimens, not only determines bone density but also is strongly correlated with key mechanical properties such as stiffness (Young's or elastic modulus), failure and yield strength (stress at failure or yield), failure and yield strain, and energy absorbed per unit bone volume (Currey 1979, 1990). This relationship, however, is not linear; stiffness and strength increase nonlinearly with mineralization to some maximum value beyond which further increases in mineral content increase the brittleness of bones, reflected in increasing stiffness but decreasing material strength and energy-absorbing capacity or toughness. For this reason, measurements of ash content may be strong indicators of mechanical properties, but further detailed mechanical tests are needed to assess full functional analysis.

We found substantial variation in the ash content of the wing bones of adult, wild-caught *Tadarida brasiliensis* (Figure 7.8); indeed, a single *T. brasiliensis* wing encompasses a greater range of mineralization levels than the entire spectrum of mammalian long bones studied to date. Variation in ash content follows a strong proximodistal gradient, with the highest levels found in the humerus, followed by the radius, metacarpals III, V, IV, and II, the proximal phalanges, and the distal phalanges in turn. This pattern is not unique to *T. brasiliensis;* to date we have documented a similar pattern of distally decreasing mineralization in *Eptesicus fuscus, Myotis lucifugus,* and *Pteropus poliocephalus* as well (Swartz, unpublished data). These values range from within the normal mammalian range of 66%–68% in the humerus and radius to virtually no ash content in the distal phalanges. This lack of measurable mineral is also seen on more detailed examination; histological analysis with staining for

calcium salts shows a normal degree of mineralization in the subchondral bone immediately underlying the articular cartilage and extending distally a few millimeters, while the remainder of the distal phalanx consists of a fibrous collagenous shell surrounding a core of cartilage (Papadimitriou et al. 1996).

Although we have yet to complete comprehensive measurements of mechanical properties of wing bones in bats, the mechanical characteristics of these bones may be substantially different from those of homologous bones of nonvolant mammals. Our preliminary studies of wing bones in *Eptesicus fuscus* have demonstrated failure strains that are within the range for other mammals. However, strength of wing bones in this bat ranges from 16 to 21 MPa, and elastic modulus (stiffness) ranges from 1.3 to 1.8 GPa, in comparison to typical values of 140–160 MPa and 15–20 GPa for mammalian limb bones (Currey 1984b, 1984c, 1987, 1988).

The observed gradient in mineralization, presumably a reflection of an equally strong or stronger gradient in mechanical properties, directly parallels the architectural variation in wing bones. Proximal elements are stiffest and strongest by virtue of both their structural geometry (thin-walled, large-diameter hollow tubes) and their mineral content (relatively high, near the maximum strength for compact bone) (Currey 1979). Likewise, distally from the carpus the geometry shifts to increasingly promote bending rather than to resist torsional or bending deflections (decreasing diameter, increasing wall thickness) and mineralization continues to drop, further decreasing stiffness and enhancing deflection under load.

Are there reasons to expect increased deflection in the wingtip? Inertial effects can be expected to produce large deflections in the wingtip because of their high accelerations (see also Norberg 1970, 1972, 1990). This phenomenon may also be enhanced by the effects of added mass, although the magnitude of these effects is not well understood at low Reynold's numbers (Norberg et al. 1993), because a relatively large volume of air is accelerated and decelerated along with the structures of the wingtip. Without large deflections of very flexible structures in this region, the large accelerations of the wingtip and its associated air mass will generate very strong vortices that will have undesirable effects on flight performance. Moreover, large distal deflections may increase the turbulence of the wing boundary layer and thereby help to maintain attached flow at higher angles of attack.

Finally, enhanced wingtip deflection may result in the top trailing the motion of the rest of the wing through the wingbeat and a failure of the wingtip to really "snap" at the top of the upstroke. As a consequence, the motion of the wingtip is smoothed, leading to a reduction of the wing's induced drag

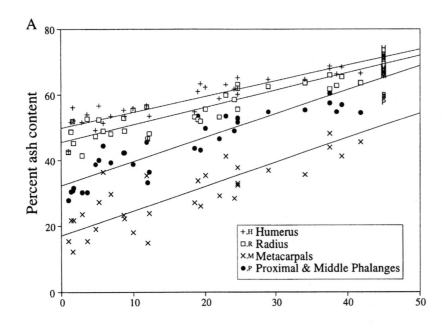

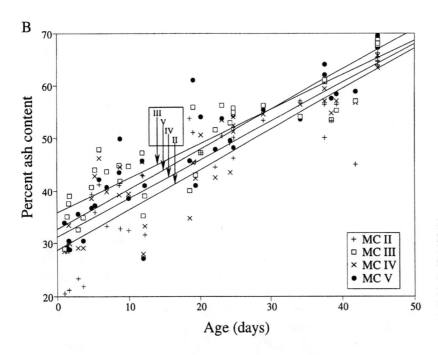

Figure 7.8. Changes in mineral content during postnatal ontogeny of *Tadarida brasiliensis*. Each data point represents a single bone of one individual. (**A**) Humerus, radius, average of all metacarpals, and average of proximal and middle phalanges of digit III. (Distal phalanx III showed no detectable mineral content.) Uppercase letters indicate adults. (**B**) Each of metacarpals (MC) II–V. (Modified from Papadimitriou et al. 1996.)

because less energy goes into the creation of wingtip vortices. In this manner, flexible wingtips may function in a manner similar to feathers, functionally increasing the span of the wing to lower induced drag as they bend to accommodate increased angle of attack.

Wing Membrane Structure and Mechanical Properties

In most vertebrates, bones provide the organism with the means to exert and resist forces and to generate locomotor propulsion. In bats, however, the structure of connective tissue of the wing membrane is, along with the bones, a primary element of the limb's structural support system, and indeed the mechanics of the bones and skin of the wing are perhaps best considered together in a unified fashion. It has long been recognized that the skin of the wing membrane is unique among mammals, particularly in its possession of an extremely reduced dermis sandwiched between dorsal and ventral layers of epidermis (Quay 1970; Sokolov 1982) and of a highly organized two-dimensional network of large, macroscopic collagen-elastin fiber bun-

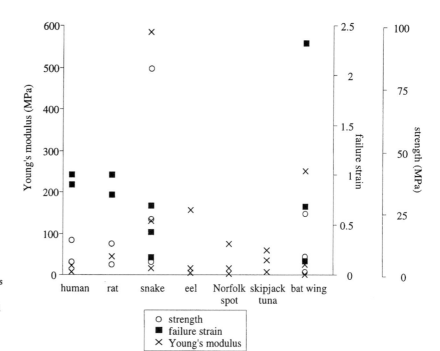

Figure 7.9. Comparison of mechanical properties of skin of a bat wing and skin of other vertebrates. Young's modulus quantifies material stiffness, failure strain quantifies length change before rupture, and strength quantifies stress developed within the skin before rupture.

dles (Holbrook and Odland 1977). However, until recently, little was known about the mechanical behavior of membrane skin.

Although it has been qualitatively recognized that the skin of the wing membrane is thin, we have for the first time placed this skin configuration in an allometric context. Our comparisons of empirically measured skin thickness to predictions based on allometric equations for a diverse range of mammalian taxa show that wing membrane skin is between 4 and 10 times thinner than would be predicted for similar-sized nonvolant mammals (Calder 1984; Swartz et al. 1996). A reduction in the thickness of skin might well imply a corresponding reduction in its ability to bear mechanical loads. To determine the magnitude of this effect, we have conducted tensile tests to assess the mechanical properties of wing membrane skin, sampling from the dactylopatagium, plagiopatagium, and uropatagium, testing the anisotropy of the membrane, and comparing mechanical behavior among a diverse group of New World microchiropterans (Swartz et al. 1996). The taxa we have sampled include both insectivores and frugivores and cover a range of body mass (7–75 g), wing loading (6.3–16.6 newtons [N] per square meter), and flight style.

The mechanical properties of wing membrane are diverse, covering, and for some characteristics also extending, the range of mechanical properties from all other vertebrate skins studied to date, often in a single wing of a single species (Figure 7.9). This wide range of strength, stiffness, failure strain, energy-absorbing ability, and load-carrying capacity arises both from gross variation in skin thickness within the wing and among taxa and from the remarkable

anisotropy of wing skin. In most mammalian skin, viewed here as a basically two-dimensional structure with some variation in thickness, mechanical properties are nearly invariant with respect to orientation within the plane of the skin (Gibson et al. 1969; Gibson 1977; Shadwick et al. 1992). In marked contrast, for example, skin from the plagiopatagium and dactylopatagium regions of a single individual may differ in strength, strain at failure, and stiffness by factors of 5.8, 18.6, and 37.2, respectively, as the result of anisotropy alone (Figure 7.10). This anisotropy is clearly related to the "skeleton" of connective tissue bundles in the wing (Holbrook and Odland 1977), but not in a simple fashion. We found that the variation in mechanical properties within a single wing region of a single species was highly correlated with the orientation of the small skin folds or corrugations that extend from fiber bundle to fiber bundle rather than with the direction of the fiber bundles themselves (Swartz et al. 1996). This result suggests that the fiber network may actually create tension in the skin tissue, and that the two-dimensional patterns of bundle intersection angles, which are primarily but not uniformly 90°, may set up a complex spatial arrangement of varying mechanical characteristics.

When we place this directional variation in mechanical properties onto the anatomical framework of the wing, the resulting pattern suggests clear functional variation (Figure 7.11). Wing skin is stiffest parallel to the wing bones, showing lower stiffness and far greater extensibility parallel to the trailing edge and perpendicular to the bony axes. This pattern may relate to shearing of the wing skin, and of the wing skin relative to its attachments to the bones, during

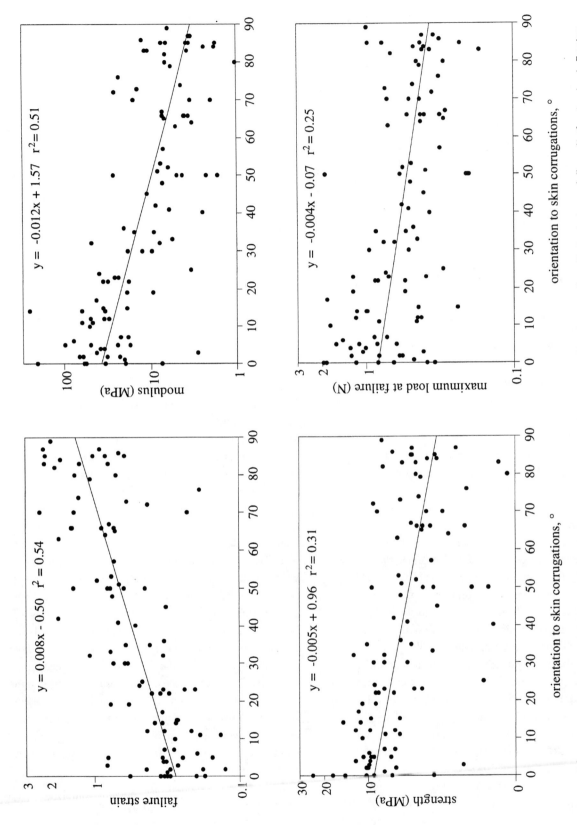

Figure 7.10. Least-squares regression plots of mechanical characteristics in relation to the corrugation angle of bat skin. Data from all wing regions and all taxa (*Artibeus jamaicensis, Eptesicus fuscus, Lasiurus cinereus, Myotis lucifugus, Pteronotus parnellii, Tadarida brasiliensis,* and *Uroderma bilobatum*) were pooled in these analyses. For all plots, the abcissa is the corrugation angle in degrees. In the absence of anisotropy, as is the case in most mammalian skin, there would be no significant regressions. (After Swartz et al. 1996.)

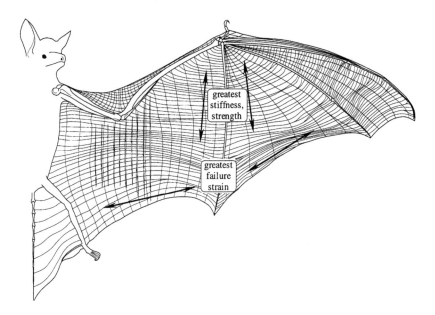

Figure 7.11. Schematic of anisotropic variation in mechanical properties of skin related to wing anatomical framework. Modulus, strength, and maximum and cross-sectional area are greatest parallel to the fifth digit; failure strain, in contrast, is greatest perpendicular to the fifth digit. (After Swartz et al. 1996.)

flapping flight. There is a large mismatch between the stiffnesses of the skin and the bones, and as aerodynamic force is exerted against the wing, this will produce both billowing of the skin in the central areas between bones (Norberg 1972) and a tendency for the skin that is attached directly along the length of the bones length to deform far more than the supporting skeleton, producing shearing forces at the skin–bone interface. To the extent that skin stiffness can be increased parallel to the bones, the destructive effect of this shearing on the interface will be minimized, and skin anisotropy may thereby contribute to maintaining the integrity of the skin–bone system.

Conversely, it is important for wing skin to be highly extensible so as to provide, simultaneously, a large wing area and the ability to retract and allow relative ease of movement when the bat wing is not in the fully extended position of mid-downstroke. This extensibility contributes most effectively to wing function if it is oriented along the proximodistal axis, contributing then to the ability of the wing to furl and extend. In comparisons among wing regions, we found substantial patterned variation in the characteristics of skin (Swartz et al. 1996). The plagiopatagium shows the greatest extensibility among wing regions. Lacking the more prehensile characteristics of the handwing, the aerodynamic function of the plagiopatagium is primarily to generate lift, and also where wing camber, a major contributor to both lift and drag, is at least partly generated and directly controlled by the billowing of the membrane.

The uropatagium, the wing area least important for the generation of lift and a critical organ for the capture of prey by insectivorous species (see Kalko and Schnitzler, Chapter 13, this volume), is by far the stiffest and strongest region. This increased stiffness likely allows better control over the

ability to propel prey from the tail membrane to the mouth, and the increased strength significantly improves puncture resistance, potentially a critical feature for a thin membrane used to catapult insects with hard cuticles at high velocity. The dactylopatagium is the wing region intermediate in its mechanical characteristics, and it is functionally involved in the generation of lift and thrust and the control of maneuverability. Its increased skin stiffness relative to the plagiopatagium may serve to minimize drag during the pronation–supination and flexion–extension movements crucial to the provision of thrust and control of maneuverability (Vaughan 1970) and may resist the larger aerodynamic forces generated in the distal wing (Norberg 1976). Altered skin characteristics may thereby maximize useful thrust and reduce the energy required for flight.

Energetic Consequences of Wing Specialization

It is indeed remarkable that bats possess such a large range of material properties within their wings. It is especially notable that this pattern parallels morphological and allometric variation among bones, functional behavior of the skeleton, and mechanical design of the skin. This multilevel variation in the mechanically important tissues of the wing may be dictated primarily by the need to adequately resist stresses induced by aerodynamic forces exerted on wings. Previous studies have documented ample interspecific variation in the functional loading of the mammalian limb skeleton, but none has observed the range of variation in material properties that we have seen in bat wing bones.

There are at least two possible explanations for the greater variation in wing structures of bats: (1) structural design is less constrained in bat wings than in other mammalian limbs, and is therefore free to vary with little functional consequence; or (2) extreme functional demands create a mechanical environment that strongly selects for mechanical properties and functional abilities beyond the range observed in nonvolant mammals. We propose that the patterned variation in multiple tissues and at several organizational levels tentatively supports the second hypothesis. We further suggest that ultimately energetic considerations are the central determinants of wing design.

Flight is an energetically expensive mode of locomotion and comprises a large portion of the total energy budget of bats (Thomas and Suthers 1972; Thomas 1975; Kurta et al. 1989; Winter et al. 1993). It is plausible, as a consequence, that selection for minimizing body mass while maintaining locomotor performance may be particularly intense in bats, particularly insectivorous species. Inertial power, the energetic cost of accelerating and decelerating the wings with respect to the animal's center of mass, will be a significant fraction of the metabolic cost of bat flight, typically between 30% and 60% of the total, unless wing inertia can be converted into useful aerodynamic work (Norberg et al. 1993; see also Norberg, Chapter 6, this volume).

Inertial power, P_{iner}, equals $16I_{\omega}\pi^2 f^3 \phi^2$, where I_{ω} is wing mass moment of inertia, f is wingbeat frequency, and ϕ is wingbeat amplitude. I_{ω} thus plays a major role in determining inertial power and is itself the sum of each mass increment multiplied by the square of its distance from the wing's center of rotation at the shoulder joint. Because of the distance squared term, I_{ω} is influenced strongly by the architecture of the dactylopatagium, far from the shoulder. It is also determined largely by skeletal mass, given that the skeleton constitutes 40%–60% of the wing's total mass, particularly in the distal wing where there is little muscle mass (Norberg 1976).

To quantitatively estimate the net effect of specialized wing bone geometry along with lowered mineralization and thus bone density, I have compared the true wing I_{ω} in a representative small insectivorous bat to estimated I_{ω} with bone geometry and density values typical of those of other mammals (Swartz 1997). In this comparison, I have shown that the actual mass moment of inertia of the bat wing is decreased by a factor of 2.3–3.0 by specializations of the wing bones. Because the reductions are most extreme in the distal wing, this effect will be enhanced in taxa with particularly elongated distal wings.

This estimate of the energetic saving from structural design may well be conservative; it does not take into consideration the effect of the enormous reduction of skin thickness in bats on the mass of the wing. Although skin typically constitutes only a small fraction of mammalian limb mass, the enormous enlargement of the surface area of the wings and the drastic reduction of muscle masses within the wing mean that skin constitutes a large portion of the nonbony wing mass. Given that skin is 4–10 times thinner, and thus presumably 4–10 times less massive, than in mammals of comparable size, this reduced skin mass may have a significant energetic effect. Hence, the modifications of the external dimensions of cortical thickness, mineralization of the wing bone, and skin design may all have important energetic as well as mechanical consequences for the evolution of bat flight (Papadimitriou et al. 1996; Swartz et al. 1996; Swartz 1997).

Conclusions

Analysis of bat wing structure using an engineering-based design suggests that the plagiopatagium, dactylopatagium, and uropatagium function as distinct but integrated elements of the flight apparatus. Evidence summarized here, drawn both from comparative studies of many taxa and from detailed analysis of single species, confirmed that wing regions differ in the geometry, material properties, and functional loading of the skeleton, as well as in the structure and material properties of the wing membrane skin. The diversity in these characteristics within a bat's wing is remarkable, often encompassing the entire range of values previously reported for a wide variety of mammals. This differentiation in characteristics of anatomically distinct wing regions may be driven by energetic considerations, and the preliminary analysis discussed here indicates that bat flight is significantly less energetically expensive as a consequence of the derived features of the wing skeleton and membranes.

Acknowledgments

Many people have contributed to the work summarized here. In vivo bone loading measurements were made and analyzed with the tireless assistance of E. Anderson, M. Bennett, D. Carrier, J. G. Chickering, L. Hall, A. Parker, and E. Mitchell. Allometric studies were made possible by access to collections of the American Museum of Natural History, the British Museum of Natural History, the Field Museum of Natural History, the Museum of Comparative Zoology at Harvard, and the U.S. Natural History Museum (Smithsonian Institution), provided by N. Simmons, P. Jenkins, B. Patterson, M. Rutzmoser, and R. Thorington. Analysis of mineralization gradients of bone during the ontogeny of *Tadarida brasiliensis* was carried out in collaboration with H. Papadimitriou and T. Kunz, with technical assistance from A. Ritter. Studies of the mechanical characteristics of wing membrane skin were done in collaboration with M. Groves, B. Walsh, and H. Kim, with technical assistance from A. Ritter, and critical review from B. Shadwick, using specimens generously provided by T. Kunz, N. Simmons, J. Simmons, J. Ryan, and W. Lancaster. My understanding of the mechanical and aerodynamic issues discussed here has been deepened through my conversations with P. Watts, D. Carrier, U. Norberg, and A. Biewener. The manuscript has been improved by suggestions from K. Dial, E. Mitchell, U. Norberg, and A. Parker. T. Freeman and E. Anderson provided encouragement and humor at several key moments. This research has been supported by the National Science Foundation (NSF) (IBN-9119143).

Literature Cited

Biewener, A. A. 1982. Bone strength in small mammals and bipedal birds: Do safety factors change with body size? Journal of Experimental Biology 98:289–301.

Biewener, A. A. 1992. *In vivo* measurement of bone strain and tendon force. *In* Biomechanics: A Practical Approach, Vol. 2, Structures, A. A. Biewener, ed., pp. 123–147. Oxford University Press, Oxford.

Biewener, A. A., and K. P. Dial. 1995. *In vivo* strain in the humerus of pigeons *(Columba livia)* during flight. Journal of Morphology 225:61–75.

Calder, W. A. III. 1984. Size, Function, and Life History. Harvard University Press, Cambridge.

Carter, D. R. 1978. Anisotropic analysis of strain rosette information from cortical bone. Journal of Biomechanics 11:199–202.

Carter, D. R., D. J. Smith, D. M. Spengler, C. H. Daly, and V. H. Frankel. 1980. Measurement and analysis of *in vivo* bone strain in the canid radius and ulna. Journal of Biomechanics 13:27–38.

Currey, J. D. 1990. Biomechanics of mineralized skeletons. *In* Skeletal Biomineralization: Patterns, Processes, and Evolutionary Trends, J. G. Carter, ed., pp. 11–25. Van Nostrand Reinhold, New York.

Currey, J. D. 1979. Mechanical properties of bone tissues with greatly differing functions. Journal of Biomechanics 12:313–319.

Currey, J. D. 1984a. Comparative mechanical properties and histology of bone. American Zoologist 24:5–12.

Currey, J. D. 1984b. Effects of differences in mineralization on the mechanical properties of bone. Philosophical Transactions of the Royal Society of London B 304:509–518.

Currey, J. D. 1984c. The Mechanical Adaptations of Bones. Princeton University Press, Princeton.

Currey, J. D. 1987. The evolution of the mechanical properties of amniote bone. Journal of Biomechanics 20:1035–1044.

Currey, J. D. 1988. The effect of porosity and mineral content on the Young's modulus of elasticity of compact bone. Journal of Biomechanics 21:131–139.

Currey, J. D., and R. M. Alexander. 1985. The thickness of the walls of tubular bones. Journal of Zoology (London) 206:453–468.

Dally, J. W., and W. F. Riley. 1978. Experimental Stress Analysis. McGraw-Hill, New York.

Gibson, T. 1977. The physical properties of skin. *In* Reconstructive Plastic Surgery: Principles and Procedures in Correction, Re-

construction, and Transplantation, J. M. Converse, ed., pp. 69–77. Saunders, Philadelphia.

Gibson, T., H. Stark, and J. M. Evans. 1969. Directional variation in extensibility of human skin *in vivo*. Journal of Biomechanics 2:201–204.

Holbrook, K. A., and G. F. Odland. 1977. A collagen and elastic network in the wing of the bat. Journal of Anatomy 126:21–36.

Keaveny, T. M., and W. C. Hayes. 1993. Mechanical properties of cortical and trabecular bone. *In* Bone Growth: B, Bone, Vol. 7, B. K. Hall, ed., pp. 285–344. CRC Press, Boca Raton, Fla.

Kurta, A., G. P. Bell, K. A. Nagy, and T. H. Kunz. 1989. Energetics of pregnancy and lactation in free-ranging little brown bats *(Myotis lucifugus)*. Physiological Zoology 62:804–818.

Norberg, U. M. 1970. Functional osteology and myology of the wing of *Plecotus auritus* Linnaeus (Chiroptera). Arkiv för Zoologi 22:483–543.

Norberg, U. M. 1972. Bat wing structure important for aerodynamics and rigidity (Mammalia, Chiroptera). Zeitschrift für Morphologie der Tiere 73:45–61.

Norberg, U. M. 1976. Aerodynamics, kinematics, and energetics of horizontal flapping flight in the long-eared bat *Plecotus auritus*. Journal of Experimental Biology 65:179–212.

Norberg, U. M. 1990. Vertebrate Flight. Springer, Berlin.

Norberg, U. M., and J. M. V. Rayner. 1987. Ecological morphology and flight in bats (Mammalia; Chiroptera): Wing adaptations, flight performance, foraging strategy, and echolocation. Philosophical Transactions of the Royal Society of London B 316: 335–427.

Norberg, U. M., T. H. Kunz, J. F. Steffensen, Y. Winter, and O. von Helverson. 1993. The cost of hovering and forward flight in a nectar-feeding bat, *Glossophaga soricina*, estimated from aerodynamic theory. Journal of Experimental Biology 182:207–227.

Nordin, M., and V. H. Frankel. 1989. Biomechanics of bone. *In* Basic Biomechanics of the Musculoskeletal System, M. Nordin and V. H. Frankel, eds., pp. 3–29. Lea and Febiger, Philadelphia.

Papadimitriou, H. M., S. M. Swartz, and T. H. Kunz. 1996. Ontogenetic and anatomic variation in mineralization of the wing skeleton of the Mexican free-tailed bat, *Tadarida brasiliensis*. Journal of Zoology (London) 240:411–426.

Pennycuick, C. J. 1967. The strength of the pigeon's wing bones in relation to their function. Journal of Experimental Biology 46: 219–233.

Perry, C. C., and H. R. Lissner. 1962. The Strain Gage Primer. McGraw-Hill, New York.

Quay, W. B. 1970. Integument and derivates. *In* Biology of Bats, Vol. II, W. A. Wimsatt, ed., pp. 2–56. Academic Press, New York.

Shadwick, R. B., A. P. Russell, and R. F. Lauff. 1992. The structure and mechanical design of rhinoceros dermal armour. Philosophical Transactions of the Royal Society of London B 337: 419–428.

Sokolov, V. E. 1982. Mammal Skin. University of California Press, Berkeley.

Swartz, S. M. 1991. Strain analysis as a tool for functional morphology. American Zoologist 31:655–669.

Swartz, S. M. 1997. Allometric patterning in the limb skeleton of bats: Implications for the mechanics and energetics of powered flight. Journal of Morphology 234:277–294.

Swartz, S. M., M. B. Bennett, and D. R. Carrier. 1992. Wing bone stresses in free flying bats and the evolution of skeletal design for flight. Nature (London) 359:726–729.

Swartz, S. M., M. B. Bennett, J. A. Gray, and A. Parker. 1993. Bones built to bend: In vivo loading in the distal wing of fruit bats. American Zoologist 33:75A.

Swartz, S. M., M. D. Groves, H. D. Kim, and W. R. Walsh. 1996. Mechanical properties of bat wing membrane skin. Journal of Zoology (London) 259:357–378.

Thomas, S. P. 1975. Metabolism during flight in two species of bats, *Phyllostomus hastatus* and *Pteropus gouldii*. Journal of Experimental Biology 63:273–293.

Thomas, S. P., and R. A. Suthers. 1972. The physiology and energetics of bat flight. Journal of Experimental Biology 57:317–335.

Vaughan, T. A. 1970. Flight patterns and aerodynamics. *In* The Biology of Bats, W. A. Wimsatt, ed., pp. 195–216. Academic Press, New York.

Winter, Y., O. von Helverson, U. M. Norberg, T. H. Kunz, and J. F. Steffensen. 1993. Flight cost and economy of nectar feeding in the bat *Glossophaga soricina* (Phyllostomidae: Glossophaginae). *In* Animal–Plant Interactions in Tropical Environments, W. Barthlott, ed., pp. 167–174. Museum Alexander Koenig, Bonn.

8
Chiropteran Muscle Biology
A Perspective from Molecules to Function

JOHN W. HERMANSON

Powered flight in bats has allowed the exploitation of a diversity of aerial feeding niches among the approximately 920 species of living bats. Functional morphologists have long focused attention on the forelimb and associated musculoskeletal features that underlie the powered downstroke. Historically, these studies focused on gross anatomy (Macalister 1872; Miller 1907) and were part of a more widespread attempt to document natural history or to assess taxonomic affiliations. Vaughan (1959) set a standard with his morphological study of three species of bats representing the families Molossidae, Phyllostomidae, and Vespertilionidae. In that study, he standardized the nomenclature of the chiropteran musculoskeletal system and made a series of astute observations on the relation between form and function in *Myotis velifer, Macrotus californicus,* and *Eumops perotis.* This study represented a starting point for work published a decade later, including Vaughan's efforts to describe the specific adaptations of molossids for high-speed flight (Vaughan 1966) and muscular specializations of mormoopids for maneuverability and fatigue-resistant flight (Vaughan and Bateman 1970).

This tradition was continued with a series of elegant anatomical studies of the interrelations between wing structures and aerodynamics on *Plecotus auritus* (Norberg 1969, 1970), *Rousettus aegyptiacus* (Norberg 1972a), and bats in general (Norberg 1972b). These studies foreshadowed her subsequent contributions on ecomorphology (Norberg and Rayner 1987) and flight energetics (Norberg 1976; Norberg et al. 1993). Studies by Altenbach (1979) on the morphology of the flight apparatus of *Desmodus rotundus* included observations on the temporal activity of the flight muscles, as revealed by electromyography, that demonstrated the remarkable terrestrial capabilities of the common vampire bat.

Strickler's (1978) monograph on the myology of the Chiroptera drew upon the diversity of bats and examined myological characters from 14 of the 18 living families. This broad-based comparative study added important new knowledge concerning the design and potential role of specific muscles in the Chiroptera. Each of the aforementioned studies contributed to our understanding of the role of the forelimb in powered flight, which is the focus in this chapter. My objective is to describe developments in the study of the chiropteran forelimb, which span from the gross level as undertaken in these germinal studies to the molecular level where we are only beginning to apply advanced biochemical or molecular approaches toward understanding the development and function of the flight-enabling musculature.

Muscular Activity during Flight

The muscular motors behind the powered downstroke of bats during flight are the pectoralis, the serratus ventralis (or "serratus anterior" of human anatomical terminology), and the subscapularis. A hypothesis of the three-part division of labor among these muscles was clearly articulated by Vaughan's early study (1959) and has been endorsed by all subsequent researchers with only minor revision. With Scott Altenbach we described the temporal activity patterns of these flight muscles in *Antrozous pallidus* (Vespertilionidae) (Hermanson and Altenbach 1981, 1983) and *Artibeus jamaicensis* (Phyllostomidae) (Hermanson and Altenbach 1985) and found several points of interest. We discovered that the pectoralis muscle exhibited electrical activity well before the onset of downstroke movements. This activity was recorded approximately 20 msec before the transition between the upstroke and downstroke, suggesting either that a significant "warm-up" time was necessary to allow the muscle to reach peak force production before the action or that the muscle experienced what physiologists refer to as an eccentric contraction.

The former interpretation, that the muscle needed time to generate force before such a movement was necessary, has been well entrenched in the literature of motor control since the studies of Engberg and Lundberg (1969) on quadrupedal locomotion in cats. However, the limb muscles of cats have a distinctly different challenge. They are not faced with repetitive activity every 100 msec nor with the enormous power requirements associated with flight in bats. We know from one brief study of *Tadarida brasiliensis* that in vivo twitch time (the time it takes from electrical stimulation to maximal twitch force production) in the pectoralis is approximately 17 msec, a remarkably rapid rate compared to muscles studied in larger animals such as cats in which 25 msec seems to be the fastest twitch contraction time (Sypert and Munson 1981; Altenbach and Hermanson 1987). By contrast, hummingbirds have a twitch time of about 14 msec (Hagiwara et al. 1968).

In an eccentric contraction, electrical activity observed in a muscle undergoing passive stretch leads to an enhancement of force produced by that muscle (Cavagna et al. 1968, 1985). This hypothesis is reasonable given that the pectoralis muscle must not only initiate adduction of the forelimbs but must also arrest the upward, abductory motion of the forelimbs that is passively imposed by residual lift generation of the bat's wings. A combination of a lag time with the idea of eccentric contraction is the most appealing hypothesis.

A similar question arises with respect to the offset (termination) of electrical activity in the pectoralis and other downstroke muscles. As observed in *Artibeus jamaicensis, Antrozous pallidus,* and *Eptesicus fuscus* (Altenbach and Hermanson 1987), the downstroke muscles turn off approximately halfway through the duration of the downstroke. The mechanical activity of the muscles apparently continues for several milliseconds after termination of the electromyographic (EMG) signal. This finding also suggests that the active phase of downstroke power is focused within the first half of the downstroke, and perhaps that wing momentum carries the wing downward through the remainder of the downstroke, which facilitates the smooth transition to an upstroke. In one of the few studies that correlated wingbeat kinematics with muscle contractile properties, Goslow and Dial (1990) reported that twitch contraction time, including rise time and relaxation time, actually exceeds the duration of the wingbeat in European starlings flying at a wingbeat frequency of 13.5 Hz.

The observations of twitch times and EMG periods pointed to the need for specific data about the time course of force production in chiropteran flight muscles. How these aerial athletes repetitively activate and relax their muscles at remarkably high frequencies and how they do so without rapidly fatiguing these muscles invites further study. Notwithstanding, downstroke muscles of bats are spectacular with respect to their speed of shortening and their ability to withstand repeated cycles of activity over the course of a long flight. Such activity would surely deplete the energy stores in the muscles of any human athlete in short order.

Is the upstroke a passive activity "powered" by lift generated by the airfoil? This question was answered in EMG studies of slow-flying bats by evidence of strong muscle activity in the upstroke muscles such as the trapezius, deltoideus, supraspinatus, and infraspinatus muscles (Hermanson and Altenbach 1983, 1985). These muscles apparently contribute to an active and controlled upstroke. Rayner (1987) utilized these data in combination with wingroot moments (the effect of lift and inertia at the shoulder joint) to postulate that the upstroke was not only active but that upstroke activity of abductory muscles must commence about halfway through the downstroke, precisely when the main adductors turn off in slow flight.

Most interesting were Rayner's predictions that at high speed, as suggested by estimates of wingroot moments, pectoralis activity would be continuous throughout the upstroke and downstroke. This temporal activity pattern is unknown in the locomotory muscles of other organisms, but it is compelling to suggest that lift and thrust could be produced continuously. These suggestions point to a need for EMG studies of bats during fast forward flight; among the EMG studies mentioned previously, all were conducted while bats were flying at a speed about 3 m/sec or slower. Although the study of low-speed flight highlights a technical limitation to observing bats in the laboratory, these observations suggest that the pectoralis muscles of bats may be designed in a fundamentally different manner from the typical appendicular muscles in other mammals.

Muscle Histochemistry Predicts Muscle Function

Although electromyography has yielded insights into temporal relations between muscle use and activity of muscles in some species, histochemistry extends this knowledge of flight musculature to the tissue and cellular level. Given the remarkable diversity of flight styles and food resources utilized by bats, it is interesting that the flight muscles are relatively homogeneous in fiber composition (Armstrong et al. 1977; Foehring and Hermanson 1984; Hermanson et al. 1991a). It should be noted that the pectoralis muscles are active for a very short burst during the late upstroke and early downstroke, a burst of approximately 40–50 msec in a bat flying slowly (3 m/sec with wingbeat frequency of about 10 Hz in bats weighing 10–45 g). During this active period the muscle must attain peak force generation, which follows a brief rise time to peak tension (17 msec in *Tadarida brasiliensis*). Thus, flight musculature must also be designed to rapidly replenish its oxygen content.

Many histochemical studies have demonstrated that insectivorous vespertilionid (Armstrong et al. 1977; Brigham et al. 1990; Hermanson et al. 1991a) and molossid bats (Foehring and Hermanson 1984) have pectoralis muscles composed of a single fiber type referred to as type IIa or type FO (fast oxidative). These nonfatigable muscle fibers are richly endowed with capillaries and mitochondria, have surface-to-volume values that enhance diffusion of oxygen and metabolic by-products into and out of the muscle fiber (Mathieu-Costello et al. 1992), and have a high percentage of the fast-twitch myosin associated with rapid muscle shortening. Mathieu-Costello et al. (1992) noted that the pectoralis of Eptesicus fuscus had an exceptionally large capillary-to-fiber surface interface, a fact that was likely more important in enhancing O_2 transfer than mere reduction of fiber size.

I have used the classification of these fibers as type IIa (after the classification of Brooke and Kaiser 1970) to be consistent with my earlier publications. However, to fully understand the literature, one must understand the several classification schemes used by histochemists and muscle physiologists (for review, see Pette and Staron 1990). Other studies conducted by Alexander and George (1982) and by Ohtsu and Uchida (1979) used different classifications but nonetheless demonstrated similar homogeneity in the pectoralis muscle of *Myotis lucifugus, Myotis macrodactylus,* and *Pipistrellus abramus.*

A histochemically homogeneous pectoralis muscle stands in contrast to the locomotory muscles in most mammals (Figure 8.1). Generally, muscles include varying proportions of type I (slow, fatigue-resistant), type IIa (fast, fatigue-resistant), and type IIb (fast, fatigable), with the proportions dependent on the specific role of the muscle. In most mammals, type I fibers are richly represented in postural muscles that are continuously active or which undergo slow, precise contractions. Type IIa fibers are most common in many muscles that undergo rapid, powerful contractions and which are not subject to fatigue. Type IIb fibers are generally associated with the most forceful, rapid contractions but are quickly susceptible to fatigue. Dogs (*Canis familiaris*), for example, lack type IIb fibers in their locomotory muscles (Snow et al. 1982) and exhibit large proportions of type IIa or special fast oxidative fibers. This may relate to the habit of domestic or wild dogs of chasing prey for long periods of time, unlike the distantly related cats, which tend to employ sprints for quick capture and killing of their prey (Armstrong et al. 1982).

Imagine the coordination of movements, from simple grooming behavior to rapid and forceful movement of the limbs during a run (or flight). Prevailing opinion is that this coordination is achieved through the graded or orderly recruitment of motoneurons and their associated muscle fibers (or muscle units). Slow, precise movements are powered by type I (generally small-diameter) and, to a lesser extent, by type IIa (small- to intermediate-diameter) fibers. High-speed movements necessitate the recruitment of larger, more powerful type IIa and type IIb muscle fibers. We have learned that the pectoralis in adult *M. lucifugus* and *T. brasiliensis* lacks type I and IIb fibers (Figure 8.2). This appears to be a derived condition because other mammals, except for some shrews (Suzuki 1990; Savolainen and Vornanen 1995), exhibit all three fiber types in locomotory muscles. Thus, the pectoralis of these bats contains a remarkably homogeneous substrate upon which to build recruitment strategies. The other downstroke muscles of these bats, including the serratus ventralis, subscapularis, and the short head of biceps brachii, contain nearly 100% type IIa fibers. Thus, the downstroke muscles exhibit monotonous homogeneity of muscle fiber design in the flight motor.

Goldspink (1977) argued that birds could not afford the

Slow, fatigue-resistant fibers

Small diameter
Many mitochondria
"Slow" myosin ATPase

Fast, fatigue-resistant fibers

Small to intermediate diameter
Many mitochondria
"Fast" myosin ATPase

Fast, fatigable fibers

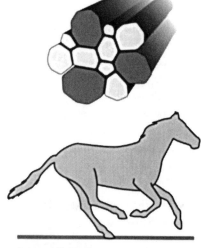

Large diameter
Few mitochondria
"Fast" myosin ATPase

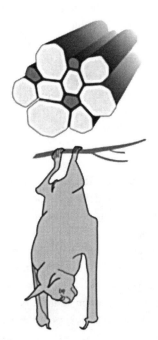

Small diameter
Many mitochondria
"Slow" myosin ATPase

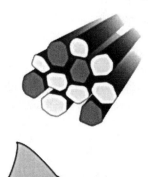

Intermediate diameter
Many mitochondria
"Fast" myosin ATPase

Intermediate diameter
Fewer mitochondria
"Fast" myosin ATPase

Figure 8.1. Schematic of the recruitment of type I, IIa, and IIb muscle fibers in terrestrial and aerial locomotion. Each fiber type is illustrated as the darkened fibers of a generalized appendicular muscle in transverse section. Type I fibers (left panels) might be most prominent in equine muscles involved in quiet standing; these fibers contract slowly and with low force levels. Type IIa fibers (middle panels) are commonplace in most appendicular muscles and are involved in most activities. Type IIb fibers (right panels), which are often but not always the largest fibers, contract rapidly and are fatigable. These muscle fibers may be most useful in short-duration, forceful activities. In roosting bats, postural slow-twitch muscle fibers may effect quiet hanging in conjunction with passive digital locking mechanisms (see Schutt, Chapter 10, this volume). The muscle illustrated for postural fibers in bats might be any of the hindlimb muscles. The bat muscle illustrated in the middle and right panels is a pectoralis muscle, which has almost no postural fibers. Type IIa fibers are critical to the pectoralis of vespertilionid bats, which is "unitypic" (consists of a single type of fibers). In contrast, the additional fast fibers (type IIb or IIe) observed in the pectoralis of several phyllostomids may fulfill the high power requirements these bats have when carrying loads or hovering.

"luxury" of multiple fiber types in the major downstroke muscles because this would require additional weight concentrated in the muscles. Our evidence from bats supports this view. However, bats are not devoid of type I fibers. Muscles associated with posture, such as hindlimb muscles (knee flexors) used in hanging, contain many type I fibers (Hermanson et al. 1991a) and express type I myosin isoforms (Schutt et al. 1994). In a sense, the decoupling of many forelimb muscles from postural (roosting) behaviors has allowed the specialization of form and function seen in the pectoralis of many bat species.

Other exceptions to the homogeneous histochemical design of downstroke muscles have been noted in both birds and bats (Figure 8.3). Several chiropteran species exhibit two fiber types in the downstroke muscles. George and Naik (1957) reported white (glycogen-loaded) and red (fat-loaded) pectoralis fibers in *Hipposideros speoris* (Hipposideridae) but did not assess the myosin adenosine triphosphatase (ATPase) system of these muscles. Strickler (1980) demonstrated two fiber types in *Pteronotus parnellii* (Mormoopidae) and *Phyllostomus hastatus* (Phyllostomidae), a majority of which he classified as SO (slow, oxidative) on the basis of their myosin ATPase activity following acidic incubation. A subsequent study of the flight muscles of *Artibeus jamaicensis* yielded similar images of myosin ATPase, but Hermanson and Foehring (1988) concluded that both types of fibers present were presumed fast-twitch based on recruitment strategies and could be classified as types IIa and IIb.

There are challenges to applying the terminology of histochemistry, which was developed in experiments on a limited number of cat limb muscles, to bats. However, the histochemistry can be nonequivocal if used judiciously and if researchers are clear in documenting their techniques so others can conduct similar experiments. Whether one assumes that a population of fibers is fast or slow, one should recognize that the fundamental difference between the phyllostomids and mormoopids on the one hand and the vespertilionids on the other is that the latter have homogeneous downstroke muscles whereas the former have two fiber types. One versus two: This is not the sort of striking difference that we might expect when searching for characters that have phylogenetic or functional significance. In this case, these differences may relate more to the relationships among the Phyllostomoidea and Vespertilionoidea and not primarily to function. One potentially conflicting observation can be seen in *Miniopterus schreibersi,* for which Ohtsu and Uchida (1979) described the presence of two fiber types. This problem needs to be clarified with myosin isoform data to fully clarify our phylogenetic perspectives on these distributions of fiber types.

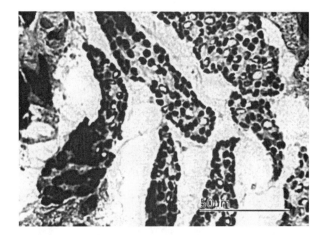
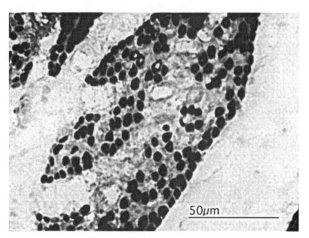
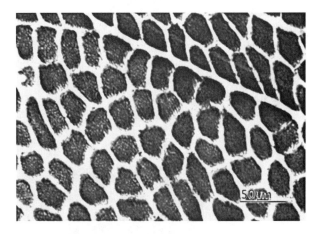

Figure 8.2. Transverse sections of the midbelly of the pectoralis muscle of fetal and adult *Myotis lucifugus* stained for myosin ATPase after incubation at pH 10.3. Dark-stained fibers are type II and presumed to be fast-twitch. **Top and middle:** Two sections from a fetus, illustrating the relative disorganization and the high concentrations of connective tissue. Fibers with central vacuoles are thought to represent an early phase of myogenesis, which was still ongoing in this specimen. **Bottom:** Section from an adult, showing fibers that have relatively invariant diameters and are uniformly type IIa. Compare this to the "black-and-white" staining pattern in the pectoralis muscles of phyllostomids.

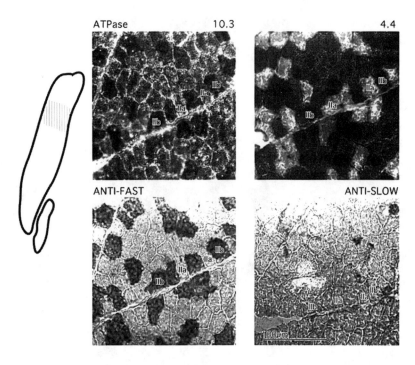

Figure 8.3. Transverse sections of the midbelly of the pectoralis muscle from adult *Artibeus lituratus* after incubation at pH 10.3 and pH 4.4 and then staining for myosin ATPase, and after reacting against anti-fast myosin antibodies and anti-slow myosin antibodies. After the alkaline incubation, type IIa, IIb, and IIe fibers stain dark. After the acidic incubation, type IIb and IIe stain dark. The fibers that appear dark in the anti-fast myosin antibody sample are fast. Type IIe fibers, which can only be discriminated with antibody reactions, are lacking in this section. This is a "bitypic" pectoralis, meaning it has two types of fibers to recruit. Hatching in schematic transverse section indicates region of the muscle from which the sample was taken.

More recently, we had the opportunity to analyze the flight muscles of the common vampire bat *Desmodus rotundus* (Hermanson et al. 1993). Building upon Altenbach's (1979) observations of the behavior and locomotion of this bat, we have found a seemingly astonishing four fiber types in the pectoralis (Figures 8.4 and 8.5). Type I fibers were distributed in specific, deep regions of the pectoralis, type IIa and IIb fibers were heterogeneously distributed throughout the entire pectoralis, and an additional fiber profile appeared throughout the superficial and middle portions of the muscle that we later called "IIe." This fiber profile did not fit any of the standard descriptions of type II fibers and, in fact, failed to react against our standard anti-fast myosin antibody.

We concluded that this might be an additional, unique fiber type associated with the pectoralis of *D. rotundus*.

Is this histochemistry to be trusted? The following section addresses this point at the molecular level. The study of *D. rotundus* was the first to use monoclonal and polyclonal antibodies as part of a fiber type paradigm for bats, and the IIe fibers would have gone unnoticed without the antibodies. Although our more recent work (Hermanson et al., unpublished data) is not yet complete, we see similarities in the presence of a IIe fiber in *Artibeus lituratus* although a standard histochemical approach would have generated a conclusion similar to that reached by Foehring and Hermanson (1984).

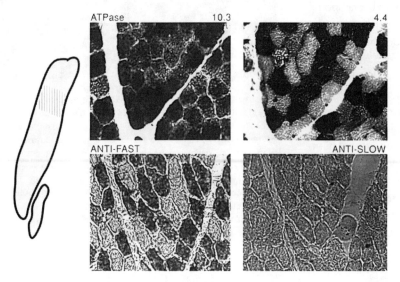

Figure 8.4. Transverse sections of superficial samples of midbelly pectoralis from *Desmodus rotundus* after staining for myosin ATPase following incubation at pH 10.3 (dark fibers are type II) and pH 4.4 (dark fibers are type IIb or IIe), and after reaction against anti-fast myosin antibodies (dark fibers are type II) and anti-slow myosin antibodies (no fibers are reactive). The IIe fibers failed to react to either anti-fast or anti-slow myosin antibodies. Hatching in schematic transverse section indicates region of the muscle from which the sample was taken. (After Hermanson et al. [1993, *Journal of Morphology*], with permission from John Wiley and Sons.)

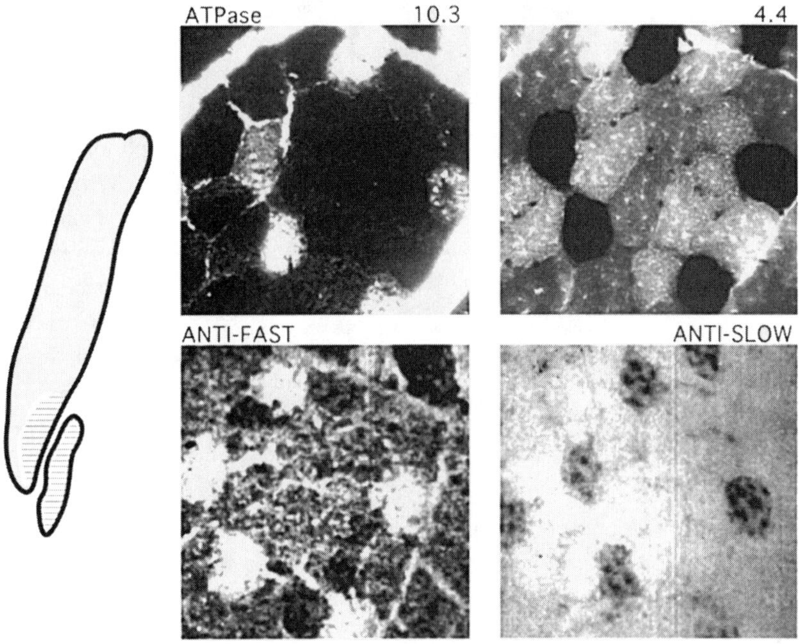

Figure 8.5. Four fiber types in *Desmodus rotundus*, demonstrated by combined histochemical and immunocytochemical analysis. These transverse sections of midbelly pectoralis from *D. rotundus*, sampled adjacent to the pectoralis abdominalis (hatched area in schematic), are shown after staining for myosin ATPase following incubation at pH 10.3 (dark fibers are type II) and pH 4.4 (dark fibers are type IIb or IIe), and after reaction against anti-fast myosin antibodies (dark fibers are type II) and anti-slow myosin antibodies (dark fibers are type I). (After Hermanson et al. [1993, *Journal of Morphology*], with permission from John Wiley and Sons.)

Correlation of the Myosin Isoforms with Structure and Function

It is clear that histochemical analysis is fraught with difficult interpretations and conflicting classification schemes. Although care has been employed in these procedures, correlation with antibody reactions and with the underlying myosin variants yields a more robust approach to the study of muscle design in bats. Myosin is the major contractile protein in muscles and it is one of the largest, although its size makes the protein difficult to isolate and to manipulate. At 300 kilodaltons (kDa), the molecule can be studied relatively intact with nondenaturing electrophoresis. Or, with denaturing and separation on sodium dodecyl–polyacrylamide gel electrophoresis (SDS-PAGE) gels, the large myosin heavy chains (MHC) can be isolated. The MHC sepa-

rates on the basis of molecular weight but provides a good correlate against speed of shortening (Reiser et al. 1985). As such, these MHC isoforms correlate with histochemical fiber types (Table 8.1). The nondenaturing gels elucidate differences among myosin isoforms on the basis of molecular weight as well as charge properties. The latter are dictated by the two pairs of light chains associated with each myosin heavy chain.

In general, our histochemical and electrophoretic studies show good agreement on the same tissues. This has been true for our studies on both horse (Hermanson et al. 1991b; Cobb et al. 1994) and bat muscle (Hermanson et al. 1991a). In the latter study, we demonstrated an exact one-to-one match between the histochemically unitypic pectoralis of *Myotis lucifugus* and single MHC isoforms and single native myosin isoforms (Figure 8.6). When the histochemistry has

Table 8.1

Bat Muscle Properties Associated with Histochemical and Electrophoretic Analyses

Fiber type	Speed of shortening	Fatigue-resistant	pH 10.3 incubation, myosin ATPase stain	Myosin heavy chain isoforms[a]	Native myosin isoforms[a]
I	Slow	Yes	Light	I	SM
IIa	Intermediate to fast	Yes	Dark	IIa	FM4, FM3, FM2
IIb	Fast	No	Dark	IIb and/or IIx	FM4, FM3
IIx	Intermediate	Yes[b]	Dark[b]	Unknown	Unknown
IIe	Fast	Yes	Dark	IIe	FM5, FM4, FM3

[a]Based on electrophoretic mobilities. Estimates are not given for type IIx fiber because it has not been identified in bats.

[b]Predictions of appearance based on rat costal diaphragm.

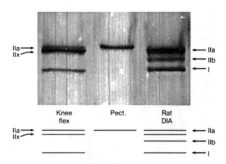

Figure 8.6. Myosin heavy-chain (MHC) isoforms of *Myotis lucifugus* isolated with 6% SDS-PAGE. **Top:** Gel showing the vertical distribution of MHC isoforms and a comparison with identified MHC isoforms from rat diaphragm. **Bottom:** Schematic of the isoform pattern. (Modified from Hermanson et al. [1991a, *Journal of Experimental Zoology*], with permission from John Wiley and Sons.)

been difficult to interpret, we have observed unique electrophoretic patterns in the corresponding electrophoretic procedure. The study of pectoralis muscle in *D. rotundus* is one such example (Hermanson et al. 1993): We noted a "black-and-white" pattern based on myosin ATPase that correlated with antibody data but was interspersed with a nonrandom pattern of fibers which were nonreactive against either fast or slow myosin antibodies. This latter population accounted for as much as 70% of the total fiber population in select regions of the pectoralis in *D. rotundus*.

However, MHC isoform analysis initially suggested the presence of small amounts of type IIb MHC isoform and large amounts of type I and IIa MHC isoforms (concentration can be estimated by band density on the electrophoretic gels).

Why is there so much type I isoform in a presumed fast-twitch muscle? Little or no type I native myosin isoform was detected on the nondenaturing gels. Indeed, our histochemistry predicted a nonrandom distribution of type I fibers with most being located in pectoralis abdominalis (a distinct "head" of the bat pectoralis) or the adjacent deep pectoralis. When we reanalyzed our tissue samples, extracted very small samples of myosin from this type I–rich region, and carefully applied minute concentrations of the extracts, we found a clear double band on MHC gels that migrated similarly to rat type I MHC isoform but which was distinct from the type I MHC isoform (Figure 8.7). This result showed that *D. rotundus* has a unique MHC associated with the otherwise "black-and-white" histochemical story. We called these fibers and this isoform type IIe, because the published terms for fast myosin already included the string IIa, IIb, IIc, IId, and IIx. Although a MHC isoform with a similar molecular weight has been observed in several phyllostomids including representative species of *Artibeus*, *Carollia*, and *Diaemus* (Hermanson et al., unpublished data),

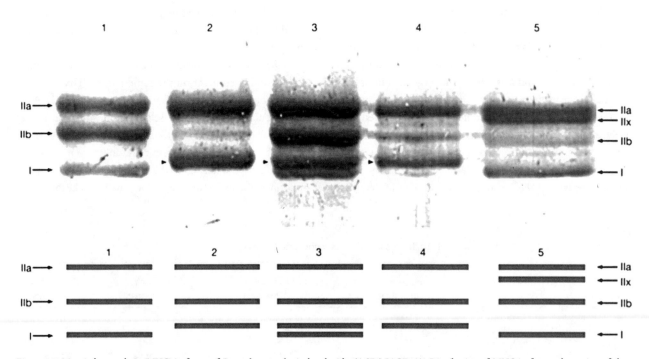

Figure 8.7. Myosin heavy-chain (MHC) isoforms of *Desmodus rotundus* isolated with 6% SDS-PAGE. (A) Distribution of MHC isoforms, the region of the muscle sampled, and comparison with identified MHC isoforms from rat diaphragm. (B) Schematic of the isoform pattern seen in a majority of gels. The sampled muscles are pectoralis abdominalis (lane 1), pectoralis cranial midbelly regions (lane 2; similar to the sample in Figure 8.4), pectoralis caudal midbelly regions (lane 3; comparable to the region depicted in Figure 8.5), and pectoralis cranial midbelly regions of another individual *D. rotundus*. For comparison, samples of rat costal diaphragm (lane 5) are shown on the same gel. (Modified from Hermanson et al. [1993], with permission from John Wiley and Sons.)

the distribution within the Chiroptera and the function of the isoform await further study.

Histochemistry Meets Ecomorphology

Where do these data fit into general interests about bat biology? I revisit Rayner's comment (1987, p. 27) that "There is considerable adaptive variation in the flight morphology and flight patterns of bats [he cites Norberg 1981, 1987], but the order as a whole shows remarkable homogeneity." I, too, am struck by how little intrinsic biological variability appears in flight muscles of bats as opposed to the great breadth of flight styles and niche selections exhibited by the Chiroptera.

For comparison, I overlaid some of our data on Norberg's ubiquitous plots of principal component analysis (PCA) comparing wing loading against aspect ratio (Norberg and Rayner 1987). Although these do not represent a formal statistical approach to this question, I predicted that the unitypic- and bitypic-pectoralis phenotypes of bats would correspond to their ecological and aerodynamic orientations (Figure 8.8). For example, I predicted that the

pectoralis of all insectivorous bats should be unitypic, as is the pectoralis of *M. lucifugus*. These bats tend to cluster in the lower-left quadrant of the PCA plots of Norberg and Rayner (1987), in part in response to the aerodynamic constraints of utilizing slower flight and enjoying high maneuverability. In addition, we have evidence from histochemistry and myosin isoform analysis suggesting that *Molossus ater, Antrozous pallidus,* and *Eptesicus fuscus* (Hermanson, unpublished data) fit the unitypic-pectoralis model.

Similarly, I predicted that the frugivorous bats would be distributed specifically in the lower-right quadrant of the PCA plots from Norberg and Rayner (1987) (cross-hatched region in Figure 8.8), a position suggested by aerodynamic constraints related to higher flight speeds for commuting flights and other factors (Norberg and Rayner 1987, p. 393). The predictions are consistent with the plot, but the observations become clouded by introducing *Pteronotus parnellii* into the sample. Strickler's data (1980) clearly showed that *P. parnellii* has a bitypic pectoralis muscle, as discussed earlier. Thus, although this bat is a good insect-eater, its flight muscle profile is distinctly like that of frugivores. *Desmodus rotundus* is clustered with frugivorous phyllo-

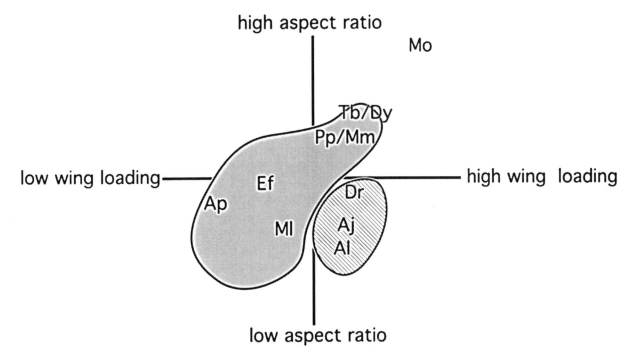

Figure 8.8. A principal component analysis (PCA) of wing shape data by Norberg and Rayner (1987), showing morphotype distribution of several vespertilionid bats (stippling), which have unitypic pectoralis muscles, frugivorous phyllostomids (hatching), which have bitypic pectoralis muscles, and two molossid bats with unitypic pectoralis muscles. Note that the plots of *Diaemus youngi* (Dy) and *Pteronotus parnellii* (Pp), both bitypic bats, overlap with the plots of unitypic bats and suggest a strong phylogenetic component to the fiber type of the pectoralis. Preliminary electrophoretic data for *Mormoops megalophylla* (Mm) corroborate the observation on *P. parnellii* (Pp) and places a bitypic bat in the midst of the PCA morphospace otherwise occupied by unitypic bats. Norberg (1990) published a comparable figure based on metabolic properties and fiber types of bat pectoralis muscles. Other species plotted: Aj, *Artibeus jamaicensis;* Al, *Artibeus lituratus;* Ap, *Antrozous pallidus;* Ef, *Eptesicus fuscus;* Ml, *Myotis lucifugus;* Mo, *Molossus ater;* Tb, *Tadarida brasiliensis;* Vh, *Vampyrops helleri.* (After Norberg and Rayner [1987, *Philosophical Transactions of the Royal Society*], with permission from the Royal Society of London.)

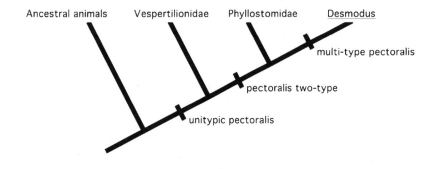

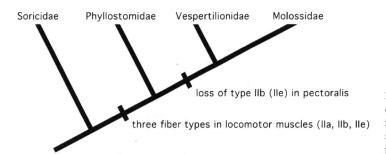

Figure 8.9. Two phylogenetic hypotheses of the evolution of the pectoralis muscle morphotypes in bats. These diagrams are simplistic but illustrate the correlation of available data with possible phylogenetic relationships.

stomids on the PCA plots of Norberg and Rayner (1987) and yet has a unique feeding niche and distinct flight requirements. *Diaemus youngi,* having a bitypic pectoralis muscle, falls quite close to *P. parnelli* in the PCA plots of Norberg and Rayner (1987) and is an outlier with respect to the PCA plots of other members of the Phyllostomidae that we studied.

Would this observation lend credibility to the hypothesis that *Diaemus* is more or less derived than *Desmodus* (see Chapter 10 by Schutt, this volume)? In all, the distinction between an insectivorous "morph" and a frugivorous or sanguinivorous morph tends to be unclear. As an alternative hypothesis, I suggest that the flight muscle phenotypes might be dictated more closely by phylogenetic history than by function. If this is true, perhaps the ancestral Phyllostomoidea expressed several myosin isoforms in the pectoralis muscles. This trend seems to be continued in all species we have studied, although we have examined only a small fraction of the living taxa of Phyllostomidae or Mormoopidae (Figure 8.9).

Several lessons can be drawn from parallel studies on avian flight muscles. Two main "pectoralis morphs" are represented by starlings *(Sturnus vulgaris)* and pigeons *(Columba livia).* The starlings have unitypic pectoralis muscles (Rosser and George 1986) that appear quite similar to those of *T. brasiliensis* and *Myotis lucifugus.* Pigeons have a complex distribution of small (type FO or IIa) and large (type FG or IIb) fibers (Kaplan and Goslow 1989). The small fibers are distributed throughout the muscle while the large

fibers are primarily found on the circumference of muscle fascicles (bundles of muscle that are obvious when viewing muscle transverse sections microscopically).

The myosin isoforms of diverse species have not been well studied; however, the data summarized here point to a dichotomy between "one-gear" and "two-gear" birds. Dial et al. (1988) used this dichotomy and their observations of electrical activity during differing modes of flight to postulate that the large type IIb fibers are primarily recruited during periods of high-intensity muscle activity. This idea was based on the observation of the presence of large-amplitude spikes of EMG activity during takeoff or landing that were not present during level forward flapping flight. Their conclusion was that the smaller type IIa fibers were sufficient to power normal flight movements.

Bats such as *Artibeus jamaicensis, A. lituratus,* and *Carollia perspicillata* are all known to hover briefly while removing fruits from a tree before transporting the fruit to a nearby "feeding" tree (see Norberg and Rayner 1987). Similarly, *D. rotundus* has obvious mechanical demands when trying to fly with a 10- to 20-g blood meal in its digestive system. Some of this problem is addressed by a unique tubular stomach and a highly specialized renal system to remove water rapidly (Wimsatt and Guerriere 1962; McFarland and Wimsatt 1969). However, bats are frequently observed in flight shortly after the feeding bout and must be able to lift significant amounts of weight greater than their normal body mass. I suggest that the bitypic or polytypic pectoralis facilitates this sort of stressful flight and that some of the fiber populations will prove to be specialized for short-duration powerful flight.

Which population might be associated with this need is not yet clear, although it is likely the IIe or IIb fibers (see Figure 8.1). Unlike the pectoralis of the pigeon in which a significant size difference was discerned between the IIa and IIb fibers, the size differences are more muted in the bats that we have studied. This size difference is the root of the biophysical principles involved in muscle recruitment that allowed Dial et al. (1988) to postulate a division of labor between the IIa and IIb fiber populations. Not only must further analysis of the phyllostomids yield a larger sample size on which to conduct a rigorous analysis, but I would expect to see how the different fiber type morphs are distributed across subfamily lines as well as across classifications of flight styles and food habits. I predict that the small population of IIb fibers (fast and glycolytic) may be involved in bursts of activity associated with terrestrial jumps unique to *Desmodus rotundus*.

Conclusions

The design of flight muscle in birds and bats provides some of the more interesting lessons about the evolution of flight in these two taxa. Both groups represent a range of flight styles and wing morphology, yet the design of the primary muscular "motor" is remarkably monotonous. Apparently, the minute modifications provided by distinct myosin isoforms or variations in the proportions of distinct histochemical fiber types (when more than one type is present) are adequate to power and coordinate the wingbeat cycle.

A dichotomy between "one-gear" and "two-gear" muscles is established in bats that places the former morph primarily in Vespertilionidae and the latter in Phyllostomidae and Mormoopidae. However, one cannot take this as a major principle until more comparative studies are completed. Such studies are needed to illustrate the distribution of the unitypic or polytypic pectoralis muscle morphs in various taxa. In addition, comparable data for the pectoralis muscles of appropriate mammalian outgroups have not been analyzed. The association of one-gear and two-gear morphs is not well correlated with feeding types or with size.

Acknowledgments

Numerous colleagues participated in or contributed to the studies discussed herein. I thank W. Schutt, W. LaFramboise, M. Daood, M. Cobb, J. Petrie, M. Lillard, and J. Ryan. Finally, I acknowledge the patience of M. Simmons, who has illustrated so many of our studies during the past eight years. Equipment and partial support accrued from funding available through the H. M. Zweig Memorial Fund. I offer a special thanks to the late D. Klingener for his encouragement and good counsel.

Literature Cited

Alexander, K. M., and J. C. George. 1982. Fibre type composition of the breast muscle of the North American little brown bat and of the Indian giant fruit bat. Journal of Animal Morphology and Physiology 29:98–102.

Altenbach, J. S. 1979. Locomotor morphology of the vampire bat, *Desmodus rotundus*. American Society of Mammalogists, Special Publication 6:1–137.

Altenbach, J. S., and J. W. Hermanson. 1987. Bat flight muscle function and the scapulo-humeral lock. *In* Recent Advances in the Study of Bats, M. B. Fenton, P. Racey, and J. M. V. Rayner, eds., pp. 100–118. Cambridge University Press, Cambridge.

Armstrong, R. B., C. D. Ianuzzo, and T. H. Kunz. 1977. Histochemical and biochemical properties of flight muscle fibers in the little brown bat, *Myotis lucifugus*. Journal of Comparative Physiology B 119:141–154.

Armstrong, R. B., C. W. Saubert IV, H. J. Seeherman, and C. R. Taylor. 1982. Distribution of fiber types in locomotory muscles of dogs. American Journal of Anatomy 163:87–98.

Brigham, R. M., C. D. Ianuzzo, N. Hamilton, and M. B. Fenton. 1990. Histochemical and biochemical plasticity of muscle fibers in the little brown bat *(Myotis lucifugus)*. Journal of Comparative Physiology B 160:183–186.

Brooke, M. H., and K. K. Kaiser. 1970. Three "myosin adenosine triphosphatase" systems: The nature of their pH lability and sulfhydryl dependence. Journal of Histochemistry and Cytochemistry 18:670–672.

Cavagna, G. A., B. Dusman, and R. Margaria. 1968. Positive work done by a previously stretched muscle. Journal of Applied Physiology 24:21–32.

Cavagna, G. A., M. Mazzanti, N. C. Heglund, and G. Citterio. 1985. Storage and release of mechanical energy by active muscle: A non-elastic mechanism? Journal of Experimental Biology 115:79–87.

Cobb, M. A., W. A. Schutt, Jr., and J. W. Hermanson. 1994. Morphological, histochemical, and myosin isoform analysis of the diaphragm of adult horses, *Equus caballus*. Anatomical Record 238:317–325.

Dial, K. P., S. R. Kaplan, G. E. Goslow, Jr., and F. A. Jenkins, Jr. 1988. A functional analysis of the primary upstroke and downstroke muscles in the domestic pigeon *(Columba livia)* during flight. Journal of Experimental Biology 134:1–16.

Engberg, I., and A. Lundberg. 1969. An electromyographic analysis of the muscular activity in the hindlimb of the cat during unrestrained locomotion. Acta Physiologica Scandinavica 75: 614–630.

Foehring, R. C., and J. W. Hermanson. 1984. Morphology and histochemistry of flight muscles in free-tailed bats, *Tadarida brasiliensis*. Journal of Mammalogy 65:388–394.

George, J. C., and R. M. Naik. 1957. Studies on the structure and physiology of the flight muscles of bats. 1. The occurrence of two types of fibres in the pectoralis major muscle of the bat (*Hipposideros speoris*), their relative distribution, nature of the fuel store, and mitochondrial content. Journal of Animal Morphology and Physiology 4:96–101.

Goldspink, G. 1977. Mechanics and energetics of muscle in animals of different sizes, with particular reference to the muscle fiber composition of vertebrate muscle. *In* Scale Effects in Animal Locomotion, T. J. Pedley, ed., pp. 37–55. Academic Press, London.

Goslow, G. E., Jr., and K. P. Dial. 1990. Active stretch-shorten contractions of the m. pectoralis in the European starling *(Sturnus vulgaris):* Evidence from electromyography and contractile properties. Netherlands Journal of Zoology 40:106–114.

Hagiwara, S., S. Chichibu, and N. Simpson. 1968. Neuromuscular mechanisms of wing beat in hummingbirds. Zeitschrift für Vergleichende Physiologie 60:209–218.

Hermanson, J. W., and J. S. Altenbach. 1981. Functional anatomy of the primary downstroke muscles in a bat, *Antrozous pallidus*. Journal of Mammalogy 62:795–800.

Hermanson, J. W., and J. S. Altenbach. 1983. The functional anatomy of the shoulder of the pallid bat, *Antrozous pallidus*. Journal of Mammalogy 64:62–75.

Hermanson, J. W., and J. S. Altenbach. 1985. Functional anatomy of the shoulder and arm of the fruit-eating bat, *Artibeus jamaicensis*. Journal of Zoology (London) 205:157–177.

Hermanson, J. W., and R. C. Foehring. 1988. Histochemistry of flight muscles in the Jamaican fruit bat, *Artibeus jamaicensis:* Implications for motor control. Journal of Morphology 196:353–362.

Hermanson, J. W., M. J. Daood, and W. A. LaFramboise. 1991a. Unique myosin isoforms in the flight muscles of little brown bats, *Myotis lucifugus*. Journal of Experimental Zoology 259:174–180.

Hermanson, J. W., M. T. Hegemann-Monachelli, M. J. Daood, and W. A. LaFramboise. 1991b. Correlation of myosin isoforms with anatomical divisions in equine Musculus biceps brachii. Acta Anatomica 141:369–376.

Hermanson, J. W., M. A. Cobb, W. A. Schutt, Jr., F. Muradali, and J. M. Ryan. 1993. Histochemical and myosin composition of vampire bat *(Desmodus rotundus)* pectoralis muscle targets a unique locomotory niche. Journal of Morphology 217:347–356.

Kaplan, S. R., and G. E. Goslow, Jr. 1989. Neuromuscular organization of the pectoralis (pars thoracicus) of the pigeon *(Columba livia):* Implications for motor control. Anatomical Record 224:426–430.

Macalister, A. 1872. The myology of the Chiroptera. Philosophical Transactions of the Royal Society of London 162:125–173.

Mathieu-Costello, O., J. M. Szewczak, R. B. Logemann, and P. J. Agey. 1992. Geometry of blood-tissue exchange in bat flight muscle compared with bat hindlimb and rat soleus muscle. American Journal of Physiology 262:R955–R965.

McFarland, W. N., and W. A. Wimsatt. 1969. Renal function and its relation to the ecology of the vampire bat, *Desmodus rotundus*. Comparative Biochemistry and Physiology 28:985–1006.

Miller, G. S., Jr. 1907. The families and genera of bats. Bulletin of the U.S. National Museum 57:1–282.

Norberg, U. M. 1969. An arrangement giving a stiff leading edge to the hand wing in bats. Journal of Mammalogy 50:766–770.

Norberg, U. M. 1970. Functional osteology and myology of the wing of *Plecotus auritus* Linnaeus (Chiroptera). Arkiv für Zoologi 22:483–543.

Norberg, U. M. 1972a. Functional osteology and myology of the wing of the dog-faced bat *Rousettus aegyptiacus* (É. Geoffroy) (Mammalia, Chiroptera). Zeitschrift für Morphologie der Tiere 73:1–44.

Norberg, U. M. 1972b. Bat wing structures important for aerodynamics and rigidity (Mammalia, Chiroptera). Zeitschrift für Morphologie der Tiere 73:45–61.

Norberg, U. M. 1976. Aerodynamics, kinematics, and energetics of horizontal flapping flight in the long-eared bat *Plecotus auritus*. Journal of Experimental Biology 65:179–212.

Norberg, U. M. 1981. Allometry of bat wings and legs and comparison with bird wings. Philosophical Transactions of the Royal Society of London B 292:359–398.

Norberg, U. M. 1987. Wing form and flight mode in bats. *In* Recent Advances in the Study of Bats, M. B. Fenton, P. Racey, and J. M. V. Rayner, eds., pp. 43–56. Cambridge University Press, Cambridge.

Norberg, U. M., and J. M. V. Rayner. 1987. Ecological morphology and flight in bats (Mammalia; Chiroptera): Wing adaptations, flight performance, foraging strategy, and echolocation. Philosophical Transactions of the Royal Society of London B 316:335–427.

Norberg, U. M. 1990. Vertebrate Flight. Springer, Berlin.

Norberg, U. M., T. H. Kunz, J. F. Steffenseon, Y. Winter, and O. Von Helversen. 1993. The cost of hovering and forward flight in a nectar-feeding bat, *Glossophaga soricina*, estimated from aerodynamic theory. Journal of Experimental Biology 182:207–227.

Ohtsu, R., and T. A. Uchida. 1979. Further studies on histochemical and ultrastructural properties of the pectoral muscles of bats. Journal of the Faculty of Agriculture, Kyushu University 24:145–155.

Pette, D., and R. S. Staron. 1990. Cellular and molecular diversities of mammalian skeletal muscle fibers. Reviews of Physiology, Biochemistry, and Pharmacology 116:1–76.

Rayner, J. M. V. 1987. The mechanics of flapping flight in bats. *In* Recent Advances in the Study of Bats, M. B. Fenton, P. Racey, and J. M. V. Rayner, eds., pp. 23–42. Cambridge University Press, Cambridge.

Reiser, P. J., R. L. Moss, G. G. Giulian, and M. L. Greaser. 1985. Shortening velocity in single fibers from adult rabbit soleus muscles is correlated with myosin heavy chain composition. Journal of Biological Chemistry 260:14403–14405.

Rosser, B. W. C., and J. C. George. 1986. The avian pectoralis: Histochemical characterization and distribution of muscle fiber types. Canadian Journal of Zoology 64:1174–1185.

Savolainen, J., and M. Vornanen. 1995. Fiber types and myosin heavy chain composition in muscles of common shrew *(Sorex araneus)*. Journal of Experimental Zoology 271:27–35.

Schutt, W. A., Jr., M. A. Cobb, J. L. Petrie, and J. W. Hermanson. 1994. Ontogeny of the pectoralis muscle in the little brown bat, *Myotis lucifugus.* Journal of Morphology 220:295–305.

Snow, D. H., R. Billeter, F. Mascarello, E. Carpenè, A. Rowlerson, and E. Jenny. 1982. No classical type IIB fibres in dog skeletal muscle. Histochemistry 75:53–65.

Strickler, T. L. 1978. Functional osteology and myology of the shoulder in the Chiroptera. Contributions to Vertebrate Evolution 4:1–198.

Strickler, T. L. 1980. Downstroke muscle histochemistry in two bats. *In* Proceedings of the Fifth International Bat Research Conference, D. E. Wilson and A. L. Gardner, eds., pp. 61–68. Texas Tech University Press, Lubbock.

Suzuki, A. 1990. Composition of myofiber types in limb muscles of the house shrew *(Suncus murinus):* Lack of type I myofibers. Anatomical Record 228:23–30.

Sypert, G. W., and J. B. Munson. 1981. Basis of segmental motor control: Motoneuron size or motor unit type? Neurosurgery (Baltimore) 8:608–621.

Vaughan, T. A. 1959. Functional morphology of three bats: *Eumops, Myotis, Macrotus.* University of Kansas, Publications of the Museum of Natural History 12:1–153.

Vaughan, T. A. 1966. Morphology and flight characteristics of molossid bats. Journal of Mammalogy 47:249–260.

Vaughan, T. A., and G. C. Bateman. 1970. Functional morphology of the forelimb of mormoopid bats. Journal of Mammalogy 51:217–235.

Wimsatt, W. A., and A. L. Guerriere. 1962. Observations on the feeding capacities and excretory functions of captive vampire bats. Journal of Mammalogy 43:17–27.

9
Form, Function, and Evolution in Skulls and Teeth of Bats

PATRICIA W. FREEMAN

Bats provide a model system for tracking change from the primitive mammalian tooth pattern to patterns indicating the more-derived food habits of carnivory, nectarivory, frugivory, and sanguinivory. Whereas microchiropteran bats show all these transitions, megachiropterans illustrate an alternative pattern concerned only with frugivory and nectarivory. In microchiropterans, it is likely that carnivory, nectarivory, frugivory, and sanguinivory are all derived from a dilambdodont insectivorous tooth pattern. Megachiropterans are troublesome because they appear as nectarivores or frugivores without a clear relationship to ancestral taxa.

The nature of the food item and how teeth respond to that item evolutionarily is an issue I have addressed previously diet by diet (Freeman 1979, 1981a, 1981b, 1984, 1988, 1995). Within the insectivorous family Molossidae, and among insectivorous microchiropteran bats in general, consumers of hard-bodied prey can be distinguished from consumers of soft-bodied prey by their more robust mandibles and crania, larger but fewer teeth, longer canines, and abbreviated third upper molars (M3; Freeman 1979, 1981a, 1981b; Strait 1993a, 1993b). Carnivorous microchiropterans have distinctive large upper molars with lengthened meta-

stylar shelves and elongated skulls with larger brain volumes and external ears than their insectivorous relatives. As in terrestrial mammals, however, there is no clear distinction between insectivorous and carnivorous species (Savage 1977; Freeman 1984). Microchiropteran nectarivores are also on a continuum with insectivores but are characteristically long-snouted with large canines and diminutive postcanine teeth (Freeman 1995). Finally among microchiropterans, frugivores differ from insectivore/carnivores and insectivore/nectarivores by having a substantially different cusp pattern on the molars. The paracone and metacone are pushed labially or buccally to become a simple, raised but sharpened ridge at the perimeter of the dental arcade (Freeman 1988, 1995).

First I examine function of differently shaped skulls and palates of bats in different dietary groups. Among Megachiroptera, frugivores are on a continuum with nectarivores, but there are characteristics of robustness that appear to be good indicators of diet that distinguish the two (Freeman 1995). Megachiropterans have several convergent characteristics in common with microchiropteran nectarivores. I believe this convergence is not only the key to explaining cranial and palatal shape and jaw function in bats but also is critical to understanding the evolution of nectarivory and frugivory in chiropterans. Associated with the shape of the palate is the way that allocation and emphasis of tooth material on the toothrow shift between suborders. The relative area that each kind of tooth occupies on the toothrow is quantified and serves as the basis for my interpretations.

A second goal is to examine function in bat teeth. Here I synthesize my past work on tooth function, particularly with regard to canines and molars, and introduce a novel way to examine function in canines. Function in more complex teeth involves a review of the principal cusps on the upper and lower molars and how cusp patterns have evolved relative to different diets. Specifically, I contrast carnivory in terrestrial mammals and bats, insectivory in insectivorous and nectarivorous species, and frugivory in mega- and microchiropterans. Finally, I suggest that the evolution of dilambdonty can be correlated with packaging and digestibility of the food item.

Study Methods

This study is based on 103 species representing 78 genera, 10 families, and two suborders of the order Chiroptera. Among microchiropterans there are 40 insectivorous, 7 carnivorous, 18 nectarivorous, 14 frugivorous, and 2 sanguinivorous species. Megachiropteran frugivores and nectarivores are represented by 11 species each (Appendix 9.1). Each species was usually represented by a single adult male skull in perfect or near-perfect condition (i.e., no broken or missing parts), although a perfect adult female skull was preferable to an imperfect male skull. There are no missing data except for naturally missing teeth in the toothrow, which are treated as missing data and not as zero; including the latter would substantially affect the average of those bats with the tooth present.

Homologies for tooth number are from Andersen (1912). Areal measurements, recalculated for this study, are from camera lucida drawings that were scanned into a Macintosh computer and taken automatically inside (teeth) or outside (palate) high-contrast occlusal outlines. Areas include upper incisors (I); upper canines (C); nonmolariform upper premolars (other PMs); fourth upper premolars (PM4); and first, second, and third upper molars (M1, M2, and M3), where found; the area of the raised stylar shelf (including PM4); and the area of the palate (as modified in Freeman 1988, 1995). Linear measurements are the same as those in Freeman (1995; but see also Freeman 1984, 1988).

This chapter is concerned with large-scale patterns. Details on variation among species can be found in earlier papers. The size character is the same as that used in previous papers (SIZE = sum of the natural logs of condylocanine length, zygomatic breadth, and temporal height; Freeman 1984, 1988, 1992, 1995).

Experimental work examining form and function in bat canine teeth involved finite-element modeling and photoelastic analysis. Shapes of cross sections in canine teeth can be edged and nonedged (Freeman 1992), and experiments with models of teeth puncturing a substance can show how these two different types of cross sections initiate different patterns of stress or toothmarks in a substance ("food"; Freeman and Weins, unpublished data). Finite-element modeling is a mathematical description of dimensional or geometric change in a structure when a force is applied to deform the structure to reveal where the most intense stresses should occur (Zienkiewicz and Taylor 1989; Rensberger 1995).

Two-dimensional models were constructed to show stresses occurring in the "food" when penetrated by teeth with 30°, 60°, and 90° angles at their edges and a circular or nonedged tooth. In three dimensions, actual stress analysis tests were performed with a metal cone and a pyramid with an edge of 90°, simulating oversized replicas of teeth (Caputo and Standlee 1987). These oversized "teeth" were loaded into plastic ("food") that had been heated to the point of being liquid and allowed to cool around the loaded forms (called stress-freezing). Cooling freezes the stress-induced patterns permanently in the plastic. The plastic is photoelastic, which means that the birefringent (refractive in two directions) patterns of stress caused by the deformation of the plastic by the different shapes can be observed under polarized light. The visual results of that experiment are presented here. Photoelasticity has been used in dentistry for several years (Guard et al. 1958; Fisher et al. 1975), but only to examine what stresses are being placed on the tooth and not how the tooth is stressing the food.

Results

Cranial and Palatal Form

Shapes of bat skulls, represented by zygomatic breadth divided by condylocanine length, vary between being as wide as they are long to being only a third of the skull length. However, skull width of most species is one-half to three-quarters the length of the skull (Figure 9.1A). Extremes are represented by the microchiropteran family Phyllostomidae, with *Centurio* and other stenodermatines on the wide end and *Musonycteris* and other glossophagines on the narrow end. Four wide-faced insectivorous species, mentioned in earlier studies (Freeman 1984), group together at 0.8 above the majority of species. On the other hand, shapes of palates of bats (breadth across the molars divided by length

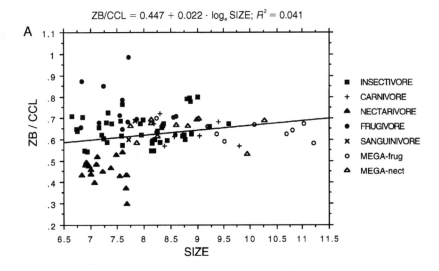

$$\text{ZB/CCL} = 0.447 + 0.022 \cdot \log_e \text{SIZE}; \ R^2 = 0.041$$

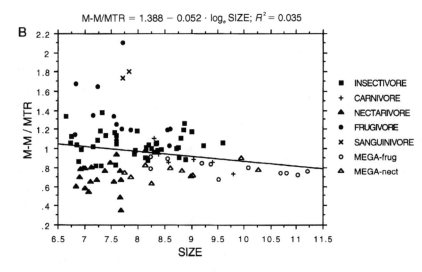

$$\text{M-M/MTR} = 1.388 - 0.052 \cdot \log_e \text{SIZE}; \ R^2 = 0.035$$

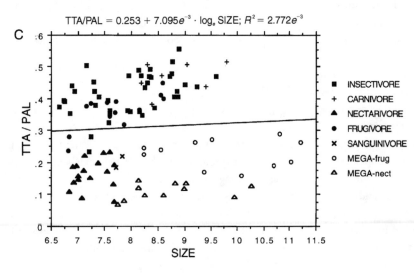

$$\text{TTA/PAL} = 0.253 + 7.095e^{-3} \cdot \log_e \text{SIZE}; \ R^2 = 2.772e^{-3}$$

Figure 9.1. Cranial and palatal features that are important in chiropterans. (**A**) Zygomatic breadth (ZB) divided by condylocanine length (CCL) regressed against SIZE (see *Methods*). (**B**) Breadth across upper molars (M-M) divided by length of maxillary toothrow (MTR) regressed against SIZE. (**C**) Total tooth area (TTA) divided by palatal area (PAL) regressed against SIZE. Open symbols denote megachiropterans; all other symbols denote microchiropterans. Megachiropterans have heads that are relatively as wide as most other chiropterans' heads (A) but have narrower palates (B). Megachiropterans (both frugivores and nectarivores, MEGA-frug and MEGA-nect) and microchiropteran nectarivores have small teeth on large palates, whereas microchiropteran insectivores, carnivores, and frugivores have large teeth on small palates (C). Two microchiropteran frugivores, *Ametrida* and *Ectophylla* (in order away from regression line), and two microchiropteran insectivores, *Lonchorhina* and *Mormoops*, have smaller teeth on larger palates than do others in these two dietary groups.

of maxillary toothrow) show a substantial downward shift from the line representing skull shape, such that microchiropteran nectarivores, megachiropterans, and several microchiropteran carnivorous species have long, narrow palates (Figure 9.1B). Again, phyllostomids show the greatest variation, but the wide-palated forms include two species of sanguinivores.

Form and Emphasis of Teeth

With few exceptions, the microchiropteran insectivores, carnivores, and frugivores have relatively large teeth on small palates, and microchiropteran nectarivores and megachiropterans have relatively small teeth on large palates (Figure 9.1C). Megachiropteran nectarivores have relatively smaller teeth on the palate than megachiropteran frugivores. These relative proportions are maintained regardless of the size of the bat (as represented here by the composite SIZE character) and presumably body mass (Freeman 1988).

The relative area of the toothrow occupied by different teeth can be compared across teeth, suborders, and feeding groups (Figure 9.2A,B). Dietary categories are further subdivided into groups of species sharing the same tooth formula (Figure 9.2C). First and second upper molars in microchiropterans occupy the greatest relative areas (23% and 22%, respectively), followed by canine area (18%) and third upper molars (10%) if present, and other PMs and incisors (6% each). In contrast, canines occupy the greatest amount of area in megachiropteran toothrows (27%), followed by other PMs, PM4s, and M1s (20% each); second upper molars occupy only 9%, and incisors, a small 4.5%.

Among microchiropterans, insectivorous and carnivorous species are similar in that nearly half the toothrow is devoted to first and second molars. In insectivores with only PM4 in the upper row (insectivore 2, Figure 9.2C), the incisors, first and second molars, and stylar shelves become larger. Carnivorous species have the most variability in teeth present on the toothrow; premolars, incisors, or both premolars and incisors can be missing. In the latter case, canines and M3s become larger. Carnivorous microchiropterans have, relatively, the largest M1s, M2s, and stylar shelves of any bat, and like insectivores with only PM4

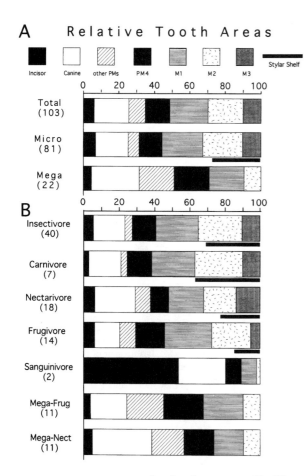

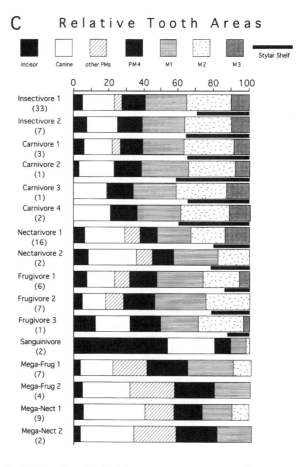

Figure 9.2. Average percentage of total tooth area occupied by different teeth: (**A**) for all bats and by suborder; (**B**) by dietary group (see Appendix 9.1); (**C**) by dietary subgroups (dietary groups subdivided further by tooth formula; see Appendix 9.1).

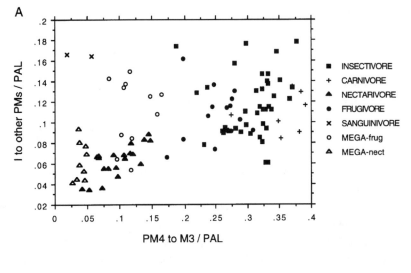

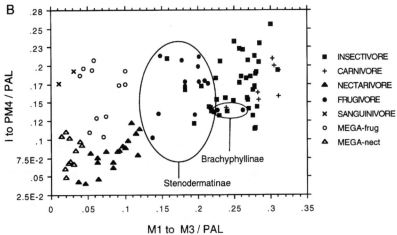

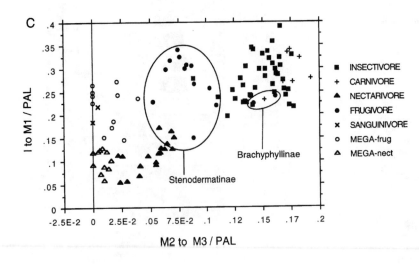

Figure 9.3. The allocation of dental material shifts anteriorly on the palate (PAL) between suborders and among different dietary groups of microchiropterans. Open symbols denote megachiropterans; all other symbols denote microchiropterans. (**A**) Area that nonmolariform incisors, canines, and other premolar teeth (I to other PMs/PAL) occupy of palatal area (not total tooth area), compared to area occupied by molariform (M) teeth (PM4 to M3/PAL). (**B**) Area that incisors, canines, other premolars, and PM4 (I to PM4/PAL) occupy of palate, compared to area occupied by molars (M1 to M3/PAL). (**C**) Area that incisors through the first molar (I to M1/PAL) occupy of palatal area, compared to area occupied by the most-posterior molars (M2 to M3/PAL).

present, have larger M2s than M1s. In fact, the smallest carnivores—*Nycteris*, *Cardioderma*, and *Trachops*—have some of the very largest M2s and stylar shelves (*Nycteris* is carnivore 2 in Figure 9.2C). Regardless of the variation anteriorly, M1 and M2 constitute more than 50% of the total tooth area in carnivorous bats.

The largest tooth on the microchiropteran nectarivore toothrow, relative to total tooth area, is the canine (23%), followed by the molars, PM4, and other PMs. In addition to large canines, nectarivores differ from insectivores by having larger M3s (13.5%) and smaller PM4s (10%) but other PMs that are larger (8.5%). In nectarivores in which M3 is absent, canines and M1s are larger (27% and 25%, respectively). Stylar shelves are smaller (22%) than in insectivores (32%) and much smaller than in carnivores (37%). In absolute size, teeth in nectarivores are considerably smaller than in other microchiropterans (see Figure 9.1C; Freeman 1995).

Frugivorous microchiropterans have relatively the largest PM4s and M1s and the smallest canines, M3s, and stylar shelves of all microchiropteran dietary groups. In those bats in which M3 is absent, M1 occupies a third of the toothrow and all the premolars occupy a third. In the one frugivore with PM4 only, incisors and canines are larger (12% and 20%, respectively; *Pygoderma* is frugivore 3, Figure 9.2C). Finally, large incisors (54%) and large canines (27%) dominate toothrows of blood-feeding microchiropterans at the expense of postcanine teeth (20%; *Desmodus* has no M2, but I did not subdivide this category further).

Megachiropterans have the greatest relative area invested in canines (27%) followed by equal proportions invested in other PMs, PM4, and M1 (20% each); however, these figures obscure the large difference in canine area between megachiropteran frugivores and nectarivores. Among frugivores, canines are smaller (21%) and PM4 and M1 are larger (23% each), whereas megachiropteran nectarivores have larger canine areas (34%) and smaller PM4s and M1s (17% each). Other PMs are larger in megachiropterans than microchiropterans. Megachiropteran frugivores that have lost M2 have larger areas for canines (27%) and for other PMs (26%); nectarivores without M2s have less area for canines (30%) but increased area for other PMs (25%), PM4 (23%), and M1 (19%). However, as in microchiropterans, the absolute size of teeth of megachiropteran nectarivores is smaller than the teeth of megachiropteran frugivores (see Figure 9.1C; Freeman 1995).

Emphasis in tooth material on the palate shifts between suborders and among dietary groups. Megachiropteran palates have relatively larger areas for nonmolar teeth while most microchiropterans have relatively larger areas for molariform teeth (Figure 9.3A). This pattern is true regardless of the size of the bat. Not surprisingly the proportion of

the palate occupied by anterior nonmolariform teeth in megachiropterans is affected by the absence of molars (M3s and M2s) in the toothrow (Figure 9.3A). The proportion of the palate occupied by all teeth anterior to the molars (I to PM4) further concentrates in the megachiropterans on the horizonal axis and starts to show the difference between microchiropteran frugivores and carnivores (Figure 9.3B). Stenodermatines have small or no M3s (but large M1s) and are different from the two brachyphyllines, which have substantial M3s. Finally, when the toothrow is divided so that M1 is included in the anterior portion of the palate (Figure 9.3C), stenodermatine frugivores are well separated from all carnivorous microchiropterans and insectivorous bats and brachyphyllines range in between.

Canine teeth have different cross-sectional shapes (Freeman 1992). Finite-element modeling in two dimensions predicts the distribution of stress in a "food" material and indicates where cracks are most likely to form. Stresses around the edged mark are greater than those surrounding the circular mark and are concentrated at the edges (Figure 9.4A). Further, stresses actually increase as the angle at the edge decreases. Initial experimentation in three dimensions at the tooth–food interface with photoelastic techniques verified that the lines or fringes of stress are concentrated at the edges of the edged tooth. In the circular or nonedged tooth, stresses are uniformly distributed (Figure 9.4B).

Discussion

Cranial and Palatal Function

NARROW SKULLS. It is not surprising that bat skulls can be short and wide or long and narrow, but it is of considerable evolutionary interest that the shortest, widest skulls and the longest, narrowest skulls are found in the same family, Phyllostomidae, and not across suborders. However, bats with long, narrow palates include microchiropteran and megachiropteran nectarivores and megachiropteran frugivores. In addition to elongated palates, these bats have in common (1) greater distances from the last lower molar to the jaw joint (Freeman 1995), (2) smaller tooth-to-palate ratios with spaces separating the teeth, (3) fused mandibles, (4) shallower, less well defined glenoid fossae (Freeman 1995), and (5) thegosed upper canines that occupy relatively large areas of the toothrow.

Possession of a long tongue explains these shared features. With the development of a large, elongated tongue, the palate lengthens, and teeth not only occupy a smaller proportion of palate but also have more space between them. In nectarivores, teeth become absolutely smaller (see Figure 9.1C). Also, teeth appear to have moved anteriorly

A

B

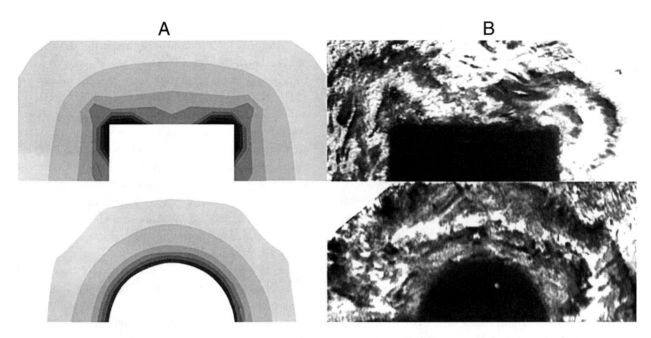

Figure 9.4. Distribution of stresses in a substance that is being penetrated by an edged tooth (top panels) and a nonedged or conical tooth (bottom panels). (**A**) Two-dimensional constructs simulated with finite-element modeling. In this schematic, the darker the area, the greater the concentration of stress (and the greater the likelihood of cracks forming there). (**B**) Verification of models using photoelastic stress analysis, a three-dimensional experiment (see *Methods*). Concentrations of stress are visible at the corners of the edged tooth; a more uniformly distributed pattern of fringes (lines of stress) can be seen around the nonedged tooth.

with elongation of the palate, resulting in the last lower molar being farther from the jaw joint than in microchiropteran insectivores, carnivores, and frugivores (Freeman 1995). Greater distance to last lower molar from the jaw joint might suggest bite strength is weak, but this is not realistic for large megachiropteran frugivores that are powerful and can crack open cocoa pods (Hill and Smith 1984). The megachiropteran palate is narrow and elongated but the skull is not (see Figure 9.1A,B). The ratio of zygomatic width to length of skull for megachiropterans is similar to that in most other bats, which suggests muscular strength is similar as well. Moreover, the origin of the masseter at the anterior base of the zygoma in megachiropteran frugivores is farther forward on the maxilla than in megachiropteran nectarivores and overlaps the last upper molar. This arrangement results in a greater mechanical advantage of the masseter (Freeman 1979, 1995). Bite force on the toothrow may be greatest at the anterior root of the zygoma because the buttressing can absorb much of the force (Crompton and Hiiemäe 1969; Werdelin 1989).

Although frugivorous megachiropterans have larger teeth than their more nectarivorous relatives, the teeth of both sit on elongated palates (see Figure 9.1C). Intertooth space is greatest in the most elongated jaws, which may be useful in determining degree of nectarivory (Freeman 1995). For example, among microchiropteran nectarivores *Musonycteris*, *Choeronycteris*, and *Choeroniscus* have the greatest space between teeth, and all three species have lost lower incisors. This loss reflects the most derived result of the protrusion and retraction of the working nectarivorous tongue (this is discussed later). Among megachiropterans, nectarivorous species have greater intertooth space and more frugivorous species have less (Freeman 1995).

Fusion of the mandibular symphysis occurs in frugivorous and nectarivorous species of both suborders and may stabilize or strengthen the anterior end of the jaws (Beecher 1979; Freeman 1988, 1995). Short-faced frugivorous phyllostomids even have chins, which reflects the strengthening of the jaws to resist vertical forces at the front of particularly short, wide jaws. Symphyseal fusion in bats with elongated jaws reflects the reinforcement needed to stabilize the mandible as the long tongue protrudes well beyond the anterior margin of the jaw for nectar-feeding. Some nectarivores have developed a sagittal, bladelike, bony reinforcement at the symphysis.

Terrestrial carnivores that must have precise occlusion of cusps on upper and lower teeth to slice meat have an unfused symphysis (Scapino 1965) associated with a snug-fitting, tapered cylindrical condyle that limits the degrees of freedom of movement of the jaw. In early creodonts the condyle was loose fitting and the symphysis was long and very well fused. Microadjustment at occlusion in these animals was thought to be accomplished by swinging the mandible sideways so that the transverse ridges on the molars guide the occlusion of the carnassial teeth (Savage 1977).

Microchiropteran insectivores and carnivores also have unfused mandibles. Their close-fitting dilambdodont teeth must fit precisely during each chewing cycle to be effective and not malocclude. Microadjustment of the teeth can be made anteriorly at the unfused symphysis and to some extent at the glenoid. Although not so tight fitting as that in terrestrial carnivores, the glenoid fossa in these bats is a well-defined platform with a well-developed postglenoid process (Freeman 1979, 1995).

Microchiropteran nectarivores and megachiropterans all have looser-fitting glenoid articulations than insectivorous and carnivorous microchiropterans (Freeman 1995) but have fused symphyses. The glenoid fossae are shallower with less distinct articular platforms and poorly developed postglenoid processes. With fusion of the symphysis, teeth either do not need to occlude precisely or, if they do, can adjust at the rear of the jaw in a looser-fitting articulation. Although small, teeth of microchiropteran nectarivores have discernible but not deeply emarginate dilambdodont cusps. As shown next, these small teeth still register with each other at the talonid–protocone contact. Several of these bats have a ventral extension of the jugal (the posterior root of the zygoma) that may limit lateral movement of the condyle, but all have narrow condyles and reduced postglenoid processes (Freeman 1995). The ventral extension may aid the registration of dilambdodont teeth in jaws with a fused symphysis.

Microchiropteran frugivores have a fused mandible as well as a well-defined glenoid articular platform with a well-developed postglenoid process. Here the dilambdodont pattern and precisely fitting transverse cusps are diminished to a raised rim, and only vertical registration is necessary for the lower dental arcade to nest inside the perimeter of the upper teeth.

Megachiropteran frugivores, which use long tongues to eat both fruit and nectar, also have fused symphyses. Indeed large, pointed tongues of "enormous protruding capacity" are a hallmark of feeding in megachiropteran frugivores (Greet and De Vree 1984). For example, to eat bananas the tongue moves forward to mash the bolus against the dorsal palatal ridges. The tongue does not stop at the margin of the mouth but protrudes beyond the mouth to curl forward over the nose. The tongue protrudes less and less with harder foods like apples. Megachiropterans have simple-basined teeth surrounded by low edges that hardly touch at occlusion. Indeed, without touching and interlocking, homologous cusps cannot be discerned (Koopman and MacIntyre 1980). Fused mandibles may limit the ability of the jaws to register the teeth precisely or at all at occlusion. If registration of teeth is inversely related to fusion at the mandible, and, if fusion evolved first because having fused mandibles meant better support of a long tongue, then

teeth in megachiropterans cannot be expected to interlock very well or at all nor could discernible cusps be expected.

WIDE SKULLS. Bats can also have wide skulls and palates (see Figure 9.1A,B). Wide skulls in microchiropteran insectivorous bats are robust with enlarged cranial crests, teeth closer to the fulcrum of the jaw joint, increased muscle mass for the more anteriorly placed masseter, and usually lengthened canines (Freeman 1984). Wider-skulled insectivores eat harder-shelled insect prey (Freeman 1981a; Strait 1993a, 1993b). However, wide skulls in microchiropteran frugivores are not necessarily robust, and the very widest, such as *Centurio,* is paedomorphic and fragile. The widest stenodermatines also have wide palates that allow more teeth to be involved with the bite, a bite which should have good mechanical advantage given the close proximity of the teeth to the jaw joint (Greaves 1985). Canines in these most extreme frugivores are small, and the cheekteeth close on food from front to back rather than from back to front as in insectivores and carnivores (Freeman 1988). In these frugivores, wide faces would be well adapted for taking plugs out of fruit and being able to secure a wide grip on fruits for transporting. Morrison (1980) observed feeding in *Artibeus* and found that this stenodermatine consumed fruit in small bites. After chewing, each bolus was then pressed against the ridges of the palate with the tongue, the juice swallowed, and dry pellets spat out. There was no discussion of the tongue extending beyond the mouth during feeding, and although it can be extended to drink (C. J. Phillips, personal communication), the tongue is not thought to be specialized for protrusible feeding (T. A. Griffiths, personal communication). These observations lead me to believe that tongues are not elongated in phyllostomid frugivores.

Dental Function

CANINES. Calculating the percentages of the toothrow that are occupied by particular teeth is a first attempt in determining the functional emphasis of a tooth. However, area indicates little about how shanks of canines and stylar shelf patterns on molars may function at the tooth–food interface. Canine teeth occupy a substantial proportion of the toothrow and have the primary function in gathering and subduing prey (Freeman 1992). I have speculated that the cross-sectional shape of insectivorous bats involved the flattening of at least one side of the tooth to form a knifelike flange that would allow the tooth to more easily pierce the exoskeleton of insect prey (Freeman 1979). In truth, canines of bats are quite diverse and cross-sectional shapes can be triangular or polygonal, with the vertices of the triangle or polygon representing edges that extend longitudinally from tip to cingulum. A reasonable assumption is that there should be differences in how cracks are propagated in the substance being penetrated on the basis of the shape of the tooth (Freeman 1992).

In recent experiments, Freeman and Weins (1997) punctured apples with casts of bat teeth to determine what sharpnesses and forces occurred at the tips of teeth. Not surprisingly, sharper tips required less force to penetrate the surface than blunter tips. More complex, however, was the investigation of how the shank of the tooth might interact with a food item. Both finite-element models and experimentation with photoelastic materials support the notion that longitudinal edges on canines would be beneficial in initiating cracks in foods (see Figure 9.4A,B). Because energy increases (indicated by higher concentration of stress) at the edges of canines, cutting through prey would be optimized much like the edge of a surgical needle, which is triangular in cross section.

The alternative shape, with a round cross section, would mean the tooth must press deeper into the prey to finally break through the surface by force. Given the elasticity of surfaces of endo- and exoskeletal prey and fruits, the latter penetration would be less efficient. Freeman (1992) identified and quantified as sharp at least one or both of two edges on bat canines, one directed toward the incisors and one directed toward the ectoloph of the postcanine teeth. Edges on canine teeth may be especially beneficial to predators and harvesters whose forelimbs are modified for flight and must eat or gather (many) items while flying. Single and multiple edges are found on all canine teeth of these bats (Freeman and Hayward, unpublished data). The relationship between the pattern of sharp and blunt edges and diet is under study.

A final feature shared by megachiropterans and microchiropteran nectarivores is that the anterior surface of the upper canines is worn by the lower canines. This phenomenon is especially noticeable in nectarivores. Tooth-on-tooth wearing and self-sharpening has been called thegosis (Every 1970) or simply attrition (Butler 1972; Osborn and Lumsden 1978). In nectarivores, the lower canines are splayed laterally during jaw closure (when viewed frontally; Figure 7 in Freeman 1995) so that they engage both upper canines well before occlusion of the cheekteeth. I believe that wear occurs because both lower canines can brace themselves simultaneously against both upper canines with the posterior pull of the jaw muscles. A bracing function would help support the lower jaw while the long tongue is being extended well beyond the anterior margin of the jaw to retrieve nectar from horizontally oriented flowers (Freeman 1995). Rapid protrusion and retraction and the mass of the tongue needed to gather nectar would create large depressive loads at the front end of the lower jaw.

The extent of wear varies in microchiropteran nectarivores from just a small patch of wear at the cingulum of the upper canine in species that have lost lower incisors to an entirely worn anterior face of the upper canine in species that retain the lower incisors (Freeman 1995). Loss of incisors occurs in the most derived nectarivores, presumably to allow an unhindered path for protrusion and retraction of the tongue during feeding. Although feeding in these extreme nectarivores could be done without opening the jaws widely or at all, the lower canines still brace at the upper cingulum as evidenced by the small but quite distinct patch of wear. In those nectarivores with lower incisors present, the entire face of the upper canine is worn because the jaws have to open wide enough to allow the tongue to move but also to avoid the lower teeth.

Longitudinal edges of canines in nectarivores (but not those in insectivores, carnivores, or frugivores) are sharpened by wear, which can affect cross-sectional shape (Freeman and Hayward, unpublished data) and may also serve some function during feeding, perhaps that of cutting into flower parts for nectar. With the additional features of relatively large canines and fusion of the mandibular symphysis, megachiropteran and microchiropteran nectarivores and megachiropteran frugivores are able to effectively support large, protrusible tongues.

Artibeus shows slight wear on the distal half of the anterolingual slope of the upper canine, which is typical of many insectivorous and carnivorous microchiropterans, and occurs when one or the other lower canine engages one or the other upper canine just before occlusion. The cross-sectional shape of these microchiropterans is not affected by wear (Freeman 1992). Further, lower canines are vertically aligned and not splayed laterally so that simultaneous contact among all four canines before occlusion of the cheek teeth is rare, if not impossible.

MOLARIFORM TEETH.   The interlocking of upper and lower teeth at occlusion is straightforward: Each lower molar occludes with two upper teeth, which is where PM4 participates and why PM4 is a functional part of the molariform row (Figure 9.5; top drawing for each bat). Understanding how the high-cusped, dilambdodont molars occlude, however, is critical to understanding not only function but also evolutionary changes in cusp patterns.

The two principal cusps on the upper molar are the anterior paracone followed by the more posterior metacone (Figure 9.6). Each has two crests radiating labially or buccally from it to give the characteristic W-shape stylar shelf or ectoloph (Butler 1941). Lingual to the paracone and across a valley is the protocone. The little diamond-shaped valley formed by the protocone and the bases of the para-

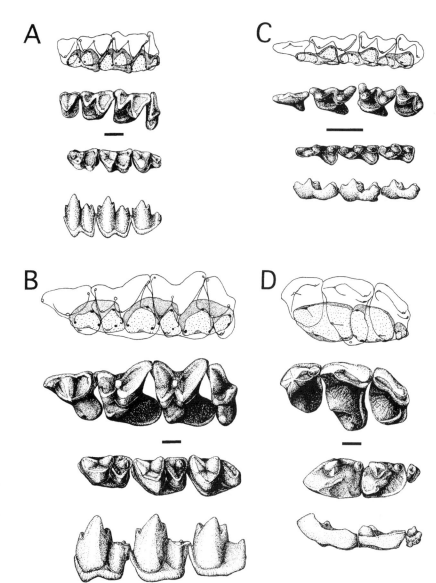

Figure 9.5. Left upper and lower molariform teeth of microchiropterans: (**A**) an insectivore, *Antrozous pallidus;* (**B**) a carnivore, *Macroderma gigas;* (**C**) a nectarivore, *Monophyllus redmani;* (**D**) a frugivore, *Artibeus jamaicensis.* Each example shows the interlocking of upper teeth and lower (stippled) teeth at occlusion, occlusal views of upper and lower teeth, and lateral view of lower teeth. Contact of the trigonid and talonid is shown by lighter stippling. The talonid is narrowest compared to the trigonid in carnivores and widest in nectarivores. Canines would be to the left, and the scale bar, which equals 1 mm, is placed lingual to the occlusal views. Names of cusps appear in Figure 9.6.

cone and metacone is the protoconal or trigon basin, and it is this basin that receives the talonid of the lower tooth. Posterior or distal to the protoconal basin is the hypocone and hypoconal basin, both of which are variable in appearance. The hypoconal basin is enormous in *Macroderma* (Figure 9.5B) and other carnivorous bats, but the actual hypocone is cryptic.

In some insectivores, there is neither hypocone nor hypoconal basin (Figure 9.5A). The nectarivore *Monophyllus* (Figure 9.5C) has a well-distinguished hypoconal area, but the most distinct hypocone is seen on M1 in the frugivore *Artibeus* (Figure 9.5D), which is lingual to the protocone and sits on a well-developed ledge. These features are clear on M1 and M2, but M3, if present, can be abbreviated from the back forward. The endpoint of the posterior arm of the W, the metastyle, is lost and the posteriormost arm or crest, the metacrista, is reduced. Microchiropteran nectarivores have

the most complete M3s, but metastyles are missing. The resulting shape is a backward N (Figure 9.5C). Further loss of the metacone and the anterior arm leading from it, the premetacrista, gives the tooth a V-shape (Figure 9.5B). The endpoint of the posterior arm of the V, the mesostyle, is lost in the most abbreviated M3s; the posterior crest leading from the paracone to the mesostyle, the postparacrista, can be much shorter than the anterior paracrista (Figure 9.5A).

The relative areas of the molariform teeth that the stylar shelves occupy compared to the relative areas of palates that teeth occupy are distinctive for microchiropterans with different diets. Generally, insectivores and carnivores have large teeth with large stylar shelves, frugivores have large teeth with the smallest stylar shelves, and nectarivores have small teeth with small to moderate stylar shelves (Figure 9.7).

The lower molars are composed of the familiar triangle of three cusps, the trigonid, with the large, buccal

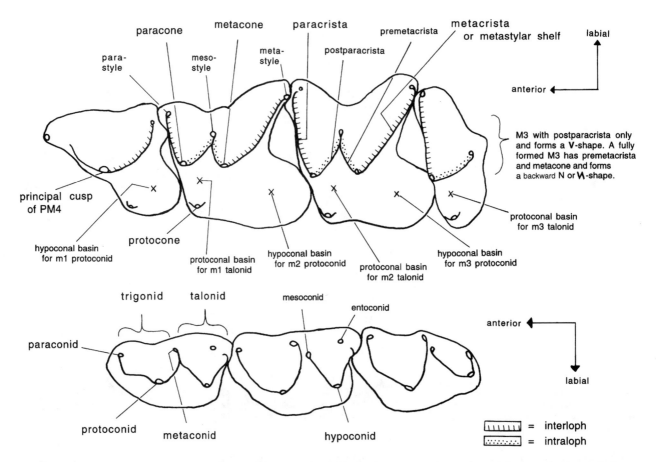

Figure 9.6. An enlarged occlusal diagram of the left upper and lower toothrows of *Macroderma gigas,* with principle cusps, basins, and cristae—including interlophs and intralophs—identified. The abbreviated cusp pattern of M3 is a more derived condition for microchiropterans. A more primitive configuration occurs in *Monophyllus* (Figure 9.5C).

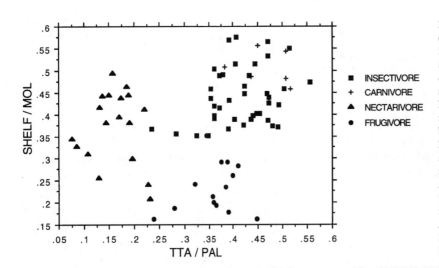

Figure 9.7. Summary of area that the stylar shelf (SHELF) occupies of molariform row (MOL, PM4 plus molars) versus the area that total tooth area (TTA) occupies of the palate (PAL) in microchiropterans. These ratios have been effective in separating insectivores and carnivores, which have the largest teeth and largest stylar shelves, from frugivores, both of which have large teeth but small stylar shelves, and nectarivores, which have small teeth and small to moderate stylar shelves. New to this study are several species that make the separation of groups less distinct. Among insectivores, *Mormoops* has the smallest teeth, and *Lonchorhina* the next smallest. Among frugivores, *Ectophylla* has the smallest teeth, then *Ametrida* and *Chiroderma; Uroderma* has the largest stylar shelf. *Phyllonycteris* and *Erophylla* have the largest teeth and the smallest stylar shelves among nectarivores.

protoconid flanked anterolingually by the paraconid and posterolingually by the metaconid (see Figure 9.6). Posterior to the trigonid is the heel of the molar or the talonid, where usually two cusps are present, the labial hypoconid and the lingual entoconid. As with M3, m3 (the third lower molar) is often abbreviated, and hypoconid and entoconid are not always distinguishable.

In general, the principal cusp on PM4 forms the anterior cutting crest of the interloph (the two-arm section of the ectoloph that is shared between upper teeth; Freeman 1984) (see Figure 9.6) with M1 to receive the trigonid of the lower molar. The protoconid itself fits into the deep valley bordered by the interloph to occlude against the lingual basin of PM4. This lingual basin is greatly expanded in *Macroderma* (Figure 9.5B) and corresponds to the hypoconal basins of M1 and M2. Functionally, these three deeply expanded basins receive the enlarged protoconids of lower molars in carnivorous bats. Posteriorly, the protoconal basin of the next posterior molar receives the talonid of the lower tooth. The first lower molar straddles the posterolingual basin of PM4 and the anterior protoconal basin of M1 (see overlying teeth, Figure 9.5).

Talonids and protoconal basins move across each other in lock-and-key fashion and appear to carry out the primary chopping and crushing action of the "pinking shears" teeth in insectivorous species. Indeed, it is this talonid–protoconal basin contact that is not only retained but even expanded in nectarivores, which have diminutive and widely spaced teeth. In insectivorous bats, the talonid is usually bigger than the trigonid and fits neatly into the deeply emarginated intraloph (that part of the ectoloph formed by the paracone, mesostyle, and metacone of the same upper tooth; Freeman 1984) (see Figure 9.6). Here the mesostyle, the middle peak of the W-shape, nearly reaches the buccal margin of the tooth. In carnivorous bats the mesostyle is more lingual, the intraloph is reduced, and the metastylar shelf (also called the metacrista) is elongated and aligned more anteroposteriorly. Talonids are narrower than the trigonids on lower molars of carnivorous bats.

The overall difference between insectivorous and carnivorous bat teeth lies in the shift of emphasis from the hypoconid in the former to the protoconid in the latter or from the talonid to the trigonid. The transverse movement of occluding teeth is kept stable and precisely guided by the transverse alignment of the transverse anterior cristae, from paracone to parastyle and from protocone to parastyle (the anterior edge of the tooth) of the upper molars and the protoconid–metaconid cutting crest of the lower teeth.

CARNIVORY. Elongation of the metastylar shelf and anteroposterior alignment on upper molars are typical in the evolution of terrestrial carnivore teeth (particularly creodonts) (Osborn 1907; Butler 1946). In the lower molars of those carnivores the trigonid starts to straighten out; that is, the paraconid and metaconid move to a plane in line with the protoconid, the metaconid disappears, the cutting crest between the paraconid and protoconid becomes a blade, and the talonid gets smaller and smaller until it disappears. As the tooth simplifies with the reduction of the protocone and metaconid, the transverse occlusal guides are lost and replaced by a more longitudinal, sagittal jaw action but one that is just as much in need of precision. This is why the tapered, cylindrical condyle fits tightly into a glenoid fossa with prominent pre- and postglenoid processes. The tight fit allows only slight lateral movement, so that microadjustment of the carnassial blades is possible only at the unfused symphysis at the anterior end of the mandibles (Scapino 1965; Savage 1977). The greatest modification of cusps occurs at the carnassial pair, which in modern carnivores is PM4–m1. The trigonid on m1 of carnivorous bats has begun to form the in-line cusp pattern in *Vampyrum*, where the paraconid–protoconid blade is quite prominent and the metaconid is reduced to a diminutive bump on the posterior crest of the protoconid. Talonids are generally smaller than the trigonids in carnivorous bats (see Figure 9.5B).

In carnivorous bats, as the metastylar shelf (and interloph) lengthens the metastyle moves lingually to shorten the intraloph. Both these modifications together simplify the complex dilambdodont pattern, which is not unlike what happens at the carnassial pair in terrestrial creodonts (particularly the series represented by *Sinopa*, *Pterodon*, and *Hyaenodon* in Hyaenodontinae, but also seen in Limnocyoninae and Machaeroidinae; Butler 1946). Here the metastyle moves lingually and the intraloph becomes shorter. The paracone and metacone move closer and closer together until no space is left between them (the intraloph disappears). Simultaneously, the protocone diminishes completely to leave only a paracone–metacone blade for the upper carnassial.

The toothrow in carnivorous bats simplifies into three large pestle-and-mortar systems and two smaller ones. Enlarged protoconids (pestles) in carnivorous bats would be good for deep penetration of endoskeletal foods and the large hypoconal basins (mortars) good for crushing bones of small mammals and birds. The basins may also shield the gums from pieces of bone, chitin, or other hard parts. Remains of prey including bones and teeth are finely chewed up by *Macroderma* (Douglas 1967). *Vampyrum* also eats bones (Peterson and Kirmse 1969; McCarthy 1987), and in captivity was observed to eat rodents and bats head first, teeth and all, but leaving a cape of skin along with the hindfeet and tail (J. S. Altenbach, personal communication). *Chrotopterus* ate all but the rostrum, wings, legs, and associated patagia of bats (McCarthy 1987). Patterns of molar microwear may correlate with the extent to which carnivorous bats masticate bony material (Strait 1993b), but more experimental data are needed. Strait did not find as much evidence of hard-item ingestion in *Vampyrum* as she did in *Macroderma* when she examined microwear at the hypoconid–protocone contact. She did not examine wear on the protoconids that contact the expanded hypoconal basin, which is likely the critical crushing area in carnivorous bats.

NECTARIVORY.   Although diminutive and not as high-crowned or emarginate as insectivores, the molars of nectarivores are still working teeth. Occlusion is particularly apparent between the wide talonid and the upper protoconal basin (see Figure 9.5C). The wear from this occlusion is obvious in even the most extreme nectarivores such as *Musonycteris*. Anterior to the talonid, the three cusps of the narrow trigonid are prominent, but the paraconid and protoconid form a small, in-line cutting crest and the especially prominent metaconid sits lingually to the protoconid. In a representative nectarivore, *Monophyllus redmani* (Figure 9.5C), there is a small upper hypoconal area against which this small cutting crest occludes. However, in nectarivores with spaces separating teeth, the crest falls between teeth, occluding longitudinally against the upper gum only, and is likely capable of processing only the softest items.

FRUGIVORY.   Teeth of frugivorous bats are completely different from those of their insectivorous relatives. Paracone and metacone have moved to the labial or buccal edge of the upper molars to form a raised rim and occupy the least area of all microchiropterans (Figure 9.5D). As a result, the stylar shelf is a wavy edge only vaguely reminiscent of the W-shape. The lingualmost face of this raised edge may not be homologous with that in insectivores, but may also include the metaconule (Slaughter 1970)—which would make my estimates of area of stylar shelves overestimates. However, the posterior crest on the primary cusps (paracones?) of the upper canine and premolar has a tendency to split to form two sharp edges. This phenomenon could also affect development of paraconids on molars and may determine what exactly comprises the homologous ectoloph (see paraconid on M1, Figure 9.5D; also see Freeman 1988).

Occlusion in frugivores is not an interlocking affair dependent on transverse crests. The upper ectoloph has simplified from a complex zigzag pattern to a single continuous, cutting edge (a cookie cutter) so that the dental arcade of lower teeth fits inside its perimeter. The outside or buccal edge of the lower molars acts as a pestle that nestles into the wide continuous mortar created by the lingual basin of PM4 and protoconal basins of M1 and M2 and has little side-to-side movement (see Figure 9.5D). The lingual parts of the teeth are broad crushing areas with small upper and lower lingual cusps adding rugosity to the surface. These small cusps fit together as loose mortars and pestles that would help crush rather than chop foods, but there is little contact between teeth lingually. When viewed laterally the dental arcade of the upper teeth, which are close-fitting teeth, is sharp and serrated.

Evolution

MEGACHIROPTERAN ANCESTOR.   Traditionally nectarivores are thought to be derived from frugivorous megachiropterans. However, I believe all megachiropterans evolved from a long-tongued ancestor, which could easily have been a nectarivore. Both nectarivores and frugivores have narrow, elongated palates with space between relatively simple, noncomplex teeth that do not register, fused mandibles, shallower, less distinct jaw joints, and thegosed upper canines. This suite of characteristics argues for a tongue-feeding ancestor that may have been either a nectarivore or a tongue-feeding frugivore or both. If this were the case, frugivorous megachiropterans became more robust with bigger teeth. Many of them reached large body masses and some of them became short-snouted (intertooth space is nonexistent in *Cynopterus* and small in *Dobsonia* and *Nyctimene*). Recent DNA evidence suggesting that nectarivorous megachiropterans are polyphyletic does not negate this possibility (Kirsch and Lapointe 1997). These authors show that more obligate nectarivores could have arisen several times from a long-tongued ancestor. Megachiropteran nectarivory and frugivory are based on a similar feeding mechanism and represent a continuum along a nectarivory–frugivory gradient.

TWO KINDS OF FRUGIVORY.   Frugivory in bats has evolved twice but has been achieved in fundamentally different ways. Microchiropteran frugivores differ from their nectarivorous confamilials in their short, wide palates and nondilambdodont, close-fitting teeth. Microchiropteran nectarivory and frugivory are not on the same continuum but represent two entirely different feeding mechanisms.

Microchiropteran and megachiropteran frugivores both mash fruit against their ridged palates with their tongues and have teeth with less discernible cusps, but they use different equipment and execution. Microchiropteran bats possess short, nonprotrusible tongues and use short, wide faces and a toothrow of serrated edges to cut into fruits with small bites. Perhaps the anecdote of why the wrinkled-face bat *Centurio* has wrinkles is true: It probably does have to bury its wide scoop-face right into a fruit to eat, and juice may well run down the wrinkles to the mouth in the absence of a big tongue. These stories also liken the nakedness of *Centurio*'s face to the baldness of the heads of vultures (Findley 1993).

In contrast, megachiropteran frugivore jaws are usually long, allowing bigger gapes and bigger bites. Processing is by teeth with blunt labial cusps and by a long, protusible tongue that reduces fruit on the ridged palate as it moves forward past the food to extend out of the mouth. Further, chewing is orthal or more vertical than chewing in a microchiropteran insectivore, which has a more lateral component to the chewing cycle (Greet and De Vree 1984). Large bites could be a function of longer jaws, teeth more anterior on the toothrow, and larger gapes (Savage 1977). The result is that tongue-feeding frugivory has evolved in megachiropterans and cookie-cutter frugivory has evolved in microchiropterans.

Modifying cookie-cutter frugivory to sanguinivory does not seem evolutionarily difficult but is not totally without obstacles. The raised stylar shelf of an ancestral frugivore could become the edge that is the cheekteeth in blood-feeders, and the shift of emphasis to the incisors at the anterior end of the palate continues the shift to the more anterior teeth seen in stenodermatine frugivores. *Pygoderma* (see Figure 9.2C) has greater incisor and canine area than consubfamilials. Sanguinivores also have relatively wide palates like stenodermatines (see Figure 9.1B), and occlusion is the ultimate in two cookie-cutter edges that shear past one another vertically without side-to-side movement. However, if derivation from stenodermatine frugivory were the case, the lingual areas of the teeth must have diminished and disappeared entirely, which would not be unlikely if the tongue were significantly larger. The anterior surfaces of the upper canines are heavily worn on the entire anterior surface and probably support a tongue that protrudes enough to lap. As in stenodermatines, the mandibles are fused completely in *Diphylla*. Fusion is less complete in *Desmodus*, which is problematical.

DILAMBDODONTY AND DIGESTIBILITY. Finally, dilambdodonty in microchiropteran bat teeth is likely to be correlated with digestibility of prey such that deeply emarginate teeth process the foods most difficult to digest. Chopping up insect prey into fine pieces may be as critical to the digestibility of insects in bats as it is for primates (Kay and Sheine 1979). The complex pinking-shears pattern allows insectivorous bats to take advantage of an abundant insect resource. How-

ever, the principal trend in the evolution of molar teeth in bats, as in terrestrial carnivores, is one of simplification, or a decrease in dilambdodonty. Bats evolved to larger sizes to take advantage of larger insects, but they also became large enough to take small vertebrates as well. Endoskeletal prey items are probably easier packages to break into and digest, especially with relatively large teeth. These teeth however would not need to be as deeply emarginate to chop and prepare food for digestion. The less-emarginate pattern in carnivorous species with their elongated interlophs, shortened intralophs, and correspondingly large protoconids seems well suited for processing meat on the bones of small prey and the bones themselves.

Teeth of nectarivorous bats have simplified by becoming diminutive and having a dilambdodont pattern that is shallower and not as high cusped. Nectar, pollen and whatever soft foods (insects? flowers?) nectarivores might be taking are surely more digestible than a diet restricted to insects. Although the enlarged tongue gathers much of the food, the talonids and protocones still have a crushing function and M3s can be fully formed. Dilambdodonty is reduced to form a dental arcade with a continuous, raised and sharpened edge in frugivores, a good design for cutting through the skin of fruit. A mouthful of the fruit's contents is crushed between broad horizontal surfaces of the teeth lingual to the rim and mashed by the tongue against the roof of the mouth to release the easily digested juice. Finally, blood-feeding bats need only vertical edges to cut through the skin of endoskeletal foods to release the liquid from within. Horizontal surfaces on the teeth and their crushing function disappear completely, and the protrusible tongue with its lapping function takes a more central role.

Conclusions

Megachiropteran frugivores, megachiropteran nectarivores, and microchiropteran nectarivores have in common craniodental characteristics that are correlated with having a long tongue. A long tongue in megachiropterans can explain why teeth do not interlock and why homologous cusp patterns cannot be discerned. It also means that megachiropterans evolved from a long-tongued ancestor that could have been a nectarivore or a tongue-feeding frugivore or both. Further, frugivory has been achieved in two different ways in bats: tongue-feeding frugivory in megachiropterans and cookie-cutter frugivory in microchiropterans. Among microchiropterans, nectarivores have functional although diminutive postcanine teeth, proportionally large-sized canines that may brace the long jaw during feeding, and molars which still function at the talonid–protocone contact.

Carnivores, in contrast, emphasize the trigonid–hypocone contact. Here, the protoconids act as large pestles that fit into deep and expanded mortars, which are the hypoconal basins and are probably useful for crushing bones of small vertebrates. Carnivores and insectivores have more dental material at the rear of the toothrow while all other bats emphasize dental material at the front of the toothrow. Not only is there a shift in allocation of dental material between suborders, there is also a shift among microchiropterans. Canine teeth in bats have edges that are nonrandomly oriented, which may aid in cutting into foods efficiently. Indeed, elaborate canines may be a necessity for aerial mammalian predation and harvesting in general.

Appendix 9.1.
Species Examined and Their Categories

Numbers to the left of the species' names refer to categories in Figure 9.2C (e.g., Insectivore 1, Insectivore 2). An asterisk indicates that the species is new to this study.

Microchiropteran Insectivores
1	Saccolaimus peli
1	Taphozous nudiventris
1	Peropteryx kappleri
1	Rhinolophus luctus
1	Rhinolophus rufus
1	Rhinolophus blasii
1	Hipposideros commersoni gigas
1	Hipposideros commersoni commersoni
1	Hipposideros lankadiva
1	Hipposideros pratti
1	Hipposideros ruber
2	Scotophilus nigrita gigas
1	Ia io
1	Myotis myotis
1	Myotis velifer
1	Nyctalus lasiopterus
2	Antrozous pallidus*
2	Eptesicus serotinus*
2	Otonycteris hemprichi*
1	Lasiurus cinereus*
1	Lasiurus borealis
2	Cheiromeles torquatus
1	Eumops perotis
1	Eumops underwoodi
1	Otomops martiensseni
1	Tadarida brasiliensis
2	Molossus molossus
2	Noctilio leporinus
1	Macrotus californicus*
1	Lonchorhina aurita*
1	Micronycteris megalotis*
1	Mimon bennettii*
1	Phylloderma stenops*
1	Phyllostomus hastatus
1	Phyllostomus elongatus*
1	Phyllostomus discolor*
1	Tonatia silvicola*
1	Carollia perspicillata

1	Mormoops megalophylla*
1	Pteronotus parnellii*

Microchiropteran Carnivores
4	Macroderma gigas
3	Megaderma lyra
4	Cardioderma cor
2	Nycteris grandis
1	Vampyrum spectrum
1	Chrotopterus auritus
1	Trachops cirrhosus

Microchiropteran Nectarivores
1	Phyllonycteris poeyi
1	Erophylla sezekorni
1	Glossophaga soricina
1	Glossophaga longirostris
1	Monophyllus plethodon
1	Monophyllus redmani
2	Lichonycteris obscura
2	Leptonycteris curasoae
1	Anoura caudifer
1	Anoura geoffroyi
1	Hylonycteris underwoodi
1	Choeroniscus godmani
1	Choeroniscus intermedius
1	Choeronycteris mexicana
1	Musonycteris harrisoni
1	Lonchophylla thomasi
1	Lonchophylla handleyi
1	Lionycteris spurrelli

Microchiropteran Frugivores
2	Artibeus jamaicensis
2	Artibeus lituratus
2	Artibeus phaeotis
2	Artibeus toltecus
1	Chiroderma villosum*
1	Uroderma bilobatum*

1	Ametrida centurio
2	Centurio senex
2	Ectophylla alba
3	Pygoderma bilabiatum
1	Sphaeronycteris toxophyllum
1	Sturnira lilium
1	Brachyphylla nana
1	Brachyphylla cavernarum

Microchiropteran Sanguinivores
	Diphylla ecaudata*
	Desmodus rotundus*

Megachiropteran Frugivores
1	Eidolon helvum*
1	Rousettus angolensis*
1	Pteropus poliocephalus*
1	Pteropus vampyrus*
1	Acerodon jubatus
1	Dobsonia moluccensis
1	Harpyionycteris whiteheadi
2	Cynopterus brachyotis*
2	Paranyctimene raptor*
2	Nyctimene draconilla
2	Nyctimene major

Megachiropteran Nectarivores
1	Pteropus scapulatus
2	Epomops buettikoferi
2	Scotonycteris zenkeri
1	Eonycteris spelaea
1	Eonycteris major
1	Megaloglossus woermanni
1	Macroglossus minimus
1	Macroglossus sobrinus
1	Syconycteris australis
1	Melonycteris melanops
1	Notopteris macdonaldi

Acknowledgments

Thanks go to the curators and staff at many museums over the years who have been very patient with my long-overdue loans. The museums include American Museum of Natural History; Field Museum of Natural History; Royal Ontario Museum; Texas A&M University, Texas Cooperative Wildlife Collection; Texas Tech University, The Museum; U.S. National Museum, Fish and Wildlife Labs; University of California–Berkeley, Museum of Vertebrate Zoology; University of Michigan, Museum of Zoology; University of Kansas, Museum of Natural History; and University of Nebraska State Museum. Pauline Denham, museum artist, helped with figures. Discussions of jaw mechanics with M. Joeckel were insightful, and the text benefited substantially from the helpful comments of U. Norberg, S. Swartz, and B. Van Valkenburgh. However, the final synthesis of ideas presented here would have suffered mightily without the constant repartee, unrelenting logic, and endless encouragement of C. Lemen. To him, as ever, I am most grateful.

Literature Cited

Andersen, K. 1912. Catalogue of the Chiroptera in the Collection of the British Museum. British Museum (Natural History), London.

Beecher, R. M. 1979. Functional significance of the mandibular symphysis. Journal of Morphology 159:117–130.

Butler, P. M. 1941. A theory of the evolution of mammalian molar teeth. American Journal of Science 239:421–450.

Butler, P. M. 1946. The evolution of carnassial dentitions in the Mammalia. Proceedings of the Zoological Society of London 116:198–220.

Butler, P. M. 1972. Some functional aspects of molar evolution. Evolution 26:474–483.

Caputo, A. A., and J. P. Standlee. 1987. Biomechanics in Clinical Dentistry. Quintessence, Chicago.

Crompton, A. W., and K. Hiiemäe. 1969. How mammalian teeth work. Discovery 5:23–34.

Douglas, A. M. 1967. The natural history of the ghost bat, *Macroderma gigas* (Microchiroptera, Megadermatidae), in western Australia. Western Australian Naturalist 10:125–138.

Every, R. G. 1970. Sharpness of teeth in man and other primates. Postilla 143:1–30.

Findley, J. S. 1993. Bats: A Community Perspective. Cambridge University Press, New York.

Fisher, D. W., A. A. Caputo, H. T. Shillingburg, and M. G. Duncanson. 1975. Photoelastic analysis of inlay and onlay preparations. Journal of Prosthetic Dentistry 33:47–53.

Freeman, P. W. 1979. Specialized insectivory: Beetle-eating and moth-eating molossid bats. Journal of Mammalogy 60:467–479.

Freeman, P. W. 1981a. Correspondence of food habits and morphology in insectivorous bats. Journal of Mammalogy 62:166–173.

Freeman, P. W. 1981b. A multivariate study of the family Molossidae (Mammalia, Chiroptera): Morphology, ecology, evolution. Fieldiana Zoology 7:1–173.

Freeman, P. W. 1984. Functional cranial analysis of large animalivorous bats (Microchiroptera). Biological Journal of the Linnean Society 21:387–408.

Freeman, P. W. 1988. Frugivorous and animalivorous bats (Microchiroptera): Dental and cranial adaptations. Biological Journal of the Linnean Society 33:249–272.

Freeman, P. W. 1992. Canine teeth of bats (Microchiroptera): Size, shape, and role in crack propagation. Biological Journal of the Linnean Society 45:97–115.

Freeman, P. W. 1995. Nectarivorous feeding mechanisms in bats. Biological Journal of the Linnean Society 56:439–463.

Freeman, P. W., and W. N. Weins. 1997. Puncturing ability of bat canine teeth: The tip. *In* Life among the Muses: Papers in Honor of James S. Findley, T. L. Yates, W. L. Gannon, and D. E. Wilson, eds., pp. 225–232. University of New Mexico Press, Albuquerque.

Greaves W. S. 1985. The generalized carnivore jaw. Zoological Journal of the Linnean Society 85:267–274.

Greet, G. De, and F. De Vree. 1984. Movements of the mandibles and tongue during mastication and swallowing in *Pteropus giganteus* (Megachiroptera): A cineradiographical study. Journal of Morphology 179:95–114.

Guard, W. F., D. C. Haack, and R. L. Ireland. 1958. Photoelastic stress analysis of buccolingual sections of class II cavity restorations. Journal of the American Dental Association 57:631–635.

Hill, J. E., and J. D. Smith. 1984. Bats: A Natural History. University of Texas Press, Austin.

Kay, R. F., and W. S. Sheine. 1979. On the relationship between chitin particle size and digestibility in the primate *Galago senegalensis*. American Journal of Physical Anthropology 50:301–308.

Kirsch, J. A. W., and F.-J. Lapointe. 1997. You aren't (always) what you eat: Evolution of nectar-feeding among Old World fruitbats (Megachiroptera: Pteropodidae). *In* Molecular Evolution and Adaptive Radiations, T. Givnish and K. Sytsma, eds., pp. 313–330. Cambridge University Press, New York.

Koopman, K. F., and G. T. MacIntyre. 1980. Phylogenetic analysis of chiropteran dentition. *In* Proceedings of the Fifth International Bat Research Conference, D. E. Wilson and A. L. Gardner, eds., pp. 279–288. Texas Tech University Press, Lubbock.

McCarthy, T. J. 1987. Additional mammalian prey of the carnivorous bats *Chrotopterus auritus* and *Vampyrum spectrum*. Bat Research News 28:1–3.

Morrison, D. W. 1980. Efficiency of food utilization by fruit bats. Oecologia 45:270–273.

Osborn, H. F. 1907. Evolution of Mammalian Molar Teeth. Macmillan, New York.

Osborn, J. W., and A. G. S. Lumsden. 1978. An alternative to "thegosis" and a re-examination of the ways in which mammalian molars work. Neues Jahrbuch für Geologie und Paläontologie Abhandlungen 156:371–392.

Peterson, R. L., and F. Kirmse. 1969. Notes on *Vampyrum spectrum*, the false vampire bat, in Panama. Canadian Journal of Zoology 47:140–142.

Rensberger, J. M. 1995. Determination of stresses in mammalian dental enamel and their relevance to the interpretation of feeding behaviors in extinct taxa. *In* Functional Morphology in Vertebrate Paleontology, J. J. Thomason, ed., pp. 151–172. Cambridge University Press, New York.

Savage, R. J. G. 1977. Evolution in carnivorous mammals. Paleontology (London) 20:237–271.

Scapino, R. P. 1965. The third joint of the canine jaw. Journal of Morphology 116:23–50.

Slaughter, B. H. 1970. Evolutionary trends of chiropteran dentitions. *In* About Bats, B. H. Slaughter and D. W. Walton, eds., pp. 51–83. Southern Methodist University Press, Dallas, Tex.

Strait, S. G. 1993a. Molar morphology and food texture among small-bodied insectivorous mammals. Journal of Mammalogy 74:391–402.

Strait, S. G. 1993b. Molar microwear in extant small-bodied faunivorous mammals: An analysis of feature density and pit frequency. American Journal of Physical Anthropology 92: 63–79.

Werdelin, L. 1989. Constraint and adaptation in the bone-cracking canid *Osteoborus* (Mammalia: Canidae). Paleobiology 15:387–401.

Zienkiewicz, O. C., and R. L. Taylor. 1989. The Finite Element Method, Vols. 1 and 2. McGraw-Hill, London.

10
Chiropteran Hindlimb Morphology and the Origin of Blood Feeding in Bats

WILLIAM A. SCHUTT, JR.

Functional morphologists are concerned with the relationship between form and function, so it is not surprising that most such studies on bats have focused on the forelimb and its exquisite modification into a wing (Vaughan 1959, 1966, 1970a, 1970b; Altenbach 1968, 1979; Norberg 1969, 1972a, 1972b, 1976, 1987; Vaughan and Bateman 1970; Findley et al. 1972; Pirlot 1977; Strickler 1978; Smith and Starett 1979; Hermanson and Altenbach 1981, 1985; Norberg and Rayner 1987). Although it plays an important role in functions such as flight, terrestrial locomotion, and hanging, the chiropteran hindlimb and its morphology have been examined by relatively few researchers.

The study by Macalister (1872) of chiropteran myology remains a principal reference for bat musculature and offers an accurate description of the unique posture of the chiropteran hindlimbs:

The position of the parts in these limbs is so remarkable that a brief review is necessary before describing the muscles. The variation in position from the usual disposition of hind limbs in Mammalia may be described (as follows): the limbs instead of having suffered a rotation forwards from their embryonic position, have been rotated backwards, and this has caused the following peculiarities: the knee joints are directed backwards and outwards, the tibial side of the leg inclines outwards and forwards, the fibular side inwards, the plantar surface of the foot is directed forwards, the outside of the femur is directed backwards and a little inwards.

Although Macalister described a rotation of the chiropteran knee joint by as much as 180°, giving bats a stance more like that of a grasshopper than a typical quadrupedal vertebrate, few researchers have extended his examination of chiropteran hindlimb morphology. Those who did, however, encountered a number of interesting problems.

In a study on the digital anatomy of hanging mammals, Schaffer (1905) described a mechanism in bat digits that apparently allows these animals to hang for extended periods of time without active contraction of the hindlimb musculature (Figure 10.1). This ratchet-like passive digital lock (or tendon locking mechanism) was further examined by Schutt (1992, 1993), Bennett (1993), and Quinn and Baumel (1993). The presence, absence, or modification of the digital lock has been used to address questions of

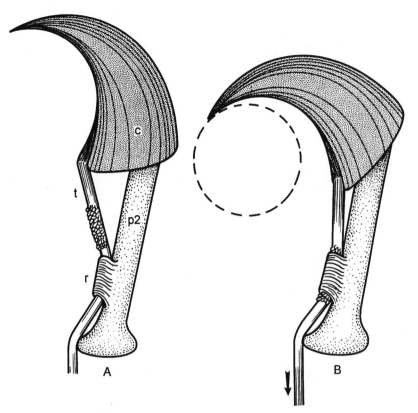

Figure 10.1. Lateral-view schematic of a typical bat hindlimb digit. (**A**) During digital extension, the scaled section of the flexor digitorum longus tendon lies distal to the retinaculum. (**B**) When the digit is flexed, the flexor digitorum longus tendon is pulled proximally (in direction of arrow) until the scales engage the plicae investing the inner surface of the retinaculum, forming the passive digital lock. p2, second or middle phalanx; t, tendon; r, retinaculum; c, claw. (Modified from Schaffer 1905.)

higher-level taxonomy within the Chiroptera (Simmons and Quinn 1994). For example, the presence of the passive digital lock in microchiropterans, megachiropterans, and dermopterans (but not in primates or scandentians) was used as yet another source of evidence supporting the monophyletic origin of bats (Simmons 1994).

Vaughan (1959) was the first investigator to draw specific parallels between adaptations of the hindlimb and their functional significance by studying the functional morphology of three North American bats, *Eumops perotis* (Molossidae), *Myotis velifer* (Vespertilionidae), and *Macrotus californicus* (Phyllostomidae). Vaughan proposed that the unique rotation of the chiropteran knee joint was related to the attachment of the plagiopatagium to the hindlimb as well as the uropatagium or interfemoral membrane, which stretches between the hindlimbs and whose form and presence vary between species. Variations in the osteology and myology of the hindlimb and pelvic girdle, according to Vaughan, "were related principally to great differences that occur in roosting habits and modes of terrestrial locomotion." The "strikingly reptilian posture" of bats was the result of the importance of the hindlimbs in both aerial and terrestrial locomotion. Because of the scanty fossil record of bats, Vaughan also stressed that the morphology of modern bats can provide important information on chirop-

teran evolution. For example, the broad, flat body shape of bats, modifications of the pelvic and pectoral girdles, and the posture of the limbs were cited to suggest that bat ancestors may have roosted in crevices.

Interfamiliar relationships within the Phyllostomidae were investigated by Walton and Walton (1970). Although their study was primarily a comparison of the osteology of the pelvic and pectoral girdles, characters generated from the hindlimb (the presence or absence of fibulae and variation in hindfoot phalanx number) were used, apparently for the first time, in phylogenetic analysis. In another study, characters such as foot and hindlimb length were used to provide phenetic evidence for a paleotropical origin for the genus *Myotis* (Findley 1972).

The unique morphology of the hindlimb bones of bats was used to propose a theory on the origin of hanging in the Chiroptera (Howell and Pylka 1977). These authors suggested that during bat evolution the hindlimb bones, especially the femur, may have undergone weight reduction as an adaptation to flight. This adaptation can be seen in birds through the evolution of pneumatic bones. In bats, however, load may have been lessened through a decrease in bone diameter relative to bone length. Because these delicate limb bones may have been more prone to buckling under the compressive loads normally associated with an upright body stance, hanging, in which tensile loads are applied to the hindlimb, made bat hindlimb bones less prone to mechanical failure.

Noctilio leporinus and *Myotis (Pizonyx) vivesi* use robust hindlimbs with large specialized feet to capture both aquatic and terrestrial prey (Reeder and Norris 1954; Bloedel 1955). Because these bats are also skilled at quadrupedal locomotion, it was proposed that well-developed hindlimbs may have been a preadaptation to capturing prey by foot (Novick and Dale 1971). Comparative hindlimb morphology of these two genera seems to support this scenario because hindlimb muscles are not especially adapted to fish-catching (Blood 1987).

Beyond the basic aspects of form and function, I have asked questions that arise when variations in hindlimb morphology can be demonstrated, especially between related taxa. As previous studies have recorded intergeneric variation in quadrupedal locomotor performance of vampire bats (Altenbach 1979, 1988; Schutt et al. 1993; Schutt 1995), I am interested in determining if these differences are reflected by variations in hindlimb morphology. Additionally, the evolution of blood feeding in bats is an intriguing subject. Because the fossil record of vampire bats offers no clues to the origin of blood feeding, I used comparative morphology, information on phyllostomid behavior and phylogeny, and South American Miocene faunal diversity to address two questions concerning vampire bat origins. Those questions are (1) how did blood feeding evolve in bats? and (2) why is blood feeding confined to only three genera of Neotropical phyllostomids? This chapter presents an overview of previous research on the bat hindlimb. Additionally, morphological data on vampire bats and other information are used to propose a new hypothesis for the evolution of blood feeding in bats. Where appropriate, I have tried to identify areas for future study.

Blood-Feeding Bats

Although approximately 920 species of bats occur worldwide, comprising one-quarter of the living mammal fauna, only 3 of these are vampire bats (Koopman 1993). Vampire bats have been described as "pinnacles of specialization and adaptation in the bat world" (Greenhall 1988). Although they are unique in that they exhibit a lifestyle in which blood is their sole source of nourishment, only *Desmodus rotundus*, the common vampire bat, has been studied extensively (for a review, see Greenhall and Schmidt 1988). Relatively little is known about the other 2 genera, *Diaemus youngi*, the white-winged vampire bat, and *Diphylla ecaudata*, the hairy-legged vampire bat. Vampire bats exhibit a fascinating assortment of morphological and behavioral specializations related to their diet. Past research on the morphology of vampire bats has examined features such as highly derived dentition (Storch 1968; Greenhall 1972; Phillips and Steinberg 1976), salivary factors that inhibit blood clotting (Hawkey 1966, 1967, 1988), the digestive system (Huxley 1865; Rouk and Glass 1970; Forman 1972; Kamiya et al. 1979), and specialized kidneys (Horst 1969; McFarland and Wimsatt 1969).

Also related to their unique feeding habits, vampire bats exhibit a degree of complexity and agility during quadrupedal locomotion that is unparalleled in bats (Altenbach 1979). Alighting near its intended prey, *Desmodus rotundus* uses a quadrupedal gait that varies between walking, spider-like scrambling, jumps into short flights, and hopping. *Desmodus* feeds primarily from the ground and from there it can make powerful jumps to initiate flight, escape predators, and avoid being stepped on by large prey. Variation exists in quadrupedal locomotor performance between *Desmodus rotundus* and *Diaemus youngi*, which is related to terrestrial versus arboreal feeding modes in these respective genera (Schutt et al. 1993; Schutt 1995). The diet of *D. youngi* consists primarily of avian blood, which is obtained from perching birds, stalked as they sleep, but these bats have also been reported to feed on mammalian blood (Goodwin and Greenhall 1961; Greenhall 1970, 1988; Uieda 1994). In captivity, *D. youngi* exhibits terrestrial feeding behavior similar to that of *Desmodus rotundus* but never shows the rapid terrestrial gait and jumping ability seen in *D. rotundus*. *Diphylla ecaudata* feeds solely on the blood of birds, stalked as they perch in trees (Uieda 1982, 1986, 1994; Uieda et al. 1992). There are no reports of *D. ecuadata* feeding from the ground or by any means other than arboreal feeding.

Vampire Bat Phylogeny

Phylogenetic assessment of vampire bats has been both difficult and controversial (Koopman 1988). Although phenetic specializations related to feeding have led some researchers (Miller 1907; Vaughan 1986) to consider vampire bats as a distinct family, phylogenetic analyses of morphological data (Machado-Allison 1967; Forman et al. 1968; Smith 1972, 1976; Griffiths 1982; Hood and Smith 1982) and immunological data (Honeycutt and Sarich 1987) indicate that vampire bats form a clade, currently recognized as the subfamily Desmodontinae within the family Phyllostomidae. Immunological, morphological, and chromosomal data now support the affiliation of the vampire bats as a basal phyllostomid lineage that originated between 2 and

10 million years ago (Honeycutt et al. 1981; Baker et al. 1988; Koopman 1988).

Koopman (1988) determined that seven of nine synapomorphies (shared-derived characters) within the Desmodontinae are shared only by *Desmodus rotundus* and *Diaemus youngi* (presence of thumb pads, reduction in calcar size, reduction in molar number, upper molar form, loss of posterior upper incisors, increased size of upper incisors and canines, and increased coronoid height compared to length of mandible). These synapomorphies likely indicate increased specialization for blood feeding and quadrupedal locomotion in *D. rotundus* and *D. youngi*. Koopman concluded that "it is therefore probable that the latest common ancestor of *Diaemus* and *Desmodus* was not also the ancestor

of *Diphylla* (which had branched off earlier)." Additionally, studies based on immunological and chromosomal evidence support the morphological data and indicate that the *Diphylla* lineage was an early offshoot from the *Desmodus* and *Diaemus* line (Honeycutt et al. 1981; Baker et al. 1988).

Hindlimb Morphology of Vampire Bats

Skeletal elements of *Desmodus rotundus* (Figure 10.2A) and *Diaemus youngi* (Figure 10.2B) are more robustly built than the same elements in *Diphylla ecaudata* (Figure 10.2C). For example, the femur and tibia in *Desmodus rotundus* and (to a lesser extent) *Diaemus youngi* are U-shaped in cross section while the same bones in *Diphylla ecaudata* do not exhibit this

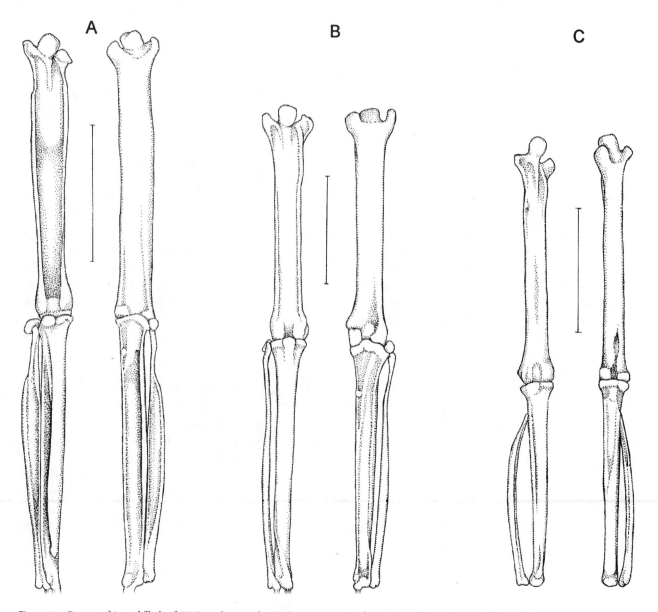

Figure 10.2. Femora, tibia, and fibula of (**A**) *Desmodus rotuindus,* (**B**) *Diaemus youngi,* and (**C**) *Diphylla ecaudata.* Scale bar = 10 mm.

grooved appearance. The shaft of the fibula in *D. rotundus* and *D. youngi* is roughly T-shaped in cross section because of lateral and medial ridges. These flangelike ridges are extremely prominent in *D. rotundus*, less prominent in *D. youngi*, and absent in *D. ecaudata*. Additionally, in *D. rotundus* and *D. youngi* the fibula spans the entire length of the tibia. The head of the fibula is in contact with the lateral condyle of the tibia. In *D. ecaudata*, the head of the fibula extends from a position 2 mm distal to the medial condyle of the tibia. Vaughan (1959) described similar morphology in the hindlimb of *Macrotus* as an "incomplete" fibula.

In bats, the semitendinosus and gracilis muscles work as a functional unit (Vaughan 1959, 1970a). With origins on the pelvis, these muscles insert together via a fused tendon onto the tibia where they function to flex the knee and adduct the femur. In *Desmodus rotundus* and *Diaemus youngi*, the fused tendon attaches near the knee joint (approximately one-fifth of the way along the tibia, on the posteromedial surface); in *Diphylla ecaudata* the tendon attaches more distally (approximately one-third of the way along the tibia).

OSTEOLOGICAL VARIATION. Howell and Pylka (1977) found that the hindlimb bones of most bat species are so delicately built that they do not fit engineering models for minimum weight support. Vampire bats appear to be exceptions with their unique locomotor habits (e.g., quadrupedal locomotion) leading to a reversal of the trend toward delicate limb bones in bats. Hindlimb bones, most notably the femora, tibia, and fibula, are more robustly built in *Desmodus rotundus* and *Diaemus youngi* then they are in *Diphylla ecaudata*. The complete fibula exhibited by *D. rotundus* and *D. youngi* is a synapomorphy for these genera relative to *D. ecaudata*. Derived conditions of the hindlimb (like those noted for cranial and postcranial characters) appear to reflect adaptations for terrestrial locomotion and feeding. For example, deeply grooved flanging of the tibia and fibula, synapomorphies shared by *D. rotundus* and *D. ecaudata*, may provide increased surface areas for the origin of muscles involved in flexion (m. plantaris and flexor digitorum fibularis) and extension (m. peroneus longus and peroneus brevis) of the foot. These muscles are important components of the terrestrial stride (Vaughan 1959, 1970a). Note that *Noctilio* (Noctilionidae) and *Pteronotus* (Mormoopidae) were used for outgroup comparison as most recent workers agree that they are the sister-taxa of the Phyllostomidae (see Simmons, Chapter 1, this volume).

MYOLOGICAL VARIATION. Proximal insertion of the gracilis and semitendinosus muscles onto the tibia is another synapomorphy linking *Desmodus rotundus* and *Diaemus youngi*.

One possible reason for the variation in tendon insertion may be related to knee flexion in the vampires. Mechanically, a more proximal insertion in *D. rotundus* and *D. youngi* translates to a shorter in-lever; this may allow the femur to be adducted through a wider angle during quadrupedal locomotion while increasing the out-velocity of the limb. This advantage in speed would be at the expense of the generation of force (see Hildebrand 1995). This observation merits further study, especially because the hindlimbs of vampire bats are not important components in the generation of force during quadrupedal locomotion (Altenbach 1979; Schutt 1995). Conversely, the distal insertion of these muscles in *Diphylla ecaudata* increases the length of the in-lever, which may permit the knee to be flexed more forcefully (at the expense of moving the limb quickly). Photographs of *D. ecaudata* climbing a wooden dowel indicate that knee flexion is an important component of branch-grasping behavior (Altenbach 1988). Additional study is required to determine if variation in the arrangement of vampire bat hindlimb muscles is related to rapid terrestrial locomotion versus climbing efficiency.

THE CALCAR. The uropatagium provides a substantial lift surface for many bat species and also functions during rapid braking and turning maneuvers (Vaughan 1959, 1970b, 1970c; Norberg 1990). In a number of genera (e.g., *Myotis*) it is used during feeding behavior to capture aerial insects before transfer to the mouth (Webster and Griffin 1962; Kalko and Schnitzler 1989). In vampire bats, the uropatagium is reduced or absent. Perhaps this is a phylogenetic constraint, if vampire ancestors also lacked a uropatagium, or a trade-off related to the dual function of the vampire bat hindlimb in quadrupedal and aerial locomotion. Researchers have suggested that a uropatagium and tail reduced in size would be "less of a hindrance in climbing and clinging among vegetation" (Norberg and Rayner 1987). It is not surprising that the support structure for the uropatagium, the calcar, is absent in *Diaemus youngi* and reduced to a tablike flap in *Desmodus rotundus* (Figure 10.3A,B). Koopman (1988) reported that calcar reduction was a synapomorphy linking these genera but described "a small but well-developed calcar" in *Diphylla ecaudata*. I found this digitiform structure extended some 3 mm past the edge of the fringelike uropatagium (Figure 10.3C). Depressor ossis styliformis and gastrocnemius muscles attach to the calcar and likely function in abduction and stabilization of the calcar, respectively.

The behavior of *D. ecaudata* in the field indicates that this bat is an arboreal hunter, stalking its prey while they sleep perched among branches (Uieda 1982, 1986, 1994; Greenhall 1988; Uieda et al. 1992; J. S. Altenbach, B. Villa-R., and

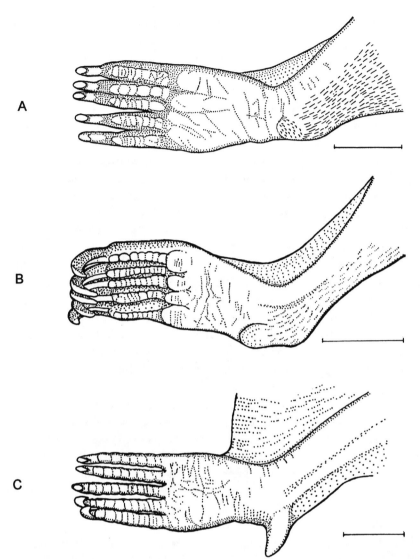

Figure 10.3. Superficial aspect of the foot of three vampire bats: (**A**) *Diaemus youngi;* (**B**) *Desmodus rotundus;* (**C**) *Diphylla ecaudata.* Note the reduction of the calcar in *Desmodus* and *Diaemus.* The calcar of *Diphylla* extends past the uropatagium and apparently functions as an opposable digit during arboreal locomotion and feeding. Scale bar = 3 mm.

A. Greenhall, personal communications). Hoyt and Altenbach (1981) maintained *Diphylla* in captivity for the first time and described its feeding behavior. In light of the anatomical specialization of the calcar, I reexamined previously unpublished photographs from their study. These photographs provide evidence that *Diphylla* uses its calcar as an opposable sixth digit to facilitate branch-grasping during arboreal locomotion (Schutt 1994; Schutt and Altenbach 1997).

The Origins of Blood Feeding in Bats

How did blood feeding originate in the ancestors of modern vampire bats? Why is blood feeding confined to three genera of New World phyllostomid bats? The very sparse vampire fossil record sheds no light on these questions, although a number of interesting hypotheses have been presented.

Ectoparasite-Feeding Hypothesis

In separate studies, Gillette (1975) and Turner (1975) suggested that the ancestors of vampire bats fed on the ectoparasites of large mammals. In this scenario, protovampires fed on blood-engorged parasites "and later directly on the mammalian blood" (Turner 1975). Seemingly, this hypothesis was founded upon the knowledge that most bat species were insectivores, combined with anecdotal evidence that vampire bats had been observed consuming parasitic moths (Lepidoptera: Noctuidae, Westermanniinae). If protovampires were ectoparasite gleaners, I propose that blood feeding might first have originated during mutual grooming behavior. Although there have been no formal reports of vampire bats feeding on ectoparasites, extant vampire genera are highly social animals and spend a considerable amount of time (mean, 5% of total time) involved in mutual grooming behavior (Greenhall 1965; Crespo et al.

1972; Wilkinson 1986). During such behavior, protovampires may have obtained their first taste of blood from members of the soft-bodied tick family Argasidae that commonly parasitize bats. McClearn (1992) reported that coatis *(Nasua narica)* on Barro Colorado Island occasionally fed on ticks groomed off tapirs *(Tapirus bairdii)*, but no extant mammals are reported to feed extensively on ectoparasites.

Fenton (1992a) suggested that the small size of ectoparasites, combined with the difficulty of locating them on another animal, made the ectoparasite-feeding hypothesis improbable. He added that because ectoparasites have a worldwide distribution, this theory did not explain the restriction of vampire bats to the New World.

Wound-Feeding Hypothesis

Fenton (1992a) proposed that blood feeding derived from bats feeding on insects present at the wound sites of large mammals. Bats feeding on these insects and their larvae would have gained additional nourishment from the blood and flesh of the wounded mammal. While early protovampires would have fed on both insects and blood, eventually a switch to feeding solely on blood would have followed. Fenton's hypothesis has four components: (1) flexibility in phyllostomid foraging behavior (e.g., a tendency to hunt for prey in different settings); (2) diverse Miocene mammal fauna present in South America at the time vampire bats made their appearance, which ensured a readily available supply of potential prey; (3) the robust nature of the upper incisor teeth in some phyllostomids, which would have been available for modification into the bladelike incisors that characterize vampire bats; and (4) wounds caused by intraspecific competition or unsuccessful attacks by predators that offered protovampires the opportunity to feed on blood.

As additional support for his theory, Fenton cited the feeding behavior of two bird species, the oxpeckers *(Buphagus)* that feed by gleaning ectoparasites from large mammals (Stutterheim and Panegis 1985; Hustler 1987) and at times feed at wounds and sores. Fenton also speculated that thermoperception may have been an effective way for protovampire bats to locate feeding sites because wounds infested by parasites (e.g., screwworm larvae) are, at times, warmer than the surrounding areas (Rubink 1987).

The first three components of Fenton's hypothesis make a strong case for the reason that blood feeding evolved in the phyllostomid lineage, but open wounds are an unpredictable food source and furthermore this theory relies on a feeding strategy not employed by any extant bat species. The wound-feeding hypothesis proposes that vampire bats and their specializations for blood feeding evolved in the face of selection pressures that would have seemingly acted against the development of such behavior. Not only would an appropriate prey need to be wounded but it would have to be of a conspicuously large size and relatively immobile. Because fresh blood contains no fat and virtually no carbohydrates, modern vampire bats cannot store energy in the manner of non-blood-feeding mammals, and they are required to consume between 50% and 100% of their body weight in blood each day to survive (Wimsatt and Guerriere 1962). Subadult vampire bats are unsuccessful in 33% of nightly feeding forays (Wilkinson 1985), and I predict that this figure would be much higher if the prey were required to have existing open wounds. It is difficult to imagine how natural selection would have led protovampires to abandon an insectivorous lifestyle for one dependent on locating large wounded mammals on a daily basis.

Moreover, in extant vampire bats thermoreceptors are accurate at a distance of 13–16 cm (Kurten and Schmidt 1982). If protovampires were similarly thermoperceptive, then thermoreception may have assisted these bats in the close range of parasite infestations but would be useless for locating feeding sites from distances greater than 16 cm.

Finally, if protovampires hunted at night, it is unlikely that vision and echolocation would have enabled them to differentiate between wounded and nonwounded animals. If the wound-feeding hypothesis is correct, acute hearing capabilities and olfaction may have led protovampires to insects swarming around wounds. The importance of olfaction in prey detection by vampire bats requires further investigation.

Dentition Hypothesis

In the dentition hypothesis, well-developed incisors used to slice through thick fruit rinds would have evolved into the bladelike teeth that characterize extant vampire bats (Slaughter 1970). Fenton (1992a, 1992b) rejected this hypothesis on the basis that "vampirism" never evolved in Old World fruit bats even though they possess robust upper incisors. This reasoning is similar to rejecting the ectoparasite-feeding hypothesis on the grounds that the worldwide distribution of ectoparasites fails to support the relatively localized distribution of vampire bats. Both arguments fall short because they suggest that the results of natural selection are predictable and invariant rather than unpredictable and largely a product of contingency (see Gould 1989).

Fenton cited a lack of flexibility in foraging behavior as the reason megachiropterans never evolved into vampires (Fenton 1992a, 1992b). It is likely that megachiropterans carried with them many characters (e.g., generally large body size, lack of echolocating ability) that may have acted as constraints to the development of blood feeding. These characters may help to explain the megachiropteran's "lack of behavioral flexibility" (Fenton 1992a), but even if ancient megachiropterans had exhibited extreme foraging flexibility, the exact set of circumstances that led to the evolution of blood feeding in phyllostomid protovampires was not present for the megachiropterans. Even had those exact circumstances existed, there would be no guarantee that blood feeding would have evolved again. It is, therefore, unreasonable to relate the lack of blood feeding among megachiropterans to the origin of microchiropteran protovampires, and similarly the worldwide distribution of ectoparasites cannot be used to rule out the ectoparasite-feeding hypothesis.

Arboreal-Feeding Hypothesis

As an alternative, I propose that protovampires foraged in much the same manner as a number of extant phyllostomids; namely, they were arboreally feeding omnivores. This argument is based on (1) comparative morphology of extant vampire bats, (2) knowledge of extant phyllostomid feeding behavior and phylogeny, and (3) Miocene faunal diversity.

Diphylla ecaudata is the only extant species of bat reported to exhibit arboreal feeding exclusively. *Diaemus youngi* typically preys on birds roosting in trees, but it can also feed from the ground (Muradali et al. 1993; Uieda 1994; Schutt 1995) and will take the blood of mammals (Goodwin and Greenhall 1961; Greenhall 1970; Greenhall and Schmidt 1988; Uieda 1994; Schutt 1995). Although *D. youngi* and *Desmodus rotundus* exhibit some morphological variation (e.g., thumb length and the number of metacarpal pads), which may explain observed differences in quadrupedal performance (Schutt et al. 1993; Schutt 1995), morphological, chromosomal, and molecular evidence indicates a close relationship between these two genera relative to *Diphylla ecaudata*. More importantly, these two genera share a number of synapomorphies related to increased specialization for blood feeding and quadrupedal locomotion. Because all the available evidence indicates that *D. ecaudata* is the most primitive of the vampire bats, and in the absence of a clear fossil record, I propose that the arboreal foraging behavior of *D. ecaudata* offers a plausible hypothesis for the origin of blood feeding in bats. As is shown, arboreal-feeding behavior, unlike wound feeding and ectoparasite gleaning, is employed by a number of bat taxa including members of the basal phyllostomid subfamily Phyllostominae.

OMNIVOROUS PHYLLOSTOMIDS.   The Neotropical phyllostomids are extremely diverse with regard to diet. Many are frugivorous while others forage for insects, pollen, nectar, or a combination of food items. Most members of the subfamily Phyllostominae are more or less omnivorous (Norberg and Raynor 1987). *Macrotus* and *Micronycteris,* for example, feed on fruit and insects while *Phyllostomus hastatus, Trachops cirrhosus,* and *Chrotopterus auritus* are known to include vertebrates in their diet (Dobson 1878; Goodwin 1946; Ruschi 1953; Goodwin and Greenhall 1961; Olrog 1973; Gardner 1977; Sazima 1978; Norberg and Raynor 1987). The diet of the largest microchiropteran, *Vampyrum spectrum,* includes birds, bats, and arboreal rodents (Goodwin and Greenhall 1961; Peterson and Kirmse 1969; Vehrencamp et al. 1977). Observations of a *V. spectrum* roost in Costa Rica revealed that foliage-roosting birds (weighing to 150 g) were the preferred prey (Vehrencamp et al. 1977). Roosting behavior of prey species was characterized as "stationary and hidden" rather than "active and exposed," and it was further suggested that *V. spectrum* locates stationary prey by olfaction rather than by vision or echolocation. Although there are no reports of the actual technique used by *V. spectrum* to capture prey in the field, in captivity this bat was reported to be "a careful, slow stalker" (Greenhall 1968). The nocturnal behavior of the prey exploited by *V. spectrum* in the field was also used to support prey capture by stalking (Vehrencamp et al. 1977).

OMNIVOROUS PROTOVAMPIRES.   The discovery of a fossil phyllostomid, *Notonycteris,* from the late Miocene of Columbia indicates that phyllostomids were established in South America by this time (Savage 1951; Smith 1976). Furthermore, *Notonycteris* is similar in morphology to *Vampyrum* and presumably had similar feeding habits, including arboreal insects, small reptiles, birds, and marsupials in its diet. It is likely that prey located among the branches were attacked and subdued by bites before being eaten. During the mid- to late Miocene, evidence suggests that South America was an island continent with rainforests existing as isolated, fragmented refugia surrounded by grasslands (Simpson and Haffer 1978; Simpson 1980). These tropical refugia may have been perfect staging grounds for evolutionary branching events such as speciation within the phyllostomids.

At this time, it is likely that omnivorous phyllostomids encountered members of an increasingly diverse arboreal and scansorial fauna. Larger marsupials, as well as primates, sloths, and procyonids, were taking up residence in the trees during the Miocene (Simpson 1980; Marshall et al. 1982) and larger forms of birds were also present (Feduccia 1980). Many of these animals may have been too large for bats to prey upon using previously existing attack strategies. Over time, an isolated population of phyllostomids may have undergone dietary and behavioral changes that enabled them to exploit larger animals as a food source. Perhaps changes in climate led to a high adaptive premium on alternative foraging strategies. Stalking their prey as they slept among the branches, protovampires may have begun biting larger species. There are several predators (e.g., the sabertooth blenny, *Aspidontus*, which mimics the cleaner wrasse, *Labroides*) that forage by taking a bite out of living prey. As in Fenton's wound-feeding hypothesis, early protovampires may have supplemented their normal diets with the flesh of the wounded animal and from licking its blood at the wound site. Selection would favor adaptations that provided opportunities for maximal nutritional payoff such as the infliction of painless bites.

It is likely that protovampires underwent evolutionary modification of previously existing anatomy (e.g., teeth and tongue) and physiology (e.g., digestive and excretory systems). Adaptations related to quadrupedal locomotion may have included a reduction in the size of the uropatagium (Norberg and Rayner 1987). Modification of the hindlimb digits was constrained by their position and function in hanging, so the calcar was retained and modified to aid in arboreal locomotion. Derived conditions for terrestrial locomotion (e.g., robust limb bones, elongated thumbs, further reduction of the calcar) may have evolved as some vampires (the *Desmodus-Diaemus* lineage) adapted their arboreal hunting techniques to exploit ground-dwelling vertebrates as a blood source.

It is possible that at one time terrestrially feeding vampires were more opportunistic in their feeding behavior (i.e., they could feed either terrestrially or arboreally). Where they existed sympatrically, however, it may have become selectively advantageous to excel in one or the other feeding niche. Character displacement may have resulted in a reduction of competition between terrestrially hunting vampire bats and a relatively recent return to primarily arboreal hunting in *Diaemus youngi*. This hypothesis would be supported if *Desmodus rotundus* and *Diaemus youngi*, which have apparently evolved into primarily terrestrial and arboreal niches, respectively, where their ranges overlap, have failed to do so or are convergent where their ranges do not overlap.

Conclusions

Comparative hindlimb morphology of three vampire bat genera has shown variation that appears to be related to arboreal versus terrestrial feeding. *Desmodus rotundus* and *Diaemus youngi* have in common derived hindlimb characters related to quadrupedal locomotion and terrestrial blood feeding. Sturdier hindlimb bones (e.g., with increased surface area for muscle attachment) and the proximal inser-

tion of the semitendinosus and gracilis muscles appear to be adaptations for terrestrial locomotion.

Diphylla ecaudata exhibits hindlimb characters related to arboreal feeding and locomotion that are primitive for vampire bats. These characters include an incomplete fibula and hindlimb bones lacking flangelike ridges (relative to the same bones in *D. rotundus* and *D. youngi*). Distal insertions of the gracilis and semitendinosus muscles suggest an important role for these muscles in forceful knee flexion. *Diphylla ecaudata* possesses a uniquely digitiform calcar that is apparently employed as an opposable sixth digit during arboreal feeding and locomotion.

Competition between *Desmodus rotundus* and *Diaemus youngi* may have resulted in the relatively recent shift of *D. youngi* to a diet of avian blood. The close phylogenetic relationship of *D. rotundus* and *D. youngi*, coupled with reports of dietary flexibility and terrestrial feeding behavior in *D. youngi*, support this scenario.

The arboreal-feeding hypothesis proposes that protovampire bats may have been arboreally feeding, omnivorous phyllostomids. It is based upon the following reasoning:

1. Unlike the hypothesized parasite-feeding and wound-feeding modes, omnivory is a feeding mode exhibited in many extant bats (including a number of phyllostomids).
2. Behavioral and morphological similarities exist between omnivorous phyllostomids and vampire bats.
3. Arboreal feeding appears to be the primitive condition in vampire bats.
4. The ability to exploit larger arboreal species as prey may have provided the selection pressure for protovampires to adopt a blood-feeding lifestyle.

Under the arboreal-feeding hypothesis, I make the following prediction: Protovampire fossils will be characterized by the presence of well-developed calcars, incomplete fibula, and hindlimb bones that lack flangelike ridges.

Acknowledgments

Very special thanks and gratitude to John W. Hermanson and Nancy B. Simmons for their time, encouragement, and support. I acknowledge contributions from the following individuals: R. Adamo, J. S. Altenbach, J. Bertram, K. Brockmann, A. Greenhall, R. Horst, K. Koopman, T. Kunz, D. Lunde, Deedra McClearn, R. McPhee, F. Muradali, J. Ryan, J. P. Schutt, W. R. Schutt, R. Tuna, and P. Wynne. This manuscript was greatly improved through the editorial contributions of S. Swartz, B. Fenton, U. Norberg, T. Kunz, and an anonymous reviewer. This project was funded in part through Cornell University, the American Museum of Natural History (Coleman Research Fellowship, Collection Study Grant, Theodore Roosevelt Memorial Award), the Society of Mammalogy, and Sigma Xi.

Literature Cited

Altenbach, J. S. 1968. The functional morphology of two bats: *Leptonycteris* and *Eptesicus*. Master's thesis, Colorado State University, Boulder.

Altenbach, J. S. 1979. Locomotor morphology of the vampire bat *Desmodus rotundus*. Special Publications of the American Society of Mammalogy 6:1–137.

Altenbach, J. S. 1988. Locomotion. *In* Natural History of Vampire Bats, A. M. Greenhall and U. Schmidt, eds., pp. 71–83. CRC Press, Boca Raton, Fla.

Baker, R. J., R. L. Honeycutt, and R. A. Bass. 1988. Genetics. *In* Natural History of Vampire Bats, A. M. Greenhall and U. Schmidt, eds., pp. 31–40. CRC Press, Boca Raton, Fla.

Bennett, M. B. 1993. Structural modifications involved in the fore and hind limb grip of some flying foxes (Chiroptera: Pteropodidae). Journal of Zoology (London) 229:237–248.

Bloedel, P. 1955. Hunting methods of fish-eating bats, particularly *Noctilio leporinus*. Journal of Mammalogy 36:390–399.

Blood, B. R. 1987. Convergent hind limb morphology and the evolution of fish-catching in the bats *Noctilio leporinus* and *Myotis (Pizonyx) vivesi*. Ph.D. dissertation, University of Southern California, Los Angeles.

Crespo, R. F., S. B. Linhart, and R. J. Burns. 1972. Behavior of the vampire bat *(Desmodus rotundus)* in captivity. Southwestern Naturalist 17:139.

Dobson, G. E. 1878. Catalogue of the Chiroptera in the Collection of the British Museum. British Museum, London.

Feduccia, A. 1980. The Age of Birds. Harvard University Press, Cambridge.

Fenton, M. B. 1992a. Wounds and the origin of blood-feeding in bats. Biological Journal of the Linnean Society 47:161–171.

Fenton, M. B. 1992b. Bats. Roundhouse, Oxford.

Findley, J. S. 1972. Phenetic relationships among bats of the genus *Myotis*. Systematic Zoology 21:31–52.

Findley, J. S., E. Studier, and D. E. Wilson. 1972. Morphologic properties of bat wings. Journal of Mammalogy 53:429–444.

Forman, G. L. 1972. Comparative morphological and histochemical studies of the stomachs of selected American bats. University of Kansas Science Bulletin 49:593–618.

Forman, G. L., R. J. Baker, and J. D. Gerber. 1968. Comments on the systematic status of vampire bats (family Desmodontidae). Systematic Zoology 17:417–425.

Gillette, D. D. 1975. Evolution of feeding strategies in bats. Tebiwa 18:39–48.

Gardner, A. F. 1977. Feeding habits. *In* Biology of the New World Family Phyllostomidae, R. J. Baker, J. K. Jones, and D. C. Carter, eds., pp. 293–350. No. 13, Special Publications, The Museum, Texas Tech University, Lubbock.

Goodwin, G. G. 1946. Mammals of Costa Rica. Bulletin of the American Museum of Natural History 68:1–60.

Goodwin, G. G., and A. M. Greenhall. 1961. A review of the bats of Trinidad and Tobago: Descriptions, rabies infection, and ecology. Bulletin of the American Museum of Natural History 122:187–302.

Gould, S. J. 1989. Wonderful Life: The Burgess Shale and the Nature of History. W. W. Norton, New York.

Greenhall, A. M. 1965. Notes on behavior of captive vampire bats. Mammalia 29:441–451.

Greenhall, A. M. 1968. Notes on the behavior of the false vampire bat. Journal of Mammalogy 49:337–340.

Greenhall, A. M. 1970. The use of a precipitin test to determine host preferences of the vampire bats *Desmodus rotundus* and *Diaemus youngi*. Bijdrogen tot de Dierkunde 40:36–39.

Greenhall, A. M. 1972. The biting and feeding habits of the vampire bat *Desmodus rotundus*. Journal of Zoology (London) 168:451–461.

Greenhall, A. M. 1988. Feeding behavior. *In* Natural History of Vampire Bats, A. M. Greenhall and U. Schmidt, eds., pp. 31–40. CRC Press, Boca Raton, Fla.

Greenhall, A. M., and U. Schmidt, eds. 1988. Natural History of Vampire Bats. CRC Press, Boca Raton, Fla.

Griffiths, T. 1982. Systematics of the New World nectar-feeding bats (Mammalia, Phyllostomidae), based on the morphology of the hyoid and lingual regions. American Museum of Natural History Novitates 2742:1–45.

Hawkey, C. M. 1966. Plasminogen activator in saliva of the vampire bat *Desmodus rotundus*. Nature (London) 211:434–435.

Hawkey, C. M. 1967. Inhibitor of platelet aggregation present in saliva of the vampire bat *Desmodus rotundus*. British Journal of Haematology 13:1014–1020.

Hawkey, C. M. 1988. Salivary antihemostatic factors. *In* Natural History of Vampire Bats, A. M. Greenhall and U. Schmidt, eds., pp. 133–143. CRC Press, Boca Raton, Fla.

Hermanson, J. W., and J. S. Altenbach. 1981. Functional anatomy of the primary downstroke muscles in the pallid bat, *Antrozous pallidus*. Journal of Mammalogy 62:801–805.

Hermanson, J. W., and J. S. Altenbach. 1985. Functional anatomy of the shoulder and arm of the fruit-eating bat *Artibeus jamaicensis*. Journal of Zoology (London) 205:157–177.

Hildebrand, M. 1995. Running and jumping. *In* Analysis of Vertebrate Structure, 4th Ed., pp. 457–481. Wiley, New York.

Honeycutt, R. L., and V. M. Sarich. 1987. Albumin evolution and subfamiliar relationships among the New World leaf-nosed bats (Family Phyllostomidae). Journal of Mammalogy 68:508–517.

Honeycutt, R. L., I. F. Greenbaum, R. J. Baker, and V. M. Sarich. 1981. Molecular evolution of vampire bats. Journal of Mammalogy 62:805–814.

Hood, C. S., and J. D. Smith. 1982. Cladistical analysis of female reproductive histomorphology in phyllostomid bats. Systematic Zoology 31:241–251.

Horst, G. R. 1969. Observations on the structure and function of the kidney of the vampire bat *(Desmodus rotundus)*. *In* Physiological Systems in Semiarid Environments, C. C. Huff and M. L. Riedesel, eds., pp. 71–84. University of New Mexico Press, Albuquerque.

Howell, D. J., and J. Pylka. 1977. Why bats hang upside-down: A biomechanical hypothesis. Journal of Theoretical Biology 69: 625–631.

Hoyt, R. A., and J. S. Altenbach. 1981. Observations on *Diphylla ecaudata* in captivity. Journal of Mammalogy 62:215–216.

Hustler, K. 1987. Host preference of oxpeckers in the Hwange National Park, Zimbabwe. African Journal of Ecology 25:241–246.

Huxley, T. H. 1865. On the structure of the stomach in *Desmodus rufus*. Proceedings of the Zoological Society of London 35: 386–391.

Kalko, E. K. V., and H.-U. Schnitzler. 1989. The echolocation and hunting behavior of Daubenton's bat, *Myotis daubentoni*. Journal of Behavioral Ecology and Sociobiology 24:225–238.

Kamiya, T., P. Pirlot, and T. Matsubara. 1979. A note on the stomach of the vampire bat *(Desmodus rotundus)*. Acta Anatomica Nipponica 54:85.

Koopman, K. F. 1988. Systematics and distribution. *In* Natural History of Vampire Bats, A. M. Greenhall and U. Schmidt, eds., pp. 7–17. CRC Press, Boca Raton, Fla.

Koopman, K. F. 1993. The Order Chiroptera. *In* Mammal Species of the World: A Taxonomic and Geographic Reference, D. E. Wilson and D. Reeder, eds., pp. 137–241. Smithsonian Institution Press, Washington, D.C.

Kurten, L., and U. Schmidt. 1982. Thermoperception in the common vampire bat, *Desmodus rotundus*. Journal of Comparative Physiology 146:223–230.

Macalister, A. 1872. The myology of the Cheiroptera. Philosophical Transactions of the Royal Society of London 162:125–171.

Machado-Allison, C. E. 1967. The systematic position of the bats *Desmodus* and *Chilonycteris*, based on host–parasite relationships (Mammalia: Chiroptera). Proceedings of the Biological Society of Washington 80:223–226.

Marshall, L. G., S. D. Webb, J. J. Sepkoski, and D. M. Raup. 1982. Mammalian evolution and the great American interchange. Science 215:1351–1357.

McClearn, D. K. 1992. The rise and fall of a mutualism? Coatis, tapirs, and ticks on Barro Colorado Island, Panamá. Biotropica 24:220–222.

McFarland, W. N., and W. A. Wimsatt. 1969. Renal function and its relation to the ecology of the vampire bat, *Desmodus rotundus*. Comparative Biochemical Physiology 28:985–1006.

Miller, G. S. 1907. The families and genera of bats. Bulletin of the U.S. National Museum 57:1–282.

Muradali, F., N. Mondol, and W. A. Schutt, Jr. 1993. Observations on feeding behavior in the white-winged vampire bat, *Diaemus youngi*. Bat Research News 34:121.

Norberg, U. M. 1969. An arrangement giving a stiff leading edge to the hand wing in bats. Journal of Mammalogy 50:766–770.

Norberg, U. M. 1972a. Functional osteology and myology of the wing of the dog-faced bat, *Rousettus aegyptiacus* (Pteropidae). Zeitschrift für Morphologie der Tiere 73:1–44.

Norberg, U. M. 1972b. Bat wing structures important for aerodynamics and rigidity. Zeitschrift für Morphologie der Tiere 73: 45–61.

Norberg, U. M. 1976. Kinematics, aerodynamics, and energetics of horizontal flapping flight in the long-eared bat *Plecotus auritus*. Journal of Experimental Biology 65:459–470.

Norberg, U. M. 1987. Wing form and flight mode in bats. *In* Recent Advances in the Study of Bats, M. B. Fenton, P. A. Racey, and J. M. V. Rayner, eds., pp. 43–56. Cambridge University Press, Cambridge.

Norberg, U. M. 1990. Vertebrate Flight. Springer, Berlin.

Norberg, U. M., and J. M. V. Rayner. 1987. Ecological morphology and flight in bats (Mammalia; Chiroptera): Wing adaptations, flight performance, foraging strategy, and echolocation. Philosophical Transactions of the Royal Society of London B 316: 335–427.

Novick, A., and B. A. Dale. 1971. Foraging behavior in fishing bats and their insectivorous relatives. Journal of Mammalogy 52: 817–818.

Olrog, C. C. 1973. Alimentación del falso vampiro *Chrotopterus auritus* (Mammalia, Phyllostomidae). Acta Zoologica Lilloana 30:5–6.

Peterson, R. L., and P. Kirmse. 1969. Notes on *Vampyrum spectrum*, the false vampire bat, in Panama. Canadian Journal of Zoology 47:140–142.

Phillips, C. J., and B. Steinberg. 1976. Histological and scanning microscopic studies of tooth structure and thegosis in the common vampire bat, *Desmodus rotundus*. Occasional Papers of the Museum, Texas Tech University 42:1–12.

Pirlot, P. 1977. Wing design and the origin of bats. *In* Major Patterns in Vertebrate Evolution, M. K. Hecht, P. C. Goody, and B. M. Hecht, eds., pp. 375–410. Plenum Press, New York.

Quinn, T. H., and J. J. Baumel. 1993. Chiropteran tendon locking mechanism. Journal of Morphology 216:197–208.

Reeder, W. G., and K. S. Norris. 1954. Distribution, habits, and type locality of the fish-eating bat, *Pizonyx vivesi*. Journal of Mammalogy 35:81–87.

Rouk, C. S., and B. P. Glass. 1970. Comparative gastric histology of five North and Central American bats. Journal of Mammalogy 51:455–472.

Rubink, W. L. 1987. Thermal ecology of the screwworm larva, *Cochliomyia hominivorax* (Diptera: Calliphoridae). Environmental Entomology 16:599–604.

Ruschi, A. 1953. Morcegos do Estado do Espírito Santo. XI. Familia Phyllostomidae, chaves analiticas para subfamilias, generos e especies representados nos Estados do E. Santo. Descrição das especies: *Trachops cirrhosus* e *Tonatia brasiliense*, com algumas observações a respeito. Boletin Museo Biolagica Professore Mello-Leitão, Santa Teresa, Zoologica 14:1–14.

Savage, D. E. 1951. A Miocene phyllostomid from Colombia, South America. University of California Publications in Geology 28:357–366.

Sazima, I. 1978. Vertebrates as food items of the woolly false vampire, *Chrotopterus auritus*. Journal of Mammalogy 59:617–618.

Schaffer, J. 1905. Anatomisch-histologische Untersuchungen über den Bau der Zehen bei Fledermäusen und einigen kletternden Säugetieren. Zeitschrift für Wissenschliche Zoologie 83:231–284.

Schutt, W. A., Jr. 1992. Preliminary anatomical studies on the chiropteran hindlimb: Does the digital flexor retinaculum help bats get the hang of it? Bat Research News 33:74–75.

Schutt, W. A., Jr. 1993. Digital morphology in the Chiroptera: The passive digital lock. Acta Anatomica 148:219–227.

Schutt, W. A., Jr. 1994. Comparative hindlimb morphology of the vampire bats (Phyllostomidae: Desmodontinae). Does *Diphylla ecaudata* possess a sixth digit? Bat Research News 35:114.

Schutt, W. A., Jr. 1995. The chiropteran hindlimb: Evolutionary, ecological, and behavioral correlates of morphology. Ph.D. dissertation, Cornell University, Ithaca, N.Y.

Schutt, W. A., Jr., and J. S. Altenbach. 1997. A sixth digit in *Diphylla ecaudata*, the hairylegged vampire bat. Mammalia 61:280–285.

Schutt, W. A., Jr., J. W. Hermanson, J. E. A. Bertram, D. Cullinane, Y. H. Chang, F. Muradali, and J. S. Altenbach. 1993. Aspects of locomotor morphology, performance, and behavior in two vampire bats: *Desmodus rotundus* and *Diaemus youngi*. Bat Research News 34:127–128.

Simmons, N. B. 1994. The case for chiropteran monophyly. American Museum of Natural History Novitates 3103:1–54.

Simmons, N. B., and T. H. Quinn. 1994. Evolution of the digital tendon locking mechanism in bats and dermopterans: A phylogenetic perspective. Journal of Mammalian Evolution 2:231–254.

Simpson, G. G. 1980. Splendid Isolation: The Curious History of South American Mammals. Yale University Press, New Haven.

Simpson, G. G., and J. Haffer. 1978. Speciation patterns in the Amazonian forest biota. Annual Review of Ecology and Systematics 9:497–518.

Slaughter, B. H. 1970. Evolutionary trends of chiropteran dentitions. *In* About Bats, B. H. Slaughter and D. W. Walton, eds., pp. 51–83. Southern Methodist University Press, Dallas, Tex.

Smith, J. D. 1972. Systematics of the chiropteran family Mormoopidae. Miscellaneous Publications, Museum of Natural History, University of Kansas 56:1–132.

Smith, J. D. 1976. Chiropteran evolution. *In* Biology of the New World Family Phyllostomidae, R. J. Baker, J. K. Jones, and J. D. Smith, eds., pp. 49–69. No. 10, Special Publications, The Museum, Texas Tech University, Lubbock.

Smith, J. D., and A. Starett. 1979. Morphometric analysis of chiropteran wings. *In* Biology of the New World Family Phyllostomidae, R. J. Baker, J. K. Jones, and J. D. Smith, eds., pp. 229–316. No. 16, Special Publications, The Museum, Texas Tech University, Lubbock.

Storch, G. 1968. Funktionsmorphologische Untersuchungen an der Kaumuskulatur und an korrelierten Schädelstrukturen der Chiropteren. Abhandlungen aus dem Gebiete de Beschreibenden Naturgeschichte von Mitgliedern der Senckenbergischen Naturforschenden Gesellschaft in Frankfurt am Main 517:1–92.

Strickler, T. L. 1978. Functional Osteology and Myology of the Shoulder in the Chiroptera. Contributions to Vertebrate Evolution, Vol. 4. S. Karger, Basel.

Stutterheim, I. M., and K. Panegis. 1985. Roosting selection and host selection of oxpeckers (Aves: Buphagidae) in Moremi Wildlife Reserve, Botswana, and eastern Caprivi, southwest Africa. South African Journal of Zoology 20:237–240.

Turner, D. C. 1975. The Vampire Bat: A Field Study in Behavior and Ecology. Johns Hopkins University Press, Baltimore.

Uieda, W. 1982. Aspectos do comportamento alimentar dos tres especies do morcegos hematófagos (Chiroptera, Phyllostomidae). Master's thesis, Instituto Biologico Universidade Estadual de Campinas, Brasil.

Uieda, W. 1986. Aspectos da morfologia lingual das três espécies de morcegos hematófagos (Chiroptera, Phyllostomidae). Review of Brasilian Biology 46:581–587.

Uieda, W. 1994. Comportamento alimentar do morcegos hematófagos ao atacar aves, caprinos e suínos, em condições de cativeiro. Ph.D. thesis, Instituto Biologico Universidade Estadual de Campinas, Brasil.

Uieda, W., S. Buck, and I. Sazima. 1992. Feeding behavior of the vampire bats *Diaemus youngi* and *Diphylla ecaudata* on smaller birds in captivity. Journal of the Brazilian Association for the Advancement of Science 44:410–412.

Vaughan, T. A. 1959. Functional morphology of three bats: *Eumops, Myotis, Macrotus*. Publications of the Museum of Natural History, University of Kansas 12:1–153.

Vaughan, T. A. 1966. Morphology and flight characteristics of molossid bats. Journal of Mammalogy 47:249–260.

Vaughan, T. A. 1970a. The muscular system. *In* Biology of Bats, W. A. Wimsatt ed., Vol. 1, pp. 139–194. Academic Press, New York.

Vaughan, T. A. 1970b. The skeletal system. *In* Biology of Bats, W. A. Wimsatt, ed., Vol. 1, pp. 97–138. Academic Press, New York.

Vaughan, T. A. 1970c. Adaptations for flight in bats. *In* About Bats, R. Slaughter and D. Walton, eds., pp. 127–143. Southern Methodist University Press, Dallas, Tex.

Vaughan, T. A. 1986. Mammalogy, 3rd Ed. Saunders, New York.

Vaughan, T. A., and G. C. Bateman. 1970. Functional morphology of the forelimb of mormoopid bats. Journal of Mammalogy 51:217–235.

Vehrencamp, S. L., F. G. Stiles, and J. W. Bradbury. 1977. Observations on the foraging behavior and avian prey of the neotropical carnivorous bat, *Vampyrum spectrum*. Journal of Mammalogy 58:469–478.

Walton, D. W., and G. M. Walton. 1970. Post-cranial osteology of bats. *In* About Bats, B. H. Slaughter and D. W. Walton, eds., pp. 93–125. Southern Methodist University Press, Dallas, Tex.

Webster, F. A., and D. R. Griffin. 1962. The role of the flight membranes in insect capture by bats. Animal Behaviour 10:332–340.

Wilkinson, G. S. 1985. The social organization of the common vampire bat. II. Mating system, genetic structure, and relatedness. Behavioral Ecology and Sociobiology 17:123.

Wilkinson, G. S. 1986. Social grooming in the common vampire bat, *Desmodus rotundus*. Animal Behaviour 34:1880–1889.

Wimsatt, G. A., and A. Guerriere. 1962. Observations on the feeding capacities and excretory functions of captive vampire bats. Journal of Mammalogy 43:17–27.

11
Interspecific and Intraspecific Variation in Echolocation Call Frequency and Morphology of Horseshoe Bats, *Rhinolophus* and *Hipposideros*

CHARLES M. FRANCIS AND JÖRG HABERSETZER

Most insectivorous bats rely mainly on echolocation to locate their prey. The design of their echolocation calls is related to their particular foraging requirements (Neuweiler 1989). The dominant frequencies in the calls influence the size of prey that the bat can detect, the distance at which prey can be detected, and the ability of certain prey items to detect and avoid the bat (Jones 1992), while the structure of the call influences the ability of the bat to detect its prey in cluttered environments. Production and processing of echolocation calls may also be constrained by the structural characteristics and dimensions of the vocal apparatus and the outer, middle, and inner ear. For example, low-frequency calls may be processed more efficiently with larger resonance cavities or by larger outer-ear structures.

Understanding the relationship between echolocation calls and certain morphological features could provide insight into the evolution of the particular morphology and behavior of each species. It could also be valuable for predicting the echolocation behavior of species or populations presently known only from their morphology, such as fossil bats (Habersetzer and Storch 1992), or extant species known only from museum specimens. Such information could be useful for locating and surveying poorly known species using bat detectors, or for identifying species that differ internally but not externally in morphology (e.g., Jones et al. 1993; Robinson 1995).

Horseshoe bats of the genera *Rhinolophus* and *Hipposideros* are particularly amenable to a comparative study of morphology and echolocation behavior because both genera have a high diversity of species, highly specialized morphology associated with both emission and reception of echolocation calls, and relatively stereotyped echolocation calls. In each genus more than 60 species are currently recognized (Corbet and Hill 1992), ranging in size from less than 5 g to more than 100 g. They have elaborate noseleaves and enlarged nasal cavities associated with transmission of the echolocation signals, and large pinnae and cochleae associated with reception of the echoes.

Species in both genera have echolocation calls that are characterized by a strong constant-frequency (CF) component, usually with a short beginning or terminal frequency-modulated (FM) component (Schnitzler 1968; see also Figure 11.1). Much of the energy of the call is contained within the CF component, and the cochleae of each species are finely tuned to the frequency emitted by the bat when it is stationary (Schuller and Pollak 1979; Vater 1987). During flight the emitted frequency is lowered to compensate for Doppler shifts induced by its flight speed, so that the returning echo is close to the resting frequency (Schnitzler 1968). As might be expected, if the frequency is constrained by anatomy, the resting frequency of the CF component remains relatively constant within individuals (Vater 1987; Heller and Helversen 1989; Jones et al. 1994), although small changes in frequency have been reported in relation to age or season (Jones and Ransome 1993).

Among species, the frequency of the CF component of the call averages lower in larger species of both *Rhinolophus* and *Hipposideros* (Heller and Helversen 1989; Jones 1995), but species in the genus *Hipposideros* usually have higher frequency calls for a given body size than species of *Rhinolophus* (Heller and Helversen 1989). Within species, the relationship between frequency and body size seems to vary among species (Jones 1995). Females emit higher frequencies than males in *Rhinolophus rouxi* (Neuweiler et al. 1987), *Rhinolophus hipposideros* (Jones et al. 1992), and *Asellia tridens* (Jones et al. 1993), but females are the same size as males in *R. rouxi*, larger than males in *R. hipposideros,* and smaller than males in *A. tridens.* In other species, no sexual difference in echolocation calls has been reported (Heller and Helversen 1989).

These relationships between body size and echolocation frequencies could result from ecological differences among species (e.g., larger species fly faster or hunt larger prey that are better detected with loud, lower-frequency calls) or from allometric relationships between sound-generating or sound-receiving apparatus and body size (e.g., larger resonant bodies may be associated with lower frequencies). In the latter case, one might expect a closer correlation between echolocation frequency and the size of the sound-processing apparatus than with overall body size. If morphology tightly constrains echolocation calls, then one might expect similar relationships within species as among species. However, if structural aspects only constrain the echolocation calls within a limited range, then different relationships may be apparent within and among species.

We have been investigating the relationships between size, morphology, and echolocation frequency among and within species of *Hipposideros* and *Rhinolophus* in southeast Asia. In this chapter, we examine the relationships between overall body size, cochlear diameter, and echolocation frequency across species, and we also examine the relationships between morphology and echolocation frequency within two species that showed marked intraspecific variation in echolocation frequencies.

Methods

This study is based on specimens captured by one of us (C. M. Francis) in Laos, peninsular Malaysia, and Sabah (north Borneo) between 1993 and 1995. Most bats were captured using four-bank harp traps (Francis 1989) set across trails or small streams in the forest understory. Traps were checked one or more times in the evening and again in the early morning. A few bats were also captured by hand within cave roosts or by using mist nets in the forest.

All captured bats were transported to a base camp in cloth bags, then removed from the bags one at a time, and handheld 4–10 cm from a microphone for recording. If necessary, bats were moved back and forth in front of the microphone to induce calling, but an attempt was made to record only calls produced when the bats were actually stationary. Most bats were recorded in the evenings, shortly after capture, but some individuals were recorded the following morning. In nearly all cases, bats appeared to be fully active when recorded. Recordings were made using the microphone of an Ultra Sound Advice S-25 Bat Detector that was fed into an Ultra Sound Advice Portable Ultra-Sound Processor (PUSP), which digitized the calls with a sampling rate of 350 kHz. The calls were then replayed with a 20-fold time expansion and recorded onto a Sony WM-D6C professional cassette recorder. The approximate frequency of the CF component of a representative series of calls for each individual was recorded from the digital zero-crossing display of the PUSP.

More precise measurements for calls of some species were made by analyzing the tapes using digital sound-processing software (Sona PC) on a Pentium-based personal computer and averaging measurements for 6–10 calls of each individual bat (Figure 11.1). If multiple harmonics were present on the recording (Figure 11.1D), the frequency was measured from the dominant harmonic. After recording, each bat was collected, measured, and preserved by being fixed initially in 10% formalin, then transferred after a few days into 70% alcohol. To measure skull parameters, the skulls were extracted from the specimens in the laboratory, cleaned, and then measured and radiographed using high-resolution x-ray equipment (Figure 11.2). The basicranial width was measured between the outermost bony margin of the semicircular canals of each ear, and the cochlear width was measured across the second half-turn of the cochlea (Figure 11.2C,D) (Habersetzer and Storch 1992).

Species were identified on the basis of descriptions by Medway (1983), Payne and Francis (1985), or Corbet and Hill (1992), with nomenclature based on Koopman (1993). Some specimens, particularly from Laos, have not yet been

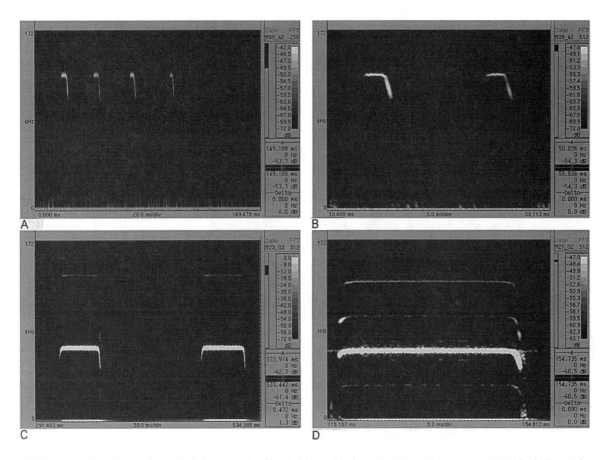

Figure 11.1. Frequency time displays of typical echolocation calls of handheld horseshoe bats: (**A, B**) *Hipposideros cervinus;* (**C, D**) *Rhinolophus creaghi.* Graphs on the left (A, C) differ in scale between the species, but graphs on right (B, D) are at an expanded scale of 5.0 msec per division for both species.

identified with certainty, either because the taxonomy requires review or because they may represent new taxa. For this chapter, "interspecific" comparisons were based either on distinct species or on populations in disjunct areas (i.e., Laos, peninsular Malaysia, or Borneo) that differed in either echolocation frequency or morphology (Table 11.1).

For statistical analysis, body measurements and call frequencies were log-transformed, but the results were essentially the same for untransformed data, and the latter have been plotted for ease of interpreting the graphs. For intraspecific comparisons, measurements were made for a sample of three to five individuals from each population or subgroup. For "interspecific" comparisons, mean values from several individuals of each species were used, although most skull measurements were based on single individuals because for most species only one skull had been prepared (Table 11.1).

Results

External measurements and recordings were obtained for at least one individual of 15 distinct populations of *Hipposi-*

deros (constituting at least 11 species) and of 19 populations of *Rhinolophus* (constituting at least 17 species; Table 11.1). Of these, skulls were prepared and measured for representatives of 11 populations of *Hipposideros* and 9 populations of *Rhinolophus.*

Across species, within each genus, the frequency of the CF component of the echolocation calls was negatively correlated with body size, as indexed by either length of forearm ($F_{1,31}$ = 19.5, $p < 0.0001$) or basicranial width ($F_{1,18}$ = 18.0, $p = 0.005$) (Figure 11.3). The slope of the relationship (on log-transformed data) did not differ significantly between the genera ($F_{1,17}$ = 0.01, $p > 0.94$), but for a given body size, the echolocation frequency of *Hipposideros* averaged about 50 kHz higher than that for *Rhinolophus* ($F_{1,18}$ = 25.0, $p < 0.0001$). Nevertheless, there was considerable scatter about the regression line. Members of some populations of *Hipposideros* (including *H. galeritus* and *H. ridleyi*) had much lower echolocation frequencies than expected on the basis of length of forearm or basicranial width (Figure 11.3). In fact, their call frequencies were similar to those of similar-sized *Rhinolophus* (note that skulls were prepared for only two of the three outlying forms in Figure 11.3A).

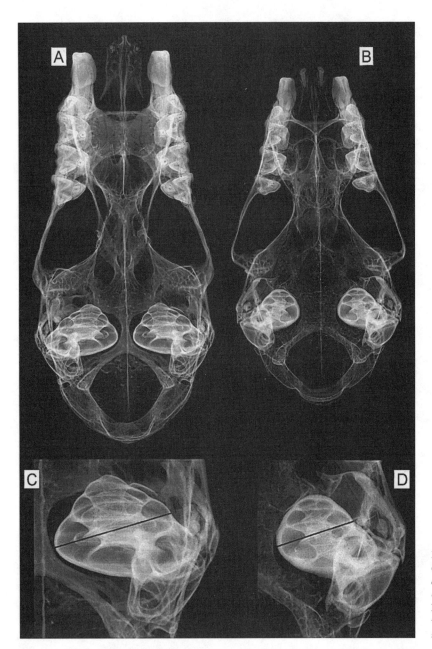

Figure 11.2. Radiographs of typical skulls: (**A, C**) *Hipposideros cervinus;* (**B, D**) *Rhinolophus creaghi.* Inset radiographs (C, D) show the enlarged details of the right cochleae; the black lines indicate where measurements of cochlear width were made. Both skulls (A, B) are to the same scale, as are the enlarged cochleae (C, D).

Because the cochlea is directly involved in processing echoes from the calls, one might expect a tighter relationship between echolocation frequency and cochlear size. Cochlear width was very closely correlated with basicranial width within both genera (*Hipposideros,* $r^2 = 0.94$, $n = 11$, $p < 0.0001$; *Rhinolophus,* $r^2 = 0.90$, $n = 10$, $p < 0.0001$), but averaged about 30% smaller for a given body size in *Hipposideros* than *Rhinolophus* (Figure 11.4A). This difference in cochlear size almost exactly matched the difference in echolocation frequencies, such that the relationship between cochlear width and frequency was nearly identical in the two genera (Figure 11.4B). On this basis, the two *Hip-*

posideros that appeared to resemble *Rhinolophus* on the basis of the body size versus frequency comparison (see Figure 11.3) are clearly outliers from both genera on cochlear size (Figure 11.4B). They have cochleae typical for their body size within *Hipposideros* and hence much smaller cochleae relative to their echolocation frequency than typical *Rhinolophus* or other species of *Hipposideros.*

We examined intraspecific variation in echolocation frequency at two levels: interpopulation variation in *Hipposideros cervinus* from peninsular Malaysia and Sabah and intrapopulation variation in *Rhinolophus creaghi* from Sabah. *Hipposideros cervinus* from peninsular Malaysia had

Table 11.1

Morphological and Echolocation Data on *Rhinolophus* and *Hipposideros* Species from Several Areas

Species[a]	Origin[b]	Length of forearm (mm)[c]	Basicranial width (mm)[c]	Cochlear width (mm)[c]	Echolocation call (kHz)[d]
R. cf. pusillus	L	35	—	—	100.0*
R. lepidus	M	41	7.90	2.79	98.0*
R. malayanus	L	41	—	—	78.0*
R. borneensis	S	43	9.03	3.16	81.8
R. cf. borneensis	M	46	8.97	3.09	92.0*
R. affinis	M	50	10.11	3.48	78.4
	L	51	—	—	73.0*
R. thomasi	L	44	—	—	76.0*
R. stheno	M	47	8.93	3.23	86.0*
R. acuminatus	S	48	—	—	89.0*
R. species A	L	45	—	—	93.0*
R. species B	L	45	—	—	77.0*
R. robinsoni	M	44	8.81	3.21	67.0*
R. shameli	L	44	—	—	76.0*
R. creaghi	S	50	9.73	3.56	68.0
R. trifoliatus	S	53	10.00	3.62	51.2
R. sedulus	S	44	—	—	76.0*
R. macrotis	L	43	—	—	51.0*
R. philippinensis	S	50	9.45	3.47	36.6
H. diadema	M	83	13.49	3.70	59.8
H. larvatus	M	60	10.65	2.57	97.7
H. galeritus	M	49	8.58	2.36	89.2
	S	49	—	—	110.0*
H. cervinus	M	50	8.83	2.25	128.8
	S	49	9.29	2.37	115.8
H. bicolor	M	45	9.12	2.30	132.0*
	S	47	—	—	135.0*
H. species A	M	45	8.89	2.22	142.0*
H. pomona	L	42	—	—	125.0*
H. ater	S	42	8.11	2.16	139.5
H. dyacorum	S	43	8.38	2.14	161.9
H. cineraceus	M	36	7.71	2.05	>154.0*
H. ridleyi	S	49	—	—	65.0*
H. species B	L	46	9.47	2.51	70.0*

[a]Species labeled "species A" or "species B" have not yet been identified; the taxonomic status of those labeled "cf." is still uncertain. All the bats included here were caught between 1993 and 1995.

[b]Origin of specimens: L, Laos; M, peninsular Malaysia; S, Sabah.

[c]Forearm measurements are either from single specimens or represent means of 2–10 individuals. Most skull measurements are from single specimens, except those for R. creaghi and H. cervinus which represent the means of several individuals.

[d]Echolocation call frequencies were recorded from hand-held bats. Those marked with an asterisk (*) were estimated from the zero-crossing display on the Portable UltraSound Processor, which has an uncertainty of at least ±2 kHz (greater for frequencies above 150 kHz). The unmarked frequencies were measured with digital sound processing software, which has an accuracy of ±0.3 kHz, and most were taken from the same individuals used for skull measurements. For R. creaghi and H. cervinus, these represent the means of several individuals.

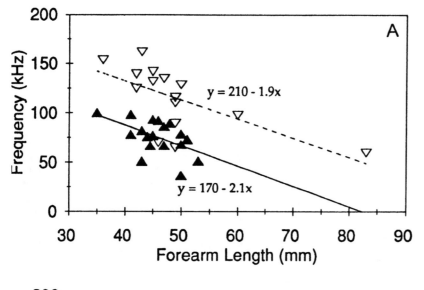

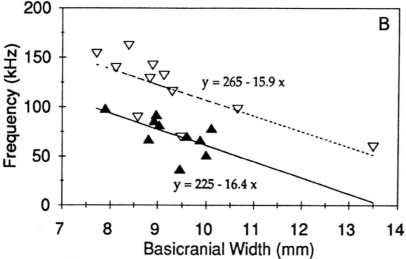

Figure 11.3. Relationships between frequency of the constant-frequency (CF) component of the echolocation call and morphological variables for *Rhinolophus* (solid triangles) and for *Hipposideros* (open triangles): (**A**) call frequency and forearm length; (**B**) call frequency and basicranial width. Each point represents a distinct population or species (see Table 11.1); the lines represent best-fit linear regressions, estimated separately for each genus.

significantly higher echolocation calls than those from Sabah ($F_{1,5} = 80.2$, $p = 0.0003$), and their basicranial width was smaller ($F_{1,5} = 30.5$, $p = 0.003$; Figure 11.5A), as would be expected from the interspecific relationships. However, their cochleae overlapped in size ($F_{1,5} = 4.7$, $p = 0.08$; Figure 11.5B), and specimens from peninsular Malaysia actually had slightly longer forearms than those from Sabah ($F_{1,5} = 11.6$, $p = 0.02$; Figure 11.5C), despite smaller skulls.

Rhinolophus creaghi showed marked sexual dimorphism in echolocation calls, with no overlap between sexes in the CF frequency of the 11 individuals recorded ($F_{1,9} = 37.9$, $p = 0.0003$), with adult males having lower-frequency calls than adult females. Although males had slightly longer forearms on average (49.6 versus 49.1), there was considerable overlap, and the differences were not significant ($F_{1,9} = 1.54$, $p = 0.24$; Figure 11.6C). However, both basicranial width ($F_{1,9}$ 31.0, $p = 0.0003$) and cochlear width ($F_{1,9}$ —

14.9, $p = 0.004$) were invariably larger in males (Figure 11.6A,B).

Discussion

Interspecific Variation

The negative correlation among species between overall body size and echolocation frequency (see Figure 11.3) is consistent with the results of previous studies (Heller and Helversen 1989; Jones 1995). However, the nearly identical relationship between cochlear width and echolocation frequency within both genera had not been noted previously (see Figure 11.4). This suggests that, despite some differences in the internal cochlear structure between *Hipposideros* and *Rhinolophus* (Kössl and Vater 1995), there may be similar functional relationships between cochlear size

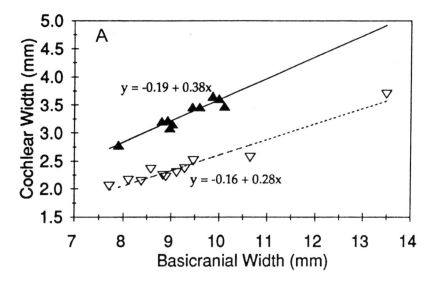

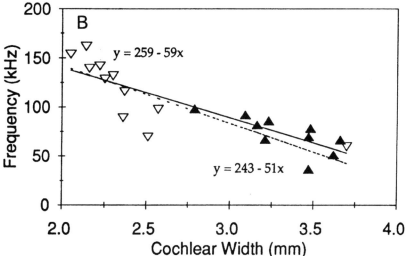

Figure 11.4. Relationships between (**A**) basicranial width and cochlear width, and (**B**) cochlear width and frequency of the CF component of the echolocation call, for *Rhinolophus* (solid triangles) and for *Hipposideros* (open triangles). Each point represents a distinct population or species (see Table 11.1); the lines represent best-fit linear regressions estimated separately for each genus.

and call frequency in both genera. The physical or physiological basis of this relationship has not yet been determined. Active resonance appears to be important in enhancing tuning in some bats (Kössl and Vater 1995), and a correlation between size and frequency might be expected on this basis.

Despite the overall close correlations, some species were clear outliers on both graphs, with lower echolocation calls than expected on the basis of body size or cochlear size. Two of these species (the lowest *Rhinolophus*, *R. philippinensis*, and the lowest, small *Hipposideros*, *H. ridleyi*; see Figure 11.4B) have exceptionally large noseleaves and ears, as might be expected for species with relatively low frequency calls. Several possibilities can be suggested why the cochleae are not similarly enlarged, at least as measured by cochlear width. The cochleae may differ in shape, being enlarged in other directions; the cochleae may compensate in other

ways, perhaps through differences in the internal structure; or the bats may process information in different ways. Of course, the possibility always remains that the similarity of the relationships between cochlear size and frequency in the two genera is a coincidence, perhaps the result of correlations with other factors related to overall body size.

Interpopulation Variation

Populations of *Hipposideros cervinus* from both peninsular Malaysia and Borneo are classified as the same subspecies, *H. c. labuanensis*, by Jenkins and Hill (1981), but they differ in echolocation calls as well as in size. These populations may prove to represent distinct subspecies, but this cannot be confirmed until additional specimens have been recorded and measured from other locations in Borneo. The smaller skull size and higher-frequency calls in peninsular

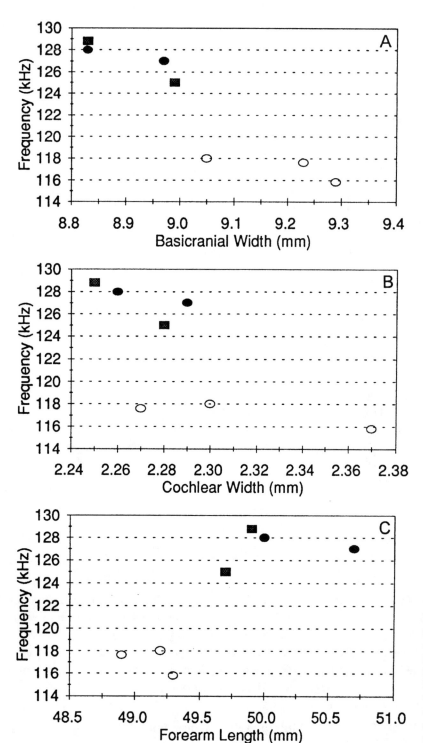

Figure 11.5. Interpopulation differences in relationships between frequency of the CF component of the echolocation call and morphological variables for *Hipposideros cervinus* from peninsular Malaysia and from Sabah, East Malaysia: (**A**) basicranial width; (**B**) cochlear width; (**C**) forearm length. Each point represents one individual (male, square; female, circle). Solid symbols represent peninsular Malaysian bats, and the open symbols denote Sabah bats.

Malaysia are consistent with expectations from interspecific relationships. However, the differences in length of forearm are opposite to that expected, highlighting the limitations of using a single measurement, such as length of forearm, as a measure of body size.

Our recordings of the closely allied species, *H. galeritus*, also indicate different calls in Borneo than in peninsular

Malaysia, but in this species the Bornean population has higher-frequency calls (see Table 11.1). We have not yet compared skull measurements of our specimens, but the Bornean populations are classified as a distinct subspecies (*H. g. insolens*), which is reported to be larger with a more robust skull and a longer and less angular interorbital region than *H. g. galeritus* from peninsular Malaysia (Jenkins

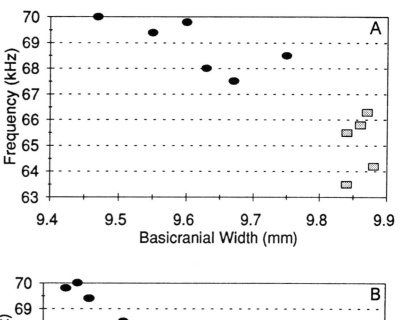

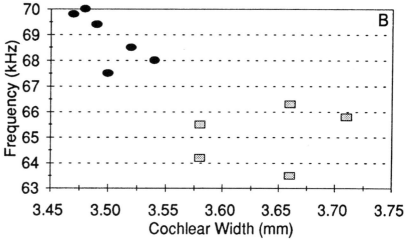

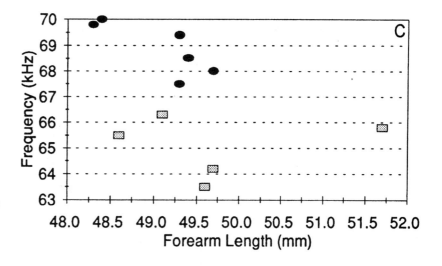

Figure 11.6. Intrapopulation variation in the relationships between frequency of the CF component of the echolocation call and morphological variables for *Rhinolophus creaghi* from Sabah, East Malaysia: (**A**) basicranial width; (**B**) cochlear width; (**C**) forearm length. Each point represents one individual (male, square; female, circle).

and Hill 1981). A difference in size contrary to that expected based on the observed relationship between call frequency and morphology within *H. cervinus* is thus suggested, but this cannot be confirmed until we have made skull measurements of the individuals that we have recorded.

Heller and Helversen (1989) suggested that sympatric species of rhinolophoid bats may be selected to have non-overlapping echolocation calls, possibly because of niche partitioning. If so, the differences in echolocation calls between peninsular Malaysia and Sabah may result from interactions with a different suite of species that differ in their echolocation calls. Alternatively, differences in the prey base or other aspects of their ecology may have selected for different frequency calls, or the differences in call frequency may be a consequence of selection on body size. These hypotheses cannot be tested until recordings and measurements are available for the complete suite of species present in both geographic areas.

The overlap in cochlear width between the two populations of *H. cervinus*, despite lack of overlap in echolocation frequency, suggests moderate plasticity in the relationship between cochlear size and the frequency of sound processed. Alternatively, other components of cochlear shape may differ between these populations. Some plasticity might be expected on the basis of observations that call frequency can change, at least slightly, with age in some *Rhinolophus* (Jones and Ransome 1993).

The geographic variation in call frequency of *H. galeritus* is particularly intriguing. Specimens from Sabah have call frequencies close to those expected for their body size on the basis of interspecific correlations with length of forearm. However, those from peninsular Malaysia have much lower call frequencies and are one of the outliers on the interspecific graphs. Further research on these populations may help to clarify the basis for the interspecific relationships between body size and echolocation frequency.

Intrapopulation Variation

The higher echolocation frequencies of female *Rhinolophus creaghi* were associated with smaller cranial and cochlear size. Cochlear measurements were taken across the bony capsule of the cochlea at the outer bony margins because these structures have much higher contrast in the radiographs compared with the cochlea diameter measured across the fluid spaces. Thus, the measured cochlea diameter is somewhat larger than the cochlea diameter across the fluid spaces, which might be more relevant for cochlear mechanics. It is possible that sexual dimorphism in cochlear size could occur because the males have thicker bony capsules than the females. However, we found no difference in

the thickness of the capsule between the sexes (Francis and Habersetzer, unpublished data). The observed differences between males and females (see Figure 11.6) are in the range expected from interspecific relationships between cochlear diameter and frequency (see Figure 11.4).

Among other species of *Rhinolophus*, females have been reported to have higher echolocation frequencies in *Rhinolophus rouxi* (Neuweiler et al. 1987) and in *Rhinolophus hipposideros* (Jones et al. 1992), but to be the same size as males in *R. rouxi* and larger than males in *R. hipposideros*. However, cranial and cochlear measurements were not compared in either study. Among 16 species of Southeast Asian *Rhinolophus* for which we recorded at least one individual of each sex, clearly distinct sexual differences in echolocation calls were noted in only 2 species: *R. creaghi* and *R. thomasi* (Francis and Habersetzer, unpublished data). In both cases, females had higher frequency calls than males, but we do not yet have data on sexual dimorphism in size of *R. thomasi*, nor on how the observed sexual dimorphism in skull characters of *R. creaghi* compares with that of other species without dimorphism in call frequencies. If call frequency is tightly correlated with size, one would expect greater size dimorphism in species that differ in echolocation frequency than in species with similar call frequencies. There were some indications of a relationship between skull dimensions and call frequencies within each sex of *R. creaghi* (Figure 11.6), but the sample size was not adequate to confirm this statistically.

Conclusions

A strong relationship overall appears to exist between cochlear size and echolocation frequency across species and, at least in some cases, within species. Nevertheless, there are several outlying species, and within species the correlations were inconsistent. This result suggests moderate plasticity in the relationship between size and frequency, perhaps through varying other components of size or shape of the cochlea or through adjustment of soft tissue.

From the perspective of predicting behavior from morphology, clearly neither body size nor cochlear size alone is adequate to make precise predictions. Other factors that might be expected to vary in relation to echolocation frequency include external features such as area or width of the noseleaf, internarial width, or size of the pinna, and internal features such as the nasal chambers in the skull. Robinson (1996) found a close relationship among species of both *Rhinolophus* and *Hipposideros* between echolocation calls and noseleaf widths. On the basis of relationships with cochlear diameter, quite likely none of these features individually can predict the precise echolocation frequencies of all bats because of subtle differences among species in the strategies employed to produce and process sound. However, the outliers in each relationship may provide insight into the mechanisms underlying each observed pattern, and a multivariate approach using a combination of factors should be tested for its ability to predict echolocation calls both within and among species.

Acknowledgments

We thank the Economic Planning Unit of the Malaysian Prime Minister's Department, the National Parks and Wildlife Department in peninsular Malaysia, and the Sabah Wildlife Department for permission to carry out field research in Malaysia, and the Lao Forestry Department for permission to carry out surveys in Laos. Y. H. Sen and D. Wells kindly provided logistic support and assisted with permits in peninsular Malaysia. The Malaysian Nature Society assisted with fieldwork during the Belum Expedition in Perak, Malaysia. We are very grateful to Juliane Altmann, who prepared the skulls. The Wildlife Conservation Society provided financial assistance for field research as well as logistic support in Laos. Additional financial assistance and equipment was provided by the Senckenberg Museum (Mammal Section).

Literature Cited

Corbet, G. B., and J. E. Hill. 1992. The Mammals of the Indomalayan Region: A Systematic Review. Oxford University Press, Oxford.

Francis, C. M. 1989. A comparison of mist nets and two designs of harp traps for capturing bats. Journal of Mammalogy 70:865–870.

Habersetzer, J., and G. Storch. 1992. Cochlea size in extant Chiroptera and Middle Eocene microchiropterans from Messel. Naturwissenschaften 79:462–466.

Heller, K.-G., and O. von Helversen. 1989. Resource partitioning of sonar frequency bands in rhinolophoid bats. Oecologia (Berlin) 80:178–186.

Jenkins, P. D., and J. E. Hill. 1981. The status of *Hipposideros galeritus* Cantor, 1846, and *Hipposideros cervinus* (Gould, 1854). Bulletin of the British Museum (Natural History), Zoology 41:279–294.

Jones, G. 1992. Bats vs moths: Studies on the diets of rhinolophid and hipposiderid bats support the allotonic frequency hypothesis. In Prague Studies in Mammalogy, I. Horácek and V. Vohralík, eds., pp. 87–92. Charles University Press, Prague.

Jones, G. 1995. Variation in bat echolocation: Implications for resource partitioning and communication. Le Rhinolophe 11:53–59.

Jones, G., and R. D. Ransome. 1993. Echolocation calls of bats are influenced by maternal effects and change over a lifetime. Proceedings of the Royal Society of London B 252:125–128.

Jones, G., T. Gordon, and J. Nightingale. 1992. Sex and age differences in the echolocation calls of the lesser horseshoe bat, *Rhinolophus hipposideros*. Mammalia 56:189–193.

Jones, G., M. Morton, P. M. Hughes, and R. M. Budden. 1993. Echolocation, flight morphology, and foraging strategies of some West African hipposiderid bats. Journal of Zoology (London) 230:385–400.

Jones, G., K. Sripathi, D. A. Waters, and G. Marimuthu. 1994. Individual variation in the echolocation calls of three sympatric Indian hipposiderid bats, and an experimental attempt to jam bat echolocation. Folia Zoologica 43:347–362.

Koopman, K. 1993. Order Chiroptera. In Mammal Species of the World: A Taxonomic and Geographic Reference, D. E. Wilson and D. M. Reeder, eds., pp. 137–242. Smithsonian Institution Press, Washington, D.C.

Kössl, M., and M. Vater. 1995. Cochlear structure and function in bats. In Hearing in Bats, A. N. Popper and R. R. Fay, eds., pp. 191–234. Springer Handbook in Auditory Research, Vol. 5. Springer-Verlag, New York.

Medway, L. 1983. The Wild Mammals of Malaya (Peninsular Malaysia) and Singapore. Oxford University Press, Kuala Lumpur, Malaysia.

Neuweiler, G. 1989. Foraging ecology and audition in echolocating bats. Trends in Ecology and Evolution 4:160–166.

Neuweiler, G., W. Metzner, U. Heilmann, R. Rübsamen, M. Eckrich, and H. H. Costa. 1987. Foraging behaviour and echolocation in the rufous horseshoe bat *(Rhinolophus rouxi)* of Sri Lanka. Behavioral Ecology and Sociobiology 20:53–67.

Payne, J. B., and C. M. Francis. 1985. A Field Guide to the Mammals of Borneo. The Sabah Society and World Wildlife Fund, Kota Kinabalu, Malaysia.

Robinson, M. F. 1995. Field identification of two morphologically similar horseshoe bats—*Rhinolophus malayanus* and *R. stheno*. Bat Research News 36:3–4.

Robinson, M. F. 1996. A relationship between echolocation calls and noseleaf widths in bats of the genera *Rhinolophus* and *Hipposideros*. Journal of Zoology (London) 239:389–393.

Schnitzler, H.-U. 1968. Die Ultraschall-Ortungslaute der Hufeisen-Fledermäuse (Chiroptera: Rhinolophidae) in verschiedenen Orientierungssituationen. Zeitschrift für Vergleichende Physiologie 57:376–408.

Schuller, G., and G. D. Pollak. 1979. Disproportionate frequency representation in the inferior colliculus of horseshoe bats: Evidence for an "acoustic fovea." Journal of Comparative Physiology 132:47–54.

Vater, M. 1987. Narrow-band frequency analysis in bats. In Recent Advances in the Study of Bats, M. B. Fenton, P. Racey, and J. M. V. Rayner, eds., pp. 200–225. Cambridge University Press, Cambridge.

Part Three
Echolocation

All the nearly 760 species of Microchiroptera have evolved sophisticated echolocation systems that they use for orientation and obtaining food. The evolution of flight and echolocation opened opportunities for a nocturnal lifestyle in largely predator-free habitats, including many unexploited food resources such as insects and arthropods, small vertebrates, blood, fruit, nectar, and pollen. These new opportunities led to extensive adaptive radiation among the Chiroptera, exhibited by an astonishing diversity. This diversity makes bats excellent models for the study of evolutionary processes.

Two components are important for understanding this extraordinary diversity in recent bats: phylogeny and environment. All these bats are related by common ancestry, but their form and function are also closely linked to the environments in which they live. If we are to understand the numerous variations in morphology, physiology, sensorimotor systems, and behavior in bats and to develop explanations for evolutionary change, we must examine both phylogenetic and environmental factors.

The principal focus of Part Three, "Echolocation," is to understand the various types of echolocation systems observed in bats. Using a comparative approach, the following chapters examine how ecological constraints are reflected in adaptations of sensory and motor systems that are associated with echolocation. The study of echolocation systems is especially well suited for evaluating how behavioral tasks imposed by ecological constraints have led to specific adaptations. The behavioral tasks solved by echolocating bats are similar to the tasks of man-made technical systems such as sonar and radar. Thus, we can apply the theoretical framework used in the design and functional analysis of man-made systems toward understanding biological echolocation systems. When we ask how bats search for and find food, how they approach and acquire food, and how the auditory periphery of bats is adapted for echolocation, we consider the operation of biological systems in a manner similar to that in which an engineer might assess the function of a technical system designed to meet specific tasks. In other words, we infer that evolution acts as an engineer to seek the optimal solutions for specific tasks, influenced by environmental constraints (adaptations) and using the available hardware provided by biological ancestors (phylogeny).

Unfortunately, we know very little about the phylogeny of echolocation and thus the following chapters are concerned mainly with the adaptational aspect. One conclusion that can be drawn from the studies presented in Part Three is that similarities in echolocation tasks are positively correlated with different echolocation systems. Thus, unrelated species subjected to similar ecological constraints have evolved rather similar echolocation systems. Many morphological, physiological, and behavioral parameters that describe echolocation systems (e.g., signal structure) are often homoplasies and cannot be used for the construction of phylogenetic trees.

Recent methodological and technical advances (sound recording and analysis, acoustical monitoring, photographic documentation, night vision devices, infrared video, and radiotelemetry) have enabled researchers to collect new information on the echolocation behavior of foraging bats in the field to address questions such as this: How do different types of habitat, foraging modes, and diet favor different kinds of echolocation systems? An important finding of comparative studies of echolocation is that the structure of search signals is intimately linked to habitat type and to foraging mode (see Schnitzler and Kalko). The signals emitted during the approach are influenced by other constraints. When approaching a target, most bats emit rather similar broadband frequency-modulated (FM) signals that permit precise localization of food or food sites. When closing in on a target, all bats reduce sound duration and pulse interval. However, maximum repetition rate differs, depending on whether the food is moving or stationary (see Kalko and Schnitzler). The high variability in signal design between and within species is another reason to be cautious when using the parameters of selected recorded signals as traits for the construction of phylogenetic trees.

Not only has the transmitter of bat echolocation systems (sound emission apparatus and echolocation signals) adapted to the ecological constraints set by different habitats, foraging mode, and diet, but also the receivers (external ear and auditory system) that process the echoes have become highly adapted. Comparative studies have demonstrated remarkable adaptations of the cochlear structure for the processing of species-specific signals, particularly in bats using long, constant-frequency (CF) signals (see Vater). Several species of bats with long CF signals adjust the frequency of sonar emissions to offset Doppler shifts introduced by their own flight velocity. A bat lowers the frequency of its sonar emissions with increased flight velocity to produce a relatively fixed echo reference frequency. This audiovocal feedback behavior is referred to as Doppler shift

compensation, and it is accompanied by high sensitivity and frequency selectivity at the echo reference frequency, which can be traced to cochlear specializations.

Species of bats that use long CF sounds are not the only ones which modify the features of their outgoing sounds in response to acoustic information carried by the echoes. Bats that use FM sonar sounds also adjust the duration, repetition rate, and bandwidth of their vocal signals with changing echo information. The sensorimotor feedback system of FM bats, which has been studied in behavioral and neurophysiological experiments (as shown by Valentine and Moss), depends on a three-dimensional system of coordinates for tracking sonar targets. Bats estimate the horizontal position of a target from interaural differences in echoes, the vertical position from spectral filtering by the external ear, and target distance from the time delay between the outgoing sound and the returning echo. It is the time delay between the sonar emission and the echo that guides range-dependent vocal behavior in bats. The neural mechanisms that support information on target-distance processing in the bat have been studied in extracellular recording experiments, and it has been postulated that single cells which show the response characteristic of echo-delay tuning may play a role in guiding range-related behaviors in bats (see Valentine and Moss; Dear). However, discrimination of target distance by FM bats that perform in laboratory experiments suggests that animals can detect changes in echo delay which are several orders of magnitude smaller than the width of the sharpest echo-delay tuning curves measured for single neurons. This discrepancy between behavioral performance and neurophysiological data challenges us to explore other characteristics of neural response that may support range processing in the central nervous system of bats (see Dear).

The echolocation system of bats has been studied extensively in the field and the laboratory. Field research permits a detailed analysis of an animal's natural sonar behavior. Laboratory research emphasizes the experimental study of echolocation and includes studies using behavioral, neurophysiological, and neuroanatomical methods. Each line of research complements the other, and representative work on echolocation in bats using these different approaches provides a glimpse of the constraints under which echolocation systems have evolved.

HANS-ULRICH SCHNITZLER AND CYNTHIA F. MOSS

12
How Echolocating Bats
Search and Find Food

HANS-ULRICH SCHNITZLER AND ELISABETH K. V. KALKO

The nearly 760 species of the suborder Microchiroptera occupy most terrestrial habitats and climatic zones and exploit a great variety of foods ranging from insects and other arthropods, small vertebrates, and blood to fruit, leaves, nectar, and pollen. Numerous morphological, physiological, and behavioral adaptations of sensory and motor systems such as echolocation and flight permit bats access to a wide range of habitats and resources at night.

Foraging bats are confronted by a multitude of problems when flying to their hunting grounds and searching for food. These problems differ depending on where the bats hunt: Those that capture insects in the open, with no obstacles in their flight paths, encounter conditions different from those confronting bats searching for prey near the edges of vegetation, in vegetation gaps or in dense forest, or near the ground. The problems differ further depending on whether bats catch insects in flight (aerial mode) or collect food from surfaces such as leaves, ground, or water (gleaning mode).

Echolocation is one of the adaptations that help bats to find their food. Echolocating bats emit ultrasonic signals and analyze the returning echoes so as to detect, characterize, and localize the reflecting objects (Schnitzler and Henson 1980). All bats use echolocation for orientation in space, that is, for determining their position relative to the echo-producing environment. Many bats (especially those which hunt for flying insects) also use echolocation to find their food (active mode). Other bats use information from other sensory systems (auditory, olfactory, visual) to perceive food-specific signals (passive mode). Some bats even randomly screen known or presumed feeding sites for food (random mode) (Figure 12.1).

The echolocation signals and hearing systems of bats are well adapted for gathering behaviorally relevant information (Schnitzler and Henson 1980; Neuweiler 1989; Fenton 1990). To understand the adaptive value of signal structure we must learn how different circumstances such as habitat type, foraging mode, and diet favor different signal types. We ask: What are the species-specific information needs of bats living under different ecological conditions and how have these needs shaped the structure of the echolocation signals and auditory systems?

Basic Perceptual Problems of Foraging Bats

When searching for food, bats must detect, classify, and localize a preferred target and distinguish it from unwanted

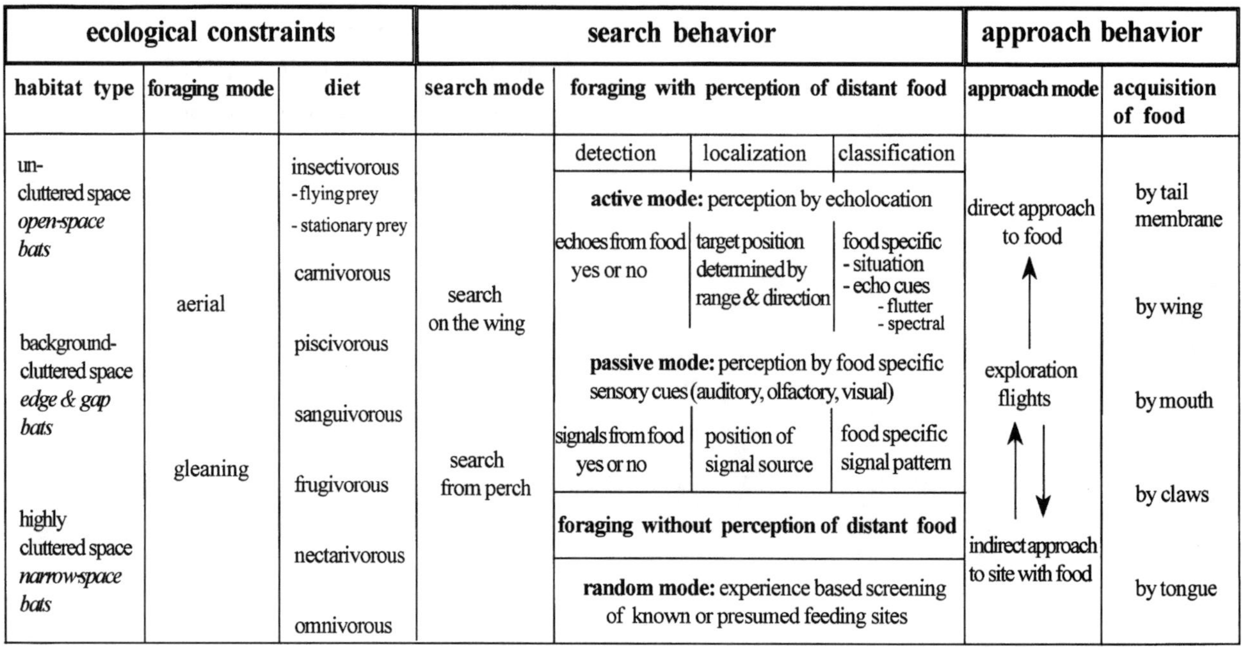

ecological constraints			search behavior				approach behavior	
habitat type	foraging mode	diet	search mode	foraging with perception of distant food			approach mode	acquisition of food
un-cluttered space *open-space bats*	aerial	insectivorous - flying prey - stationary prey carnivorous	search on the wing	detection	localization	classification	direct approach to food	by tail membrane
				active mode: perception by echolocation				

				echoes from food yes or no	target position determined by range & direction	food specific
						- situation
						- echo cues
						  - flutter
						  - spectral
								by wing

background-cluttered space *edge & gap bats*		piscivorous sanguivorous		**passive mode:** perception by food specific sensory cues (auditory, olfactory, visual)			exploration flights	by mouth
	gleaning	frugivorous	search from perch	signals from food yes or no	position of signal source	food specific signal pattern		by claws
highly cluttered space *narrow-space bats*		nectarivorous		**foraging without perception of distant food**			indirect approach to site with food	by tongue
		omnivorous		**random mode:** experience based screening of known or presumed feeding sites				

Figure 12.1. Ecological constraints on the search and approach behavior of echolocating bats.

or clutter targets such as tree branches, foliage, or the ground. For many bats, echolocation delivers all the information they need. However, some bats rely partly or completely on other sensory systems to find their food or randomly screen for prey at known or presumed feeding sites. Independent of their specific foraging conditions, all bats must solve the following basic perceptual problems.

Detection

An echolocating bat must decide whether or not an echo of its own echolocation signal is present. If bats use other sensory systems for target localization, they also must decide whether there are relevant sensory signals from the target. However, it is difficult to conceptualize detection independent of classification and localization.

Classification

Bats categorize targets according to echo features that reveal their characteristics (Ostwald et al. 1988) or according to other specific sensory cues. Target properties such as size, form, material, depth and angular extension, and texture are encoded in the complex temporal and spectral parameters of an echo. Furthermore, rhythmical amplitude and frequency modulations in the echo reveal characteristic movements of a target such as the beating wings of

a fluttering insect (Schnitzler 1987). In some bats, prey-specific sensory signals such as odor or prey-generated sounds are used for classification.

Localization

The position of a target is defined by its range and by its horizontal and vertical spatial angles. Range is encoded in the time delay between an emitted signal and a returning echo; the horizontal angle is determined from binaural echo cues and the vertical angle from monaural echo cues. Bats that use other sensory cues to find their food must localize the source of the food-specific sensory signal such as the position of an odor or sound source.

Interfering factors such as internal and external noise, clutter echoes, and signals from other bats may complicate the echolocation process involved with detecting, identifying, and localizing a preferred target. Interference between target echo and clutter, and between target echo and emitted signal, may be important in limiting the processing of relevant information.

No one signal is ideal for the extraction of all desired information with maximal accuracy and for the avoidance of all interfering factors. A signal well suited for one task may not work for another. Thus, it is important to know what kind of information can be carried by individual signal elements found in the echolocation signals of bats.

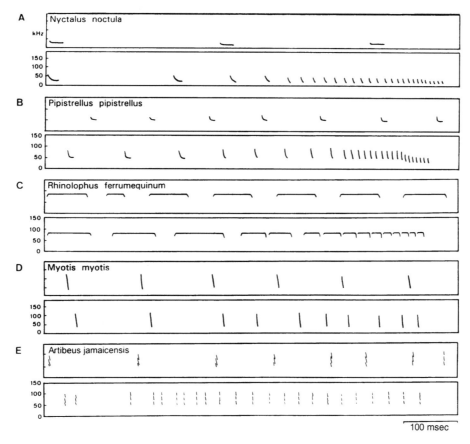

Figure 12.2. Search and approach signals of foraging bats. (**A–C**) Signals of bats that captured a flying insect at the end of the sequence. (**D, E**) Signals of bats that gleaned their food—an insect (D) and a fig (E)—from a surface at the end of the echolocation sequence. In all sequences, the increase in repetition rate and the reduction of sound duration indicate the switching from the search phase to the approach phase. Note the distinct terminal phase in sequences where the bats caught flying insects (A–C) and the lack thereof in sequences where the bats gleaned their food from surfaces (D, E).

Suitability of Signal Elements for Specific Echolocation Tasks

Bats use a wide variety of signal types. The structure of echolocation signals is generally species specific, and each species varies depending on the echolocation task confronting the bat. For example, the structure of signals emitted by a bat when it searches for food differs from that produced when a bat approaches food (Figure 12.2). Most echolocation signals of Microchiroptera consist of constant-frequency (CF), quasi-constant frequency (QCF), or downward frequency-modulated (FM) elements or combinations of these (Figure 12.2; see also Figure 12.5). These signal elements differ in absolute frequency, bandwidth, harmonic structure, duration, and sound pressure level (SPL), creating the wide variety of signal types found in echolocating bats (reviewed in Pye 1980; Schnitzler and Henson 1980; Simmons and Stein 1980; Neuweiler 1989; Fenton 1990).

The information that can be extracted from the echoes of different signal elements depends on their physical struc-

ture and on the performance of the bat's auditory system. A correlation between signal structure and auditory performance allows the following predictions concerning the suitability of signal elements for basic echolocation tasks.

Narrowband signals or signal elements (bandwidth of only a few kilohertz, kHz) such as QCF and CF signals are well suited for detection because the signal energy is concentrated during the entire echo within the neuronal filters tuned to the corresponding frequency band. However, narrowband signals are less suited for precisely localizing a target when bats must accurately measure range and the horizontal and vertical angle. Range is encoded in the time delay between emitted signal and returning echo. For accurate range determination, bats must determine the exact instant of sound emission and echo reception.

Narrowband signals are rather imprecise time markers because they stay within the corresponding neuronal filters for a long time and activate only a few auditory channels, thus diminishing range accuracy. The horizontal angle is encoded in binaural echo cues and the vertical angle in monaural echo cues; narrowband signals with their small frequency range activate only a few channels that deliver such cues, thus reducing a precise angle determination. Narrowband signals can be used for target classification if bats evaluate the amplitude and frequency modulations in the echoes arising from characteristic target movements. In particular, when a narrowband signal hits a fluttering insect at the favorable instant when the insect's wings are perpendicular to the impinging sound wave, a short and very prominent amplitude peak in the echo, a so-called acoustic glint, reveals a fluttering insect (see Figure 12.4B). This glint, which can be as much as 20–30 dB stronger than the echo from the body of the insect, also increases the probability for detection (Kober and Schnitzler 1990; Moss and Zagaeski 1994). The probability of receiving such a glint depends on the duty cycle (percentage of time in which signals are emitted) and the wingbeat rate of the insect. For instance, a bat with a duty cycle of 10% perceives an average of 6 glints/sec from a moth with a wingbeat rate of 60 Hz.

Signals of broad bandwidth such as uniharmonic and multiharmonic FM signals are less suited for the detection of weak echoes. These signals sweep rapidly through the tuning areas of the corresponding neuronal filters so that each channel receives only a small amount of energy. FM signals of broad bandwidth are well suited for exact target localization in which range and angle must be measured accurately. These signals activate each filter only for a very short period, producing the discrete time markers needed for an exact determination of the time delay that encodes the range. Large signal bandwidths activate more neuronal filters, thus improving the accuracy of range determination if, as is indicated by behavioral experiments, the range information is averaged over all activated channels (Moss and Schnitzler 1995).

To determine target angle, bats evaluate binaural cues. For determination of horizontal angle, bats measure time and intensity differences between corresponding left and right ear channels with similar best frequencies (Schnitzler and Henson 1980). FM signals activate many channels, so that the accuracy of angular determination improves with increasing bandwidth. FM signals of broad bandwidth also deliver spectral cues that can be used for target classification. Target features such as texture, which affects the absorption of sound at different frequencies, or depth structure, which causes an interference pattern from the overlapping multiwavefront echoes, are somehow reflected in the echo spectrum, thus encoding information about the character of a target (Ostwald et al. 1988).

In the laboratory, bats have learned to use such spectral differences to discriminate different targets and so it has been assumed that broadband FM signals allow for spectral characterization of prey (Neuweiler 1989, 1990). However, this is true only if the spectral signature of echoes from an insect is so specific that, independent of aspect angle, it is possible for the echo from an insect to be distinguished from clutter echoes.

Long CF signals in combination with Doppler shift compensation and a specialized hearing system serve to detect and classify fluttering insects in a cluttered environment. The beating wings of insects produce a rhythmical pattern of amplitude and frequency modulations that encode wing-beat rate, wing size, and other species-specific information. The most prominent flutter features are the very short and strong amplitude peaks or "acoustic glints" produced when the wings are perpendicular to the impinging sound waves (Kober and Schnitzler 1990). Bats use CF signals to detect fluttering insects in clutter and to classify insects according to the prey-specific pattern of echo modulations (von der Emde and Schnitzler 1990). The transmitters and receivers of the echolocation systems of CF bats are especially adapted for this kind of information processing (Schnitzler and Ostwald 1983; Neuweiler 1990).

By lowering their emission frequency, CF bats compensate for Doppler shifts caused by their own flight movement. Thus, the frequency of the CF component of insect echoes is kept within an "expectation window." A corresponding "analysis window" is established in the hearing system by a specialized cochlea with a highly expanded frequency representation in the range of the insect echoes. This acoustic fovea leads to an overrepresentation of sharply tuned neurons with special response characteristics throughout the auditory pathway. With these specific adaptations, CF bats are able to discriminate the modulated insect echo from overlapping unmodulated clutter echoes and to classify insects according to their specific modulation pattern.

Narrowband signals are ideal for target detection but less well suited for target localization. Broadband FM signals, however, are good for localization but less well suited for detection. This tradeoff between detectability and accuracy of localization is reflected in the design of search signals in some bats. In situations in which bats must solve several tasks simultaneously, they combine suitable signal elements; QCF-FM, FM-QCF, and CF-FM signals are such combinations that allow good detection and localization. Another combination is found in mouse-eared bats (*Myotis*, Vespertilionidae). When flying in the open, they produce broadband uniharmonic FM signals with changing steepness. Within a single sound, a steep initial part is followed by a more shallow part that ends in a steep segment. We assume that the steeper FM components are good for localization and that the more shallow component improves detection by introducing more signal energy into the corresponding neuronal filters.

The very short multiharmonic broadband FM signals of other bats such as New World leaf-nosed bats (Phyllostomidae) also can be interpreted as a compromise between detectability and localization accuracy. These signals have the advantage of a rather wide bandwidth, but each harmonic is still sufficiently shallow to introduce enough energy into the corresponding neuronal filters despite the short signal duration. This improves detectability without losing too much accuracy for determination of range and angle.

Thus, narrowband signals are well suited for target detection and classification of targets by flutter cues but less suited for precise target localization. Broadband FM signals, however, are less suited for detection but allow more precise target localization. FM signals can be used only for prey classification if the spectral signature of food echoes is specific enough so that bats can learn to discriminate food from clutter. The long CF signals of Doppler-compensating bats are especially adapted to detect fluttering insects in clutter and to classify insects according to their specific flutter pattern.

Limitations for Echolocating Bats

Not all the problems facing foraging bats can be solved by echolocation alone, one reason being the limited range of echolocation. The SPL of echoes decreases sharply with increasing target distance because of the geometric and atmospheric attenuation of sound traveling in air. If we assume a temperature of 20°C, a relative humidity of 50%, and a realistic detection threshold of about 15 dB, a bat with a signal SPL of 112 dB 40 cm in front of its head and with a signal frequency of 20 kHz will detect a fluttering insect with a wing length of 25 cm at a detection range of not more than 10.5 m. Under similar conditions, a sphere with a diameter of 2.5 cm cannot be detected beyond 7 m. The maximum detection distance decreases still more with increasing signal frequency, humidity, temperature, and decreasing prey size (Kober and Schnitzler 1990). Thus, echolocation represents a sonar system that works only over short distances. For long-distance orientation, bats must use other sensory systems such as vision.

The separation of the target echo from interfering signals is another important task facing echolocating bats. The evaluation of a sonar target is hampered when the neuronal activity evoked by clutter echoes and by the bat's own emitted signal interferes with the activity evoked by the target echo. Interfering signals that precede the target echo (such as the emitted signal) produce a forward masking effect. Interfering signals that follow the target echo (such as clutter echoes) produce a backward masking effect. Depending on the signal type, several strategies are used to avoid masking. Field and laboratory studies indicate that bats using QCF or FM signals avoid an overlap of the target echo with clutter echoes and also with their own emitted signal (Kalko and Schnitzler 1989, 1993). This avoidance of overlap suggests that these signal types are overlap sensitive. Because of overlap interference the ability of bats to evaluate insect echoes depends on the position of an insect relative to the bat and to clutter targets. When an insect flies so close to a bat that the returning echo overlaps with the still-emitted signal, forward masking effects interfere with the evaluation of the insect echo.

We assume that forward masking reduces the probability of detection in the zone in front of the bat where overlap occurs. The width of this signal-overlap zone depends on signal duration (Figure 12.3). For example, with a signal duration of 10 msec the overlap zone is 1.70 m wide. If undisturbed detection is only possible beyond this signal-overlap zone, signal duration sets a minimum detection distance. Each millisecond of signal duration adds 17 cm to this minimum detection distance. When an insect flies so close to clutter targets that its echoes overlap the clutter echoes, backward masking reduces the probability of detection in this clutter-overlap zone. The width of this zone is also determined by signal duration. Only insects flying so far away from the bat and from the clutter-producing background that no overlap occurs can be detected without interference. In our definition, these insects fly in an overlap-free window (Figure 12.3).

In bats with long CF signals, the CF component of the emitted signal and that of the returning echo often overlap (Figure 12.4B). This overlap produces no masking effect because Doppler shift compensation separates the target echo from the CF component of the emitted signal. The target echo is kept in the range of the extremely sharply tuned neurons of the acoustical fovea, whereas the emitted signal is lower in frequency and falls in a range where the auditory threshold is high (Schnitzler and Ostwald 1983; Neuweiler 1990). Therefore, long CF components of Doppler-compensating bats are not vulnerable to overlap and are thus overlap insensitive.

Foraging Habitats

Comparative studies reveal that foraging and echolocation behavior, especially the design of search signals, strongly

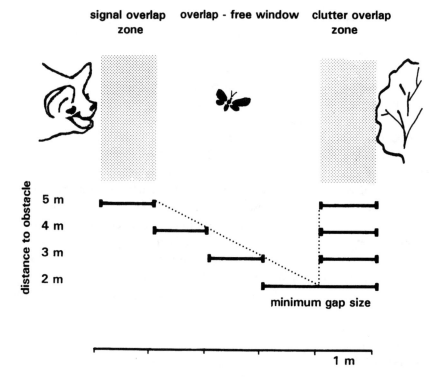

signal overlap   overlap - free window   clutter overlap
      zone   zone

distance to obstacle

5 m
4 m
3 m
2 m

minimum gap size

1 m

Figure 12.3. Schematic display of the clutter situation of a bat foraging near vegetation and emitting signals with a duration of 6 msec. The prey echo overlaps with the emitted signal when the insect flies in the signal-overlap zone and overlaps with the clutter echoes when it flies in the clutter-overlap zone. No overlap occurs when the insect flies in the overlap-free window. At a distance of 2 m, the overlap-free window is closed and the bat has reached, for the given signal duration, the minimum gap size, at which an overlap-free echolocation is no longer possible.

depends on the clutter situation encountered by a bat in its specific habitat. As the clutter situation is the most important ecological constraint, we define the foraging habitats of bats with respect to the proximity of the desired prey item to clutter targets such as vegetation, ground, or water surfaces. Such clutter targets represent a perceptual as well as a mechanical challenge for bats (Fenton 1990). Perceptually, bats are constrained by their sensory capacities (e.g., echolocation, vision, olfaction, passive listening) to detect, classify, and locate food in the vicinity of clutter targets. Mechanically, bats are constrained by their motor capacities, such as flight physiology (Norberg and Rayner 1987). For example, bats that forage near clutter targets need special maneuvering abilities (e.g., adaptations in wing morphology) to intercept insects while also avoiding collisions with obstacles. Here we discuss only the perceptual problem.

Bats foraging in various habitats encounter different clutter conditions. These differences have led us to define three habitat types (Figure 12.4).

Uncluttered Space

For bats that search for insects in the open, high above the ground and far away from vegetation, clutter echoes from the background are so far apart from the emitted signal and

target echo that they play no role in the echolocation process (Figure 12.4B). For these "open-space bats," a returning echo most often indicates a flying insect. In our terminology, these bats forage in "uncluttered space."

Background-Cluttered Space

When bats search for insects near the edges of vegetation, in vegetation gaps, or near the ground or water surfaces, the pair of pulse and echo is followed by clutter echoes from the background (Figure 12.4B). These "edge and gap bats" must solve two problems. First, they must recognize the insect echo and separate it from the background clutter. Second, they must determine the position of clutter targets to avoid collision. These bats are considered to hunt in "background-cluttered space."

Highly Cluttered Space

Bats that forage for insects very close to surfaces such as leaves, ground, or water and bats gleaning stationary food (sitting insects, other animals, fruit, nectar, blood) from surfaces encounter two situations. For gleaning bats that use short FM signals the echoes from food items are buried in clutter (Figure 12.4B), and for insectivorous bats using long CF signals the overlapping pulse–echo pair also over-

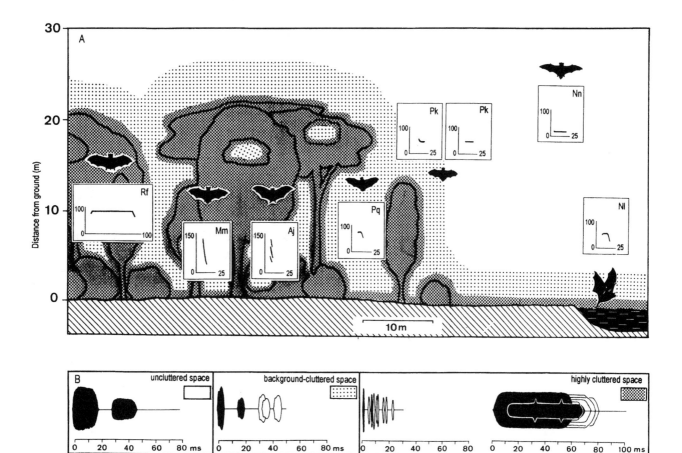

Figure 12.4. Schematic displays of bats foraging in different clutter situations. (**A**) Foraging habitats of bats, and flight silhouettes and search signals of representative species according to the clutter situation. The depicted species (from left: Rf, *Rhinolophus ferrumequinum*; Mm, *Myotis myotis*; Aj, *Artibeus jamaicensis*; Pq, *Pipistrellus quadridens*; Pk, *Pipistrellus kuhlii*; Nn, *Nyctalus noctula*; Nl, *Noctilio leporinus*) do not all occur sympatrically. Search signals (inset graphs) are reported in kilohertz (*y*-axes) and milliseconds (*x*-axes). In uncluttered space, prey is far away from vegetation and the ground. In background-cluttered space, prey is near obstacles, such as edges of vegetation, near the ground or water surfaces, and in gaps between and in vegetation. In highly cluttered space, prey is either close to or on vegetation or the ground. The border between uncluttered and background-cluttered space is defined by the echolocation behavior of bats. When entering the uncluttered space from background-cluttered space, bats switch from broadband signals to narrowband signals (e.g., *Pipistrellus kuhlii*), and vice versa. In this depiction, the border is about 5 m from vegetation and the ground, as described for pipistrelle bats. The border between background-cluttered and highly cluttered space is determined by the beginning of the clutter-overlap zone in which insect echoes overlap with clutter echoes. (**B**) Input into the auditory system of bats foraging in different clutter situations. The emitted pulse and the returning insect echo are depicted by solid black symbols. In uncluttered space, the pulse-echo pair is far from clutter; in background-cluttered space, the pulse-echo pair is followed by clutter echoes, depicted by open symbols. In highly cluttered space, the target echo is buried in overlapping clutter echoes. Sound duration and envelope correspond to search signals typical for the different spaces: uncluttered space, QCF signal of an open-space bat; background-cluttered space, broadband FM-QCF signal of an edge-and-gap bat; highly cluttered space, short and broadband FM signal of a narrow-space FM bat (left), and long CF-FM signal of a narrow-space CF bat (right). The echo of the long CF-FM signal shows amplitude modulations or glints created by the beating wings of an insect.

laps clutter echoes (Figure 12.4B). These "narrow-space bats" must discriminate between echoes from the food item and overlapping clutter echoes. Moreover, to avoid collision they must control their position relative to the clutter targets. These bats considered to forage in "highly cluttered space."

The border between uncluttered and background-cluttered space is defined by the echolocation behavior of the bats (Kalko and Schnitzler 1993). In uncluttered space,

bats use relatively long search signals of narrow bandwidth, which indicates that they do not locate the background clutter targets. In background-cluttered space, bats use search signals with FM components of broad bandwidth, indicating that echoes from clutter targets in addition to echoes of insects also guide the bats' behavior. When crossing the border between the two spaces, bats change the structure of their search signals.

The border between background-cluttered space and

highly cluttered space is defined by the relation between insect echoes and masking clutter echoes. A bat forages in highly cluttered space when the prey is situated in the clutter-overlap zone where clutter echoes overlap prey echoes. The width of the clutter-overlap zone is determined by signal duration. We are aware that bats flying parallel to clutter targets can reduce the masking effect of clutter echoes by spatial clutter rejection. However, to facilitate comparison between species, we use signal duration to define the width of the clutter-overlap zone independent of the bat's flight direction.

Search Behavior in Various Habitats

We postulate that habitat, foraging mode, and diet pose ecological constraints that have shaped the evolution of foraging and echolocation in bats. To test this hypothesis, we compared the foraging and echolocation behavior of bats in similar habitats with similar foraging modes and diets (see Figure 12.1).

To describe the habitat we use the three clutter spaces as previously defined. Foraging modes define the way bats obtain their food. Here we distinguish between bats that capture insects in the air (aerial mode) and those which glean insects and other food from surfaces (gleaning mode). According to the preferred food of various species, we categorize bats as insectivores, carnivores, piscivores, sanguivores, frugivores, nectarivores, and omnivores. To describe foraging behavior, we distinguish between the two search modes: search on the wing and search from a perch (see Figure 12.1). Furthermore, we address the problem of how bats find distant food. Here we discriminate between three modes. In the active mode bats find their food by echolocation alone; in the passive mode, they use food-specific sensory cues (auditory, olfactory, visual); and in the random mode, they rely on previous experience and screen known or presumed feeding sites.

In foraging bats we recognize two main classes of signal types: search and approach signals. Search signals are emitted when bats search for prey, or move from one place to another, and do not approach a specific target. Bats that approach a target or landing site emit a sequence of approach signals. These signals have the function of guiding the bat to the chosen target or site. The sequence of approach signals is dominated less by the habitat type than by the movement of the chosen target. Moving targets pose a more difficult task than stationary targets. Thus, in this chapter we focus our discussion on the search behavior of bats and consider the approach behavior of bats in a companion chapter (see Kalko and Schnitzler, Chapter 13 this volume).

Search Behavior in Uncluttered Space

Only insectivorous bats forage in uncluttered space. When they search for insects they have no masking problem so long as the emitted signal does not overlap the returning insect echo (Figure 12.4). However, bats often experience the problem that their rather small prey is sparsely distributed in a large space. Thus, the bats must cover a large search area to find an insect. This fact and the rather low SPL of insect echoes make it difficult to detect potential prey. Thus, echolocation signals should be optimized for detection.

The search signals of bats hunting for insects in uncluttered space are overlap-sensitive QCF signals or narrowband (<0.5 octave) FM-QCF signals of rather long duration (~8–25 msec) and low (<30 kHz) to medium (~30–60 kHz) terminal frequency. The signals are emitted at a rather low repetition rate correlated with the bat's wingbeat. Often these bats make two, three, or more wingbeats without pulse emission, resulting in very long pulse intervals (≥450 msec). Pure QCF signals of low frequency and long duration are emitted mainly by bats that only forage in uncluttered space (Figures 12.2A and 12.5 A_1). QCF or narrowband FM-QCF signals of higher frequency and shorter duration are used by bats that sometimes leave background-cluttered space to forage in uncluttered space (Figure 12.5 B_1). Moreover, a number of bats that usually forage in highly cluttered space and emit FM signals sometimes switch into uncluttered space where they emit broadband FM signals with a dominating shallow component (Figure 12.5 D_1).

Relatively long narrowband signals are adapted for long-range detection of insects in open space and may also deliver some flutter information. A glint in an echo not only improves the chances of detection but also indicates the presence of a fluttering insect. However, this may not be important, because in open space an echo from any moving target is a typical food-specific situation and indicates a flying insect. Bats switch to broadband signals when they change to the approach flight because these signals allow the precise localization necessary for successful insect pursuit.

An increase in signal duration should improve the chances of detecting an insect, but there seems to be a limit for such an increase. With increasing signal duration the signal-overlap zone gets wider and, if undisturbed detection is only possible beyond this signal-overlap zone, the minimum detection distance becomes larger (see Figure 12.3). Furthermore, signal duration and the corresponding minimum detection distance influence the minimum size of the prey that can be detected. At long signal durations, and therefore large minimum detection distances, only rela-

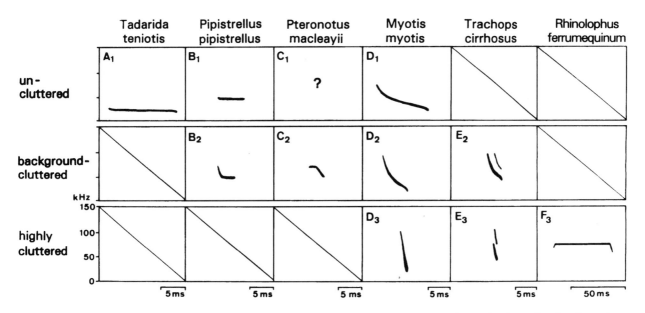

Figure 12.5. Examples of various search signals and their associated habitat types, by species. Bats may leave their preferred space for a less-cluttered space but not for a more-cluttered space.

tively large insects yield echoes that are strong enough to be detected.

Detection is also influenced by the frequency of a signal. Signals with low frequencies cover a wide search volume because they have low directionality and low atmospheric attenuation; with a large wavelength, they have the disadvantage of being less suited for the detection of small insects. Signals with higher frequencies cover a smaller search volume, having a higher directionality and a greater atmospheric attenuation; the shorter wavelength means they are better suited for the detection of small targets.

From the detection limits set by signal duration and signal frequency, we suggest a size-filtering hypothesis for the narrowband search signals of open-space bats: The longer the signals and the lower the frequency, the larger is the just-detectable prey. This concept suggests that bats with long signals of low frequency are adapted for detection of large insects at long distances, while bats with shorter signals and higher frequency can detect insects that are smaller and nearer. Several field studies have confirmed this trend (Barclay 1985, 1986). According to this hypothesis, an open-space bat hunting for small insects should produce shorter signals than when hunting for large insects.

Relatively long pulse intervals should be no problem for open-space bats because their maximum detection range is still larger than their own travel distance between consecutive signals, thus minimizing the possibility of overlooking an important target. The pulse intervals are far too large for setting a maximum detection distance for prey that might arise from masking by the next pulse of the returning echo of the preceding one. A pulse interval of only 150 msec

(which is small in open space) would set the maximum detection distance at 25.5 m. This distance is far too great for an echo produced from an insect to be audible to a bat. We thus assume that detection distances greater than 10 m are very unlikely.

Open-space bats with narrowband search signals are found among free-tailed bats (Molossidae), mouse-tailed bats (Rhinopomatidae), some sheath-tailed bats (Emballonuridae; e.g., *Peropteryx, Diclidurus, Taphozous*), and some evening bats (Vespertilionidae; e.g., *Nyctalus, Lasiurus*). All open-space bats forage continuously on the wing.

Search Behavior in Background-Cluttered Space

Nearly all bats that forage in background-cluttered space are insectivorous. Very few bats find their food (e.g., a pendulous fruit, a flower) so exposed that it produces an echo returning ahead of clutter echoes (e.g., fruits of *Gurania spinolosum* offered to *Phyllostomus hastatus;* Kalko and Condon 1998). Bats that search for insects in background-cluttered space must detect and identify insects flying near clutter-producing background and determine the position of large clutter targets and avoid them. This presents a detection and a classification problem with the rather weak echoes from the flying insects and a localization problem with the clutter echoes.

The search signals of species that hunt for insects in background-cluttered space are overlap sensitive, broadband (bandwidth > 0.5 octave) FM-QCF (Figures 12.2B and 12.5 B_2) or QCF-FM (Figure 12.5 C_2) signals of intermediate duration (~3–10 msec) and medium frequency (QCF,

~30–60 kHz). Broadband FM signals with a distinct shallow component are emitted by some narrow-space FM bats that forage in background-cluttered space (Figure 12.5 D_2, E_2). There is a clear tendency for reduced signal duration with decreasing distance to the clutter-producing background. The pulse interval is shorter than in open-space bats. Bats normally emit one signal per wingbeat, resulting in pulse intervals of about 80–120 msec.

The structure of search signals indicates that bats that forage in edge and gap habitats pursue a mixed strategy. The narrowband signal elements are adapted for medium-range detection of insects flying in front of clutter-producing background and for delivering some flutter information. The broadband FM components are adapted for precise localization of the clutter-producing background that must be avoided by the bats. In some bats, such as *Pipistrellus* and *Eptesicus* (Vespertilionidae), the FM component precedes the QCF component (FM-QCF) (Figure 12.5 B_2). In other species, such as some leaf-chinned bats (Mormoopidae), the FM component follows the QCF component (QCF-FM) (Figure 12.5 C_2).

Bats that forage in edge and gap habitats must cope with the forward masking effects of the emitted search signal on target-echo detection when insects fly in the signal-overlap zone. In addition, backward masking by strong clutter echoes presumably affects detection when the prey flies in the clutter-overlap zone. Therefore, detection without interference is only possible when insects fly in the overlap-free window between the two zones (Figure 12.3). The shorter the signal, the smaller are the overlap zones and the larger is the overlap-free window. On the other hand, we know that detectability increases with signal duration. This leaves the bats in a situation in which they have to find a compromise between detectability and width of the overlap-free window. The signals should be long enough for optimum detection but short enough to minimize overlap. An appropriate strategy would be to reduce signal duration with decreasing distance to the clutter-producing background, thereby keeping the overlap-free window open. Field studies show that this strategy is used by bats that come close to clutter targets (Kalko and Schnitzler 1993). Bats that forage in edge and gap habitats are often found hunting in gaps inside the forest, for example, between the canopy and subcanopy layers (Kalko 1995). These bats require a certain gap size to be able to search for insects in an overlap-free window. The minimum gap size at which the overlap-free window closes is defined by sound duration (Figure 12.3).

Bats that forage in edge and gap habitats with broadband FM-QCF signals are found among the evening bats (Vespertilionidae; *Eptesicus* and *Pipistrellus*) (Figure 12.5 B_L) Sometimes these bats also hunt for insects in uncluttered space

where they emit QCF or narrowband FM-QCF signals characteristic of this type of clutter situation (Figure 12.5 B_1) (Kalko and Schnitzler 1993). Some leaf-chinned bats (Mormoopidae) are examples of species with QCF-FM signals (Figure 12.5 C_2). Most bats that forage in edge and gap habitats do so continuously on the wing.

The border between uncluttered and background-cluttered space is defined by changes in the echolocation behavior of bats. For example, pipistrelle bats switch from narrowband signals that are typical for uncluttered space to broadband signals which are typical for background-cluttered space, and vice versa, at a transition distance of about 5 m from the clutter-producing background (Kalko and Schnitzler 1993). When flying away from clutter targets pipistrelles do not perceive clutter echoes from ahead. Nonetheless, they do not switch from the broadband to the narrowband mode before they have crossed the border between the two spaces at the transition distance of about 5 m. This indicates that the border between the two spaces does not depend on the perceived clutter situation alone. We assume that bats use some kind of a spatial memory to keep track of the background and that the border between uncluttered and background-cluttered space is marked on a three-dimensional cognitive space map (or spatial memory). Some preliminary data from the big brown bat *(Eptesicus fuscus)* and the noctule bat *(Nyctalus noctula)* have indicated that the transition distance is species specific (H.-U. Schnitzler and E. Kalko, unpublished observations).

Search Behavior in Highly Cluttered Space

Bats that forage for food in the clutter-overlap zone forage in highly cluttered space. Some of these narrow-space bats are insectivorous and either capture flying insects in the aerial mode or take insects that sit on clutter sources in the gleaning mode. Other bats are carnivorous, piscivorous, sanguivorous, nectarivorus, or omnivorous and glean their food from vegetation, the ground, the water, and other clutter sources. These bats must cope with a situation in which echoes from their food items are buried in background clutter (Figure 12.4B). Only if the echoes are so characteristic that they can be distinguished from the clutter echoes are these bats able to detect, classify, and localize the prey by echolocation alone. Moreover, these bats face the problem that the space for flight maneuvers is very restricted, which means that they must know the exact position to avoid the clutter-producing background.

Two strategies have evolved to solve these problems. Some species use signals consisting of a rather long CF component followed by a FM component (long CF-FM signals; Figures 12.2C and 12.5 F_3). Others use rather short broadband FM signals, often of low intensity, with either a uni- or a multiharmonic structure (broadband uniharmonic FM and broadband multiharmonic FM) (Figures 12.2D,E, and 12.5 D_3, E_3). In view of these very different echolocation systems we consider separately the "CF bats," which use long CF-FM signals, and the "FM bats," which use short FM signals.

NARROW-SPACE CF BATS.   Narrow-space CF bats are insectivorous and search for fluttering insects in highly cluttered space. They use long-duration ($\sim$10–100 msec), medium to high (CF $>$ 30 kHz) CF-FM signals with Doppler shift compensation (Figures 12.2C and 12.5 F_3). The CF component is overlap insensitive, but the FM component is not. Depending on the distance to the clutter targets these bats emit either one search signal per wingbeat or groups of two and more signals per wingbeat.

The very long CF-FM signals of narrow-space CF bats create such large signal- and clutter-overlap zones that they do not encounter an overlap-free window. The bats always face the problem of the target echo overlapping the emitted signal and the clutter echoes. Nonetheless, these bats are able to detect and classify prey echoes in strong clutter by evaluating the specific flutter signature of the CF component. Adaptations of the transmitter and receiver of the echolocation system, such as Doppler shift compensation and the auditory fovea in the cochlea, enable these bats to distinguish the modulated insect echo from the overlapping emitted signal and from overlapping clutter echoes, neither of which are modulated. As in other bats, the FM component is used to determine the position of targets. Bats that approach a target prevent overlap between the FM components of the emitted signal and the target echo. Thus, the CF component of the long CF-FM signals of these bats is adapted for the evaluation of Doppler shifts that encode relative movements and flutter information. The FM component is used for the localization of targets. With these adaptations, the signals are especially suited to forage for fluttering insects in highly cluttered environments.

Typical narrow-space CF bats are all species of horseshoe and Old World leaf-nosed bats (Rhinolophidae and Hipposideridae) and the New World mustached bat (Mormoopidae; *Pteronotus parnellii*). These bats search for fluttering insects in the forest close to vegetation or on the ground with two search modes. Some CF bats forage only on the wing (e.g., *Pteronotus parnellii*), and others also hunt from perches for passing insects (e.g., *Rhinolophus*). The insects are captured mainly in the aerial mode, but sometimes fluttering insects are gleaned from surfaces. The two species of noctilionid bats also evaluate flutter information

when hunting for insects with their CF or CF-FM signals. When gleaning fish from the water, *Noctilio leporinus* evaluates the prey-specific pattern of glints in echoes of jumping fish (Schnitzler et al. 1994).

NARROW-SPACE FM BATS.   Narrow-space FM bats that search for food in highly cluttered space use short-duration ($\sim$1–3 msec), broadband ($\geq$0.5 octave), uni- or multiharmonic, overlap-sensitive FM signals (Figures 12.2D,E, and 12.5 D_3, E_3), often with very low SPL (whispering bats). Depending on their distance from clutter targets, these bats emit either one signal per wingbeat or groups of two or more signals.

All narrow-space FM bats are gleaners. The echoes from their food items are buried in clutter (Figure 12.4), and the bats can only recognize these items when they have a food-specific signature. FM bats in the laboratory have learned to discriminate various targets according to spectral differences in the echoes (reviewed in Ostwald et al. 1988), and it has been postulated that these bats use spectral cues to distinguish echoes of sitting insects from clutter (Simmons and Stein 1980; Neuweiler 1990). However, no one to date has been able to demonstrate that spectral cues in echoes from typical food items (e.g., from an insect or a fruit) are sufficiently specific for bats to recognize them in a highly cluttered situation. We assume that under normal field conditions broadband FM signals do not allow the recognition of food items in clutter. An exception may be plants presenting flowers or fruits in such a way that they act as strong reflectors producing loud echoes that can be discriminated from the weaker echoes of the background (e.g., flowers of *Cardon* cactus offered to the nectar-drinking *Leptonycteris curasoae* [Phyllostomidae]; Schnitzler and Kalko, unpublished observation).

Narrow-space FM bats that feed on animals mainly use prey-generated acoustic cues for detection, classification, and localization (Tuttle and Ryan 1981; Belwood and Morris 1987; Faure and Barclay 1994). Many of these bats have very large ears especially adapted for passive localization of targets with acoustic cues. Gleaners that feed on fruit and nectar often use olfactory cues to find their food (Kalko and Condon 1998). In spite of their mostly passive localization of food, narrow-space FM bats nonetheless emit echolocation signals when they search for food. We assume that most gleaners use echolocation for determining the position of the site where the food is located, but not for detecting, classifying, and locating the food item itself. The low SPL of the echolocation signals may help to prevent overloading the hearing system with loud clutter echoes.

Narrow-space FM bats are represented among the ghost-faced bats (Megadermatidae), slit-faced bats (Nycteridae), New World leaf-nosed bats (Phyllostomidae), and evening bats (Vespertilionidae). Many of the mouse-eared bats (*Myotis*) belong to this group. Two search modes have been described for narrow-space FM bats: They either fly continuously along vegetation or close to the ground and search for acoustical signals signifying animal prey or for olfactory signals indicating fruits or flowers, or they hunt from perches for noisy prey. Particularly attractive are insects that crash into vegetation. Some of the insectivorous narrow-space FM bats have been found foraging in the aerial mode in background-cluttered and in uncluttered space. There, they emit longer FM signals, often with a distinct, shallow modulated component (Figure 12.5 D_1, D_2, E_2).

The use of food-specific cues for the detection, localization, and classification of food (passive mode) does not exclude the possibility that a narrow-space FM bat under favorable conditions will use echolocation to find its food (active mode). If the food produces such a loud echo that it can be discriminated from clutter, a bat has a chance to find it by echolocation (e.g., *Leptonycteris curasoae* feeding on *Cardon* flowers). Furthermore, learning of a specific situation may play an important role. For example, if we offer a *Myotis myotis* a noisy insect sitting on a screen, it first will use the prey-generated sounds for passive localization and approach and capture the insect. After some experience the bat learns that on this screen echolocation cues from a protruding target indicate a sitting insect and thus will also approach silent insects. Bats make such a transfer to echolocation only if some kind of echolocation cues can be associated with prey at a specific place. At other places, they do not react to similar echolocation cues.

Bat Guilds

Comparative studies on echolocation and foraging behavior of bats confirm the prediction that similarities in echolocation tasks correlate with signal design. In particular, the design of search signals is intimately linked to habitat type and foraging mode. Thus, design of search signals indicates the type of habitat in which a bat forages. Given this correlation, how can we use this relationship to address questions of community structure and diversity in bats? To understand how sensory and motor adaptations in bats promote species diversity we need to compare species assemblages at different localities.

In this context, assigning assemblages of species into guilds, that is, groups of species that live under similar ecological conditions, has proven to be a useful approach. We suggest that bats can be subdivided into guilds (similar to the original guild concept sensu Root 1967) characterized by habitat type, foraging mode, and diet. To provide

a standardized definition of these categories, we further define habitat type and foraging mode in relation to problems that must be solved by foraging bats. This definition characterizes ecological constraints as a set of tasks imposed on sensory as well as on motor systems. For example, to avoid clutter targets bats need sensory adaptations such as specific echolocation signals as well as motor adaptations such as specific wing shapes. From our considerations we propose the following guilds:

Habitat type	Feeding mode	Diet
Uncluttered space	Aerial	Insectivore
Background-cluttered space	Aerial	Insectivore
Highly cluttered space	Aerial	Insectivore
Highly cluttered space	Gleaning	Insectivore
Highly cluttered space	Gleaning	Carnivore
Highly cluttered space	Gleaning	Piscivore
Highly cluttered space	Gleaning	Sanguivore
Highly cluttered space	Gleaning	Frugivore
Highly cluttered space	Gleaning	Nectarivore
Highly cluttered space	Gleaning	Omnivore

The usefulness of the guild concept is often disputed, in particular for bats. As outlined in detail by Fenton (1990), bats can be highly variable in their foraging behavior, making it difficult to clearly define boundaries between guilds and to assign individual species to a particular guild. Some bats hunt in more than one of the defined habitats, or use the aerial as well as the gleaning mode, or feed on more than one of the defined diets. Some species that mainly capture insects in highly cluttered space are also able to forage for prey in background-cluttered space and perhaps even in uncluttered space (e.g., *Myotis myotis*, Vespertilionidae; Figure 12.5 D_1–D_3). Many bats that mainly forage in background-cluttered space also search for insects in uncluttered space (e.g., *Pipistrellus pipistrellus*, Vespertilionidae; Figure 12.5 B_1, B_2).

However, there are limits to this behavioral plasticity. Bats that are especially adapted for hunting in uncluttered space (e.g., *Tadarida*; Molossidae) are usually restricted to this habitat and cannot search for insects in background- and highly cluttered space. Bats that are adapted for background-cluttered space do not exploit highly cluttered space. Thus, the access of bats from their specific space to a less-cluttered space is possible, but not the reverse. Fenton (1990) explained this restriction with perceptual (ability to detect prey in clutter) and mechanical (ability to fly in close vicinity to clutter targets) problems.

We assume that limitations of the motor system largely prevent access to habitats with a more difficult clutter situation. For instance, long pointed wings (high aspect ratio and low wing loading) allow only fast agile flight with reduced maneuverability, thus restricting these bats to open spaces. By contrast, short broad wings (low aspect ratio and low wing loading) allow the slow maneuverable flight that is needed in obstacle-rich environments (Norberg and Rayner 1987). The sensory abilities, or at least the echolocation abilities, should be less restrictive, because all bats are able to produce the short FM signals that are necessary in cluttered situations (see approach sequences in Figure 12.2).

Despite the behavioral variability found in some species of bats, we suggest that bats can be classified into guilds according to their dominant sensory and motor adaptations. With this approach we have divided bats into 10 guilds. Bats that only forage for insects in open space belong to the guild of uncluttered-space aerial insectivores. When searching for prey, these bats emit rather long QCF signals of low frequency that are adapted for long-range detection of insects. Bats that search for insects mainly near edges, in gaps, or near the ground belong to the guild of background-cluttered aerial insectivores. When hunting in their preferred habitat, these bats emit mixed search signals of medium duration consisting of a QCF component of medium frequency and a FM component of wide bandwidth (FM-QCF or QCF-FM signals). These signals are adapted for medium-range detection of insects as well as for the localization of clutter targets.

Bats that search for insects very close to clutter targets or on the ground and capture them when using echolocation, mainly in the aerial mode, belong to the guild of highly cluttered space aerial insectivores. In this complex-clutter situation only bats with very long CF-FM signals are able to discriminate prey echoes from clutter by evaluating flutter information. Bats of the seven other guilds glean their food from surfaces in the highly cluttered space. Except for those taking food from water surfaces, bats mainly use prey-generated acoustic cues (e.g., animal-eating bats) or olfactory cues (e.g., frugivorous or nectarivorous bats) to detect, locate, and classify their prey. While searching for prey these bats emit broadband uni- or multiharmonic FM signals of short duration and low SPL that are used for the short-range localization of clutter targets. Piscivorous bats deviate in their behavior from other gleaners. For example, *Noctilio leporinus* uses prey-specific modulations in the CF component of echoes caused mainly by distortions of the water surface to recognize jumping fish.

Using the proposed guild concept we should be able to compare bat communities in a standardized way to gain a better understanding of the diversity of bats ecologically and in overall species richness (Kalko et al. 1996).

Acknowledgments

This chapter is based in part on a review that is to be published in *Bioscience* (Schnitzler and Kalko, submitted). Some of the ideas of this chapter have been stimulated by the work of people who often could not be cited because of space limitations. We are aware of this fact and thank all of them. In particular, we would like to acknowledge the work of D. Griffin, B. Fenton, G. Neuweiler, and their co-workers. We also thank I. Kaipf for technical assistance in the laboratory and in the field. This article has benefited greatly from the comments of C. Moss, C. Handley, A. Herre, and J. Ostwald. Our fieldwork has been supported by grants from the Deutsche Forschungsgemeinschaft (SFB 307 and SPP Mechanismen der Aufrechterhaltung tropischer Diversität).

Literature Cited

Barclay, R. M. R. 1985. Long- versus short-range foraging strategies of hoary (*Lasiurus cinereus*) and silver-haired (*Lasionycteris noctivagans*) bats and the consequences for prey selection. Canadian Journal of Zoology 63:2507–2515.

Barclay, R. M. R. 1986. The echolocation calls of hoary (*Lasiurus cinereus*) and silver-haired (*Lasionycteris noctivagans*) bats and the consequences for prey selection. Canadian Journal of Zoology 64:2700–2705.

Belwood, J. J., and G. K. Morris. 1987. Bat predation and its influence on calling behavior in neotropical katydid. Science 238: 64–67.

Faure, P. A., and R. M. R. Barclay. 1994. Substrate-gleaning versus aerial-hawking: Plasticity in the foraging and echolocation behaviour of the long-eared bat, *Myotis evotis*. Journal of Comparative Physiology A 174:651–660.

Fenton, M. B. 1990. The foraging behavior and ecology of animal eating bats. Canadian Journal of Zoology 68:411–422.

Kalko, E. K. V. 1995. Echolocation signal design, foraging habitats and guild structure in six Neotropical sheath-tailed bats (Emballonuridae). Symposia of the Zoological Society of London 67: 259–273.

Kalko, E. K. V., and M. Condon. 1998. Echolocation, olfaction, and fruit display: How bats find fruit of flagellichorous cucurbits. Functional Ecology 12:364–372.

Kalko, E. K. V., and H.-U. Schnitzler. 1989. The echolocation and hunting behavior of Daubenton's bat, *Myotis daubentoni*. Behavioral Ecology and Sociobiology 24:225–238.

Kalko, E. K. V., and H.-U. Schnitzler. 1993. Plasticity in echolocation signals of European pipistrelle bats in search flight: Implications for habitat use and prey detection. Behavioral Ecology and Sociobiology 33:415–428.

Kalko, E. K. V., C. O. Handley, and D. Handley. 1996. Organization, diversity and long-term dynamics of a neotropical bat community. *In* Long-Term Studies in Vertebrate Communities, M. Cody and J. Smallwood, eds., pp. 503–553. Academic Press, Los Angeles.

Kober, R., and H.-U. Schnitzler. 1990. Information in sonar echoes of fluttering insects available for echolocating bats. Journal of the Acoustical Society of America 87:882–896.

Moss, C. F., and H.-U. Schnitzler. 1995. Behavioral studies of auditory information processing. *In* Hearing by Bats, A. N. Popper and R. R. Fay, eds., pp. 87–145. Springer Handbook of Auditory Research, Vol. 5. Springer, Heidelberg.

Moss, C. F., and M. Zagaeski. 1994. Acoustic information available to bats using frequency modulated echolocation sounds for the perception of insect prey. Journal of the Acoustical Society of America 95:2745–2756.

Neuweiler, G. 1989. Foraging ecology and audition in echolocating bats. Trends in Ecology and Evolution 6:160–166.

Neuweiler, G. 1990. Auditory adaptations for prey capture in echolocating bats. Physiological Reviews 70:615–641.

Norberg, U. M., and J. M. V. Rayner. 1987. Ecological morphology and flight in bats (Mammalia, Chiroptera): Wing adaptations, flight performance, foraging strategy and echolocation. Philosophical Transactions of the Royal Society of London Series B 316:335–427.

Ostwald, J., H.-U. Schnitzler, and G. Schuller. 1988. Target discrimination and target classification in echolocating bats. *In* Animal Sonar Systems, P. Nachtigall, ed., pp. 413–434. Plenum Press, New York.

Pye, J. D. 1980. Adaptiveness of echolocation signals in bats. Flexibility in behaviour and in evolution. Trends in Neurosciences 3: 232–235.

Root, R. B. 1967. The niche exploitation pattern of the blue-gray gnatcatcher. Ecological Monographs 37:317–350.

Schnitzler, H.-U. 1987. Echoes of fluttering insects: Information for echolocating bats. *In* Advances in the Study of Bats, B. Fenton, P. A. Racey, and J. M. V. Rayner, eds., pp. 226–243. Cambridge University Press, Cambridge.

Schnitzler, H.-U., and O. W. Henson, Jr. 1980. Performance of airborne animal sonar systems: I. Microchiroptera. *In* Animal Sonar Systems, R. G. Busnel and J. F. Fish, eds., pp. 109–181. Plenum Press, New York.

Schnitzler, H.-U., and J. Ostwald. 1983. Adaptations for the detection of fluttering insects by echolocating bats. *In* Advances in Vertebrate Neuroethology, J. P. Ewert, R. R. Capranica, and D. J. Ingle, eds., pp. 801–827. Plenum Press, New York.

Schnitzler, H.-U., E. K. V. Kalko, I. Kaipf, and A. D. Grinnell. 1994. Fishing and echolocation behavior of the greater bulldog bat, *Noctilio leporinus*, in the field. Behavioral Ecology and Sociobiology 35:327–345.

Simmons, J. A., and R. A. Stein. 1980. Acoustic imaging in bat sonar: Echolocation signals and the evolution of echolocation. Journal of Comparative Physiology A 135:61–84.

Tuttle, M. D., and M. J. Ryan. 1981. Bat predation and the evolution of frog vocalizations in the neotropics. Science 214:677–678.

von der Emde, G., and H.-U. Schnitzler. 1990. Classification of insects by echolocating greater horseshoe bats. Journal of Comparative Physiology A 167:423–430.

13
How Echolocating Bats
Approach and Acquire Food

ELISABETH K. V. KALKO AND HANS-ULRICH SCHNITZLER

In the preceding chapter (Schnitzler and Kalko, "How Echolocating Bats Search and Find Food"), we described basic perceptual tasks that confront echolocating bats searching for food in various habitats. We discussed echolocation and other sensory systems (olfactory, visual, auditory) that bats use in searching for and finding food. Here, we present a brief overview of our current understanding of how bats approach and acquire various kinds of food. The perceptual and motor tasks that must be solved by foraging bats vary in difficulty depending on the feeding mode (see Figure 12.1). Bats that pursue and capture moving targets such as flying insects (aerial mode) or collect stationary food in highly cluttered environments (gleaning mode) face very different problems.

Approach Behavior by Aerial Insectivorous Bats

Perceptual Tasks

Bats that capture insects in the air face the problem that their prey constantly moves in three-dimensional space. Consequently, to successfully pursue and capture an insect,

a bat must continuously monitor its position. Aerial insectivorous bats rely on echolocation to find and track their food and approach it directly after detection (Griffin et al. 1960). In our definition, these bats forage in the active mode and make a direct approach (see Figure 12.1).

Transition from Search to Approach

Photographic sequences of aerial captures of insects by bats in the field and in the laboratory, synchronized with sound recordings, strongly suggest that the bats switch from search flight to target-oriented approach flight immediately after detecting prey (Griffin et al. 1960; Kalko and Schnitzler 1989; Kalko 1995a). During approach flight, the head and ears of a bat are pointed toward the target (Figure 13.1). All detailed studies of aerial insectivorous bats hunting in open space, edges, and gaps have shown that the beginning of pursuit coincides with simultaneous changes in echolocation behavior (Figure 13.1). The bat switches from the search phase to the approach sequence (Griffin et al. 1960; Schnitzler et al. 1987, 1994; Kalko and Schnitzler 1989; Kalko 1995a). The suggestion that the big brown bat (*Eptesicus fuscus*) may detect and pursue a target at first

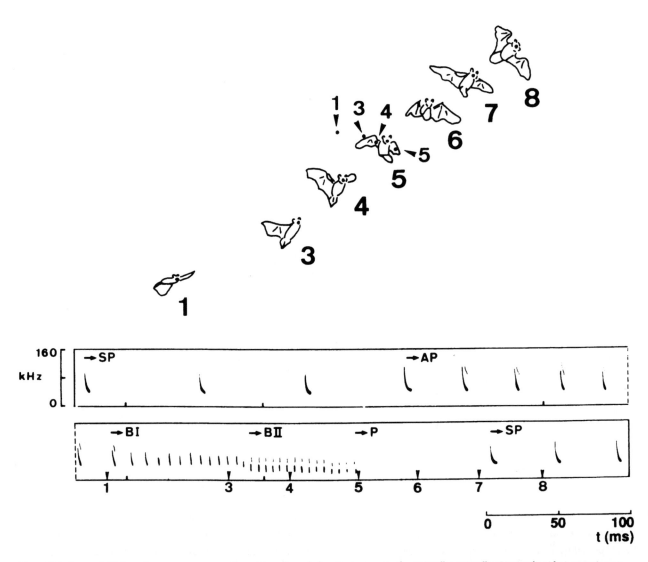

Figure 13.1. Approach flight and capture maneuver and corresponding echolocation sequence of a *Pipistrellus pipistrellus* (Vespertilionidae) capturing an insect (redrawn from a multiflash sequence). Numbers indicate the sequential order of the flashes and the corresponding synchronization pulses of the echolocation sequence; small numerals (with arrowheads) indicate the position of the insect. At 1–4, the bat is in approach flight; at 5 it extends one wing toward the insect and moves its tail into capture position (tail down); at 6, the bat bends its head into the tail membrane pouch to retrieve the insect (head down); at 7, the bat straightens its body and resumes search flight. In the sequence of echolocation signals (bottom panel), arrows indicate the following phases in echolocation behavior: SP, search phase; AP, beginning of approach sequence; BI, buzz I; BII, buzz II; P, pause.

without a distinct change in echolocation behavior (Kick 1982; Kick and Simmons 1984) is not supported by field data. However, hipposiderid and rhinolophid bats, which produce long constant-frequency, frequency-modulated (CF-FM) signals while searching for insects in highly cluttered environments and while hunting in flycatcher-style from a perch, may be an exception. These bats continue to emit typical search signals for some time after they detect their prey (Schnitzler et al. 1985).

Approach Sequence

The approach sequence of aerial insectivorous bats is characterized by distinct changes in signal structure and signal pattern. Typically, the bandwidth of the signals and the repetition rate increase, and pulse interval and sound duration decrease, with approach to the target (Griffin et al. 1960; Schnitzler and Henson 1980; Kalko and Schnitzler 1989; Kalko 1995a). Except for rhinolophids and hipposiderids, which produce CF-FM signals, aerial insectivorous bats emit broadband FM signals during the approach sequence (Figure 13.1). Species that have narrowband components (quasi-constant-frequency [QCF], shallow-modulated FM) in their search signals eliminate them during the first few signals of the approach sequence (Rydell 1990; Surlykke et al. 1993; Kalko 1995a).

In contrast to the various types of search signals used by aerial insectivorous bats, the signals in approach sequences are remarkably similar (see Figure 12.2). All bats that approach and close in on a flying insect face a similar challenge in localizing this target. The broadband, short FM signals of the approach sequence are well suited for this task (Schnitzler and Henson 1980; Simmons and Stein 1980; Neuweiler 1989, 1990). The increase in repetition rate enhances the information flow, enabling the bats to pinpoint the exact position of the insect, to position themselves optimally for capture, and to compensate for last-instant changes in prey position.

Field studies of aerial insectivorous bats have demonstrated that the bats reduce signal duration as they approach a target. Consequently, overlap of target echoes and the emitted signal and overlap of target echoes and clutter echoes are avoided. In this way the target is kept in an overlap-free window (Kalko and Schnitzler 1993; Kalko 1995a). In both search phase and approach sequence, an overlap might mask important information from the bat. Consequently, we assume that both search and approach signals, with the exception of those of CF-FM bats, are overlap sensitive. CF-FM bats foraging in highly cluttered space maintain a CF component that overlaps the CF component of the target echo while they approach the target. However, this has no masking effect because Doppler shift compensation, in combination with an auditory fovea with many sharply tuned neurons in the frequency range of the CF component of the echoes, prevents masking between the CF components of emitted signal and echoes. Hence, we conclude that the CF components of the approach signals, like those of search signals, are overlap insensitive. Further, we assume that the bats still need the CF component during approach to distinguish the fluttering target from the background. During approach, however, bats reduce the duration of the FM component following the CF part of the signal and thus avoid overlap with the overlap-sensitive FM component (Tian and Schnitzler 1997).

Terminal Phase

The approach sequence of aerial insectivorous bats ends in a distinct terminal phase, or final buzz, emitted just before capture of aerial prey (see Figures 12.2 and 13.1). In pipistrelle bats (Vespertilionidae), for instance, the onset of the terminal phase occurs at distances between 30 and 70 cm before the bat reaches its target (Kalko 1995a). The terminal phase is characterized by a series of short FM signals, mostly produced in one but sometimes in two or more groups (Simmons et al. 1979; Schnitzler and Henson 1980; Kalko and Schnitzler 1989; Kalko 1995a). The repetition rate of the terminal phase may reach 150–200 Hz.

In many vespertilionids the terminal phase is divided into two parts: buzz I and buzz II (Figure 13.1). Sound duration and pulse interval are continuously reduced throughout buzz I. Sound duration decreases from about 3–4 msec to about 1 msec and pulse interval from 30–40 to 10 msec. Buzz II is characterized by a distinct drop in frequency, sound duration of less than 1 msec, and a constant pulse interval about 4–5 msec. Photographic sequences synchronized with sound recordings show that the onset of buzz II coincides with preparations for the capture (Kalko and Schnitzler 1989; Kalko 1995a).

During approach, with the exception of the terminal phase, sound emission is coupled with wingbeat and respiratory cycle, and groups of two or more signals per wingbeat are emitted. The number of pulses per group increases as the bat comes closer to the insect (Schnitzler and Henson 1980; Kalko 1994). In the terminal phase a bat makes several wingbeats while it emits a continuous train of signals. In prolonged pursuits, sound emission in the terminal phase may be interrupted briefly while the bat presumably takes a quick breath. During insect pursuits involving steep dives coupled with sharp turns, a bat may glide briefly (Kalko 1994). In contrast to gliding flights in search flight, when signal emission is often stopped, the bat continues to emit signals during gliding intervals in pursuits.

Approach Strategies

Our field and laboratory studies of the foraging behavior of aerial insectivorous bats of five families (Emballonuridae, Molossidae, Mormoopidae, Noctilionidae, and Vespertilionidae) have permitted reconstruction of the flight paths of bats and prey in three-dimensional space. These studies demonstrated that aerial insectivorous bats employ a variety of pursuit strategies (Griffin et al. 1960; Webster and Brazier 1965; Kalko 1995b), depending largely on the movement of their prey. A bat that pursues a slow-flying insect flies straight to the prey and captures it. Evidence to date indicates that a bat in pursuit of a fast-flying insect, a moth, for example, is able to predict its future position and intercept it. By comparing simulated pursuits with real flight courses of bats and insects, we found that the interception course can be achieved if a bat keeps the tracking angle between itself and the prey constant. This action leads the bat onto a collision course with the insect so long as the insect does not change its flight speed and direction.

Our multiflash photographic sequences indicated that a bat usually approaches an insect from behind (Kalko 1995a). After detecting an insect a bat will often make a wide turn to position itself behind the prey, making it easier to correct for changes in the insect's flight path. Before capture, the bat typically moves to an optimal position for capture, just above the insect (see Figure 13.1).

Duration of insect pursuit is highly variable and depends on factors such as flight speed and maneuverability of both bat and insect; it ranges from a few hundred milliseconds to several seconds. Our photographs show that aerial insectivorous bats with short, broad wings adapted for slow flight and high maneuverability often follow evasive insects tightly and repeatedly attack them. Bats with long, pointed wings, such as free-tailed bats (Molossidae), attack insects at high flight speeds. When they miss the insect on the first attempt they often do not resume pursuit.

Flight Speed

Compared with its speed during search flight, an aerial insectivorous bat reduces flight speed considerably during pursuit and capture of an insect, sometimes coming almost to a standstill (Griffin et al. 1960; Webster and Brazier 1965; Jones and Rayner 1988; Kalko 1995a). Presumably reduction in flight speed, combined with an approach from behind, reduces the closing speed between the bat and an insect and gives the bat better control over the capture maneuver and adjustments for last-instant changes of the insect's flight path. This strategy also might help to reduce potential air turbulence that could push the insect off the capture membrane.

Acquisition of Food

Immediately before most captures, a bat tilts its body slightly upward and curls its tail membrane downward, perpendicular to its body—the tail-down stage. Insects may be captured with the tail membrane (tail scoop), or with a wing (wing capture), or be funneled onto the interfemoral membrane with a wing (wing capture with tail scoop) (Griffin et al. 1960; Webster and Griffin 1962; Trappe and Schnitzler 1982; Schnitzler et al. 1987; Kalko 1995a; see Figure 13.1; also see Figure 12.1).

There is only anecdotal evidence of bats capturing insects directly in their mouth (Griffin et al. 1960; Surlykke et al. 1993). Because a mouth capture requires a high degree of accuracy in prey localization, it must be more efficient to utilize the large membrane area of tail or wing to maximize the rate of capture success. Moreover, use of the large membrane areas for capture facilitates last-instant corrections to compensate for changes in prey position. In pipistrelle bats, for example, an estimated wing and tail membrane area as large as 170–270 cm^2 can be used in captures in contrast to only about 1 cm^2 for captures by mouth

(Griffin et al. 1960; Kalko 1995a). Multiflash photographs synchronized with sound recordings show that aerial insectivorous bats detect and track individual insects rather than simply flying through dense swarms of insects with their mouth open (Kalko 1995a; unpublished data). The 1-cm^2 area of the open mouth is too small to ensure a capture success rate high enough per unit of pursuit time to satisfy the energy needs of a bat.

After capturing an insect, a bat retrieves it immediately by lifting its tail membrane forward and bending its head into the pouch formed by the membrane (head-down stage) to retrieve the insect (see Figure 13.1). Horseshoe bats (Rhinolophidae) that capture an insect with a wing retrieve it directly from the wing (Trappe and Schnitzler 1982). Most insects are consumed in flight. Small insects are eaten whole; inedible parts such as wings and legs of moths and other large insects may be discarded while a bat circles briefly in a "parking position." Usually no echolocation signals are emitted (silent period or pause) during the head-down stage (see Figure 13.1). After retrieving and eating an insect, aerial insectivorous bats resume search flight and emission of search signals. Very large insects may be transported to a feeding roost for consumption.

Attack Rate

The attack rate of aerial insectivorous bats, particularly of bats feeding on small insects, can be very high. A complete pursuit maneuver, consisting of the approach flight, capture, and retrieval of an insect, may take only 200–500 msec. As many as two or three capture maneuvers per second have been documented. In such instances the bats emitted only one or two calls in the search and approach phase before starting the next terminal phase (Griffin et al. 1960; Fenton 1990; Kalko 1995a).

A question remains whether the duration and number of terminal phases and the duration of the silent period following a terminal phase can be reliably linked to capture success. Successful capture attempts frequently are accompanied by longer silent periods, indicating that the bat is retrieving the prey from its tail pouch, while shorter silent intervals are often associated with unsuccessful capture attempts (Schnitzler et al. 1987; Kalko 1995a). However, loss of prey may also lead to longer pauses and rapid retrieval of prey from the tail pouch to short pauses. Moreover, longer pursuit maneuvers following complex evasive maneuvers of prey that hear ultrasound, difficulties in capturing and handling prey, a longer time to eat large insects, and low insect density all prolong the interval between capture attempts (Acharya and Fenton 1992; Kalko 1995a).

Approach Behavior by Gleaning Bats

Perceptual Tasks

Echolocation tasks for gleaning bats differ from those of aerial insectivores because their food is stationary or moves little in two-dimensional space. Because gleaning bats forage in highly cluttered space, echolocation gives them only little information. Echoes from the target overlapping clutter echoes and the emitted signal create masking effects. Thus, gleaning bats detect and find their food largely by means of food-specific sensory cues (olfactory, visual, and auditory) rather than by echolocation (Neuweiler 1989; Fenton 1990). In our definition these bats forage in the passive mode (see Figure 12.1). Echolocation serves them primarily to keep track of surrounding obstacles and to control their approach to the food site.

However, a few gleaners do use echolocation to detect and find stationary food, including bats of the subgenera *Leuconoe* and *Pizonyx* of *Myotis*, bulldog bats (Noctilionidae) that take prey from the water surface, and several species which feed on exposed fruits, flowers, and small vertebrates (Jones and Rayner 1988, 1991; Kalko and Schnitzler 1989; Kalko and Condon 1993; Schnitzler et al. 1994; Marimuthu et al. 1995). Some gleaners use neither echolocation nor other sensory cues to find food or a site with food; rather, they screen known or presumed feeding sites on the basis of previous foraging experience (Schnitzler et al. 1994; von Staden 1995). We classify this as a random mode (see Figure 12.1). For example, we have observed fish-eating greater bulldog bats *(Noctilio leporinus)* capture fish by raking their feet through water in areas with no cues indicating the presence of the fish.

Approach Sequence

If the information from acoustic, olfactory, or visual cues does not allow exact localization of the food, gleaning bats will approach the food site rather than the food itself. In our definition this behavior is an indirect approach to a site with food (see Figure 12.1). If the food can be localized unambiguously, the bats approach it directly. According to our definition this is a direct approach to food.

After detecting food-specific signals, gleaning bats fly toward the signal source, switch in the vicinity of the food to approach flight, and start the approach sequence. Those attracted by acoustical cues often approach the source directly, sometimes with a brief hovering flight, before landing and grasping the prey. Bats attracted by scent seem to have more difficulty in localizing the food. Often they fly repeatedly toward the target and sometimes hover briefly (exploratory flights), before making a final approach to the food (see Figure 12.1). Gleaning bats reduce flight speed, as aerial insectivorous bats do, when approaching mostly stationary targets.

The switch from search signals to the approach sequence is characterized by a reduction in sound duration and pulse interval, indicating that the approach is guided by echolocation. Characteristically, approach signals of gleaning bats are broadband, uniharmonic or multiharmonic FM signals, resembling approach calls of the aerial insectivores (see Figure 12.2).

With decreasing distance to the food or food site, pulse interval and pulse duration are reduced to avoid signal overlap. The signals are arranged in groups of two or more signals, correlated with wingbeat and respiratory cycle. More signals are emitted per wingbeat as the bats come closer to the food or food site. A distinct terminal phase is lacking in bats that glean food from surfaces (Barclay et al. 1981; Neuweiler 1989; Fenton 1990; Schumm et al. 1991; Marimuthu et al. 1995). This leads to a lower repetition rate at the end of the approach sequence than aerial insectivorous bats use. Gleaning bats that take prey from the water surface are exceptions (Suthers 1965; Suthers and Fattu 1973; Jones and Rayner 1988, 1991; Kalko and Schnitzler 1989; Schnitzler et al. 1994).

During exploration flights, gleaning bats may change from olfactory localization (passive mode) to echolocation (active mode). For instance, some frugivorous New World leaf-nosed bats (Phyllostomidae) switch from a primarily scent-guided approach to ripe fruit to a primarily echolocation-guided final approach (Thies 1993; Kalko and Condon 1998).

Acquisition of Food

Gleaning insectivorous and carnivorous bats fly directly toward prey and pounce on it, or they hover briefly before swooping down upon it. Sometimes they land close to the prey site and crawl toward it, searching for the prey by scanning the surroundings with constant head movements. The papillae surrounding the chin of the frog-eating fringe-lipped bat, *Trachops cirrhosus,* may harbor chemoreceptors that help discriminate between poisonous and nonpoisonous frogs. As soon as the bat reaches its prey, it uses its mouth, wings, and tail membrane to subdue and secure it. To immobilize the prey, the bat aims its first bite at the neck, head, or thorax of the victim.

Gleaning phyllostomid frugivores use their mouths and sharp teeth to pluck fruit while in flight, or they may land briefly (Morrison 1978; Handley et al. 1991; Kalko et al. 1996; see Figure 12.1, this volume). Echolocation information is needed at close range to successfully grasp fruit in flight. In contrast to flying foxes, gleaning phyllostomids do not stay in the fruit tree to eat a fruit but instead fly off with it to a more distant feeding roost (Handley et al. 1991).

Nectarivorous gleaning phyllostomids (Glossophaginae and Lonchophyllinae) ingest liquid food by tongue-lapping. These nectar drinkers have long, extensible tongues with lateral grooves or feather-like papillae that increase surface area. The bat hovers momentarily or lands briefly, sticks its head into the corolla of the flower, and laps the nectar with rapid movements of its tongue (Dobat and Peikert-Holle 1985; Helversen 1993).

The sanguivorous phyllostomids (Desmodontinae), the vampire bats *(Desmodus rotundus, Diaemus youngi,* and *Diphylla ecaudata),* approach prey mainly on foot and use heat sensing to locate bare skin with blood vessels close to the surface (see Schutt, Chapter 10, this volume). There they make a small incision with razor-sharp incisors and lap the blood that trickles from the wound. A strong anticoagulant prevents the blood from clotting.

The fish-eating greater bulldog bat *(Noctilio leporinus)* and other bats that glean prey from water surfaces are characterized by large feet. They use their feet and tail membrane together to capture prey (Jones and Rayner 1988, 1991; Kalko and Schnitzler 1989; Schnitzler et al. 1994). The strategy of *Noctilio leporinus* is to swoop down briefly to the water surface and gaff its prey with its large hooklike claws, or to drag its feet through the water (rake) for as long as several meters (Suthers 1965; Schnitzler et al. 1994). Laterally compressed claws are adapted to minimize drag while the bat is raking (Fish et al. 1991). During rakes the long, strong calcars and large interfemoral membrane are folded forward between the legs to avoid contact with water. After a successful capture, *Noctilio leporinus* unfolds its large tail membrane and transfers the prey from its feet to its mouth with the support of its interfemoral membrane.

In contrast to aerial insectivorous bats, which mostly ingest their prey on the wing, gleaning insectivorous, carnivorous, and frugivorous bats transport their food to a more distant feeding roost (Belwood 1988; Fenton 1990; Handley et al. 1991; Kalko et al. 1996). Bulldog bats (Noctilionidae) have well-developed cheek-pouches in which to store their prey before thoroughly chewing and finally swallowing it (Murray and Strickler 1975).

At the feeding roost, gleaning bats hold food items between the thumb claws and manipulate them with their teeth. Small vertebrates are usually eaten whole. Gleaning insectivores discard chitinous material from insects such as wings, legs, mandibles, elytra, and ovipositors. Gleaning carnivores drop inedible parts such as feathers or tails from their prey. Frugivorous gleaners take bites from a fruit, and in some cases also from the leaves, chew them, press the pulp with their strong tongues against a sculptured palate, swallow the juice, and spit out the "dry" pellets (Handley et al. 1991; Kunz and Diaz 1995; Kalko et al. 1996); alternatively, they may open a fruit and lick and swallow its content. Large seeds are discarded undamaged at the feeding site. Small seeds are swallowed, usually undamaged, and are later dispersed in the feces of the bat.

Conclusions

Structure and pattern of approach signals are strongly influenced by the perceptual tasks that must be solved by bats that approach food or a specific food site. In contrast to the various signal types found in the search phase, signal structure in approach is rather uniform with broadband, overlap-sensitive, uni- or multiharmonic FM signals. Only CF-FM bats maintain an overlap-insensitive CF component throughout the approach sequence. The similarity of the structure of approach signals reflects a similar problem for all bats: precise localization of food or food site. All echolocating bats reduce sound duration and pulse interval when closing in on food, using either a direct or indirect approach. However, maximum repetition rate differs, depending on whether the food is moving or stationary. The difficult task of tracking a moving target is reflected in distinct terminal groups with high maximum repetition rates and hence increased information flow. Gleaners do not produce distinct terminal groups. A lower repetition rate is sufficient to make a controlled approach to a more or less stationary target.

To conclude, this chapter and the preceding chapter by Schnitzler and Kalko (Chapter 12, this volume) have clearly demonstrated that comparative field studies of bats living under different ecological conditions are essential to better understand the adaptive value of the wide range of signal structures that have evolved in bats. Only with a thorough understanding of the constraints influencing signal structure can we recognize which differences in signal structure are likely to reflect adaptations to the environment and which differences may mirror phylogenetic relationships among species.

Acknowledgments

As was the case with our companion chapter in this volume, this chapter is based on a review that is to be published in *Bioscience* (Schnitzler and Kalko, submitted). We thank C. Moss, C. Handley,

A. Herre, and J. Ostwald for many valuable comments and suggestions on the manuscript, C. Handley for correcting the English, and I. Kaipf for excellent technical assistance in the field and in the laboratory.

Literature Cited

Acharaya, L., and M. B. Fenton. 1992. Echolocation behavior of vespertilionid bats *Lasiurus cinereus* and *Lasiurus borealis* attacking airborne targets including arctiid moths. Canadian Journal of Zoology 70:1292–1298.

Barclay, R. M. R., M. B. Fenton, M. D. Tuttle, and M. J. Ryan. 1981. Echolocation calls produced by *Trachops cirrhosus* (Chiroptera: Phyllostomatidae) while hunting for frogs. Canadian Journal of Zoology 59:750–753.

Belwood, J. J. 1988. Foraging behavior, prey selection, and echolocation in phyllostomine bats (Phyllostomidae). *In* Animal Sonar, P. E. Nachtigall and P. W. B. Moore, eds., pp. 601–605. NATO ASI Series, Vol. 156. Plenum Press, New York.

Dobat, K., and T. Peikert-Holle. 1985. Blüten und Blumenfledermäuse. Bestäubung durch Fledermäuse und Flughunde (Chiropterphilie). Waldemar Kramer, Frankfurt am Main.

Fenton, M. B. 1990. The foraging behavior and ecology of animal-eating bats. Canadian Journal of Zoology 86:411–422.

Fish, F. E., B. R. Blood, and B. D. Clark. 1991. Hydrodynamics of the feet of fish-catching bats: Influence of the water surface on drag and morphological design. Journal of Experimental Zoology 258:164–173.

Griffin, D. R., F. A. Webster, and C. R. Michael. 1960. The echolocation of flying insects by bats. Animal Behaviour 8:141–154.

Handley, C. O., Jr., A. L. Gardner, and D. E. Wilson. 1991. Demography and natural history of the common fruit bat, *Artibeus jamaicensis*, on Barro Colorado Island, Panamá. Smithsonian Contributions to Zoology, Vol. 511. Smithsonian Institution Press, Washington, D.C.

Helversen, O. von. 1993. Adaptations of flowers to the pollination by glossophagine bats. *In* Animal–Plant Interactions in Tropical Environments, W. Barthlott, C. M. Naumann, K. Schmidt-Loske, and K. L. Schuchmann, eds., pp. 41–60. Zoologisches Forschungsinstitut und Museum Alexander Koenig, Bonn, Germany.

Jones, G., and J. M. V. Rayner. 1988. Flight performance, foraging tactics, and echolocation in free-living Daubenton's bats *Myotis daubentoni* (Chiroptera: Vespertilionidae). Journal of Zoology (London) 215:113–132.

Jones, G., and J. M. V. Rayner. 1991. Flight performance, foraging tactics, and echolocation in the trawling insectivorous bat, *Myotis adversus* (Chiroptera: Vespertilionidae). Journal of Zoology (London) 225:393–412.

Kalko, E. K. V. 1994. Coupling of sound emission and wingbeat in naturally foraging European pipistrelle bats (Microchiroptera: Vespertilionidae). Folia Zoologica 43:363–376.

Kalko, E. K. V. 1995a. Foraging behavior, capture techniques, and echolocation in European pipistrelle bats. Animal Behaviour 50: 861–880.

Kalko, E. K. V. 1995b. Predator–prey interactions: Evidence for predictive pursuit strategies in naturally foraging aerial insectivorous bats. American Zoologist 35(5): 40A.

Kalko, E. K. V., and M. Condon. 1998. Echolocation, olfaction, and fruit display: How bats find fruit of flagellichorous cucurbits. Functional Ecology 12:364–372.

Kalko, E. K. V., and H.-U. Schnitzler. 1989. The echolocation and hunting behavior of Daubenton's bat, *Myotis daubentoni*. Behavioral Ecology and Sociobiology 24:225–238.

Kalko, E. K. V., and H.-U. Schnitzler. 1993. Plasticity in echolocation signals of European pipistrelle bats in search flight: Implications for prey detection and habitat use. Behavioral Ecology and Sociobiology 33:415–428.

Kalko, E. K. V., E. A. Herre, and C. O. Handley, Jr. 1996. Relation of fig fruit characteristics to fruit-eating bats in the New and Old World tropics. Journal of Biogeography 23:565–576.

Kick, S. A. 1982. Target-detection by the echolocating bat, *Eptesicus fuscus*. Journal of Comparative Physiology 145:431–435.

Kick, S. A., and J. Simmons. 1984. Automatic gain control in the bat's sonar receiver and the neuroethology of echolocation. Journal of Neuroscience 4:2725–2737.

Kunz, T. H., and C. A. Diaz. 1995. Folivory in fruit-eating bats, with new evidence from *Artibeus jamaicensis* (Chiroptera: Phyllostomidae). Biotropica 27:106–120.

Marimuthu, G., J. Habersetzer, and D. Leippert. 1995. Gleaning from the water surface by the vampire bat. Ethology 99:61–74.

Morrison, D. W. 1978. Foraging ecology and energetics of the frugivorous bat, *Artibeus jamaicensis*. Ecology 59:716–723.

Neuweiler, G. 1989. Foraging ecology and audition in echolocating bats. Trends in Ecology and Evolution 4:160–166.

Neuweiler, G. 1990. Auditory adaptations for prey capture in echolocating bats. Physiological Reviews 70:615–641.

Murray, P. F., and T. Strickler. 1975. Notes on the structure and function of cheek pouches within the Chiroptera. Journal of Mammalogy 56:637–676.

Rydell, J. 1990. Behavioral variation in echolocation pulses of the northern bat, *Eptesicus nilssonii*. Ethology 85:103–113.

Schnitzler, H.-U., and O. W. Henson, Jr. 1980. Performance of airborne animal sonar systems. I. Microchiroptera. *In* Animal Sonar Systems, R. G. Busnel and J. F. Fish, eds., pp. 109–181. Plenum Press, New York.

Schnitzler, H.-U., H. Hackbarth, U. Heilmann, and H. Herbert. 1985. Echolocation behavior of rufous horseshoe bats hunting for insects in the flycatcher style. Journal of Comparative Physiology 157:39–46.

Schnitzler, H.-U., E. Kalko, L. A. Miller, and A. Surlykke. 1987. The echolocation and hunting behavior of the bat, *Pipistrellus kuhli*. Journal of Comparative Physiology 161:267–274.

Schnitzler, H.-U., E. K. V. Kalko, I. Kaipf, and A. D. Grinell. 1994. Fishing and echolocation behavior of the greater bulldog bat, *Noctilio leporinus*, in the field. Behavioral Ecology and Sociobiology 35:327–345.

Schumm, A., D. Krull, and G. Neuweiler. 1991. Echolocation in the notch-eared bat, *Myotis emarginatus*. Behavioral Ecology and Sociobiology 28:255–261.

Simmons, J. A., and R. A. Stein. 1980. Acoustic imaging in bat sonar: Echolocation signals and the evolution of echolocation. Journal of Comparative Physiology 135:61–84.

Simmons, J. A., M. B. Fenton, and M. J. O'Farrell. 1979. Echolocation and pursuit of prey by bats. Science 203:6–21.

Surlykke, A., L. A. Miller, B. Mohl, B. B. Andersen, J. Christensen-Dalsgaard, and M. B. Jorgensen. 1993. Echolocation in two very small bats from Thailand: *Crasenonycteris thonglongyai* and *Myotis siligorensis*. Behavioral Ecology and Sociobiology 33: 1–12.

Suthers, R. A. 1965. Acoustic orientation by fish-catching bats. Journal of Experimental Zoology 158:3119–3148.

Suthers, R. A., and J. M. Fattu. 1973. Fishing behavior and acoustic orientation by the bat *Noctilio labialis*. Animal Behaviour 21: 661–666.

Thies, W. 1993. Echoortung und Futtersuche bei zwei frugivoren Fledermausarten *(Carollia perspicillata* und *Carollia castanea)* im Flugkaefig. Diplomarbeit, University of Tuebingen, Tuebingen, Germany.

Tian, B., and H.-U. Schnitzler. 1997. The design of echolocation signals of the greater horseshoe bat *(Rhinolophus ferrum-equinum)* during transfer flight and landing. Journal of the Acoustical Society of America 101:2347–2364.

Trappe, M., and H.-U. Schnitzler. 1982. Doppler shift compensation in insect-catching horseshoe bats. Naturwissenschaften 69:193–194.

von Staden, D. 1995. Das Jagd- und Echoortungsverhalten der Bechsteinfledermaus *(Myotis bechsteini;* Kuhl 1818). Diplomarbeit, University of Tuebingen, Tuebingen, Germany.

Webster, F. A., and O. B. Brazier. 1965. Experimental studies on target detection, evaluation, and interception by echolocating bats. AD 628055. Aerospace Medical Research Laboratory, Wright-Patterson Air Force Base, Ohio.

Webster, F. A., and D. R. Griffin. 1962. The role of flight membranes in insect captures by bats. Animal Behaviour 10:332–342.

14
Computational Strategies in the Auditory Cortex of the Big Brown Bat, *Eptesicus fuscus*

STEVEN P. DEAR

Field observations (Griffin et al. 1965; Schnitzler and Henson 1980; Neuweiler 1990) and behavioral studies (Griffin 1958; Webster and Brazier 1968; Simmons 1971, 1973) show that echolocating bats simultaneously perceive the shape and correct spatial locations of multiple objects in the environment as acoustic images derived from returning echoes. An important aspect of bat echolocation is the determination of the distance (range) of multiple targets. One well-documented cue for range determination is the time or echo delay between an emitted bat biosonar vocalization or pulse and the returning echo (Simmons 1971, 1973, 1989).

Cortical neurons specialized for encoding echo delay have been found in the four species of echolocating bats thus far studied neurophysiologically: *Pteronotus parnellii, Rhinolophus rouxi, Myotis lucifugus,* and *Eptesicus fuscus* (Feng et al. 1978; Suga et al. 1978; O'Neill and Suga 1979, 1982; Sullivan 1982; Wong and Shannon 1988; Berkowitz and Suga 1989; Schuller et al. 1991; Dear et al. 1993a, 1993b; Paschal and Wong 1994). These neurons, typically referred to as delay-tuned neurons, respond maximally to specific values of echo delay known as the best echo delay (BD). Populations of cortical delay-tuned neurons in two of these species, *P. parnellii* and *R. rouxi,* exhibit an additional organizational feature. The BDs of cortical neurons in these species are systematically arrayed across the cortical surface, forming a topographic map of echo delay or target range (Suga et al. 1978, 1983; O'Neill and Suga 1979; Suga and O'Neill 1979; Suga and Horikawa 1986; Edamatsu et al. 1989; Schuller et al. 1991). Cortical target-range maps are important to range perception in *P. parnellii* because focal inactivation of cortical delay-tuned neurons disrupts the bat's behavioral discrimination of range (Riquimaroux et al. 1991).

In contrast to the extraordinary map organization exhibited by *P. parnellii* and *R. rouxi,* in *M. lucifugus* and *E. fuscus* the populations of cortical delay-tuned neurons are not organized into topographical maps of BD (Wong and Shannon 1988; Dear et al. 1993b; Paschal and Wong 1994). Consequently, the question remains: How is range information conveyed into perception in these species of FM bats? Two newly discovered cortical computational strategies for encoding target-range information employed by *E. fuscus* are discussed in this chapter. Unlike computational strategies based on spatially organized computational maps on the cortical surface, these new cortical computational strategies utilize organization in time.

Temporal Considerations for Multiple Target Representation

Eptesicus fuscus typically hunts insects in open spaces with few conspecifics present (Geggie and Fenton 1985; Furlonger et al. 1987) but will chase insects into foliage. To keep track of insect prey when hunting near foliage, presumably its perceptual machinery is able to distinguish insect echoes from foliage echoes. Because the velocity of sound imposes a temporal sequence on the order of echo arrival to a bat's ears (echo delay), it is tempting to equate the process of perceptual segmentation solely with the resolution of temporally separated echoes. However, the process of perceptual segmentation is more complicated than just resolving a sequence of temporally segregated echoes. Because the scattering of wideband biosonar pulses off a single, rough, three-dimensional target such as an insect or branch can give rise to multiple, temporally overlapping echoes, environments containing multiple targets may return to the bat more temporally simultaneous or nearly simultaneous echoes than those containing single targets. For multiple targets, perceptual segmentation must link together or integrate the different echoes scattered from each target, as well as resolve the range differences between targets. The process of linking or integrating multiple echoes into a single target shares, with general imaging processes, an additional requirement for a temporally concurrent or simultaneous representation of different target ranges (Marr 1982).

Cumulative, Concurrent Neural Representation of Target Range

In the case of bat biosonar, time can be defined as the temporal interval following the start of each biosonar vocalization. For neuronal responses in the bat's auditory system, time is equated with the time interval or latency between the beginning of an acoustic stimulus and the first neuronal action potential elicited by the stimulus. For delay-tuned neurons, the acoustic stimulus consists of a pair of sounds simulating the particular species biosonar vocalization or pulse and a returning echo. Time intervals measured from the start of the sound pulse to the neurophysiological responses of delay-tuned neurons are referred to as the pulse facilitation latencies (PFLs). Although neuronal PFLs measured in passive listening experiments have not been compared to the neural response intervals following an active biosonar emission in a freely vocalizing *E. fuscus*, a related comparison of BDs measured during passive and active listening by *P. parnellii* yielded similar BDs (Kawasaki et al. 1988). By implication, PFL is assumed to be equivalent to the neural response time after a biosonar vocalization.

Both the midbrain (Dear and Suga 1995) and auditory cortex (Dear et al. 1993a, 1993b) of *E. fuscus* contain delay-tuned neurons for encoding target range. These delay-tuned neurons exhibit two additional response properties specialized for temporal coding. First, unlike delay-tuned neurons in other species, midbrain and cortical delay-tuned neurons discharge with an average of about one action potential in response to an optimal acoustic stimulus pair. Second, the PFLs of these delay-tuned neurons exhibit low temporal jitter to repeated presentations of the same acoustic stimulus. Thus, delay-tuned neurons in *E. fuscus* fire phasically, permitting each neuron to encode the range of a target as a single point in time.

Given that delay-tuned neurons in *E. fuscus* can encode target range as discrete points in time, the temporal organization of these neurons can be examined in midbrain and cortex by plotting the PFL of each neuron as a function of its target range or best echo delay (BD). As can be seen from Figure 14.1, the temporal organization of BDs of a population of midbrain delay-tuned neurons (Figure 14.1A; $n = 33$) is different from a population of cortical delay-tuned neurons (Figure 14.1B; $n = 99$). For the midbrain population, PFL is proportional to BD with very little scatter. This temporal organization is consistent with the sequential nature of echo arrival times at the bat's ears, added to a time delay before neuronal firing. However, cortical delay-tuned neurons exhibit additional complexity. The PFL of a neuron can be nearly as short as its BD or can extend to a maximum of about 40 msec (Figure 14.1B).

As *P. parnellii* and *R. rouxi* exhibit beautiful cortical topographic maps for values of neuronal BD (Suga et al. 1978, 1983; O'Neill and Suga 1979; Suga and O'Neill 1979; Suga and Horikawa 1986; Edamatsu et al. 1989; Schuller et al. 1991) and as cortical delay-tuned neurons in *E. fuscus* fire phasically with complex temporal organization, is the PFL topographically organized in the cortex of *E. fuscus* instead of the BD? Figure 14.2 shows a composite map representing the spatial distribution of PFLs of cortical delay-tuned neurons from eight *E. fuscus* individuals. The auditory cortex consists of two subdivisions, the variable and the tonotopic areas (Dear et al. 1993a, 1993b). Neither area exhibits global organization of PFLs. Thus, neither PFLs nor BDs of delay-tuned neurons are spatially organized in the *E. fuscus* cortex.

Although response times of cortical delay-tuned neurons in *E. fuscus* are not spatially organized, temporal organization does exist. The key to understanding this temporal order is illustrated in Figure 14.3. Figure 14.3A shows the cortical data illustrated in Figure 14.1B. Interestingly, variable-area delay-tuned neurons (squares to left of curved line plot) exhibit temporal organization similar to that of midbrain delay-tuned neurons (PFL proportional to BD). However, tonotopic-area delay-tuned neurons exhibit PFLs over the range from the neuron BD to about 40 msec. This

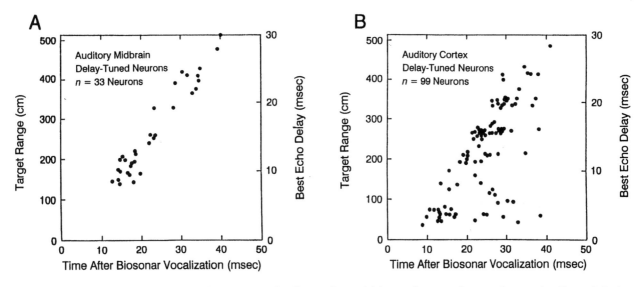

Figure 14.1. Best echo delays (BD) and response latencies (PFL) of midbrain and cortical delay-tuned neurons of *Eptesicus fuscus*. Each millisecond of echo delay corresponds to 17.3 cm of target range, so the ordinate is expressed in both range and echo delay. (A) PFLs and BDs of an ensemble of midbrain delay-tuned neurons. PFL is proportional to BD for these neurons. This reflects the sequential arrival of echoes to a bat's ears. (B) PFLs and BDs of an ensemble of cortical delay-tuned neurons. The systematic interaction of PFL with BD in cortical neurons transforms the sequential pattern of echo arrival times at sequentially longer delays into a simultaneous temporal representation of multiple targets at different delays.

property allows neural representation of short-range targets to be delayed until neurons that are tuned to longer ranges, but have shorter latencies, can discharge.

The additional complexity in temporal organization exhibited by tonotopic-area delay-tuned neurons creates a concurrent, cumulative representation of target range (Dear et al. 1993a). Figure 14.3B illustrates this temporal organization with three subsets of tonotopic-area data sampled from Figure 14.3A at PFLs of about 11, 25, and 38 msec. The PFLs are intended to reflect the neural representation of range information unfolding in the tonotopic area of the cortex as a function of time after each biosonar vocalization. Tonotopic-area delay-tuned neurons in the first subset discharge about 12 msec after the pulse to encode ranges of 50 to 140 cm, representing objects relatively near the bat. At about 25 msec after the pulse, different neurons (subset 2) discharge to encode ranges of about 275 cm, accumulating new ranges. Also, neurons with short BDs and long PFLs discharge simultaneously with the longer-BD neurons to concurrently encode ranges from 50 to 275 cm. This systematic trend is continued in subset 3 with new delay-tuned neurons simultaneously encoding ranges from 50 to 413 cm. Thus, the phasic discharges of tonotopic-area delay-tuned neurons integrate temporally dispersed echoes into a cumulative, concurrent neural representation of range (Dear et al. 1993a).

Temporal Representation of Target-Range Acuity

The topographic maps of cortical BD in *P. parnellii* and *R. rouxi* (O'Neill and Suga 1979; Suga and O'Neill 1979; Suga et

al. 1978, 1983; Suga and Horikawa 1986; Edamatsu et al. 1989; Schuller et al. 1991) suggest that the spatial location of a target in the environment corresponds to the location of the activity of delay-tuned neurons responding at their BD on the cortical surface. However, delay-tuned neurons also respond to nonoptimal values of echo delay as evidenced by their delay-tuning curves. How does a bat resolve two targets of similar range and direction? Target-range acuity of a single delay-tuned neuron refers to its ability to discriminate or resolve two echoes with similar echo delays. A metric for quantifying range acuity has been defined (Figure 14.4A) as the $Q_{50\%BD}$ (Dear et al. 1993a; Dear and Suga 1995). $Q_{50\%BD}$ values for cortical (variable- and tonotopic-area) delay-tuned neurons in *E. fuscus* (Figure 14.4B) vary by a factor of 12.8.

By examining PFL as a function of $Q_{50\%BD}$ for tonotopic-area delay-tuned neurons, the temporal organization of range acuity can be studied. Tonotopic-area delay-tuned neurons (Figure 14.5A) exhibit two overlapping patterns of $Q_{50\%BD}$ temporal organization, suggesting a division into two subpopulations. The smaller subpopulation, clustered around the horizontal dashed linear regression line, exhibits an almost constant $Q_{50\%BD}$ as a function of PLF, indicating constant fractional range acuity. The larger subpopulation, clustered around the sloping regression line, possesses the interesting property that the fractional range acuity of these neurons is proportional to their response latency. The relationship between $Q_{50\%BD}$ and neuronal BD for this subpopulation (Figure 14.5B) shows that the subpopulation with $Q_{50\%BD}$ proportional to PFL contains neurons with all values of BD.

Together, Figures 14.4 and 14.5 show that (1) the tempo-

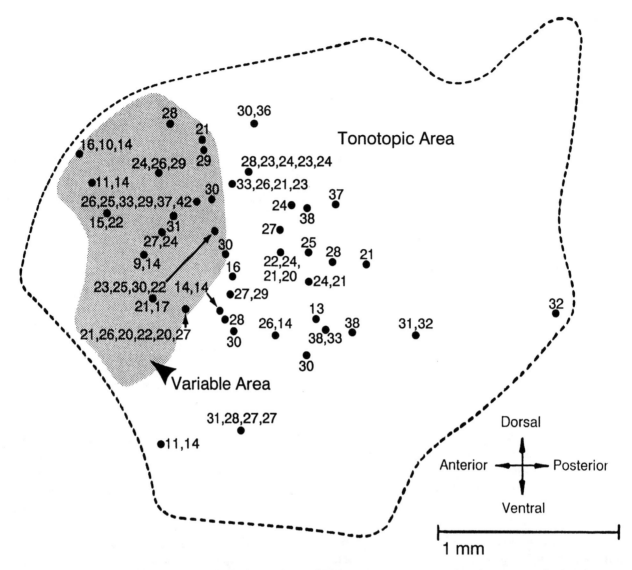

Figure 14.2. A composite spatial map representing the distribution of PFLs of delay-tuned neurons ($n = 84$) in the auditory cortices of eight E. *fuscus*. Numbers superimposed on the topographic map represent PFLs in milliseconds; multiple numbers separated by commas represent PFLs of sequential delay-tuned neurons encountered in a single penetration, in order of increasing depth. Cortical delay-tuned neurons were found in both the variable area (shaded, $n = 38$) and the tonotopic area ($n = 48$). PFLs are not uniform within a single orthogonal electrode penetration, nor is the global organization of PFL topographic. Consequently, response time is not spatially organized in the cortex.

ral organization of the tonotopic-area subpopulation of delay-tuned neurons with proportional $Q_{50\%BD}$ exhibits a concurrent, cumulative representation of target range; (2) delay-tuned neurons discharging at any instant of time after the pulse share about the same fractional range acuity; and (3) the fractional range acuity is progressively greater for neurons firing progressively later after the pulse. To gether these three response properties correspond to a neural implementation of a well-established class of computational algorithms used in image and signal processing—a type of pyramidal decomposition (Burt and Adelson 1983) or multiresolution decomposition (Mallat 1989). These computational algorithms share a common architecture of

groups of parallel filter banks. The filters in each bank share a common central frequency but differ in their individual bandwidths. For E. *fuscus*, the "signal" corresponds to the target-range or target-distance axis, and each biological "filter" corresponds to an individual cortical delay-tuned neuron. Individual "filter banks" correspond to delay-tuned neurons with the same BD but different $Q_{50\%BD}$ values.

For biologists, an important criteria for evaluating computational strategies is to determine the perceptual or behavioral utility of the proposed computational strategy to the animal. For bats, pyramidal or multiresolution algorithms can be understood in the context of more familiar examples of topographical cortical range maps (Suga et al.

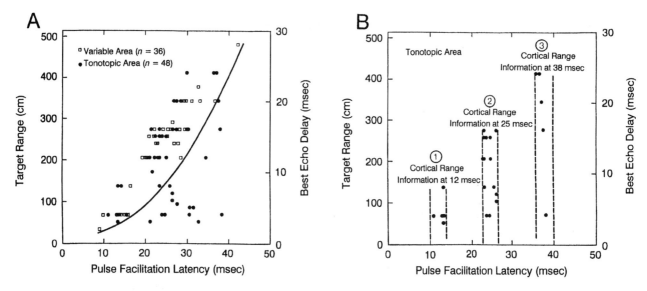

Figure 14.3. Temporal organization exhibited by delay-tuned neurons in the tonotopic area of the auditory cortex of *E. fuscus*. (**A**) PFLs and BDs of delay-tuned neurons (*n* = 84) from the variable and tonotopic areas of the auditory cortices of eight bats (see Figure 14.2). PFL is proportional to BD for variable-area delay-tuned neurons, much as it is for midbrain delay-tuned neurons (see Figure 14.1A). In the tonotopic area, many neurons tuned to short ranges have long PFLs (15 dots below the curved line plot); this retards the neural representation of their short-range estimates until neurons tuned to longer ranges, but with shorter latencies, can discharge. Thus, a concurrent, cumulative representation of range is generated exclusively by delay-tuned neurons in the tonotopic area. (**B**) Three illustrative subsets of tonotopic-area delay-tuned neurons from (A). About 12 msec after the simulated vocalization in subset 1, neurons discharge to encode ranges of 50–140 cm, representing objects relatively near the bat. About 25 msec after simulated vocalization, different neurons (subset 2) discharge to encode ranges up to 275 cm, thereby accumulating new ranges and concurrently encoding ranges from 50 to 275 cm. This systematic trend is continued in subset 3, with new delay-tuned neurons simultaneously encoding ranges from 50 to 413 cm. The phasic discharges of tonotopic-area delay-tuned neurons integrate temporally dispersed echoes into a cumulative, concurrent neural representation of range.

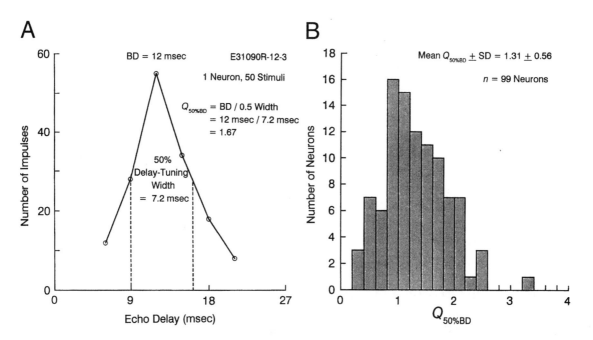

Figure 14.4. Quantification of range acuity of delay-tuned neurons. (**A**) Delay-tuning curve for a single neuron of *E. fuscus*. Note that delay-tuned neurons respond to nonoptimal values of echo delay. A quality factor called the $Q_{50\%BD}$ (Dear and Suga 1995) is a measure of the sharpness of delay tuning or range acuity for a delay-tuned neuron, quantified by dividing the BD by the width of the delay-tuning curve at half the maximum response (50% width). The cortical delay-tuned neuron represented here had a 12-msec BD with a 50% width of 7.2 msec, and thus a $Q_{50\%BD}$ of 1.67. (**B**) Distribution of $Q_{50\%BD}$ values for cortical delay-tuned neurons of *E. fuscus*. $Q_{50\%BD}$ values ranged from 0.26 to 3.33 (i.e., by a factor of 12.8).

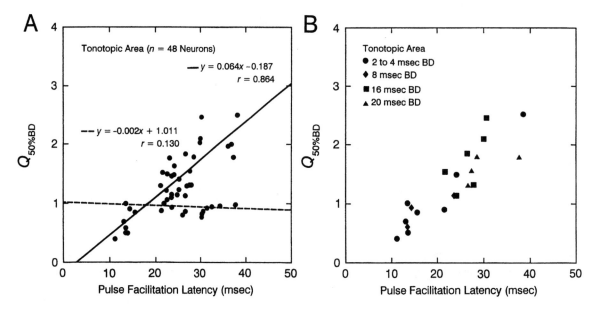

Figure 14.5. Temporal organization of range acuity ($Q_{50\%BD}$) for tonotopic-area delay-tuned neurons of *E. fuscus*. (**A**) Scatterplot of range acuity in tonotopic-area delay-tuned neurons against PFLs suggests a segregation into two subpopulations of neurons. The horizontal dashed line represents a linear regression fit to the smaller subpopulation ($n = 20$); $Q_{50\%BD}$ represented by this subpopulation is nearly constant, suggesting a constant fractional range acuity even though this subpopulation encompasses all values of BD. The sloping solid line represents a linear regression fit to the larger subpopulation ($n = 37$). $Q_{50\%BD}$ is correlated with PFL ($r = 0.86$), suggesting that range acuity of neurons in this subpopulation improves as a function of firing time after a biosonar vocalization. (**B**) Tonotopic area delay-tuned neurons belonging to the increasing range acuity subpopulation encompass all values of BD. The $Q_{50\%BD}$ values of neurons with BDs from 2 to 4 msec (circles) systematically increase by a factor of 6.2 with increasing PFL. Also, $Q_{50\%BD}$ values of neurons with BDs of 8, 16, and 20 msec increase by factors of 1.9, 3.2, and 1.3, respectively.

1978, 1983; O'Neill and Suga 1979; Suga and O'Neill 1979; Suga and Horikawa 1986; Edamatsu et al. 1989; Schuller et al. 1991). With topographic range maps, the underlying assumption often made is that a point in space is represented to perception by localized activity of neurons on a corresponding point or region of the cortical map. While this assumption seems plausible for single targets, it does not hold for multiple targets because most or all regions of the cortical map are active. Pyramidal decomposition or multiresolution analysis schemes circumvent this difficulty by representing range in a highly redundant fashion at different resolutions.

Consequently, these schemes greatly simplify subsequent "complex" imaging tasks such as edge detection or stereoscopic vision to processes of comparing neural activity across the multiresolution population. Neural implementation of a multiresolution representation of target range in bat cortex suggests the first computational step of sonar "image processing" in the bats and provides physiological support for the long-held hypothesis that *E. fuscus* processes biosonar echoes into acoustic images (Dear et al. 1993b).

Sensitivity of Cortical Neuronal Ensembles to Biosonar Phase Shifts

One subtle implication of pyramidal or multiresolution computational strategies is that neural information about

fine target-range features may reside within a distributed population of neurons as opposed to individual "filter" neurons. This possibility seems likely because the best target-range resolution of individual delay-tuned neurons in *E. fuscus* is of the order of milliseconds (Dear et al. 1993a, 1993b; Dear and Suga 1995) while the behaviorally determined target-range resolution is of the order of tens of nanoseconds to microseconds (for reviews, see Simmons 1993; Moss and Schnitzler 1995). For neurophysiologists, these behavioral data raise an interesting question: Where is the corresponding neural representation of fine target-range information in *E. fuscus*?

One hypothesis under active study in the visual system is that temporal correlation of neuronal discharges may bind distributed neuronal activity into unique representations (Engel et al. 1992). In the spirit of this hypothesis, Simmons and co-workers have been looking for evidence of neural temporal coding of fine target-range information in multi-unit discharges of *E. fuscus* inferior colliculus and auditory cortex neurons (Ferragamo et al. 1992; Haresign et al. 1993a, 1993b; Simmons et al. 1993, 1996). Their method of extracting temporal correlation of neuronal discharges consists of averaging multiunit extracellular potentials that were recorded from microelectrodes and elicited by repeated presentations of acoustic stimuli containing fine target-range information. This methodology (Glaser and

Ruchkin 1976) assumes that (1) neural spike trains contain some action potentials that are temporally synchronized to features in the acoustic stimulus and others which are not temporally synchronized; (2) the temporal synchronized firing pattern is similar for each repetition of the acoustic stimulus; and (3) the nontemporally synchronized action potentials can be considered as statistically independent samples of a random process.

Using acoustic stimuli containing fine target-range information, interaural time-delay differences, and echo phase differences, Simmons and co-workers (Ferragamo et al. 1992; Haresign et al. 1993a, 1993b; Simmons et al. 1993, 1996) found that (1) temporally synchronized responses can be extracted from multiunit spike trains by averaging potentials from inferior colliculus or auditory cortex neurons; (2) small time changes in the acoustic stimuli caused by phase shifts or interaural time delays resulted in larger time changes in the response latency of averaged inferior colliculus and auditory cortex potentials (time expansion); and (3) interaural time-delay differences up to 50 μsec evoke

proportional-response latency shifts of averaged potentials up to several milliseconds in the inferior colliculus. Thus, the response latency of collicular-averaged potentials appears to encode interaural time delay. One of the many questions remaining from these studies is whether the response latency of averaged cortical potentials actually encodes phase shifts or merely exhibits phase sensitivity. To address this question, the following experiment was performed.

Figure 14.6 illustrates the experimental procedures used to collect averaged cortical extracellular potentials along with representative examples. The acoustic stimuli for these experiments is based on the pursuit sequence (Figure 14.6A), which consists of 54 pairs of characteristic sounds simulating the biosonar emissions (pulses) and returning echoes during the pursuit and capture of an insect. The first sound in each pair is the pulse and the second sound simulates the echo at a later time (echo delay). The pursuit sequence is a dynamic stimulus because the echo delay, repetition rate, amplitude, and harmonic content are changing between successive pairs, a key feature in the experimental design.

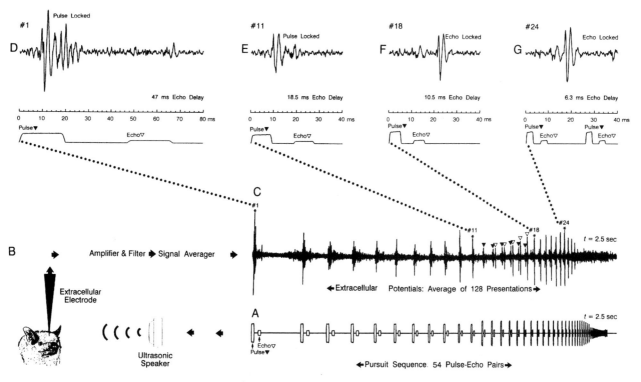

Figure 14.6. Experimental procedure used to collect averaged cortical multiunit extracellular potentials in *E. fuscus.* (**A**) A 2.5-sec temporal sequence (pursuit sequence) of 54 sound pairs simulating emitted biosonar vocalizations (pulses; solid triangle) and returning echoes (open triangle) during the pursuit and capture of an insect by the bat. The pursuit sequence is a dynamic acoustic stimulus in the sense that the echo delay, repetition rate, duration, and harmonic content vary throughout the sequence. The pursuit sequence is amplified and broadcast every 4 sec to an awake bat through an ultrasonic transducer. (**B**) Extracellular potentials elicited by the pursuit sequence are recorded from ensembles of cortical neurons from a single, high-impedance, 10-μm-diameter, carbon fiber microelectrode. Cortical potentials are amplified, bandpass filtered (300–8,000 Hz), and averaged 64 times. (**C**) An averaged cortical extracellular potential sequence elicited by the pursuit sequence. The extracellular potentials are time-locked to the pulse for the first 11 sequence sound pairs. For pairs #13 through #18, distinct averaged potentials are elicited by pulses (filled triangles) and echoes (open triangles). For pairs #19 through #28, averaged potentials are time-locked to the echo. (**D–G**) Time-expanded display of four averaged potentials from (C) evoked by the underlying sound pairs #1, #11, #18, and #24. Different sound pairs elicit different averaged potentials even though the recording electrode is in the same cortical location.

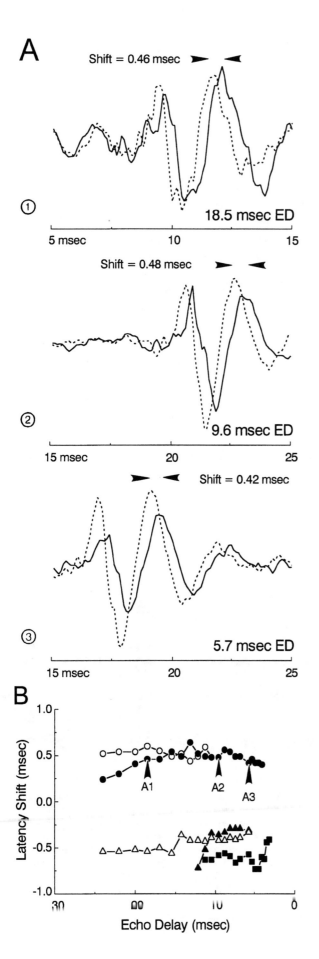

Thus, phase manipulations on either all the pursuit sequence pulses or all the pursuit sequence echoes that elicit a consistent shift in the averaged cortical response latency to most or all pursuit sequence sound pairs should reflect the encoding of the phase manipulations as opposed to responses to specific combinations of pursuit sequence echo delay, repetition rate, amplitude, or harmonic content.

Figure 14.7 illustrates the phase sensitivity of averaged cortical extracellular potentials recorded in different locations. Figure 14.7A compares potentials recorded at one cortical location; potentials elicited by sound pairs of a normal pursuit sequence are compared with potentials elicited by a phase-inverted pursuit sequence in which the phase of all the pulses (but not the echoes) has been inverted. Compared with the normal pursuit sequence, inverting the pulse phase by 180° causes the same latency difference of about 0.5 msec for each of the three sound pairs, even though the shape of the averaged potential remains about the same. Thus, the latency difference is most likely the result of the phase inversion.

Figure 14.7B compares the latency differences elicited by pulse-phase inversion for five different cortical locations. Although the latency differences are consistent for each cortical location, the shift may be in either the positive or negative direction with an amplitude of about 0.5 msec. While this result may seem puzzling as a neural code, the result is consistent with the predictions of a cross-correlation model. Shifting the phase of either the pulse or echo by 180° inverts the cross-correlation function, resulting in peaks with a symmetrical positive and negative time shift as compared to the single peak at time zero for the normal cross-correlation function. However, the cross-correlation model predicts a time shift of ±14 μsec while the cortical shifts are ±0.5 msec, a time-expansion factor of about 35. Because bats behaviorally perceive the phase of an echo relative to a pulse (Simmons 1993; Moss and Schnitzler 1995), these data suggest a link between cortical extracellular potentials and behavior.

←

Figure 14.7. Phase inversion of either all the pulses or all the echoes in the pursuit sequence elicits consistent forward or backward latency shifts in the averaged cortical potentials of E. fuscus. (A) Three plots of averaged extracellular responses (solid lines) recorded from one cortical location to three different pursuit sequence sound pairs (#11, #19, and #25) consisting of a 9.25-, 4.8-, or 2.85-msec pulse, followed by an echo 18.5, 9.6, or 5.7 msec later (echo delay, ED). Inverting all the pursuit sequence pulses by 180° (multiplying the electrical pulse signal by −1) causes a consistent shift (see arrowheads) in the averaged potential response latency (dashed lines). (B) Differences in averaged potential response latencies between normal and 180° inverted-phase pursuit sequences recorded from five different cortical locations. Arrowheads mark the data from (A). Depending on cortical location, either a positive or negative latency shift can occur in response to inverting the pursuit sequence pulses. The positive and negative latency shifts are consistent with a cross-correlation representation of target range.

The Temporal Patterns of Averaged Cortical Potentials May Carry Information

In the analysis of time shifts in fine target-range biosonar cues and corresponding neural response latency time shifts of averaged potentials, the specific pattern or patterns of averaged potentials have previously been ignored. The question remains whether the pattern or patterns of the averaged cortical potentials convey stimulus-specific information per se or merely serve as carriers of a relative latency code irrespective of their underlying pattern. For the remainder of this chapter, I provide preliminary evidence that the patterns of averaged cortical potentials may covey stimulus-specific information. The mechanism of information that is conveyed strongly resembles, in part, a computational equivalent of an inverse wavelet transform. Explanation, discussion, and implications of this evidence are based on three sources: (1) the correspondence of the shape of averaged cortical potentials to a particular class of compactly supported wavelets; (2) the temporal scaling of averaged cortical potentials; and (3) the temporal translation properties of forward and inverse wavelet transforms. Although each of these sources is necessary for neural implementation of a wavelet transform, only the combination of all three is sufficient.

Averaged Cortical Potentials Closely Resemble Wavelets

Evidence that ensembles of action potentials may carry fine target-range information per se starts with the observation that the shapes of averaged action potentials are not random but fall into distinct patterns. Figure 14.8 shows four examples of averaged cortical potentials and their associated time scales. Averaged potentials sd722 #11 and sd720 #12 were recorded at the same cortical location but were elicited by different sound pairs from the pursuit sequence. Potentials sd393 #18 and sd391 #21 were recorded at a different cortical location and were also elicited by different sound pairs from the pursuit sequence. However, potentials sd393 #18 and sd722 #11 closely resemble one another except for temporal scaling. In total, 57 different averaged potentials that correspond to the pattern illustrated by sd393 #18 and sd722 #11 were elicited by different sound pairs and recorded from different cortical locations. Averaged potentials sd391 #21 and sd720 #12 also resemble each other, but they represent a different pattern from the previous two examples. The pattern represented by averaged potentials sd391 #21 and sd720 #12 is less common, with a total of 25 examples recorded in auditory cortex.

The possible connection between the shapes or patterns of averaged potentials and an information-carrying capabil-ity is suggested by a close correspondence between these averaged potentials and the symlet family of wavelets (Daubechies 1992). In general, wavelets correspond to the impulse response of bandpass filters. The particular family members that these averaged potentials resemble are the S20 and S18 symlets (Daubechies 1992). Figure 14.9 compares the mathematical S20 symlet wavelet with the averaged cortical potential sd393 #18. As can be seen, the correspondence is remarkable. Figure 14.10 compares the mathematical S18 symlet wavelet with the averaged cortical potential sd720 #12. While the correspondence of averaged potential sd720 #12 to the wavelet is not quite so good as the previous example, any correspondence is interesting because the symlet family of wavelets is the result of a recursive mathematical calculation (Daubechies 1992) and the averaged potentials result from temporally synchronized action potentials from ensembles of cortical neurons.

The symlet family of wavelets are one of the few examples of wavelets that exhibit both good time and frequency resolution. Thus, a neural implementation of these particular wavelets seems logical given the signal-processing requirements of biosonar in the time and frequency domains. However, because the frequencies associated with the averaged potentials are in the range of hundreds of hertz, it seems unlikely that wavelets convey information about biosonar signals in the tens of kilohertz directly. If frequency information on biosonar was modulated down to the frequency range of the averaged potentials by the bat brain, then time information should be expanded by a corresponding factor. This prediction is consistent with the neural time expansions noted by Simmons and co-workers (Ferragamo et al. 1992; Haresign et al. 1993a, 1993b; Simmons et al. 1993, 1996) and the results of the cortical phase manipulation experiments noted in the previous section.

Temporal Scaling of Averaged Cortical Extracellular Potentials

The ability to carry information is not an intrinsic property of wavelets. Rather, wavelets carry information about signals or images when used as basis functions in wavelet transforms much in the way that cosines are used as basis functions in Fourier transforms (Daubechies 1992). One important difference between wavelet transforms and Fourier transforms is that the "time window" associated with a wavelet transform "scales" with frequency whereas the time window in a Fourier transform is constant (Rioul and Vetterli 1991). Consequently, wavelets carrying information in a wavelet transform context should exhibit temporal scaling. Figure 14.11A shows four examples of averaged cortical potentials resembling the S20 symlet recorded from four different cortical locations. Amplitudes have been normalized

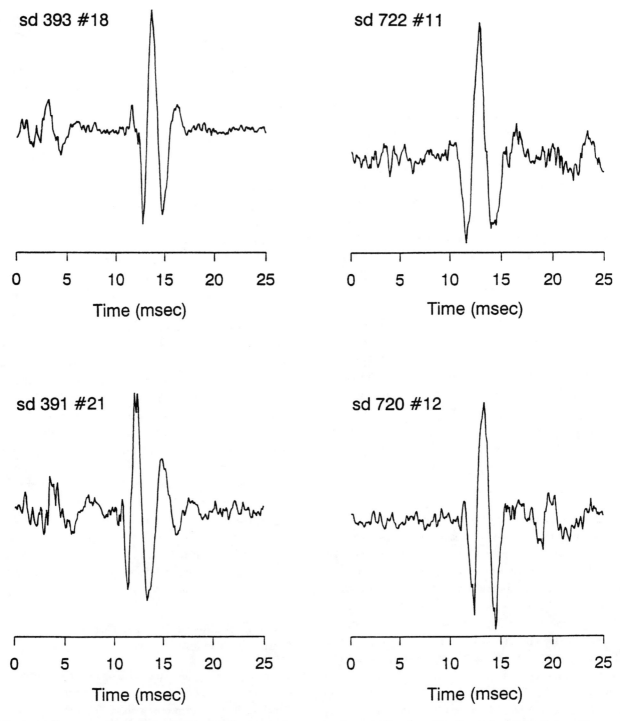

Figure 14.8. Four averaged extracellular potentials elicited by the pursuit sequence and recorded from ensembles of cortical neurons in *E. fuscus*. The two potentials in the top row strongly resemble Daubechies S18 symlet wavelet (see Figure 14.10), and the bottom row potentials strongly resemble Daubechies S20 symlet wavelet (see Figure 14.9). These results suggest that the extracellular potentials represent fine information using the computational equivalent of an inverse wavelet transform.

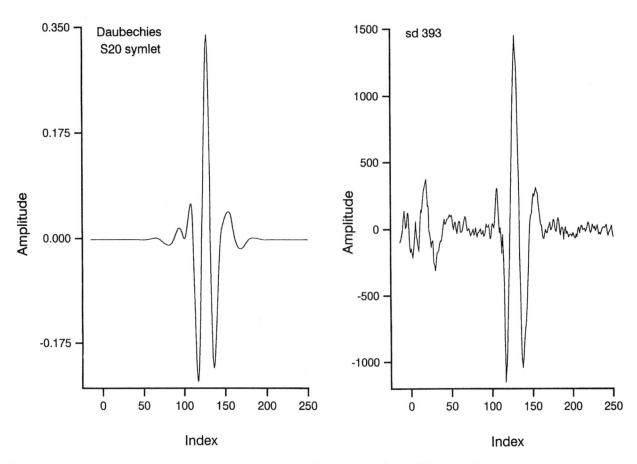

Figure 14.9. Correspondence between a Daubechies S20 symlet wavelet (**left**) and an averaged extracellular potential, sd393, that a pursuit sequence elicited in *E. fuscus* (**right**).

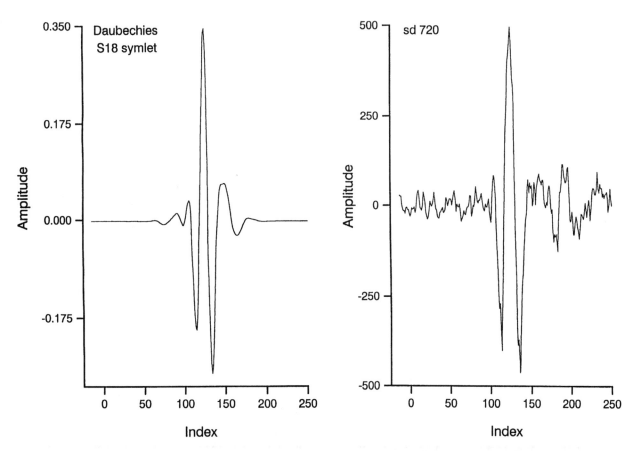

Figure 14.10. Correspondence between a Daubechies S18 symlet wavelet (**left**) and an averaged extracellular potential, sd720, that a pursuit sequence elicited in *E. fuscus* (**right**).

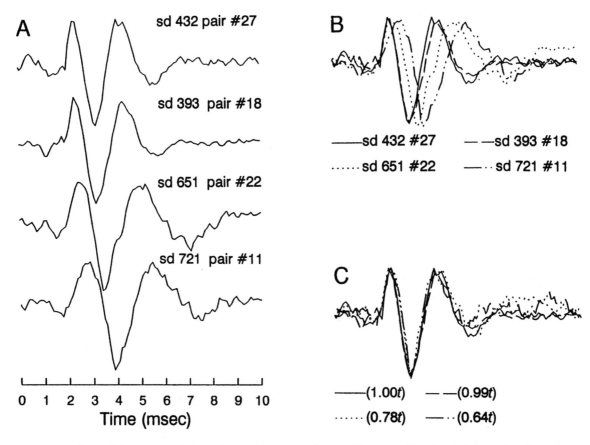

Figure 14.11. Averaged extracellular potentials elicited by the pursuit sequence and recorded from ensembles of cortical neurons in *E. fuscus* are time-scaled replicas of one another. **(A)** Solid lines represent four averaged potentials with normalized amplitudes from four different cortical recording sites. Each averaged potential was evoked by a different pursuit sequence sound pair: #27, #18, #22, or #11. **(B)** The averaged potentials from (A) are aligned on the same time axis. **(C)** Compressing the time scales of the last three potentials by factors of 0.99, 0.78, and 0.64, respectively, causes all the potentials to superimpose, which demonstrates that these oscillations are time-scaled replicas of the same basic shape.

for the sake of comparison, but time scales are preserved. By superimposing the four potentials (Figure 14.11B), the potentials appear to be time-dilated versions with the same shape. With proper time scaling, the potentials superimpose upon one another (Figure 14.11C). Thus, the averaged potentials recorded at different locations are time-scaled replicas of one another.

Representation of Fine Target-Range Information by Neural Wavelets

The results of the previous sections—latency shifts in response to fine stimulus timing manipulations, resemblance of averaged potentials to wavelets, and temporal scaling of averaged potentials—suggest that extracellular cortical potentials may represent fine target-range information to perception as the computational equivalent of an inverse wavelet transform. The relationships between each of these results and the computational aspects of wavelet transforms are illustrated graphically in Figure 14.12. Like a Fourier transform, the purpose of a wavelet transform is to take a

signal such as a Gaussian pulse (left side of Figure 14.12) and decompose it to a sequence of coefficients at different levels of resolution (d1, d2, . . . , s6). For those familiar with Fourier transforms, the transform coefficients represent the frequency content of the input signal. The levels in a wavelet transform (d1, d2, . . . , s6; Figure 14.12) can be thought of in terms of frequency such that level d1 represents the highest frequency and level s6 represents the lowest. The frequency associated with each d level (d1, . . . , d6) is the result of a temporally expanded or contracted prototype or mother wavelet.

In addition, all the levels of a wavelet transform are divided into a sequence of time increments. Thus, a wavelet transform is a time-frequency transform analogous to a spectrogram representation (Rioul and Vetterli 1991). This analogy is the basis of the suggestion that averaged multi-unit potentials recorded in the different frequency lamina in the inferior colliculus of *E. fuscus* correspond to a wavelet-like transform because the duration of collicular-averaged potentials varies across the frequency lamina (Haresign et al. 1993c). However, this variation of duration is not equiva-

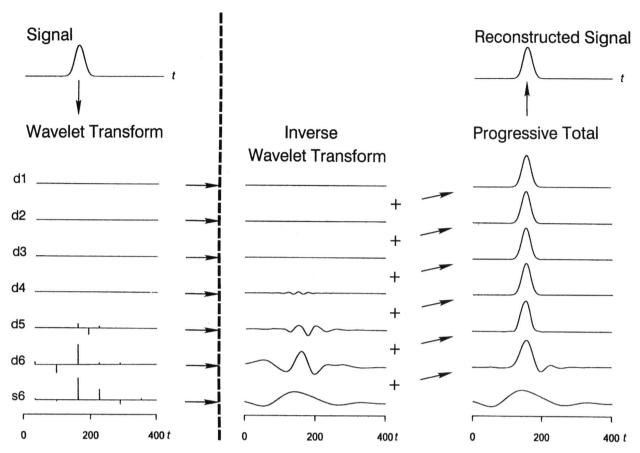

Figure 14.12. Graphical representation of the computational aspects of wavelet transforms: wavelet decomposition (to the left of the vertical partition) and wavelet synthesis (to the right). A Gaussian pulse is used as an example of a time domain input signal. A wavelet transform converts the input signal into temporal sequences of coefficients at different resolution levels: d1, d2, . . . , d6, and s6. Coefficients at each d level are the result of convolutions of the input signal with an analyzing wavelet (time expansions or contractions of a mother wavelet; a Daubechies S6 symlet for this example). The levels are organized so that the analyzing wavelet is longest at d6 and shrinks by a time factor of 2 at successive levels. Thus, level d1 contains the shortest analyzing wavelet. A signal can be represented by very few nonzero coefficients. Coefficients at the s level are the result of convolutions of the input signal with an analyzing scaling function (father wavelet) corresponding to the mother wavelet. The input signal can be reconstructed from the wavelet transform coefficients by the process of an inverse wavelet transform. At each level, the wavelet coefficients are multiplied with the corresponding analyzing wavelet or scaling function and added together, resulting in the graphs at each level. As the levels are successively added together, starting with level s6, the original signal is obtained.

lent to temporal scaling nor have the shapes of the averaged potentials been shown to resemble wavelets.

One property of a wavelet transform is that it can be inverted to recover the input signal. The right side of Figure 14.12 illustrates the first step in the inverse wavelet transform. Coefficients from each level of the wavelet transform are multiplied by the expanded or contracted decomposing wavelet, resulting in a temporal sequence of overlapping wavelets at each level. The original signal is recovered by starting at the lowest level, s6, and progressively summing the levels together.

The theoretical properties of the first stage of an inverse wavelet transform match the empirical response properties exhibited by the averaged cortical extracellular potentials on the following points. (1) Actual wavelets that appear in this part of the wavelet transform process correspond to

the averaged potentials. Thus, the single wavelet filter associated with each "d" level in Figure 14.12 is probably represented by many different ensembles at different cortical locations. (2) Temporally scaled wavelets associated with the different d levels correspond to the temporally scaled averaged potentials. (3) A time shift in the input signal time shifts all the wavelets, corresponding to the latency shifts of averaged potentials elicited by pulse or echo phase shifts. It is in this context, then, that the following hypothesis is proposed: Analog, parallel, distributed multiunit potentials convey biosonar information per se as part of the computational equivalent of an inverse wavelet transform.

Theoretically, this method of representing temporal information offers three computational advantages. (1) Because wavelets have a zero mean, the parallel sum of temporally random potentials will be zero unless an exact temporal relationship is maintained between the analysis of biosonar signals and the resulting inverse representation. Thus, a neural representation can be manipulated by controlling the neural timing. (2) By maintaining several different types of wavelet basis functions such as S18 and S20, only a few wavelets are needed to represent a time-varying signal (dynamic data compression). (3) Nonlinear wavelet transform methods can improve the neural signal-to-noise ratio close to the performance of a matched filter (Donoho 1995). For the neurophysiologist, the difficulty raised by this type of computational scheme is that the relationship between biosonar cues and neural responses has become less apparent and more complex. The wavelet computational hypothesis, if correct, suggests that wavelets are a common language repesenting all types of fine target structure.

Conclusions

Echolocating bats navigate and hunt in the dark by emitting biosonar vocalizations or pulses and analyzing returning echoes. Neurons specialized for the tasks of echolocation have been found in the brains of every species of echolocating bat thus far studied. Typically, the response properties of these neurons are systematically arrayed on the cortical surface in the form of computational maps. Are computational maps sufficient to explain all echolocation behavior in all species?

Single-unit and multiunit data recorded from the auditory cortex of the big brown bat *(Eptesicus fuscus)* supply evidence for two computational strategies providing target-range information to perception about complex targets without computational maps. The first computational strategy corresponds to a neural implementation of a common class of image analysis used in computational vision—pyramidal decomposition or multiresolution analysis. A general feature of this signal-processing class is that parallel banks of filters are employed with filters in each bank sharing a common central frequency but differing in their individual bandwidths. For the bat, the signal corresponds to the range axis of the bat and the filters correspond to individual cortical delay-tuned neurons. A bank of filters corresponds to groups of delay-tuned neurons with the same BD but different $Q_{50\%BD}$ values. It is important to note that although pyramidal decomposition or multiresolution analysis schemes appear to represent information in highly redundant fashion, they greatly simplify subsequent complex imaging tasks such as edge detection or stereoscopic vision. Neural implementation of a multiresolution representation of target range in the bat cortex is suggestive of the first computational steps of sonar image processing in the bat.

The second strategy suggests a wavelet representation of fine target structure. Specifically, the hypothesis proposed that analog, parallel, distributed averaged potentials convey fine target information per se as part of the computational equivalent of an inverse wavelet transform. However, unlike the preceding computational strategy, individual wavelet filters correspond to the ensemble activity of a group of neurons. Together, both computations form a multiresolution representation of target range.

Acknowledgments

This work was supported by Office of Naval Research grants no. N00014-95-10565 and N00014-97-0365.

Literature Cited

Berkowitz, A., and N. Suga. 1989. Neural mechanisms of ranging are different in two species of bats. Hearing Research 41:255–264.

Burt, P. J., and E. H. Adelson. 1983. The laplacian pyramid as a compact image code. IEEE Transactions on Communications 31:532–540.

Daubechies, I. 1992. Ten Lectures on Wavelets. Society for Industrial and Applied Mathematics, Philadelphia.

Dear, S. P., and N. Suga. 1995. Delay-tuned neurons in the midbrain of the big brown bat. Journal of Neurophysiology (Bethesda) 73:1084–1100.

Dear, S. P., J. A. Simmons, and J. Fritz. 1993a. A possible neuronal basis for representation of acoustic scenes in auditory cortex of the big brown bat. Nature (London) 364:620–623.

Dear, S. P., J. Fritz, T. Haresign, M. Ferragamo, and J. A. Simmons. 1993b. Tonotopic and functional organization in the auditory cortex of the big brown bat, *Eptesicus fuscus*. Journal of Neurophysiology (Bethesda) 70:1988–2009.

Donoho, D. L. 1995. De-noising by soft-thresholding. IEEE Transactions on Information Theory 41:613–627.

Edamatsu, H., M. Kawasaki, and N. Suga. 1989. Distribution of combination sensitive neurons in the ventral fringe area of the auditory cortex of the mustached bat. Journal of Neurophysiology (Bethesda) 61:202–207.

Engel, A. K., P. Konig, A. K. Kreiter, T. B. Schillen, and W. Singer. 1992. Temporal coding in the visual cortex: New vistas on integration in the nervous system. Trends in Neurosciences 15:218–226.

Feng, A. S., J. A. Simmons, and S. A. Kick. 1978. Echo detection and target-ranging neurons in the auditory system of the bat, *Eptesicus fuscus*. Science 202:645–648.

Ferragamo, M., T. Haresign, and J. A. Simmons. 1992. Neural responses to dichotic FM sweeps in the auditory cortex of *Eptesicus fuscus*. Proceedings of the Third International Congress of Neuroethology Abstracts 3:217.

Furlonger, C. L., H. J. Dewar, and M. B. Fenton. 1987. Habitat use by foraging insectivorous bats. Canadian Journal of Zoology 65:284–288.

Geggie, J. F., and M. B. Fenton. 1985. A comparison of foraging by *Eptesicus fuscus* (Chiroptera: Vespertilionidae) in urban and rural environments. Canadian Journal of Zoology 63:263–267.

Glaser, E. M., and D. S. Ruchkin. 1976. Principles of Neurobiological Signal Analysis. Academic Press, New York.

Griffin, D. R. 1958. Listening in the Dark. Yale University Press, New Haven.

Griffin, D. R., J. H. Friend, and F. A. Webster. 1965. Target discrimination by the echolocation of bats. Journal of Experimental Zoology 158:155–168.

Haresign, T., M. Ferragamo, and J. A. Simmons. 1993a. Averaged multiunit activity in the inferior colliculus of *Eptesicus fuscus* shows sensitivity to phase and microsecond time shifts in echoes. Association for Research in Otolaryngology Abstracts 16:271.

Haresign, T., M. Ferragamo, and J. A. Simmons. 1993b. Neural sensitivity to phase and microsecond time shifts at ultrasonic frequencies in *Eptesicus fuscus*. Society for Neuroscience Abstracts 19:273.

Haresign, T., J. A. Simmons, and M. Ferragamo. 1993c. Wavelet-like transforms in the auditory system of the bat. Journal of the Acoustical Society of America Abstracts 93:2290.

Kawasaki, M., D. Margolish, and N. Suga. 1988. Delay-tuned combination-sensitive neurons in the auditory cortex of the vocalizing mustached bat. Journal of Neurophysiology (Bethesda) 59:623–635.

Mallat, S. 1989. Multifrequency channel decompositions of images and wavelet models. IEEE Transactions on Acoustics, Speech, and Signal Processing 37:2091–2110.

Marr, D. 1982. Vision. Freeman, San Francisco.

Moss, C. F., and H.-U. Schnitzler. 1995. Behavioral studies of auditory information processing. *In* Hearing by Bats, A. N. Popper and R. R. Fay, eds., pp. 87–145. Springer Handbook of Auditory Research, Vol. 5. Springer-Verlag, New York.

Neuweiler, G. 1990. Auditory adaptations for prey capture in echolocating bats. Physiology Reviews 70:615–641.

O'Neill, W. E., and N. Suga. 1979. Target-range-sensitive neurons in the auditory cortex of the mustached bat. Science 203: 69–73.

O'Neill, W. E., and N. Suga. 1982. Encoding of target range information and its representation in the auditory cortex of the mustached bat. Journal of Neuroscience 2:17–31.

Paschal, W. G., and D. Wong. 1994. Frequency organization of delay-sensitive neurons in the auditory cortex of the FM bat, *Myotis lucifugus*. Journal of Neurophysiology (Bethesda) 72: 366–379.

Rioul, O., and M. Vetterli. 1991. Wavelets and signal processing. IEEE Signal Processing Magazine 8:14–38.

Riquimaroux, H., S. J. Gaioni, and N. Suga. 1991. Cortical computational maps control auditory perception. Science 251:565–568.

Schnitzler, H.-U., and O. W. Henson. 1980. Performance of airborne animal sonar systems: I. Microchiroptera. *In* Animal Sonar Systems, R. G. Busnel and J. F. Fish, eds., pp. 109–181. Plenum Press, New York.

Schuller, G., S. Radtke-Schuller, and W. E. O'Neill. 1991. Processing of paired biosonar signals in the cortices of *Rhinolophus rouxi* and *Pteronotus parnellii*: A comparative neurophysiological and neuroanatomical study. *In* Animal Sonar Systems, P. E. Nachtigal and P. W. B. Moore, eds., pp. 259–264. Plenum Press, New York.

Simmons, J. A. 1971. The sonar receiver of the bat. Annals of the New York Academy of Sciences 188:161–174.

Simmons, J. A. 1973. The resolution of target range by echolocating bats. Journal of the Acoustical Society of America 54:157–173.

Simmons, J. A. 1989. A view of the world through the bat's ear: The formation of acoustic images in echolocation. Cognition 33:155–199.

Simmons, J. A. 1993. Evidence for perception of fine echo delay and phase by the FM bat, *Eptesicus fuscus*. Journal of Comparative Physiology A 172:533–547.

Simmons, J. A., T. Haresign, and M. Ferragamo. 1993. Reconstruction of the delay and phase of echo waveforms from neural responses in the inferior colliculus of the echolocating bat, *Eptesicus fuscus*. Association for Research in Otolaryngology Abstracts 16:270.

Simmons, J. A., P. A. Saillant, M. J. Ferragamo, T. Haresign, S. P. Dear, J. Fritz, and T. A. McMullen. 1996. Auditory computations for biosonar target imaging in bats. *In* Auditory Computations, A. N. Popper and R. R. Fay, eds., pp. 401–468. Springer Handbook of Auditory Research, Vol. 6. Springer-Verlag, New York.

Suga, N., and J. Horikawa. 1986. Multiple time axes for representation of echo delays in the auditory cortex of the mustached bat. Journal of Neurophysiology (Bethesda) 55:776–805.

Suga, N., and W. E. O'Neill. 1979. Neural axis representing target range in the auditory cortex of the mustached bat. Science 206:351–353.

Suga, N., W. E. O'Neill, and T. Manabe. 1978. Cortical neurons sensitive to particular combinations of information-bearing elements of biosonar signals in the mustached bat. Science 200:778–781.

Suga, N., W. E. O'Neill, K. Kujirai, and T. Manabe. 1983. Specificity of combination sensitive neurons for processing of complex biosonar signals in the auditory cortex of the mustached bat. Journal of Neurophysiology (Bethesda) 49:1573–1626.

Sullivan, W. E. 1982. Neural representation of target distance in auditory cortex of the echolocating bat *Myotis lucifugus*. Journal of Neurophysiology (Bethesda) 48:1011–1032.

Webster, F. A., and O. G. Brazier. 1968. Experimental studies on echolocation mechanisms in bats. AMRL-TR-67-192. Clearinghouse for Federal Scientific and Technical Information, Springfield, Va.

Wong, D., and S. L. Shannon. 1988. Functional zones in the auditory cortex of the echolocating bat *Myotis lucifugus*. Brain Research 453:349–352.

15
Sensorimotor Integration in Bat Sonar

DOREEN E. VALENTINE AND CYNTHIA F. MOSS

As an echolocating bat pursues its insect prey, it actively probes the environment with ultrasonic signals and listens to the echoes that return from reflecting targets (Griffin 1958). Information carried by the echoes is processed by the bat's sonar receiver to determine the direction and distance of insect prey, and this spatial information is then used to guide adjustments in the position of the head and pinnae, in the activity of the muscle groups controlling the flight path, and in the production of subsequent sonar signals (Busnel and Fish 1980; Nachtigall and Moore 1986; Fay and Popper 1995). The bat's biosonar system thus requires a sensorimotor interface in which spatial information computed by the central nervous system directs motor commands for appropriate orientation behaviors. The emphasis of this chapter is on the neural mechanisms that support this sensorimotor feedback system in the echolocating bat *Eptesicus fuscus*.

Anatomical and neurophysiological data suggest that many interconnecting neural structures support sensorimotor integration in bat sonar; these include the midbrain superior colliculus, cerebellum, and pontine reticular formation, as well as other brain regions (Huffman and Henson 1990; Covey and Cassedy 1995). Our particular focus

has been on the superior colliculus (SC) of the bat, and we here present data from behavior–lesion studies, extracellular recording, and microstimulation experiments that suggest biologically relevant specializations in the SC of the bat used for acoustic orientation by sonar. We also present a model that describes the bat SC as part of a complex interactive system which coordinates vocalizations and orientation maneuvers to the reception of echoes.

The bat's auditory system receives echoes and other sounds in its environment and processes this information for spatial perception using acoustic cues. However, the bat essentially has a "standard mammalian auditory system" (Suga 1988; Covey and Cassedy 1995). Many of the same cues used by other mammalian species to localize sound and to process complex patterns of acoustic information are exploited by the bat for orientation and perception by sonar. Binaural cues for sound localization are used by the bat to estimate the azimuthal position of an acoustic target. Monaural cues are important to bats for assigning a location in azimuth, but these also are considered essential for determining the elevation of a sound in space (Heffner and Heffner 1992).

The bat's two pinnae and tragi produce changes in the

spectrum of incoming echoes, creating patterns of interference that it uses to estimate target elevation (Grinnell and Grinnell 1965; Lawrence and Simmons 1982; Mogdans et al. 1988; Wotton et al. 1995). Interaural spectral cues produced by the directionality of the two ears may provide additional information for determining target angle in the vertical plane (Grinnell and Grinnell 1965). The third spatial dimension, target range, is conveyed by temporal cues produced by echo reflections of the bat's sonar emissions. The delay between an outgoing vocalization and a returning echo corresponds to the distance between the bat and its target (Hartridge 1945; Simmons 1973). During target pursuit, the bat tailors its vocalizations and regulates its hearing as distance and echo amplitude change (Cahlander et al. 1964; Hartley 1992).

The bat also maneuvers its flight, aims its head, and focuses its ears as it pursues its target. Changes in the position of the bat's head and pinnae may enhance the information carried by these binaural and monaural cues for sound localization, just as a person gazes at an object to gain more detailed visual information. The directionality and motility of the bat's auditory receiver, relative to the external auditory field, allow the animal to actively sharpen its perception. The characteristics and temporal patterning of its echolocation sounds constrain the acoustic information available to the bat's sonar receiver.

A neural mechanism for orienting by sonar toward an acoustic target would likely involve a three-dimensional representation of auditory space that includes information about the current and desired position of the head and ears and about the timing and features of each vocal emission. Converging lines of evidence from comparative anatomical, physiological, and behavioral studies suggest that the midbrain SC links auditory spatial perception with motor pathways for acoustic orientation (Jen et al. 1984; Poussin and Schlegel 1984; Wong 1984; Covey et al. 1987; Cassedy et al. 1989). The mammalian SC contains multimodal sensory inputs that are topographically organized in spatial registration, and motor circuitry which controls orientation behavior (Stein and Meredith 1993). We hypothesized that this midbrain structure may reveal distinct functional specializations in the bat that are important for its acoustic orientation.

The SC has connections with many structures in the auditory brainstem and midbrain, receiving multiple inputs at different stages of processing. It receives inputs from cortical areas that may be involved in processing target-range information and in controlling vocal behavior (Casseday et al. 1989). The SC also receives input from the accessory or extralemniscal central acoustic tract. A fast route to thalamocortical structures, the central acoustic tract by-

passes the central nucleus of the inferior colliculus (ICC) (Papez 1929; Cassedy et al. 1989), and thus forms an ascending auditory pathway in parallel with the main auditory tract. Moreover, the extensive connections of the SC with pontine and reticular structures, and with the periaqueductal gray, trigeminal, and olivocerebellar systems, support its role in acoustic orientation, acoustic reflex production, and audiovocal control. In both the SC and the descending acousticomotor system, microstimulation reliably elicits vocal behavior in echolocating bats (Schuller and Radtke-Schuller 1988, 1990; Metzner 1993; see following). We have investigated the role of the SC in a sensorimotor feedback system that coordinates the bat's active motor control over the acoustic features of reflected echoes. Because the directional control of echolocation behavior is closely linked to changes in target position (Kick and Simmons 1984), a system of sensorimotor feedback in the SC must include spatioacoustic information in azimuth, elevation, and range.

Sensorimotor Feedback in Bat Echolocation Behavior

Field and laboratory observations have documented adaptive behaviors exhibited by echolocating bats in the pursuit of insect prey (Griffin 1958; Griffin et al. 1960; Webster 1963a, 1963b; Webster and Brazier 1965). High-speed motion pictures show that a bat approaching an insect locks its head onto the target, even when the bat is unable to maneuver its body to the target's exact position (Webster 1963a, 1963b). As a bat flies toward a target, changes in the repetition rate, bandwidth, and duration of its sonar emissions have been used to divide the bat's insect pursuit sequence into different phases: search, approach, and terminal (Griffin et al. 1960; Webster 1963a, 1993b; Schnitzler and Kalko, Chapter 12, this volume).

Laboratory Studies of Insect-Capture Behavior

We conducted studies of bats capturing insects in a large laboratory flight room to detail adaptive motor behaviors in an acoustic orienting task and to provide a behavioral assay to anchor our neurophysiological studies of sensorimotor integration in bats. In these studies, insects were tethered to a string attached to a rotating turntable mounted on the ceiling of a large flight room, and blinded bats were trained to intercept these moving insects on the wing while their behaviors were recorded on high-speed video (Eastman Kodak, 500 frames/sec) and audio tape (Racal Store 4, 30 inches/sec).

The bat's success rate and motor behaviors required for the task were measured during a 3- to 4-week period to document baseline performance. Our video analyses showed that the bat typically positioned its head 3–4 cm above the tethered insect at the point of capture, locking the position of its head with an accuracy of approximately 5°. The bat then scooped the insect into its tail membrane, transferred the prey item to its mouth, and continued in flight. Sonar sounds recorded from the animal showed a systematic increase in repetition rate, decrease in duration, and change in overall bandwidth with decreasing target distance (Figure 15.1). This pattern of sound production resembles that recorded from *Eptesicus fuscus* foraging in the field (Griffin 1958).

Our analysis of sounds produced by the bat in this capture task revealed that the sound repetition rate of *E. fuscus* during target pursuit does not change continuously over time but rather remains stable for fixed intervals before increasing. During the approach phase, the sound repeti-

tion rate may plateau at about 30 Hz for time periods as long as 150 msec (Iannucci 1993; Moss et al. 1996) (see Figure 15.1). A review of the literature uncovered this pattern of vocal production in data collected more than 40 years ago from *E. fuscus;* however, this pattern of stable repetition rates was never explicitly described (see Griffin 1953, Plate 2). In other species of bats, the pattern of vocal production is tied to respiration and wingbeat cycles (Schnitzler and Henson 1980), and this also appears to occur in *E. fuscus* (see Figure 15.1C). Note that the sound amplitude is modulated in the terminal phase of insect pursuit without the discrete breaks in sound production that appear in the approach phase.

The stable periods of sound repetition rate produced by *E. fuscus* in the approach phase of insect pursuit may provide insight into the bat's ability to process sequences of spatioacoustic information. For example, the duration of

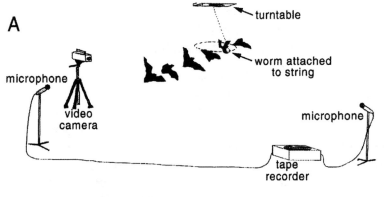

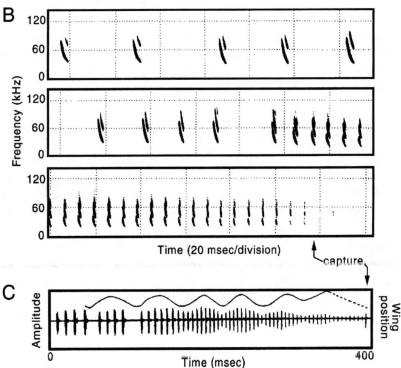

Figure 15.1. Observation of a bat's motor behaviors in an acoustic-orienting task (insect capture). (**A**) Setup for laboratory observation (video and sound recording) of a bat engaged in an insect-capture task. (**B**) Spectrograms of sound sequence produced by a bat during insect-capture task. The three panels represent a continuous sequence. (**C**) Time waveforms of sounds produced by a bat during the final 400 msec of an insect-capture sequence and the corresponding wingbeat cycle during this period.

these stable periods and the number of echolocation sounds contained in each may reveal behavioral response latencies for processing of distance information because the bat adjusts its repetition rate appropriately for particular echo delays (Cahlander et al. 1964).

Insect-Capture Behavior Following Lesions in the Superior Colliculus

We conducted behavioral studies to evaluate the role of the midbrain SC in an insect-capture task in the laboratory. After establishing baseline performance in the task based on an approximately 85% successful capture rate, we electrolytically lesioned the SC in two animals. The bat's postlesion performance was carefully studied for 6–8 weeks, using high-speed video and sound recordings to document the bat's behavior (Sheen et al. 1995; Valentine et al. 1995). We found that insect capture was disrupted by the lesions, and the magnitude of the bat's deficit appeared to be related to the size of the lesions, which were confirmed histologically and confined to the SC. During a period of 4–6 weeks, recovery of performance was observed (Figure 15.2). Obstacle avoidance and landing behavior were not affected by the lesions. Sonar sounds also appeared unimpaired by the lesions and resembled those of an unlesioned bat (such as in Figure 15.1B).

Wenstrup and Suthers (1981) studied wire-avoidance behavior in *E. fuscus* that were subjected to lesions of the SC and reported similar observations. Sonar vocalizations were unimpaired, and flight behavior could not be distinguished from prelesion measurements. Three animals with large lesions that included the periaqueductal gray (PAG) exhibited a small but statistically reliable decrease in wire-avoidance performance. Five other experimental animals with lesions restricted to the caudal SC showed no change in postlesion performance. Our work indicates that lesions restricted to the SC disrupt goal-directed behaviors such as insect capture in the bat, a finding consistent with the idea that the SC supports species-specific orienting behaviors, such as echolocation.

Neurophysiological Studies of the Bat Superior Colliculus

The effect of SC lesions on the sensorimotor behavior required for insect capture in *E. fuscus* suggests that there may be specialization in this neural structure for acoustic orientation by sonar. To explore this hypothesis, we conducted studies of the neural response characteristics and motor organization of the bat SC.

Extracellular Recording

Using extracellular recording methods and auditory stimulation under free-field conditions, we studied the responses of SC neurons in *E. fuscus* for evidence of spatial selectivity in three dimensions (Valentine and Moss 1993, 1997). Ninety-eight isolated units were studied in the awake animal using computer-generated frequency-modulated (FM) sounds characteristic of those produced by *E. fuscus* during the approach phase of insect pursuit (Webster and Brazier 1965). Azimuthal sensitivity was studied by recording re-

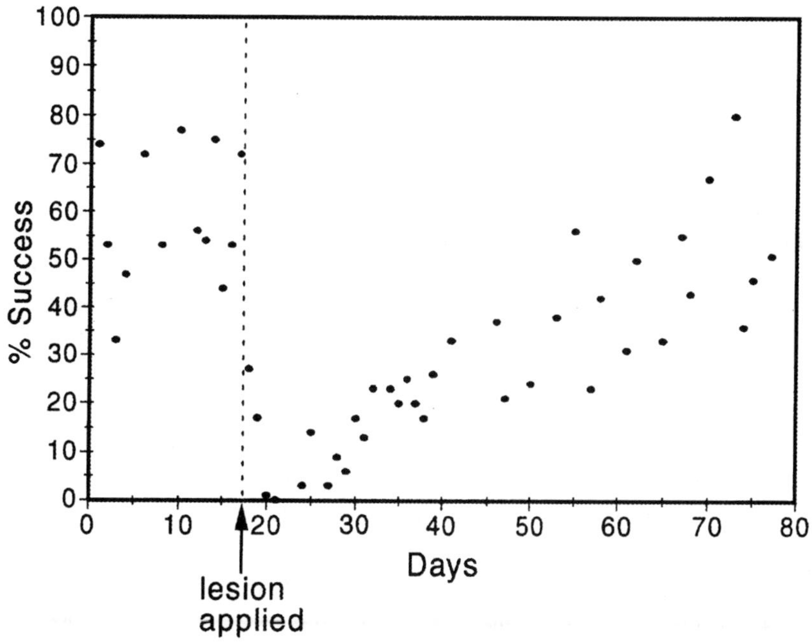

Figure 15.2. Behavioral performance of a single bat in an insect-capture task. Baseline data (i.e., from before the lesioning of the superior colliculus) are to the left of the vertical dashed line.

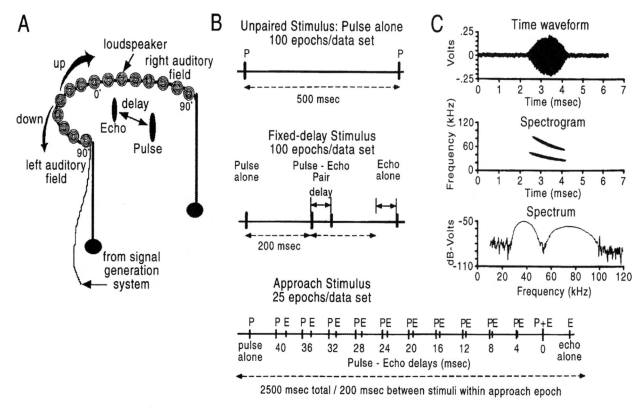

Figure 15.3. Experimental setup for determining spatial selectivity (in terms of azimuth and elevation) of neurons of the bat superior colliculus (SC). (A) Equipment for manipulating the direction and distance of acoustic stimulation. (B) Stimulus epochs (unpaired sound, fixed delay, and approach sequence) of stimuli broadcast through loudspeakers. The pulse and echo were 2-msec frequency-modulated sounds, with the echo 20 dB weaker than the pulse. (C) Sound characteristics (time waveform, spectrogram, and spectrum) of the broadcast stimuli.

sponses to acoustic stimuli broadcast through each of 15 loudspeakers arranged in a frontal hemifield around the bat (Figure 15.3A). The speaker hoop was rotated to move the sound source along the vertical axis to test sensitivity to changes in elevation. Target range was simulated using pairs of computer-generated FM bat sounds separated by particular delays (Suga and O'Neill 1979). The pulse (P), intended to mimic the bat's own sonar emission, and the echo (E), intended to mimic a target reflection, were synthesized in two separate channels and could be modified independently. The echo was attenuated 20 dB relative to the pulse.

Two neuronal populations were distinguished by their selectivity to synthetic echoes and to the spatial location of auditory stimulation. Two-thirds of the population (66/98) responded to auditory stimuli arriving predominantly from a central region of space. These cells did not show facilitation to synthetic P-E pairs in which the delay between the sounds in the pair corresponded to target range; rather, their spatial response area was defined as two dimensional (2-D). A second class of cells (33%; 32/98) were distinguished by a facilitated response to the paired acoustic stimulus (Figure 15.4A) that was coupled to spatial selectivity in azimuth and elevation (Figure 15.4B). That is, echo sensitivity in this

population depended on both the time interval separating P and E signals and the location of the sounds, indicating these cells may encode the spatial locus of an acoustic object in three-dimensional (3-D) coordinates.

The range of best delays recorded from the population of 3-D neurons was 4–20 msec (mean, 13.5 ± 8.1 msec), corresponding to a target distance of about 0.68–3.40 m. On average the number of spikes fired per stimulus presentation to the paired stimulus at the P-E delay value eliciting the maximum discharge was 7.2 times greater than the response to a single-sound stimulus at any sound level. These facilitated responses cannot be explained by the summation of stimulus energy from the paired P-E stimuli, as the auditory integration time of the bat E. fuscus is only about 2 msec (Surlykke and Bojesen 1996). Presenting a sequence of P-E stimuli such that the time interval between the paired sounds progressively decreased from longer to shorter delays (closing distance) or reversing the order to present short-delay pairs before long-delay pairs (increasing distance) evoked the same delay-selective responses.

Responses of 3-D neurons showed selectivity to echo delay that was tagged to the azimuth (and elevation) of stimulation (Figure 15.4B). In 68% of these cells (19/28), facilitation was exhibited along the delay axis only from a

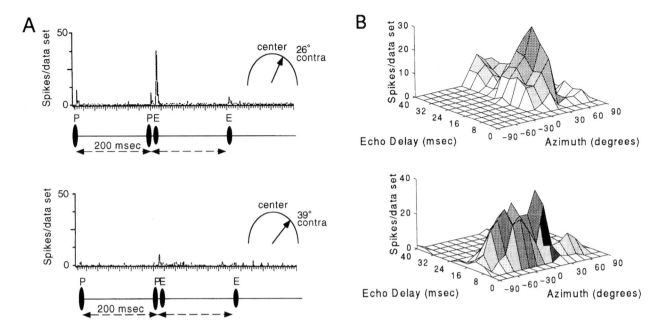

Figure 15.4. Spatial selectivity of bat SC neurons. (A) Response of a single neuron stimulated with a pulse alone (P), a pulse-echo pair with a 12-msec delay (P-E), and echo alone (E) broadcast through a loudspeaker positioned 26° and 39° off the bat's midline and contralateral to the recording site. These plots illustrate that echo-delay facilitation depends upon the azimuth of stimulation. (B) Three-dimensional response profiles of two different neurons to stimuli varying in azimuth and pulse-echo delay. The top plot illustrates the spatial response profile of a neuron with relatively narrow azimuthal tuning and relatively broad echo-delay tuning; the bottom plot illustrates the spatial response profile of a neuron with relatively broad azimuthal tuning and relatively narrow echo-delay tuning.

restricted azimuth, whereas in 21% (6/28), selective delay tuning was observed across a broad region of auditory space. No map of best echo delay was identified in the SC. This finding is consistent with neurophysiological observations on the auditory cortex of this species, which also appears to lack a topography of echo delay (Dear et al. 1993a, 1993b). The organization of best azimuth in 2-D and 3-D cells also did not reveal an orderly topography, a result that supports earlier work in the bat SC (Jen et al. 1984; Poussin and Schegel 1984; Wong 1984). Further, we found that the 2-D and 3-D populations were not segregated although 3-D cells tended to be found in the anterior and midregions of the SC.

Microstimulation

Acoustic information about target location might be expected to coordinate appropriate orienting responses in the bat, for example, head and pinna movements, as well as sonar vocalizations. To study the motor organization of the bat SC, we conducted microstimulation experiments in 14 adult animals, and responses were studied under both head-fixed and head-free conditions (Valentine et al. 1994; Valentine 1995). Single trains of constant-current, electrically isolated twin pulses were generated at a pulse rate of 100–200 Hz by a Grass S48 stimulator and stimulus isolation unit (model PSIU6). The twin-pulse stimulus proved

sufficient to activate premotor circuitry for producing orienting responses. Motor and vocal responses were simultaneously recorded on video (Canon, 30 frames/sec; Redlake, 300 frames/sec) and audio tape (Racal Store-4).

Microstimulation of the bat SC elicited pinna, head, and body movements similar to those reported in other species (Stein and Clamman 1981), revealing a motor map of the bat SC similar to that which has been observed in other mammals (Figure 15.5A,B). Microstimulation also elicited sonar vocalizations, a motor behavior specific to the bat's acoustic orientation by sonar (Figure 15.6A,B; see also Schuller and Radtke-Schuller 1988, 1990). When microstimulation was carried out in the same animal under head-fixed and head-free conditions using the same stimulation parameters, 50% (5/10) of the stimulation sites that elicited vocal-motor responses did so only when the bat's head was free to move. This result suggests a coupling may exist between the occurrence of a head movement and the output of the bat's orienting vocalization system.

The properties of the pinna movements under both head-fixed and head-free conditions were consistent with the placement of the electrode at the same site under the head-fixed and head-free testing conditions. The placement of the electrode in the tissue was confirmed histologically: The site was approximately 500 μm from the periaqueductal gray, a midbrain region known to play a role in vocalization (Jürgens and Pratt 1979). Thus our results provide

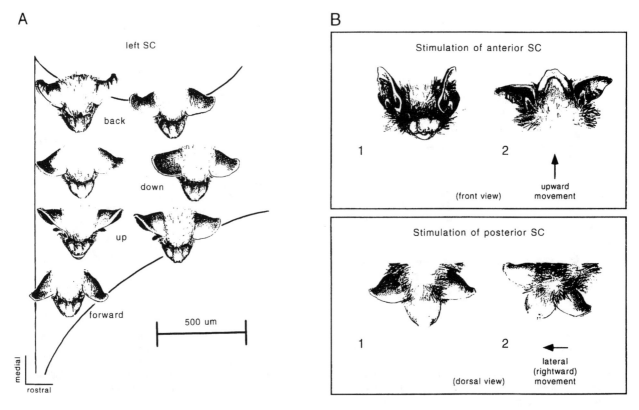

Figure 15.5. Movements elicited by microstimulation of the SC of a bat. (A) Map of pinna movements on dorsal surface of the SC. (B) Head movements with respect to initial position: upward when stimulus is applied to the anterior SC, and lateral when it is applied to the posterior SC.

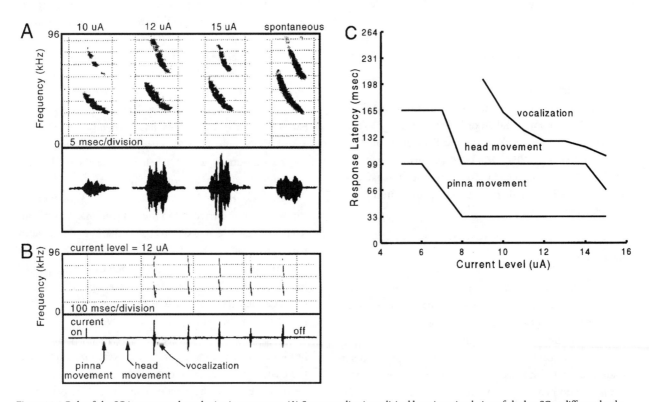

Figure 15.6. Role of the SC in motor and vocal orienting responses. (A) Sonar vocalizations elicited by microstimulation of the bat SC at different levels of current. Spontaneous vocalization recorded from a bat outside of the microstimulation experiment is also shown. (B) Temporal relation between pinna movements, head movements, and sonar vocalizations elicited by SC microstimulation. (C) Latency to initiation of vocalization, pinna movement, and head movement as a function of current strength.

convincing evidence for the role of the SC in vocal-motor orienting responses.

The current threshold for producing a detectable response was determined using a single train of twin pulses that were presented at a rate of 100 Hz. The stimulus pulse duration was 0.3 msec. The criterion for determining stimulus threshold in this study was a twitch of the pinna or neck muscle detectable on the video image. Vocalization threshold was defined a single vocal pulse emitted in response to a single train of the stimulating current. *Eptesicus fuscus* did not spontaneously vocalize during the testing, and vocalizations were observed only as a consequence of stimulation. Stimulus threshold was lowest for pinna movements and highest for vocalizations, although the difference in the amount of stimulation required to produce a vocalization was small.

The relationship between current strength and the time to initiate a behavioral response decreased as the current strength increased above the minimum threshold and showed saturation at suprathreshold current levels. In agreement with DuLac and Knudsen (1990), who studied orienting gaze behavior mediated by the tectum (SC) in the barn owl, we found that the direction of the evoked movement depended on the site of stimulation, while the metrics of the motor response (e.g., amplitude and latency) were determined by both the site and the magnitude of the stimulating current. Response latencies for eliciting head or pinna movements and sonar vocalizations in the same trial are shown in Figure 15.6C. Each curve in this plot follows a similar relationship to increasing levels of electrical stimulation. In this example, a fine measurement of response latency for the movements was not available, but by counting the number of frames from the onset of the stimulating current to the first twitch of the neck or ear flap, the latency could be determined within 33.3 msec (1 frame). In experiments carried out for two bats, a high-speed digital video system was used (Redlake, 300 frames/sec). The results of these experiments suggest the shortest latency to a twitch of the contralateral ear was of the order of 16–21 msec. This plot also emphasizes the order in which motor responses were executed. Pinna movements typically were initiated first, followed by a movement of the head. Vocalizations occurred with the longest latency, lagging behind the head movement by tens of milliseconds.

A Model of Superior Colliculus Function in the Echolocating Bat

The bat orients toward targets using mechanisms common to other species: Two-dimensional spatioacoustic cues processed by SC neurons sensitive to target locations in azi-

muth and elevation may function to direct the aim of the head and pinnae to a desired position. As diagrammed in Figure 15.7A, a signal of motor error derived from the current position of the head and pinnae and from information about the 2-D location of a target (desired position of the head and pinnae) is processed by SC circuitry and relayed to premotor and motor centers that encode the appropriate commands to the muscles. A "moving hill" of neuronal activity across the SC is thought to form the mechanism by which a population response is translated into discrete and specific movements of the head and pinnae (Munoz et al. 1991; Guitton 1992).

Although 2-D cells may coordinate orientation responses appropriate for the bat's position in 2-D space, the population of 3-D cells may direct echolocation behavior that is also coupled to target range. We suggest that the activity of 2-D and 3-D cells determines a pattern of integrated sensory information tagged with spatial coordinates in azimuth, elevation, and range, shown in Figure 15.7B as a facilitated population response. As echoes return from different distances, the temporal and spatial pattern of neural discharges in the SC may encode echo-derived spatial information as a signal of dynamic motor error for the control of acoustic orientation by sonar. The signal of a dynamic relationship between the current position of the bat with an estimate of its desired position relative to the source of salient echoes would guide appropriate head and pinna movements and vocalizations appropriate for tracking the target.

Our single-unit data, combined with microstimulation experiments, suggest that the bat SC may play an important role in the approach phase of insect pursuit (Table 15.1). Specifically, the features of the sonar vocalizations elicited by microstimulation resemble those produced by the bat during target approach. In addition, the best delays of 3-D neurons correspond to the operating range of this behavioral phase of insect capture.

Conclusions

The results of our experiments suggest that the sensorimotor function of the bat SC follows a general mammalian plan while it also supports specializations that may be important for acoustic orientation by sonar. The sensorimotor specializations we have found in our studies may play an important role in coordinating acoustic information about the position of a target with premotor and motor circuitry that permits a bat to adjust the position of its head and pinna and the features of its sonar vocalizations in response to spatial information contained in sonar echoes. It has been our goal in these studies to apply the findings from a specialized animal system to broaden our understanding of

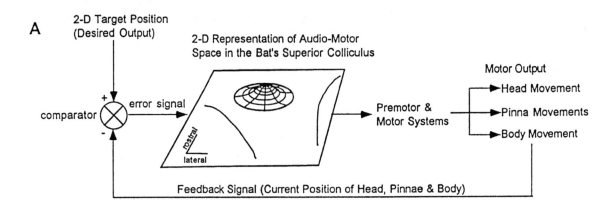

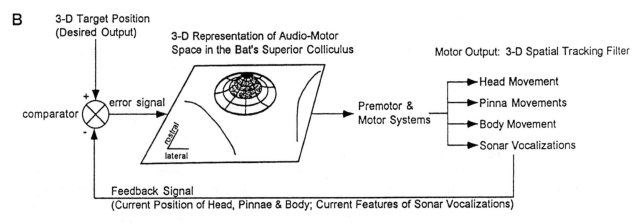

Figure 15.7. Model diagramming the role of the bat SC in the translation of spatioacoustic information into signals that produce adaptive changes in echolocation behavior: (**A**) translation of two-dimensional information; (**B**) translation of three-dimensional information.

Table 15.1

Role of the Bat Superior Colliculus (SC) in the Approach Phase of Insect Pursuit

Aspect of echolocation	SC function and properties[a]	Behavior of echolocating bat[b]
Operating range	For 3-D neurons of SC: 2.30 ± 1.38 m (0.68–3.40 m), with delay of 13.5 ± 8.1 msec (4–20 msec)	Bat is 0.5–3 m from the insect it is pursuing
Resolution of image	Coarse delay (range) tuning	After detection, the bat may pursue a moving target without fine image resolution until it is at close range
Directional control	Electrical microstimulation of SC evokes orienting movements of head and pinna SC may have a role in orienting flight musculature 3-D neurons of SC are involved in coupling of range axis to directional hearing mechanisms	As the bat flies in pursuit, the aim of its head and the axis of its directional hearing are maintained on the moving insect target
Vocal control	Electrical microstimulation of SC elicits vocalizations coupled with head and pinna movements	The bat processes echo information, which then guides vocalization properties and head and pinna movements
Sonar emissions:		
Pulse duration	2.7 ± 1.0 msec (1.11–7.00 msec)	Typically 2–5 msec
Repetition rate	15.4 ± 5.5 sounds/sec (8.1–56.9 sounds/sec)	10–50 sounds/sec
Frequency of first harmonic	Sweeping from 50 kHz (±4.3 kHz) to 23.7 kHz (±4.8 kHz)	Sweeping from 50 kHz to 25 kHz

[a]Within this column, data are given as mean ± 3D.

[b]Behavioral information is from Webster and Brazier (1965) and Kick and Simmons (1984).

the general principles guiding the functional organization of the vertebrate midbrain.

Acknowledgments

This work was supported by a Whitehall Foundation Grant and an NSF Young Investigator Award to C.F.M. and a Sackler Programme Fellowship to D.E.V. We thank R. Iannucci for research support, and A. Grossetête, A. Huang, R. Iannucci, D. Rios, G. Sheen, L. Taft, J. Wadsworth, and M. Zagaeski for technical assistance with the experiments summarized in this chapter.

Literature Cited

Busnel, R. G., and J. F. Fish, eds. 1980. Animal Sonar Systems. Plenum Press, New York.

Cahlander, D. A., J. J. G. McCue, and F. A. Webster. 1964. The determination of distance by echolocation. Nature (London) 201:544–546.

Casseday, J. H., J. B. Kobler, S. F. Isbey, and E. Covey. 1989. Central acoustic tract in an echolocating bat: An extralemniscal auditory pathway to the thalamus. Journal of Comparative Neurology 287:247–259.

Covey, E., and J. H. Cassedy. 1995. The lower brainstem auditory pathways. *In* Hearing by Bats, R. R. Fay and A. N. Popper, eds., pp. 235–295. Springer Handbook of Auditory Research, Vol. 5. Springer-Verlag, New York.

Covey, E., W. C. Hall, and J. B. Kobler. 1987. Subcortical connections of the superior colliculus in the mustache bat, *Pteronotus parnellii.* Journal of Comparative Neurology 263:179–197.

Dear, S. P., J. A. Simmons, and J. Fritz. 1993a. A possible neuronal basis for representation of acoustic scenes in auditory cortex of the big brown bat. Nature (London) 364:620–622.

Dear, S. P., J. Fritz, T. Haresign, M. Ferragamo, and J. A. Simmons. 1993b. Tonotopic and functional organization in the auditory cortex of the big brown bat, *Eptesicus fuscus.* Journal of Neurophysiology (Bethesda) 70:1988–2009.

Du Lac, S., and E. I. Knudsen. 1990. Neural maps of head movement vector and speed in the optic tectum of the barn owl. Journal of Neurophysiology (Bethesda) 63:131–146.

Fay, R. R., and A. N. Popper, eds. 1995. Hearing by Bats. Springer Handbook of Auditory Research, Vol. 5. Springer-Verlag, New York.

Griffin, D. R. 1953. Bat sounds under natural conditions, with evidence for echolocation of insect prey. Journal of Experimental Zoology 123:435–465.

Griffin, D. 1958. Listening in the Dark. Yale University Press, New Haven.

Griffin, D. R., F. A. Webster, and C. R. Michael. 1960. The echolocation of flying insects by bats. Animal Behaviour 8:141–154.

Grinnell, A. D., and V. S. Grinnell. 1965. Neural correlates of vertical localization by echolocating bats. Journal of Physiology (London) 181:830–851.

Guitton, D. 1992. Control of eye–head coordination during orienting gaze shifts. Trends in Neurosciences 15:174–179.

Hartley, D. J. 1992. Stabilization of perceived echo amplitudes in echolocating bats. II. The acoustic behaviour of the big brown bat, *Eptesicus fuscus,* when tracking moving prey. Journal of the Acoustical Society of America 91:1133–1149.

Hartridge, H. 1945. Acoustic control in the flight of bats. Nature (London) 156:490–494.

Heffner, R. S., and H. E. Heffner. 1992. Evolution of sound localization in mammals. *In* The Evolutionary Biology of Hearing, D. B. Webster, R. R. Fay, and A. N. Popper, eds., pp. 691–715. Springer-Verlag, New York.

Huffman, R. F., and O. W. Henson, Jr. 1990. The descending auditory pathway and acousticomotor systems: Connections with the inferior colliculus. Brain Research Reviews 15:295–323.

Iannucci, R. 1993. Behavioral and physiological characterization of sensorimotor integration in the big brown bat, *Eptesicus fuscus.* Undergraduate honors thesis, Harvard University, Cambridge.

Jen, P. H.-S., X. Sun, T. Kamada, S. Zhang, and R. Shimozawa. 1984. Auditory response properties and spatial response areas of superior colliculus neurons of the FM bat *Eptesicus fuscus.* Journal of Comparative Physiology A 154:407–413.

Jürgens, U., and R. Pratt. 1979. Role of the periaqueductal grey in vocal expression of emotion. Brain Research 167:367–378.

Kick, S. A., and J. A. Simmons. 1984. Automatic gain control in the bat's sonar receiver and the neuroethology of echolocation. Journal of Neuroscience 4:2725–2737.

Lawrence, B. D., and J. A. Simmons. 1982. Echolocation in bats: The external ear and perception of the vertical positions of targets. Science 218:481–483.

Metzner, W. 1993. An audio-vocal interface in echolocating horseshoe bats. Journal of Neuroscience 13:1899–1915.

Mogdans, J., J. Ostwald, and H.-U. Schnitzler. 1988. The role of pinna movement for the localization of vertical and horizontal wire obstacles in the greater horseshoe bat, *Rhinolophus ferrumequinum.* Journal of the Acoustical Society of America 84:1676–1679.

Moss, C. F., R. Iannucci, and W. W. Wilson. 1996. Spatial tracking of moving targets by the echolocating bat *Eptesicus fuscus:* Perceptual consequences of vocal-motor behavior. Society for Neuroscience Abstracts 22:161.19.

Munoz, D. P., D. Pélisson, and D. Guitton. 1991. Movement of neural activity on the superior colliculus motor map during during gaze shifts. Science 251:1358–1360.

Nachtigall P. E., and P. W. B. Moore, eds. 1986. Animal Sonar: Processes and Performance. Plenum Press, New York.

Papez, J. W. 1929. Central acoustic tract in cat and man. Anatomical Record 42:60.

Poussin, C., and P. Schlegel. 1984. Directional sensitivity of auditory neurons in the superior colliculus of the bat *Eptesicus fuscus* using free field sound stimulation. Journal of Comparative Physiology A 154:253–261.

Schnitzler, H.-U., and W. Henson, Jr. 1980. Performance of airborne animal sonar systems. I. Microchiroptera. *In* Animal Sonar Systems, R. G. Busnel and J. F. Fish, eds., pp. 109–181. Plenum Press, New York.

Schuller, G., and S. Radtke-Schuller. 1988. Midbrain areas as candidates for audio-vocal interface in echolocating bats. *In* Animal Sonar Systems, R. G. Busnel and J. F. Fish, eds., pp. 93–98. Plenum Press, New York.

Schuller, G., and S. Radtke-Schuller. 1990. Neural control of vocalization in bats: Mapping of brainstem areas with electrical microstimulation eliciting species-specific echolocation calls in the rufous horseshoe bat. Experimental Brain Research 79: 192–206.

Sheen, G. C., L. D. Taft, and C. F. Moss. 1995. Effect of superior colliculus lesions on prey-capture behavior in the FM-bat *Eptesicus fuscus*. Tenth International Bat Research Conference, Boston (abstract).

Simmons, J. A. 1973. The resolution of target range by echolocating bats. Journal of the Acoustical Society of America 54:157–173.

Stein, B. E., and H. P. Clamann. 1981. Control of pinna movements and sensorimotor register in cat superior colliculus. Brain Behavior and Evolution 19:180–192.

Stein, B. E., and M. A. Meredith. 1993. Merging of the Senses. MIT Press, Cambridge.

Suga, N. 1988. Auditory neuroethology and speech processing: Complex-sound processing by combination sensitive neurons. *In* Auditory Function: Neurobiological Bases of Hearing, G. M. Edelman, W. E. Gall, and W. M. Cowan, eds., pp. 679–720. Wiley, New York.

Suga, N., and W. E. O'Neill. 1979. Neural axis representing target range in the auditory cortex of the mustache bat. Science 206: 351–353.

Surlykke, A., and O. Bojesen. 1996. Integration time for short broadband clicks in echolocating FM-bats *(Eptesicus fuscus)*. Journal of Comparative Physiology A 178:235–241.

Valentine, D. E. 1995. Spatial perception and sensorimotor integration: The role of the superior colliculus in the echolocating bat *Eptesicus fuscus*. Ph.D. dissertation, Harvard University, Cambridge.

Valentine, D. E., and C. F. Moss. 1993. 3-D receptive field properties of neurons in the superior colliculus of the echolocating bat *Eptesicus fuscus*. Society for Neuroscience Abstracts 19:237.15.

Valentine, D. E., and C. F. Moss. 1997. Spatially-selective auditory responses in the superior colliculus of the echolocating bat. Journal of Neuroscience 17:1720–1733.

Valentine, D. E., R. Iannucci, and C. F. Moss. 1994. Sensorimotor integration of spatial information in the superior colliculus of the echolocating bat. Society for Neuroscience Abstracts 23: 75.12.

Valentine, D. E., G. C. Sheen, and C. F. Moss. 1995. Sensorimotor integration in echolocation: The role of the superior colliculus in acoustic orientation by the FM-bat *Eptesicus fuscus*. Association for Research in Otolaryngology 18:256.

Webster, F. A. 1963a. Active energy radiating systems: The bat and ultrasonic principles. II. Acoustical control of airborne interceptions by bats. Proceedings of the International Congress on Technology and Blindness, Vol. I, pp. 49–135. New York.

Webster, F. A. 1963b. Bat-type signals and some implications. *In* Human Factors and Technology, E. Bennett, J. Degan, and J. Spiegel, eds., pp. 378–408. McGraw-Hill, New York.

Webster, F. A., and O. B. Brazier. 1965. Experimental studies on target detection, evaluation, and interception by echolocating bats. AD673373. Aerospace Medical Research Laboratory, Wright-Patterson Air Force Base, Ohio.

Wenstrup, J., and R. A. Suthers. 1981. Do lesions of the superior colliculus affect acoustic orientation in echolocating bats? Physiology and Behavior 27:835–839.

Wong, D. 1984. Spatial tuning of auditory neurons in the superior colliculus of the echolocating bat *Myotis lucifugus*. Hearing Research 16:261–270.

Wotton, J. M., T. Haresign, and J. A. Simmons. 1995. Spatially-dependent acoustic cues generated by the external ear of *Eptesicus fuscus*. Journal of the Acoustical Society of America 98: 1423–1445.

16
Adaptation of the Auditory Periphery of Bats for Echolocation

MARIANNE VATER

In the field of auditory neurobiology, echolocating bats represent excellent models to study adaptations of the peripheral auditory system for processing species-specific vocalizations. Analysis of echolocation call structure in relation to hunting habitats serves as a guideline for posing specific questions: (1) Which cochlear mechanisms generate sensitivity for ultrasonic frequencies and how do they conform with or differ from cochlear mechanisms established in nonecholocating species? (2) How is the common mammalian cochlear design modified to specifically enhance sensitivity and tuning in particular frequency ranges and to create expanded representations of the frequency ranges that are biologically most important?

Numerous physiological and anatomical studies have been devoted to these issues and summarized in recent reviews (Vater 1988; Kössl and Vater 1995). Rather than presenting an extensive review of the elegant work done in the past three decades, this chapter highlights specializations of cochlear structure and function in some selected species of echolocating bats representing different evolutionary and adaptive traits, particularly emphasizing recent advances obtained within the last 3 years.

Model Cases: Echolocation Signals and Hunting Habitat

The Microchiroptera are characterized by a remarkable diversity and adaptive radiation (Neuweiler 1990; Fenton 1995; also see Schnitzler and Kalko, Chapter 12, this volume). A recently proposed classification scheme that integrates the biophysical properties of the echolocation signals and ecological aspects, such as hunting behavior and habitat, defines 11 different guilds (Schnitzler and Kalko, Chapter 12, this volume). Detailed information on the structure and function of the peripheral auditory system is so far available only for a few species of bats, and even for these the information is far from complete. The species chosen as the central characters of this review belong to two nonrelated genera that have evolved Doppler-sensitive sonar systems in convergent evolution: the mustached bat *(Pteronotus parnellii)* and the horseshoe bat *(Rhinolophus rouxi)*. They are compared with other bat species that employ broadband sonar: the big brown bat *(Eptesicus fuscus)*, the Indian false vampire bat *(Megaderma lyra)*, and the Mexican free-tailed bat *(Tadarida brasiliensis)*. Each of these spe-

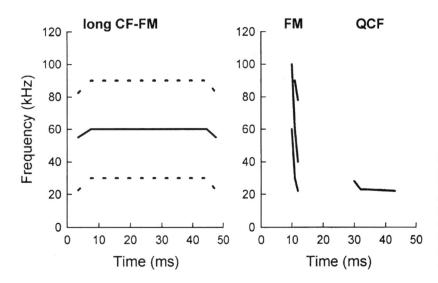

Figure 16.1. Highly schematic illustration of different types of echolocation signals. **Left:** Long constant-frequency, frequency-modulated (CF-FM) signal of *Pteronotus parnellii*. **Right:** broadband FM signal emitted by *Eptesicus fuscus* during the approach phase; quasi-constant-frequency (QCF) call emitted by many FM bats as a search signal while hunting in open space. (After Fenton 1995.)

cies is a model subject in behavioral and neurophysiological research. Furthermore, to demonstrate the evolutionary plasticity of cochlear design within the same genus, data are presented on the cochlea of *Pteronotus quadridens*.

The Doppler-sensitive echolocation systems of horseshoe bats and the mustached bat are thought to represent adaptations for hunting in dense vegetation. This environment poses severe constraints on the echolocation system because the bats must detect relevant echoes from a highly cluttered background. The echolocation signals are composed of long constant-frequency (CF) components starting and terminating with brief frequency modulations (FM) (Figure 16.1); these species are therefore called CF-FM bats. These bats compensate for frequency shifts of the echo CF component that are generated by their own flight speed by lowering the frequency of the emitted call in proportion to the Doppler shift (Schnitzler 1968, 1970). The returning echoes are thus held at a constant reference frequency, and if reflected from fluttering insect prey, the CF echo component contains small periodic modulations in frequency and amplitude. These periodic modulations stand out as rhythmic acoustic glints from the randomly structured background echoes and thus serve as clues for prey detection and identification (Neuweiler et al. 1980; Schnitzler and Henson 1980; Neuweiler 1990). The FM component is used for target ranging and determination of target texture.

As far as we know, this Doppler-sensitive sonar is used by all species of the Old World genera *Rhinolophus* and the related *Hipposideros* (Schnitzler 1968; Schuller 1980; Habersetzer et al. 1984) but by only one species of the New World genus *Pteronotus* (Schnitzler 1970). The frequency of the dominant CF signal component is characteristic for the species and, within species, characteristic for the individual (Neuweiler 1990; Kössl and Vater 1995).

The term FM bat is used for all other bat species described in this chapter, although this terminology is not quite correct. The big brown bat and the Mexican free-tailed bat are characterized by a flexible repertoire of echolocation calls. Both employ quasi-constant-frequency (QCF) signals of about 15–20 msec in duration and rather low frequency (~25 kHz) for long-range detection while hunting in open space and switch to the use of brief downward FM sweeps with a broad bandwidth while closing in on the target (see Figure 16.1) (Simmons et al. 1979; Simmons and Stein 1980; Fenton 1995).

The Indian false vampire bat employs brief FM signals for echolocation, but as a gleaning species it relies heavily on passive listening to sounds produced by its prey (Fiedler 1979). *Pteronotus quadridens* uses multiharmonic FM signals that occasionally start with a brief CF component (Kössl et al. 1997).

Narrowband and Broadband Cochlear Tuning

The pioneering studies of cochlear physiology in bats were based on recordings of cochlear microphonics and compound action potentials of the auditory nerve. These investigations highlighted the essential distinguishing features between bats employing narrowband, long CF-FM calls and those that use broadband FM calls: The cochlea of Doppler shift compensating bats is exceptionally sharply tuned to the dominant CF component of the biosonar signal, whereas cochlear tuning in FM bats is nonspecialized and matches that of nonecholocating mammals (*P. parnellii*: Pollak et al. 1972, 1979; Suga et al. 1975; Suga and Jen 1977; Henson et al. 1985; *Rhinolophus*: Schnitzler et al. 1976; Suga et al. 1976; *Eptesicus*: Dalland 1965; for a more detailed review see Kössl and Vater 1995).

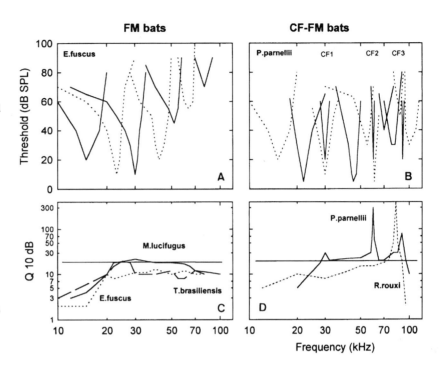

Figure 16.2. Comparison of tuning curves and Q_{10} dB values of single cochlear nucleus neurons in FM bats (left) and CF-FM bats (right). (**A**) Tuning curves of *Eptesicus fuscus* show typical mammalian shape and bandwidth. (**B**) In *Pteronotus parnellii*, exceptionally narrow tuning curves are found in frequency regions that correspond to the harmonics of the CF signal component (CF$_1$, CF$_2$, CF$_3$). (**C**) Maximal Q_{10} dB values for different species of FM bats are typically less than 20 (solid line, *Myotis;* dashed line, *Tadarida;* dotted line, *Eptesicus*). (**D**) Maximum Q_{10} dB values of CF-FM bats reach 400 at the frequency of the dominant CF$_2$ signal component. Tuning is also enhanced at the CF$_1$ and CF$_3$ signal in *Pteronotus parnelli* (solid line), but there is only one region of enhanced tuning at CF$_2$ in *Rhinolophus rouxi* (dotted line). (Data sources: *Eptesicus,* after Haplea et al. 1994; *Myotis,* after Suga 1964; *Tadarida,* after Vater and Siefer, unpublished; *Pteronotus parnellii,* after Kössl and Vater 1995; *Rhinolophus,* after Suga et al. 1976.)

Direct measurements of the tuning curves of hair cells and auditory nerve fibers are not available for different species of bats. However, the sharpness of tuning curves of single neurons at the first central station of the auditory pathway, the cochlear nucleus, can provide a first estimate of frequency-specific cochlear tuning (Figure 16.2). In *E. fuscus* (Haplea et al. 1994), all tuning curves are V-shaped with no evidence for particularly broad or sharp tuning in specific frequency ranges (Figure 16.2A). The Q_{10} dB values (i.e., the quotient of the best frequency, the frequency of the most sensitive threshold, and the bandwidth 10 dB above minimum threshold) are typically less than 20. Thus, they conform to results obtained in other bats employing broadband calls such as *Myotis* (Suga 1964) and *Tadarida* (Figure 16.2C) or nonecholocating species (e.g., *Felis catus:* Evans 1975). In the mustached bat (Figure 16.2B), there are three frequency regions characterized by enhanced tuning that correspond to the first (CF$_1$), second (CF$_2$), and third (CF$_3$) harmonic of the CF component (Suga et al. 1975; Kössl and Vater 1990). Q_{10} dB values reach a maximum of 400 in the CF$_2$ signal range in both the mustached bat and the horseshoe bat (Figure 16.2D).

The application of noninvasive techniques to measure otoacoustic emissions of the bat cochlea (CF-FM bats: Henson et al. 1985; Kössl and Vater 1985a; Kössl 1992a, 1994a, 1994b; *Megaderma lyra* and *Carollia perspicillata:* Kössl 1992b; *P. quadridens* and *Mormoops blainvilli:* Kössl et al. 1997) provides a valuable database for interpretation of specialized cochlear function and for assessing the hearing characteristics of those species that cannot routinely be kept in the laboratory. Otoacoustic emissions are sound waves generated by the cochlea that can be measured in the outer ear canal with sensitive microphones. When stimulated with two tones (f_1 and f_2), the cochlea generates distortion-product otoacoustic emissions that are most prominent at the frequency of $2 f_1 - f_2$ (for details of the technique, see Kössl 1992a). Threshold curves of distortion products give a close approximation of the audiogram. In the CF-FM bat *Pteronotus parnellii,* the distortion-product audiogram shows narrowly tuned regions of maximum and minimum thresholds and thus reflects the sharp tuning properties of the cochlea. In contrast, the distortion-product audiogram of the closely related FM bat *P. quadridens* is broadly tuned with sensitivity extending beyond 120 kHz (Kössl et al. 1997).

Synchronous evoked otoacoustic emissions can be measured by stimulation with a slow tone sweep and typically occur in frequency regions of maximal sensitivity (Kemp 1978). The cochlea of *P. parnellii* emits a strong evoked otoacoustic emission at a single frequency. This emission is present in all individuals and in both ears, and its frequency of about 61 kHz resides a few hundred hertz above the frequency of the emitted call. This corresponds to the audiogram region of sharply tuned minimum thresholds and is exactly the frequency to which the bat adjusts the CF$_2$ echo frequency during Doppler shift compensation (Henson et al. 1990). In some individuals, strong spontaneous otoacoustic emissions occur at the same frequency (Kössl and Vater 1985a; Kössl 1994a, 1994b). Evoked otoacoustic emissions are not a robust phenomenon in *Rhinolophus:* They were faint and were recorded in only a few ears (Kössl

1994a, 1994b). This finding implies a causal relationship between strong resonance phenomena observed in the cochlear microphonic response of *P. parnellii* (Suga and Jen 1977) and the presence of otoacoustic emissions at the same frequency (Henson et al. 1985; Kössl 1994a, 1994b). Resonance phenomena in the cochlear microphonic response of *Rhinolophus* are present but far less pronounced than in *P. parnellii* (Schnitzler et al. 1976; Henson et al. 1985), which suggests that the mechanisms producing exceptionally sharp cochlear tuning in these species may differ in certain aspects. The most crucial difference may reside in the damping of the sharply tuned cochlear resonator (Henson et al. 1985; Kössl 1994b).

The most direct evidence for the cochlear origin of enhanced tuning was obtained by measurements of the frequency response of the basilar membrane with a laser diode interferometer that allows recordings of vibrations in the nanometer range (Kössl and Russell 1995). The basilar membrane in the basal cochlear turn of *P. parnellii* is sharply tuned to the frequency range of the evoked otoacoustic emission with quality factors exceeding those of neuronal tuning (Q_{10} dB of 600). The mechanism of exceptional sharp tuning appears implemented through an interaction of passive resonant properties of the cochlear partition and active mechanical feedback from the outer hair cells. Sensitive sharp tuning is physiologically labile (Pollak et al. 1979; Henson et al. 1985; Kössl and Russell 1995) and shifts with body temperature in *P. parnellii* (Huffman and Henson 1991, 1993a, 1993b).

Altogether, the physiological data show that one of the most salient adaptations for processing the species-specific calls, namely the creation of strongly enhanced tuning to the biologically most important frequency bands found in certain species of bats, is already established within the receptor organ. The hydromechanical processing in the cochlea provides the basic filter design for frequency analysis. Central auditory processing has been shown to change the filter shape by mechanisms of convergence (Covey and Casseday 1995) and lateral inhibition (Pollak and Park 1995).

A further adaptive feature for processing important echolocation signal ranges, first noted in Doppler shift compensating bats, is the vast overrepresentation of single auditory neurons tuned to the CF_2 component at all levels of the auditory pathway within an otherwise typically mammalian tonotopic representation of the receptor surface (Feng and Vater 1985; Rübsamen et al. 1988; Covey and Casseday 1995; Pollak and Park 1995). This feature originates in an expanded frequency representation in the cochlea (Bruns 1976a; Vater et al. 1985; Kössl and Vater 1985b), and recent data as summarized here indicate that it may be more common than previously assumed.

Cochlear Frequency Representation

The construction of cochlear frequency maps is essential for correlations between structure and function, which in turn are the prerequisite to investigate the mechanisms of specialized cochlear tuning observed in different bat species. Such maps can be obtained by tracing the cochlear origin of auditory nerve fibers of defined frequency response characteristics and are available for several species of bats (Kössl and Vater 1985b; Vater et al. 1985; Vater and Siefer 1995), several nonecholocating mammals (Liberman 1982; Müller 1991; Müller et al. 1992), and the barn owl (Köppl et al. 1993) (Figure 16.3). The cochlear frequency maps for horseshoe bats and mustached bats feature an expanded representation of a narrow frequency band encompassing the species-specific CF_2 component of the echolocation signal in the basal turn of the cochlea. This is shown by the shallow course of the function relating frequency to its representation place on the basilar membrane. The calculated mapping coefficients within this region amount to more than 50 mm per octave. The steepness of the slopes in other frequency ranges closely conforms to those found in nonspecialized mammals such as the cat and rat (~3 mm/

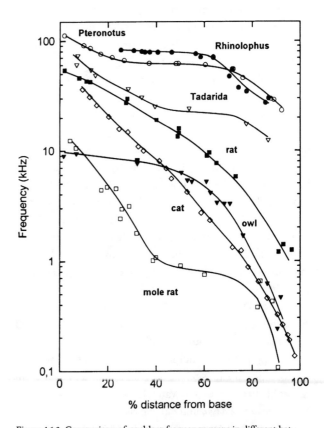

Figure 16.3. Comparison of cochlear frequency maps in different bat species, nonecholocating mammals, and the barn owl. (Data sources: *Rhinolophus rouxi*, after Vater et al. 1985; *Pteronotus parnellii*, after Kössl and Vater 1995; rat, after Müller 1991; cat, after Liberman 1982; mole rat, after Müller et al. 1992; barn owl, after Köppl et al. 1993.)

octave). In the horseshoe bat, mapping techniques based on loud, pure-tone exposure likewise demonstrated an expansion of the CF_2 signal range (Bruns 1976a), but the map is shifted toward more basal cochlear locations (for discussion, see Vater et al. 1985).

Distinct but less pronounced overrepresentations of certain frequency ranges (5–6 mm/octave) have been found in the only FM bat that has been studied, *Tadarida brasiliensis* (Vater and Siefer 1995); in one nonecholocating mammal, the African mole rat *Cryptomys hottentus* (Müller et al. 1992); and in the cochlea of the barn owl (Köppl et al. 1993). The expanded frequency representation in *Tadarida* covers the terminal frequencies of the FM signal and the range of the QCF long-range detection signal (Vater and Siefer 1995), and the barn owl cochlea devotes increased space for processing noise generated by prey (Köppl et al. 1993).

The term acoustic fovea was used to denote the specialized cochlear region in the horseshoe bat that is characterized by expanded frequency mapping and exceptionally sharp tuning properties (Schuller and Pollak 1979; Bruns and Schmieszek 1980). In contrast to the visual fovea, there is no increase in receptor cell density in the acoustic fovea relative to neighboring areas; however, because of the expanded mapping itself (Bruns and Schmieszek 1980) and an additional increase in afferent innervation density (Vater et al. 1985; Zook and Leake 1989), there is a clear overrepresentation of neurons coding for the respective frequency range. Analogies to the visual system are emphasized by the interpretation of Doppler shift compensation as a means of tracking echoes (Schuller and Pollak 1979). Doppler shift compensation keeps the image of interest (echoes from flying prey insects) focused on the foveal region of the cochlea, and as a consequence the ability to detect and resolve signal features is enhanced (Schuller and Pollak 1979). However, it is far from clear whether the audiovocal feedback systems are similarly organized as visuomotor systems.

Thus, in Doppler shift compensating bats, the analogy originally not only refers to the existence of an expanded representation of a certain frequency band on the receptor surface but also includes specializations of brain pathways that finally generate a motor output. In later studies, however, the term acoustic fovea or auditory fovea was widely used to simply denote a specialized segment of the cochlea that is characterized by expanded frequency mapping even in nonecholocating species (Müller et al. 1992; Köppl et al. 1993). In regions with expanded mapping, the mechanical frequency resolution (mm/octave) is high. Therefore, one would expect a corresponding enhancement of frequency tuning. Tuning sharpness of single neurons is, however, not significantly increased in the foveal frequency ranges of *Tadarida,* mole rat, and barn owl (Müller et al. 1992; Köppl

et al. 1993; Vater and Siefer 1995). In the mustached bat, the frequency expansion covers basilar membrane regions of enhanced tuning to the CF_2 signal range and adjacent regions with normal tuning properties (Kössl and Vater 1985b). Furthermore, in the mustached bat, enhanced tuning is found in the CF_3 signal range that is not expandedly mapped (Kössl and Vater 1985b). Therefore, cochlear frequency expansion and enhanced tuning are two distinct phenomena, and additional hydromechanical mechanisms seem to be crucial to produce the enhanced tuning.

What is the function of an acoustic fovea if it does not necessarily increase sensory acuity (tuning sharpness)? In the most general sense, the establishment of acoustic foveae in the cochlea can provide the origin of an enlarged neuronal substrate for parallel processing of different aspects of auditory signals in the frequency ranges biologically most relevant and thus have a selective advantage for species with very different ecological adaptations.

Structural and Functional Correlations

The macromechanics and micromechanics of the cochlea set the stage for species-characteristic frequency mapping and filter mechanisms (Békésy 1960; Zwislocki 1986). Therefore, the frequency maps of the cochlea obtained in different species need to be correlated with basoapical changes in structure and dimensions of the components of the organ of Corti. It must be emphasized that cochlear tuning is the outcome of a concerted action of active and passive elements of the cochlea, and specializations of several structural parameters appear to play a role. Evidence for foveal representations without enhanced tuning presented earlier emphasizes the need to distinguish structural features that relate to expanded mapping from those which relate to enhanced tuning.

The cochlea of echolocating bats follows the basic mammalian design, and the morphology of the organ of Corti is illustrated in Figure 16.4. As outlined in the following paragraphs, the main targets of adaptive evolutionary change appear to be the extrasensory components of the cochlear partition, the basilar membrane and the tectorial membrane. The receptor cells, in particular the outer hair cells, show general adaptations for high-frequency hearing and a rather conservative organization across differently specialized bat species.

Specialized Basilar Membrane Morphology

The elastic properties of the basilar membrane play a crucial role in implementation of the cochlear frequency place map (Békésy 1960). Basilar membrane stiffness can be in-

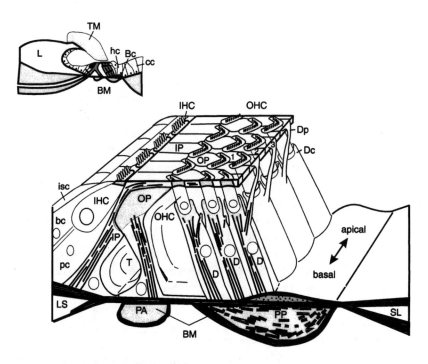

Figure 16.4. Schematic illustration of the organ of Corti in the basal turn of an echolocating bat (sparsely innervated zone of *Pteronotus parnellii*), constructed from transmission and scanning electron micrographs, emphasizing hydromechanically important features. Innervation of the receptor cells is omitted. Hensen cells (hc), Boettcher cells (Bc), and Claudius cells (cc) are shown only in the top graph. Basic organization follows the common mammalian scheme, but both tectorial membrane (TM) and basilar membrane (BM) are of specialized shape. Outer hair cells (OHC) and their stereovilli are small. Supporting cells (inner and outer pillar cells, IP, OP; Deiters cells, D) are massive and contain an abundance of microtubules. The basal poles of OHCs are tightly held in the specialized Deiters cups (Dc). Full extent of the phalangeal processes (DP) of the Deiters cells is only indicated for the outermost (third) row of Deiters cells. The portions of the Deiters cell processes of all three rows within the reticular lamina are labeled with numbers (1–3). Other labeled structures: bc, border cell; IHC, inner hair cell; isc, inner sulcus cell; L, spiral limbus; LS, spiral lamina; PA, pars arcuata of the basilar membrane; pc, phalangeal cell; PP, pars pectinata of the basilar membrane; SL, spiral ligament; T, tunnel of Corti.

ferred from morphological measurements of basoapical changes in basilar membrane width and thickness (Békésy 1960). Such measurements are now available for several species of bats, some of which are illustrated in Figure 16.5 (for further information, see Kössl and Vater 1995) and support the notion originally proposed by Bruns (1976a, 1976b) that specializations of basilar membrane morphology play an integral role in creating specialized cochlear tuning.

Most FM bats show the typical mammalian gradients with a gradual decrease in basilar membrane thickness and increase in basilar membrane width toward the cochlear apex. Regardless of differences in hunting strategy and echolocation call repertoire, the patterns are similar in the gleaning bat *Megaderma lyra* (Fiedler 1983), *Eptesicus fuscus* (Plath and Vater, unpublished data), and *Pteronotus quadridens* (Vater 1997). The basilar membrane of the FM bat *Tadarida brasiliensis* (Vater and Siefer 1995), however, features con-

stant thickness and width throughout most of the cochlea and thus presents a unique pattern among bats and even among mammals. Within the region of "constant" basilar membrane thickness there is, however, a change in basilar membrane shape and ultrastructure that is expected to influence cochlear micromechanics (see following). The cochleae of the CF-FM bats *Rhinolophus* and *P. parnellii* differ in morphological detail (Kössl and Vater 1995), but both show a stepwise change in basilar membrane thickness just basal to the foveal representation site of the CF_2 signal range and constant basilar membrane thickness within the biologically relevant frequency range of the fovea (Kössl and Vater 1985b; Vater et al. 1985).

In all species that possess foveal representations (CF-FM bats, *Tadarida*, mole rat, barn owl), the frequency expansion is located within regions which are characterized by almost "constant" values of basilar membrane thickness and width that are expected to create a very shallow stiffness gradient.

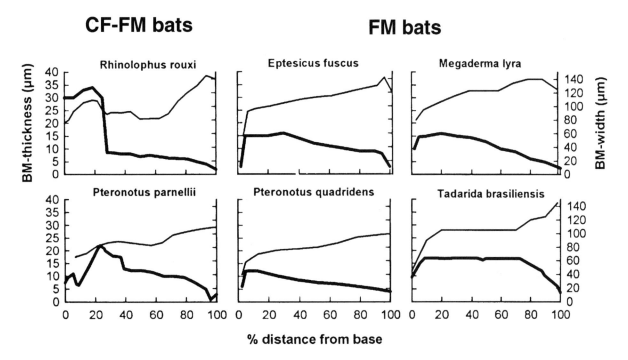

Figure 16.5. Basilar membrane dimensions in CF-FM bats and FM bats. BM thickness (thick lines) and BM width (thinner lines) are plotted versus cochlear length given in percent distance from base. (Data sources: *Rhinolophus*, after Bruns 1976a; *Pteronotus parnellii*, thickness after Kössl and Vater 1985b, and width after Vater 1996; *Eptesicus fuscus*, Plath and Vater, unpublished; *Pteronotus quadridens*, after Vater 1996; *Megaderma lyra*, after Fiedler 1983; *Tadarida*, after Vater and Siefer 1995.)

Constant basilar membrane morphology thus appears to correlate with expanded frequency representation (Kössl and Vater 1985b; Vater et al. 1985; Müller et al. 1992). Stepwise changes in basilar membrane morphology are only found in CF-FM bats with enhanced foveal tuning. These may represent reflection points for traveling waves that are important for resonant reverberations which lead to sharp tuning (Duifhuis and Vater 1985; Kössl 1994a, 1994b).

Simple measurements of basilar membrane thickness and basilar membrane width certainly present only very indirect estimates of basilar membrane stiffness. Furthermore, they neglect place-specific and species-specific differences in basilar membrane shape and ultrastructure that very likely influence the vibration modes (Kolston et al. 1989). Some of these features are illustrated in Figure 16.6. The basilar membrane of the basal cochlear turn of *Eptesicus* (Figure 16.6A) features two distinct thickenings in the radial direction separated by a thinner hinge region. The lateral thickening (pars pectinata) is composed of the addition of a homogenous substance that is packed with radially directed filaments. This appearance is also typical for the basal cochlear turn of many rodents with good hearing in the ultrasonic frequency range, for most bats with broadband echolocation signals (e.g., Pye 1966), and for nonspecialized cochlear regions of CF-FM bats.

The highly specialized, thickened basilar membrane

basal to the foveal representation site of the CF_2 frequencies in *P. parnellii* (Vater, unpublished data) and *Rhinolophus* (Bruns 1976b, 1980) is characterized by two features: The pars pectinata is densely packed with interwoven filaments, which are mainly oriented radially, and there is a further thickening of the basilar membrane on the side that faces the organ of Corti (the vestibular thickening) that is predominantly composed of longitudinal filaments (Figure 16.6B). This specialization probably creates an increased longitudinal coupling (Kössl and Vater 1985b). The basilar membrane within the fovea of *Tadarida* features a continuous radial thickening that is mainly created by the addition of a homogenous ground substance (Figure 16.6C) and which probably serves as a mass load (Vater and Siefer 1995).

The shape of the basilar membrane in the lower and upper basal cochlear turns is compared for different bats in Figure 16.7. *Eptesicus* and *Megaderma* are the typical nonspecialized cases. In these bats, the thickenings of the basilar membrane gradually diminish toward the cochlear apex. In horseshoe bats and mustached bats, there is a specialized vestibular basilar membrane thickening in cochlear regions basal to the fovea and a nonspecialized basilar membrane shape in the fovea. In *Tadarida*, the basilar membrane basal to the fovea is nonspecialized, whereas the basilar membrane within the fovea is thickened and not radially seg-

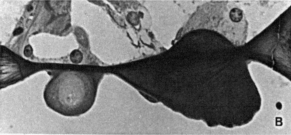

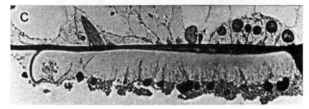

Figure 16.6. Morphology of the basilar membrane of different bat species as seen in semithin sections (2 μm). (**A**) Nonspecialized BM in the basal turn of *Eptesicus fuscus*. (**B**) Specialized BM from the lower basal turn of *Rhinolophus rouxi*. (**C**) Specialized BM in the fovea of *Tadarida brasiliensis*. (Photos by M. Vater.)

mented. This pattern represents an evolutionary line clearly separate from CF-FM bats and other bats that have broadband echolocation systems.

The micromechanical properties of the basilar membrane that are an essential part of the cochlear frequency analyzer are not fully understood. Clearly, there is a need for more thorough modeling of the frequency response of the cochlea that, in addition to estimated stiffness, takes into account the specialized geometry of the basilar membrane. Further studies are needed on specialized aspects of basilar membrane attachment sites such as the secondary

spiral lamina (Bruns 1980), and the possibility that the basilar membrane may be under tension that is generated by specialized cell types of the spiral ligament (Henson and Henson 1988).

The Tectorial Membrane as a Resonant Structure

Textbooks generally describe the tectorial membrane as an anchoring device for hair cell stereovilli. Only relatively recently has the tectorial membrane been attributed a role in frequency analysis by acting as a second filter superimposed on basilar membrane mechanics (Zwislocki and Kletzky 1979; Allen and Fahey 1993) that improves the tip-to-tail ratio of neuronal tuning curves. Recent anatomical findings (Henson and Henson 1991; Vater and Kössl 1996) indeed suggest that the tectorial membrane of the mustached bat contributes to specialized aspects of cochlear tuning. The tectorial membrane shows pronounced regional-specific differences in shape and cross-sectional area and in its attachment site to the spiral limbus (Figure 16.8A), which are largest within the expanded representation place of the CF_2 frequencies (Henson and Henson 1991; Vater and Kössl 1996).

Within the more basally located region that is characterized by maximally thickened basilar membrane (Kössl and Vater 1985b) and low innervation density (the sparsely innervated [SI] zone: Henson 1973; Zook and Leake 1989), both cross-sectional area and limbal attachment site of the tectorial membrane are significantly reduced (Figure 16.8A). The transitions in basilar membrane and tectorial membrane morphology occur at the same place and are steep. In the hook region of the cochlea, the attachment site of the tectorial membrane is again significantly enlarged (not shown; see Vater and Kössl 1996). Such patterns are not seen in the cochlea of the related FM bat *Pteronotus quadridens* (Figure 16.8B) or other bats with broadband echolocation signals (*Tadarida*: Vater and Siefer 1995; *Eptesicus*: Vater, un-

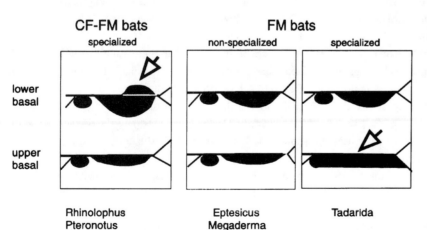

Figure 16.7. Schematic illustration of different types of basilar membrane specializations (arrows) in CF-FM and FM bats.

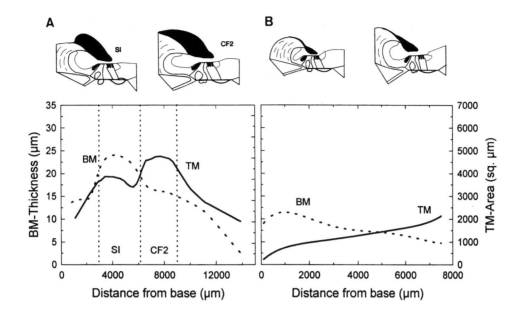

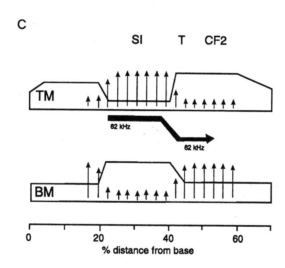

Figure 16.8. Specializations of the tectorial membrane (TM) and basilar membrane (BM) in the CF-FM bat *Pteronotus parnellii*. (**A, B**) Cross-sectional area and shape of the BM and TM of *P. parnelli* (A), in comparison to those of the related FM bat *P. quadridens* (B). Note the regional specific differences in (A), in contrast to the continous gradients in (B). (After Vater 1996.) (**C**) Working hypothesis of specialized function of tectorial membrane in *P. parnelli*. It is assumed that most of the resonant energy to 62 kHz is carried by the tectorial membrane of the sparsely innervated zone (SI) and transferred to the basilar membrane at the transition (T) in TM and BM morphology, thus enhancing vibration amplitudes within the CF$_2$ region. The thick arrow in the middle represents this energy transfer. Upward arrows indicate relative amount of vibration. (After Kössl and Vater 1996a, 1996b.)

published data) nor in the cochlea of nonecholocating mammals (e.g., *Rattus;* Roth and Bruns 1992). A steep transition in tectorial membrane morphology at the transition in basilar membrane morphology is likewise observed in the horseshoe bat (Vater 1997), which indicates that it represents a common feature in cochlear design for sharp tuning in CF-FM bats.

These observations are relevant in the context of cochlear models that view the tectorial membrane as a second resonator to improve neural tuning (Zwislocki and Ce-

faratti 1989; Allen and Fahey 1993). The resonance frequency of the tectorial membrane is determined by the quotient of mass (cross-sectional area) and stiffness (material properties and attachment to the spiral limbus). Frequency maps of the tectorial membrane as inferred from specific measurement paradigms of distortion-product otoacoustic emissions (Gaskill and Brown 1990; Allen and Fahey 1993) indicate that in nonspecialized mammals the tectorial membrane tuning is about half an octave lower than the basilar membrane tuning at the same cochlear position. It has been postulated that the tectorial membrane shunts energy back into the cochlear fluids at its resonance frequency, thereby isolating stereovilli from low-frequency motion and improving the tip-to-tail ratios of neuronal tuning curves (Allen and Fahey 1993).

Similar measurements in *P. parnellii* (Kössl and Vater 1996a, 1996b) show that inferred tectorial membrane tuning and basilar membrane tuning are almost identical within cochlear regions representing the second and third harmonic of the CF signal. This indicates either that no second filter is present or, more likely, that the second filter is tuned to frequencies almost matching those of the basilar membrane. Furthermore, within the SI zone tectorial membrane tuning stays almost constant at the CF_2 signal frequency. Within this region, basilar membrane thickness is maximal and the cross-sectional area and limbal attachment of the tectorial membrane are reduced, suggesting that here the tectorial membrane is capable of strong oscillatory movements and may carry most of the energy of cochlear resonance. This energy is transferred into a basilar membrane movement more apically, as illustrated in the phenomenological model of Figure 16.8C.

Hair Cells

General information on hair cell arrangements and patterns of innervation in the bat cochlea was presented in a review by Kössl and Vater (1995). Here, the main focus is on the outer hair cells (OHCs), which in general auditory research are regarded as active elements in cochlear tuning. OHCs perform both forward and reverse mechanical trans-duction; the latter appears to be linked to their capacity of fast and slow motile responses (Brownell et al. 1985; Zenner et al. 1985; Ashmore 1987; Dallos and Corey 1991; Kalinec et al. 1992). The presence of fast motility has been demonstrated for isolated OHCs of *Carollia perspicillata* (Kössl et al. 1993) and is in the range reported for other mammals in both speed and amplitude. Motility as measured in vitro may represent an epiphenomenon, and in situ the associated stiffness changes may be more important for active OHC function, in particular at very high frequencies.

Certain aspects of OHC morphology in bats correlate with the ability of hearing in the ultrasonic range. The dimensions of both OHC body and sterovilli throughout the cochlea are small and basoapical gradients are far less pronounced than in mammals with good hearing abilities in both the low- and high-frequency ranges (Figure 16.9). The size of the OHCs is thus adapted to process high frequencies and does not reach dimensions typically associated with good low-frequency hearing. The pronounced differences in arrangement and size of the OHC sterovilli bundles are exemplified by the comparison of extreme cases: an OHC from the basal cochlea of *Tadarida* and an OHC from the apical cochlea of a macaque (Figure 16.10). Morphological features such as opening angle of the sterovilli bundles, size and inclination of sterovilli, and pattern of bundle arrangement in the three rows of OHC follow basoapical gradients. Many of these parameters have been determined for several bat species (Dannhof and Bruns 1991; Vater and Lenoir 1992; Vater and Siefer 1995; Vater et al. 1997) and for the cochlea of some nonecholocating mammals (Wright 1984; Lim 1986; Pujol et al. 1992).

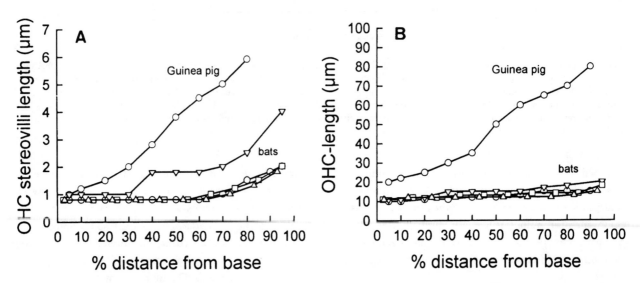

Figure 16.9. Comparison of outer hair cell (OHC) morphology in different bat species and in the guinea pig. (A) Dimensions of OHC stereovilli. (Data sources: guinea pig, after Wright 1984; *Tadarida brasiliensis*, after Vater and Siefer 1995; *Hipposideros speoris*, after Dannhof and Bruns 1991; *Rhinolophus rouxi*, after Vater and Lenoir 1992; *Pteronotus parnellii*, after Vater and Kössl 1996.) (B) Dimensions of the OHC body (guinea pig data from Pujol et al. 1992; bat species and data sources as in A).

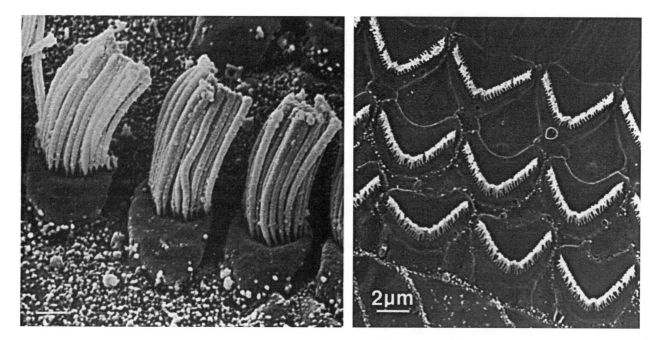

Figure 16.10. Comparison of stereovilli arrangements in low-frequency and high-frequency cochlear turns. Notice the differences in length of stereovilli and bundle morphology. These scanning electron micrographs were taken at the same magnification. **Left:** Apical cochlear turn (low frequency) of the monkey *Macacua mulatta*. **Right:** Basal cochlear turn (high frequency) of the bat *Eptesicus fuscus*. (Photos by M. Vater.)

There is clearly a need to incorporate these parameters into micromechanical models. Both the size of OHC bodies and OHC stereocilia found in the basal turn of the bat cochlea seem to have reached minimal dimensions that still guarantee function: The limiting values appear to be 0.7 μm for the length of stereocilia and 11–12 μm for the length of OHC bodies in the cochlea of *P. parnellii*, but it would be of interest to extend these observations further on bat species known to emit orientation calls at frequencies close to 200 kHz.

To date, no morphological features have been identified that distinguish OHCs from the sharply tuned foveal regions of CF-FM bat from those in adjacent frequency ranges or from OHCs of bats with broadly tuned sonar systems (Vater and Lenoir 1992; Vater and Kössl 1996). Significantly, bat OHCs have conserved the intricate submembranous cytoskeleton that is important in the context of active motility (Pujol et al. 1996). This suggests that OHC function is quite conservative across different species of bats and that the salient mechanisms creating enhanced cochlear tuning reside in specializations of the passive mechanical system.

There are, however, profound species-specific differences in the organization of the medial olivocochlear system that provides efferent synapses to the OHCs. Throughout the cochlea of the mustached bat, each OHC is contacted by a large efferent terminal (Bishop and Henson 1988; Xie et al. 1993). In horseshoe bats and related hippossiderids, how-

ever, the medial olivocochlear system is completely lacking (Bruns and Schmieszek 1980; Aschoff and Ostwald 1987; Bishop and Henson 1988; Vater et al. 1992).

The medial olivocochlear system is in a strategic position to modulate electrical and mechanical properties of the OHCs. By influencing the gain of electromechanical feedback, it controls sensitivity and frequency selectivity of the cochlea (Mugarusa and Russell 1996). Because its action is mainly inhibitory it suppresses basilar membrane displacement, thus leading to a suppression of inner hair cell (IHC) receptor potentials and afferent spike activity (Brown and Nuttall 1984). In *P. parnellii*, stimulation of the medial olivocochlear efferents reduces the decay time of the transient-evoked cochlear microphonic ringing responses at the CF_2 frequency and produces shifts in resonance frequency (Henson et al. 1995). Thus, the damping of the oscillatory response of the cochlear partition is increased by efferent activity, and the resonance frequency and tonotopy along the cochlear partition are altered. A damping effect can serve as a protection against overdrive, but as stated by Henson et al. (1995): "It is difficult to envision any perceptual benefit of rapid shifts in a system that must continuously monitor a very complex, constantly changing acoustic environment." Consequently, the loss of medial efferents in the subfamily Rhinolophoidea was interpreted to represent an advantageous feature because it adds stability to the system (Henson et al. 1995), although it is not confined to the fovea.

Conclusions

Comparative studies have demonstrated remarkable adaptations of cochlear structure and function for processing the CF components of the biosonar signal in nonrelated species of CF-FM bats, *Pteronotus parnellii* (Phyllostomidae) and *Rhinolophus rouxi* (Rhinolophoidea). Both species show foveal representations and enhanced cochlear tuning to narrow frequency ranges encompassing the CF signal component of the second harmonic. Although cochlear design differs to a certain extent in these bats, specializations of both tectorial membrane and basilar membrane appear necessary for implementation of exceptionally sharply tuned cochlear filter mechanisms.

Diversity in cochlear design of FM bats is greater than previously assumed. Foveal representations of certain frequency ranges are also found in the cochlea of molossid bats *(Tadarida brasiliensis)*. They are less pronounced than in CF-FM bats and not paralleled by enhanced cochlear tuning. The cochleae of other FM bats studied to date appear nonspecialized with only general adaptations for high-frequency hearing.

The mechanisms of profound adaptive plasticity at the level of the auditory receptor organ are not fully understood. Nor do we know whether there are further adaptive traits in cochlear design within the approximately 760 species of Microchiroptera; only about half a dozen of these species have been studied.

Acknowledgments

I thank G. Neuweiler and H.-U. Schnitzler for their continuous support and many fruitful discussions, and M. Kössl for helpful comments on the manuscript.

Literature Cited

Allen, J. B., and P. F. Fahey. 1993. A second cochlear-frequency map that correlates distortion product and neural measurements. Journal of the Acoustical Society of America 94:809–816.

Aschoff, A., and J. Ostwald. 1987. Different origins of cochlear efferents in some bat species, rats, and guinea pigs. Journal of Comparative Neurology 264:56–72.

Ashmore, J. F. 1987. A fast motile response in guinea-pig outer hair cells: The cellular basis of the cochlear amplifier. Journal of Physiology (Cambridge) 288:323–347.

Békésy, G. von. 1960. Experiments in Hearing. McGraw-Hill, New York.

Bishop, A. L., and O. W. Henson, Jr. 1988. The efferent auditory system in Doppler-shift compensating bats. *In* Animal Sonar, P. E. Nachtigall and P. W. B. Moore, eds., pp. 307–311. Plenum Press, New York.

Brown, M. C., and A. L. Nuttall. 1984. Efferent control of cochlear inner hair cell responses in the guinea pig. Journal of Physiology (Cambridge) 354:625–646.

Brownell, W. E., C. R. Bader, D. Bertrand, and D. E. Ribaupierre. 1985. Evoked mechanical responses of isolated cochlear outer hair cells. Science 227:194–196.

Bruns, V. 1976a. Peripheral auditory tuning for fine frequency analysis by the CF-FM bat *Rhinolophus ferrumequinum*. II. Frequency mapping in the cochlea. Journal of Comparative Physiology 106:87–97.

Bruns, V. 1976b. Peripheral auditory tuning for fine frequency analysis by the CF-FM bat *Rhinolophus ferrumequinum*. I. Mechanical specializations of the cochlea. Journal of Comparative Physiology 106:77–86.

Bruns, V. 1980. Basilar membrane and its anchoring system in the cochlea of the greater horseshoe bat. Anatomy and Embryology 161:29–51.

Bruns, V., and E. Schmieszek. 1980. Cochlear innervation in the greater horseshoe bat: Demonstration of an acoustic fovea. Hearing Research 3:27–43.

Covey, E., and J. H. Casseday. 1995. The lower brainstem auditory pathways. *In* Hearing by Bats, A. N. Popper and R. R. Fay, eds., pp. 235–296. Springer Handbook of Auditory Research, Vol. 5. Springer-Verlag, New York.

Dalland, J. I. 1965. Hearing sensitivity in bats. Science 150:1185–1186.

Dallos, P., and M. E. Corey. 1991. The role of outer hair cell motility in cochlear tuning. Current Opinion in Neurobiology 1:215–220.

Dannhof, B. J., and V. Bruns. 1991. The organ of Corti in the bat *Hipposideros bicolor*. Hearing Research 53:253–268.

Duifhuis, H., and M. Vater. 1985. On the mechanics of the horseshoe bat cochlea. *In* Peripheral Auditory Mechanisms, J. B. Allen, J. L. Hall, A. Hubbard, S. T. Neely, and A. Tubis, eds., pp. 89–96. Springer, Berlin.

Evans, E. F. 1975. Cochlear nerve and cochlear nucleus. *In* Auditory Systems, W. D. Keidel and W. D. Neff, eds., pp. 1–108. Handbook of Sensory Physiology, Vol. 5/2. Springer, Berlin.

Feng, A.-S., and M. Vater. 1985. Functional organization of the cochlear nucleus of rufous horseshoe bats *(Rhinolophus rouxi)*: Frequencies and internal connections are arranged in slabs. Journal of Comparative Neurology 235:529–553.

Fenton, M. B. 1995. Natural history and biosonar signals. *In* Hearing by Bats, A. N. Popper and R. R. Fay, eds., pp. 37–87. Springer Handbook of Auditory Research, Vol. 5. Springer-Verlag, New York.

Fiedler, J. 1979. Prey catching with and without echolocation in the Indian false vampire *(Megaderma lyra)*. Behavioral Ecology and Sociobiology 6:155–160.

Fiedler, J. 1983. Vergleichende Cochlea-Morphologie der Fledermausarten *Molossus ater, Taphozous nudiventris kachhensis* und *Megaderma lyra*. Ph.D. dissertation, University of Frankfurt, Germany.

Gaskill, S. A., and A. M. Brown. 1990. The behavior of the acoustic distortion product, f_1–f_2, from the human ear and its relation to auditory sensitivity. Journal of the Acoustical Society of America 88:821–839.

Habersetzer, J., G. Schuller, and G. Neuweiler. 1984. Foraging behavior and Doppler shift compensation in echolocating bats *Hipposideros bicolor* and *Hipposideros speoris*. Journal of Comparative Physiology A 155:559–567.

Haplea, S., E. Covey, and J. H. Casseday. 1994. Frequency tuning and response latencies at three levels in the brainstem of the echolocating bat *Eptesicus fuscus*. Journal of Comparative Physiology A 174:671–683.

Henson, M. M. 1973. Unusual nerve-fiber distribution in the cochlea of the bat *Pteronotus p. parnellii* (Gray). Journal of the Acoustical Society of America 53:1739–1740.

Henson, M. M., and O. W. Henson, Jr. 1988. Tension fibroblasts and the connective tissue matrix of the spiral ligament. Hearing Research 35:237–258.

Henson, M. M., and O. W. Henson, Jr. 1991. Specializations for sharp tuning in the mustached bat: The tectorial membrane and spiral limbus. Hearing Research 56:122–132.

Henson, O. W., Jr., G. Schuller, and M. Vater. 1985. A comparative study of the physiological properties of the inner ear in Doppler shift compensating bats *(Rhinolophus rouxi, Pteronotus parnellii)*. Journal of Comparative Physiology A 157:587–597.

Henson, O. W., Jr., P. A. Koplas, A. W. Kating, R. F. Huffman, and M. M. Henson. 1990. Cochlear resonance in the mustached bat: Behavioral adaptations. Hearing Research 50:259–274.

Henson, O. W., Jr., D. H. Xie, A. W. Keating, and M. M. Henson. 1995. The effect of contralateral stimulation on cochlear resonance and damping in the mustached bat: Role of the medial efferent system. Hearing Research 86:111–124.

Huffman, R. F., and O. W. Henson, Jr. 1991. Cochlear and CNS tonotopy: Normal physiological shifts in the mustached bat. Hearing Research 56:79–85.

Huffman, R. F., and O. W. Henson, Jr. 1993a. Labile cochlear tuning in the mustached bat. I. Concomitant shifts in biosonar emission frequency. Journal of Comparative Physiology A 171:725–734.

Huffman, R. F., and O. W. Henson, Jr. 1993b. Labile cochlear tuning in the mustached bat. II. Concomitant shifts in neural tuning. Journal of Comparative Physiology A 171:735–748.

Kalinec, F., C. M. Holley, K. H. Iwasa, D. J. Lim, and B. Kachar. 1992. A membrane-based force generation mechanism in auditory sensory cells. Proceedings of the National Academy of Sciences of the United States of America 89:8671–8675.

Kemp, D. T. 1978. Stimulated acoustic emissions from within the human auditory system. Journal of the Acoustical Society of America 64:1386–1391.

Kolston, P. J., M. A. Viergever, E. de Boer, and R. J. Diependaal. 1989. Realistic mechanical tuning in a micromechanical model. Journal of the Acoustical Society of America 86:133–140.

Köppl, C., G. A. Manley, and O. Gleich. 1993. An auditory fovea in the barn owl cochlea. Journal of Comparative Physiology A 171:695–704.

Kössl, M. 1992a. High frequency two-tone distortions from the cochlea of the mustached bat *Pteronotus parnellii* reflect enhanced cochlear tuning. Naturwissenschaften 79:425–427.

Kössl, M. 1992b. High frequency distortion products from the ears of two bat species, *Megaderma lyra* and *Carollia perpicillata*. Hearing Research 60:156–164.

Kössl, M. 1994a. Otoacoustic emissions from the cochlea of the "constant frequency" bats *Pteronotus parnellii* and *Rhinolophus rouxi*. Hearing Research 72:59–72.

Kössl, M. 1994b. Evidence for a mechanical filter in the cochlea of the "constant frequency" bats *Rhinolophus rouxi* and *Pteronotus parnellii*. Hearing Research 72:73–80.

Kössl, M., and I. J. Russell. 1995. Basilar membrane resonance in the cochlea of the mustached bat. Proceedings of the National Academy of Sciences of the United States of America 92:276–279.

Kössl, M., and M. Vater. 1985a. Evoked acoustic emissions and cochlear microphonics in the mustache bat, *Pteronotus parnellii*. Hearing Research 19:157–170.

Kössl, M., and M. Vater. 1985b. The cochlear frequency map of the mustache bat, *Pteronotus parnellii*. Journal of Comparative Physiology A 157:687–697.

Kössl, M., and M. Vater. 1990. Resonance phenomena in the cochlea of the mustache bat and their contribution to neuronal response characteristics in the cochlear nucleus. Journal of Comparative Physiology A 166:711–720.

Kössl, M., and M. Vater. 1995. Cochlear structure and function in bats. *In* Hearing by Bats, A. N. Popper and R. R. Fay, eds., pp. 191–235. Springer Handbook of Auditory Research, Vol. 5. Springer-Verlag, New York.

Kössl, M., and M. Vater. 1996a. A tectorial membrane fovea in the cochlea of the mustached bat. Naturwissenschaften 2:89–92.

Kössl, M., and M. Vater. 1996b. Further studies on the mechanics of the cochlear partition in the mustached bat. II. A second cochlear frequency map derived from acoustic distortion products. Hearing Research 94:78–87.

Kössl, M., G. Frank, M. Faulstich, and I. J. Russell. 1997. Acoustic distortion products as indicator of cochlear adaptations in Jamaican mormoopid bats. *In* Diversity in Auditory Mechanics, E. R. Lewis, G. R. Long, R. F. Lyon, P. M. Narins, C. R. Steele, and E. Hecht-Poinar, eds., pp. 42–49. World Scientific, Singapore.

Kössl, M., G. Reuter, W. Hemmert, S. Preyer, U. Zimmermann, and H.-P. Zenner. 1993. Electromotility of outer hair cells from the cochlea of the echolocating bat *Carollia perspicillata*. Journal of Comparative Physiology A 175:449–455.

Liberman, M. C. 1982. The cochlear frequency map for the cat: Labeling auditory-nerve fibers of known characteristic frequency. Journal of the Acoustical Society of America 72:1441–1449.

Lim, D. J. 1986. Functional structure of the organ of Corti: A review. Hearing Research 22:117–146.

Müller, M. 1991. Frequency representation in the rat cochlea. Hearing Research 51:247–254.

Müller. M., B. Laube, H. Burda, and V. Bruns. 1992. Structure and function of the cochlea in the African mole rat *(Cryptomys hottentottus)*: Evidence for a low frequency acoustic fovea. Journal of Comparative Physiology A 171:469–476.

Mugarusa, E., and I. J. Russell. 1996. The effect of efferent stimulation on basilar membrane displacement in the basal turn of the guinea pig cochlea. Journal of Neuroscience 16:325–332.

Neuweiler, G. 1990. Auditory adaptations for prey capture in echolocating bats. Physiological Reviews 70:615–641.

Neuweiler, G., V. Bruns, and G. Schuller. 1980. Ears adapted for the detection of motion, or how echolocating bats have exploited the capacities of the mammalian auditory system. Journal of the Acoustical Society of America 68:741–753.

Pollak, G. D., and T. J. Park. 1995. The inferior colliculus. In Hearing by Bats, A. N. Popper and R. R. Fay, eds., pp. 296–368. Springer Handbook of Auditory Research, Vol. 5. Springer-Verlag, New York.

Pollak, G. D., O. W. Henson, Jr., and A. Novick. 1972. Cochlear microphonic audiograms in the "pure tone" bat *Chilonycteris parnellii parnellii*. Science 176:66–68.

Pollak, G. D., O. W. Henson, Jr., and R. Johnson. 1979. Multiple specializations in the peripheral auditory system of the CF-FM bat *Pteronotus parnellii*. Journal of Comparative Physiology A 131:255–266.

Pujol, R., A. Forge, T. Pujol, and M. Vater. 1996. Specializations of the outer hair cell lateral wall and at the junction with the Deiter's cell in bats. In Abstracts of the Nineteenth Midwinter Research Meeting of the Association for Research in Otolaryngology, St. Petersburg, Fla., February 4–8, 1996, p. 60.

Pujol, R., M. Lenoir, S. Ladrech, F. Tribillac, and G. Rebillard. 1992. Correlation between the length of outer hair cells and the frequency coding of the cochlea. In Auditory Physiology and Perception, Y. Cazals, L. Demany, and K. Horner, eds., pp. 45–52. Advances in Biosciences. Pergamon Press, New York.

Pye, A. 1966. The structure of the cochlea in Chiroptera. A selection of Microchiroptera from Africa. Journal of Zoology (London) 162:335–343.

Roth, B., and V. Bruns. 1992. Postnatal development of the rat organ of Corti. I. General morphology, basilar membrane, tectorial membrane, and border cells. Anatomy and Embryology 185:559–569.

Rübsamen, R., G. Neuweiler, and K. Sripathi. 1988. Comparative collicular tonotopy in two bat species adapted to movement detection, *Hipposideros speoris* and *Megaderma lyra*. Journal of Comparative Physiology A 163:271–285.

Schnitzler, H.-U. 1968. Die Ultraschall-Ortungslaute der Hufeisen-Fledermäuse (Chiroptera: Rhinolophidae) in verschiedenen Orientierungssituationen. Zeitschrift für Vergleichende Physiologie 57:376–408.

Schnitzler, H.-U. 1970. Echoortung bei der Fledermaus *Chilonycteris rubiginosa*. Zeitschrift für Vergleichende Physiologie 68:25–39.

Schnitzler, H.-U., and O. W. Henson, Jr. 1980. Performance of airborne animal sonar systems: I. Microchiroptera. In Animal Sonar Systems, R. G. Busnel and J. F. Fish, eds., pp. 109–181. Plenum Press, New York.

Schnitzler, H.-U., N. Suga, and J. A. Simmons. 1976. Peripheral auditory tuning for fine frequency analysis in the CF-FM bat *Rhinolophus ferrumequinum*. Journal of Comparative Physiology A 106:99–110.

Schuller, G. 1980. Hearing characteristics and Doppler shift compensation in South Indian CF-FM bats. Journal of Comparative Physiology A 139:349–356.

Schuller, G., and G. D. Pollak. 1979. Disproportionate frequency representation in the inferior colliculus of Doppler-compensating greater horseshoe bats: Evidence for an acoustic fovea. Journal of Comparative Physiology A 132:47–54.

Simmons, J. A., and R. Stein. 1980. Acoustic imaging in bat sonar: Echolocation signals and the evolution of echolocation. Journal of Comparative Physiology A 135:61–84.

Simmons, J. A., M. B. Fenton, and M. J. O'Farrell. 1979. Echolocation and pursuit of prey by bats. Science 203:16–21.

Suga, N. 1964. Single unit activity in cochlear nucleus and inferior colliculus of echolocating bats. Journal of Physiology (Cambridge) 172:449–474.

Suga, N., and P.-H. S. Jen. 1977. Further studies on the peripheral auditory system of CF-FM bats specialized for fine frequency analysis of Doppler shifted echoes. Journal of Experimental Biology 69:207–232.

Suga, N., J. A. Simmons, and P.-H. S. Jen. 1975. Peripheral specialization for fine analysis of Doppler shifted echoes in the auditory system of the CF-FM bat *Pteronotus parnellii*. Journal of Experimental Biology 63:161–192.

Suga, N., G. Neuweiler, and J. Möller. 1976. Peripheral auditory tuning for fine frequency analysis by the CF-FM bat *Rhinolophus ferrumequinum*. Journal of Comparative Physiology A 106:111–125.

Vater, M. 1988. Cochlear physiology and anatomy in bats. In Animal Sonar, P. E. Nachtigall and P. W. B. Moore, eds., pp. 225–242. Plenum Press, New York.

Vater, M. 1997. Evolutionary plasticity of cochlear design in echolocating bats. In Diversity in Auditory Mechanics, E. R. Lewis, G. R. Long, R. F. Lyon, P. M. Narins, C. R. Steele, and E. Hecht-Poinar, eds., pp. 49–55. World Scientific, Singapore.

Vater, M., and M. Kössl. 1996. Further studies on the mechanics of the cochlear partition in the mustached bat. I. Ultrastructural observations on the tectorial membrane and its attachments. Hearing Research 94:63–78.

Vater, M., and M. Lenoir. 1992. Ultrastructure of the horseshoe bat's organ of Corti. I. Scanning electron microscopy. Journal of Comparative Neurology 318:367–379.

Vater, M., and W. Siefer. 1995. The cochlea of *Tadarida brasiliensis*: Specialized functional organization in a generalized bat. Hearing Research 91:178–195.

Vater, M., A. S. Feng, and M. Betz. 1985. An HRP-study of the frequency place map of the horseshoe bat cochlea: Morphological correlates of the sharp tuning to a narrow frequency band. Journal of Comparative Physiology A 157:671–686.

Vater, M., M. Lenoir, and R. Pujol. 1992. Ultrastructure of the horseshoe bat's organ of Corti. II. Transmission electron microscopy. Journal of Comparative Neurology 318:380–391.

Vater, M., M. Lenoir, and R. Pujol. 1997. Development of the organ of Corti in horseshoe bats: Scanning and transmission electron microscopy. Journal of Comparative Neurology 377: 520–534.

Wright, A. 1984. Dimensions of the stereocilia in man and the guinea pig. Hearing Research 13:89–98.

Xie, D. H., M. M. Henson, A. L. Bishop, and O. W. Henson, Jr. 1993. Efferent terminals in the cochlea of the mustached bat: Quantitative data. Hearing Research 66:81–90.

Zenner, H.-P., U. Zimmermann, and U. Schmidt. 1985. Reversible contraction of isolated cochlear hair cells. Hearing Research 18:127–133.

Zook, J. M., and P. A. Leake. 1989. Connections and frequency representation in the auditory brainstem of the mustache bat, *Pteronotus parnellii*. Journal of Comparative Neurology 290: 243–261.

Zwislocki, J. J. 1986. Analysis of cochlear mechanics. Hearing Research 22:155–169.

Zwislocki, J. J., and L. K. Cefaratti. 1989. Tectorial membrane. II. Stiffness measurements in vivo. Hearing Research 42:211–228.

Zwislocki, J. J., and E. J. Kletzky. 1979. Tectorial membrane: A possible effect on frequency analysis in the cochlea. Science 204:639–641.

Part Four
Conservation Biology

The chapters included in Part Four, "Conservation Biology," represent eight geographic regions and document a growing international concern for the conservation status of bats. There are specific threats in particular countries or regions, and there are also common themes. All identified threats, be they local or global, can be linked to human activities and serve as a painful reminder that thanks to our anthropocentric pursuits there are few places left on this planet that are "wild and free." Much of the earth is now a managed landscape. As our human populations expand and our technological advances allow more efficient and thorough resource extraction, many organisms, including bats, pay an increasingly high price. Extinction rates for bats have, however, been lower than for many other organisms.

Vagility, a willingness to roost in anthropogenic structures, and broad diets have afforded at least some species a buffer against human alterations of natural habitat. Nevertheless, the available data suggest that human impacts are resulting in reduced diversity and shifts in relative abundance. Those species with the most limited distributions or most restricted habitat requirements are declining or disappearing as a result of human impacts. Those species most tolerant of such impacts may remain stable or even increase in some areas, but they are frequently at increased risk of direct mortality from pest control programs, vandalism, and loss of roosts. Relatively few species persist in totally altered landscapes, particularly urbanized and intensely cultivated areas.

Loss of native forest habitat to timber production, agricultural conversion, or human infrastructures is noted by all authors as a primary threat to the long-term persistence of many bat species. The specific habitats differ: for example, the broadleaf woodlands of Europe, the savannah woodlands of Africa, the mixed deciduous and coniferous forests of North America, the coastal rainforests of Australia, the Neotropics, the Philippines, and Pacific oceanic islands. Yet in all localities a high proportion of the bat species, including many of those that now appear to be declining, rely heavily on tree roosts (i.e., cavities, foliage, and loose bark).

Although the diet of most insectivorous forest-dwelling bats is not known, many frugivorous and nectarivorous species rely on native plants for food and play an important role in pollination, seed dispersal, and forest regeneration. Cave-dwelling bats appear to be at high risk wherever they occur. The threats to these bats differ regionally and include recreational caving in Europe and North America, vampire

247

bat control in Middle and South America, and killing for food or because of misconceptions regarding disease risks in Africa, in Asia, and on Pacific islands. The tendency of cave-dwelling bats to form large aggregations and be highly visible, however, renders them particularly vulnerable to human-induced disturbance or mortality.

Conservation efforts are frequently hampered by insufficient knowledge of the distribution and natural history of most species. Distributional maps are generally a better indication of where studies have been conducted than where the bats occur: New distribution records are still emerging, even in regions that have been intensively studied. Additionally, the specific ecological information sought by land managers is generally unavailable—information such as the number of dead roost trees per hectare that should be retained in timber operations, the microclimatic conditions that determine roost selection, the flexibility of a species in its roosting or foraging requirements, and the extent to which habitat alteration affects longevity or reproductive success.

Unresolved taxonomic issues can also inhibit the formulation of species-specific conservation strategies. From both the biological and the political perspective it is necessary to define the taxonomic unit for which a policy is recommended. A taxon may be assigned different conservation priorities depending on whether it is viewed as a subspecies or a distinct species, and it can be overlooked altogether if its correct taxonomic status is not recognized. New species are still being discovered, identified, or redefined even in areas such as Europe and North America, which have been subject historically to relatively close taxonomic scrutiny.

Public misconceptions regarding bats have thwarted conservation efforts everywhere, but the source of misunderstanding can differ regionally. In North America it may be fear of rabies, in Middle and South America fear of vampire bats, or in Australia concern over predation in fruit orchards. The result, however, is the same: bats are often treated as vermin, subjected to eradication campaigns, and overlooked or undervalued by land managers.

Most of the following chapters recognize two basic approaches to conservation: protection of individual species and preservation of habitat. Identified risk factors for individual species include small population size or rarity, evidence of population declines, limited distribution, restricted habitat requirements, and particular vulnerability to human-induced impacts. The highest priority is usually granted to endemic and monotypic species. A species-focused approach has been and likely will be critical for the persistence of certain taxa. For example, in areas where bats are intensively hunted or subjected to systematic eradication, habitat protection does not suffice.

An additional approach is to protect habitats or habitat features critical to the survival of bat communities. Although habitat-oriented conservation efforts generally focus on areas of high biodiversity, this is not sufficient for bats. Because of the very pronounced latitudinal gradient in species diversity, if species richness were the only criterion many habitats important to bats would be overlooked. Because bat distribution is frequently more closely linked to structural rather than botanical characteristics of the landscape, conservation strategies based solely on botanical communities may not account for the habitat features most needed by many bat species (e.g., caves or rock formations). In tropical regions, low-elevation and coastal forests have been most vulnerable to conversion for human development. The loss of dry forests, which do not receive the same conservation attention as rainforests, is also of serious concern.

The authors have generally recognized that conservation strategies for any species or habitat must include protection of both roosting and foraging areas. Acknowledgment of roosting requirements generally leads to the recognition that caverns (caves, old mines, and other anthropogenic structures) and mature native forests are critical to the protection of both numbers of species and overall diversity. Foraging needs are, in general, less well understood than roosting requirements, but unpolluted water sources, intact riparian zones, complex forest structure, and substantial tracts of native habitat are mentioned by a number of authors as important for the maintenance of suitable foraging areas.

While the most effective conservation actions are often based on grass-roots efforts and require the dedication of many people at a local level, national and regional policies protecting bat populations are also necessary. The Australian Bat Action Plan (Richards and Hall), the establishment of federal conservation units in Brazil (Marinho-Filho and Sazima), the European Bats Agreement (Racey), the United States Endangered Species Act (Pierson), or CITES (Convention on International Trade in Endangered Species of Wild Fauna and Flora) regulations (Rainey) serve as cases in point.

Part Four represents a significant benchmark as the first symposium to focus on the conservation biology of bats. The inclusion of this topic in a volume of international scope signals the intention of bat biologists to provide a scientific basis for the conservation of their study animals, so as to inform policy makers and managers, in the hope that bats will continue to be major contributors to mammalian biodiversity worldwide.

ELIZABETH D. PIERSON AND PAUL A. RACEY

17
Ecology of European Bats in Relation to Their Conservation

PAUL A. RACEY

Before bat detectors enabled field identification of foraging bats (Ahlén 1981; Baagøe 1986, in manuscript), the majority of studies of European species involved counting individuals and describing their activity and roosting habits in maternity colonies and hibernacula. Great disparities were often seen between summer and winter counts, particularly for vespertilionids, with winter populations accounting for a small proportion of those counted in summer and vice versa, depending on species. Although the most reliable estimates of population size have generally been derived from studies of maternity colonies (Gaisler 1975, 1978; Speakman et al. 1991a; Entwistle 1994), there remains considerable difficulty in establishing reliable population trends in European bat species.

Conservation legislation in many European countries has protected bats but not their roosts and habitats (Stebbings 1988). More recently, however, the European Bats Agreement 1992 (part of the 1979 Bonn Convention on the Conservation of Migratory Species of Wild Animals) and the European Union (EU) Habitats and Species Directive 1992 have directed attention toward the preservation of foraging habitats. Although the miniaturization of radiotransmitters and the application of radiotracking in re-

cent years has aided the study of bat habitat requirements, generalizations applicable on a nationwide scale have not hitherto been possible in any European country.

The U.K. National Bat Habitat Survey

The first national bat habitat survey was initiated in the United Kingdom in 1990, and the results have been published by Walsh et al. (1995) and Walsh and Harris (1996a, 1996b). The survey adopted a random stratified sampling system (Magurran 1988) based on a land classification that assigns every 1-km square in Britain to 1 of 32 land classes. Squares in each land class have a similar climate, physiogeography, and pattern of land use (Bunce et al. 1981a, 1981b, 1983). Within each land class a sample of 1-km squares was selected at random, to avoid observer bias in the selection of sites and to ensure a standard sampling effort in different landscape types.

Fieldwork was carried out during three consecutive summers from 1990 to 1992 and involved both professional and amateur bat biologists, the latter drawn mainly from Britain's 90 bat groups. Each volunteer was allocated one or more 1-km squares and walked a transect in each square four times during predetermined periods in summer, avoiding nights when weather conditions were unfavorable to bats. Volunteers carried tunable bat detectors (mainly QMC Mini-2) set at 45 kHz to maximize the range of species encountered, and noted the total number of bat passes and feeding buzzes in each square and within each habitat type. The more experienced surveyors were able to identify some bats to species or species groups.

Analysis of the data revealed relationships between bat activity and habitat variables within and across the 32 land classes, which for the purposes of the analysis were combined into seven major groups: three arable, two pastoral, marginal upland, and upland. Avoidance or selection of habitat types was examined by constructing Bonferroni confidence intervals around the observed use of each habitat type (Neu et al. 1974), and regression analyses were used to evaluate habitats of critical importance in determining high bat activity in each land-class group.

Of the 1,030 surveyed 1-km squares, 910 provided data suitable for analysis, involving 2,700 hours of search effort with 30,000 bat passes recorded in the 9,000 km walked. Observers identified 24% of bat passes to a particular species or species group, and 71% of these were *Pipistrellus pipistrellus*, 17% *Myotis* spp., 8% *Nyctalus noctula*, 3% *Plecotus* spp., and 2% *Eptesicus serotinus*. Because a similar proportion of the unidentified bat passes were probably *Pipistrellus pipistrellus*, habitat preferences (and their implications for conservation) apply to the Vespertilionidae as a whole and to *P. pipistrellus* in particular.

Land class was demonstrated to be a highly significant factor influencing the incidence of bat passes, with the greatest bat activity occurring in pastoral land classes. Across all land-class groups, bats tended to forage selectively in edge and linear habitats and avoided more open and intensively managed habitat types. They showed a far stronger preference for all bodies of water and woodland edges than for any other habitat type, emphasizing the importance of these habitats as key foraging sites. In the woodland category, edges were more strongly selected than openings, and seminatural broad-leaved woodland was more strongly selected than either mixed or coniferous woodland. Urban areas, which included villages, small towns, and the suburbs of some large cities, were selected in three land-class groups. Linear vegetation corridors, particularly tree-lined hedgerows and covered ditches, were also selected by bats, emphasizing the importance of landscape connectivity. Sandy, shingle, or rocky beaches and estuarine coastal marsh were also significantly selected as foraging sites. Habitats strongly and consistently avoided in all land-class groups were those that were more exposed and more intensively managed, including moorland, improved grassland, upland, and arable land. The only grassland type not consistently avoided was lowland, unimproved grassland.

Bats foraged preferentially in habitats that were comparatively rare within each land-class group. For example, the percentage availability in each land-class group of the preferred habitats of woodland edge, tree lines, hedgerows, and bodies of water ranged from 14% to 31%, with a mean of 25%. In contrast, the availability of habitats that were consistently avoided (stone walls, moorland, arable, and most grassland categories) ranged from 40% to 54%, with a mean of 47%. Optimal habitats tend to be at the perimeters of other habitat types or linear strips, and thus in comparison with contiguous blocks of pasture or arable land, for instance, their area is proportionally smaller. Bodies of water generally represent less than 1% of the available habitat, and broad-leaved woodland edge ranges from 3% to 4%. Because the selection patterns were consistent between individual land classes, the results of the analysis by Walsh and Harris (1996a, 1996b) can be summarized by habitat type rather than by land class (Table 17.1).

The primary aim of the U.K. National Bat Habitat Survey was to provide a means of assessing the significance of habitat change for bat populations. By expanding the scale of previous studies of habitat preferences (reviewed in Walsh et al. 1995) and distribution surveys (Ahlén 1980–1981; Baagøe 1986; Rydell 1986; Jüdes 1989) to a national level and using a land classification system widely accepted in the United Kingdom, Walsh et al. (1995) and Walsh and Harris (1996a, 1996b) have developed a potential method of detecting change, establishing its direction and measuring its magnitude using a protocol for simultaneously monitoring bat activity and habitat use. Analysis of habitat factors affecting high levels of bat activity resulted in equations with high predictive power and of particular value for forecasting the effects of changes in land use on bats. Although vespertilionids use a diversity of habitats, the regression models identify riparian and woodland habitats as particularly important. Once more numerous and widespread, woodland habitats are now patchily distributed and further habitat fragmentation may reduce the value of those that remain (Bright 1993). Thus, for conservation purposes the relative magnitude of such an effect may be predicted using data collected in the survey.

The principal caveat in interpreting the results of the survey is that they reflect the habitat preferences of the pipistrelle, the most abundant bat in the United Kingdom (Harris et al. 1995). In future U.K. surveys, it is hoped that volunteers will be better able to identify bats to species or species groups. To achieve this goal, training courses in the use of bat detectors are being held throughout the country

Table 17.1

Habitat Types Selected, Used in Proportion to Availability, and Avoided by Bats in Britain

GOOD HABITATS ◄				► POOR HABITATS
Selected in all land classes	Selected in some land classes; never avoided	Selected in some land classes and avoided in others	Avoided in some land classes; never selected	Avoided in all land classes
Treeline	Hedgerow	Open ditch	Improved grassland	Arable
Broad-leaved woodland edge	Stream	Covered ditch	Semi-improved grassland	Moorland
Lake or reservoir	Coniferous woodland edge	Stone wall	Lowland unimproved grassland	Upland unimproved grassland
	Mixed woodland edge	Coniferous woodland opening		
	Broad-leaved woodland opening	Scrub		
	Mixed woodland opening	Park land		
	Felled woodland	Urban land		
	River or canal			
	Pond			

Notes: The results interpreted here are from Walsh and Harris (1996a, 1996b). "Land classes" refers to the 19 discrete land classes in their study.

(Catto 1994), following a successful initiative in the Netherlands. The results of a nationwide bat distribution survey in that country, also involving volunteers (Kapteyn 1991), have been recently published (Limpens et al. 1997). A detector survey of bat distribution also has been completed in Denmark (Baagøe, in manuscript).

Landscape Ecology: From the National to the Local Scale

In a major study of habitat use in the province of Uppland, Sweden (59° N), de Jong (1994) found that relatively open deciduous forests and adjacent shallow eutrophic lakes were the only habitats attracting large numbers and a high diversity of bats during early summer, as a result of high chironomid productivity (de Jong and Ahlén 1991). Bats foraged in more diverse habitats later in summer, although lakes, wetlands, and deciduous forest remained important. In a landscape mosaic, patch size is important, and the number of species foraging in a patch of favorable habitat significantly decreased when the area was less than 30 ha and was mainly affected by the abundance of deciduous forest in the patch and by the extent of its isolation from other patches.

Species that avoid open areas (*Myotis* spp., *Plecotus auritus*) are less likely to use such patches, but in some cases patches are only used by *Nyctalus noctula*. The composition of the mosaic also affects species number, and fewer species were found in patches isolated by open fields and clearcuts. Dense closed-canopy forest was avoided by all species. Observed preferences were for relatively open coniferous forest by *Myotis brandti* and *M. nattereri*, for deciduous for-

est by *Pipistrellus pipistrellus,* and for forest edge and glades by *P. pipistrellus* and *Eptesicus nilssoni*. Corridors of trees that connect feeding patches are important, particularly for *M. nattereri.*

The corridor concept has an intuitive appeal and has been widely adopted by ecologists and land-use planners in advance of formal proof of its validity. Recent reviews (Hill et al. 1993; Spellerberg and Gaywood 1993; Dawson 1994) concluded that, with few exceptions, relevant studies lacked statistical rigor. The use of linear landscape elements by bats has been demonstrated in the Netherlands, where in the open landscape, lanes, avenues of trees, hedgerows, and canals are used as flight paths (Limpens et al. 1989; Limpens and Kapteyn 1991). In areas where linear landscape elements are abundant, bat detector surveys revealed a dense network of flight paths and foraging areas. In more open areas devoid of such elements, few flight paths and bats are recorded. Some species, such as *Myotis daubentonii*, make detours along hedgerows rather than cross an open area on the shortest route to a feeding habitat.

In comparing bats foraging in arable land and in an adjacent nature reserve with abundant woodland, Gaisler and Kolibac (1992) noted that the density of foraging bats was lower in the farmland than in the reserve, by an order of magnitude for *Nyctalus noctula* and by a factor of two for the remaining species. In farmland, the bats often flew along windbreaks, which attract and provide shelter for insects, and bats frequently travel on their leeward sides (Racey and Swift 1985; Limpens and Kapteyn 1991). Windbreaks may also reduce the risk of predation (Schofield 1996) and may be used by bats for orientation by echolocation (Limpens and Kapteyn 1991). *Pipistrellus nathusii* is thus observed migrating southwest in autumn along Dutch polders (P. H. C. Lina, personal communication). In a river valley mosaic of woodland and pasture, Racey and Swift (1985) showed that *P. pipistrellus* used a regular nightly route and flew directly between foraging areas.

Pipistrelle bats roosting in bat boxes were studied in southernmost Sweden (57° N) for eight consecutive breeding seasons in two contrasting situations: a 150-ha pine plantation adjoining a lake and a 16-ha park dominated by deciduous trees of varying ages but surrounded by intensive farming and industry (Gerell and Lundberg 1993). The population in the latter area declined over the study period, while that in the former increased. This difference is not attributable to movement of the populations, as Lundberg and Gerell (1986) have shown high roost site fidelity in this species. Body mass indices ($\log_e$ mass/$\log_e$ forearm^3) were used to estimate fat reserves, and these were significantly lower in 3 of 4 years in September in the area adjacent to intensive farmland. Although the levels of organochlorine residues and cadmium were also higher in the population living close to farmland, Gerell and Lundberg (1993) considered that the proximate cause of the decline in the pipistrelle population in the farmland and industrial area was the deterioration in feeding conditions caused by drainage and water pollution, which result in decreased aquatic insect populations.

Because of the continued preference of the noctule bat *Nyctalus noctula* for roosting in cavities in deciduous trees rather than in houses and the loss of mature and postmature deciduous trees from managed landscapes, concerns have been expressed about the long-term survival of this species in western Europe (Hutson 1993). *Nyctalus noctula* does, however, make use of other anthropogenic factors, and two studies (Cranbrook and Barrett 1965; Kronwitter 1988) have shown that it forages on house crickets (*Acheta domestica*) over domestic refuse dump sites.

Bats and Street Lamps

There have been a number of studies in Europe and North America of bats foraging around street lamps (Rydell and Racey 1995). The bluish-white light of mercury-vapor street lamps, which emit ultraviolet radiation, attracts insects whose density can be determined by flash photography. In contrast, low-pressure sodium lamps, which emit monochromatic orange light, do not attract insects, and high-pressure sodium lamps that include some mercury vapor and hence emit some ultraviolet radiation are intermediate in terms of insect attraction. Rydell (1992), using a bat detector fixed to a moving car (Ahlén 1980–1981; Jüdes

1989), detected northern bats (*Eptesicus nilssoni*) at relatively high densities (2–5/km) in southern Sweden near white street lamps, compared with 0.1–0.4 bats per kilometer of unlit road. Means of 3.2 and 3.1 pipistrelle bats (*Pipistrellus pipistrellus*) were recorded per kilometer of lit road in England and Scotland, respectively (Blake et al. 1994; Rydell et al., unpublished data). The gross energy intake of *E. nilssoni* foraging around street lamps was more than twice as high (0.5 kJ/min) as that recorded in woodland (0.2 kJ/min) and comparable to that over pasture where the bats foraged on dung beetles (0.6 kJ/min). Street lamps may allow some bat species to increase their energy intake and may account for the frequent occurrence of these species in built-up urban areas (Rydell 1992).

A radiotracking study showed that noctules (*Nyctalus noctula*) spent most of their foraging time (~65%) either over a lake or in a town, hunting over an asphalt surface (a car park and road junction) illuminated by strong lights (Kronwitter 1988). Adjacent woods and farmland were used only occasionally. Typically, the bats fed over the lake at dusk and later, after it was fully dark, moved on to feed over the lights in the town. During this second period, 75% of the foraging time was spent over the lights. Similar observations of noctules in England feeding over a well-lit railway yard, which subsequently became an equally well-lit prison, have been made by A. J. Mitchell-Jones (personal communication).

In a contrasting study, Rachwald (1992) used bat detectors to monitor noctules in Bialowieza primeval forest in Poland where there are several villages but no street lamps. Bat activity was consistently highest over water but small forest clearings, maintained for traditional farming, were also exploited. Bats did not prefer the villages to other open areas. In a radiotracking study of serotines (*Eptesicus serotinus*) in England, Catto et al. (1995) showed that second to cattle pastures that provided an abundant source of *Aphodius* dung beetles, roads with white street lamps were the most frequently used feeding sites. In southern Switzerland, Haffner and Stutz (1985–1986) monitored the activity of *Pipistrellus kuhlii* and *P. pipistrellus* for 3 years over about 500 km of road using a bat detector on the roof of a moving car; 94% of *P. kuhlii* and 45% of *P. pipistrellus* were observed near street lamps.

The frequent use of the area around street lamps by foraging bats has obvious implications for conservation. In contrast to many other bat habitats, the illumination of streets, roads, and private properties is increasing and is likely to benefit at least some bat species. Potentially less beneficial has been a recent tendency to replace mercury vapor lamps with sodium lamps, which use less energy (and a less hazardous element) but are of less value to bats.

The urban areas of large cities have depleted insect faunas (Frankie and Ehler 1978; Taylor et al. 1978), and some studies in North America have shown that few bats can survive in this kind of habitat (Geggie and Fenton 1987; Kurta and Terramino 1992). However, a major survey of bats in the London area in 1985–1986 revealed 137 active summer roosts, 75 of which were pipistrelles *(Pipistrellus pipistrellus)*. A total of 430 summer roosts was recorded during a 36-year period; 450 possible feeding sites were surveyed and bats were recorded at 397 of these. Pipistrelles were the most abundant species, although *Nyctalus noctula* and *N. leisleri* were also widespread. In contrast, only 28 of 106 possible hibernacula surveyed were used by bats, and in 70% of these fewer than six individuals were found (Mickleburgh 1987). Gaisler (1979) showed that the number of foraging *P. pipistrellus* increased from the suburbs to the city center of Brno in the Czech Republic and reviewed the common occurrence of this species in several European cities, which he attributed to the presence of street lights.

Different species of bats are not likely to be equally affected by the presence of street lamps. Those most likely to benefit are aerial hawking species such as *Nyctalus*, *Vespertilio*, *Eptesicus*, and *Pipistrellus*. With the exception of *Nyctalus noctula*, there is little clear evidence that any of the species in these genera is presently threatened (Hutson 1993). On the other hand, several European species of *Myotis*, *Plecotus*, and *Rhinolophus*, which do not take advantage of street lamps, have suffered population declines and are endangered, at least in some countries (Stebbings 1988). A possible negative effect of street lights is that they may attract moths that then become unavailable to bat species which are adapted to gleaning.

The echolocation pulses of most *Myotis* and *Plecotus* species lack the narrowband component necessary for the long-range detection of insects in open air (Neuweiler 1989). It may be that these bats are less efficient at exploiting insects in open situations than in cluttered environments. Alternatively, the predation risk in brightly lighted conditions in combination with open situations may be too high for slow-flying bats. Evidence from high latitudes in Scandinavia and Finland suggests that bats, particularly *Myotis* spp., tend to avoid open habitats such as lakes in the ambient light conditions prevailing around midsummer, but return to such areas as the nights become darker later in the year (Nyholm 1965; Rydell 1992). Although these findings suggest that the movements of *Myotis* spp. may be restricted by illuminated motorways and other lit areas, Krull et al. (1991) found that radiotagged *M. emarginatus* made detours along a motorway to reach underpasses leading to foraging areas.

National Roost Surveys

The first national standardized survey of bat roosts in England—The National Bats in Churches Survey—was organized by the Bat Conservation Trust from 1992 to 1994 and involved visits to 538 churches and chapels (Sargent 1995). Visits showed that 90% of 132 churches used as roosts in the late 1960s were still occupied by bats. The most important factor influencing the likelihood of bat occupancy was age of the building, but when the relative effects of all the recorded factors were taken into account, including aspects of the building related to age, the latter decreased in its level of importance, and a specific combination of other factors described more variation in the data than age alone.

The four factors having a significant effect on the presence of bats in churches and chapels were roof type, wall material, geographic location, and the level of building development around the church. Bats were more likely to occupy churches with limestone walls and lead roofing, features that probably indicate other properties which attracts bats such as temperature and roost sites. In general bats occupied churches and chapels surrounded by pre-1800 buildings and avoided those with modern neighboring buildings. Isolated rural and village churches were also favored by bats rather than those in urban areas.

Autecological Studies

The Greater Horseshoe Bat

The greater horseshoe bat *(Rhinolophus ferrumequinum)* is categorized as an endangered species throughout much of Europe, with the current British population of this species estimated at about 4,000 individuals divided among about 14 major maternity colonies (Mitchell-Jones 1995). Radiotracking studies have revealed that bats usually forage within 4 km of the roost, and the conservation of foraging habitats within this radius is therefore particularly important (Jones and Morton 1992; Jones et al. 1995). During spring, bats forage in ancient seminatural woodland, but during late summer they feed mainly over pasture. The ambient temperature in woodland is generally higher than that over pasture, and this differential widens as temperature decreases. Because insect abundance increases rapidly above 6°–10°C, it is likely to be more profitable for bats to forage in woodland in spring.

The shift to pastures during the summer is associated with the dominance of *Aphodius* dung beetles in the diet as cattle dung accumulates. The abundance of such beetles may be threatened by the use of Avermectin antihelminthics in cattle, which because of their persistence reduce the insect fauna associated with dung (Strong 1992). Because juvenile bats forage independently of their mothers both before and after weaning, prime foraging habitat (cattle-grazed permanent pasture close to ancient woodland) adjacent to the maternity roost is likely to be important to juvenile survival. The use of Avermectin in cattle pastured in the vicinity of such roosts could be particularly detrimental to juvenile bats (Duvergé and Jones 1994).

A similarly detailed picture of the characteristics of the hibernacula of greater horseshoe bats has been assembled by Ransome (1968, 1971, 1990). These hibernacula contain completely dark areas with a relative humidity in excess of 95%, a range of ambient temperatures between 5° and 10°C, and regions of slow air flow resulting from two or more entrances or a sloping entrance. Close access to sheltered, often south-facing winter foraging areas is also important, as is freedom from repeated human disturbance.

As a result of the protection of maternity roosts and hibernacula, the decline in numbers of greater horseshoe bats (Stebbings and Arnold 1987; Stebbings 1988) has been halted. The way in which knowledge of the autecology of horseshoe bats is being applied to their conservation in England and Wales has been detailed by Mitchell-Jones (1995) and includes designating key roosts as sites of Special Scientific Interest managed by the relevant statutory conservation agencies (English Nature and The Countryside Council for Wales). The ambient temperature of maternity roosts is increased by the use of heaters (Mitchell-Jones 1995), and gates have been installed at the entrances to hibernacula to prevent disturbance (Ransome 1990). Attention is presently being focused on the ways in which key feeding sites around maternity roosts can be protected (Mitchell-Jones 1995).

The Lesser Horseshoe Bat

Historically, the lesser horseshoe bat *(Rhinolophus hipposideros)* roosted all year around in caves (Horacek 1984), and the most marked change in its roosting behavior has been the adoption of buildings as summer roosts. In a survey of the characteristics of 156 breeding roosts in the United Kingdom (Schofield 1996), *R. hipposideros* selected predominantly nineteenth-century buildings (77%; $n = 61$) with stone walls (81%; $n = 82$) and slate roofs (88%; $n = 77$). Their roosting sites were usually located in roof spaces of attic rooms (95%; $n = 76$). Access to roosts was commonly through large openings (>0.5 m²), such as the open doorways or windows that often characterize derelict or semiderelict buildings. The volume of breeding roosts was frequently greater than 250 m³. Similar findings are re-

ported from other countries in Europe (Gaisler 1963; Stutz and Haffner 1984; McAney and Fairley 1989).

Comparisons of availability between 19 roost buildings and 20 randomly selected buildings in the same geographic area indicate that buildings used as roosts are located closest to stands of broadleaf and mixed woodland. Buildings containing roosts are more likely to be connected to foraging areas by continuous linear landscape features such as hedgerows or stands of trees.

A Bonferroni pairwise comparison (Byers et al. 1984) between the land classes established (or identified) by the Institute of Terrestrial Ecology (ITE) (Bunce and Howard 1991) in 1-km squares containing *R. hipposideros* roosts and those for England and Wales as a whole indicate that this species selects for those land classes associated with areas of gentle rolling and undulating countryside, often enclosed by hedgerows and tree lines. Land classes associated with flat open countryside with intensive agriculture or exposed upland areas are avoided. Details of land cover (as distinct from land class) obtained from satellite data showed that *R. hipposideros* selects deciduous woodland, and bats were found foraging as much as 2 km from breeding roosts in stands of broadleaf and mixed woodland, riparian trees, and hedgerows.

Bats commuted to and from foraging areas and their roosts (both breeding and night roosts) along continuous linear landscape features (particularly well-grown hedgerows). In addition to providing foraging areas, the use of these features may reduce predation. The ambient light level and height at which bats flew across a 5-m gap in one of the hedgerows were recorded using an infrared video camera. At ambient light levels greater than 1 lux, bats flew at a height of less than 1 m from the ground: at less than 1 lux, the height of flight increased to 4 m, suggesting avoidance of avian predators. Conservation strategies and recovery programs for *R. hipposideros* must take into account both the roosting and landscape requirements of this species.

The Brown Long-Eared Bat

BUILDING ROOSTS.   In a study of the roosting ecology of the brown long-eared bat *(Plecotus auritus)* in northeastern Scotland (57° N), Entwistle et al. (1997) located 56 roosts in buildings and compared their characteristics with a randomly chosen sample of buildings from the same area. This approach revealed that this species preferentially roosted in buildings that were older, higher, and had more roof compartments (which in Scotland are fully lined with wood). Brown long-eared bats were found in such roosts between May and October, and the typical group size within the roof space was 15–20 bats.

The mean temperature within roosts was 17.9°C, and roosts were significantly warmer than a random sample of buildings. When captured in their roosts, bats were generally active, and warmer roost temperatures may have reduced their dependence on torpor. The temperature inside the roost was positively correlated to the frequency of occupancy and also with the size of the bat's forearms, with larger individuals being captured in warmer roosts.

The buildings used as roosts were closer to trees and water and had a larger area of woodland within 0.5 km than a random sample of buildings. Brown long-eared bats are foliage gleaners (Anderson and Racey 1991, 1993), and radiotracking revealed that bats foraged mainly in deciduous woodland in the vicinity of the roost, using a series of feeding sites to which they frequently returned and that were occasionally shared with other bats from the same roosts. Females spent most of their foraging time within 0.5 km of the roost, whereas males traveled further. Bats returned to the main roost on 77% of mornings, but also used alternative sites, which had cooler internal temperatures, following colder nights (Entwistle et al. 1996).

The characteristics of the roost and its location were related to the ecology and behavior of its occupants. Across the different roost sites examined, the area of deciduous woodland within 0.5 km of the roost was correlated with foraging patterns, colony size, and the progress of the male reproductive cycle (Entwistle 1994). Although large-scale distribution patterns for this species are probably linked to areas of woodland, local abundance of bats may be affected by the availability of suitable roost sites. The implications of the work for conservation and management are clear. Building roosts are important and need protection, but so also do the adjacent woodlands, which should be maintained and their quality improved where appropriate. But if building roosts of brown long-eared bats have a suite of characteristics that distinguish them from other roof spaces, why are bat boxes attractive to this species?

BAT BOXES. Bat boxes have been widely used for many years throughout Europe (Mayle 1990), and their occupancy rate is highest when they are used in large numbers in coniferous plantations devoid of natural roost sites, although bird boxes are also used by bats in deciduous forests (Schlapp 1990). In the 1970s, a scheme involving 3,000 bat boxes was initiated at six sites across a north–south transect in the United Kingdom by R. E. Stebbings. From data collected at one of these sites, Thetford Forest, Norfolk (53° N), one of the few comprehensive studies of the contribution of such boxes to the population ecology of bats was conducted. Boyd and Stebbings (1989) analyzed the brown long-eared bat occupancy of 480 wooden boxes attached to 100 trees around the perimeter of a 7-ha rectangle in the center of a plantation of mature Corsican (*Pinus nigra* Arnold) and Scots pine (*Pinus sylvestris* Linn). The boxes were checked two to four times a year for 10 years, and a total of 219 females and 182 males were captured and individually marked. Following the establishment of a population of bats in the boxes, immigration probably accounted for a small proportion of the total recruitment, the remainder coming from reproduction within the population. The mean number of young born per female per year was 0.55, and the total population increased during the study from 74 to 140 bats, giving a doubling time of 10 years. The annual survival rate was 0.78–0.86 for females, depending on the method of estimation used, and 0.60 for males, similar to rates found in populations of brown long-eared bats roosting in houses in northeast Scotland (Entwistle 1994) and on the south coast of England (Stebbings 1970a).

Benzal (1991) studied brown long-eared bats occupying 520 bird boxes in a 130-ha *Pinus sylvestris* forest at the comparatively high altitude of 1,400 m in central Spain. The boxes were widely used in the first summer after installation; a total of 197 individuals was found in 3% of box inspections and droppings were found in an additional 8%. Although the bats arrived in the study area in the first half of May and stayed until the beginning of November, they left the boxes at the end of May to form maternity colonies in small caves and attics and returned to them in mid-July with flying young. These studies provide ecological data that support the promotion of bat boxes as alternative roosts for bats, particularly in areas devoid of such roosts, although the Spanish study suggests that boxes may not always be appropriate for breeding.

Long-term studies are now needed to compare the population dynamics of bats occupying different types of boxes, particularly those made from sawdust and cement, which are often preferred to those made only of wood (Taake and Hildenhagen 1989; Mayle 1990). Despite the positive conservation values of bat boxes, they should not be considered as a substitute for hollow trees, the loss of which is a major threat to less synanthropic species such as *Nyctalus noctula*.

The Serotine Bat

The serotine bat *(Eptesicus serotinus)* is widely distributed in mainland Europe even at more northerly latitudes (Schober and Grimmberger 1993) where it is associated with highly exploited landscapes and appears to be increasing its range (Baagøe 1986). By contrast, in Britain this species is largely restricted to southern England (Hutson 1991). Radiotracking studies have revealed that serotines in Britain generally forage within 3 km of their roosts, using as many as five foraging sites a night in a wide range of habitats that include chalk grassland, scrub, pasture, and areas around white street lamps. Serotines located and exploited temporary feeding sites such as recently mown grass from which summer chafers *(Amphimallon solstitialis)* emerged (Catto et al. 1995). The increasing incidence of *Aphodius* beetles in the diet of serotines as summer progressed, reaching a peak of 85% of identifiable fragments in the feces in August, confirmed the importance of cattle pasture as feeding habitat for this species (Catto et al. 1994). Catto et al. (1995) concluded that serotines are well adapted to an anthropogenic environment. They are strongly philopatric to their roosts in houses that are located close to a range of feeding sites where they can take advantage of current farming practice and street lamps.

Daubenton's Bat

Daubenton's bat *(Myotis daubentonii)* forages almost exclusively over water and around riparian vegetation (Jones and Rayner 1988; Kalko and Schnitzler 1989), feeding opportunistically on insects that swarm in such situations, principally chironomids (Swift and Racey 1983; Kokurewicz 1995). Increased numbers of this species of bat, sometimes of several hundred percent, have been reported from counts of hibernating individuals in the Netherlands (Daan 1980; Voûte et al. 1980; Weinrich and Oude Voshaar 1992), Czech and Slovak republics (Barta et al. 1981; Cerveny and Bürger 1990), and southwest Germany (von Helversen et al. 1987).

Kokurewicz (1995) postulated that the increase in Daubenton's bat is attributable to eutrophication and canalization of waterways throughout Europe, factors which increase the numbers of chironomids. This hypothesis was recently tested in northeast Scotland by comparing the activity of pipistrelles and Daubenton's bats in two rivers with sharply contrasting nitrate levels, one of which may be designated as a nitrate-sensitive area by the European Union (Racey 1998). This study provides some support for Kokurewicz's hypothesis and is one of the few examples of a beneficial effect of aquatic pollution on bat populations.

Hidden Biodiversity

Many of the recent additions to European national faunas have been bats. *Plecotus austriacus* was first described in the United Kingdom by Stebbings in 1967. Although Topal (1958) first questioned the status of subspecies of *Myotis mystacinus*, it was not until more than a decade later that Hanak (1971) eventually distinguished *Myotis brandti* as a separate sympatric species. These two sibling species are now widely recognized throughout Europe (Baagøe 1973; Corbet and Harris 1991), and recent molecular work by Nemeth and von Helversen (1993) has pointed to the existence of a third European species in the *M. mystacinus* group.

Jones and van Parijs (1993) described two phonic types of *Pipistrellus pipistrellus,* and evidence is amassing that these are separate species (Barratt et al. 1995). It also raises the question of whether additional cryptic bat species remain to be recognized.

Pipistrellus nathusii is a migratory species (Strelkov 1969) and moves southwest from the Baltic states into the low countries in autumn. The first breeding colony of *P. nathusii* was recently recorded in the Netherlands (Kapteyn and Lina 1994), and since its first appearance in Britain in 1969 (Stebbings 1970b) it has been recorded there with increasing frequency in autumn (Speakman et al. 1991b) and summer (Rydell and Swift 1995; Barlow and Jones 1996).

Conclusions

The first national bat habitat survey of a European country concluded that the main foraging habitats of vespertilionids are associated with broad-leaved woodland and water. Many autecological studies indicate that these conclusions are widely applicable throughout Europe, and one of the benefits of the European Union is that initiatives to restore woodland and to maintain the quality of rivers apply in all member states.

For example, the long decline in total woodland cover in the United Kingdom continued until 1920 when woodland occupied only 5% of the U.K. land surface. Since then, a large reforestation program has increased total woodland cover to 10%. Although conifers account for most of this increase, the proportion of broadleaf trees planted each year has increased 10-fold since 1985 and now exceeds conifer planting, so that a third of the 2.4×10^6 ha of woodland in the United Kingdom includes broadleaf trees. This change has been largely driven by the provision of generous financial incentives for broadleaf planting, not only because of its aesthetic value but because of a wide appreciation of the value of such trees to many faunal groups (Osborne and Krebs 1981; Kennedy and Southwood 1984). Similar grant aid has been available for the restoration of ponds, 75% of which have been lost in Britain since 1880 (Alstrop and Biggs 1993). Such schemes have the added value of providing good foraging habitat for bats and countering the trend toward habitat fragmentation. Unfortunately, the rate of loss of hedgerows, which are important landscape elements for bats, still exceeds the rate of their replacement (Barr et al. 1994).

The widespread use of agricultural pesticides and remedial timber treatments with organochlorides may have had major deleterious effects on bat populations in the 1950s and 1960s. Restrictions on the most persistent and toxic chemicals may have resulted in the reported increase in numbers of some bat species, similar to the recovery that has occurred in raptors. Pollution is also being reduced in many rivers, although eutrophication may have a beneficial effect for those species that feed over water.

The principal way in which the value of conifer plantations is being enhanced for bats is by the provision of roosting boxes, and viable and self-sustaining populations of bats become established in such boxes when large numbers are provided in relatively small areas. Further research is needed on the population dynamics of bats inhabiting boxes made of sawdust and cement, which are more attractive to them than wooden ones.

Recent studies have detailed with some statistical rigor those physical features associated with buildings used by bats as roosts. Several vespertilionid and rhinolophid species forage within a relatively short distance of such roosts so that the conservation and management of adjacent woodland is particularly important. One of the most valuable farmland habitats for bats is unimproved pasture, where bats forage on insects with subterranean larval stages, such as chafers, or on dung beetles. Farming practices such as the use of antihelminthics that reduce the insect fauna associated with dung or which reduce the availability of dung such as zerograzing (transporting cut grass to cattle) may adversely effect several bat species. Bats generally forage on a variety of insect groups, and we need to know whether the loss of certain taxa can be accommodated by a shift toward others. In some cases, however, particularly for highly endangered species such as horseshoe bats, financial incentives are needed to stimulate management of farmlands near roosts.

Bats are among the most synanthropic mammals, relying on human-made structures for roosting in both rural and urban areas. They feed opportunistically on concentrations of insects, such as those around white street lights and domestic refuse dumps. The use of white street lights may have some conservation value because the energy intake of such bats is enhanced. Unfortunately, those species exploiting this feeding opportunity are less threatened than those that do not.

In many European countries, bats are the most important contributors to mammalian biodiversity, and the use of bat detectors and application of the powerful techniques of molecular genetics continues to add bat species to national faunal lists. This is potentially important because at the 1992 Rio Convention on Biological Diversity many participating politicians pledged to halt the worldwide loss of animal and plant species and genetic resources.

The first conference of the parties to the European Bats Agreement held in the United Kingdom in 1995 agreed to a wide-ranging conservation and management plan, which involves survey and monitoring of populations, the identification and protection of important roosts and foraging habitats, and the promotion of public awareness about bats. The majority of these goals will be achieved by a strong voluntary sector willing and able to undertake surveys and monitoring, underpinned by the work of ecologists in identifying and characterizing important roosts and foraging habitats.

Acknowledgments

I am grateful to the following colleagues who commented on various drafts of this chapter: H. Baagøe, J. Burton, C. M. C. Catto, M. B. Fenton, J. Gaisler, A. M. Hutson, G. Jones, S. Mickleburgh, J. Rydell, G. Sargent, H. W. Schofield, and A. L. Walsh. I thank L. Young for her valiant efforts at the keyboard.

Literature Cited

Ahlén, I. 1980–1981. Field identification of bats and survey methods based on sounds. Myotis 18–19:128–136.

Ahlén, I. 1981. Identification of Scandinavian bats by their sounds. Report 6, Department of Wildlife Ecology, The Swedish University of Agricultural Sciences, Uppsala.

Alstrop, C., and J. Biggs. 1993. Proceedings of the Conference on Protecting Britain's Ponds. Wildfowl and Wetland Trust and Pond Action, Slimbridge, U.K.

Anderson, M. B., and P. A. Racey. 1991. Feeding behaviour in captive brown long-eared bats *Plecotus auritus*. Animal Behaviour 42:489–493.

Anderson, M. B., and P. A. Racey. 1993. Discrimination between fluttering and non-fluttering moths by brown long-eared bats *Plecotus auritus*. Animal Behaviour 46:1151–1155.

Baagøe, H. J. 1973. Taxonomy of two sibling species of bats in Scandinavia: *Myotis mystacinus* and *Myotis brandti* (Chiroptera). Videnskabelige Meddelelser fra Dansk Naturhstorisk Forening 136:191–216.

Baagøe, H. J. 1986. Summer occurrence of *Vespertilio murinus* (Linné 1758) and *Eptesicus serotinus* (Schreber 1780) (Chiroptera, Mammalia) on Zealand, Denmark, based on records of roosts and registration with bat detectors. Annalen de Naturhistorischen Museums in Wien Serie B 88–89:281–291.

Barlow, K. E., and G. Jones. 1996. *Pipistrellus nathusii* (Chiroptera: Vespertilionidae) in Britain in the mating season. Journal of Zoology (London) 24:767–773.

Barr, C., M. Gillespie, and D. Howard. 1994. Hedgerow Survey 1993. Institute of Terrestrial Ecology, Grange-over-Sands, U.K.

Barratt, E. M., M. W. Bruford, T. Burland, G. Jones, P. A. Racey, and R. K. Wayne. 1995. Characterisation of mitochondrial DNA variability within the microchiropteran genus *Pipistrellus*: Approaches and applications. Symposia of the Zoological Society of London 67:291–307.

Barta, Z., J. Cerveny, J. Gaisler, P. Hanák, I. Horacek, L. Hurka, P. Miles, M. Nevrly, Z. Rumler, J. Sklenár, and J. Zalman. 1981. Results of winter census of bats in Czechoslovakia 1969–1979. Sbornik Oresniho Muzea v Moste 3:71–116.

Benzal, J. 1991. Population dynamics of the brown long-eared bat (*Plecotus auritus*) occupying bird boxes in a pine forest plantation in central Spain. Netherlands Journal of Zoology 41:241–249.

Blake, D., A. M. Hutson, P. A. Racey, J. Rydell, and J. R. Speakman. 1994. Use of lamplit roads by foraging bats in southern England. Journal of Zoology (London) 234:453–462.

Boyd, I. L., and R. E. Stebbings. 1989. Population changes of brown long-eared bats (*Plecotus auritus*) in bat boxes at Thetford Forest. Journal of Applied Ecology 26:101–112.

Bright, P. 1993. Habitat fragmentation and predictions for British mammals. Mammal Review 230:101–111.

Bunce, R. G. H., and D. C. Howard. 1991. The distribution and aggregation of ITE land classes. Report to the Department of the Environment. The Institute of Terrestrial Ecology, Grange-over-Sands, U.K.

Bunce, R. G. H., C. J. Barr, and H. A. Whitaker. 1981a. Land Classes in Great Britain: Preliminary Descriptions for Users of the Merlewood Method of Land Classification. The Institute of Terrestrial Ecology, Grange-over-Sands, U.K.

Bunce, R. G. H., C. J. Barr, and H. A. Whitaker. 1981b. An integrated system of land classification. *In* The Institute of Terrestrial Ecology Report for 1980, pp. 29–33. The Institute of Terrestrial Ecology, Cambridge, U.K.

Bunce, R. G. H., C. J. Barr, and H. A. Whitaker. 1983. A stratified system for ecological sampling. *In* Ecological Mapping from Ground, Air, and Space, R. M. Fuller, ed., pp. 39–46. The Institute of Terrestrial Ecology, Abbots Ripton, U.K.

Byers, C. R., R. K. Steinhorst, and P. R. Krausman. 1984. Clarification of the technique for analysis of utilization–availability data. Journal of Wildlife Management 48:1050–1053.

Catto, C. 1994. Bat Detector Manual. Bat Conservation Trust, London.

Catto, C. M. C., A. M. Hutson, and P. A. Racey. 1994. The diet of *Eptesicus serotinus* in southern England. Folia Zoologica 43:307–314.

Catto, C. M. C., A. M. Hutson, P. A. Racey, and P. J. Stephenson. 1995. Foraging behaviour and habitat use of the serotine bat (*Eptesicus serotinus*) in southern England. Journal of Zoology (London) 238:623–633.

Cerveny, J., and P. Bürger. 1990. Changes in bat population sizes in the Sumava Mts (south-west Bohemia). Folia Zoologica 39:213–226.

Corbet, G. B., and S. Harris. 1991. The Handbook of British Mammals. Blackwell, Oxford, U.K.

Cranbrook, the Earl of, and H. G. Barrett. 1965. Observations on noctule bats *Nyctalus noctula* captured while feeding. Proceedings of the Zoological Society of London 144:1–24.

Daan, S. 1980. Long-term changes in bat populations in the Netherlands. Lutra 22:95–105.

Dawson, D. G. 1994. Are habitat corridors conduits for animals and plants in a fragmented landscape? A review of the scientific evidence. English Nature Research Report 74:1–89.

de Jong, J. 1994. Distribution patterns and habitat use by bats in relation to landscape heterogeneity, and consequences for conservation. Ph.D. dissertation, Swedish University of Agricultural Sciences, Uppsala.

de Jong, J., and I. Ahlén. 1991. Factors affecting the distribution patterns of bats in Uppland, central Sweden. Holarctic Ecology 14:92–96.

Duvergé, P. L., and G. Jones. 1994. Greater horseshoe bats: Activity, foraging behaviour, and habitat use. British Wildlife 6:69–77.

Entwistle, A. 1994. The roost ecology of the brown long-eared bat *Plecotus auritus*. Ph.D. thesis, University of Aberdeen, U.K.

Entwistle, A. C., P. A. Racey, and J. R. Speakman. 1996. Habitat exploitation by a gleaning bat, *Plecotus auritus*. Philosophical Transactions of the Royal Society of London B 351:921–931.

Entwistle, A. C., P. A. Racey, and J. R. Speakman. 1997. Roost selection by the brown long-eared bat (*Plecotus auritus*). Journal of Applied Ecology 34:399–408.

Frankie, G. W., and L. E. Ehler. 1978. Ecology of insects in urban environments. Annual Review of Entomology 23:367–387.

Gaisler, J. 1963. The ecology of the lesser horseshoe bat, *Rhinolophus hipposideros hipposideros* (Bechstein, 1800), in Czechoslovakia. Part I. Vestník Ceskoslovenské Spolecnosti Zoologické 27:211–233.

Gaisler, J. 1975. A quantitative study of some populations of bats in Czechoslovakia (Mammalia: Chiroptera). Acta Scientiarum Naturalium Academiae Scientiarum Bohemoslovacae Brno 9:1–44.

Gaisler, J. 1978. Tentative estimates of the population densities of some European bats. *In* Proceedings of the Fourth International Bat Research Conference, R. J. Olembo, J. B. Castelino, and F. A. Mutere, eds., pp. 283–285. Kenya Literature Bureau, Nairobi.

Gaisler, J. 1979. Results of bat census in a town (Mammalia: Chiroptera). Vestník Ceskoslovenské Spolecnosti Zoologické 43:7–21.

Gaisler, J., and J. Kolibac. 1992. Summer occurrence of bats in agrocoenoses. Folia Zoologica 41:19–27.

Geggie, J. F., and M. B. Fenton. 1987. A comparison of foraging by *Eptesicus fuscus* (Chiroptera: Vespertilionidae) in urban and rural environments. Canadian Journal of Zoology 63:263–267.

Gerell, R., and K. G. Lundberg. 1993. Decline of a bat, *Pipistrellus pipistrellus*, population in an industrialised area in southern Sweden. Biological Conservation 65:153–157.

Haffner, M., and H. P. Stutz. 1985–1986. Abundance of *Pipistrellus pipistrellus* and *Pipistrellus kuhlii* foraging at street lamps. Myotis 23–24:167–171.

Hanak, V. 1971. *Myotis brandtii* Eversmann, 1845 (Vespertilionidae, Chiroptera) uider Tschechoslowakei. Vestník Ceskoslovenské Spolecnosti Zoologické Praha 35:175–185.

Hill, M. O., P. D. Carey, B. C. Eversham, H. R. Arnold, C. D. Preston, M. G. Telfer, N. J. Brown, N. Veitch, R. C. Welch, G. W. Elmes, and A. Buse. 1993. The role of corridors, stepping stones, and islands for species conservation in a changing climate. English Nature Research Report 75:1–112.

Horacek, I. 1984. Remarks on the causality of population decline in European bats. Myotis 21–22:138–146.

Hutson, A. M. 1991. The serotine. *In* The Handbook of British Mammals, G. B. Corbet and S. Harris, eds., pp. 112–116. Blackwell, London.

Hutson, A. M. 1993. Action Plan for the Conservation of Bats in the United Kingdom. The Bat Conservation Trust, London.

Jones, G., and M. Morton. 1992. Radio-tracking studies on habitat-use by greater horseshoe bats *(Rhinolophus ferrumequinum)*. *In* Wildlife Telemetry, Remote Monitoring, and Tracking of Animals, I. G. Priede and S. M. Swift, eds., pp. 521–537. Ellis Horwood, Chichester, U.K.

Jones, G., and J. M. V. Rayner. 1988. Flight performance, foraging tactics, and echolocation in free-living Daubenton's bats *Myotis daubentonii* (Chiroptera: Vespertilionidae). Journal of Zoology (London) 215:113–132.

Jones, G., and S. M. van Parijs. 1993. Bimodal echolocation in pipistrelle bats: Are cryptic species present? Proceedings of the Royal Society of London B 251:119–125.

Jones, G., P. L. Duvergé, and R. D. Ransome. 1995. Conservation biology of an endangered species: Field studies of greater horseshoe bats. Symposia of the Zoological Society of London 67:309–324.

Jüdes, U. 1989. Analysis of the distribution of flying bats along line-transects. *In* European Bat Research, V. Hanák, I. Horácek, and J. Gaisler, eds., pp 311–318. Charles University Press, Prague.

Kalko, E., and H.-U. Schnitzler. 1989. The echolocation and hunting behavior of Daubenton's bat, *Myotis daubentoni*. Behavioral Ecology and Sociobiology 24:225–238.

Kapteyn, K. 1991. Intraspecific variation and echolocation. *In* Proceedings of the First European Bat Detector Workshop, K. Kapteyn, ed., pp. 45–57. Netherlands Bat Research Foundation, Amsterdam.

Kapteyn, K., and P. H. C. Lina., 1994. Eerste Vondst van een Kraamkolonie van Nathusius' dwergvleermuis *Pipistrellus nathusii* in Nederland. Lutra 37:106–109.

Kennedy, C. E. J., and T. R. E. Southwood. 1984. The numbers of species of insects associated with British trees: A reanalysis. Journal of Animal Ecology 53:455–478.

Kokurewicz, T. 1995. Increased population of Daubenton's bat *Myotis daubentonii* (Kuhl 1819) (Chiroptera: Vespertilionidae) in Poland. Myotis 32–33:155–161.

Kronwitter, F. 1988. Population structure, habitat use, and activity patterns of the noctule bat, *Nyctalus noctula* Schreb., 1974 (Chiroptera: Vespertilionidae) revealed by radiotracking. Myotis 26:23–85.

Krull, D., A. Schumm, W. Metzner, and G. Neuweiler. 1991. Foraging areas and foraging behavior in the notch-eared bat, *Myotis emarginatus* (Vespertilionidae). Behavioral Ecology and Sociobiology 28:247–253.

Kurta, A., and J. A. Terramino. 1992. Bat community structure in an urban park. Ecography 15:257–261.

Limpens, H. J. G. A. 1991. Dutch survey project. *In* Proceedings of the First European Bat Detector Workshop, K. Kapteyn, ed., pp. 105–111. Netherlands Bat Research Foundation, Amsterdam.

Limpens, H. J. G. A., and K. Kapteyn. 1991. Bats, their behavior, and linear landscape elements. Myotis 29:39–48.

Limpens, H. J. G. A., W. Helmer, A. Van Winden, and K. Mostert. 1989. Vleermuizen (Chiroptera) en lintvormige landschapselementen. Lutra 32:1–20.

Limpens, H., K. Mostert, and W. Bongers. 1997. Atlas van de Nederlandse vleermuizen. Onderzoek naar verspreiding en ecologie, p. 260. Koninklijke Nederlandse Natuurhistorische, Vereniging, Utrecht.

Lundberg, K., and R. Gerell. 1986. Territorial advertisement and mate attraction in the bat *Pipistrellus pipistrellus*. Ethology 71: 115–124.

Magurran, A. E. 1988. Ecological Diversity and Its Measurement. Croom Helm, London.

Mayle, B. A. 1990. A biological basis for bat conservation in British woodlands: A review. Mammal Review 20:159–195.

McAney, C. M., and J. S. Fairley. 1989. Analysis of the diet of the lesser horseshoe bat *Rhinolophus hipposideros* in the west of Ireland. Journal of Zoology (London) 217:491–498.

Mickleburgh, S. 1987. Distribution and status of bats in the London area. London Naturalist 66:41–91.

Mitchell-Jones, A. J. 1995. The status and conservation of horseshoe bats in Britain. Myotis 32–33:271–284.

Nemeth, A., and O. von Helversen. 1993. The phylogeny of the *Myotis mystacinus* group: A molecular approach. *In* Sixth European Bat Research Symposium, Evora, Portugal, August 22–27, 1993 (abstract).

Neu, C. W., C. R. Byers, and J. M. Peek. 1994. A technique for analysis of utilization–availability data. Journal of Wildlife Management 38:541–545.

Neuweiler, G. 1989. Foraging ecology and audition in echolocating bats. Trends in Ecology and Evolution 4:160–166.

Nyholm, E. S. 1965. Zur Ökologie von *Myotis mystacinus* (Leisl) und *Myotis daubentoni* (Leisl) (Chiroptera). Annales Zoologici Fennici 2:77–123.

Osborne, L., and J. Krebs. 1981. Replanting after Dutch elm disease. New Scientist 29:212–215.

Racey, P. A. 1998. The importance of the riparian environment as a habitat for British bats. *In* Behaviour and Ecology of Riparian Mammals, N. Dunstone and M. L. Gorman, eds., Symposia of the Zoological Society of London, No. 71, pp. 69–91.

Racey, P. A., and S. M. Swift. 1985. Feeding ecology of *Pipistrellus pipistrellus* (Chiroptera: Vespertilionidae) during pregnancy and lactation. 1. Foraging behaviour. Journal of Animal Ecology 54:205–215.

Rachwald, A. 1992. Habitat preference and the activity of the noctule bat *Nyctalus noctula* in the Bialowieza Primeval Forest. Acta Theriologica 37:413–422.

Ransome, R. D. 1968. The distribution of the greater horseshoe bat, *Rhinolophus ferrum-equinum*, during hibernation, in relation to environmental factors. Journal of Zoology (London) 154: 77–112.

Ransome, R. D. 1971. The effect of ambient temperature on the arousal frequency of the hibernating greater horseshoe bat *Rhinolophus* in relation to site selection and the hibernation state. Journal of Zoology (London) 164:353–371.

Ransome, R. D. 1990. The Natural History of Hibernating Bats. Christopher Helm, London.

Rydell, J. 1986. Foraging and diet of the northern bat *Eptesicus nilssoni* in Sweden. Holarctic Ecology 14:203–207.

Rydell, J. 1992. Exploitation of insects around street lamps by bats in Sweden. Functional Ecology 6:744–750.

Rydell, J., and P. A. Racey. 1995. Street lamps and the feeding ecology of insectivorous bats. Symposia of the Zoological Society of London 67:291–307.

Rydell, J., and S. M. Swift. 1995. Observations of Nathusius' pipistrelle *Pipistrellus nathusii* in northern Scotland. Scottish Bats 3:6–7.

Sargent, G. 1995. The Bats in Churches Project. The Bat Conservation Trust, London.

Schlapp, G. 1990. Populationsdichte und Habitatansprüche der Bechstein Fledermaus *Myotis bechsteini* (Kuhl, 1818) in Steigerwald (Forstamt Ebrach). Myotis 28:39–57.

Schober, W., and E. Grimmberger. 1993. Bats of Britain and Europe. Hamlyn, London.

Schofield, H. W. 1996. The ecology and conservation biology of *Rhinolophus hipposideros*, the lesser horseshoe bat. Ph.D. thesis, University of Aberdeen, Scotland.

Speakman, J. R., P. A. Racey, C. M. C. Catto, P. I. Webb, S. M. Swift, and A. M. Burnett. 1991a. Minimum summer populations and densities of bats in N.E. Scotland, near the northern borders of their distribution. Journal of Zoology (London) 225:327–345.

Speakman, J. R., P. A. Racey, A. M. Hutson, P. I. Webb, and A. M. Burnett. 1991b. Status of Nathusius pipistrelle *(Pipistrellus nathusii)* in Britain. Journal of Zoology (London) 225:685–690.

Spellerberg, I. F., and M. Gaywood. 1993. Linear features: Linear habitats and wildlife corridors. English Nature Research Report 60:1–74.

Stebbings, R. E. 1967. Identification and distribution of bats of the genus *Plecotus*. Journal of Zoology (London) 150:53–75.

Stebbings, R. E. 1970a. A comparative study of *Plecotus auritus* and *Plecotus austriacus* (Chiroptera: Vespertilionidae) inhabiting the same roosts. Bijdragen tot de Dierkunde 40:91–94.

Stebbings, R.E. 1970b. A new bat to Britain, *Pipistrellus nathusii*, with notes on its identification and distribution in Europe. Journal of Zoology (London) 161:282–286.

Stebbings, R. E. 1988. The Conservation of European Bats. Christopher Helm, London.

Stebbings, R. E., and H. R. Arnold. 1987. Assessment of trends in size and structure of a colony of the greater horseshoe bat. Symposia of the Zoological Society of London 58:7–24.

Strelkov, P. P. 1969. Migratory and stationary bats (Chiroptera) of the European part of the Soviet Union. Acta Zoologica Cracoviensia 14:393–440.

Strong, L. 1992. Avermectins: A review of their impact on insects in cattle dung. Bulletin of Entomological Research 82:265–274.

Stutz, H. P., and M. Haffner. 1984. Arealverlust und Bestandsrückgang der Kleinen Hufeisennase *Rhinolophus hipposideros* (Bechstein, 1800) (Mammalia: Chiroptera) in der Schweiz. Jahresbericht Naturforschende Gesellschaft Fraubuendens 101:169–178.

Swift, S. M., and P. A. Racey. 1983. Resource partitioning in two species of vespertilionid bats (Chiroptera) occupying the same roost. Journal of Zoology (London) 200:249–259.

Taake, K-H., and H. Hildenhagen. 1989. Nine years' inspections of different artificial roosts for forest-dwelling bats in northern Westfalia: Some results. *In* European Bat Research 1987, V. Hanák, I. Horácek, and J. Gaisler, eds., pp. 487–493. Charles University Press, Prague.

Taylor, L. R., R. A. French, and L. P. Woiwood. 1978. The Rothamsted insect survey and the urbanization of land in Great Britain. *In* Perspectives in Urban Entomology, G. W. Frankie and C. S. Koehler, eds., pp. 31–65. Academic Press, New York.

Topal, G. 1958. Morphological studies on the penis of bats in the Carpatian Basin. Annals of the National Natural History Museum of Hungary 50 (n.s. 9): 331–342.

von Helversen, O., M. Esche, F. Kretzschmar, and M. Boschert. 1987. Die Fledermäuse Südbaden. Mitteilungen des Badischen Landesvereins für Naturkunde und Naturschutz E V Freiburg im Breisgau 14:409–475.

Voûte, A. M., J. W. Sluiter, and P. F. van Heerdt. 1980. De vleermuizen stand in einige zuidlimburgse groeven sedert 1942. Lutra 22:18–34.

Walsh, A. L., and S. Harris. 1996a. Foraging habitat preferences of vespertilionid bats in Britain. Journal of Applied Ecology 33: 508–518.

Walsh, A. L., and S. Harris. 1996b. Factors determining the abundance of vespertilionid bats in Britain: Geographic, land class, and local habitat relationships. Journal of Applied Ecology 33: 519–529.

Walsh, A. L., S. Harris, and A. M. Hutson. 1995. Abundance and habitat selection of foraging vespertilionid bats in Britain: A landscape-scale approach. Symposia of the Zoological Society of London 67:325–344.

Weinrich, J. A., and J. H. Oude Voshaar. 1992. Population trends of bats hibernating in Marl Caves in the Netherlands (1943–1987). Myotis 30:75–84.

18
Impacts of Ignorance and Human and Elephant Populations on the Conservation of Bats in African Woodlands

M. BROCK FENTON AND I. L. RAUTENBACH

The Ethiopian zoogeographic region (sub-Saharan Africa) covers an area of 23,426,000 km², of which less than 10% historically was rainforest (Keast 1972). Although the region is best known for its large mammals, the Chiroptera include more species than any other order of mammals. Approximately 174 species of bats live mainly in the Ethiopian region, showing a high level of endemism (24 of 41 genera) with the ranges of only about 6 species (3%) extending beyond this area (Hayman and Hill 1971; Nowak 1994). Approximately 24 other species whose ranges extend into Africa occur mainly outside the Ethiopian region, making a total of about 198 species of bats known from Africa (Hayman and Hill 1971).

The purpose of this chapter is to consider the conservation challenges posed by bats in African woodlands. We selected these habitats for two reasons: First, our collective experience is focused there, and second, the woodlands dominate the zoogeographic region (Figure 18.1) and often overlap with areas of high human population density. In this chapter, we use the nomenclature for bats presented in Nowak (1994).

The General Situation

The human population in sub-Saharan Africa is projected to increase five-fold in a little more than a hundred years, from 0.6 billion in 1990 to 2.8 billion by 2100 (Bongaarts 1994). In 1989, the human population density in Africa ranged from 2 to 265 (mean ± SD, 41.8 ± 51.5) persons per square kilometer (Stuart et al. 1990). The potential impact of the expanding human population for the conservation of African biota cannot be overstated. For example, since 1900, the human population in Zimbabwe has increased from 0.5 to more than 10 million (Cumming 1991), and in the past 40 years the rate of land clearance in the Sebungwe District of Zimbabwe has been about 4% per annum (D. H. M. Cumming, personal communication). Changes of this magnitude are expected to have profound effects on habitats and thus pose a major threat to the survival of the African fauna (Martin and de Meulenaer 1988). The direct and indirect effects of an expanding human population are expected to be the principal factor causing the extinction of many organisms, including bats.

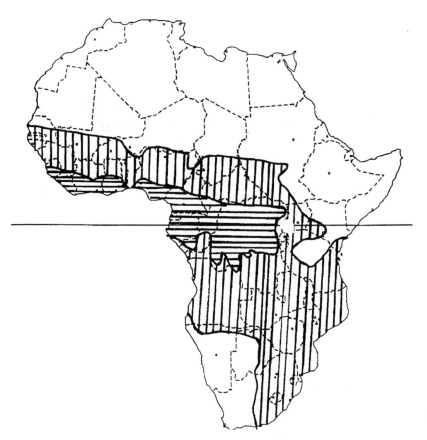

Figure 18.1. The distribution of woodlands (vertical hatching) and rainforests (horizontal hatching) in sub-Saharan Africa. (After Smithers 1983.)

Table 18.1

The Orders of Insects in the Diets of Some Common and Widespread African Bats

Species	N_b	N_p	Main food item	Diptera	Source
Hipposideros caffer	18	?	Lepidoptera (72%)	Yes	Fenton 1985
	90[a]	0	Lepidoptera (70%)	<5%	Whitaker and Black 1976
Nycteris thebaica	33[a]	0	Coleoptera (45%)	<5%	Whitaker and Black 1976
Eptesicus capensis	29	115	Coleoptera (65.3%)	No	Fenton et al. 1998
	21	88	Coleoptera (72.6%)	No	Fenton et al. 1998
Nycticeius schlieffeni	11	55	Coleoptera (47.7%)	Yes	Fenton et al. 1998
Scotophilus borbonicus	40	170	Coleoptera (80.5%)	No	Fenton et al. 1998
	32	130	Coleoptera (76.6%)	No	Fenton et al. 1998
Scotophilus viridis	14	52	Coleoptera (80.0%)	No	Fenton et al. 1998
	12	51	Coleoptera (74.6%)	No	Fenton et al. 1998
Scotophilus leucogaster	48	0	Hemiptera (47%)	<1%	Barclay 1985
Miniopterus schreibersii	10	?	Coleoptera (60%)	No	Fenton 1985
	23[a]	0	Isoptera (34%)	<2%	Whitaker and Black 1976
Tadarida condylura	5	25	Coleoptera (74.0%)	No	Fenton et al. 1998
Tadarida pumila	7	31	Coleoptera (52.9%)	No	Fenton et al. 1998

Note: N_b is the number of bats in the sample; N_p is the number of pellets analyzed.

[a]Stomach contents, rather than pellets, were analyzed.

Bats offer some special conservation concerns in Africa and elsewhere. For example, a lack of interest in and concern about African bats, combined with the public perception that they may be dangerous because of their role in the epidemiology of ghastly diseases (e.g., those caused by filoviruses; Preston 1994) does not make bats strong candidates for attracting public support for the conservation of African woodlands.

In promoting the conservation of bats in areas where malaria poses an important threat to human health (Collins and Besansky 1994; Godal 1994), it is tempting to identify bats as valuable because of their role as consumers of mosquitoes (Tuttle 1988). However, the available evidence, particularly for bat species that are widespread and occur in large populations in Africa, indicates that Coleoptera, Lepidoptera, and Hemiptera (beetles, moths, and bugs) are the most common food items of bats (Table 18.1). These data suggest that the bats rarely eat flies (Diptera), let alone mosquitoes. Ellis (1995) found that flies were rarely part of the diets of seven species of bats that she studied in the northern Transvaal (South Africa).

Diseases and parasites, however, can be positive conservation factors. For example, through their impact on domestic animals and humans, parasites such as some species of *Trypanosoma* and their insect vectors *(Glossina morsitans)* are directly responsible for preserving large tracts of woodland habitat in Africa by making it unsuitable for maintaining cattle and horses (McKelvey 1973).

Roosts and Food

Plans to conserve bats must consider two vital resources, roosts and food, although the role of either as a limiting factor usually remains unclear (Findley 1993). Although many authors have proposed that populations of bats are limited by the availability of roosts and food, there is little direct evidence to support this view (Brosset 1966; Findley 1993).

How do bats use day roosts? Radiotracking studies have revealed at least four basic patterns of day roost use by bats that are relevant to the assessment of the impact of roost availability (Lewis 1995). The impact of roosts on bats depends on the type of roosts, the fidelity of bats to specific roosts, and the propensity of bats to exploit new roosts.

Like bats elsewhere (Kunz 1982), some African species consistently use the same roosts (buildings) day after day (e.g., *Mops midas*) while others change roosts (tree hollows) almost daily (e.g., *Scotophilus borbonicus;* Fenton and Rautenbach 1986). Still others which roost in foliage use different sites in the same general area day after day (e.g., *Epomophorus wahlbergi;* Fenton et al. 1985). A cavity-roosting species, *Nycteris grandis,* abandons preferred roosts after disturbance, returning there several days later (Fenton et al. 1993). Knowledge of patterns of roost use is a prerequisite to appreciating the impact of roost availability on bat populations. For tree-roosting species, the destruction of woodlands has a significant impact on distribution and survival.

As elsewhere in the world, the conservation of bats must involve the protection of cave roosts. In South Africa, for example, 24 of 76 species roost in caves, perhaps because they require specific conditions of temperature and humidity. The importance of particular roosts for some species is exemplified by one cave in the De Hoop Nature Reserve in South Africa. This cave harbors 300,000 bats that are estimated to consume annually more than 100 tons of insects during the 8 months of summer (McDonald et al. 1990). The community in the cave is dominated by *Miniopterus schreibersi* but includes 4 other species (McDonald et al. 1990). In other locations in South Africa, for example, around Pretoria, caves that harbored bats 50 years ago now have none, apparently reflecting the impact of a variety of human disturbances.

What impact does food availability have on bat populations? The appearance of ripe fruits influences the patterns of movement and reproduction in African pteropodids (Thomas 1983; Thomas and Marshall 1984). In the northern part of Kruger National Park, the seasonal activity of insectivorous bats is strongly influenced by prey availability. Nonflying insects are more consistently available than flying ones through the dry season (Rautenbach et al. 1988). In some areas, permanently flowing rivers such as the Zambezi may be crucial to the survival of some species by providing access to food throughout the year (e.g., *Nycteris grandis;* Fenton et al. 1993), but other species are not so tied to water (e.g., *Pipistrellus zuluensis* and *Sauromys petrophilus;* Roer 1970). Still other species, such as *Lavia frons,* use the same roosts and habitat from season to season, adjusting their foraging patterns according to seasonal patterns of prey availability (Vaughan and Vaughan 1986). At a range of sites in Kruger National Park in South Africa, the activity of echolocating bats hunting flying insects is significantly associated with insect activity (Rautenbach et al. 1995), suggesting that food availability affects bat activity and habitat use.

The availability of food and roosts influences the distribution and abundance of bats, but it appears that conservation efforts may be more appropriately focused on roosts which, for some species, can be clearly associated with specific structures.

Bats and People

As usual, human activities produce both lethal risks and opportunities for bats. Humans eat some larger species in some countries (Brosset 1966; Mickleburgh et al. 1992). Unlike the situation on some South Pacific Islands (Craig et al. 1994), human predation pressure currently appears to pose a minimal threat to African bats (Mickleburgh et al. 1992). Changes in the availability of suitable habitat, however, combined with increased human populations could change this situation dramatically (see Rainey, Chapter 23, this volume). The importance of habitat destruction may be reflected by the situation on offshore islands. In the Seychelles, for example, a larger species (*Pteropus seychellensis*) appears to be numerous even though subject to human predation, and a smaller one (*Coleura seychellensis*) is endangered (Racey 1979; Mickleburgh et al. 1992).

Many species of African bats exploit human changes in the environment. The most obvious examples are species that roost in tree hollows and have occupied artificial structures such as mines, buildings, and culverts. Some crevice-roosting species also roost in bridges and buildings. In Africa, we know of no data concerning the relative incidence of bats roosting in artificial versus natural structures. With decreasing availability of natural roosts, often arising from the destruction of woodland and forest, more and more bat populations may come to depend on artificial structures. In this context, agricultural practices that leave larger trees should have less impact on roost availability for bats than practices requiring extensive clear-cutting of woodlands. Forest preservation, with its implications for conservation in general, remains a prime concern (Myers 1995), integrally associated with the conservation of soils (Pimentel et al. 1995). Huston (1993) noted that conserving biodiversity does not conflict with agriculture because habitats with high biodiversity are often relatively infertile. In Africa this means that Miombo woodlands, for example, do not offer good agricultural potential and should be left undeveloped.

As elsewhere (e.g., Rydell and Racey 1995), some African insectivorous bats often feed in concentrations of insects around lights (Kingdon 1974; Ellis 1995). Insect density near lights is often statistically related to bat activity as reflected by echolocation calls. Ellis (1995) found behavioral evidence that some bats responded initially to the lights rather than to the insects.

In many parts of Africa the risk to humans and livestock posed by insect-borne diseases (e.g., malaria [Godal 1994] or sleeping sickness [McKelvey 1973]) means that pesticides such as DDT are still being used. For example, until recently in Zimbabwe, DDT was used to control tsetse flies, and by comparing sprayed and unsprayed areas McWilliam (1994) showed that the spraying increased mortality in some species. Some bat species were more at risk from DDT sprayed in tsetse control operations than others, with roosting habits and foraging behavior affecting their vul-

nerability. In sprayed areas, bats such as *Nycteris thebaica* that roost in large tree hollows and glean prey from the ground were more at risk than those such as *Scotophilus* species which roost in small hollows and take airborne prey (McWilliam 1994).

The State of Our Knowledge

One overriding problem facing those who would develop conservation, preservation, or recovery plans for African bats is our lack of knowledge about most species. At a basic level, taxonomic problems complicate the issue to some extent. Even in the common and widespread genera such as *Eptesicus* and *Pipistrellus,* there are questions about the numbers of species and their distribution. For example, in 1994, five species listed as *Eptesicus* by Hayman and Hill (1971) had been reassigned to *Pipistrellus* (Nowak 1994). If the situation mirrors the one associated with *Pipistrellus pipistrellus* in Europe (Jones and van Parijs 1993), namely the discovery of an apparent cryptic species, then we may see more species added to the list of African bats. This "hidden biodiversity" may complicate the development of plans to conserve the bats of African woodlands.

The 1994 IUCN (International Union for Conservation of Nature and Natural Resources) Red List recognized that lack of data is a major problem for the development of conservation plans. Excluding populations of some species on some offshore islands, we found no evidence that any species of bats in Africa could be classified as "endangered" as defined by the IUCN 1994 standards. Most species of African bats fall squarely in the data-deficient (DD) category.

Changes in the technology available for studying bats have profoundly affected the kind of data that are potentially available about these animals. Radiotracking and portable equipment for monitoring the echolocation calls of bats are particularly relevant.

Migrations

Although some African bats are migratory (Rosevear 1965; Smithers 1983; Happold 1987), we lack detailed knowledge about the generalities, let alone the specifics, of seasonal movements. One exception is provided by Thomas' work (1983) in Ivory Coast, where he documented the migrations and reproductive patterns of several pteropodids and associated these with the seasonal availability of fruit. Elsewhere, bat migrations are indicated by regular seasonal movements of bats ranging from *Eidolon helvum* to *Scotophilus* species and some *Miniopterus* spp. and may reflect the availability of food.

For most species, we have no information about possible

Figure 18.2. The range of *Nycteris grandis,* a species largely confined to the rainforests of sub-Saharan Africa. (After Smithers 1983.)

migratory habits. It is probably true that the smaller ($\leq$5 g) species are less likely to be migratory than larger ones, but the long-distance migrations of hummingbirds (Baker 1980) and small European bats (Racey, Chapter 17, this volume) oblige us to keep an open mind on this topic.

Six Specific Examples

The following six examples reflect the diversity of situations facing individuals concerned about the survival and conservation of bats in African woodlands.

Some General Knowledge and a Well-Documented Distribution: *Nycteris grandis*

For some African species, there are general knowledge and various detailed data but no published studies of the overall conservation problems facing bats. For example, *Nycteris grandis* is the largest species in the genus (forearm length, 55–65 mm; body mass, 25–30 g). While this is primarily a bat of the rainforest, there are two outlying populations in woodlands (Figure 18.2). In Mana Pools National Park in Zimbabwe at the southern edge of its range, *N. grandis* takes a wide range of prey, from large insects to frogs, birds, bats, and fish. There the species roosts in tree hollows,

foliage, disused military bunkers, and the attics of buildings, and rarely are taken outside these situations (Fenton et al. 1993). The data from Mana Pools have been used to predict that the survival of this species at the southern edge of its range is related to the Zambezi River, its water, and the associated array of prey. In Mana Pools National Park and other locations in Zimbabwe, this species has readily accommodated itself to anthropogenic activities. Indeed, it was their use of buildings as day roosts and feeding roosts that drew attention to the range of food this species consumed (Smithers and Wilson 1979).

Our knowledge of *N. grandis* comes from distribution records that provide an indication of the extent of its geographic range, anecdotal observations, and more detailed data from one site in Zimbabwe. This database does not permit an assessment of the size of or trends in its populations, let alone the elements essential to a conservation plan. The plasticity of its diet and roosting habits suggests that this species does not need a conservation plan.

Little General Knowledge and Scattered Distribution Records: *Chaerephon chapini*

Many, if not most, African species are known from some portion of their geographic ranges. For example, *Chaerephon chapini,* one of the smallest of the African molossids (fore-

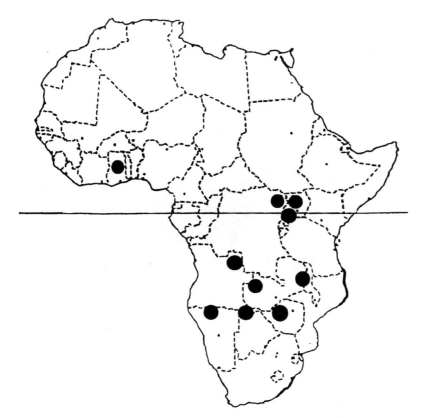

Figure 18.3. The distribution records for *Chaerophon chapini* in sub-Saharan Africa, which, although scattered, do suggest a woodland distribution.

arm length, 34–40 mm; body mass, 8–10 g), was first described from a specimen found in the crop of a bat hawk (*Macheirhamphus alcinus*) shot in Uganda. Subsequently, it has been taken at a progressively wider range of localities suggesting a pattern of distribution typical of woodland bats (Figure 18.3). Virtually nothing is known of its roosting habits, and most specimens appear to have been captured in mist nets set over water. While uncommon in collections, this species may prove to be both widely distributed and locally abundant. In the Sengwa Wildlife Research Area in Zimbabwe, this bat is often the most numerous molossid captured in mist nets.

How widely distributed is *C. chapini*? Is this species restricted to woodland situations by virtue of its roosting or foraging behavior?

Common, Commensal, and Widespread: *Chaerephon pumila*

A few African species are common and widespread, apparently readily adapting to human-induced environmental changes. Their propensity to roost in artificial structures accounts for most of our current knowledge. For example, *Chaerephon pumila* is widespread (Figure 18.4) and is one of the most abundant bats in Africa, commonly establishing large populations in roosts in bridges and the attics of build-

ings. This small, monotocous molossid (forearm length, 36–39 mm; body mass, 10–14 g) is remarkably prolific for a bat. In artificial roosts (buildings) near the equator, females appear to produce five litters annually (McWilliam 1987). Further to the south in Kruger National Park, this species roosts in buildings and bridges, where it apparently produces three litters each year (Van der Merwe et al. 1986).

Buildings and bridges can have a strong impact on the populations of these and other bat species. For example, at the Skukuza townsite in the Kruger National Park, as many as 85% of the buildings are used by *C. pumila*. This raises two important questions. What impact do roosts have on populations of the bats in Africa? Do the large populations of *C. pumila* adversely affect other species of bats and bat communities near their roosts? Both questions are particularly relevant for species such as *C. pumila* that occur in high numbers where buildings provide spacious roosting opportunities.

Unresolved Taxonomic Questions: *Scotophilus* spp.

For several genera of African bats, unresolved taxonomic issues could complicate the development of sound conservation plans even for commonly encountered species. One excellent example is provided by the African species of *Scotophilus* (Schlitter et al. 1980). The name changes reflect the situation. Since 1971, three different names—*Scotophilus*

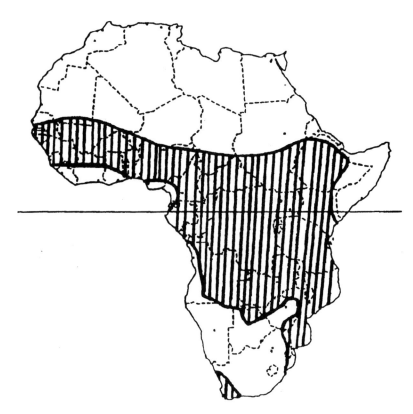

Figure 18.4. The widespread distribution of *Chaerophon pumila* in sub-Saharan Africa, covering areas of woodlands and rainforests. (After Smithers 1983.)

leucogaster, S. viridis, and *S. borbonicus*—have been associated with the smaller species in Zimbabwe and South Africa.

In southern Africa, we typically have encountered two species of *Scotophilus*, the smaller *Scotophilus borbonicus* (forearm length, 43–50 mm) with a whitish venter, and the larger *Scotophilus dingani* (forearm length, 50–60 mm) with a bright yellowish-orange venter. At any site we hoped to catch the much larger *Scotophilus nigrita* (forearm length, 75–80 mm) (=*Scotophilus gigas* in Hayman and Hill 1971), which has been reported once from Zimbabwe (Smithers and Wilson 1979). Elsewhere in African woodlands, the situation for *Scotophilus* is more complex (e.g., in West Africa [Rosevear 1965] or Nigeria [Happold 1987]). In 1971 a conservative view held that there were at least three species of *Scotophilus* in sub-Saharan Africa (Hayman and Hill 1971). By 1994 (Nowak 1994), this number was eight.

Smithers (1983) commented that size rather than color was the most accurate way to distinguish *S. dingani* from *S. borbonicus*. In 1992 and 1993 in Kruger National Park in South Africa, however, we noticed two color variants of the smaller species (both whitish and yellowish venters), presumably *S. borbonicus*, as well as the larger *S. dingani*. This situation also prevailed in the north-central part of Zimbabwe in 1994. Are there three species of *Scotophilus* in southern Africa in addition to the very large *S. nigrita*? If the answer is yes, our view of which of these species is most common may have to be reassessed. Would one conserva-

tion plan equally serve African species in the genus *Scotophilus*? Resolving the specific identities of the bats is important because of differences in roost use. While some southern African *Scotophilus* roost in buildings (Smithers 1983), others use hollow trees (Fenton et al. 1985) or abandoned mines (Wilkinson 1995). Is the selection of roost sites species specific?

Habitat Associations of Bats

What are the habitat associations of African bats? In woodland areas, some species are most often captured in riverine habitats rather than in drier woodlands further from rivers. This situation could reflect the impact of the structural diversity of habitats, as is known from the United States (Humphrey 1975). Apparently the difference in bat diversity and community structure in Africa does not reflect the dramatic change in trophic roles that prevails in Central America (Wilson 1974), because almost all the African woodland species eat either fruit or insects (Smithers 1983, although *N. grandis* also eats small vertebrates including bats, birds, fish, and frogs.

Do captures reflect the actual patterns of habitat preference? Not necessarily, because radiotracking (Fenton and Rautenbach 1986) has shown that species such as *Rhinolophus hildebrandti* roost in woodland and forage in riverine forest, while others (e.g., *Scotophilus* spp.) may roost in woodlands and forage widely, including along rivers. Patterns of habitat use also can be complicated by reproductive state. In North America, for example, patterns of habitat use by foraging bats change during pregnancy, reflecting different aerodynamic considerations (Kalcounis and Brigham 1995).

To explore the habitat preferences of bats, we simultaneously sampled bats at nine pairs of riverine and woodland sites along the long (north–south) axis of Kruger National Park (Rautenbach et al. 1995). The capture data supported the prediction that bats were more abundant in riverine woodlands, but did not indicate that the diversity (α of Fisher et al. 1943) or evenness of the bat communities differed significantly between riverine and dry woodland. The same species dominated the catch at sites in riverine and drier woodland. This pattern also prevailed at sites in northern Zimbabwe where we typically captured *Scotophilus* (two or three species, as noted earlier), *Nycticeinops schlieffeni*, *Eptesicus* (as many as four species), *Pipistrellus* (to four species), and a variety of molossids, including *Mops condylura*, *Chaerephon pumila*, *Tadarida aegyptiaca*, *Mops midas* in more southerly locations, and others (*Chaerephon nigeriae*, *C. chapini*, and *C. bivitatta* further north; Fenton 1985). In the northern part of Kruger National Park (22° S), a number of species are netted occasionally, for example, *Kerivoula* (two species) and *Laephotis botswannae*. Farther north, in the Sengwa Wildlife Research Area (18° S) in Zimbabwe, species of *Kerivoula* are present but rarely caught, while *L. botswannae* is more consistently taken, usually in mopane (*Colophospermum mopane*) woodlands. In both areas and elsewhere, species such as the spectacular *Myotis welwitschii* are captured occasionally, too rarely to provide any clear indication of their habitat preferences.

For a few species, the patterns of distribution are more obviously associated with roost sites. Along the Luvuvhu River in Kruger National Park, for example, the incidence of *Rousettus aegyptiacus* decreases with increasing distance from a known cave roost (Braack 1989). Some of the *Rhinolophus* species are most often caught in harp traps set at the entrances to underground hollows (such as caves, mines, or tunnels used by warthogs [*Phacochoerus aethiopicus*] or ant bears [*Orycteropus afer*]), and this also is true of hipposiderids such as *Hipposideros caffer* or *Cloeotis percivali*. Some rhinolophids and *Hipposideros caffer* and other species, such as *Nycteris thebaica*, also roost in tree hollows.

The Kruger transect data also demonstrated that monitoring bat activity by echolocation calls did not provide the same picture of habitat use by bats as mist netting (Rautenbach et al. 1995). This study and others using echolocation calls (McWilliam 1991) showed that molossids are often active but less often captured.

Conflicting Conservation Priorities

Protection for some species can imperil others, including bats. In parts of Zimbabwe and elsewhere in southern Africa where elephants (*Loxodonta africana*) occur at high density, they have a strong impact on the habitat (Owen-Smith 1989; Spinage 1994; Cumming et al. 1997). In Zimbabwe, the population of elephants has increased from fewer than 4,000 to more than 60,000 since 1900, despite heavy culling in the 1980s (Cumming 1991). The changes from closed-canopy Miombo woodland to grassland and bushland (Starfield et al. 1993) affect bats. When we compared paired elephant-disturbed and undisturbed sites in northern Zimbabwe, we captured significantly more bats in undisturbed woodlands where bats were more active (as indicated by echolocation calls). Although the bat species composition did not differ significantly between disturbed and undisturbed sites, the diets of some species differed significantly. Bats of the genus *Scotophilus* and *Pipistrellus capensis* ate different insects relative to their availability in the two types of sites. We never captured *Epomophorus* spp. or other fruit-eating bats in disturbed woodland sites, although they were captured in undisturbed woodland. The differences in bats between disturbed and undisturbed woodlands were paralleled by differences in the numbers of species of birds, indicating that the effect of habitat destruction was not limited to bats. It is important to realize that the changes in elephant population density in Zimbabwe, and elsewhere, reflect the expanding human population, which means that less habitat is available for wildlife.

Conclusions

The conservation of bats poses the same kinds of problems as the conservation of other organisms. The principal threat to most organisms and habitats in Africa is the impact of the expanding human population. However, so long as our knowledge about the habits, classification, and distribution of bats remains incomplete, it is difficult to use these animals to document the impact of expanding human populations on biodiversity. The picture is further complicated by the fact that bats are often viewed with suspicion by many and further clouded by those species that prosper from human-wrought changes to the environment. The conservation of some animals, notably elephants, can imperil habitats and the other organisms that live there.

What can biologists who work with bats do to further the conservation of African bats? It is obvious that further research can provide much-needed data about almost all aspects of the biology of these animals, from the roosts that they use to the habitats in which they forage. Colleagues making (relatively) short field expeditions can, often in the pursuit of more theoretical questions, provide some basic information. The onus is on those of us engaged in such research to highlight the conservation implications of our findings. The longer term studies needed to increase our knowledge of populations certainly will fall more to local researchers. The fiscal situation of many African countries and the worldwide economic conditions will, however, make it difficult to fund long-term research. Studies of the role of bats in the epidemiology of filoviruses may be the most promising source of financial support for bat research in Africa, although bat biologists are apt to be unenthusiastic about data linking bats to these public scourges.

Acknowledgments

We thank L. Acharya, J. J. Belwood, D. H. M. Cumming, E. R. Fenton, L. Packer, P. A. Racey, and B. Stutchbury for critically examining earlier drafts of this manuscript, and the director of the Department of National Parks and Wildlife in Zimbabwe and the warden of the Kruger National Park for permitting us to study bats in areas under their jurisdiction. M.B.F.'s research on bats has been supported by the Natural Sciences and Engineering Research Council of Canada. I.L.R.'s research has been supported by a Core Programme Rolling Research Support Grant from the Foundation for Research Development (South Africa).

Literature Cited

Baker, R. 1980. The Mystery of Migration. Wiley and Sons, Toronto.

Barclay, R. M. R. 1985. Foraging behavior of the African insectivorous bat, *Scotophilus leucogaster.* Biotropica 17:65–70.

Bongaarts, J. 1994. Population policy options in the developing world. Science 263:771–776.

Braack, L. E. O. 1989. Arthropod inhabitants of a tropical cave "island" environment provisioned by bats. Biological Conservation 48:77–84.

Brosset, A. 1966. La Biologie des Chiroptères. Masson et Cie, Paris.

Collins, F. H., and N. J. Besansky. 1994. Vector biology and the control of malaria in Africa. Science 264:1874–1875.

Craig, P. C., P. Trail, and P. E. Morrell. 1994. The decline of fruit bats in American Samoa due to hurricanes and overhunting. Biological Conservation 69:261–266.

Cumming, D. H. M. 1991. Wildlife products and the market place: A view from southern Africa. *In* Wildlife Production: Conservation and Sustainable Development, L. A. Renecker and R. J. Hudson, eds., pp. 11–25. AFES Miscellaneous Publications 91-6. University of Alaska, Fairbanks.

Cumming, D. H. M., M. B. Fenton, I. L. Rautenbach, J. S. Taylor, G. S. Cumming, M. S. Cumming, J. Dunlop, M. D. Hovorka, D. S. Johnston, M. C. Kalcounis, Z. Mahlanga, and C. V. Portfors.

1997. Elephant impacts on biodiversity of Miombo woodlands in Zimbabwe. South African Journal of Science 93:231–236.

Ellis, S. 1995. Seasonal responses of South African bats to insect densities and lights. Master's thesis, York University, North York, Ontario.

Fenton, M. B. 1985. The feeding behaviour of insectivorous bats: Echolocation, foraging strategies, and resource partitioning. Transvaal Museum Bulletin 21:5–16.

Fenton, M. B., and I. L. Rautenbach. 1986. A comparison of the roosting and foraging behaviour of three species of African insectivorous bats. Canadian Journal of Zoology 64:2860–2867.

Fenton, M. B., R. M. Brigham, A. M. Mills, and I. L. Rautenbach. 1985. The roosting and foraging areas of *Epomphorus wahlbergi* (Pteropodidae) and *Scotophilus viridis* (Vespertilionidae) in Kruger National Park, South Africa. Journal of Mammalogy 66:461–468.

Fenton, M. B., I. L. Rautenbach, D. Chipese, M. B. Cumming, M. K. Musgrave, J. S. Taylor, and T. Volpers. 1993. Variation in foraging behavior, habitat use, and diet of large slit-faced bats *(Nycteris grandis).* Zeitschrift für Saeugetierkunde 58:65–74.

Fenton, M. B., D. H. M. Cumming, I. L. Rautenbach, G. S. Cumming, M. S. Cumming, J. Dunlop, M. D. Hovorka, D. S. Johnston, C. V. Portfors, M. C. Kalcounis, and Z. Mahlanga. 1998. Bats and the loss of tree canopy in African woodlands. Conservation Biology 12:399–407.

Findley, J. S. 1993. Bats: A Community Perspective. Cambridge University Press, Cambridge.

Fisher, R. A., A. S. Corbet, and C. B. Williams. 1943. The relation between the number of species and the number of individuals in a random sample of an animal population. Journal of Animal Ecology 12:42–58.

Godal, T. 1994. Fighting the parasites of poverty: Public research, private industry, and tropical diseases. Science 264:1864–1866.

Happold, D. C. D. 1987. The Mammals of Nigeria. Oxford Science Publications. Clarendon Press, Oxford.

Hayman, R. W., and J. E. Hill. 1971. Chiroptera. *In* The Mammals of Africa: An Identification Manual, J. Meester and H. W. Setzer, eds., pp. 1–73. Smithsonian Institution Press, Washington, D.C.

Humphrey, S. R. 1975. Nursery roosts and community diversity of Nearctic bats. Journal of Mammalogy 56:321–346.

Huston, M. 1993. Biological diversity, soils, and economics. Science 262:1676–1679.

IUCN. 1994. IUCN Red Book Categories. Prepared by the IUCN Species Survival Commission. Approved by the 40th Meeting of the IUCN Council, Gland, Switzerland.

Jones, G., and S. M. van Parijs. 1993. Bimodal echolocation in pipistrelle bats: Are cryptic species present? Proceedings of the Royal Society of London B 251:119–125.

Kalcounis, M. C., and R. M. Brigham. 1995. Intraspecific variation in wing-loading affects habitat use by little brown bats *(Myotis lucifugus).* Canadian Journal of Zoology 73:89–95.

Keast, A. 1972. Introduction: The southern continents as backgrounds for mammalian evolution. *In* Evolution, Mammals, and Southern Continents, A. Keast, F. C. Erk, and B. Glass, eds., pp. 19–22. State University of New York Press, Albany.

Kingdon, J. 1974. Mammals of East Africa: An Atlas of Evolution, Vol. 2A. Academic Press, New York.

Kunz, T. H. 1982. Roosting ecology of bats. In Ecology of Bats, T. H. Kunz, ed., pp. 1–56. Plenum Press, New York.

Lewis, S. E. 1995. Roost fidelity of bats: A review. Journal of Mammalogy 76:481–496.

Martin, R. B., and T. de Meulenaer. 1988. Survey of the status of the leopard (Panthera pardus) in sub-Saharan Africa. Secretariat of the Convention on International Trade in Endangered Species of Wild Flora and Fauna, Lausanne, Switzerland.

McDonald, J. T., I. L. Rautenbach, and J. A. J. Nel. 1990. Foraging ecology of bats observed at the De Hoop Provincial Nature Reserve, southern Cape Province. South African Journal of Wildlife Research 20:133–145.

McKelvey, J. J., Jr. 1973. Man Against Tsetse: Struggle for Africa. Cornell University Press, Ithaca.

McWilliam, A. N. 1987. Polyoestry and postpartum oestrus in Tadarida (Chaerephon) pumila (Chiroptera: Molossidae) in northern Ghana. Journal of Zoology (London) 213:735–739.

McWilliam, A. N. 1994. Nocturnal animals. In DDT in the Tropics: The Impact on Wildlife in Zimbabwe of Ground-spraying for Tsetse Fly Control, J. R. Douthwaite and C. C. D. Tingle, eds., pp. 103–133. Natural Resources Institute, Chatham, U.K.

Mickelburgh, S. P., A. M. Hutson, and P. A. Racey. 1992. Old World fruit bats: An action plan for conservation. IUCN/SSC Chiroptera Specialist Group. IUCN, Gland, Switzerland.

Myers, N. 1995. The world's forests: Need for a policy appraisal. Science 268:823–824.

Nowak, R. W., ed. 1994. Walker's Bats of the World. Johns Hopkins University Press, Baltimore.

Owen-Smith, N. 1989. Megaherbivores. Cambridge University Press, Cambridge.

Pimentel, D., C. Harvey, P. Resosudarmo, K. Sinclair, D. Kurz, M. McNair, S. Crist, L. Shpritz, L. Fitton, R. Saffouri, and R. Blair. 1995. Environmental and economic costs of soil erosion and conservation benefits. Science 267:1117–1122.

Preston, R. 1994. The Hot Zone, a Terrifying True Story. Random House, New York.

Racey, P. A. 1979. Two bats in the Seychelles. Oryx 15:148–152.

Rautenbach, I. L., A. C. Kemp, and C. H. Scholtz. 1988. Fluctuations in availability of arthropods correlated with microchiropteran and avian predator activities. Koedoe 30:77–90.

Rautenbach, I. L., M. B. Fenton, and M. J. Whiting. 1995. Bats in riverine forests and woodlands: A longitudinal transect in southern Africa. Canadian Journal of Zoology 74:312–322.

Roer, H. 1970. Zur Wasserversorgung der Microchiropteren Eptesicus zuluensis vansoni (Vespertilionidae) und Sauromys petrophilus erongensis (Molossidae) in der Namibwúste. Bijdragen tot de Dierkunde 40:71–73.

Rosevear, D. R. 1965. The Bats of West Africa. British Museum of Natural History, London.

Rydell, J., and P. A. Racey. 1995. Streetlamps and the feeding ecology of insectivorous bats. Symposia of the Zoological Society of London 67:291–308.

Schlitter, D. A., I. L. Rautenbach, and D. A. Wolhuter. 1980. Karyotypes and morphometrics of two species of Scotophilus in South Africa (Mammalia, Vespertilionidae). Annals of the Transvaal Museum 32:231–239.

Smithers, R. H. N. 1983. Mammals of the Southern African Subregion. University of Pretoria Press, Pretoria.

Smithers, R. H. N, and V. J. Wilson. 1979. The Mammals of Zimbabwe-Rhodesia. National Museums and Monuments of Zimbabwe-Rhodesia, Salisbury.

Spinage, C. 1994. Elephants. Poyser Natural History, London.

Starfield, A. M., D. H. M. Cumming, R. D. Taylor, and M. S. Quadling. 1993. A frame-based paradigm for dynamic ecosystem models. AI Applications 7:1–13.

Stuart, S. N., R. J. Adams, and M. D. Jenkins. 1990. Biodiversity in sub-Saharan Africa and its islands: Conservation, management, and sustainable use. No. 6, Occasional Papers of the IUCN Species Survival Commission, Gland, Switzerland.

Thomas, D. W. 1983. The annual migrations of three species of West African fruit bats (Chiroptera: Pteropodidae). Canadian Journal of Zoology 61:2266–2272.

Thomas, D. W., and A. G. Marshall. 1984. Reproduction and growth in three species of West African fruit bats. Journal of Zoology (London) 202:265–281.

Tuttle, M. D. 1988. America's Neighborhood Bats. University of Texas Press, Austin.

Van der Merwe, M., I. L. Rautenbach, and W. J. van der Colf. 1986. Reproduction in females of the little free-tailed bat, Tadarida (Chaerophon) pumila, in the eastern Transvaal, South Africa. Journal of Reproductive Fertility 77:355–364.

Vaughan, T. A., and R. M. Vaughan. 1986. Seasonality and the behavior of the African yellow-winged bat. Journal of Mammalogy 67:91–102.

Whitaker, J. O., Jr., and H. L. Black. 1976. Food habits of cave bats of Zambia. Journal of Mammalogy 57:199–204.

Wilkinson, G. S. 1995. Evolution of infant vocalizations in CF and FM bats. In Tenth International Bat Research Conference, Boston, August 6–11, 1995 (abstract).

Wilson, J. W. III. 1974. Analytical zoogeography of North American mammals. Evolution 28:124–140.

19
Conservation Biology
of Australian Bats
Are Recent Advances Solving Our Problems?

GREGORY C. RICHARDS AND LESLIE S. HALL

Even though its colonial economy—and the related ecological pressures—began only a little more than 200 years ago, Australia has one of the worst conservation records of any country in the world, being responsible for half of all the mammalian extinctions worldwide in this time period (Short and Smith 1994). This high extinction rate has resulted primarily from loss of habitat as a result of agricultural development, but several introduced predators and herbivores also have had a major influence.

Australia is a diversely vegetated continent, offering a wide range of habitats for a relatively large bat fauna. Although not all of them have been described, at least 85 bat taxa have been identified, constituting about one-third of the mammal fauna found on this continent. Several years ago the Australian government supported the preparation of an action plan for bat conservation (Richards and Hall 1994). Its main intent was to set conservation priorities during the next decade as a guide for state governments that have final responsibility for species conservation. This chapter evaluates the current status of bat conservation in Australia,[1] and

assesses the likelihood that identifiable problems can be rectified, given the recent scientific advances that have been made.

Historical Perspective

Little has been published about the conservation of Australian bats. Ride (1970) was the first to comment on the status of bats when he listed Australian mammals that were represented by fewer than 20 specimens in collections or recorded from fewer than 10 locations. Frith (1973) used a similar approach for categorizing Australian bats and also listed a number of factors considered to be threats to this fauna. It is interesting to note that more than two decades ago identifiable threats included habitat clearance, roost destruction, and environmental contaminants. Frith (1973) also predicted that further research would change the status of many Australian bats; at that stage only about 50 species were recognized. Hamilton-Smith (1974) was the first to focus specifically on bat conservation in this country. Later, he evaluated the endangered and threatened species known at the time (Hamilton-Smith 1978) and reviewed the current threats (Hamilton-Smith 1980).

More recent research, including a survey of rare bats, changed the known status of several species for the better

[1]The data analyzed came from Richards and Hall (1994). Slight changes in the species identified as endangered may have occurred as a result of the revision of assessment criteria by the International Union for Conservation of Nature and Natural Resources (IUCN) in November 1994, after the submission of the draft action plan.

(Hall 1983). Richards and Tidemann (1988) discussed the effects of habitat disturbance and alteration; Lunney (1989) listed priorities for bat conservation based upon a survey of bat researchers at the Eighth International Bat Research Conference in Sydney; and Richards (1990, 1991, 1992) addressed the problems of bat conservation in the Australian wet tropics. The need for improved taxonomic knowledge, first noted by Frith (1973) as basic to resolving problems in Australian bat conservation, was again emphasized by Parnaby (1991) and further summarized by Hall (1990) and Richards (1991).

Current Problems

The Need for Conservation Action

We cannot make adequate conservation plans for many species because we lack biological information about them. Even the distribution patterns and habitat requirements of many species are essentially unknown, particularly for many of those 39 species that are classified in IUCN terms as being threatened to some extent. However, from a list of more than 80 taxa, the draft bat action plan of Richards and Hall (1994) identified a broad range of threats that needed to be addressed and the complement of species which should be placed in IUCN threat categories. These threats are summarized as follows.

Habitat loss and habitat modification are a problem where loss relates to total removal of habitat such as through clear-cutting of forests or the clearing of woodland for agricultural development, and modification that encompasses selective logging and partial clearance for real estate development.

Colony disturbance and roost destruction are involved where bats within a roost are either adversely affected by human visitation and disturbance, or where the roost itself is totally removed by human activities such as excavation during the mining of previously abandoned sites.

Clarification of species taxonomy becomes a threat when a species complex is known to contain unnamed taxa with highly restricted distributions and populations that may be threatened, but the taxon is lumped with an abundant, widespread species.

Predation relates to contact with introduced predators such as feral cats and cane toads that capture bats as they emerge from subterranean roosts.

Direct killing occurs during, for example, culls of flying-foxes (fruit bats) in orchards.

Lack of biological information prevents adequate conservation planning, for example, if a species is recognized to be specialized in its ecology (confirmed by morphological characters) but major aspects of its ecology are unknown.

Environmental pollution threats include the ingestion of lead from petrol fumes by flying-foxes, the absorption of fluoride by insectivorous bats around smelters, and the ingestion of DDT via the food chain of bats feeding in areas of intensive agriculture.

Isolation of populations becomes a threat when the threats just listed are suspected but not confirmed, and the restriction of the entire population of the species to a sensitive area dictates the application of the Precautionary Principle.

New parasites have been shown to be a threat to one flying-fox species (Pteropus conspicillatus) that now feeds closer to the ground, within the range of the paralysis tick, which causes deaths.

Poisoning is inflicted on flying-foxes when they eat commercial fruit baited with strychnine.

Starvation has been shown as a problem when lactating flying-foxes are killed in orchards and their non-volant young die where they have been left behind at the maternity roost.

The action plan also indicated, through a coarse species-by-species analysis, the extent to which each identifiable threat had to date been ameliorated in some way. From this research we can score the current level of conservation action for the bat fauna overall. For example, if a species is suffering from habitat loss, habitat modification, or colony disturbance, and part of the total population lives within a national park (where such threats would be ameliorated), then the estimated proportion of the population or proportion of distributional range so protected can be calculated. If we have been unable to adequately analyze a species distribution or use of habitat because of poor taxonomic resolution that has recently been corrected, then this effort can also be scored as a value, and so on.

Some species face only one of these threats, others endure several different ones. A species that faced three threats would be given a score of 3.0, but if one threat were resolved it would then be scored with 2.0, giving a degree of threat resolution of 33.3%. Furthermore, if 10 species suffered in the past from habitat loss and this problem was resolved for half of them, then for that threat the degree of resolution would be 50%. Overall calculations can then be made using a species × threats matrix.

Results of such analyses are summarized in Tables 19.1–19.3, which show that bat conservation in Australia

Table 19.1

Degree of Resolution of Conservation Problems for Each Species Allocated to
IUCN Threat Categories by Richards and Hall (1994)

IUCN status	Taxon	No. of problems	Degree of resolution	% resolution
EX(p)	*Nyctophilus howensis*	1.0	0.0	0
	Pteropus brunneus	2.0	0.0	0
END	*Murina florium*	2.0	1.0	50
	Taphozous australis	3.0	0.6	20
	Taphozous troughtoni	2.0	0.0	0
VUL	*Chalinolobus dwyeri*	3.0	0.2	70
	Hipposideros cervinus	4.0	0.4	10
	Hipposideros diadema	2.0	0.4	20
	Kerivoula papuensis	2.0	0.4	20
	Macroderma gigas	3.0	0.9	30
	Nyctophilus timoriensis (eastern form)	3.0	0.9	30
	Pteropus conspicillatus	3.0	0.1	3
	Pteropus poliocephalus	6.0	0.4	7
RARE	*Chalinolobus picatus*	2.0	0.0	0
	Dobsonia moluccensis	2.0	0.0	0
	Hipposideros (*inornatus* form)	3.0	2.3	77
	Hipposideros semoni	4.0	0.4	10
	Hipposideros stenotis	0.0	0.0	0
	Nyctimene vizcaccia	0.0	0.0	0
	Nyctimene (*cephalotes* form)	2.0	0.3	15
	Pteropus macrotis	0.0	0.0	0
	Pteropus sp. nov.	2.0	0.1	5
	Rhinolophus philippinensis	4.0	0.2	5
	Rhinolophus (*maros* form)	3.0	1.2	40
	Saccolaimus mixtus	2.0	0.2	10
	Saccolaimus saccolaimus	2.0	0.4	20
IK	*Chalinolobus morio* (cave form)	3.0	0.7	23
	Falsistrellus mckenziei	1.0	0.0	0
	Mormopterus (*setirostra* form)	2.0	0.3	15
	Murina sp. (from Iron Range)	3.0	0.5	17
	Myotis sp. (from Kimberley)	2.0	0.9	45
	Nyctophilus timoriensis (central form)	2.0	0.5	25
	Nyctophilus timoriensis (western form)	2.0	0.5	25
	Nyctophilus (*daedelus* form)	3.0	1.0	33
	Taphozous kapalgensis	1.0	0.0	0
	Vespadelus douglasorum	2.0	0.0	0

Notes: In this analysis, we estimated the current level of resolution of each problem for each species: fully resolved = 100% (1.0); half resolved = 0.5; not at all resolved = 0.0. IUCN categories: EX(p), extinct (presumed); END, endangered; VUL, vulnerable; IK, insufficiently known. Number of species in quartiles based upon the degree of resolution of their identified conservation problems: first quartile (0–25% resolution), 28 species; second quartile, 6; third quartile, 1; fourth quartile, 1.

falls seriously short of providing needed species protection. Nothing has been done to ease the problems of conservation of 11 of the 36 threatened bat species in Australia; there has been 25% or less resolution of the problems of an additional 17 taxa, and we are halfway or better toward ameliorating threats to only 3 other species (Table 19.1).

Most attention (value, 23%) has been given to the problems of the Endangered group and least to the Vulnerable group (value, 12%), but when we consider the total of 85 threats to the 36 sensitive species, fewer than 15% of the problems have been addressed (Table 19.2). Habitat problems comprise four of the five major types of threat, involving as

Table 19.2

Analysis of Degrees of Resolution of Threats to Species in IUCN Categories

Status	Threat	No. of species in class	Score	Degree of resolution (%)
EX(p)	Direct killing	1	0.0	0
	Habitat loss	1	0.0	0
	Predation	1	0.0	0
	All	3	0.0	0
END	Colony disturbance	1	0.2	20
	Habitat loss	2	0.7	35
	Habitat modification	1	0.5	50
	No biological information	1	0.0	0
	Roost destruction	2	0.2	10
	All	7	1.6	23
VUL	Colony disturbance	4	0.2	5
	Direct killing	2	0.0	0
	Environmental pollution	1	0.1	10
	Habitat loss	7	1.2	17
	Habitat modification	4	0.8	20
	New parasites	1	0.0	0
	Poisoning	1	0.0	0
	Predation	1	0.0	0
	Roost destruction	3	0.9	30
	Starvation	1	0.0	0
	Taxonomic problems	1	0.5	50
	All	26	3.7	12
RARE	Colony disturbance	6	0.6	10
	Direct killing	2	0.0	0
	Habitat loss	5	0.8	16
	Habitat modification	3	0.9	30
	Isolation of population	1	0.1	10
	Predation	2	0.0	0
	Roost destruction	6	1.1	18
	Taxonomic problems	3	1.9	63
	All	28	5.4	18
IK	Colony disturbance	2	0.7	35
	Habitat loss	5	1.0	20
	Habitat modification	2	0.2	10
	No biological information	3	0.0	0
	Predation	2	0.0	0
	Taxonomic problems	7	1.9	27
	All	21	3.8	15
All	All	85	14.5	15

Notes: IUCN status categories (as defined by Richards and Hall 1994): EX(p), extinct (presumed), END, endangered; VUL, vulnerable; IK, insufficiently known Number of conservation-problem types (combination of IUCN status and threat type) in quartiles based upon the degree of resolution: first quartile (0–25% resolved), 25 types of conservation problems; second quartile, 7; third quartile, 1; fourth quartile, 0.

Table 19.3

Occurrence and Current Resolution of Each Threat Irrespective of IUCN Category, Ranked by Number of Species in Threat Class

Threat	No. of species in threat category	Total score	Degree of resolution (%)
Habitat loss	19	3.7	19
Colony disturbance	13	0.2	2
Roost destruction	11	2.2	20
Taxonomy needs clarification	11	4.3	39
Habitat modification	10	2.4	24
Predation	5	0.0	0
Direct killing	4	0.0	0
No biological information	4	0.0	0
Environmental pollution	1	0.1	10
Isolation of population	1	0.1	10
New parasites	1	0.0	0
Poisoning	1	0.0	0
Starvation	1	0.0	0

many as 19 species, but less than 20% of the required effort to counteract them has been expended (Table 19.3). These analyses quantify and emphasize priorities for the aforementioned action plan and are a severe indictment of the lack of support for research and conservation in the past.

Habitat Loss and Modification and Forest Management

Major habitat problems include the selective logging of forests for timber, clear-cutting of forests for woodchip-paper production, the total clearance of forests and woodlands for agricultural crops or pasturelands, and dieback of large areas of woodland as a consequence of groundwater salinity, fungal infection, or insect attack. About half of Australia's forests have been cleared since settlement in 1788 (Lunney 1991; Graetz et al. 1992). Removal of large tracts of woodland and forest for agriculture has had deleterious effects on other faunal groups, particularly after World War II when machinery replaced the axe and land clearing became more rapid and efficient. The Resource Assessment Commission (1991) concluded that if logging continued at the current rate, by 2015 there would be very little unlogged forest remaining in timber production tenures. It is highly likely that this would have a deleterious impact on the bat fauna overall, but particularly on flying-foxes that depend on large areas of blossom from mature eucalyptus (Eby 1991a). To place our forest issue into an international perspective, the area that is under discussion is approximately 400,000 km², about 80,000 km² of which is in conservation reserves (Lunney 1991).

Habitat loss results not only in fewer foraging areas for bats but also in fewer tree hollows for roosts. Bennett et al. (1994) showed that in woodland the general abundance of tree hollows in grazing areas is strongly influenced by land management practices. These authors expressed concern that the grazing of seedlings and resultant lack of recruitment of sapling trees will reduce the future tree-hollow resource. Lumsden et al. (1995) showed that remnant patches of woodland within cleared pastoral areas retained high numbers of local species, and proposed that only their ability to fly enabled the bats to utilize other resources in this patchy landscape, thereby giving them an advantage over other mammal groups.

We do not yet know however if the reduction of foraging area and probable reduction in prey levels as a consequence of habitat loss have had any subtle and suppressive effects on reproductive output and recruitment. We suspect that fast-flying forest bats with high aspect ratio wings may be restricted to roosting only in emergent trees. To launch into horizontal flight after exit, a drop of 2–3 m is necessary before normal flight speed is achieved, particularly for large emballonurids and molossids (Richards, unpublished data); such bats may not be able to exit into uncluttered flight space unless the roost is elevated above foliage. If this hypothesis is correct, then the removal of emergent trees in logging operations may have been more disadvantageous than previously thought.

There are no quantitative data to show the effects of land clearance on insectivorous bats, but the effects on populations of the grey-headed flying-fox (*Pteropus poliocephalus*) indicate the likely problems. This species is distributed along the east coast of the southern half of Australia and favors eucalypt forests for foraging and roosting. Maps produced by the Australian Surveying and Land and Information Group show that before European settlement forest habitat was apparently continuous whereas now it is fragmented, resulting in discontinuous migration routes. Of the coastal eucalypt forest types that are most suitable for flying-foxes (i.e., having a canopy height exceeding 10 m) within the distributional range of *P. poliocephalus*, Lunney (1991) showed that only 52% remains. Further, when tracts that have been used as traditional roost sites are removed for urban or agricultural development, colonies break up into smaller camps (Nelson 1965; Hall and Richards 1991). The social and genetic effects of these problems are unknown.

Based on numbers estimated in the late 1920s (Ratcliffe 1931), declines have occurred in Australian flying-fox populations since then and are likely to have been concomitant with the extensive clearing of forest for pastoral activities.

The problem for these bats has been exacerbated by natural phenomena such as the failure of normal flowering in native forest caused by drought, or through loss of nectar from rain, and remaining populations periodically experience severe food stress. This appears to drive them to feed on less nutritious commercial fruit (Ratcliffe 1931; Nelson 1965; Fleming and Robinson 1987; Tidemann and Nelson 1987). Moreover, where forest reduction and fragmentation continues this situation will worsen.

Foresters also manage the intensity of wildfires by "controlled burning" in seasons such as winter when fires will burn at low intensity and simply reduce fuel loads that could otherwise ignite during dangerous periods. These practices appear to rapidly modify habitat for bats and can have a variety of effects, given that the density and height of the understory influence habitat utilization by some forest bat species (Richards 1994).

The extraction of logs during selective logging of tropical rainforest can have a more serious effect than the felling operation itself (Crome et al. 1991). The opening up of tropical rainforest to gain access to logging areas can cause the introduction of non-rainforest bat species (Crome and Richards 1988), although in temperate eucalypt forests, which apparently sustain a greater harvesting pressure than does tropical rainforest, the effects are apparently mixed (Richards 1994).

Flying-Foxes

Flying-foxes have been culled in New South Wales under permits issued to orchardists in a haphazard fashion, with little regard to life cycles and population dynamics, without calculating the proportion that can be safely removed, and without ongoing monitoring of the population status. Because flying-foxes need 2 years to reach sexual maturity and females bear only one young per year, Pierson and Rainey (1992) estimated that annual culling of even 10% of a population will almost halve it in 5 years and can reduce it to approximately one-fifth of its original size after 20 years. On the basis of the number of licenses allocated to orchardists for culling, Wahl (1994) estimated that during the years 1986 to 1992, inclusive, at least 240,000 flying-foxes could have been killed in the coastal area between Sydney and the Queensland border.

Using the model of Pierson and Rainey (1992), and by averaging Wahl's data to a culling of 30,000 bats per year, we can predict the long-term effects. We estimated that if the base population is 2 million animals (40 camps of 50,000), and if it is culled at 10% per year, then fewer than 800,000 animals will remain in 10 years and fewer than 100,000 in about 30 years. This is an untenable situation, and gives credence to the listing of *P. poliocephalus,* the primary target in culling, as Vulnerable. However, the problem is being addressed to some extent by the relevant management agency, which has recently published a formal management plan that addresses these issues (Eby 1995).

We need information about the interaction between species of *Pteropus,* their food sources, and forest phenological and successional cycles. It is essential to recognize that flying-foxes have "pest" status in some areas, particularly in Queensland. Passive exclusion and other nondestructive management techniques are essential and urgently required. One area of needed research is the development of models to enable the prediction of influxes. Richards (1995) showed that we can predict migrations of *Pteropus scapulatus* to fruit-growing areas in north Queensland by monitoring rainfall patterns within the core range of this species. Hall (1994) noted that *P. scapulatus* undergoes a regular cycle of annual movement in southeast Queensland, contrary to the pattern in north Queensland, but again allowing influxes to be predicted.

It is almost impossible to design a reserve system or a coherent management strategy for flying-foxes on the east coast of Australia because (1) they are highly mobile and seasonally migratory or nomadic, traveling as much as 50 km per night and 1,000 km seasonally (Eby 1991b), and (2) the Australian climate is so variable and the fragmentation of food resources is so great. We stress that different management plans need to be devised for different regions. For example, populations of the same species in southeast Queensland, north Queensland, and Torres Strait all behave differently in their movements and should be managed accordingly.

Cave Bat Populations

No state or national advisory group has the authority to implement a cave protection system, and there is currently no way to effectively manage the large number of recreational cavers who frequently, if unwittingly, interfere with nursery colonies and hibernating bats. Many cave colonies have completely disappeared since the 1980s (Richards and Hall 1994); most of these were close to urban areas. A survey of known maternity sites of the large bentwing bat (*Miniopterus schreibersi*) and the eastern horseshoe bat (*Rhinolophus megaphyllus*) in eastern Victoria during the 1994–1995 breeding season showed that the population sizes were similar to those estimated in the 1960s (E. Hamilton-Smith, personal communication). However, this sample was from a small area; a more extensive survey, covering a wider area and encompassing more species, is needed.

Abandoned mines play an important role by providing a suitable alternative to natural underground roosts for bats. Australia as a continent is not endowed with a large number of natural caves, and in some areas it is obvious that the presence of a particular bat species is a result of the construction of mineshafts. It is also possible that some isolated cave areas (such as at Chillagoe, Queensland) were not colonized until they had been linked to occupied caves by the interstitial provision of artificial roosts in the form of mines.

Of 36 species in IUCN threat categories, 17 of them or 47% inhabit subterranean roosts, which is quite disproportionate to the 35% of the national bat fauna that roosts underground (Table 19.4). Recent advances in processing techniques allow gold to be extracted from lower-grade ore than was possible in the past century, so that reworking of old mines is becoming commonplace. In turn, this places more urgent pressure on bat biologists to produce management strategies for cave-dwelling bats. Many old mines have a crucially important role in bat conservation, particularly as maternity sites and as roosts for endangered species. The type locality of Troughton's sheathtail bat (*Taphozous troughtoni*) was Native Bee Mine, near Mt. Isa, Queensland, but the mineshaft was bulldozed during mining operations. This endangered species is probably found in nearby mines, but if such roosts do exist then they may also suffer the same fate. The type locality of the large-eared pied bat (*Chalinolobus dwyeri*), described in the 1960s, was a mine that was flooded after construction of a reservoir in the 1970s.

The largest colony of the vulnerable and endemic ghost bat (*Macroderma gigas*) roosts in an abandoned gold mine at Pine Creek in the Northern Territory. The conservation of this colony, comprising approximately 1,500 individuals of a known national population of less than 10,000 (Phillips

Table 19.4

Proportion of Species Using Subterranean Roosts According to the National Species Inventory of Richards and Hall (1994)

Family	Total no. of taxa	No. of taxa roosting underground	Underground taxa threatened No.	%
Pteropodidae	13	1	1	100
Megadermatidae	1	1	1	100
Rhinolophidae	3	3	2	67
Hipposideridae	7	7	5	71
Emballonuridae	8	5	2	40
Molossidae	11	0	0	0
Vespertilionidae	42	13	6	46
All	85	30	17	57

1990), is obviously crucial. The colony uses the site throughout the year and is the largest known maternity aggregation of this species. Recently, however, the area was considered as part of an extensive open-pit mining operation and the future of the colony looked precarious. Only the low value of the ore reprieved the site, and the roost was retained intact. This is a situation, nevertheless, that could be easily reversed in the future.

At present there is no national policy for the management of bat populations occupying old mines. Any abandoned mines that are considered unsafe and dangerous to the public are bulldozed, and some collapse before their formal destruction. A complex problem therefore develops in that some roost sites either are deteriorating and require some human intervention to keep them open for bats or are being destroyed during or after mineral extraction operations.

Taxonomic Problems

Many of the threatened taxa addressed in the draft of the Australian bat action plan (Richards and Hall 1994) are newly recognized and undescribed species distinguished by pseudonyms while they await formal description. This situation results partly from the lack of chiropteran taxonomists and partly from the inaccessibility of type material held in museums overseas. Access to many of these specimens is essential for resolving many taxonomic questions of identity; 72 holotypes (84%) are lodged in overseas museums and can only be accessed by visiting, which is a prohibitively expensive and time-consuming exercise for Australians. Consequently, although we can identify and classify new species, applying the correct nomenclature is a difficult task. Unfortunately, until new species can be formally named for reference, our conservation efforts are thwarted, particularly in the definition of distribution and habitat utilization patterns and the preparation of environmental impact statements.

Lack of Basic Biological Information

Even though our knowledge of the basic natural history of some common species is gradually increasing, primarily in localities where environmental studies are legally required, there have been very few comprehensive biological studies of single species. Virtually nothing is known about the habitat preferences of Australian bats, and little information is available concerning their food preferences as well (Vestjens and Hall 1977). If we are to plan bat conservation effectively, we need more information about patterns of distribution and habitat utilization, maternity site requirements for tree dwellers, and the sizes of home ranges and foraging areas. We now know something of the roosting

biology of common forest species through the efforts of Lunney et al. (1985, 1988), Tidemann and Flavel (1987), and Taylor and Savva (1988).

Using geographic analyses, Richards and Hall (1994) determined the level of past field survey efforts and estimated that 63% of Australia has no records for bats and a further 28% of the continent has fewer than one record per 3,000 km^2. In terms of land area, this total of 91% equates to 5 million km^2, primarily in the arid zone.

Introduced Predators, Competitors, and Pollutants

There is circumstantial evidence that introduced predators caused the extinction of the Lord Howe Island bat (*Nyctophilus howensis*) in the early 1900s (Richards and Hall 1994). The inadvertent introduction of rats (*Rattus norvegicus*) from a shipwreck on this small island appears to have started the demise of this species, which was apparently later completed by owls introduced to control the rats. Because this species is now extinct it is no longer regarded as a current conservation problem, but it does serve as an example of the sensitivity of some species of bats, particularly those with restricted distributions, to introduced predators.

Predators of Australian bats include introduced species such as the European fox (*Vulpes vulpes*), feral cats (*Felis catus*), and cane toads (*Bufo marinus*) that hunt at cave entrances or within caves (Dwyer 1964; Hall, unpublished data). There is evidence that English starlings (*Sturnus vulgaris*) evict species such as Gould's wattled bat (*Chalinolobus gouldii*) from tree hollows just as they do possums and native birds.

In southeast Queensland, deaths of flying-foxes have followed the accumulation of high levels of lead in their tissues (Sutton and Hariano 1987); the source of the lead is suspected to be vehicle exhaust vapors that settle on vegetation. High levels of fluoride have been found in the body tissues of insectivorous bats captured in the vicinity of large aluminum smelters (Hoye 1994). Hoye suggested that as a consequence older individuals may be prematurely removed from the population.

A survey for the presence of DDT and its derivatives in bent-wing bats (*Miniopterus schreibersi*) showed high levels in body tissues, even in young bats that had not left the maternity roost (Dunsmore et al. 1974). DDT buildup was the suspected cause of several mass die-offs involving many thousands of this species. Now that DDT has been phased out as an agricultural chemical, such poisoning should no longer pose a problem; nevertheless, this exemplifies the potential danger of insecticides and other toxins in the food web of bats.

Table 19.5

Classification of Australian Fruit Bats in an Ecological Framework

Size class (body weight)	Frugivorous specialist	Nectarivorous specialist	Dietary generalist
Large species (>300 g)	Pteropus conspicillatus*	Pteropus scapulatus*	Pteropus alecto*
	Pteropus sp. nov.	Pteropus macrotis	Pteropus poliocephalus
	Dobsonia moluccensis		Pteropus brunneus
Small species (<60 g)	Nyctimene robinsoni*	Syconycteris australis*	
	Nyctimene vizcaccia	Macroglossus minimus	
	Nyctimene (cf. cephalotes)		

Notes: These classifications are based on dental morphology and diet. Specialists are considered to be those species with approximately 90% of their food as one type; generalists have a broader diet. Species that are abundant and widespread are indicated by an asterisk (*). *Pteropus brunneus* is considered to be extinct.

Recent Advances

Conservation Planning

For a nation that has devoted few resources to bat biology in the past, the development of a draft bat action plan (Richards and Hall 1994) presented a unique set of problems. Recent taxonomic revisions have increased the number of species from 55 known in the early 1980s to approximately 70. Because other species are still in the process of being described, the action plan covers not only known species but also other conservation units, such as undescribed species and subspecies that are likely to be elevated to species rank. As a result, the plan concerns 85 taxa. Action plan analyses were carried out by linking Geographic Information System software to a large database of distribution records. Analyses of regional species richness confirmed that Cape York Peninsula supports more species than elsewhere. Closer analysis of this region shows that rainforest areas are the most vital to conserve. Nationwide, 11 areas have been identified as key conservation sites on the basis of species richness.

Another set of what would be valuable national reserves for bats has been elucidated by making a Critical Areas analysis, based on species endemic to Australia and prioritized to endemics with highly restricted distributions. Recovery plans for the 36 species allocated to IUCN threat categories have been developed.

Amelioration of Habitat Loss

The logging of native forests, a contentious issue, has been ameliorated to some extent. Recent federal policy (July 1995) requires that 15% of each forest type present at the time of Caucasian settlement be excluded from logging and that no further old growth forests are to be logged. However, this policy is not enshrined in legislation and cannot

be implemented without state cooperation. The state government of New South Wales enforces a licensing procedure for the forestry industry whereby areas for logging must be surveyed for endangered species and prescriptions enacted for their conservation at a local level. The Queensland Department of Primary Industries (Forestry) has introduced a manual for use by forest managers to ensure that areas of forest are retained for rare and threatened species (Borsboom 1994).

We are now in a position to make the best use for bats of national initiatives such as the One Billion Trees Program, which is intended to restore deforested areas. The dietary components of some of the threatened species of *Pteropus* are becoming well known, so it should be possible to advise this program of which trees to plant that are primary food in areas where they are now depauperate, particularly along migration routes.

Flying-Fox Biology

Recent research has shown that Australia's pteropodid fauna of 13 species can be separated into a 2 × 2 matrix of large and small species that are generalist or specialist feeders; the specialists can be further subdivided into fruit- or nectar-feeding species (Table 19.5) (Richards 1995). For conservation planning and management, the position of a species in this matrix will indicate the likely impact of particular threats. We anticipate that specialists will experience greater impacts than generalists, given the same level of threat.

Conservation and Management of Forest Bats

In surveys of 24 species over 222 sites in eastern Australia, Richards (1994) showed that the microchiropteran forest bat fauna can be divided, by species, into habitat generalists and habitat specialists. As a rule, the former constitute genera endemic to Australia, with the latter being extralimital genera. The use of habitat by generalist species appears to be unaffected by logging, whereas specialists are limited largely to unlogged forests. This information has allowed the development of models showing the types of habitat that need to be included in reserve networks.

Richards (1994) also showed that each habitat specialist occupied a unique position within a gradient between habitats that have low foliage-nutrient concentrations (primarily foliage potassium) and open forest structure and habitats which have high foliage-nutrient concentrations and dense forest structure. This information allows further "fine tuning" of the reserve selection process in eastern forests to accommodate endangered species and could have application elsewhere.

Resolution of Taxonomic Problems

Because type material of already described taxa is so inaccessible, a system using computerized video imagery as a taxonomic research tool has been developed (Richards, unpublished data). Most of the holotypes lodged in non-Australian museums can now be accessed via computerized images, allowing taxonomists to make comparisons with Australian specimens without visiting overseas. Although our resources may have increased and many taxonomic problems can now be resolved, the number of chiropteran taxonomists in Australia unfortunately has declined rather than increased. However, although classical descriptive taxonomy may be suffering from delays, molecular research (particularly in the area of bat genetics) is increasing. DNA analysis has now identified a new form of tube-nosed bat *(Nyctimene)* from Cape York (Richards and Collet, unpublished data). A genetic study of the ghost bat *(Macroderma gigas),* a threatened species that is restricted to approximately 10 disjunct maternity sites, has shown that for the purposes of management each population should be treated as a separate entity (Worthington Wilmer et al. 1994).

Cave Bat Populations

Interactive software has been proposed so that geologists and mining companies can determine whether subterranean roosts are present in lease areas. It is hoped that this proposal will stop some of the present inadvertent destruction of bat roosts in mines. The software will use a database containing more than 60,000 distribution records that is linked to Geographic Information System (GIS) software.

Conclusions

Current threats to bats in Australia are wide ranging: habitat loss associated with land clearance and timber extraction from forests, the culling of flying-foxes in orchards, the disturbance or destruction of colonies in underground roosts, starvation, environmental pollution, introduced predators and competitors, poisoning, and new parasites. Complicating the resolution of many of these threats is a plethora of taxonomic problems and a lack of basic biological information for many species, which make broad-scale conservation planning quite difficult.

However, because of recent advances in research methods and the development of a national action plan, the range of threats can now be specified; we are now more optimistic that we can quickly respond to, and resolve, the problems that have been identified. Unfortunately, without better funding conserving Australia's bat fauna will continue to be difficult to accomplish and is probably beyond the capacity of the small core of researchers currently available. To resolve these problems we suggest it may be prudent to establish a national Co-operative Research Center for Bat Biology. Such centers have become a successful government initiative in Australia, and, because they draw together the skills of scientists and students in universities and government agencies, they become a major source for funding and can focus research directions.

Acknowledgments

This chapter is not only the work of the two authors. It summarizes the efforts of many dedicated Australians who for many years have strived for the conservation of a faunal group that has not had as much public acceptance as the "cuddlier" members of the Australian fauna, and it is to these people that we dedicate this chapter. We also thank A. Milligan and R. Schodde for their valuable comments and criticisms.

Literature Cited

Bennett, A. F., L. F. Lumsden, and A. O. Nicholls. 1994. Tree hollows as a resource for wildlife in remnant woodlands: Spatial and temporal patterns across the northern plains of Victoria. Pacific Conservation Biology 1:222–235.

Borsboom, A. 1994. Development of a fauna and flora information system for rare, threatened, and potentially threatened species within the Queensland forest estate. Establishment report. Queensland Department of Primary Industries, Forestry, Brisbane.

Crome, F. H. J., and G. C. Richards. 1988. Bats and gaps: Microchiropteran community structure in a Queensland rain forest. Ecology 69:1960–1969.

Crome, F. H. J., L. A. Moore, and G. C. Richards. 1991. A study of logging damage in upland rainforest in north Queensland. Forest Ecology and Management 49:1–29.

Dunsmore, J. D., L. S. Hall, and K. H. Kottek. 1974. DDT in the bent-winged bat in Australia. Search (Carlton) 5:110–111.

Dwyer, P. D. 1964. Fox predation on cave bats. Australian Journal of Science 26:397.

Eby, P. 1991a. "Finger-winged night workers": Managing forests to conserve the role of grey-headed flying foxes as pollinators and seed dispersers. *In* Conservation of Australia's Forest Fauna, D. Lunney, ed., pp. 91–100. Royal Zoological Society of New South Wales, Mosman.

Eby, P. 1991b. Seasonal movements of grey-headed flying-foxes, *Pteropus poliocephalus* (Chiroptera: Pteropodidae), from two maternity camps in northern New South Wales. Australian Wildlife Research 18:547–559.

Eby, P. 1995. The biology and management of flying foxes in New South Wales. Species Management Report No. 18. National Parks and Wildlife Service of New South Wales, Sydney.

Fleming, P. J. S., and D. Robinson. 1987. Flying fox (Chiroptera: Pteropodidae) on the north coast of New South Wales: Damage to stonefruit crops and control methods. Australian Mammalogy 10:143–145.

Frith, H. J. 1973. Wildlife Conservation. Angus and Robertson, Sydney.

Graetz, R. D., R. P. Fisher, and M. A. Wilson. 1992. Looking back. The changing face of the Australian continent, 1972–1992. CSIRO, Canberra, Australia.

Hall, L. S. 1983. A study of the status, habitat requirements, and conservation of rare and endangered Australian bats. Report to World Wildlife Fund Australia, Sydney, June 1983.

Hall, L. S. 1990. Bat conservation in Australia. Australian Zoologist 26:1–4.

Hall, L. S. 1994. Predicting flying fox movements: When, why, and where. *In* Bird and Bat Control Seminar Proceedings, pp. 30–37. Queensland Department of Primary Industry, Nambour.

Hall, L. S., and G. C. Richards. 1991. Flying fox camps. Wildlife Australia 28:19–22.

Hamilton-Smith, E. 1974. The present knowledge of Australian Chiroptera. Australian Mammalogy 1:95–108.

Hamilton-Smith, E. 1978. Endangered and threatened Chiroptera of Australia and the Pacific region. *In* The Status of Endangered Australian Wildlife, M. J. Tyler, ed., pp. 85–91. Royal Zoological Society of South Australia, Adelaide.

Hamilton-Smith, E. 1980. The status of Australian Chiroptera. *In* Proceedings, Fifth International Bat Research Conference, D. E. Wilson and A. L. Gardner, eds., pp. 199–205. Texas Tech University, Lubbock.

Hoye, G. 1994. Background and elevated bone fluoride concentrations in the little forest bat *Vespadelus vulturnus*. Abstracts, Sixth Australasian Bat Research Conference, January 1994, Southern Cross University, Lismore, Australia.

Lumsden, L. F., A. F. Bennett, S. P. Krasna, and J. E. Silins. 1995. The conservation of insectivorous bats in rural landscapes of northern Victoria. *In* People and Nature Conservation: Perspectives on Private Land Use and Endangered Species Recovery, A. Bennett, G. Backhouse, and T. Clark, eds., pp. 142–148. Royal Zoological Society of New South Wales, Mosman.

Lunney, D. 1989. Priorities for bat conservation: Analysis of the response to a questionnaire in July 1989 by the participants of the Eighth International Bat Research Conference. Australian Zoologist 25:71–78.

Lunney, D. 1991. The future of Australia's forest fauna. *In* Conservation of Australia's Forest Fauna, D. Lunney, ed., pp. 1–24. Royal Zoological Society of New South Wales, Mosman.

Lunney, D., J. Barker, and D. Priddel. 1985. Movements and day roosts of the chocolate wattled bat *Chalinolobus morio* (Gray) (Microchiroptera: Vespertilionidae) in a logged forest. Australian Mammalogy 8:313–317.

Lunney, D., J. Barker, D. Priddel, and M. O'Connell. 1988. Roost selection by Gould's long-eared bat *Nyctophilus gouldi* Tomes (Microchiroptera: Vespertilionidae) in a logged forest on the south coast of New South Wales. Australian Wildlife Research 15:375–384.

Nelson, J. E. 1965. Movements of Australian flying foxes (Pteropodidae: Megachiroptera). Australian Journal of Zoology 13: 53–73.

Parnaby, H. 1991. A sound species taxonomy is crucial to the conservation of forest bats. *In* Conservation of Australia's Forest Fauna, D. Lunney, ed., pp. 101–120. Royal Zoological Society of New South Wales, Mosman.

Phillips, W. R. 1990. Priorities for Ghost Bat (*Macroderma gigas*) Conservation and Management. Australian National Parks and Wildlife Service, Canberra.

Pierson, E. D., and W. E. Rainey. 1992. The biology of flying foxes of the genus *Pteropus*: A review. *In* Pacific Island Flying Foxes: Proceedings of an International Conservation Conference, D. E. Wilson and G. L. Graham, eds., pp. 1–17. Biological Report 90(23). U.S. Fish and Wildlife Service, Washington, D.C.

Ratcliffe, F. N. 1931. The flying fox (*Pteropus*) in Australia. Bulletin of the Council for Scientific and Industrial Research Australia 53:1–80.

Resource Assessment Commission. 1991. Forest and timber inquiry. Draft report, Vols. 1 and 2. Australian Government Publishing Service, Canberra.

Richards, G. C. 1990. Rainforest bat conservation: Unique problems in a unique environment. Australian Zoologist 26: 44–46.

Richards, G. C. 1991. The conservation of forest bats in Australia: Do we really know the problems and solutions? *In* Conservation of Australia's Forest Fauna, D. Lunney, ed., pp. 81–90. Royal Zoological Society of New South Wales, Mosman.

Richards, G. C. 1992. The conservation status of the rainforest bat fauna of north Queensland. *In* The Rainforest Legacy: Australian National Rainforests Study. Vol. 2: Flora and Fauna of the Rainforests, G. L. Werren and A. P. Kershaw, eds., pp.177–186. Special Australian Heritage Publications Series No. 7(2). Australian Government Publishing Service, Canberra.

Richards, G. C. 1994. Relationships of forest bats with habitat structure and foliage nutrients in south-eastern Australia. Abstracts, Sixth Australasian Bat Research Conference, January 1994, Southern Cross University, Lismore, Australia.

Richards, G. C. 1995. A review of ecological interactions of fruit bats in Australian ecosystems. Symposia of the Zoological Society of London 68:79–96.

Richards, G. C., and L. S. Hall. 1994. An action plan for the conservation of bats in Australia (draft). Australian Nature Conservation Agency, Canberra.

Richards, G. C., and C. R. Tidemann. 1988. Habitat. The effects of alteration and disturbance upon bat communities in Australia. Australian Science Magazine 2:40–45.

Ride, W. D. L. 1970. A Guide to the Native Mammals of Australia. Oxford University Press, Melbourne.

Short, J., and A. Smith. 1994. Mammal decline and recovery in Australia. Journal of Mammalogy 75:288–297.

Sutton, R. H., and B. Hariano. 1987. Lead poisoning in flying foxes. Australian Mammalogy 10:125–126.

Taylor, R. J., and N. M. Savva. 1988. The use of roost sites by four species of bats in state forest in south-eastern Tasmania. Australian Wildlife Research 15:637–645.

Tidemann, C. R., and S. C. Flavel. 1987. Factors affecting choice of roost site by tree-hole bats (Microchiroptera) in southeastern Australia. Australian Wildlife Research 14:459–473.

Tidemann, C. R., and J. E. Nelson. 1987. Flying-foxes (Chiroptera: Pteropodidae) and bananas: Some interactions. Australian Mammalogy 10:133–135.

Vestjens, W. J. M., and L. S. Hall. 1977. Stomach contents of forty-two species of bats from the Australasian region. Australian Wildlife Research 4:25–35.

Wahl, D. E. 1994. The management of flying foxes (*Pteropus* spp.) in New South Wales. Master's thesis, University of Canberra, Australia.

Worthington Wilmer, J., C. Moritz, L. Hall, and J. Toop. 1994. Extreme population structuring in the threatened ghost bat, *Macroderma gigas:* Evidence from mitochondrial DNA. Proceedings of the Royal Society of London B 257:193–198.

20
Brazilian Bats and Conservation Biology
A First Survey

JADER MARINHO-FILHO AND IVAN SAZIMA

Brazil, with an area of 8,511,996 km², accounts for 48% of the total area of South America and contains five of the major biomes in the continent: Amazonia, the Atlantic Forest, the Caatinga, the Cerrado, and the Pantanal, four of which are exclusively Brazilian. More than 60% of the area of Amazonia, the only exception, is within Brazil's territory. This huge country, with its impressive geographic and physiognomic variety, bears an astonishing biological diversity, as indicated by the highest number of flowering plants (approximately 55,000 species), primates (68 species), psittacine birds (70 species), amphibians (518 species), terrestrial vertebrates (3,022 species), and freshwater fishes (~3,000 species) in the world, placing Brazil at the top of the world's six megadiversity countries: Brazil, Colombia, Mexico, Zaire, Madagascar, and Indonesia (Mittermeyer 1988; Mittermeyer et al. 1992).

As a biodiversity leader, Brazil not only contains a major portion of the world's total biological diversity but also has an even greater fraction of the world's diversity at risk (Mittermeyer 1988). As are other megadiversity countries, Brazil is going through a rapid environmental change and is facing severe socioeconomic problems that make it very difficult to develop the conservation programs needed to protect its biological diversity. Recent conservation efforts and campaigns have had an important impact on the conservation of some charismatic animal species and also have led to a general increase of conservation awareness. In spite of this, the official list of Brazilian mammal species threatened with extinction (see Bernardes et al. 1990) contains 58 species, which corresponds to 12% of the Brazilian mammal fauna. This figure is considered as very conservative and it is believed that at least twice that number, that is, 1 in each 4 Brazilian mammal species, are currently threatened (Fonseca et al. 1994).

Bats represent approximately one-third of the Brazilian land mammal fauna. Because they are so speciose, abundant, and biologically diverse, bats play a number of different and important ecological roles in the tropical communities in Brazil and in South America. However, bats are not considered as charismatic species for conservation purposes, and in Brazil they are seen by common people as vermin or even as harmful creatures associated with blood-feeding vampires. Possibly because of this poor public image, bats have been largely neglected in conservation and environmental education programs in Brazil and elsewhere in South America.

Herein we examine the distributional patterns of Brazilian bats at the biome level, analyzing their endemism and rarity, as well as the effect of habitat modifications and the main threats for bats in Brazil. Because the guidelines to South America's bat biogeography were mostly established by Koopman (1982), the biome-level analysis seems appropriate for two reasons: (1) understanding of faunal relationships at a regional scale is enhanced; and (2) land use and natural resource utilization in Brazil are clearly associated with habitat physiognomy and vegetation types.

Methods

We followed Koopman (1993) as an authority for the taxonomic status and distribution of bats in Brazil and South America. However, the distributional ranges of a great number of bat species are far from being well defined. We established the database on the distribution of Brazilian bats from a number of different sources that include comprehensive works such as Vieira (1942), Cabrera (1957), and Koopman (1982, 1993); published studies on local or regional faunas, such as Handley (1967), Piccinini (1974), Reis and Schubart (1979), Taddei and Reis (1980), Mok et al. (1982), Schaller (1983), Willig (1983), Willig and Mares (1989), Reis (1984), Trajano (1984), and Peracchi and Albuquerque (1993); masters' theses and doctoral dissertations such as those by Marinho-Filho (1992), Pedro (1992), Aguiar (1994), Fazzolari-Corrêa (1995), and Lima (1995); recent papers describing species, setting new distributional limits, or reviewing the taxonomic status of some species such as Taddei and Pedro (1993), Pedro et al. (1994), Fazzolari-Corrêa (1994), and Zortéa and Taddei (1995); and unpublished original information from scientific collections at the Universidade de Brasília, the Museu de História Natural da Universidade de Campinas, and the authors' personal observations.

This database generated a checklist of bat species for the entire country as well as for each of the major Brazilian biomes (Appendix 20.1). Because many species of bats present great distributional ranges, for the analysis of endemism we verified the distribution of each species for the whole neotropical region. There are still many gaps to be filled, controversies about the status of some species, and species to be described, but even with these potential sources of error the general picture seems adequate for our purposes.

The limits of each biome are those defined by the Brazilian Institute of Geography and Statistics (IBGE 1993), with minor modifications. The limits of non-Brazilian Amazonia and southwestern borders of the Atlantic Forest were adapted from Hueck (1972). The maps showing these limits were produced by the Biodiversity Conservation Data Center of the Fundação Biodiversitas (Belo Horizonte, Brazil). The areas of the whole country and of each biome were calculated by means of a PC-based Geographical Information System (CISIG), developed by Conservation International.

To investigate the faunal affinities between areas, we used Whittaker's index for beta diversity (Whittaker 1960), which gives an idea of the species turnover from one sample to another. It is calculated from the formula

$$W = (s/a) - 1,$$

where s is the regional diversity (in our case the number of species combined for the two biomes considered) and a is mean alpha diversity (here the average richness of the two biomes). This index ranges from 0 (complete similarity) to 1 (highest species turnover or complete dissimilarity).

Brazilian Biomes

The five major Brazilian biomes are defined on a vegetational and physiognomic basis (Hueck 1972; Ab'Saber 1977), and therefore they do not necessarily correspond to well-defined zoogeographic units. There are minor enclaves of neighbor formations within all biomes, and the real figure tends to be more patchy than the representation on Figure 20.1. However, for our purposes we consider the whole of Amazonia and the Atlantic Forest as major habitats of tropical forest and the Caatinga, Cerrado, and Pantanal as habitat formations representing an open corridor between the two great forest systems.

Brazilian Amazonia, with 3,940,449 km^2, is a land of dense wet forests with the canopy about 40 m or more high (Murça-Pires and Prance 1985), an annual rainfall from 2,000 to 4,000 mm, well drained by large rivers, with an impressive biological diversity. The Atlantic Forest corresponds to 2,052,533 km^2 mainly along the seashore, entering to the west in the southern part of the country, covering the mountains (Figure 20.2a) that form the edge of the so-called Brazilian Central Plateau at altitudes to 2,500 m. Some parts of this region may receive as much as 3,000–3,500 mm annual rainfall, drained to the sea through an extense web of rapid mountain streams and small rivers. This forest formation is now reduced to approximately 5% of its original area (Fonseca 1985). The Caatinga, with 770,442 km^2 in the Brazilian semiarid region, is a complex of thorn forests, with small trees and shrubs (Figure 20.2b), and open formations growing over dry and poor soils (Willig 1983). Most of the area receives no more than 500 mm of annual rainfall. In places where rivers and lakes persist during the dry season, one can find mesic vegetation such as gallery forests. The Cerrado contains 2,052,533 km^2 and

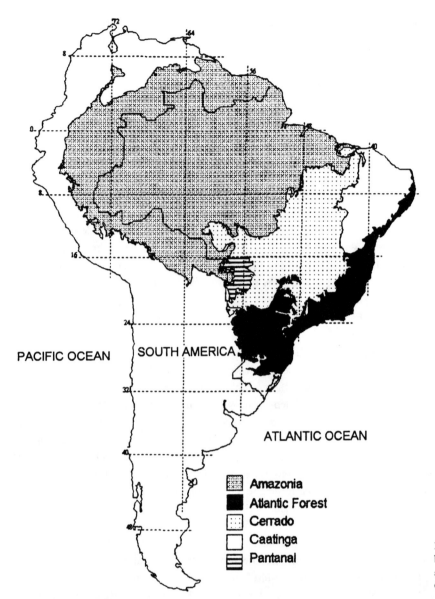

PACIFIC OCEAN

SOUTH AMERICA

ATLANTIC OCEAN

▦ Amazonia
■ Atlantic Forest
☐ Cerrado
☐ Caatinga
▤ Pantanal

Figure 20.1. Map of the five major Brazilian biomes. (From IBGE 1993; GIS treatment and analysis by the Fundação Biodiversitas / Conservation International.)

is the greatest and richest Neotropical savannah (Goodland and Ferri 1979; Sarmiento 1983). It covers aproximately 20% of the country's area and lies over the Central Brazilian Plateau, at altitudes from 600 to 1,400 m. A number of different types of habitat ranging from open grasslands to semideciduous forests may occur side by side (Figure 20.2c). Gallery forests along the margins of the rivers seem to play a crucial role in the maintenance of high levels of diversity (Redford and Fonseca 1986). The Pantanal is a great floodplain of 143,046 km², seasonally flooded by the waters of the Paraguay River. It is a mixture of grasslands, forest patches, and even cerrados on the higher parts of the terrain (Prance and Schaller 1982) that become true islands during the rainy season (Figure 20.2d). A more complete characterization of each major Brazilian biome can be found in Hueck (1972).

Brazilian Bat Fauna

Species Richness, Endemicity, Species–Area Relationships, and Faunal Affinities

Brazil harbors 137 of the 190 South American bat species, thus 72% of the total bat fauna (Appendix 20.1). Brazilian bat species are distributed among nine families and 56 genera. Phyllostomids, with 77 species, are by far the most speciose group, followed by molossids (18 species), vespertilionids (18 species), and emballonurids (15 species). Amazonia is the greatest Brazilian biome and has the richest bat fauna (Table 20.1). The Cerrado is the second largest biome in total area but the remnants of the Atlantic Forest hold a greater number of species. The other two open formations show less diversity than the forested areas and the Cerrado (Table 20.1).

Figure 20.2. Photos of four of the five major Brazilian biomes: (**a**) remnants of the Atlantic Forest, which are found mostly over rugged terrain; (**b**) a dry riverbed (foreground) of the Caatinga in northeastern Brazil, illustrative of the region's dry climate; (**c**) Cerrado (far background) and grasslands with termite mounds, which are found mostly in central Brazil; (**d**) the Pantanal flatlands in southwestern Brazil, which are seasonally flooded by the Paraguay River.

Endemism for bats is relatively low in Brazil. The relative frequencies of endemics for the seven South American zoogeographic areas defined by Koopman (1982) range from 3% to 39%, and most of them were between 4% and 13%. In Brazil, 19 species of bats (13.8% of the total fauna) are endemic. Amazonia accounts for 13 endemic species, the Atlantic Forest has 5, and the Cerrado only 1 (see Table 20.1). The two forest biomes support 95% of the Brazilian

endemic bat species. It is important to note that, despite the loss of more than 90% of its original area, the Atlantic Forest remnants still support high levels of species richness and endemicity (Table 20.1).

There is a positive and statistically significant relationship between the size and number of species per biome (Figure 20.3). The two forest formations are above the regression line, suggesting a higher than expected number of species. This correlation may indicate that Brazilian bats are primarily associated with forested areas. Even in the open formations, the gallery forests and forest patches seem to play a crucial role for the maintenance of bat populations (Marinho-Filho and Reis 1989).

The beta-diversity indexes indicate that the three open formations are very similar in species composition. Although it may appear that there is a corridor between the two forested areas, they are more similar to each other than to any forested area. Amazonia and the Atlantic Forest are very similar also, although separated by the open corridor and considered by Koopman (1982) as two distinct zoogeographic provinces. This is only an apparent contradiction—the relatively low endemism for bats in South America

Table 20.1

Bat Species Richness and Endemism in the Five Major Brazilian Biomes

Biome	Area (km²)	No. of species	Endemic species No.	Endemic species %
Amazonia	3,940,449	117	13	11.1
Atlantic Forest	1,272,539	96	5	5.2
Cerrado	2,052,533	80	1	1.2
Caatinga	770,442	69	0	0
Pantanal	143,046	61	0	0

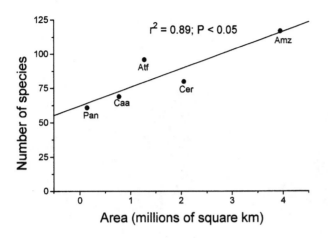

Figure 20.3. Linear regression of bat species richness and area of the five major Brazilian biomes: Amz, Amazonia; Atf, Atlantic Forest; Caa, Caatinga; Cer, Cerrado; Pan, Pantanal.

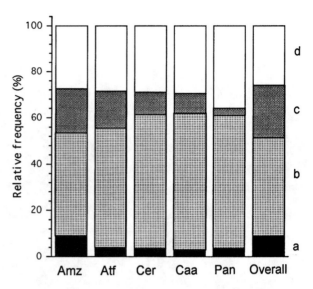

Figure 20.4. Relative frequencies of Brazilian bat species in the four categories of rarity defined by Arita (1993): (a) locally abundant with restricted distribution; (b) locally abundant and widespread; (c) locally scarce with restricted distribution; (d) locally scarce and widespread. Biomes: Amz, Amazonia; Atf, Atlantic Forest; Caa, Caatinga; Cer, Cerrado; Pan, Pantanal.

reduces the species turnover between areas, resulting in great similarity.

Distribution and Abundance

Another important concept related to conservation biology is rarity. We followed Arita (1993), who classified the number of Neotropical bats into four categories, considering their distributional ranges and relative abundance from data in the literature: (1) species locally abundant with restricted distribution; (2) species locally abundant and widespread; (3) species locally scarce with restricted distribution; and (4) species locally scarce and widespread (Figure 20.4). Even considering the variation in numbers that populations of species with wide distribution may present, Arita's classification seems adequate enough for most Brazilian bat species and therefore helps to delineate a general picture. Although there is no information for all species in each biome, the sample is quite representative. Approximately 42% of the Brazilian bat species may be considered as common and widespread, 47% of the species appear rare, and at least 21% may deserve special attention; they seem rare and have restricted distribution (Figure 20.4).

Amazonia has the highest number of rare species and also the highest proportion of rare species with restricted distribution. The forest biomes show a higher proportion of rare or restricted species than the open ones (Figure 20.4).

Conservation of Bats in Brazil

State of the Art

The Brazilian official list of endangered mammal species (see Bernardes et al. 1990) contains 58 species threatened

with extinction: 25 primates, 13 carnivores, 4 xenarthrans, 2 sirenians, 3 cetaceans, 7 rodents, and 3 artiodactyls, which represent roughly 12% of the whole Brazilian mammal fauna. The high number of endangered species of primates and carnivores in the list reflects the impact of hunting and deforestation, especially in the coastal zone, but also reflects the fact that primatology is one of the best developed branches of mammalogy in Brazil and thus has the greatest input.

For Brazilian bats, data on basic natural history are lacking for most if not all species. Available information is concentrated on a few and widespread species such as *Artibeus lituratus, Desmodus rotundus, Molossus ater,* and *Tadarida brasiliensis* (Sazima 1978; Marques 1986; Uieda 1992; Marques and Fabián 1994; I. Sazima et al. 1994). For phyllostomids, for instance, studies on bat–plant interactions predominate and are concerned with flower pollination (Carvalho 1960; Sazima and Sazima 1978; Gribel and Hay 1993; M. Sazima et al. 1994), frugivory, folivory, and fruit resource sharing (Uieda and Vasconcellos-Neto 1985; Marinho-Filho 1991; Zortéa and Mendes 1993; Marinho-Filho and Vasconcellos-Neto 1994). There are few studies on reproduction and activity patterns (Taddei 1976; Reis 1984; Marinho-Filho and Sazima 1989). Community studies are still scarcer but may give some hints on conservation biology (Trajano 1985; Pedro 1992; Aguiar 1994; Reis and Muller 1995). However, most data available from field studies are fragmentary and insufficient to even document the process of declining populations.

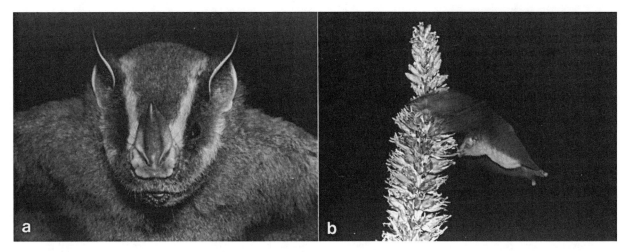

Figure 20.5. Two vulnerable Brazilian bats: (**a**) the stenodermatine *Chiroderma doriae;* and (**b**) the glossophagine *Lonchophylla bokermanni* (shown here taking nectar from a bromeliad inflorescence). Both species have a patchy distribution over a restricted area in southeastern Brazil.

The absence of bats in the official list of endangered species reflects more the difficulties in assessing the conservation status of bat species in Brazil than the nonexistence of risk to them. Recently a group of Brazilian chiropteran specialists in a meeting held at the Museu de Biologia Prof. Mello Leitão, Santa Teresa, Espírito Santo, recognized nine bat species as being "vulnerable," and recommended these be included in the Brazilian Official List of Species Threatened with Extinction: one emballonurid, *Saccopteryx gymnura;* five phyllostomids, *Chiroderma doriae, Lichonycteris obscura, Lonchophylla bokermanni, Platyrrhinus recifinus,* and *Vampyrum spectrum;* and three vespertilionids, *Lasiurus ebenus, Lasiurus egregius,* and *Myotis ruber* (Figure 20.5).

Threats to Bat Populations

Direct threats to bat populations caused by roost disturbance or pest species management do exist in Brazil but seem to be less important than habitat destruction or fragmentation. Roost disturbance is particularly important for cave-dwelling species in the karstic region in the southeast, where limestone exploitation is an important economic activity and tourism and cave visitation are not adequately controlled (Trajano 1995).

Inadequate vampire control may represent a great risk to other bat species populations. Ignorance and prejudice against bats in general may lead to extermination of bat colonies in refuges and roosting sites. However, this mismanagement represents isolated individual actions that are not encouraged by any Brazilian governmental agency.

There is only one published paper about the effects of habitat fragmentation on bat populations in Brazil. Reis and Muller (1995) sampled three forest patches of different sizes, remnants of the southern Atlantic Forest isolated by

intense agricultural activity, in a developed region of the Parana State, and the university campus in the urban area. They compared species composition and abundance, estimated by the frequency of capture of each bat species (Table 20.2). There was a trend of reduction in species

Table 20.2

Frequency of Mist-Net Captures of Bats, by Species, in Forest Remnants of Different Sizes and in an Urban Area in Southeastern Brazil

Species	680-ha forest	60-ha forest	6.2-ha forest	Urban area
Micronycteris megalotis	6	2	2	0
Phyllostomus hastatus	0	5	0	0
Chrotopterus auritus	8	0	1	0
Anoura caudifer	1	0	0	0
Carollia perspicillata	58	67	9	1
Sturnira lilium	42	41	7	28
Platyrrhinus lineatus	1	1	2	36
Vampyressa pusilla	6	0	0	0
Chiroderma doriae	0	0	1	0
Artibeus jamaicensis	13	7	1	0
Artibeus lituratus	96	234	51	351
Pygoderma bilabiatum	3	4	1	1
Desmodus rotundus	0	0	12	0
Myotis nigricans	6	14	1	0
Myotis ruber	3	0	0	0
Eptesicus furinalis	1	0	0	2
Eptesicus diminutus	1	0	0	0
Histiotus velatus	2	0	0	0
Lasiurus borealis	1	0	0	0
All species	15	11	11	6

Note: Data represent number of bats and are from Reis and Muller (1995).

richness and also in the abundance of most species, associated with the reduction of the size of the forest remnants and urbanization. It is important to note however that while some phyllostomids such as *Carollia perspicillata* seem to be negatively affected by forest habitat degradation, others such as *Sturnira lilium, Artibeus lituratus,* and *Platyrrhinus lineatus* apparently thrive in the urban environment. The two latter are generalist and versatile species that roost in trees or human settlements and forage on the fruits of native or introduced trees within the city and thus may even be favored by the urban environment (Sazima et al. 1994).

Aguiar (1994) studied three isolated remnants of the Atlantic Forest in eastern Brazil. These fragments were approximately the same size but were subject to different levels of disturbance. Despite the relatively small sample sizes, there was also a trend of reduction in species richness and in the abundance of most bat species, related to quality of the habitat. Again, *C. perspicillata* seemed to suffer an important reduction in abundance in the degraded habitat, whereas *S. lilium* and the vespertilionid *Myotis nigricans* were favored by the same condition.

Conservation Status of Brazilian Biomes

In a large country such as Brazil, environmental problems are historically and geographically dissimilar, depending on the different ecological and socioeconomic settings of each region.

Amazonia represents the greatest humid tropical forest area in the world, and it is the planet's largest biological reservoir. Although this may not be in accordance with the concepts and ideas broadcast by the communication media, which emphasize deforestation and environmental degradation, Amazonia remains in a relatively pristine stage. Estimates of the percentage of forest cover eliminated and degraded are still disputed, but it is accepted that not more than 12% of its original area in Brazil has been altered by humans, and at least 7.6% is legally protected by federal conservation units (CIMA 1991; Rylands 1991). However, it is known from previous experience in the states of Para and Rondonia that large-scale logging and colonization projects based on agriculture and cattle ranching have eliminated significant portions of native forest in just a few years (Fearnside 1986; Mahar 1989).

The Cerrado is the second largest Neotropical biome. It is also the main colonization frontier in the country, the native vegetation giving place to grain crops and cattle ranching (Klink et al. 1993). Virtually untouched until the first half of this century, this biome has experienced very high demographic and economic development, especially in the past 30 years. Statistical projections indicate that by the year 2000 about 50% of the total Cerrado area will be modified by human activities (Alho and Martins 1995). This situation is critical because only 1% of the Cerrado area is included in federal conservation units (Dias 1992).

The Caatinga stands as a dramatic scenario of poverty and deleterious use of natural resources, its fauna and flora included. The region, colonized long ago, bears a miserable and unassisted population who, without alternatives, have been destroying the environment and overexploiting the available resources for the last three centuries. The widespread gathering of firewood and hunting for food still represent an important threat to the biological diversity of the Caatinga. Impoundments along the main perennial rivers for hydroelectric plants are a recent additional threat to gallery forests, caves, and other native habitats associated with the river edges. Inadequate irrigation projects and traditional practices of slash-and-burn agriculture lead to a very high risk of desertification (CIMA 1991). To make things worse, only 0.1% of the Caatinga is included in federal conservation units (CIMA 1991).

Because of the difficulties in reaching its hinterland and the limits imposed by a peculiar hydric cycle, the main economic activity in the Pantanal since the beginning of the eighteenth century has been low-density, extensive cattle ranching (CIMA 1991). This practice never produced high economic standards in the region, but did provide a fairly good level of biological conservation associated with an established, sustainable economic activity. However, governmental policies aiming for acceleration of regional economic development, with support from international financial agencies, indicate that this figure is about to change radically. Nowadays the expansion of agriculture through the drainage of wet terrain and irrigation on the surrounding plateaus, and gold mining, as well as the establishment of a fluvial exportation corridor in the Paraguay River basin, are among the most important environmental threats to the Pantanal (Alho et al. 1987). Only 0.7% of this biome is protected in federal conservation units (CIMA 1991).

The Atlantic Forest is the region in which the initial colonization took place in Brazil and where, for more than three centuries, almost all economic activity was concentrated. This biome changed into an agricultural and industrial center. It is the most densely populated portion of the country and includes 2 of the 10 largest cities in the world, São Paulo and Rio de Janeiro. The forest cover is reduced to about 5% of its original area and only about 2% of these forest remnants is protected under federal conservation jurisdiction (CIMA 1991). Demographic pressure with accelerated growth of urban centers, expansion of agriculture and cattle farming, plus expansion of industrial areas and the associated pollutants are the current main threats to this biome. Most of the remaining forest lies in southeastern Brazil, whereas the greatest remnant of continuous forest in the northeast, the Biological Reserve of Una, has only 11,500 ha (Conservation International 1995). Although few, these remnants of the Atlantic Forest still bear an impressive biological diversity which, in the case of bats, corresponds to 71% of the Brazilian bat species.

Conclusions

From the major Brazilian biomes, Amazonia has the highest number of species and also most of the endemic or rare bat species. Because of its relative isolation and small human population it may be considered as the least disturbed biome in Brazil. The open formations (Cerrado, Caatinga, and Pantanal) have less species richness and fewer endemicity indicators, but it is important to note that this relative paucity refers to Brazilian standards and that the 59 species found in the Pantanal represent only one-third of all

South American bat species. These three biomes are under intense anthropogenic pressure and have an insignificant representation in the national conservation units system. The Atlantic Forest is undoubtedly the most threatened biome in Brazil and stands among those most imperiled on Earth (Mittermeyer 1988). In spite of the tremendous reduction in its original area and the current threats, this biome supports more than one-half the South American bat species.

It is clear that direct measures to counteract the loss of Brazilian bat diversity are unlikely to be realized in the short term. Although a list of vulnerable species may attract attention to this mammal group and perhaps contribute to a more general conservation awareness, the protection of native areas probably is more productive and cost effective than species-oriented strategies. The trends that we indicate for the Brazilian bat fauna and its conservation probably extend to most of the tropical and subtropical regions in South America.

Appendix 20.1.
Distribution of Brazilian Bat Species in the Five Major Biomes of Brazil

Endemic species are identified by an asterisk (for Amazonia), a dagger (for Atlantic Forest), or a double dagger (for Cerrado).

Taxon	Amazonia	Atlantic Forest	Cerrado	Caatinga	Pantanal
Family Emballonuridae					
(7 genera, 15 species)					
Centronycteris maximiliani	–	+	+	+	–
Cormura brevirostris	+	–	–	–	–
Cyttarops alecto	+	–	–	–	–
Diclidurus albus	+	+	–	+	–
Diclidurus ingens	+	–	–	–	–
* *Diclidurus isabellus*	+	–	–	–	–
* *Diclidurus scutatus*	+	–	–	–	–
Peropteryx kappleri	+	+	+	–	+
Peropteryx leucoptera	+	+	–	–	–
Peropteryx macrotis	+	+	+	+	+
Rhynchonycteris naso	+	+	+	+	+
Saccopteryx bilineata	+	+	+	+	+
Saccopteryx canescens	+	–	–	–	–
* *Saccopteryx gymnura*	+	–	–	–	–
Saccopteryx leptura	+	+	+	+	+
Family Noctilionidae					
(1 genus, 2 species)					
Noctilio albiventris	+	+	+	+	+
Noctilio leporinus	+	+	+	+	+
Family Mormoopidae					
(1 genus, 3 species)					
Pteronotus gymnonotus	+	–	+	–	+
Pteronotus parnellii	+	–	+	+	+
Pteronotus personatus	+	–	+	+	+

Taxon (continued)	Amazonia	Atlantic Forest	Cerrado	Caatinga	Pantanal
Family Phyllostomidae					
(33 genera, 77 species)					
Chrotopterus auritus	+	+	+	−	+
Lonchorhina aurita	+	+	+	+	+
Macrophyllum macrophyllum	+	+	+	+	+
Micronycteris behnii	+	−	+	−	+
Micronycteris brachyotis	+	+	−	−	−
Micronycteris hirsuta	+	+	−	−	−
Micronycteris megalotis	+	+	+	+	−
Micronycteris minuta	+	+	+	+	+
Micronycteris nicefori	+	+	−	−	−
* Micronycteris pusilla	+	−	−	−	−
Micronycteris schmidtorum	+	−	−	−	−
Micronycteris sylvestris	+	+	−	+	−
Mimon bennettii	+	+	+	+	−
Mimon crenulatum	+	+	+	+	−
Phylloderma stenops	+	+	+	−	−
Phyllostomus discolor	+	+	+	+	+
Phyllostomus elongatus	+	+	+	+	+
Phyllostomus hastatus	+	+	+	+	+
Tonatia bidens	+	+	+	+	+
Tonatia brasiliense	+	+	+	+	−
* Tonatia carrikeri	+	−	−	−	−
* Tonatia schulzi	+	−	−	−	−
Tonatia silvicola	+	+	+	+	+
Trachops cirrhosus	+	+	+	+	+
Vampyrum spectrum	+	−	−	−	+
Lionycteris spurrelli	+	−	−	−	−
Lonchophylla bokermanni	−	+	+	−	−
‡ Lonchophylla dekeyseri	−	−	+	−	−
Lonchophylla mordax	−	+	−	+	−
Lonchophylla thomasi	+	−	−	−	−
Anoura geoffroyi	+	+	+	+	+
Anoura caudifer	+	+	+	+	+
Choeroniscus intermedius	+	−	−	−	−
Choeroniscus minor	+	+	+	−	−
Glossophaga commissarisi	+	−	−	−	−
Glossophaga longirostris	+	−	−	−	−
Glossophaga soricina	+	+	+	+	+
Linchonycteris obscura	+	+	−	−	−
* Scleronycteris ega	+	−	−	−	−
Carollia brevicauda	+	+	−	+	+
Carollia castanea	+	−	−	−	−
Carollia perspicillata	+	+	+	+	+
* Rhinophylla fischerae	+	−	−	−	−
Rhinophylla pumilio	+	+	+	+	−
Ametrida centurio	+	−	−	−	−
Artibeus anderseni	+	−	−	−	−
Artibeus cinereus	+	+	+	+	+
Artibeus concolor	+	+	+	+	−
Artibeus fimbriatus	−	+	−	−	+
Artibeus glaucus	+	+	−	−	+
Artibeus jamaicensis	+	+	+	+	+
Artibeus lituratus	+	+	+	+	+
Artibeus obscurus	+	+	−	+	+

Taxon (continued)	Amazonia	Atlantic Forest	Cerrado	Caatinga	Pantanal
Artibeus planirostris	+	+	+	–	–
† Chiroderma doriae	–	+	–	–	–
Chiroderma trinitatum	+	–	+	–	–
Chiroderma villosum	+	+	+	–	+
Mesophylla macconnelli	+	–	–	–	–
Platyrrhinus brachycephalus	+	–	–	–	–
Platyrrhinus helleri	+	–	+	–	+
Platyrrhinus infuscus	+	–	–	–	–
Platyrrhinus lineatus	+	+	+	+	+
† Platyrrhinus recifinus	–	+	–	–	–
Pygoderma bilabiatum	–	+	–	–	+
Sphaeronycteris toxophyllum	+	–	–	–	–
* Sturnira bidens	+	–	–	–	–
Sturnira lilium	+	+	+	+	+
Sturnira tildae	+	+	+	+	–
Uroderma bilobatum	+	+	+	+	+
Uroderma magnirostrum	+	+	+	+	–
Vampyressa bidens	+	–	–	–	–
* Vampyressa brocki	+	–	–	–	–
Vampyressa pusilla	–	+	+	–	–
Vampyrodes caraccioli	+	–	–	–	–
Desmodus rotundus	+	+	+	+	+
Diaemus youngi	+	+	+	+	+
Diphylla ecaudata	+	+	+	+	+
Family Natalidae					
(1 genus, 1 species)					
Natalus stramineus	+	+	–	+	–
Family Furipteridae					
(1 genus, 1 species)					
Furipterus horrens	+	+	+	+	–
Family Thyropteridae					
(1 genus, 2 species)					
Thyroptera discifera	+	–	–	–	–
Thyroptera tricolor	+	+	–	–	–
Family Vespertilionidae					
(5 genera, 18 species)					
Eptesicus brasiliensis	+	+	+	+	+
Eptesicus diminutus	–	+	+	+	–
Eptesicus furinalis	+	+	+	+	+
† Histiotus alienus	–	+	–	–	–
Histiotus montanus	–	+	–	–	–
Histiotus velatus	–	+	+	–	–
Lasiurus borealis	+	+	+	+	+
Lasiurus cinereus	–	+	+	–	+
† Lasiurus ebenus	–	+	–	–	–
Lasiurus ega	+	+	+	+	+
Lasiurus egregius	–	+	–	–	–
Myotis albescens	+	+	+	+	+
Myotis levis	–	+	–	–	–
Myotis nigricans	+	+	+	+	+
Myotis riparius	+	+	+	+	+
† Myotis ruber	–	+	–	–	–
Myotis simus	+	+	–	–	–
Rhogeessa tumida	+	+	+	+	–

Taxon *(continued)*	Amazonia	Atlantic Forest	Cerrado	Caatinga	Pantanal
Family Molossidae					
(6 genera, 18 species)					
Eumops auripendulus	+	+	+	+	+
Eumops bonariensis	+	+	+	+	+
Eumops glaucinus	+	+	+	+	+
Eumops hansae	+	+	+	–	–
Eumops perotis	+	+	+	+	+
Molossops abrasus	+	–	+	–	+
Molossops greenhalli	+	+	–	+	–
Molossops matogrossensis	+	+	+	+	–
* *Molossops neglectus*	+	–	–	–	–
Molossops planirostris	+	+	+	+	+
Molossops temminckii	–	+	+	+	+
Molossus ater	+	+	+	+	+
Molossus molossus	+	+	+	+	+
Nyctinomops aurispinosus	+	+	+	+	–
Nyctinomops laticaudatus	+	–	+	+	+
Nyctinomops macrotis	+	+	+	+	+
Promops nasutus	+	+	+	+	–
Tadarida brasiliensis	–	+	+	–	–
Total (56 genera, 137 species)	117	96	80	69	61

Acknowledgments

We thank R. Machado, the Fundação Biodiversitas, and Conservation International for the GIS treatment and analysis of the Brazilian biomes map. Marina Anciães kindly helped with data compilation. CNPq provided financial support (grant 300992/79) to I.S.

Literature Cited

Ab'Saber, A. N. 1977. Os domínios morfoclimáticos na América do Sul: Primeira aproximação. Geomorfologia 52:1–21.

Aguiar, L. M. S. 1994. Comunidades de Chiroptera em três áreas de Mata Atlântica em diferentes estádios de sucessão, Estação Biológica de Caratinga, Minas Gerais. Master's thesis, Universidade Federal de Minas Gerais, Belo Horizonte, Brasil.

Alho, C. J. R., and E. S. Martins, eds. 1995. De Grão em Grão o Cerrado Perde Espaço. World Wildlife Fund and Sociedade de Pesquisas Ecológicas do Cerrado, Brasília.

Alho, C. J. R., T. E. Lacher, Jr., and H. C. Gonçalves. 1987. Environmental degradation in the Pantanal ecosystem. Bioscience 38: 164–171.

Arita, H. T. 1993. Rarity in Neotropical bats: Correlations with phylogeny, diet, and body mass. Ecological Applications 3:500–517.

Bernardes, A. T., A. B. M. Machado, and A. Rylands. 1990. Fauna Brasileira Ameaçada de Extinção. Fundação Biodiversitas, Belo Horizonte, Brasil.

Cabrera, A. 1957. Catalogo de los Mamíferos de America del Sur. Revista del Museo Argentino de Ciencias Naturales "Bernardino Rivadavia" 4:1–307.

Carvalho, C. T. 1960. Das visitas de morcegos às flores (Mammalia, Chiroptera). Anais da Academia Brasileira de Ciências 32: 359–377.

CIMA. 1991. Subsídios Técnicos para Elaboração do Relatório Nacional do Brasil para a CNUMAD. IBAMA, Brasília.

Conservation International. 1995. Prioridades para Conservação da Biodiversidade da Mata Atlântica do Nordeste (map). Conservation International, Fundação Biodiversitas, and Sociedade Nordestina de Ecologia, Washington, D.C.

Dias, B. F. S. 1992. Alternativas de Desenvolvimento dos Cerrados: Manejo e Conservação dos Recursos Naturais Renováveis. FUNATURA/IBAMA, Brasília.

Fazzolari-Corrêa, S. 1994. *Lasiurus ebenus,* a new vespertilionid bat from southeastern Brazil. Mammalia 58:119–123.

Fazzolari-Corrêa, S. 1995. Aspectos sistemáticos, ecológicos e reprodutivos de morcegos na Mata Atlântica. Doctoral dissertation, Universidade de São Paulo, São Paulo, Brasil.

Fearnside, P. M. 1986. Human Carrying Capacity of the Brazilian Rain Forest. Columbia University Press, New York.

Fonseca, G. A. B. 1985. The vanishing Brazilian Atlantic forest. Biological Conservation 34:17–34.

Fonseca, G. A. B., C. M. R. Costa, R. B. Machado, and Y. L. R. Leite. 1994. Livro Vermelho dos Mamíferos Brasileiros Ameaçados de Extinção. Fundação Biodiversitas, Belo Horizonte, Brasil.

Goodland, R., and M. G. Ferri. 1979. Ecologia do Cerrado. Editora Itatiaia, Belo Horizonte, Brasil.

Gribel, R., and J. D. Hay. 1993. Pollination ecology of *Caryocar brasiliense* (Caryocaraceae) in central Brazil cerrado vegetation. Journal of Tropical Ecology 9:199–211.

Handley, C. O. 1967. Bats of the canopy of an Amazonian forest. Atas do Simpósio Sobre a Biota Amazônica (Zoologia) 5:211–215.

Hueck, K. 1972. As Florestas da América do Sul. Editora Polígono, São Paulo, Brasil.

IBGE. 1993. Mapa de Vegetação do Brasil. Fundação Instituto Brasileiro de Geografia e Estatística, Brasília.

Klink, C. A., A. G. Moreira, and O. T. Solbrig. 1993. Ecological impact of agricultural development in the Brazilian cerrado. *In* The World Savannas, M. D. Young and O. T. Solbrig, eds., pp. 259–282. Parthenon Publishing, New York.

Koopman, K. F. 1982. Biogeography of the bats of South America. *In* Mammalian Biology in South America, M. A. Mares and H. H. Genoways, eds., pp. 273–302. Pymatuning Laboratory of Ecology, University of Pittsburgh, Linesville, Pa.

Koopman, K. F. 1993. Order Chiroptera. *In* Mammal Species of the World: A Taxonomic and Geographic Reference, D. E. Wilson and D. M. Reeder, eds., pp. 137–241. Smithsonian Institution Press, Washington, D.C.

Lima, M. D. 1995. Estrutura morfométrica da taxocenose de morcegos do Brasil central. Master's thesis, Universidade de Brasília, Brasília.

Mahar, D. J. 1989. Government Policies and Deforestation in Brazil's Amazon Region. World Wildlife Fund, The Conservation Foundation, International Bank for Reconstruction and Development, and The World Bank, Washington, D.C.

Marinho-Filho, J. 1991. The coexistence of two frugivorous bat species and the phenology of their food plants in southeastern Brazil. Journal of Tropical Ecology 7:59–67.

Marinho-Filho, J. 1992. Ecologia e história natural das interações entre palmeiras, epífitas e frugívoros na região do Pantanal Matogrossense. Doctoral dissertation, Universidade Estadual de Campinas, Campinas, Brasil.

Marinho-Filho, J., and M. L. Reis. 1989. A fauna de mamíferos associada às matas de galeria. *In* Anais do Simpósio Sobre Mata Ciliar, L. M. Barbosa, ed., pp. 43–60. Fundação Cargill, Campinas, Brasil.

Marinho-Filho, J., and I. Sazima. 1989. Activity patterns of six phyllostomid bat species in southeastern Brazil. Revista Brasileira de Biologia 49:777–782.

Marinho-Filho, J., and J. Vasconcellos-Neto. 1994. A dispersão de sementes de *Vismia cayennensis* (Guttiferae) por morcegos na região de Manaus. Acta Botanica Brasilica 8:87–96.

Marques, S. A. 1986. Activity cycle, feeding, and reproduction of *Molossus ater* (Chiroptera, Molossidae) in Brazil. Boletim do Museu Paraense Emilio Goeldi (Zoologia) 2:159–180.

Marques, R. V., and M. E. Fabián. 1994. Ciclo reprodutivo de *Tadarida brasiliensis* (I. Geoffroy, 1924) (Chiroptera Molossidae) em Porto Alegre, Brasil. Iheringia Série Zoologia 77:45–56.

Mittermeyer, R. A. 1988. Primate diversity and the tropical forest: Case studies from Brazil and Madagascar and the importance of the megadiversity countries. *In* Biodiversity, E. O. Wilson, ed., pp. 145–154. National Academy Press, Washington, D.C.

Mittermeyer, R. A., J. M. Ayres, T. Werner, and G. A. B. Fonseca. 1992. O país da megadiversidade. Ciência Hoje 14:20–27.

Mok, W. Y., D. E. Wilson, L. A. Lacey, and R. C. C. Luizão. 1982. Lista atualizada de quirópteros da Amazônia Brasileira. Acta Amazonica 12:817–823.

Murca-Pires, J., and G. T. Prance. 1985. The vegetation types of the Brazilian Amazon. *In* Amazonia, G. T. Prance and T. E. Lovejoy, eds., pp. 109–145. Pergamon Press, Oxford.

Pedro, W. A. 1992. Estrutura de uma taxocenose de morcegos da Reserva do Panga (Uberlândia, MG.), com ênfase nas relações tróficas em Phyllostomidae (Mammalia, Chiroptera). Master's thesis, Universidade Estadual de Campinas, Campinas, Brasil.

Pedro, W. A., C. A. Komeno, and V. A. Taddei. 1994. Morphometrics and biological notes on *Mimon crenulatum* (Chiroptera, Phyllostomidae). Boletim do Museu Paraense Emilio Goeldi 10:107–112.

Peracchi, A. L., and S. T. Albuquerque. 1993. Quirópteros do município de Linhares, estado do Espírito Santo, Brasil (Mammalia, Chiroptera). Revista Brasileira de Biologia 53:575–581.

Piccinini, R. S. 1974. Lista provisória dos quirópteros da coleção do Museu Paraense Emílio Goeldi (Chiroptera). Boletim do Museu Paraense Emílio Goeldi Zoologia 77:1–32.

Prance, G. T., and G. B. Schaller. 1982. Preliminary study of some vegetation types on the Pantanal, Mato Grosso, Brazil. Brittonia 34:228–251.

Redford, K. H., and G. A. B. Fonseca. 1986. The role of gallery forests in the zoogeography of the cerrado non-volant mammalian fauna. Biotropica 18:126–135.

Reis, N. R. 1984. Estrutura de comunidades de morcegos na região de Manaus, Amazonas. Revista Brasileira de Biologia 44:247–254.

Reis, N. R., and M. F. Muller. 1995. Bat diversity of forests and open areas in a subtropical region of south Brazil. Ecologia Austral 5:31–36.

Reis, N. R., and H. O. R. Schubart. 1979. Notas preliminares sobre os morcegos do Parque Nacional da Amazônia (Médio Tapajós). Acta Amazonica 9:507–515.

Rylands, A. B. 1991. The status of conservation areas in the Brazilian Amazon. World Wildlife Fund, Washington, D.C.

Sarmiento, G. 1983. The savannas of tropical America. *In* Tropical Savannas, F. Bourliére, ed., pp. 245–288. Elsevier, New York.

Sazima, I. 1978. Aspectos do comportamento alimentar do morcego hematófago *Desmodus rotundus*. Boletim de Zoologia da Universidade de São Paulo 3:97–119.

Sazima, I., W. A. Fischer, M. Sazima, and E. A. Fischer. 1994. The fruit bat *Artibeus lituratus* as a forest and city dweller. Ciência e Cultura (Sao Paulo) 46:164–168.

Sazima, M., and I. Sazima. 1978. Bat pollination of the passion flower *Passiflora mucronata* in southeastern Brazil. Biotropica 10:100–109.

Sazima, M., I. Sazima, and S. Buzato. 1994. Nectar by day and night: *Siphocampylus sulphureus* (Lobeliaceae) pollinated by hummingbirds and bats. Plant Systematics and Evolution 191:237–246.

Schaller, G. B. 1983. Mammals and their biomass on a Brazilian ranch. Arquivos de Zoologia 31:1–36.

Taddei, V. A. 1976. The reproduction of some Phyllostomidae (Chiroptera) from the northwestern region of the State of São Paulo. Boletim de Zoologia da Universidade de São Paulo 1:313–330.

Taddei, V. A., and W. A. Pedro. 1993. A record of *Lichonycteris* (Chiroptera, Phyllostomidae) from northeast Brazil. Mammalia 57:454–456.

Taddei, V. A., and N. R. Reis. 1980. Notas sobre alguns morcegos da Ilha de Maracá, Território Federal de Roraima (Mammalia, Chiroptera). Acta Amazonica 10:363–368.

Trajano, E. 1985. Ecologia de populações de morcegos cavernícolas em uma região cárstica do sudeste do Brasil. Revista Brasileira de Zoologia 2:255–320.

Trajano, E. 1995. Protecting caves for the bats or bats for the caves? Chiroptera Neotropical 1(2): 19–22.

Uieda, W. 1992. Período de atividade alimentar e tipos de presa dos morcegos hematófagos (Phyllostomidae) no sudeste do Brasil. Revista Brasileira de Biologia 52:563–573.

Uieda, W., and J. Vasconcellos-Neto. 1985. Dispersão de *Solanum* spp. (Solanaceae) por morcegos, na região de Manaus, AM, Brasil. Revista Brasileira de Zoologia 2:449–458.

Vieira, C. O. C. 1942. Ensaio monográfico sobre os quirópteros do Brasil. Arquivos de Zoologia (Sao Paulo) 3:219–471.

Whittaker, R. M. 1960. Vegetation of the Siskiyou Mountains, Oregon and California. Ecological Monographs 30:279–338.

Willig, M. R. 1983. Composition, microgeographic variation, and sexual dimorphism in Caatinga and Cerrado bat communities from northeast Brazil. Bulletin of the Carnegie Museum of Natural History 23:1–131.

Willig, M. R., and M. A. Mares. 1989. Mammals from the Caatinga: An updated list and summary of recent research. Revista Brasileira de Biologia 49:361–367.

Zortéa, M., and S. L. Mendes. 1993. Folivory in the big fruit-eating bat, *Artibeus lituratus* (Chiroptera, Phyllostomidae), in eastern Brazil. Journal of Tropical Ecology 9:117–120.

Zortéa, M., and V. A. Taddei. 1995. Taxonomic status of *Tadarida espiritosantensis* Ruschi, 1951 (Chiroptera, Phyllostomidae). Boletim do Museu de Biologia Mello Leitão (nova série) 2:15–21.

21
The Middle American Bat Fauna
Conservation in the Neotropical–Nearctic Border

HÉCTOR T. ARITA AND JORGE ORTEGA

When the Spaniards arrived more than 500 years ago at what is now Middle America, they were looking for gold and other treasures. They did not realize that one of the major fortunes of the region was everywhere around them: the impressive biodiversity. Middle America has, in fact, one of the richest and most distinct faunas of the world because it constitutes the convergence zone of two of the major zoogeographical realms—the Neotropical and the Nearctic regions.

Bats are an important component of the impressive fauna of Middle America. In this chapter, we analyze the chiropteran fauna of the region, briefly explore its diversity, and identify the major threats to this biological wealth. One of our main objectives is to identify important elements for the conservation of the bat fauna at the species and the regional levels. Accordingly, this chapter is organized into four main sections: a description of Middle America, a summary of the bat fauna of the region, an analysis at the species level, and a survey at the regional scale.

Middle America

Middle America, as defined here, includes Mexico and the seven countries of Central America—Belize, Guatemala, El Salvador, Honduras, Nicaragua, Costa Rica, and Panama. Anthropologists have used the term Mesoamerica to refer to the area that was the seat of the great pre-Columbian civilizations, including parts of Mexico and about half of Central America. Cultural geographers use the term Middle America to include Mexico, Central America, and the Caribbean islands. In this chapter, we use this latter definition, but we restrict our analysis to the mainland portion of the region.

Bisected by the tropic of Cancer, the region extends from about 7° to 32° N latitude and totals approximately 2.5 million square kilometers in area. From the U.S.–Mexico border, where the distance between the Pacific and Atlantic Oceans is about 2,500 km, the region narrows gradually until reaching a width of only 150 km in the Darien region of Panama. This funnel-shaped morphology, the geographic position, and the complicated topography determine most of the climatic, biological, and even social features of Middle America.

Heterogeneity is a major characteristic of this region. Middle America is part of the so-called circum-Pacific Ring of Fire, and is one of the most active tectonic areas of the world. Highlands predominate over lowlands, and a high percentage of the land is located in mountainous areas. In the north, the Mexican plateau, a high-altitude and relatively flat area, is bordered by the eastern and western Sierras Madre. From central Mexico southward, volcanic rims dominate the landscape. Major features include the Mexican volcanic belt, the sierras of Oaxaca and northern Chiapas, and the Central American volcanic belt that extends from southern Chiapas to Panama and which forms the watershed between the Pacific and the Caribbean versants along most of Central America.

The two coastal plains feature contrasting characteristics. The Atlantic coastal plain extends with no major interruption from the U.S. border to Guatemala, Belize, and the Yucatan peninsula. In Central America, the plain forms the extensive mosquito coast in parts of Honduras, Nicaragua, and Costa Rica. The Pacific lowlands, in contrast, form a very narrow and often interrupted strip that in no place extends to more than a hundred kilometers in length.

Climate shows considerable variation. North of the tropic of Cancer, with the exception of small patches in the mountains, sites are temperate and dry, with annual rainfall averaging less than 500 mm. In the intertropical portion of Middle America, climate is determined by altitude, rather than by latitude. Lands located below 1,500 m are hot (mean temperature higher than 22°C), whereas highlands above 2,500 m are cold (mean annual temperature less than 15°C). Zones located between 1,500 and 2,500 m, where the bulk of the human population is concentrated, present a mild climate (mean temperature between 16° and 22°C).

In general, the Pacific lowlands are much drier than the Atlantic plains. In most areas along the Gulf of Mexico and Caribbean coasts, annual rainfall exceeds 2,000 mm, and the dry season is barely apparent. In contrast, along the Pacific lowlands annual rainfall ranges from 1,000 to 2,000 mm, and a distinct dry season is the rule, normally between November and April.

An intricate mosaic of vegetation types mirrors the complexity of the topography and climate of Middle America. The use of fine-grained classification systems is difficult for such a complex pattern. For example, at least 12 of Holdrige's life zones occur in Costa Rica alone (Holdrige 1967; Hartshorn 1983). For the purposes of this chapter, we use the simpler system proposed by Rzedowski (1978), who classified the vegetation of Mexico into 10 main types: evergreen tropical forest, subdeciduous tropical forest, deciduous tropical forest, thorn forest, grassland, xerophilic brushland, oak forest, coniferous forest, cloud forest, and aquatic vegetation.

A noticeable characteristic of Middle America is the pronounced variation in vegetation types, even over short distances. In a 60-km transect from Mexico City to Cuernavaca, for example, it is possible to observe aquatic vegetation, xerophilic brushland, coniferous forests, alpine grasslands, a relict cloud forest, oak forests, and tropical deciduous forest. In a similar transect across Central America at almost any point, it would be possible to encounter most tropical vegetation types, from deciduous forests on the Pacific Coast to rainforests in the Atlantic versant.

The fauna of Middle America is characterized by a mixture of Nearctic and Neotropical components. Central America was separated from South America until about 3 M.Y.B.P. (millions of years before present), so most Neotropical terrestrial animals are recent additions to the fauna of what is now Middle America. Furthermore, because dry conditions predominated during most of the Pleistocene, it has been argued that rainforest has existed in Middle America as a major vegetation type only since the end of the Pleistocene (Rich and Rich 1983).

The intricate topography and mixture of vegetation types have promoted the isolation of animal populations and the eventual formation of full species with small distributional ranges. As a result, approximately 2,500 vertebrate species are endemic to Middle America (Flores-V. and Gerez 1994).

Middle America is also a complex mosaic of human groups. The population totals about 120.5 million people, but density varies greatly, from 8.9 persons per square kilometer in Belize to 262.2 in El Salvador. In Central America, human populations concentrate at middle altitudes, particularly along the Pacific versant. In Mexico, larger cities are located in the northern part of the country, particularly in the Mexican plateau. Lowlands are more sparsely populated and, in general, are less developed.

The Middle American Bat Fauna

The Middle American chiropteran fauna has received considerable attention in the past 15 years. The basic reference for the bat fauna of the region is still Hall (1981). Recent changes in nomenclature, as well as new records for the entire Middle American bat fauna, were summarized by Jones et al. (1988). Monographs and checklists for individual countries include Villa-R. (1966), Ramírez-P. et al. (1986), and Cervantes et al. (1994) for Mexico; Jones (1966) for Guatemala; McCarthy (1987) for Belize; Owen et al. (1991) for El Salvador; Jones and Owen (1986) for Nicaragua; Wilson (1983) for Costa Rica; and Eisenberg (1989) for Panama.

Koopman (1993) presented a comprehensive review of the nomenclature of chiropteran species. To maintain consistency with the rest of this volume, we adopted his usage

Table 21.1

Species Richness and Endemism for Ten Zoogeographic Regions of the New World

Region	No. of genera	No. of species	Endemic species No.	Endemic species %	Data source
(1) Patagonia	27	55	4	7.3	Koopman 1982; Redford and Eisenberg 1992
(2) Eastern Brazil	45	90	5	5.6	Adapted from Koopman 1982
(3) Amazon Basin	53	142	20	14.1	Adapted from Koopman 1982
(4) Northern Andes	29	63	5	7.9	Adapted from Koopman 1982
(5) Northern Coast	50	121	6	5.0	Koopman 1982; Eisenberg 1989
(6) Peruvian Coast	16	19	7	36.8	Adapted from Koopman 1982
(7) Northwest Coast	40	69	6	8.7	Koopman 1982; Eisenberg 1989
(8) Antilles	29	49	23	46.9	Koopman 1989
(9) Middle America	67	166	28	16.9	This chapter
(10) North America	18	40	6	15.0	Jones et al. 1992

Notes: Regions 1–7 are South American regions as defined by Koopman (1982). Taxonomic nomenclature follows Koopman (1993). Because some of the data sources used different nomenclatures, the numbers given here may not exactly match those in the original sources.

of names as the basis for the analyses presented in this chapter. It is important to mention, however, that the list presents major discrepancies with the nomenclature proposed by other authorities on New World bat taxonomy.

The bat fauna of Middle America consists of a rich assemblage of 166 species, representing 67 genera in nine families. These figures imply that 38% of the 177 genera and 18% of the approximately 925 bat species of the world occur in a region that represents less than 1.8% of the emergent area of the planet.

Compared with other regions of the New World, Middle America stands out as an important center of bat diversity because it harbors the richest fauna in terms of both number of species and number of genera (Table 21.1). Even the assemblages of larger areas, such as North America (Canada and the United States), the Amazon Basin, or Patagonia, are no match for the impressive richness of the Middle American bat fauna.

Middle America represents the transition between the Nearctic and the Neotropical realms. Although the border between the two regions is drawn at different latitudes by different authors, most authorities agree that the limit is located somewhere in Middle America. The bat fauna of the region shows this transitional nature in the frequency distribution of species among families (Figure 21.1). Although the Nearctic assemblage is dominated by vespertilionid species, the Neotropics are characterized by a high percentage of phyllostomids and the presence of such tropical families as the Emballonuridae, Noctilionidae, Natalidae, Thyropteridae, and Furipteridae. The Middle American fauna pre-

sents a proportion of vespertilionids and phyllostomids that is intermediate between the Nearctic and the Neotropical assemblages, but the presence of all tropical families demonstrates that a major portion of Middle America shows a clear Neotropical affinity.

Another important characteristic of the Middle American bat fauna is its high level of endemism. In all, 28 chiropteran species, representing 12 genera, are endemic to the region (Table 21.2). Additionally, 2 subspecies *(Myotis nigricans carteri* and *Molossus molossus aztecus)* that have been considered by some authors as full species (Bogan 1978; Dolan 1989) are also endemic to the region. In absolute numbers, no region of the New World harbors a higher number of exclusive species than Middle America. In terms of percentages, only the Antilles and the dry coastal plain of Peru show a higher degree of endemism (see Table 21.1).

Two genera *(Hylonycteris* and *Musonycteris)* are exclusive to the region. In addition, *Baeurus,* a subgenus of *Antrozous* considered by most authorities as a full genus (Engstrom and Wilson 1981; Wilson 1991), is also endemic to Middle America.

Conservation at the Species Level

One possible strategy for the conservation of a given fauna is to focus efforts on particular species that are considered to be of foremost importance or which are thought to be more susceptible to extinction. A main goal of such strategies is the objective identification of taxa that (1) for some

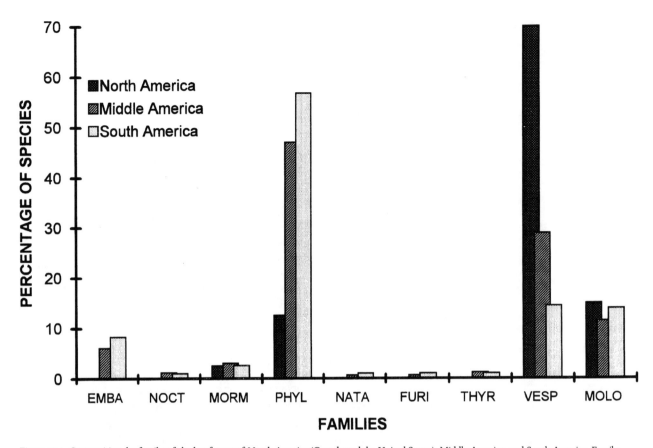

Figure 21.1. Composition, by family, of the bat faunas of North America (Canada and the United States), Middle America, and South America. Family names are abbreviated as follows: EMBA, Emballonuridae; FURI, Furipteridae; MOLO, Molossidae; MORM, Mormoopidae; NATA, Natalidae; NOCT, Noctilionidae; PHYL, Phyllostomidae; THYR, Thyropteridae; VESP, Vespertilionidae.

Table 21.2

Endemic Species of Chiroptera in Middle America

Emballonuridae	Vespertilionidae
Balantiopteryx io	*Antrozous dubiaquercus*[b]
	Myotis cobanensis
Phyllostomidae	*Myotis elegans*
Artibeus aztecus	*Myotis findleyi*
Artibeus hirsutus	*Myotis fortidens*
Artibeus inopinatus	*Myotis milleri*
Artibeus toltecus	*Myotis peninsularis*
Carollia subrufa	*Myotis planiceps*
Glossophaga leachii	*Myotis vivesi*
Glossophaga morenoi	*Plecotus mexicanus*
Hylonycteris underwoodi[a]	*Rhogeessa aeneus*[c]
Musonycteris harrisoni[a]	*Rhogeessa alleni*
Sturnira mordax	*Rhogeessa genowaysi*
Tonatia evotis	*Rhogeessa gracilis*
	Rhogeessa mira
	Rhogeessa parvula

[a]Endemic genus.

[b]Considered a distinct genus *(Bauerus)* by Engstrom and Wilson (1981). Under this arrangement, the genus would be endemic to Middle America.

[c]Not listed in Koopman (1993). Proposed as a distinct species by Audet et al. (1993).

reason have a higher conservation value and (2) are more susceptible to extinction.

Assigning Conservation Values

The conservation value of an animal species can be estimated using different criteria. Direct economic value, either by consumption or by the sale of the animal, its parts, or its products, is negligible in the case of Middle American bats. No chiropteran species is consumed or sold on a regular basis in Latin America, except for very special purposes, such as witchcraft at a very local scale. Similarly, the sale of bat guano is no longer profitable, although it is still used in some localities as a fertilizer. In the case of New World bats, direct economic importance cannot be a decisive criterion for assigning conservation value.

Some bat species could be assigned an indirect economic value. Insectivores that destroy tons of insects, or nectarivores and frugivores that pollinate or disperse plants of economic importance, surely contribute a great deal to the economy of a nation. However, no study has yet determined the exact monetary value of these activities. A negative example is the detrimental effect of vampire bats (Des-

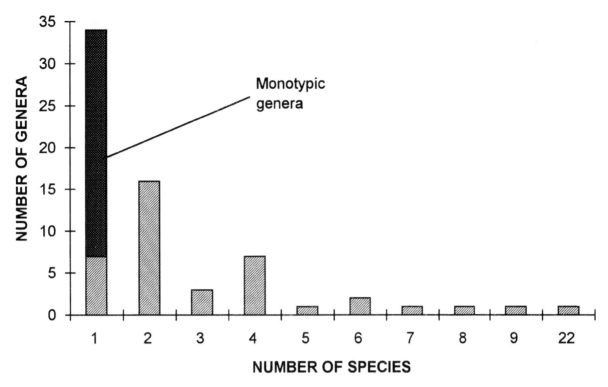

Figure 21.2. Frequency distribution of species among genera for the Middle American bat fauna. The black part of first bar denotes genera that include only one species at the global level.

modontinae) on the cattle industry in tropical America. Although the importance of these positive or negative values cannot be ignored, their use in a conservation campaign is limited by the lack of adequate estimates of the real impact on the economy.

In contrast, some bat species can have a high ecological value. In particular, many nectarivores and frugivores can be considered keystone species (sensu Gilbert 1980) because they function as efficient pollinators and seed dispersers. Moreover, because some of these species move along latitudinal or altitudinal lines, they can serve also as mobile links, binding together habitats that are spatially separated and would be totally unconnected otherwise. Long-nosed bats (*Leptonycteris curasoae*), for example, are migratory in northwestern Mexico, and they pollinate a wide range of tropical and subtropical plants along their migratory route.

In the same vein, frugivores such as epauletted (*Sturnira* spp.) and short-tailed (*Carollia* spp.) bats, among others, are key factors in the regeneration of tropical forests because they disperse the seeds of pioneer plants to areas that have been cleared. Unfortunately, the precise role of bats in Neotropical ecosystems has not been evaluated adequately, and it is difficult to assign conservation values to particular species.

Cultural values are also difficult to assign to particular species of bats because most people are unable to distinguish between them. A possible exception is the attraction

of tourists to caves or other roosting sites to watch large concentrations of bats (for example, the Mexican free-tailed bat, *Tadarida brasiliensis*) as they exit caves. The use of caves as tourist sites in Middle America is uncommon, however, and in almost no case are bats featured as an attraction.

From an evolutionary point of view, it is possible to assign conservation value to different species. All other things being equal, species that are sole representatives of a phylogenetic line should be given higher conservation values than species which share their evolutionary history with others. For some groups of Middle American bats, phylogenetic relationships are comparatively well known and conservation-oriented analyses of phylogeny are feasible. For example, such an analysis has been carried out for the nectar-feeding bats of Mexico (Arita and Santos del Prado, unpublished data).

Knowledge of the evolutionary relationships among Middle American bats, however, is incomplete for most groups, so a detailed phylogenetic analysis for the entire fauna is not possible. Instead, the principle of assigning higher values to species that are unique, in the evolutionary sense, can be maintained by giving special attention to species which represent monotypic genera or, even better, monotypic families.

In Middle America, 34 genera are represented by a single species each (Figure 21.2). Of these, 27 genera are truly monotypic, meaning that they do not have additional species outside Middle America. From an evolutionary perspective, one of these species should have higher conservation value than, for example, 1 of the 22 *Myotis* species of the region.

An alternative procedure for assigning special value to "unique" species is the selection of endemic species for conservation campaigns. Under many circumstances, endemic species are what defines the essence of a zoogeographic area. If an endemic species becomes extinct, the region loses a part of its identity. Also, the species is lost forever, because no populations exist anywhere else.

In addition to the 28 true endemic species listed in Table 21.2, other species can be considered "quasi-endemic" and should be given special value. These species are bats that have most of their distributional range within the limits of Middle America and barely penetrate to the United States or to South America. Examples include the California leaf-nosed *(Macrotus californicus),* the Mexican long-nosed *(Leptonycteris nivalis)* and hog-nosed *(Choeronycteris mexicana)* bats in North America, and the white *(Ectophylla alba)* and hairy-tailed *(Lasiurus castaneus)* bats, which were considered Middle American endemics until reported from Colombia by Cuervo et al. (1986).

Four endemic or quasi-endemic Middle American bats constitute monotypic genera: Underwood's long-tongued *(Hylonycteris underwoodi),* trumpet-nosed *(Musonycteris harrisoni),* hog-nosed *(C. mexicana),* and white *(E. alba)* bats. In any conservation strategy, these four species should be given special consideration.

Assessing Vulnerability to Extinction

Several physical and ecological features of plants and animals correlate with a higher probability of extinction (Terborgh 1974; Pimm et al. 1988). As with the criteria to establish conservation values, a precise determination of extinction probabilities is not feasible for Middle American bats because detailed demographic parameters are not known. In this chapter, we use rarity and additional indirect criteria to estimate the relative vulnerability of species.

There are three ways in which a species can be considered rare (Rabinowitz et al. 1986): It can be locally scarce, it can have a restricted distribution, or it can be a habitat specialist. Two of these aspects of rarity (local abundance and distributional range) have been used to assess the vulnerability of Neotropical mammals and other taxa (Arita et al. 1990; Arita 1993a; Gaston 1994). A detailed discussion of the relationship between rarity and probability of extinction can be found in those papers. The conclusion in both cases was that initial rarity can be, in most cases, a reliable and easily measurable indicator of vulnerability.

For Middle American bats, we assessed rarity using various indirect measures. To estimate local abundance, we started with the rank values of local density calculated by Arita (1993a) from data on 16 localities in the Neotropics. The database included 150 Neotropical species, many of which are found in Middle America. To complete the list, we interpolated in the rank series the Middle American species not included in the original study by comparing their relative local densities with species that were already in the list. Because this procedure involves a great deal of subjectivity and potential sources of error, we used only two categories to classify the species: locally rare (species located below the median in the rank series) and locally abundant (the rest of the species).

To estimate area of range, we drew distributional maps for all species. The starting point was the maps of Hall (1981), but most nomenclatural and distributional changes published up to the end of 1993 were reviewed to edit the maps. (A complete list of the references consulted is available from the first author.) We overlaid the maps with a grid of $0.5° \times 0.5°$ quadrats and tallied the number of quadrats that covered the distribution of each species. Then, area of range was measured as the number of quadrats occupied by each species. Because meridians converge toward the poles, quadrats are smaller at higher latitudes, but the maximum difference (7.5%) is negligible at the scale we used.

We ranked species according to the area of their distributional range and classified them in three categories: restricted in distribution (the lower quartile of the distribution), widespread (the upper quartile), and intermediate (the remaining species). As with other groups and geographic places, the frequency distribution of area of range follows a log-normal distribution (Figure 21.3). For this reason, species in the lower quartile (species with restricted distributions) are all included in the first bar of Figure 21.3.

We did not quantify the third dimension of rarity, habitat specialization, because data are scarce or nonexistent for most species. Instead, we simply identified those species that are characterized by any form of specialization. For instance, we gave special attention to species found only in particular vegetation types. Examples include the Honduran fruit-eating bat *(Artibeus inopinatus),* which is found only in tropical lowland thorn-scrub habitats (Dolan and Carter 1979), and Miller's myotis *(Myotis milleri),* known only from the coniferous forests of the San Pedro Martir mountains in Baja California, Mexico.

We also gave special consideration to species with particular roosting preferences. The fish-eating bat *(Myotis vivesi),* for example, roosts in crevices or under piles of stones in coastal areas of northwestern Mexico. The disk-winged bat *(Thyroptera tricolor)* shelters in the curled leaves

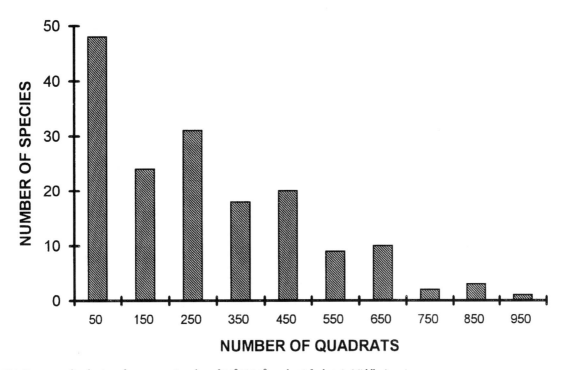

Figure 21.3. Frequency distribution of range area (number of 0.5° × 0.5° quadrats) for bats in Middle America.

of platanillo plants (*Heliconia* spp.) or other Musaceae. Cave species, particularly those that have been found in association with the common vampire bat *(Desmodus rotundus),* were also given particular attention because caves are one of the most threatened bat roosting sites in the region (Arita 1993b, 1996; Arita and Vargas 1995).

Specialized diet was an additional criterion to identify potentially vulnerable bat species. Species such as the fishing *(M. vivesi)* and bulldog *(Noctilio leporinus)* bats, which depend on crustaceans and fish for food, are very susceptible to modifications of their habitats. Bats that feed on vertebrates are of special concern because some of the life history characters that are associated with a carnivorous diet make them more vulnerable to extinction. For example, the false vampire bats *(Chrotopterus auritus* and *Vampyrum spectrum)* feed on birds and small mammals and both are large, exist at low population densities, have low reproductive rates, and form complex social groups, in this case small families that remain together for several years. The extreme dietary specialization among bats, feeding on blood, is shown by the common vampire bat *(D. rotundus),* and especially by the white-winged *(Diaemus youngi)* and hairy-legged *(Diphylla ecaudata)* vampires, which feed exclusively on the blood of birds.

A cross-tabulation between area of geographic range and local abundance identified 25 species that are scarce and restricted in distribution (Table 21.3). This list includes, however, some species that are common or widespread in North or South America. For example, the silver-haired bat

(Lasionycteris noctivagans) is known in Middle America only from a single locality in Tamaulipas, Mexico, but the species is widespread in the United States and Canada. Similarly, the dog-faced bat *(Molossops planirostris)* has a restricted distribution in Middle America but is widespread in South America. After deleting such species, the list was reduced to 11 that we consider to be of critical concern in Middle America (Table 21.4).

Fourteen additional species are included in Table 21.4. These bats do not constitute critical cases but show traits that cause them to be of particular importance for conservation or susceptible to perturbation of their habitats. In-

Table 21.3

Numbers of Species in the Six Categories of Rarity Defined in the Text

	Distribution		
Local density	Restricted	Intermediate	Wide
Abundant	17	40	25
Scarce	25	44	15

Notes: Locally abundant species are above the median for local density in the database; scarce species are below this median. Species with restricted distribution are in the lower quartile of a rank series based on area of distribution; species with wide distribution are in the upper quartile; the intermediate category includes the rest of the species.

Table 21.4

Middle American Bat Species of Critical and Special Concern

STATUS and taxon	Comments
CRITICAL	
Phyllostomidae	
Artibeus inopinatus	Endemic; very restricted range; specialized habitat
Musonycteris harrisoni	Endemic; monotypic genus; restricted distribution; found only in dry tropical forests; roosts in caves; pollinator
Vampyressa nymphaea	Restricted distribution; low population levels
Vespertilionidae	
Lasiurus castaneus	Quasi-endemic; very restricted distribution; locally rare
Lasiurus egregius	—
Myotis cobanensis	Endemic; known from only one specimen; probably extinct
Myotis findleyi	Endemic; insular
Myotis milleri	Endemic; locally rare; habitat specialist
Myotis planiceps	Endemic; very restricted distribution; extremely rare; known from three localities
Rhogeessa genowaysi	Endemic; cryptic species with very restricted distribution
Rhogeessa mira	Endemic; smallest bat of the region; habitat specialist; very restricted distribution; known from two localities
SPECIAL	
Balantiopteryx io	Endemic; restricted range
Artibeus hirsutus	Endemic; restricted range; habitat specialist; roosts in caves
Choeronycteris mexicana	Quasi-endemic; locally scarce; monotypic genus; pollinator
Ectophylla alba	Quasi-endemic; monotypic genus; specialized roosting site
Hylonycteris underwoodi	Endemic; monotypic genus
Leptonycteris nivalis	Quasi-endemic; migratory; pollinator; roosts in caves; evidence of population declines
Sturnira mordax	Endemic; restricted range
Tonatia evotis	Endemic; very low population levels
Antrozous dubiaquercus	Endemic; very low population levels; fragmented area of distribution
Myotis elegans	Endemic; very low population levels; found only in tropical rainforest
Myotis peninsularis	Endemic; relict distribution; roosts in caves
Myotis vivesi	Endemic; specialized diet and habitat; habitat being transformed rapidly
Rhogeessa alleni	Endemic; restricted range
Rhogeessa gracilis	Endemic; restricted range

cluded in this category are species that are endemic or quasi-endemic to Middle America and which show one or more of the following characteristics: (1) widespread but with very low population levels (e.g., Van Gelder's bat, *Antrozous dubiaquercus*); (2) habitat specialists (e.g., little yellow bats, *Rhogeessa* spp.); (3) ecological specialists (e.g., fishing bat, *Myotis vivesi*); (4) migratory (e.g., long-nosed bat, *Leptonycteris nivalis*); (5) potential keystone species (e.g., hog-nosed bat, *Choeronycteris mexicana*); or (6) roost in caves used by vampire bats (e.g., hairy fruit-eating bat, *Artibeus hirsutus*).

A conservation strategy at the species level should focus on the taxa listed in Table 21.4. Immediate recommended actions are the protection of areas in which the species occur and the management of roosting and feeding habitats. Because very little is known about the ecology of the bats in Table 21.4, a logical second step is to collect data on the natural history of the species. A third step is to undertake detailed population studies, especially population and habitat viability analyses (PHVAs), to determine the real status and the probabilities of survival for these species.

Conservation at the Regional Level

Besides focusing attention at particular species, conservation biologists use the alternative strategy of protecting sites with particularly high biological diversity. Traditionally, large-scale biodiversity has been measured only in terms of species richness, that is, the number of species found in a given region. Recently, however, several researchers have proposed alternative criteria for the quantification of diversity.

Phylogenetic criteria, for instance, have been used to assign conservation value to species or sites (Erwin 1991; Vane-Wright et al. 1991; Pressey et al. 1993). Similarly, the presence of rare, endemic, or endangered species constitutes an important criterion for identifying critical areas (Sisk et al. 1994). Finally, biological integrity has been proposed as an additional criterion for the characterization of conservation areas (Angermeir and Karr 1994).

To identify priority areas for the protection of Middle American bats, we used a combination of criteria. First, we employed the traditional measure of diversity, that is, species richness. Subsequently, we explored alternative estimates that included an index which measures the rarity of species in terms of their area of distribution. Finally, we analyzed patterns of presence and number of rare, endemic, and specialized species, particularly when representing monotypic genera, to identify zones of particularly high diversity.

Species Richness

We calculated species richness for each of the 1,054 quadrats ($0.5° \times 0.5°$) previously used to document the distribution of species. The number of species in each quadrat varied widely, from 5 to 101. As a consequence of the method used to gather the data, these figures should be interpreted as the number of species potentially present in a given quadrat and as overestimates of actual species richness.

To provide an indication of species richness at the local level, we also gathered information from the literature for 19 sites in Middle America (Figure 21.4). Because most sites are incompletely sampled, real figures for species richness are probably between the values shown in Figure 21.4 and those obtained using the quadrats.

Corroborating the results of previous studies (Wilson 1974; McCoy and Connor 1980; Kaufman 1995), we found a clear latitudinal gradient for species richness of bats in Middle America (Figures 21.4 and 21.5). At the local level, assemblages near the U.S. border harbor few species (7–18), whereas communities in the tropical rainforests of southern Mexico and Central America consist of as many as 65 species (Timm et al. 1989; Medellín 1994).

At a larger scale, richness is greater than 60 bat species for all quadrats below 16° (corresponding to Central America), whereas no quadrat above the tropic of Cancer contains more than 50 species. We ranked quadrats according to their species richness and found that the 20% with the top values are all located below 19°, coinciding with the limit of the Neotropical realm in Central Mexico. These results support the paradigm that tropical bat communities consist of many more species than temperate zone assemblages (Findley 1993).

Because bat species richness is highly correlated with latitude, its use as the only criterion for determining priority areas for conservation is of limited value. By this criterion alone, temperate zone communities, which are comparatively poor in species, would be excluded from any conservation strategy. We concluded from this analysis that southern Mexico and Central America support assemblages that are important because of their high species richness, but additional criteria of diversity are needed to obtain a more complete picture of the geographic patterns of bat diversity in Middle America. Some of these criteria are discussed in the following sections.

Area of Distribution

Area of range has been used to design indices that assess the relative importance of areas for conservation (Kershaw et al. 1994; Arita et al. 1997). These indices measure, for a given site, the average "rarity" of the species that occur there by quantifying the degree of restriction of their total distribution. The index proposed by Arita et al. (1997), for example, is expressed as

$$I = \left(\sum_{i=1}^{S_c} \frac{1}{A_i} \right) \div S_c,$$

where A_i is the area of distribution of species i and S_c is the number of species in quadrat c. The smaller the range of a species, the higher the value of the index.

For the Mexican mammal fauna, this index is useful for the identification of priority areas for nonvolant species. In the case of bats, however, the usefulness of the index is limited by its high correlation with species richness (Arita et al. 1997). In Mexico, areas that harbor species with restricted distributional ranges are also the richest in species. Therefore, the index is redundant with species richness and cannot be considered an independent measure of diversity.

We extended the analysis of Arita et al. (1997) to the bat fauna of Middle America and reached basically the same conclusion. Quadrats with high values of the index are located within zones of high species richness, such as the areas covered with tropical rainforest in Central America and southern Mexico. Rarity, as measured by the area of distributional range, is of little value for identifying conservation areas for Middle American bats.

Presence of Species with Critical and Special Status

We identified areas that are important for the protection of species listed in Table 21.4. For each quadrat, we tallied the

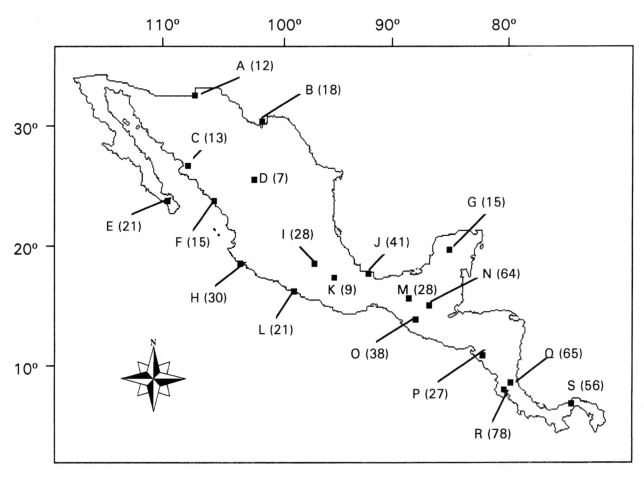

Figure 21.4. Estimates of species richness (number of species, in parentheses) in 19 Middle American bat communities, from the literature: A, Rancho San Francisco, Chihuahua, Mexico (Anderson 1972); B, Big Bend National Park, Texas (Easterla 1973); C, El Fuerte, Sinaloa, Mexico (Jones et al. 1972); D, Matalotes, Durango, Mexico (Baker and Greer 1962); E, Sierra de la Laguna (Woloszyn and Woloszyn 1982); F, Escuinapa, Sinaloa, Mexico (Jones et al. 1972); G, Loltun, Yucatan, Mexico (Arroyo-C. 1992); H, Chamela, Jalisco, Mexico (Ceballos and Miranda 1986); I, Mexico City, D.F., Mexico (Sánchez-H. et al. 1989); J, Los Tuxtlas, Veracruz, Mexico (Coates-E. and Estrada 1986); K, Tequistlan, Oaxaca, Mexico (Goodwin 1969); L, Tecpan, Guerrero, Mexico (Ramírez-P. et al. 1977); M, San Cristobal de las Casas, Chiapas, Mexico (Alvarez-C. and Alvarez 1991); N, Chajul, Chiapas, Mexico (Medellín 1994); O, El Triunfo, Chiapas, Mexico (Alvarez-C. and Alvarez 1991); P, Chinandenga, Nicaragua (Jones et al. 1971); Q, La Selva, Costa Rica (Timm et al. 1989); R, Guanacaste, Costa Rica (Wilson 1983); S, Barro Colorado Island, Panama (Handley et al. 1991).

number of species of concern and ranked the squares according to this count. Values varied from zero to eight species. We selected quadrats with five or more species of concern to identify priority areas. Additionally, we included zones that, despite harboring comparatively few species, are important for the protection of particular species. We identified 10 areas that stand out as important centers for the conservation of critical and special species (Figure 21.6). These areas are listed below. Areas 5–10 are of importance for particular species

1. The Pacific plain and adjacent portions of the Sierra Madre Occidental in the Mexican states of Nayarit, Jalisco, Colima, Michoacan, and Guerrero: Several species of concern occur in this area, including the trumpet-nosed (*Musonycteris harrisoni*), the hairy fruit-

eating (*Artibeus hirsutus*), and the miniature yellow (*Rhogeessa mira*) bats. The yellow bat is of particular importance, because it is known from only two localities in the area of El Infiernillo, in the border between Michoacan and Guerrero.

2. The Balsas Basin and the Volcanic Belt in Central Mexico: The Balsas Basin constitutes an intrusion of the Pacific lowlands into Central Mexico. Together with the volcanic belt, it forms one of the clearest transitions between the Nearctic and the Neotropical realms. Typical species of this transitional area are the long-nosed bat (*Leptonycteris nivalis*) and the highland fruit-eating bat (*Artibeus aztecus*).

3. The lower Yucatan peninsula, including the Peten area of Guatemala and Belize and the lowlands of eastern Chiapas: Besides holding the richest bat communities

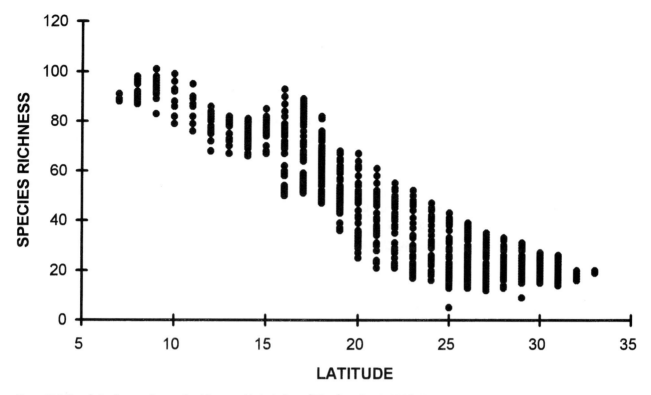

Figure 21.5. Correlation between bat species richness and latitude for 0.5° × 0.5° quadrats in Middle America.

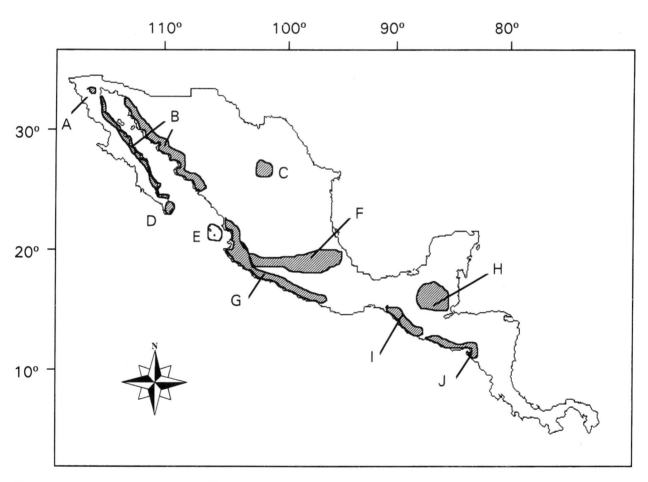

Figure 21.6. Priority areas for the conservation of bat species of critical and special status in Middle America: A, San Pedro Mountains, Baja California, Mexico; B, coastal areas of the Gulf of Cortez; C, corner of the Mexican states of Coahuila, Nuevo Leon, and Zacatecas; D, tip of the Baja California peninsula; E, Tres Marias Islands, Mexico; F, Balsas Basin and volcanic belt; G, Pacific coast of western Mexico; H, lower Yucatan peninsula; I, coastal plain of Chiapas and Guatemala; J, Pacific plain of El Salvador, Honduras, and Nicaragua.

of the region, the zone harbors populations of species such as Thomas' sac-winged bat *(Balantiopteryx io)*, the Guatemalan and elegant myotis *(Myotis cobanensis* and *M. elegans)*, the round-eared bat *(Tonatia evotis)*, and Van Gelder's bat *(Antrozous dubiaquercus)*.

4. The Pacific Coast and adjacent mountainous areas in Chiapas and Guatemala: This area forms a mosaic of diverse vegetation types, from mangrove stands to tropical subdeciduous and cloud forests. An important species is Genoways' little yellow bat *(Rhogeesa genowaysi)*, found only in a small area of the state of Chiapas.

5. The San Pedro Martir mountains in Baja California: The mountains are a virtual island of coniferous forests surrounded by xerophilic brushland. These forests are the habitat of Miller's myotis *(Myotis milleri)*.

6. The coastal areas and islands of the Gulf of Cortez: This is the habitat of the fishing bat *(Myotis vivesi)*.

7. The southern tip of the Baja California peninsula, particularly the Sierra de la Laguna: The peninsular bat *(Myotis peninsularis)* occurs only in this area, where it roosts mainly in caves (Woloszyn and Woloszyn 1982).

8. The Tres Marias islands: Findley's black myotis *(Myotis findleyi)* occurs only in these islands. Additionally, the largest known populations of Van Gelder's bat *(Antrozous dubiaquercus)* are found here (Wilson 1991).

9. The highlands in the corner of the Mexican states of Coahuila, Nuevo Leon, and Zacatecas: The flat-headed myotis *(Myotis planiceps)* is endemic to this small area (Matson and Baker 1986).

10. The Pacific plain of El Salvador, Honduras, and Nicaragua: The only known habitat of the Honduran fruit-eating bat *(Artibeus inopinatus)* is the tropical thorn forests of this dry area.

Conclusions

Middle America, with its impressively rich fauna of vertebrates, presents a difficult challenge for any conservation strategy. In the case of bats, such a strategy should be aimed at both the protection of particular species and the management of critical areas.

Because one of the main traits of the region is the high percentage of endemism, species with restricted distributions should be given first priority for protection. Species listed in Table 21.4 represent the essence of what distinguishes Middle America from other areas of the New World in terms of the bat faunas. A comprehensive strategy, perhaps at the continental level, should include additional species that are particularly important for the North and South American bat faunas.

Conservation in Middle America cannot be focused on the protection of a few areas. Because the region is a complex mosaic of habitats, vertebrate faunas are completely different from place to place. The relatively high number of areas identified here as priority sites is a clear indication that conservation in Middle America must be planned at the regional level (Arita et al. 1997).

Even a large system of protected areas would be insufficient to encompass the diversity of habitats and faunas of the region. This limitation is particularly true in the case of very vagile species, such as bats and birds. The unavoidable conclusion is, therefore, that conservation in Middle America should be viewed as a program that must include both an efficient system of reserves and adequate strategies to guarantee the conservation of bats outside protected areas.

Literature Cited

Alvarez-C., S. T., and T. Alvarez. 1991. Los murciélagos de Chiapas. Escuela Nacional de Ciencias Biológicas, Instituto Politécnico Nacional, Cuidad de México.

Anderson, S. 1972. Mammals of Chihuahua: Taxonomy and distribution. Bulletin of the American Museum of Natural History 148:151–410.

Angermeir, P. L., and J. R. Karr. 1994. Biological integrity versus biological diversity as policy directives. Bioscience 44:690–697.

Arita, H. T. 1993a. Rarity in Neotropical bats: Correlations with phylogeny, diet, and body mass. Ecological Applications 3:506–517.

Arita, H. T. 1993b. Conservation biology of the cave bats of Mexico. Journal of Mammalogy 74:693–702.

Arita, H. T. 1996. Conservation of the cave bats of Yucatan, Mexico. Biological Conservation 76:177–185.

Arita, H. T., and J. Vargas. 1995. Natural history, interspecific association, and incidence of the cave bats of Yucatan, Mexico. Southwestern Naturalist 40:29–37.

Arita, H. T., J. G. Robinson, and K. H. Redford. 1990. Rarity in Neotropical forest mammals and its ecological correlates. Conservation Biology 4:181–192.

Arita, H. T., F. Figueroa, A. Frisch, P. Rodríguez, and K. Santos del Prado. 1997. Geographical range size and the conservation of Mexican mammals. Conservation Biology 11:92–100.

Arroyo-C., J. 1992. Sinopsis de los murciélagos fósiles de México. Revista de la Sociedad Mexicana de Paleontología 5:1–14.

Audet, D., M. D. Engstrom, and M. B. Fenton. 1993. Morphology, karyology, and echolocation calls of *Rhogeessa* (Chiroptera: Vespertilionidae) from the Yucatan Peninsula. Journal of Mammalogy 74:498–502.

Baker, R. H., and J. K. Greer. 1962. Mammals of the Mexican state of Durango. Publications of the Museum Michigan State University 2:29–154.

Bogan, M. A. 1978. A new species of *Myotis* from the Islas Tres

Marías, Nayarit, México, with comments of variation in *Myotis nigricans*. Journal of Mammalogy 59:519–530.

Ceballos, G., and A. Miranda. 1986. Los mamíferos de Chamela, Jalisco. Instituto de Biología, Universidad Nacional Autónoma de México, México City, México.

Cervantes, F. A., A. Castro-C., and J. Ramírez-Pulido. 1994. Mamíferos terrestres nativos de México. Anales del Instituto de Biología, Universidad Nacional Autónoma de México Serie Zoología 65:177–190.

Coates-E., R., and A. Estrada. 1986. Manual de identificación de campo de los mamíferos de la Estación de Biología "Los Tuxtlas." Instituto de Biología, Universidad Nacional Autónoma de México, México City, México.

Cuervo, A., J. Hernández-Camacho, and A. Cadena. 1986. Lista actualizada de los mamíferos de Colombia: Anotaciones sobre su distribución. Caldasia 25:471–501.

Dolan, P. G. 1989. Systematics of Middle American mastiff bats of the genus *Molossus*. Special Publications, The Museum, Texas Tech University 23:1–71.

Dolan, P. G., and D. C. Carter. 1979. Distributional notes and records for Middle America Chiroptera. Journal of Mammalogy 60:644–649.

Easterla, D. A. 1973. Ecology of the 18 species of Chiroptera at Big Bend National Park, Texas. Part I. Northwest Missouri State University Studies 34:1–53.

Eisenberg, J. F. 1989. Mammals of the Neotropics. The Northern Neotropics. University of Chicago Press, Chicago.

Engstrom, M. D., and D. E. Wilson. 1981. Systematics of *Antrozous dubiaquercus* (Chiroptera: Vespertilionidae), with comments on the status of *Bauerus* Van Gelder. Annals of the Carnegie Museum 50:371–383.

Erwin, T. L. 1991. An evolutionary basis for conservation strategies. Science 253:750–752.

Findley, J. S. 1993. Bats: A Community Perspective. Cambridge University Press, Cambridge.

Flores-V., O., and P. Gerez. 1994. Biodiversidad y Conservación en México: Vertebrados, Vegetación y Uso del Suelo, 2nd Ed. Universidad Nacional Autónoma de México, Cuidad de México.

Gaston, K. J. 1994. Rarity. Chapman and Hall, London.

Gilbert, L. E. 1980. Food web organization and the conservation of Neotropical diversity. *In* Conservation Biology: An Evolutionary-Ecological Perspective, M. E. Soulé and B. A. Wilcox, eds., pp. 11–33. Sinauer, Sunderland, Mass.

Goodwin, G. G. 1969. Mammals from the state of Oaxaca, Mexico, in the American Museum of Natural History. Bulletin of the American Museum of Natural History 141:1–269.

Hall, E. R. 1981. The Mammals of North America. Wiley, New York.

Handley, C. O., Jr., D. E. Wilson, and A. L. Gardner. 1991. Demography and natural history of the common fruit bat, *Artibeus jamaicensis*, on Barro Colorado Island, Panamá. Smithsonian Contributions to Zoology 511:1–173.

Hartshorn, G. S. 1983. Introduction to plants. *In* Costa Rican Natural History, D. H. Janzen, ed., pp. 118–157. University of Chicago Press, Chicago.

Holdrige, L. R. 1967. Life Zone Ecology. Tropical Science Center, San José, Costa Rica.

Jones, J. K., Jr. 1966. Bats from Guatemala. University of Kansas Publications of the Museum of Natural History 16:439–472.

Jones, J. K., Jr., and R. D. Owen. 1986. Checklist and bibliography of Nicaraguan Chiroptera. Occasional Papers of the Museum, Texas Tech University 106:1–13.

Jones, J. K., Jr., J. Arroyo-C., and R. D. Owen. 1988. Revised checklist of bats (Chiroptera) of Mexico and Central America. Occasional Papers of the Museum, Texas Tech University 120:1–34.

Jones, J. K., Jr., J. R. Choate, and A. Cadena. 1972. Mammals from the Mexican state of Sinaloa. II. Chiroptera. Occasional Papers of the Museum of Natural History, University of Kansas 6:1–29.

Jones, J. K., Jr., J. D. Smith, and R. W. Turner. 1971. Noteworthy records of bats from Nicaragua, with a checklist of the chiropteran fauna of the country. Occasional Papers of the Museum of Natural History, University of Kansas 2:1–35.

Jones, J. K., Jr., R. S. Hoffmann, D. W. Rice, C. Jones, R. J. Baker, and M. D. Engstrom. 1992. Revised checklist of North American mammals north of Mexico, 1991. Occasional Papers of the Museum, Texas Tech University 146:1–23.

Kaufman, D. M. 1995. Diversity of New World mammals: Universality of the latitudinal gradients of species and bauplans. Journal of Mammalogy 76:322–334.

Kershaw, M., P. H. Williams, and G. M. Mace. 1994. Conservation of Afrotropical antelopes: Consequences and efficiency of using different site selection methods and diversity criteria. Biodiversity and Conservation 3:354–372.

Koopman, K. F. 1982. Biogeography of the bats of South America. *In* Mammalian Biology in South America, M. A. Mares and H. H. Genoways, eds., pp. 273–302. University of Pittsburgh, Linesville, Pa.

Koopman, K. F. 1989. A review and analysis of the bats of the West Indies. *In* Biogeography of the West Indies: Past, Present, and Future, C. A. Woods, ed., pp. 635–644. Sandhill Crane Press, Gainesville, Fla.

Koopman, K. F. 1993. Order Chiroptera. *In* Mammal Species of the World, D. E. Wilson and D. M. Reeder, eds., pp. 137–232. Smithsonian Institution Press, Washington, D.C.

Matson, J. O., and R. H. Baker. 1986. Mammals of Zacatecas. Special Publications, The Museum, Texas Tech University 24:1–88.

McCarthy, T. J. 1987. Distributional records of bats from the Caribbean lowlands of Belize and adjacent Guatemala and Mexico. Fieldiana Zoology (new series) 39:137–162.

McCoy, E. D., and E. F. Connor. 1980. Latitudinal gradients in the species diversity of North American mammals. Evolution 34:193–203.

Medellín, R. A. 1994. Mammal diversity and conservation in the selva Lacandona, Chiapas, México. Conservation Biology 8:780–799.

Owen, R. D., J. K. Jones, Jr., and R. J. Baker. 1991. Annotated checklist of land mammals of El Salvador. Occasional Papers of the Museum, Texas Tech University 139:1–17.

Pimm, S. L., H. L. Jones, and J. Diamond. 1988. On the risk of extinction. American Naturalist 132:757–785.

Pressey, R., C. Humphries, C. Margules, R. Vane-Wright, and P. Williams. 1993. Beyond opportunism: Key principles for systematic reserve selection. Trends in Ecology and Evolution 8: 124–128.

Rabinowitz, D., S. Cairns, and T. Dillon. 1986. Seven forms of rarity and their frecuency in the flora of British Isles. *In* Conservation Biology: The Science of Scarcity and Diversity, M. E. Soulé, ed., pp. 182–204. Sinauer, Sunderland, Mass.

Ramírez-P., J., A. Martínez, and G. Urbano. 1977. Mamíferos de la Costa Grande de Guerrero, México. Anales del Instituto de Biología, Universidad Nacional Autónoma de México, Serie Zoología 48:243–292.

Ramírez-P., J., M. C. Britton, A. Perdomo, and A. Castro-C. 1986. Guía de los mamíferos de México. Universidad Autónoma Metropolitana, México City, México.

Redford, K. H., and J. F. Eisenberg. 1992. Mammals of the Neotropics, the Southern Cone. University of Chicago Press, Chicago.

Rich, P. V., and T. H. Rich. 1983. The Central American dispersal route: Biotic history and paleogeography. *In* Costa Rican Natural History, D. H. Janzen, ed., pp. 12–34. University of Chicago Press, Chicago.

Rzedowski, J. 1978. Vegetación de México. Editorial Limusa, México City, México.

Sánchez-H., O., G. López-O., and R. López-W. 1989. Murciélagos de la ciudad de México y sus alrededores. *In* Ecología Urbana, R. Gío-Argáez, I. Hernández-R., and E. Sáinz-H., eds., pp. 141–165. Sociedad Mexicana de Historia Natural, México City, México.

Sisk, T. D., A. E. Launer, K. R. Swittky, and P. R. Ehrlich. 1994. Identifying extinction threats. Bioscience 44:592–604.

Terborgh, J. 1974. Preservation of natural diversity: The problem of extinction-prone species. Bioscience 24:715–722.

Timm, R. M., D. E. Wilson, B. L. Clauson, R. K. LaVal, and C. S. Vaughan. 1989. Mammals of the La Selva–Braulio Carrillo complex, Costa Rica. North American Fauna, No. 75. U.S. Fish and Wildlife Service, Washington, D.C.

Vane-Wright, R. I., C. J. Humphries, and P. H. Williams. 1991. What to protect? Systematics and the agony of choice. Biological Conservation 55:235–254.

Villa-R., B. 1966. Los murciélagos de México. Instituto de Biología, Universidad Nacional Autónoma de México, México City, Mexico.

Wilson, D. E. 1983. Checklist of mammals. *In* Costa Rican Natural History, D. H. Janzen, ed., pp. 443–447. University of Chicago Press, Chicago.

Wilson, D. E. 1991. Mammals of the Tres Marías Islands. Bulletin of the American Museum of Natural History 206:214–250.

Wilson, J. W. III. 1974. Analytical zoogeography of North American mammals. Evolution 28:124–140.

Woloszyn, D., and B. W. Woloszyn. 1982. Los mamíferos de la Sierra de La Laguna, Baja California Sur. Consejo Nacional de Ciencia y Tecnología, México City, México.

22
Tall Trees, Deep Holes, and Scarred Landscapes
Conservation Biology of North American Bats

ELIZABETH D. PIERSON

Concern for organismal welfare has been a consistent theme in the North American bat research community for almost 100 years. Recognition that bats were frequently maligned and misunderstood inspired biologists to call for their protection and with poetic zeal to delineate their ecological, economic, and mythological significance (Grinnell 1918; Bailey 1925; Allen 1940). In the 1950s and 1960s, researchers recognized the vulnerability of highly aggregated bat populations and the first cave protection measures were proposed (Mohr 1952, 1953; Graham 1966). Since the 1970s data have been accumulating on population declines and habitat threats for a number of North American species. As a consequence, seven species or subspecies are now recognized as threatened or endangered under the U.S. Endangered Species Act (ESA), and many are given special consideration under state or provincial jurisdictions. It was not until recently, however, that researchers have had the technology (e.g., miniature radiotransmitters, portable ultrasound detectors, and chemiluminescent markers) (Kunz 1988a; Wilson et al. 1996) to address some of the most pressing conservation-related ecological questions concerning species-specific habitat requirements.

Wildlife policy in North America has historically been oriented toward exploited game species, with a more recent secondary focus on declining nongame species. Although conservation strategies were developed for a small number of bat species officially recognized as threatened or endangered, the policy perspective on most has been set by wildlife nuisance and vector control regulations, instituted after the discovery of rabies in the 1950s. While there remains a significant tension in public and institutional perspectives on bats, attitudes are changing. With an emerging emphasis on "ecosystem management," resource agencies are now moving rapidly to include bats in their deliberations. The purpose of this chapter is to provide a conceptual context for the conservation biology of North American bats and to outline some of the most pressing issues facing these animals.

Patterns of Diversity and Distribution

There are 45 bat species in Canada and the United States (Table 22.1). All 45 species, from 19 genera and four families, occur in the United States; 21 of these, from 8 genera and two families, also occur in Canada. There is a rapid decline in number of species with increasing latitude; for

Table 22.1

Relative Range Sizes and Latitudinal Limits for Bat Species of North America (Canada and the Continental United States)

Taxon	Relative range[a] in: Canada	U.S.	Maximum latitude	Reference
Mormoopidae				
Mormoops megalophylla	—	L	31°37′	Hall 1981
Phyllostomidae				
Choeronycteris mexicana	—	L	36°11′	Constantine 1987
Diphylla ecaudata	—	L	29°41′	Hall 1981
Leptonycteris curasoae	—	L	33°32′	Hall 1981
Leptonycteris nivalis	—	L	31°35′	Hoyt et al. 1994
Macrotus californicus	—	L	36°56′	Hall 1981
Vespertilionidae				
Antrozous pallidus	L	I	49°19′	Hall 1981
Corynorhinus rafinesquii[b]	—	L	40°25′	Hall 1981
Corynorhinus townsendii[b]	L	I	52°08′	Hall 1981
Eptesicus fuscus	B	B	59°00′	Hall 1981
Euderma maculatum	L	I	52°27′	Roberts and Roberts 1993
Idionycteris phyllotis	—	L	37°??′	Hall 1981
Lasionycteris noctivagans	B	B	61°07′	Hall 1981
Lasiurus blossevillii[c]	L	L	49°19′	Nagorsen and Brigham 1993
Lasiurus borealis	I	I	57°15′	Hall 1981
Lasiurus cinereus	B	B	64°20′	Hall 1981
Lasiurus ega	—	L	27°48′	Schmidly 1991
Lasiurus intermedius	—	L	40°39′	Hall 1981
Lasiurus seminolus	—	I	42°27′	Hall 1981
Lasiurus xanthinus[c]	—	L	34°04′	Hall 1981
Myotis auriculus	—	L	35°13′	Hall 1981
Myotis austroriparius	—	L	39°01′	Hall 1981
Myotis californicus	L	I	54°54′	Hall 1981
Myotis ciliolabrum[d]	L	I	51°56′	Garcia et al. 1995
Myotis evotis	L	I	54°47′	Nagorsen and Brigham 1993
Myotis grisescens	—	L	39°27′	Hall 1981
Myotis keenii	L	L	57°55′	Hall 1981
Myotis leibii	L	I	47°32′	Hall 1981
Myotis lucifugus	B	B	66°34′	Hall 1981
Myotis occultus[e]	—	L	35°25′	Hoffmeister 1986
Myotis septentrionalis[d]	B	I	61°25′	van Zyll de Jong 1985
Myotis sodalis	—	I	44°16′	Hall 1981
Myotis thysanodes	L	I	51°54′	Rasheed et al. 1995
Myotis velifer	—	L	37°39′	Hall 1981
Myotis volans	L	I	59°35′	Hall 1981
Myotis yumanensis	L	I	52°57′	Hall 1981
Nycticeius humeralis	L	I	42°17′	Hall 1981
Pipistrellus hesperus	—	I	35°03′	Hall 1981
Pipistrellus subflavus	L	I	47°20′	Hall 1981
Molossidae				
Eumops glaucinus	—	L	26°55′	Belwood 1981
Eumops perotis	—	L	41°32′	W. E. Rainey, unpublished observation
Eumops underwoodi	—	L	31°48′	Hall 1981
Nyctinomops femorosaccus	—	L	34°17′	Hall 1981
Nyctinomops macrotis	L	L	49°12′	Hall 1981
Tadarida brasiliensis	—	I	42°19′	Hall 1981

Note: Nomenclature follows Koopman (1993) unless otherwise noted.

[a]Relative range: B, broad; I, intermediate; L, limited; —, not applicable.

[b]Nomenclature follows Frost and Timm (1992) and Tumlison and Douglas (1992).

[c]Nomenclature follows Baker et al. (1988) and Morales and Bickham (1995).

[d]Nomenclature follows van Zyll de Jong (1985).

[e]Nomenclature follows Schmidly (1991).

example, there are 40 species at 30° and only 5 above 60°. Although the negative correlation of number of mammalian species with increasing latitude has been repeatedly documented, the reasons for it are not well understood (Rohde 1992; Findley 1993; Kaufman 1995). This correlation is, however, particularly strong for bats, and may be explained by a latitude-related factor such as temperature (Willig and Selcer 1989).

While there is an overriding latitudinal gradient, there are also biogeographic factors that help explain differences in species density at comparable latitudes. The greatest number of species (24–31) are found in the southwestern United States, in those states that share a border with Mexico. By contrast, those states that border water (i.e., the Gulf

of Mexico) have significantly fewer species (11–16). A number of essentially Mexican species reach the northern limit of their ranges between about 29° and 36° N in the southwestern United States. For example, *Mormoops megalophylla*, and 5 phyllostomid, 5 vespertilionid, and 3 molossid species, are confined to this part of the country (Table 22.2).

Although Canada has 21 species, 6 of these have their primary distribution farther south and occur only in south-central British Columbia (Table 22.2), where an unusually moderate climate occurs in the Okanagan River drainage. Except for a single record of the molossid *Nyctinomops macrotis*, all Canadian species belong to the Vespertilionidae (see Table 22.1).

Bats tend to have broad distributions relative to nonvolant

Table 22.2

Distributional Characteristics of Bat Species of North America
(Canada and the Continental United States)

Family	Distribution		
	Eastern	Western	Both
Mormoopidae	—	* *Mormoops megalophylla*	—
Phyllostomidae	—	* *Choeronycteris mexicana*	—
		* *Diphylla ecaudata*	
		* *Leptonycteris curasoae*	
		* *Leptonycteris nivalis*	
		* *Macrotus californicus*	
Vespertilionidae	*Corynorhinus rafinesquii*	† *Antrozous pallidus*	† *Corynorhinus townsendii*
	Lasiurus borealis[a]	† *Euderma maculatum*	*Eptesicus fuscus*
	Lasiurus intermedius	* *Idionycteris phyllotis*	*Lasionycteris noctivagans*
	Lasiurus seminolus	† *Lasiurus blossevillii[a]*	*Lasiurus cinereus*
	Myotis austroriparius	* *Lasiurus ega*	*Myotis lucifugus*
	Myotis grisescens	* *Lasiurus xanthinus*	
	Myotis leibii[b]	* *Myotis auriculus*	
	Myotis septentrionalis[c]	*Myotis californicus*	
	Myotis sodalis	† *Myotis ciliolabrum[b]*	
	Nycticeius humeralis	*Myotis evotis*	
	Pipistrellus subflavus[d]	*Myotis keenii[c]*	
		* *Myotis occultus*	
		† *Myotis thysanodes*	
		Myotis velifer	
		Myotis volans	
		Myotis yumanensis	
		Pipistrellus hesperus[d]	
Molossidae	*Eumops glaucinus*	* *Eumops perotis*	*Tadarida brasiliensis*
		* *Eumops underwoodi*	
		* *Nyctinomops femorosaccus*	
		Nyctinomops macrotis	

Notes: An asterisk (*) marks species with North American distribution confined to the southwestern United States. A dagger (†) denotes bats with Canadian distribution confined to southern British Columbia. Species with matching superscript letters are eastern–western species pairs.

mammals of comparable size, with the smallest range for a North American species covering at least 5×10^5 km^2 (calculated from Hall 1981). Nevertheless, there are distinct patterns of distribution. The central plains of the United States and Canada separate two species assemblages (see Table 22.2). While this division is not absolute, there are a number of taxa that occur only in the east or the west, with a number of eastern and western species pairs (see Table 22.2).

Conservation Implications of Distributional Patterns

The fact that most bat species have broad distributions has profound implications for conservation policy, because there is a tendency to focus conservation efforts on species with limited distributions. For example, the U.S. Forest Service is considering putting all species through a "distribution filter" (using categories established by the Nature Conservancy) as a first step in consideration for protective status, excluding all species of broad distribution (D. MacFarlane, U.S. Forest Service, personal communication). Although organisms of broad distribution are generally less vulnerable to species-level extinction than those of limited distribution (e.g., island endemics such as the endangered Hawaiian hoary bat, *Lasiurus cinereus semotus*), local or regional populations can be severely threatened or eliminated if critical habitat factors change. Thus, knowledge of overall distribution is necessary but not sufficient for determining the status of a species, and a number of additional factors, as are now discussed, need to be considered.

PATCHY DISTRIBUTION.   While marginal records for a species may suggest a large geographic range, actual distribution is often patchy. Humphrey (1975) showed that the distribution and abundance of colonial Nearctic bats were determined largely by the availability of suitable roosts and that, as a consequence, both species number and community diversity varied in a geographically patchy rather than clinal fashion.

This phenomenon is best documented for cave-dwelling taxa. For example, the range of *Myotis grisescens* covers much of the southeastern United States, yet this obligate cave-dwelling species is confined to karst areas within its range (Tuttle 1976; Rabinowitz and Tuttle 1980). Although *Myotis austroriparius* occurs throughout the entire southeastern United States, most of the known maternity caves (18 of 21) occur in Florida (Gore and Hovis 1992). The nearly obligate cliff-dwelling species *Euderma maculatum* extends from southern British Columbia throughout the western United States to central Mexico, yet records are rare and it appears to be limited to areas with significant

rock features (Easterla 1973; Leonard and Fenton 1983; Storz 1995; Pierson and Rainey 1998).

SEASONAL DISTRIBUTION.   Many species undergo seasonal shifts in distribution. For some, this may mean only short-distance movements from a site that offers suitably warm summer temperatures for a nursery roost to one which offers winter temperatures cool and stable enough for hibernation. For several eastern species, populations consolidate in the winter at very few hibernating caves, placing these species at especially high risk. For example, 95% of all known *M. grisescens* hibernate in nine caves (Tuttle 1986); 87% of all known *Myotis sodalis* can be accounted for at seven hibernating sites (Humphrey 1978). Many of the New England populations of *Myotis lucifugus* aggregate in a few caves and mines in western Vermont and eastern New York State (Davis and Hitchcock 1965), and 75% of all known Virginia populations hibernate in only eight caves (Dalton 1987).

There are also a number of species that appear to undergo long-distance migrations, which in some cases cross international boundaries (Cockrum 1969). For these species, conservation efforts require international cooperation. For example, populations of *Tadarida brasiliensis* and both species of *Leptonycteris* migrate seasonally between the southern United States and Mexico, facing conservation challenges on both sides of the border (Barbour and Davis 1969; Wilkins 1989; Fleming et al. 1993; McCracken et al. 1994). Shared concerns have led to the establishment of the Programa Para la Conservacíon de Murciélagos Migratorios de Mexico y Estados Unidos de Norteamérica (PCMM), an ambitious program that includes conservation action, education, and research (Walker 1995).

While altitudinal zonation during the summer activity season has been examined in some areas (Grinnell and Storer 1924; Jones 1965), altitudinal migration, well documented for birds, has not been explored in any detail for North American bats. Little is known about the winter distribution of bats in areas where altitudinal migrations would be most likely, the extensive mountain ranges (Rockies and Sierra Nevada) of the West. In fact, winter distributions are very poorly understood for the majority of North American species. For example, winter records exist for only 7 of the 16 species known to occur in the summer in British Columbia (Nagorsen et al. 1993). In western North America, no winter aggregations comparable in size to summer colonies have been found for species such as *Myotis yumanensis* or *Antrozous pallidus* (Barbour and Davis 1969). While a few large winter colonies of *Corynorhinus townsendii* have been located (Pearson et al. 1952; Altenbach and Pierson 1995), they account for only a small fraction of known summer populations.

RANGE CONTRACTION.   There are a few documented cases of range contraction. The Ozark big-eared bat, *C. townsendii ingens,* which was known from four states including Missouri, no longer occurs in Missouri (Harvey 1980; Hensley and Scott 1995). *Myotis occultus,* known originally in California from a single colony along the Colorado River (Stager 1943), has not been seen in this state since 1967 (K. Stager, personal communication), despite significant survey efforts in the area (P. E. Brown, personal communication; D. Constantine, personal communication). *Macrotus californicus,* once distributed across southern California and common near the Salton Sea (Howell 1920), is now limited within that state primarily to mountain ranges in the Colorado River basin (P. E. Brown, personal communication). *Eumops glaucinus,* apparently once common along the east coast of Florida, has been seen in that area only once since 1967 (Robson et al. 1989; Belwood 1992), with one additional record from western Florida in 1979 (Belwood 1981).

LOCAL POPULATION DECLINES.   Bat species typically maintain overall distribution and undergo local population declines or extirpations. This has been best documented for colony-forming cave-dwelling species, whose large aggregations can be monitored over time. Population declines of cave-dwelling bats in the eastern United States were noted by Mohr (1952, 1953, 1972) as early as 1952, and led to the protection, under the U.S. Endangered Species Act, of *Myotis grisescens* (Tuttle 1979; Rabinowitz and Tuttle 1980), *Myotis sodalis* (Humphrey 1978), and the two eastern subspecies of *Corynorhinus townsendii* (Tolin 1994; Hensley and Scott 1995). Although each of these species maintained their historical distributions, censuses conducted in the 1970s documented declines of 76%–89% at *M. grisescens* maternity caves (Tuttle 1979; Rabinowitz and Tuttle 1980) and of as much as 73% at *M. sodalis* hibernacula (Humphrey 1978). In Florida, where *Myotis austroriparius* was known from 15 maternity caves with an estimated past population of 400,000, in 1991 only 5 caves were occupied, and the population appeared to have declined by at least 50% (Gore and Hovis 1992).

INCOMPLETE DISTRIBUTIONAL INFORMATION.   Although some parts of the United States and Canada have been the focus of considerable research, others are poorly known. Consequently, distributional information is deficient for most species, and all currently available distribution maps should be viewed as incomplete. Whenever survey efforts focus on a previously unstudied area, new records emerge. For example, recent surveys in British Columbia extended the known range of several species, most notably *Antrozous pallidus, C. townsendii, Myotis evotis,* and *Myotis thysanodes* (Nagorsen

and Brigham 1993; Roberts and Roberts 1993; Rasheed et al. 1995). In the United States, the use of acoustic monitoring devices has extended the known range of *Eumops perotis* into northern Arizona (Arizona Game and Fish Department, personal communication) and to just south of the Oregon border in California (W. E. Rainey, personal communication).

UNRESOLVED TAXONOMIC ISSUES.   It is difficult to formulate conservation strategies for species for which taxonomic status is unresolved. For example, there are differing views on whether *Lasiurus blossevillii* and *Lasiurus xanthinus* should be granted specific status, or included, respectively, in *Lasiurus borealis* and *Lasiurus ega* (Baker et al. 1988; Koopman 1993; Morales and Bickham 1995). Likewise, *Myotis occultus* and *Myotis septentrionalis,* considered separate species by some (van Zyll de Jong 1985; Schmidly 1991), are treated as subspecies, respectively, of *Myotis lucifugus* and *Myotis keenii* by Koopman (1993).

Roosting and Foraging Habitats: A Conservation Perspective

Conservation strategies for North American bat species must address both seasonal roosting and foraging requirements. All species need summer roosts with appropriate microclimatic conditions for raising young, winter sites that are disturbance free and do not experience freezing temperatures (which for many are hibernacula), and foraging sites within commuting distance of the roost. Fall roosts, particularly if used as mating sites or migratory stopovers, may also be important. Protection of only one resource is not adequate. For example, declines in the endangered species *M. sodalis* were initially attributed to disturbance at critical cave hibernacula. Although these sites have been protected, populations have continued to decline, apparently because summer foraging and tree-roosting requirements are not yet adequately understood (Clawson 1987; Missouri Department of Conservation 1992).

Considerable evidence suggests that roosts are limiting for many bat species (Humphrey 1975; McCracken 1988), particularly for cave-dwelling taxa. Yet, no species can survive without also having suitable foraging habitat. *Macrotus californicus* offers a salient example. In California, this nonhibernating species appears to be currently limited in its distribution by the availability of geothermally heated mines for winter roosts (Bell et al. 1986). Nevertheless, a colony that had long been resident in one mine declined dramatically when renewed mining removed adjacent desert-wash vegetation, shown by radiotracking to be the required winter foraging habitat (Brown et al. 1994b).

At the heart of all conservation issues facing North American bats are alterations of the natural landscape caused by expanding human populations or resource use. Urban and suburban sprawl has eliminated both roosting and foraging habitat, resulting in dramatic declines in diversity and abundance (Geggie and Fenton 1985; Kurta and Teramino 1992). Resource extraction (e.g., mining and timber harvest) is causing other problems (see following). Some alterations have created habitat. For example, most bat species across North America will roost in anthropogenic structures such as buildings, mines, and bridges. Water impoundments, while they flood roosting and foraging habitat for some species, may create foraging habitat and structural refuges for others. However, as detailed here, most alterations likely have had negative effects on diversity.

Roosting Habitats

Roosting habits of bats are often partitioned by roost type (Kunz 1982), and can be divided into "natural roosts" (e.g., caves, trees, rock crevices) and "anthropogenic roosts" (e.g., bridges, buildings, mines). Historically, research efforts have focused on those roosts most readily investigated. Thus, cave and building roosts have received relatively more research attention than tree, mine, or cliff roosts.

CAVES.   Natural caves provide some of the most important maternity and hibernating sites for bats. They are used regularly by at least 21 species and occasionally by most others (Barbour and Davis 1969; Dalton 1985, 1987). Declines in cave-dwelling bats attributable to human disturbance have been repeatedly documented, and led to protection under the U.S. Endangered Species Act for *Myotis grisescens, Myotis sodalis,* and the two eastern subspecies of *Corynorhinus townsendii.* More recently, two other predominantly cave-dwelling species, *Leptonycteris curasoae* and *Leptonycteris nivalis,* have been added to this list.

The installation of metal gates, specifically designed to exclude humans and allow free passage to bats (Tuttle 1977; Dalton and Dalton 1995), has allowed population recoveries at a number of sites (Clawson 1987; Pierson et al. 1991; Decher and Choate 1995; C. Stihler, personal communication). The importance of these conservation measures is reinforced by the recent finding of Thomas (1995) that even nontactile disturbance can cause bats to arouse from hibernation, thus compromising critical fat reserves. Gate design, however, is critical. Improperly installed gates can alter the cave environment by impeding air circulation, causing further population declines (Tuttle 1977; Richter et al. 1993). Also, gate designs that are suitable for some species may not be accepted by others. The unwillingness of

M. grisescens maternity colonies to accept full gates has been a major obstacle in protecting this species (Tuttle 1986). Thus it is critical that gating efforts be monitored and that additional inexpensive closure methods that will accommodate more species be developed.

ANTHROPOGENIC STRUCTURES.   Anthropogenic structures frequently provide ideal refuges for bats. They are used to some extent by most species, and extensively by some, such as *Myotis lucifugus* and *Eptesicus fuscus.* Conflicts between bats and humans in anthropogenic structures have long been recognized as an important conservation issue (Grinnell 1918; Greenhall 1982; Tuttle 1988). Exaggerated fears regarding rabies combined with other misconceptions about health risks have resulted in deliberate and unnecessary eradication of bats (Constantine 1970; Tuttle and Kern 1981; Brass 1994). Termite control and remodeling often cause inadvertent destruction. Uncovered chimneys and industrial exhaust stacks can also be a significant source of mortality when bats (and birds) get trapped inside these structures (Cantor 1995; W. E. Rainey, personal communication).

There are, however, numerous examples of successful efforts on the part of individuals and organizations to preserve bat colonies in human structures and to use the protected bat colonies as an educational tool. For example, M. B. Fenton and local citizens combined efforts to help the Chautauqua Institute in New York State live with their fairly numerous bat colonies (Fenton 1992); Bat Conservation International, through educational efforts, has converted a bat colony at a Texas bridge from a nuisance into a major tourist attraction (Murphy 1990).

Artificially constructed roosts have been viewed as a conservation tool since 1913 (Campbell 1913) and have proved invaluable for both research and conservation in Europe (Krzanowski 1991). There has been a recent resurgence of interest in North America, particularly on the part of the conservation-minded public, in providing bat habitat through small, commercially available bat houses. Bat Conservation International is engaged in a long-term research effort to evaluate the success of various bat house designs (Tuttle and Hensley 1993). The smaller "bat boxes" have been the least successful (Fenton 1992; Neilson and Fenton 1994). Because spatial and thermal choice appear to be critical (Hall 1962; Licht and Leitner 1967; Tuttle 1977; Pierson et al. 1991), artificial roosts that involve structural modifications to larger, already existing, thermally buffered structures (e.g., the undersides of bridges, attics of barns, interiors of dams) have worked better, particularly for crevice-dwelling species (Tuttle and Hensley 1993; S. Reynolds, personal communication; R. Wisecarver, personal communication). Artificial roosts may become increasingly important as buildings with appropriate structural features become more scarce. Yet, as currently designed, they are most readily occupied by crevice-dwelling species and are least likely to benefit those species at greatest risk, the more roost limited taxa that typically occupy large cavities.

Although bats make extensive use of bridges as both day and night roosts, these structures have received relatively little attention from the research community (Davis and Cockrum 1963) and their value, as relatively cryptic sites with generally low levels of human disturbance, has not been adequately recognized (Perlmeter 1996; Pierson et al. 1996). From a conservation perspective, there are opportunities to elicit support from transportation authorities and develop management policies for protecting and enhancing bridge roosts. The Texas Department of Transportation, working in cooperation with Bat Conservation International, is researching bat-friendly bridge design (Bat Conservation International 1994), and the California Department of Transportation has recently instituted an agency-wide policy to protect bat roosts beneath bridges (G. Erickson, personal communication).

MINES. The dependence of a number of bat species on abandoned mines has received increasing attention (Belwood and Waugh 1991; Brown and Berry 1991; Tuttle and Taylor 1994; Altenbach and Pierson 1995; H. D. Bryan and J. R. MacGregor, personal communication; R. Currie, personal communication). Of the 45 North American bat species, 28 roost in old mines (Pierson et al. 1991); for many, mines provide critical habitat. For example, the only known roosts for *Leptonycteris curasoae* in the United States are in mines.

Although underground mines have been part of the landscape for more than 100 years and have become an increasingly important resource for bats, their numbers are declining by several processes other than structural failure. Current open-pit mining operations, often located in historic mining districts, frequently destroy or are forced by regulatory agencies to reclaim old underground workings used by bats (Brown 1995). Additionally, before their value as wildlife habitat was recognized, many abandoned mines on both private and public lands were barricaded, backfilled, or dynamited shut for safety and liability reasons. This was frequently done without prior biological survey, resulting in entombment of bat colonies.

Increased awareness has resulted in a number of projects involving the mining industry, public agencies, local residents, and bat researchers that have led to protection (primarily through gating) of significant mine roosts (Pierson et al. 1991; Saugey 1991; Brown et al. 1993; Thomas 1993; Dutko 1994; Tuttle and Taylor 1994; D. C. Dalton and V. M.

Dalton, personal communication). Additionally, there have been several efforts to address this issue on a regional or national scale. The Colorado Division of Wildlife and Division of Mines and Geology, through a volunteer-based survey of abandoned mines, have identified and protected numerous previously unknown bat roosts (Navo et al. 1995). A cooperative program between the New Mexico Abandoned Minelands Bureau and the University of New Mexico has provided roost protection and highlighted the significance of deep shafts as important hibernating sites (J. S. Altenbach, personal communication).

When the state of Nevada was faced with the daunting task of making management decisions regarding as many as 300,000 abandoned mines, a workshop involving resource managers, industry representatives, and scientists was convened by the Biological Resources Research Center at the University of Nevada, Reno. From this meeting came comprehensive guidelines on how to evaluate abandoned mines and resolve potential conflicts (Riddle 1995). Bat Conservation International, with support from various federal agencies, has offered workshops on bat/mine management and published guidelines for bat/mine assessment (Tuttle and Taylor 1994).

FORESTS. Although tree cavities have been recognized for some time as important roost sites for Old World and Neotropical bat species (Kunz 1982), it is only within the last few years that their importance for Nearctic bats has been documented (Barclay and Brigham 1996; Wunder and Carey 1996). With the exception of the foliage-roosting lasiurines and *Lasionycteris noctivagans*, few if any North American species are obligate tree roosters. Nevertheless, in forested landscapes a number of species may be essentially tree dependent, roosting primarily in the cavities and under flaking bark of trees, mostly those of large diameter that are dead or dying. Even those bat species most commonly associated with anthropogenic structures rely heavily on tree roosts in some areas (e.g., *Eptesicus fuscus* [Brigham 1991; Betts 1996; Vonhof 1996; M. C. Kalcounis, personal communication], *Myotis lucifugus* [Crampton and Barclay 1996; Kalcounis and Hecker 1996], and *M. yumanensis* [Pierson and Rainey, unpublished observations]). Tree roosts may also be important for some species generally associated with caves, such as *M. austroriparius* (Gore and Hovis 1992) and *Corynorhinus rafinesquii* (M. K. Clark, personal communication).

Recent information suggests that rather than showing loyalty to a particular tree in the way colonies show high fidelity to cave, mine, and building roosts (Lewis 1995), tree-roosting bats move frequently among a number of trees, generally within a relatively small area (Barclay and Brigham 1996). Thus, in evaluating the habitat requirements of tree-dwelling bat species, the local abundance and spatial distribution of snags may be as important as the microhabitat features of particular roost trees.

Current silvicultural practices, which favor even-age monospecific stands, short rotation times, and selective removal of dead and dying trees, leave little roosting habitat for most tree-dwelling species. There is compelling evidence that bat densities are greater in old-growth stands, where structural diversity provides more roosting options (Perkins and Cross 1988; Crampton and Barclay 1996; Hayes and Adam 1996; Parker et al. 1996; Parks and Shaw 1996). While some snags are generally retained in timber harvest, the numbers are likely too low to accommodate the needs of cavity-dwelling wildlife (J. M. Perkins, personal communication). Additionally, there is frequently no consideration given to snag recruitment (e.g., the retention of green trees of varied sizes to serve as future snags). Legislation passed in 1995 by the U.S. Congress (Rescission Bill, Public Law 104-19), popularly known as the "salvage logging rider," has had particularly devastating effects. This legislation exempted timber sales on U.S. federal lands from compliance with federal environmental laws, permitted selective removal of dead and dying trees, and allowed aggressive timber harvest in previously protected old-growth stands (Durbin 1996).

Foraging Habitat

While there has been considerable discussion in the literature about whether insectivorous bats forage opportunistically or selectively and an acknowledgment that availability of prey, from the bat's perspective, is difficult to determine (Kunz 1988b; Whitaker 1994), it does appear that many species have diets that vary seasonally and geographically. Nevertheless, there are significant differences in foraging style, foraging habitat, and dietary preference among species. Additionally, habitat structure likely influences the composition of the bat community and partially determines what prey is available to the bats (Bradshaw 1996). For example, investigators have found significantly greater amounts of bat activity in unlogged versus logged riparian habitats (Hayes and Adam 1996), in old versus young or mature forest stands (Perkins and Cross 1988; Crampton and Barclay 1996), and above versus below the forest canopy (M. Kalcounis, personal communication).

Many North American bat species forage in close association with surface water. Some species, such as *Myotis austroriparius*, *M. grisescens*, *M. lucifugus*, and *M. yumanensis*, feed directly over water (Anthony and Kunz 1977; LaVal et al. 1977; Fenton and Barclay 1980; Jones and Manning 1989; Brigham et al. 1992; Findley 1993). Others, such as *M. evotis*,

M. sodalis, or *M. velifer*, feed in a variety of habitats, but are often associated with riparian vegetation (Stager 1939; Kunz 1974; Manning and Jones 1989). Even species such as *Antrozous pallidus*, which feeds on large ground-dwelling arthropods (Hermanson and O'Shea 1983), or *Corynorhinus townsendii*, which gleans moths off foliage (Kunz and Martin 1982; V. M. Dalton, personal communication; Pierson, unpublished observations), will follow stream or river corridors and forage within a broad riparian zone (Rainey and Pierson 1996). With the exception of some desert species (Bell et al. 1986), even those species that do not feed in association with water will come to water to drink.

While most species including the "beetle strategists" (sensu Freeman 1979) consume moths (Warner 1985) and a number of species feed primarily on moths, many species may be more selective than has previously been appreciated. For example, Sample and Whitmore (1993) showed that *C. townsendii* fed preferentially on forest-associated lepidopteran species in West Virginia, and Brown et al. (1994a) found that radiotagged *C. townsendii* in California ignored lush nonnative vegetation close to a roost site and traveled 5 km to forage in native ironwood forest.

To assess potential impacts of various land-use alterations, we need to understand the habitat associations of insect prey. The disappearance of formerly abundant populations of *M. velifer* from the Colorado River basin in California is correlated with the conversion of the cottonwood riparian forest to agriculture (K. Stager, personal communication). Decreased diversity in urban areas is more likely the result of loss of foraging habitat than roosting habitat, particularly for those species that accept building roosts, although predation by human commensals such as cats or rats may play a role (Gillette and Kimbrough 1970; G. Fellers, personal communication).

Recent radiotracking studies indicate that many species show remarkable loyalty to foraging sites, with individuals returning night after night, and even year to year, to the same area (Brigham 1991; Rainey and Pierson 1996; V. M. Dalton, personal communication). Thus, depending on the spatial distribution of foraging sites, landscape alterations could have profound impacts on bat populations.

Pesticide use in agriculture and forestry is also cause for concern (Henny et al. 1982; Clark 1988). Although persistent chlorinated hydrocarbons (now banned) historically caused severe problems for bats (Clark 1981, 1988; Clawson and Clark 1989; Clawson 1991), little field research has been conducted on the effects of the now widely used poisons with shorter half-lives (organophosphates), which have been documented to cause problems for birds, notably raptors (Wilson et al. 1991). The broadcast application of other agents with low vertebrate toxicity (e.g., Bt [*Bacillus thur-*

ingiensis] used in lepidopteran control) may still reduce the prey base for certain bat species, particularly the moth specialists. One study conducted in West Virginia, however, suggested that no moth genera preyed upon by *C. townsendii* were significantly less abundant after *Bt* application (Reardon et al. 1994).

Polluted or toxic water associated with resource extraction facilities poses a serious threat to foraging bats. Modern gold mining often uses cyanide solution to extract gold. This practice can involve the creation of large, toxin-laden ponds, often located in arid areas with little surface water. These impoundments have been responsible for substantial wildlife mortality (Clark and Hothem 1991). Mining operations on U.S. public lands are now generally required to cover collection ponds with protective netting. It is far more difficult, however, to control wildlife access to "heap leach pads," where a cyanide solution is sprayed over crushed ore and often pools on the pad surface. Similarly, significant wildlife mortality, involving particularly birds and bats, has been documented at exposed waste oil pits in a number of western states (Flickinger and Bunck 1987; Esmoil and Anderson 1995).

Ecosystem Role of Bats

Recognition of bats as primary predators of nocturnal insects has long provided an economic rationale for bat preservation in anthropogenic landscapes (Grinnell 1918). The idea that bats could control populations of disease-carrying insects such as mosquitoes motivated the first large-scale experimentation with artificial roosts (Campbell 1913). The mostly anecdotal recognition that bats feed on agricultural pests is now being supported by research showing bats to be significant predators on cucumber beetles (*Diabrodica* spp.) and corn earworm moths (*Heliothis zea*) (Whitaker 1993; McCracken 1996). Also, recent studies suggesting that vertebrate insectivores (birds and rodents) may play a role in limiting pest outbreaks in forested settings (Crawford and Jennings 1989; Ostfeld et al. 1996) have important implications for bats. Most major forest pests are moths, and some, at least, are known to be consumed by bats; for example, *Myotis volans* and *M. evotis* both feed on the spruce budworm moth (*Choristoneura fumiferana*) (J. M. Perkins, personal communication), and light-tagged *Corynorhinus townsendii* were observed at several localities feeding on oak moths (*Phryganidia californica*) during a recent California outbreak (Pierson, unpublished data).

Bats may also play a significant role in nutrient transfer in natural ecosystems. These organisms have an extremely high throughput (e.g., *Tadarida brasiliensis* ingests 50%–70% of its body mass per night [Kunz et al. 1995] and *M. lucifugus*

as much as 100% [Kurta et al. 1989]) and produce nitrogen-rich guano, which has been used for centuries as a fertilizer (Bailey 1925; Keleher 1996). Because bats frequently defecate on the wing, and can travel considerable distances between foraging areas and roosts, it seems possible that they act as "nutrient pepper shakers," redistributing nutrients over the landscape and creating nutrient "hot spots" within roosts, particularly large hollow trees (Janzen 1976; Kunz 1982; W. E. Rainey, personal communication). The nutrient pepper shaker function is likely to be most relevant in boreal forests, where bats tend to forage in highly productive areas (generally associated with water) and travel to roosts in a nutrient-limited setting (Bonan and Shugart 1989; Chapin et al. 1987). Because temperature is one of the prime factors in roost selection (Hall 1962; Tuttle and Stevenson 1978), tree roosts are often selected for maximum sun exposure, and thus are located in trees that reach above the surrounding canopy or occur on ridge tops, generally at some distance from water-associated foraging areas (Barclay and Brigham 1996). The need to find roosts with suitable temperature conditions, in some settings, likely dictates the distance that bats travel to foraging areas.

An ecologically distinctive characteristic of many North American bat species is the relatively large distance traveled between roosting and foraging areas (Table 22.3). Most small rodents, even the insectivorous species, travel very short distances from their nests, and those insectivorous birds most noted for their flying skills (e.g., swifts and swallows) travel a relatively short distance from their home base as compared to bats. Most passerines, for example, form seasonal pair bonds, and nesting pairs are separated by territories that include both nesting and foraging areas (Lack 1968; Newton 1989). The only insectivorous birds that travel as far as bats are the much larger, nocturnal caprimulgids (Brigham 1988), which are convergent in their foraging strategies with bats (Brigham and Fenton 1991). While it is not surprising that those bats that are adapted for long-distance flight (e.g., the molossids and lasiurines) generally travel farther than other species, the distances covered are noteworthy. Also, it is especially striking that small *Myotis* and other broad-winged species (e.g., *Euderma maculatum* and *C. townsendii*), presumably adapted for short-distance maneuverable flight (Norberg 1987; Norberg and Rayner 1987), will travel one-way distances as great as 16 km per night to feed. The flight distances undertaken by bats are more comparable to those of larger columbids, seabirds, and raptors than to those of the smaller insectivorous birds. Although long-distance movements are not observed in all North American forest-dwelling species (e.g., *M. evotis*) (Pierson, unpublished observations), and do not seem to be undertaken by some European species (see

Table 22.3

Maximum One-Way Distance Traveled from Nest or Roost to Foraging Area for Various Bats and Animals of Similar Size

Species	Distance (km)	Method[a]	Reference
Rodents			
Stephen's kangaroo rat (*Dipodomys stephensi*)	0.04	a	Price et al. 1994
Northern flying squirrel (*Glaucomys sabrinus*)	0.30	a	Witt 1992
Southern grasshopper mouse (*Onychomys torridus*)	0.12	b	Chew and Chew 1970
Birds			
Vaux's swift (*Chaetura vauxi*)	5.4	a	Bull and Beckwith 1993
Common nighthawk (*Chordeiles minor*)	12.0	a	Brigham 1988
Red-cockaded woodpecker (*Dendrocopos borealis*)	1.30	c	Ligon 1970
Cliff swallow (*Hirundo pyrrhonata*)	0.96	c	Brown et al. 1992
Bats			
Townsend's big-eared bat (*Corynorhinus townsendii*)	12.8	a	V. Dalton, personal communication
	7.0	a	Clark et al. 1993
Spotted bat (*Euderma maculatum*)	10.0	a	Wai-Ping and Fenton 1989
	>6.0	a	Pierson and Rainey 1995
Western mastiff bat (*Eumops perotis*)	>15.0	a	Pierson and Rainey 1995
	24.0	d	Vaughan 1959
Silver-haired bat (*Lasionycteris noctivagans*)	17.4	a	Rainey and Pierson 1996
Hoary bat (*Lasiurus cinereus*)	20.0	a	R. M. R. Barclay, personal communication
	6.3	a	Barclay 1989
Sanborn's long-nosed bat (*Leptonycteris curasoae*)	30.0	a	Sahley et al. 1993
Gray bat (*Myotis grisescens*)	39.0	e	LaVal et al. 1977
Long-eared myotis (*Myotis evotis*)	0.6	a	Pierson, unpublished observation
Yuma myotis (*Myotis yumanensis*)	7.0	a	Pierson and Rainey, unpublished observation
	16.2	b	Rainey and Pierson 1996

[a]Methods: a, radiotracking; b, mark-recapture; c, direct observation; d, measuring distance to nearest suitable roost habitat; e, light-tagging.

Racey, Chapter 17, this volume), they are an important factor in the ecology of some North American taxa.

Sustaining the ecosystem role of bats may require a shift in attitudes toward conservation. Most conservation efforts to date have focused on rare and endangered taxa. While these efforts have generally been successful (Brady et al. 1982; Tuttle 1986) and are likely essential to the survival of these taxa, they overlook the potentially critical ecological role played by more numerous species. From an ecosystem perspective, the most important taxa may be the most widespread and locally abundant, such as *Eptesicus fuscus*, *Lasionycteris noctivagans*, *Myotis lucifugus*, *M. yumanensis*, and *Tadarida brasiliensis*.

Status Assessment

Assessment of population trends is critical in determining a species status. The only species for which historical population estimates are available are some of the cave-dwelling taxa. Sometimes these are based on actual recorded population counts, in other cases, on extrapolations based on the size of guano deposits or ceiling staining in roosting areas (Pearson et al. 1952; Mohr 1972; Humphrey 1978; Tuttle 1979; Rabinowitz and Tuttle 1980; Clawson 1987; Gore and Hovis 1992; Missouri Department of Conservation 1992; Stihler and Brack 1992; Pierson and Rainey 1994; C. W. Stihler and J. S. Hall, personal communication). Declines are also inferred when bats no longer occupy traditional roosts (Dalton 1987).

For those species that do not aggregate at readily accessible traditional roost sites, population trends are more difficult to evaluate. In many cases, particularly for solitary or forest-dwelling species, we must rely on anecdotal information. For example, the red bat, *Lasiuris borealis*, now rarely seen in the northeastern United States, was described by Mearns in 1898 as occurring in "great flights" that could be observed "during the whole day" during fall migration in New York State. In another example, numerous museum records support the conclusion that the Los Angeles basin was, earlier in this century, a significant center of distribution for the mastiff bat, *Eumops perotis*. Although the species still occurs there, it is now rare, and appears to have been extirpated from many areas where it was previously found (Pierson and Rainey 1995).

Concern may be warranted for rare and geographically restricted taxa such as *Euderma maculatum*, *Idionycteris phyllotis*, or *Myotis keenii*, even in the absence of historical population data, because rarity alone confers vulnerability. For example, limited range and currently small population size formed the basis for federal endangered species status for the two eastern subspecies of *Corynorhinus townsendii*. Species more likely to be overlooked in a species status evaluation are those that, in the absence of historical data, would be viewed as abundant. For example, *T. brasiliensis* forms the largest known mammalian aggregations, with some colonies in Texas currently reaching 20 million (McCracken 1986). In many habitats of the southwestern United States it appears to be the most abundant and pervasive species. Historical data indicate, however, that its current population levels may be only a fraction of what they once were. At some known roost caves, populations declined in the 1950s and 1960s by as much as 99% (Mohr 1972).

One problem with comparing past and present abundance of particular bat species is that sampling methods, and consequently detection rates, have changed as new technologies have become available (Thomas and LaVal 1988; Kunz et al. 1996). Although it has always been possible to estimate numbers at roosts (by counting individuals or calculating the number of individuals per unit area), the availability of night vision equipment, video cameras, and automated monitoring devices have made it possible to obtain more precise roost counts. Historically, sampling for bats in foraging areas relied on the use of a shotgun. Kellogg (1916), in reference to a mammal and bird survey of northwestern California, commented "there is nothing quite so wasteful of ammunition as shooting at bats." On an extensive 6-month expedition, only 6 specimens of 3 species were collected. Recently, in this same region, by using mist nets as many as 130 individuals and 9 of the 17 known species have been captured in a single night (Rainey and Pierson 1996).

Conclusions

Most North American bat species have broad distributions and thus at a species level are not so vulnerable to extinction as some other taxa. Yet, particular features of bat distribution are critical to conservation and need to be considered. While marginal records may suggest a species occurs over a large area, distribution may vary seasonally and may be quite patchy, particularly for those species that rely on rare geomorphic features for roosting sites. The tendency of some species to aggregate in relatively few large colo-

nies, combined with their need to find maternity and hibernating sites with fairly restrictive microclimatic conditions, put many at high risk regardless of overall distribution. Population declines have been documented even for some of the most widely distributed species.

The often substantial geographic separation between roosting and foraging areas requires a dual focus when considering the habitat needs of these organisms. While threats to roost sites in caves and human structures have been recognized for a long time, more recent research has documented the importance of trees and abandoned mines. Nevertheless, woefully little information is available on species-specific roosting requirements in any of these settings. Even less is known regarding foraging requirements. The fact that most bat species appear to be relatively opportunistic in their selection of prey may obscure other important constraints, such as requirements for particular habitats or habitat features.

Declines in bat abundance and diversity could have serious consequences for ecosystem function. Insectivorous bats may be unique among small vertebrates in the distances traveled between roosting and foraging areas. Because they frequently forage in high productivity environments (in proximity to water), roost in nutrient-limited environments, and have unusually high throughput (consuming up to 100% of their body mass per night), they may play a significant role in nutrient transfer by acting as nutrient pepper shakers, particularly in conifer forests. This function however may be difficult to demonstrate at current levels of abundance.

Historical data on bat numbers are generally lacking, and changes in sampling methodology make it difficult to compare historic and current abundance for many species. It is, nevertheless, almost certain that current bat densities are well below historical levels. Although we lack the kind of long-term records available for birds (Marshall 1988), we have well-substantiated evidence of serious population declines for many species, particularly cave-dwelling taxa. Although there is a tendency to view current systems as though they were in equilibrium, chances are they are nowhere close to equilibrium. In this context, it is not far fetched to suggest that the fate of *Tadarida brasiliensis*, still one of North America's most abundant species, could follow the same trajectory as that of the passenger pigeon or Carolina parakeet.

Too often, insufficient information and exaggerated concerns regarding liability and health risks guide management decisions unfavorable to bats. Granting bats an equitable place in wildlife policy deliberations is an essential first step but must be supported by conservation-related research. Because management decisions are often made in haste, there is a tendency to assume that an action appropriate for one species will benefit all. This frequently is not the case. Because virtually all North America is now a managed landscape, it is critical for the long-term viability of bat populations that researchers provide land managers with sufficient information that the diverse ecological needs of the bat community can be met.

Acknowledgments

I am grateful to T. H. Kunz for having introduced me to bats; to all those colleagues who share a commitment to bat conservation—most particularly those who have been valued field companions—W. E. Rainey, P. E. Brown, and V. M. Dalton; to the California Department of Fish and Game and the National Park Service, who have supported much of my work; to the McLaughlin Mine for developing a model bat–mine project in California; and to Holohil Systems for making reliable radiotransmitters for bats. I thank R. M. Brigham, D. C. Dalton, V. M. Dalton, and W. E. Rainey for critical review of this manuscript.

Literature Cited

Allen, G. M. 1940. Bats. Harvard University Press, Cambridge.

Altenbach, J. S., and E. D. Pierson. 1995. The importance of mines to bats: An overview. *In* Inactive Mines as Bat Habitat: Guidelines for Research, Survey, Monitoring, and Mine Management in Nevada, B. R. Riddle, ed., pp. 7–18. Biological Resources Research Center, University of Nevada, Reno.

Anthony, E. L. P., and T. H. Kunz. 1977. Feeding strategies of the little brown bat, *Myotis lucifugus*, in southern New Hampshire. Ecology 58:775–786.

Bailey, V. 1925. Bats of the Carlsbad Cavern. National Geographic 48:320–330.

Baker, R. J., J. C. Patton, H. H. Genoways, and J. W. Bickham. 1988. Genic studies of *Lasiurus* (Chiroptera: Vespertilionidae). Occasional Papers of the Museum, Texas Tech University 117:1–15.

Barbour, R. W., and W. H. Davis. 1969. Bats of America. University of Kentucky Press, Lexington.

Barclay, R. M. R. 1989. The effect of reproductive condition on the foraging behavior of female hoary bats, *Lasiurus cinereus*. Behavioral Ecology and Sociobiology 24:31–37.

Barclay, R. M. R., and R. M. Brigham, eds. 1996. Bats and Forests Symposium, October 19–21, 1995, Victoria, British Columbia. Working Paper 23/1996. Research Branch, Ministry of Forests, Victoria.

Bat Conservation International. 1994. Research begins on bat-friendly bridge designs. Bats 12(4): 18–19.

Bell, G. P., G. A. Bartholomew, and K. A. Nagy. 1986. The roles of energetics, water economy, foraging behavior, and geothermal refugia in the distribution of the bat *Macrotus californicus*. Journal of Comparative Physiology B 156:441–450.

Belwood, J. J. 1981. Wagner's mastiff bat, *Eumops glaucinus flori-danus* (Molossidae), in southwestern Florida. Journal of Mammalogy 62:411–413.

Belwood, J. J. 1992. Florida mastiff bat, *Eumops glaucinus floridanus*. *In* Rare and Endangered Biota of Florida: Mammals, S. R. Humphrey, ed., pp. 216–223. University Press of Florida, Gainesville.

Belwood, J. J., and R. J. Waugh. 1991. Bats and mines: Abandoned does not always mean empty. Bats 9(3): 13–16.

Betts, B. 1996. Roosting behaviour of silver-haired bats *(Lasionycteris noctivagans)* and big brown bats *(Eptesicus fuscus)* in northeast Oregon. *In* Bats and Forests Symposium, October 19–21, 1995, Victoria, British Columbia, pp. 55–61. Working Paper 23/1996. Research Branch, Ministry of Forests, Victoria.

Bonan, G. B., and H. H. Shugart. 1989. Environmental factors and ecological processes in boreal forests. Annual Review of Ecology and Systematics 20:1–28.

Bradshaw, P. A. 1996. The physical nature of vertical forest habitat and its importance in shaping bat species assemblages. *In* Bats and Forests Symposium, October 19–21, 1995, Victoria, British Columbia, pp. 199–212. Working Paper 23/1996. Research Branch, Ministry of Forests, Victoria.

Brady, J. T., T. H. Kunz, M. D. Tuttle, and D. E. Wilson. 1982. Gray Bat Recovery Plan. U.S. Fish and Wildlife Service, Washington, D.C.

Brass, D. A. 1994. Rabies in Bats: Natural History and Public Health Implications. Livia Press, Ridgefield, Conn.

Brigham, R. M. 1988. The influence of wing morphology, prey detection system, and availability of prey on the foraging strategies of aerial insectivores. Ph.D. dissertation, York University, Downsview, Ontario.

Brigham, R. M. 1991. Flexibility in foraging and roosting behaviour by the big brown bat *(Eptesicus fuscus)*. Canadian Journal of Zoology 69:117–121.

Brigham, R. M., and M. B. Fenton. 1991. Convergence in foraging strategies by two morphologically and phylogenetically distinct nocturnal aerial insectivores. Journal of Zoology (London) 223: 475–490.

Brigham, R. M., H. D. J. N. Aldridge, and R. L. Mackey. 1992. Variation in habitat use and prey selection by yuma bats, *Myotis yumanensis*. Journal of Mammalogy 73:640–645.

Brown, C. R., M. B. Brown, and A. R. Ives. 1992. Nest placement relative to food and its influence on the evolution of avian coloniality. American Naturalist 139:205–217.

Brown, P. E. 1995. Impacts of renewed mining in historic districts and mitigation for impacts on bat populations. *In* Inactive Mines as Bat Habitat: Guidelines for Research, Survey, Monitoring, and Mine Management in Nevada, B. R. Riddle, ed., pp. 138–140. Biological Resources Research Center, University of Nevada, Reno.

Brown, P. E., and R. D. Berry. 1991. Bats: Habitat, impacts, and mitigation. *In* Issues and Technology in Management of Impacted Wildlife, April 8–10, 1991, pp. 26–30. Thorne Ecological Institute, Boulder, Colo.

Brown, P. E., R. D. Berry, and C. Brown. 1993. Bats and mines: Finding solutions. Bats 11:12–13.

Brown, P. E., R. D. Berry, and C. Brown. 1994a. Foraging behavior of Townsend's big-eared bats *(Plecotus townsendii)* on Santa Cruz Island. *In* Fourth California Islands Symposium: Update on the Status of Resources, W. L. Halvorson and G. J. Maender, eds., pp. 367–370. Santa Barbara Museum of Natural History, Santa Barbara, Calif.

Brown, P. E., R. D. Berry, and C. Brown. 1994b. The California leaf-nosed bat *(Macrotus californicus)* and American Girl Mining joint venture: Impact and solutions. *In* Thorne Ecological Institute Proceedings VI: Issues and Technology in the Management of Impacted Wildlife, pp. 54–56. Thorne Ecological Institute, Boulder, Colo.

Bull, E. L., and R. C. Beckwith. 1993. Diet and foraging behavior of Vaux's swifts in northeastern Oregon. Condor 95:1016–1023.

Campbell, C. A. R. 1913. Eradication of mosquitoes by the cultivation of bats. Monthly Bulletin of Agricultural Intelligence and Plant Diseases 4:1175–1181.

Cantor, D. 1995. Mobil goes to bat for birds and bats: Inexpensive exhaust stack caps save lives. Business and Society Review 92:34–35.

Chapin, F. S., A. J. Bloom, C. B. Field, and R. H. Waring. 1987. Plant responses to multiple environmental factors. Bioscience 37:49–57.

Chew, R. M., and A. E. Chew. 1970. Energy relationships of the mammals of a desert shrub *(Larrea virdentata)* community. Ecological Monographs 40:1–21.

Clark, B. S., D. M. Leslie, and T. S. Carter. 1993. Foraging activity of adult Ozark big-eared bats *Plecotus townsendii ingens* in summer. Journal of Mammalogy 74:422–427.

Clark, D. R., Jr. 1981. Bats and environmental contaminants: A review. U.S. Fish and Wildlife Service Special Scientific Report 1:1–27.

Clark, D. R., Jr. 1988. Environmental contaminants and the management of bat populations in the United States. U.S. Forest Service General Technical Report 166:409–413.

Clark, D. R., Jr., and R. L. Hothem. 1991. Mammal mortality at Arizona, California, and Nevada gold mines using cyanide extraction. California Fish and Game 77:61–69.

Clawson, R. L. 1987. Indiana bats: Down for the count. Endangered Species Technical Bulletin 12:9–11.

Clawson, R. L. 1991. Pesticide contamination of endangered gray bats and their prey in Boone, Franklin, and Camden counties, Missouri. Transactions of the Missouri Academy of Science 25:13–19.

Clawson, R. L., and D. R. Clark, Jr. 1989. Pesticide contamination of endangered gray bats and their food base in Boone County, Missouri, 1982. Bulletin of Environmental Contamination and Toxicology 42:431–437.

Cockrum, E. L. 1969. Migration in the guano bat, *Tadarida brasiliensis. In* Contributions in Mammalogy, No. 51, J. K. Jones, Jr., ed., pp. 304–336. Museum of Natural History, University of Kansas, Lawrence.

Constantine, D. G. 1970. Bats in relation to the health, welfare, and economy of man. *In* Biology of Bats, Vol. II, W. A. Wimsatt, ed., pp. 319–449. Academic Press, New York.

Constantine, D. G. 1987. Long-tongued bat and spotted bat at Las Vegas, Nevada. Southwestern Naturalist 32:392.

Crampton, L. H., and R. M. R. Barclay. 1996. Habitat selection by bats in fragmented and unfragmented aspen mixed wood stands of different ages. *In* Bats and Forests Symposium, October 19–21, 1995, Victoria, British Columbia, pp. 238–259. Working Paper 23/1996. Research Branch, Ministry of Forests, Victoria.

Crawford, H. S., and D. T. Jennings. 1989. Predation by birds on spruce budworm *Choristoneura fumiferana:* Functional, numerical, and total responses. Ecology 70:152–163.

Dalton, D. C., and V. M. Dalton. 1995. Mine closure methods including a recommended gate design. *In* Inactive Mines as Bat Habitat: Guidelines for Research, Survey, Monitoring, and Mine Management in Nevada, B. R. Riddle, ed., pp. 130–135. Biological Resources Research Center, University of Nevada, Reno.

Dalton, V. M. 1985. Cave bats: Their ecology, identification, and distribution. *In* Proceedings, Cave Management Symposia, H. Thornton and J. Thornton, eds., pp. 36–44. American Cave Conservation Association, Richmond, Va.

Dalton, V. M. 1987. Distribution, abundance, and status of bats hibernating in caves in Virginia. Virginia Journal of Science 38:369–379.

Davis, R., and E. L. Cockrum. 1963. Bridges utilized as day-roosts by bats. Journal of Mammalogy 44:428–430.

Davis, W. H., and H. B. Hitchcock. 1965. Biology and migration of the bat, *Myotis lucifugus,* in New England. Journal of Mammalogy 46:296–313.

Decher, J., and J. R. Choate 1995. *Myotis grisescens.* Mammalian Species 510:1–7.

Durbin, K. 1996. Lawless logging. Defenders 71:15–24.

Dutko, R. 1994. Protected at last: The hibernia mine. Bats 12:3–5.

Easterla, D. A. 1973. Ecology of the 18 species of Chiroptera at Big Bend National Park, Texas. Part II. Northwest Missouri State University Studies 34:54–165.

Esmoil, B. J., and S. H. Anderson. 1995. Wildlife mortality associated with oil pits in Wyoming. Prairie Naturalist 27:81–88.

Fenton, M. B. 1992. Bats. Facts on File, New York.

Fenton, M. B., and R. M. R. Barclay. 1980. *Myotis lucifugus.* Mammalian Species 142:1–8.

Findley, J. S. 1993. Bats: A Community Perspective. Cambridge University Press, Cambridge.

Fleming, T. H., R. A. Nunez, and L. D. S. L. Sternberg. 1993. Seasonal changes in the diets of migrant and non-migrant nectarivorous bats as revealed by carbon stable isotope analysis. Oecologia (Berlin) 94:72–75.

Flickinger, E. L., and C. M. Bunck. 1987. Number of oil-killed birds and fate of bird carcasses at crude oil pits in Texas. Southwestern Naturalist 32:377–381.

Freeman, P. W. 1979. Specialized insectivory: Beetle-eating and moth-eating molossid bats. Journal of Mammalogy 60:467–479.

Frost, D. R., and R. M. Timm. 1992. Phylogeny of plecotine bats (Chiroptera: Vespertilionidae): Summary of the evidence and proposal of a logically consistent taxonomy. American Museum Novitates 3034:1–16.

Garcia, P. F. J., S. A. Rasheed, and S. L. Holroyd. 1995. Status of the western small-footed myotis in British Columbia. Wildlife Working Report WR-74:1–14. Ministry of Environment, Lands, and Parks, Victoria, British Columbia.

Geggie, J. F., and M. B. Fenton. 1985. A comparison of foraging by *Eptesicus fuscus* (Chiroptera: Vespertilionidae) in urban and rural environments. Canadian Journal of Zoology 63:263–266.

Gillette, D. D., and J. D. Kimbrough. 1970. Chiropteran mortality. *In* About Bats, B. H. Slaughter and D. W. Walton, eds., pp. 262–281. Southern Methodist University Press, Dallas, Tex.

Gore, J. A., and J. A. Hovis. 1992. The southeastern bat: Another cave-roosting species in peril. Bats 10:10–12.

Graham, R. E. 1966. Observations on the roosting habits of the big-eared bat, *Plecotus townsendii*, in California limestone caves. Cave Notes 8:17–22.

Greenhall, A. M. 1982. House bat management. Resource Publication 143. U.S. Fish and Wildlife Service, Washington, D.C.

Grinnell, H. W. 1918. A synopsis of the bats of California. University of California Publications in Zoology 17:223–404.

Grinnell, J., and T. I. Storer. 1924. Animal Life in the Yosemite. University of California Press, Berkeley.

Hall, E. R. 1981. The Mammals of North America. Wiley, New York.

Hall, J. S. 1962. A Life History and Taxonomic Study of the Indiana Bat, *Myotis sodalis*. Scientific Publication 12, Public Museum and Art Gallery, Reading, Pa.

Harvey, M. J. 1980. Status of the endangered bats, *Myotis sodalis*, *M. grisescens*, and *Plecotus townsendii ingens*, in the southern Ozarks. *In* Proceedings, Fifth International Bat Research Conference, D. E. Wilson and A. L. Gardner, eds., pp. 221–223. Texas Tech University Press, Lubbock.

Hayes, J. P., and M. D. Adam. 1996. The influence of logging riparian areas on habitat utilization by bats in western Oregon. *In* Bats and Forests Symposium, October 19–21, 1995, Victoria, British Columbia, pp. 228–237. Working Paper 23/1996. Research Branch, Ministry of Forests, Victoria.

Henny, C. J., C. Maser, J. O. Whitaker, Jr., and T. E. Kaiser. 1982. Organochlorine residues in bats after a forest spraying with DDT. Northwest Science 56:329–337.

Hensley, S., and C. Scott. 1995. Ozark big-eared bat, *Plecotus townsendii ingens* (Handley): Revised recovery plan. U.S. Fish and Wildlife Service, Region 2, Albuquerque, N.Mex.

Hermanson, J. W., and T. J. O'Shea. 1983. *Antrozous pallidus*. Mammalian Species 213:1–8.

Hoffmeister, D. F. 1986. The Mammals of Arizona. University of Arizona Press, Tucson.

Howell, A. B. 1920. Some Californian experiences with bat roosts. Journal of Mammalogy 1:169–177.

Hoyt, R. A., J. S. Altenbach, and D. J. Hafner. 1994. Observations on long-nosed bats *(Leptonycteris)* in New Mexico. Southwestern Naturalist 39:175–179.

Humphrey, S. R. 1975. Nursery roosts and community diversity of Nearctic bats. Journal of Mammalogy 56:321–346.

Humphrey, S. R. 1978. Status, winter habitat, and management of the endangered Indiana bat, *Myotis sodalis*. Florida Scientist 41:65–76.

Janzen, D. H. 1976. Why tropical trees have rotten cores. Biotropica 8:110.

Jones, C. J. 1965. Ecological distribution and activity periods of bats of the Mogollon Mountains area of New Mexico and adjacent Arizona. Tulane Studies in Zoology 12:93–100.

Jones, C., and R. W. Manning. 1989. *Myotis austroriparius*. Mammalian Species 332:1–3.

Kalcounis, M. C., and K. R. Hecker. 1996. Intraspecific variation in roost-site selection by little brown bats *(Myotis lucifugus)*. *In* Bats and Forests Symposium, October 19–21, 1995, Victoria, British Columbia, pp. 81–90. Working Paper 23/1996. Research Branch, Ministry of Forests, Victoria.

Kaufman, D. M. 1995. Diversity of new world mammals: Universality of the latitudinal gradients of species and bauplans. Journal of Mammalogy 76:322–334.

Keleher, S. 1996. Guano: Bats' gift to gardeners. Bats 14:15–17.

Kellogg, L. 1916. Report upon mammals and birds found in portions of Trinity, Siskiyou, and Shasta counties, California, with description of a new *Dipodomys*. University of California Publications in Zoology 12:335–398.

Koopman, K. E. 1993. Order Chiroptera. *In* Mammal Species of the World: A Taxonomic and Geographic Reference, D. E. Wilson and D. M. Reeder, eds., pp. 137–241. Smithsonian Institution Press, Washington, D.C.

Krzanowski, A. 1991. Bibliography of bat roosting boxes. Lubuski Przeglad Przyrodniczy 2:65–96.

Kunz, T. H. 1974. Feeding ecology of a temperate insectivorous bat *(Myotis velifer)*. Ecology 55:693–711.

Kunz, T. H. 1982. Roosting ecology of bats. *In* Ecology of Bats, T. H. Kunz, ed., pp. 1–55. Plenum Press, New York.

Kunz, T. H., ed. 1988a. Ecological and Behavioral Methods for the Study of Bats. Smithsonian Institution Press, Washington, D.C.

Kunz, T. H. 1988b. Methods of assessing the availability of prey to insectivorous bats. *In* Ecological and Behavioral Methods for the Study of Bats, T. H. Kunz, ed., pp. 191–210. Smithsonian Institution Press, Washington, D.C.

Kunz, T. H., and R. A. Martin. 1982. *Plecotus townsendii*. Mammalian Species 175:1–6.

Kunz, T. H., J. O. Whitaker, Jr., and M. D. Wadanoli. 1995. Dietary energetics of the insectivorous Mexican free-tailed bat *(Tadarida brasiliensis)* during pregnancy and lactation. Oecologia (Berlin) 101:407–415.

Kunz, T. H., D. W. Thomas, G. C. Richards, C. R. Tidemann, E. D. Pierson, and P. A. Racey. 1996. Observational techniques for bats. *In* Measuring and Monitoring Biological Diversity: Standard Methods for Mammals, D. E. Wilson, F. R. Cole, J. D. Nichols, R. Rudran, and M. Foster, eds., pp. 105–114. Smithsonian Institution Press, Washington, D.C.

Kurta, A., G. P. Bell, K. A. Nagy, and T. H. Kunz. 1989. Energetics of pregnancy and lactation in free-ranging little brown bats *(Myotis lucifugus)*. Physiological Zoology 62:804–818.

Kurta, A., and J. A. Teramino. 1992. Bat community structure in an urban park. Ecography 15:257–261.

Lack, D. L. 1968. Ecological Adaptations for Breeding in Birds. Metheun Press, London.

LaVal, R. K., R. L. Clawson, M. L. LaVal, and W. Caire. 1977. Foraging behavior and nocturnal activity patterns of Missouri bats, with emphasis on the endangered species *Myotis grisescens* and *Myotis sodalis.* Journal of Mammalogy 58:592–599.

Leonard, M. L., and M. B. Fenton. 1983. Habitat use by spotted bats (*Euderma maculatum,* Chiroptera: Vespertilionidae): Roosting and foraging behavior. Canadian Journal of Zoology 61: 1487–1491.

Lewis, S. E. 1995. Roost fidelity of bats: A review. Journal of Mammalogy 76:481–496.

Licht, P., and P. Leitner. 1967. Behavioral responses to high temperatures in three species of California bats. Journal of Mammalogy 48:52–61.

Ligon, J. D. 1970. Behavior and breeding biology of the red-cockaded woodpecker. Auk 87:255–278.

Manning, R. W., and J. K. Jones, Jr. 1989. *Myotis evotis.* Mammalian Species 329:1–5.

Marshall, J. T. 1988. Birds lost from a giant sequoia forest during fifty years. Condor 90:359–372.

McCracken, G. F. 1986. Why are we losing our Mexican free-tailed bats? Bats 3:1–4.

McCracken, G. F. 1988. Who's endangered and what can we do? Bats 6:5–9.

McCracken, G. F. 1996. Bats aloft: A study of high-altitude feeding. Bats 14:7–10.

McCracken, G. F., M. K. McCracken, and A. T. Vawter. 1994. Genetic structure in migratory populations of the bat *Tadarida brasiliensis mexicana.* Journal of Mammalogy 75:500–514.

Mearns, E. A. 1898. A study of the vertebrate fauna of the Hudson Highlands, with observations on the Mollusca, Crustacea, Lepidoptera, and the flora of the region. Bulletin of the American Museum of Natural History 10:303–352.

Missouri Department of Conservation. 1992. Management plan for the Indiana bat and the gray bat in Missouri. Missouri Department of Conservation, St. Louis.

Mohr, C. E. 1952. A survey of bat banding in North America, 1932–1951. National Speleological Society Bulletin 14:3–13.

Mohr, C. E. 1953. Possible causes of an apparent decline in wintering populations of cave bats. National Speleological Society News 11:4–5.

Mohr, C. E. 1972. The status of threatened species of cave-dwelling bats. National Speleological Society Bulletin 34:33–47.

Morales, J. C., and J. W. Bickham. 1995. Molecular systematics of the genus *Lasiurus* (Chiroptera: Vespertilionidae) based on restriction-site maps of the mitochondrial ribosomal genes. Journal of Mammalogy 76:730–749.

Murphy, M. 1990. The bats at the bridge. Bats 11:21–23.

Nagorsen, D. W., and R. M. Brigham. 1993. Bats of British Columbia. University of British Columbia Press, Vancouver.

Nagorsen, D. W., A. A. Bryant, D. Kerridge, G. Roberts, A. Roberts, and M. J. Sarell. 1993. Winter bat records for British Columbia. Northwestern Naturalist 74:61–66.

Navo, K., J. Sheppard, and T. Ingersoll. 1995. Colorado's bats/inactive mines project: The use of volunteers in bat conservation. *In* Inactive Mines as Bat Habitat: Guidelines for Research, Survey, Monitoring, and Mine Management in Nevada, B. R. Riddle, ed., pp. 49–54. Biological Resources Research Center, University of Nevada, Reno.

Neilson, A. L., and M. B. Fenton. 1994. Responses of little brown myotis to exclusion and to bat houses. Wildlife Society Bulletin 22:8–14.

Newton, I., ed. 1989. Lifetime Reproduction in Birds. Academic Press, London.

Norberg, U. M. 1987. Wing form and flight mode in bats. *In* Recent Advances in the Study of Bats, M. B. Fenton, P. Racey, and J. M. V. Rayner, eds., pp. 46–56. Cambridge University Press, Cambridge.

Norberg, U. M., and J. M. V. Rayner. 1987. Ecological morphology and flight in bats (Mammalia; Chiroptera): Wing adaptations, flight performance, foraging strategy, and echolocation. Philosophical Transactions of the Royal Society of London B 316: 335–427.

Ostfeld, R. S., C. G. Jones, and J. O. Wolff. 1996. Of mice and mast. Bioscience 46:323–330.

Parker, D. I., J. A. Cook, and S. W. Lewis. 1996. Effects of timber harvest on bat activity in southeastern Alaska's temperate rainforests. *In* Bats and Forests Symposium, October 19–21, 1995, Victoria, British Columbia, pp. 277–292. Working Paper 23/1996. Research Branch, Ministry of Forests, Victoria.

Parks, C. G., and D. C. Shaw. 1996. Death and decay: A vital part of living canopies. Northwest Science 70:46–53.

Pearson, O. P., M. R. Koford, and A. K. Pearson. 1952. Reproduction of the lump-nosed bat (*Corynorhinus rafinesquei*) in California. Journal of Mammalogy 33:273–320.

Perkins, J. M., and S. P. Cross. 1988. Differential use of some coniferous forest habitats by hoary and silver-haired bats in Oregon. Murrelet 69:21–24.

Perlmeter, S. I. 1996. Bats and bridges: Patterns of night roost use by bats in the Willamette National Forest. *In* Bats and Forests Symposium, October 19–21, 1995, Victoria, British Columbia, pp. 132–150. Working Paper 23/1996. Research Branch, Ministry of Forests, Victoria.

Pierson, E. D., and W. E. Rainey. 1994. The distribution, status, and management of Townsend's big-eared bat (*Corynorhinus townsendii*) in California. Bird and Mammal Conservation Program Report 94-8. California Department of Fish and Game, Sacramento.

Pierson, E. D., and W. E. Rainey. 1995. Distribution, habitat associations, status, and survey methodologies for three molossid bat species (*Eumops perotis, Nyctinomops femorosaccus, Nyctinomops macrotis*) and the vespertilionid (*Euderma maculatum*). Contract FG2328WM. Report to California Department of Fish and Game, Sacramento.

Pierson, E. D., and W. E. Rainey. 1998. Distribution of the spotted bat, *Euderma maculatum,* in California, with notes on habitat associations and status. Journal of Mammalogy 79: (in press).

Pierson, E. D., W. E. Rainey, and D. M. Koontz. 1991. Bats and mines: Experimental mitigation for Townsend's big-eared bat at the McLaughlin Mine in California. *In* Thorne Ecological Institute Proceedings V: Issues and Technology in the Management of Impacted Wildlife, pp. 31–42. Thorne Ecological Institute, Boulder, Colo.

Pierson, E. D., W. E. Rainey, and R. M. Miller. 1996. Night roost sampling: A window on the forest bat community in northern California. *In* Bats and Forests Symposium, October 19–21, 1995, Victoria, British Columbia, pp. 151–163. Working Paper 23/1996. Research Branch, Ministry of Forests, Victoria.

Price, M. V., P. A. Kelly, and R. L. Goldingay. 1994. Distances moved by Stephen's kangaroo rat (*Dipodomys stephensi* Merriam) and implications for conservation. Journal of Mammalogy 75:929–939.

Rabinowitz, A., and M. D. Tuttle. 1980. Status of summer colonies of the endangered gray bat in Kentucky. Journal of Wildlife Management 44:955–960.

Rainey, W. E., and E. D. Pierson. 1996. Cantara spill effects on bat populations of the upper Sacramento River, 1991–1995. Contract FG2099R1. Report to California Department of Fish and Game, Redding.

Rasheed, S. A., P. F. J. Garcia, and S. L. Holroyd. 1995. Status of the fringed myotis in British Columbia. Wildlife Working Report WR-73. Ministry of Environment, Lands, and Parks, Victoria, British Columbia.

Reardon, R., N. Dubois, and W. McLane 1994. *Bacillus thuringiensis* for Managing Gypsy Moth: A Review. FHM-NC-01-94, Forest Health Management. U.S. Forest Service, Washington, D.C.

Richter, A. R., S. R. Humphrey, J. B. Cope, and V. Brack, Jr. 1993. Modified cave entrances: Thermal effect on body mass and resulting decline of endangered Indiana bats (*Myotis sodalis*). Conservation Biology 7:407–415.

Riddle, B. R., ed. 1995. Inactive Mines as Bat Habitat: Guidelines for Research, Survey, Monitoring, and Mine Management in Nevada. Biological Resources Research Center, University of Nevada, Reno.

Roberts, G., and A. Roberts. 1993. Biodiversity in the Caribo-Chilcotin grasslands. Report to Ministry of Environment, Lands, and Parks, Wildlife Branch, Williams Lake, British Columbia.

Robson, M. S., F. J. Mazzotti, and T. Parrott. 1989. Recent evidence of the mastiff bat in southern Florida. Florida Field Naturalist 17:81–82.

Rohde, K. 1992. Latitudinal gradients in species diversity: The search for the primary cause. Oikos 65:514–527.

Sample, B. E., and R. C. Whitmore. 1993. Food habits of the endangered Virginia big-eared bat in West Virginia. Journal of Mammalogy 74:428–435.

Sahley, C. T., M. A. Horner, and T. H. Fleming. 1993. Flight speeds and mechanical power outputs of the nectar-feeding bat, *Leptonycteris curasoae* (Phyllostomidae: Glossophaginae). Journal of Mammalogy 74:594–600.

Saugey, D. A. 1991. U.S. national forests: Unsung home to America's bats. Bats 9:3–6.

Schmidly, D. J. 1991. The Bats of Texas. University of Texas Press, Austin.

Stager, K. 1939. Status of *Myotis velifer* in California, with notes on its life history. Journal of Mammalogy 20:225–228.

Stager, K. E. 1943. Remarks on *Myotis occultus* in California. Journal of Mammalogy 24:197–199.

Stihler, C. W., and V. Brack, Jr. 1992. A survey of hibernating bats in Hellhole Cave, Pendleton County, West Virginia. Proceedings of the West Virginia Academy of Sciences 64:34–35.

Storz, J. F. 1995. Local distribution and foraging behavior of the spotted bat, *Euderma maculatum*, in northwestern Colorado and adjacent Utah. Great Basin Naturalist 55:78–83.

Thomas, D. W. 1993. Bats, mines, and politics. Bats 11:10–11.

Thomas, D. W. 1995. Hibernating bats are sensitive to nontactile human disturbance. Journal of Mammalogy 76:940–946.

Thomas, D. W., and R. K. LaVal. 1988. Survey and census methods. *In* Ecological and Behavioral Methods for the Study of Bats, T. H. Kunz, ed., pp. 77–89. Smithsonian Institution Press, Washington, D.C.

Tolin, W. A. 1994. Virginia big-eared bat (*Plecotus townsendii virginianus*) recovery plan. U.S. Fish and Wildlife Service, Region 5, Hadley, Mass.

Tumlison, R., and M. E. Douglas. 1992. Parsimony analysis and the phylogeny of the plecotine bats (Chiroptera: Vespertilionidae). Journal of Mammalogy 73:276–285.

Tuttle, M. D. 1976. Population ecology of the gray bat (*Myotis grisescens*): Philopatry, timing and patterns of movement, weight loss during migration, and seasonal adaptive strategies. Occasional Papers of the Museum of Natural History, University of Kansas 54:1–38.

Tuttle, M.D. 1977. Gating as a means of protecting cave dwelling bats. *In* National Cave Management Symposium Proceedings, 1976, T. Ailey and D. Rhodes, eds., pp. 77–82. Speleobooks, Albuquerque, N.Mex.

Tuttle, M. D. 1979. Status, causes of decline, and management of endangered gray bats. Journal of Wildlife Management 43:1–17.

Tuttle, M. D. 1986. Endangered gray bat benefits from protection. Bats 4:1–4.

Tuttle, M. D. 1988. America's Neighborhood Bats: Understanding and Learning to Live in Harmony with Them. University of Texas Press, Austin.

Tuttle, M. D., and D. Hensley. 1993. Bat houses: The secrets of success. Bats 11:3–14.

Tuttle, M. D., and S. J. Kern. 1981. Bats and public health. Milwaukee Public Museum, Contributions in Biology and Geology 48:1–11.

Tuttle, M. D., and D. E. Stevenson. 1978. Variation in the cave environment and its biological implications. *In* Proceedings, National Cave Management Symposium, R. Zuber, J. Chester, S. Gilbert, and D. Rhodes, eds., pp. 108–121. Adobe Press, Albuquerque, N.Mex.

Tuttle, M. D., and D. A. R. Taylor. 1994. Bats and Mines. Resource Publication No. 3. Bat Conservation International, Austin, Tex.

van Zyll de Jong, C. G. 1985. Handbook of Canadian Mammals. National Museum of Natural Sciences, Ottawa.

Vaughan, T. A. 1959. Functional morphology of three bats: *Eumops, Myotis, and Macrotus.* University of Kansas Museum of Natural History Publications 12:1–153.

Vonhof, M. J. 1996. Roost-site preferences of big brown bats *(Eptesicus fuscus)* and silver-haired bats *(Lasionycteris noctivagans)* in the Pend d'Oreille Valley in southern British Columbia. *In* Bats and Forests Symposium, October 19–21, 1995, Victoria, British Columbia, pp. 62–80. Working Paper 23/1996. Research Branch, Ministry of Forests, Victoria.

Wai Ping, V., and M. B. Fenton. 1989. Ecology of spotted bat *(Euderma maculatum):* Roosting and foraging behavior. Journal of Mammalogy 70:617–622.

Walker, S. 1995. Mexico–U.S. partnership makes gains for migratory bats. Bats 13:3–5.

Warner, R. M. 1985. Interspecific and temporal dietary variation in an Arizona bat community. Journal of Mammalogy 66:45–51.

Whitaker, J. O., Jr. 1993. Bats, beetles, and bugs. Bats 11:23.

Whitaker, J. 1994. Food availability and opportunistic versus selective feeding in insectivorous bats. Bat Research News 35:75–77.

Wilkins, K. T. 1989. *Tadarida brasiliensis.* Mammalian Species 331: 1–10.

Willig, M. R., and K. W. Selcer. 1989. Bat species density gradients in the New World: A statistical assessment. Journal of Biogeography 16:189–195.

Wilson, B. W., M. J. Hooper, E. E. Littrell, P. J. Detrich, M. E. Hansen, C. P. Weisskopf, and J. N. Seiber. 1991. Orchard dormant sprays and exposure of red-tailed hawks to organophosphates. Bulletin of Environmental Contamination and Toxicology 47:717–724.

Wilson, D. E., F. R. Cole, J. D. Nichols, R. Rudran, and M. Foster, eds. 1996. Measuring and Monitoring Biological Diversity: Standard Methods for Mammals. Smithsonian Institution Press, Washington, D.C.

Witt, J. W. 1992. Home range and density estimates for the northern flying squirrel, *Glaucomys sabrinus,* in western Oregon. Journal of Mammalogy 73:921–929.

Wunder, L., and A. B. Carey. 1996. Use of the forest canopy by bats. Northwest Science 70:79–86.

23
Conservation of Bats on Remote
Indo-Pacific Islands

WILLIAM E. RAINEY

Oceanic island biotas have played key roles in both the theory and practice of conservation biology. They provided the basis for the theory of equilibrium island biogeography (MacArthur and Wilson 1967), which has since been widely applied to other discontinuous habitats. They furnish some of the best-documented examples of anthropogenic extinctions and introductions (Simberloff 1995), and offer sites of manageable scale where experimental intervention, such as translocation or predator extermination, can be undertaken on behalf of endangered species (Clout and Craig 1995).

While research and management efforts on remote-island vertebrates have emphasized endemic land birds and seabirds, the less-diverse bat communities have recently attracted attention, driven by concerns about biodiversity, ecosystem function, and the maintenance of traditional human resources (Falanruw 1988; Cox et al. 1991). Three recent compilations (Mickleburgh et al. 1992; Wilson and Graham 1992; Flannery 1995) covered many topics relevant to bat conservation on paleotropical islands, including systematics, biogeography, human utilization, and the status of species. Additionally, Mickleburgh et al. (1992) have offered a program of conservation action for pteropodids.

The following discussion emphasizes the special circumstances of the small, largely endemic bat faunas of remote Indo-Pacific islands but draws on the more-diverse faunas from Southeast Asia and Australia to examine specific themes. Bats from remote islands are largely pteropodids of the wide-ranging genus *Pteropus*, most of whose members are island endemics (Rainey and Pierson 1992).

Evolutionary and Ecological Context

Among mammals, bats are outliers in terms of maximum life span in relation to body size (Austad and Fisher 1991). Long mass-specific life spans in birds and volant mammals link flight and low mortality which, in turn, are correlated with low annual reproductive output (Holmes and Austad 1995). While constraints of flight may have shaped the reproductive specializations of bats (Hayssen and Kunz 1996), populations are adapted to low predation and low stochastic adult mortality. Evolutionary responses of vertebrate populations to altered selection pressures on islands can be rapid. The most numerous examples involve morphological change (Smith et al. 1995), but Austad (1993) demonstrated that an island population of Virginia opossums *(Didelphis virginiana),* with low predation for about 5,000 years, had both reduced litter size and reduced rates of senescence relative to mainland populations. Assuming evolutionary scope for even lower reproductive rate in island-dwelling bats, such adaptations could heighten the risk of extinction with the abrupt increase in predation that accompanies human colonization of islands.

While trends in body size versus island area vary across bat taxa (Krzanowski 1967), McNab (1994) demonstrated a general trend toward smaller body size and lower mass-specific metabolic rates for some species of *Pteropus* on small islands. He noted that the long-term persistence of small populations is highly sensitive to population size and that decreased individual requirements permit larger populations of resource-limited species. Ornithologists, in particular, have intensively explored (without reaching consensus on cause) an "insular syndrome" of increased densities, broadened habitat or dietary niches, and, occasionally, reduced territoriality in birds on islands relative to the same or related taxa on larger adjacent land masses (Thiollay 1993).

Interactions among different species are simplified in the depauperate communities of small islands. In what are typically evolutionarily asymmetrical relationships, small guilds of generalists, including frugivorous and nectarivorous bats, pollinate or disperse seeds of numerous plants (Woodell 1979; Cox et al. 1991). There are notably few vertebrate dispersers of the often large-seeded fruits of island canopy trees. Thus, declines or extinctions in this guild raise questions about long-term impacts on forest regeneration (Rainey et al. 1995), and about cascading effects on other species, including humans, whose traditional subsistence systems may exploit large-seeded forest trees (Wiles and Fujita 1992). Nocturnal aerial insectivores on remote islands are few, and the ecosystem role of insectivorous bats in these locales is little known.

Extinctions and Extirpations of Bats on Islands

Archaeological Perspectives on Human Impact

Accumulating archaeological data (Flannery 1995; Steadman 1995) make it clear that human colonization of smaller oceanic islands in the Pacific was accompanied by numerous extinctions of endemic vertebrates (largely land birds) and local extirpations of indigenous species. As human populations grew on many of these islands, faunas were increasingly influenced by widespread introductions of domesticated and commensal vertebrates and by alterations in vegetation structure and composition, especially in lowland forests (Steadman and Rolett 1996). Identifying taxa found in

archeological deposits as anthropogenic extinctions requires careful interpretation. Bones may be prehistorical imports, vagrants (Wragg 1995), or taxa extirpated by environmental change (Weisler and Gargett 1993). Nonetheless, models of biogeographic processes that treat human-influenced islands as isolated exceptions (Lack 1976; Adler 1992) must be reviewed carefully against evidence of pervasive human colonization and extensive defaunation of islands in the tropics and subtropics. Then, as now, exploitation of island wildlife emphasized subsistence, and easily harvested forms declined first (e.g., colonial ground-nesting seabirds, large pigeons, and terrestrial or flightless species). Additionally, cultural preferences sometimes emphasized particular taxa (e.g., the red feathers of the Rimatara lorikeet, *Vini kuhlii*) and contributed to local declines (Watling 1995).

Archeological evidence for reductions, extirpations, and extinctions of bat populations on Indo-Pacific islands are summarized in Table 23.1. Although archaeological surveys have revealed numerous extinctions of bird taxa, there is only one record of bat extinction, an undescribed microchiropteran from Maui, Hawaii (James et al. 1987). The record does, however, suggest or document declines and local extirpation for a number of other taxa. Remains of bats from the two East Pacific islands are not from human-occupied sites, but radiocarbon dates, stratigraphy, and records for other fauna are correlated with the disappearance or extirpation of these species with human colonization (Steadman 1986; James et al. 1987, personal communication). The apparent pattern of extirpation and extinction may be biased against detection of small vertebrates, including microchiropteran bats, particularly at human occupation sites. Small vertebrates are less likely to be harvested, and their remains are less likely to survive food processing, consumption, and extended burial. Additionally, coarser screens, more commonly used in the past for sampling archaeological deposits, have selected against recovery (Flannery et al. 1988).

The only island archaeofauna showing loss of several bat species is on ʻEua, Tonga (Koopman and Steadman 1995). The extirpated species—*Pteropus samoensis, Notopteris macdonaldi,* and *Chaerephon jobensis*—are currently unknown elsewhere in the Tongan archipelago, although all are extant in Fiji (Flannery 1995). *Notopteris macdonaldi* and *C. jobensis* are present in Vanuatu and *P. samoensis* in Samoa. *Pteropus tonganus* and *Emballonura semicaudata* have been recovered from prehuman strata on ʻEua, and both species persist to the present (Koopman and Steadman 1995). The prehistoric extirpation of *N. macdonaldi* and *C. jobensis* may be attributed to the high vulnerability of relatively large-bodied, aggregated, cave-roosting species to overharvesting and disturbance. Flannery (1995) reported that *C. jobensis* is currently collected for food from caves in Fiji. Although in

Table 23.1

Archaeological Evidence of Reductions in Indo-Pacific Populations of Island Bats

Locality	Species	Impact	Local distribution	Comments	Source
Rodrigues	*Pteropus niger*	Prehistoric extirpation?	Persists on Mauritius	Skull in cave with extinct taxa	Cheke and Dahl 1981
Round Island, Mauritius	*Pteropus rodricensis*	Prehistoric extirpation?	Persists on Rodrigues	Skull in bone fissure fill	Cheke and Dahl 1981
Okinoerabu, Ryukyu Island	*Pteropus dasymallus*	Prehistoric extirpation?	Extant elsewhere in archipelago	Occupation site; limited remains	Nishinakagawa et al. 1994
Rota, Marianas Island	*Emballonura semicaudata*	Prehistoric extirpation	Extant elsewhere in archipelago	Occupation site; limited remains	Steadman 1992
'Eua, Tonga	*Pteropus samoensis* *Notopteris macdonaldi* *Chaerephon jobensis*	Prehistoric extirpation	Absent elsewhere in Tonga; present in adjacent archipelagoes	Raptor prey deposits and occupation sites; two other bat species persist to present	Koopman and Steadman 1995
Mangaia, Cook Island	*Pteropus tonganus*	Declining harvest through time	Present in island interior	Human occupation site	Kirch et al. 1992
Aitutaki, Cook Island	*Pteropus tonganus*	Prehistoric extirpation	Extant elsewhere in archipelago	Occupation site; limited remains	Steadman 1991
Ma'uke, Cook Island	*Pteropus tonganus*	Prehistoric extirpation	Extant elsewhere in archipelago	Occupation site	Walter 1990 (in Koopman and Steadman 1995)
Maui, Hawaiian Island	Undescribed vespertilionid	Prehistoric extinction	Remains also recovered elsewhere in archipelago	Cave trap; bones of extant *Lasiurus cinereus semotus* also present in deposit	James et al. 1987; H. F. James, personal communication
Floreana (=Santa Maria), Galapagos Island	*Lasiurus borealis*	Historic extirpation	Extant elsewhere in archipelago	Barn owl prey deposit; owl extirpated mid-1800s	Steadman 1986

Samoa today *P. samoensis* is much less numerous than *P. tonganus* (Craig et al. 1994), the extirpation of one of the two large (0.3–0.5 kg) canopy-roosting bats on 'Eua is difficult to explain without invoking cultural prey preference or severe prehistorical reduction in primary forest (on which the regional endemic *P. samoensis* appears more dependent). Behavioral observations (Wilson and Engbring 1992; Pierson et al. 1996) of *P. samoensis* and *P. tonganus* do not suggest differential vulnerability to hunting methods traditionally used by Polynesians.

Historical and Recent Extinctions and Extirpations

Several bat species, collected on Indo-Pacific islands after the onset of European exploration, are either extinct or locally extirpated (Table 23.2). For those first collected in the nineteenth century, the record is frequently enigmatic. There are usually one or a few specimens with uncertain provenance, and natural history descriptions provide no clues on the course and causes of decline. Similar events in the twentieth century are sometimes better documented, so that ecological traits which made them (and presumably other extant species) vulnerable can be identified.

The nineteenth-century demise of *Pteropus subniger* on both Mauritius and Réunion (with the larger *Pteropus niger* extirpated on Réunion, but still extant on Mauritius) is well documented (Cheke and Dahl 1981). Both species were initially so common that plans were made to export rendered bat oil. This study suggested that the extinction of *P. subniger* was linked to its unusual roosting habits. While most *Pteropus* species for which roosting habits are known cling externally to tree branches (Pierson and Rainey 1992), *P. subniger* roosted in aggregations as many as 400 individuals in tree cavities and rock crevices, where it was highly vulnerable to collection (Cheke and Dahl 1981). Risks to this species likely intensified as more forest was cleared.

The fate of *P. subniger* suggests we should pay particular attention to other species with similar roosting habits. Review of museum records and recent field studies indicates that cavity roosting may be more common than was previously realized among smaller, little-known *Pteropus*. For example, *Pteropus vetulus* on New Caledonia is one of several extant species that resembles *P. subniger* in having short, furred ears which barely protrude above a dense, woolly pelage. Information on specimen tags at the American Museum of Natural History indicates several *P. vetulus* were captured from a hollow tree limb at 1,200 m elevation.

Table 23.2

Historical Extirpations and Extinctions of Indo-Pacific Island Bats

Locality	Species	Date of extinction or extirpation	Probable cause	Comments	Source[a]
Mauritius, Reunion	*Pteropus subniger*	ca. 1870, 1860	Hunting; habitat alteration	Aggregated tree roosts vulnerable	Cheke and Dahl 1981
Reunion	*Pteropus niger*	Extirpated before 1801	Hunting; habitat alteration	Persists on Mauritius	Cheke and Dahl 1981
Reunion	*Scotophilus borbonicus*	Late 19th century?	Unknown; habitat alteration extensive	Taxonomic status uncertain; possible species persists; two other extant Microchiroptera	Cheke and Dahl 1981
Okinawa, Ryukyu Island	*Pteropus mariannus loochoensis*	After 1849	Unknown; habitat alteration extensive; locality error possible	Two specimens; *P. dasymallus* persists	Andersen 1912; K. Koopman, personal communication; H. Ohta, personal communication
Palau	*Pteropus pilosus*	After 1874	Unknown; habitat relatively intact	Two specimens; large bat; *P. m. pelewensis* persists	Andersen 1912; Wiles 1997
Tobi, Palau	*Pteropus mariannus pelewensis*	After 1908	Hunting?	Persists elsewhere in Palau	Wiles 1997
Nendo, Solomon Island	*Nyctimene sanctacrucis*	ca. 1900	Unknown; possibly habitat alteration	One specimen; no contemporary natural history data	Flannery 1995
Panay, Philippine Island	*Acerodon lucifer*	After 1892	Habitat alteration; hunting	Persisted in periphery of agricultural landscapes	Utzurrum 1992
Addu Atoll, Maldives	*Pteropus hypomelanus maris*	After 1922	Unknown; possibly a waif or locality error	One specimen; no contemporary natural history data; *P. giganteus ariel* persists	Hill 1958; Holmes 1994
Guam	*Pteropus tokudae*	After 1968	Uncertain; habitat alteration; bat hunting	Three specimens	Wiles 1987a
Negros, Philippine Island	*Dobsonia chapmani*	After 1964	Foraging habitat clearance; hunting; roost disturbance	Mortality from guano mining and hunting in the caves	Heaney and Heideman 1987
Kashoto Island, Taiwan	*Pteropus dasymallus formosus*	After 1986	Hunting; habitat alteration	—	H. Ohta, personal communication

[a]Species accounts for all except *Scotophilus borbonicus* are in Mickleburgh et al. (1992).

In contrast to earlier reports (Sanborn and Nicholson 1950) in which *Pteropus ornatus* was common and *P. vetulus* rare, Flannery (1995) recently found that *P. vetulus* was common in subcanopy mist net captures, despite intensive local bat hunting. Supposing that cavity roosting is common for this species, a key difference from the Mascarene species may be that New Caledonian bats have recently been hunted by spotlighting and shooting, often at night from vehicles (A. Bauer, personal communication). With this method, cryptic aggregated roosts and strict nocturnality may in fact reduce vulnerability to hunters. Flannery (1995) also reported a tree-cavity colony of about 30 *Pteropus ad-*

miralitatum on Malaita, Solomon Islands, and noted the Moluccan *Pteropus caniceps* roosts as pairs in tree hollows.

Flannery (1995) also suggested that all members of the little-studied genus *Pteralopex* may roost in tree cavities. He reported that an unnamed *Pteralopex* in lowland New Georgia and Vangunu, Solomon Islands, roosts communally in cavities of large trees and is easily captured by hand. Flannery also described how this species was observed flying away from areas of logging on Kolombangara and has not been seen there since the mid-1970s. In lowland forests now subject to intensifying timber management or other activities that truncate the age structure of trees, bats which depend on large hollow trees are certainly more vulnerable than those that roost in canopy foliage.

Acerodon lucifer, from Panay, Philippines, was last collected in 1892 from the margins of agricultural areas (Utzurrum 1992). The clearance of land on Panay has now reduced forest cover to 10%, with residual forests confined primarily to ridges. Extensive hunting and habitat loss have presumably led to the extinction of *A. lucifer* (Mickleburgh et al. 1992). *Dobsonia chapmani,* an aggregated cave-roosting species, was still common on southern Negros, Philippine Islands, in 1964 (Heaney and Heideman 1987). In later surveys these authors concluded that rapid clearing of foraging habitat in lowland forest near cave roosts, along with disturbance and mortality from guano mining and hunting in the caves, caused extinction before 1981.

Reasons for the recent decline and presumed extinction of *Pteropus tokudae* on Guam are obscure (Wiles 1987a; Mickleburgh et al. 1992). When discovered, this small species was very rare relative to the sympatric *Pteropus mariannus mariannus.* Two specimens were collected in 1930, and a third was 1 of 100 bats shot during the hunting season of 1968. Remaining forest habitat on Guam, although heavily altered, could probably still support many more than the several hundred *P. m. ariannus* currently on the island. Rapid population increase of *Boiga irregularis,* the predatory arboreal snake introduced after World War II, postdates the disappearance of *P. tokudae* (Wiles 1987a; Rodda et al. 1992). This species is not known from the adjacent, less disturbed island of Rota or elsewhere in the Marianas, despite repeated surveys (G. Wiles and D. Worthington, personal communication).

Pteropus dasymallus formosus, which persisted until at least 1986 on Kashoto, an offshore island of Taiwan, is apparently the most recent extinction of *Pteropus.* Several other subspecies of *P. dasymallus* in the Ryukyus barely survive (Mickleburgh et al. 1992; H. Ohta, personal communication). Forest cover on Kashoto has been greatly reduced by agricultural conversion, but the ultimate cause of extinction is likely hunting for food. Kuroda (1933) concluded that historical specimens of *P. d. formosus* from localities on Taiwan (Andersen 1912) originated elsewhere, but no zooarchaeological surveys are available to test this hypothesis.

Inventory, Status Assessment, and Population Monitoring

Inventory

Scientific knowledge of many bat species, particularly in the more diverse faunas of the southwest Pacific, has often been limited to their taxonomic description based on a few specimens. Early bat collections were frequently incidental to bird sampling, and thus inventories were incomplete. With accelerating habitat alteration in recent decades, the continued existence of species on small islands, especially those dependent on lowland forest areas, has been an open question. Partly from concern for the potential loss of biodiversity, the museum tradition of faunal inventories has been renewed, especially by the Australian Museum, the Western Australian Museum, and regional collaborators.

From these surveys Flannery (1995) has reported recent collections of many of the little-known *Pteropus* from the southwest Pacific (e.g., *P. chrysoproctus, P. fundatus, P. mahaganus, P. melanopogon, P. nitendiensis, P. ocularis, P. pohlei, P. rayneri cognatus,* and *P. temmincki*). These surveys also yielded two new species (one discussed earlier) of *Pteralopex* from the Solomon Islands (Flannery 1991, 1995). On Guadalcanal, *Pteralopex pulchra* is unusual in that it appears restricted to high, mossy, montane forest, where it replaces the (now declining) lowland species *Pteralopex atrata.* The wet, high-elevation habitat is analogous to that of the rare Fijian endemic, *Pteralopex acrodonta* (Hill and Beckon 1978; also recollected by Flannery 1995). In Fiji, however, no lowland *Pteralopex* species is known.

Status Assessments and Population Monitoring

Several surveys that have assessed the status of bat populations postdate the megachiropteran compilation by Mickleburgh et al. (1992). A 1991 survey of bats in Palau found that *Pteropus mariannus pelewensis,* formerly very heavily hunted for commercial export to the Marianas, was common to abundant at 40% of 54 evening census localities and that *Pteropus pilosus* is almost certainly extinct (Wiles et al. 1997). Grant (1994) observed *Pteropus tonganus* on the isolated island of Niue, and obtained local estimates that 1,200–5,000 individuals are taken annually by shooting, suggesting a much larger population than previously reported. In reviewing the distribution and current status of bats known from Tonga, Koopman and Steadman (1995) reported at least 2,400 *P. tonganus* on 'Eua and several thousand at Kolovai, Tongatapu, in 1988. Holmes (1994) described *Pteropus giganteus ariel* in the Maldives as widespread, but not numerous, and subject to lethal control efforts intended to reduce fruit depredation.

Surveys conducted over several years for *Pteropus livingstonii,* the rarer of the two *Pteropus* species in the Comores, yielded minimum population estimates of 380 on Anjouan and 60 on Moheli (W. Trewhella, personal communication). Reason and Trewhella (1994) noted that the primary threat to this species is habitat loss as forest fragments are converted to agricultural use to support a rapidly growing human population. *Pteropus livingstonii* currently roosts in the canopy of montane forest patches on steep slopes and relies more extensively on native fruits for food than the other *Pteropus* species, *Pteropus seychellensis comorensis*. Netting efforts to collect *P. livingstonii* for an ex situ captive breeding program incidentally captured more than 150 *Rousettus obliviosus*, a third little-known pteropodid (Reason et al. 1994).

In more affluent American Samoa, human population growth is comparably rapid, deforestation of limited lowlands is well advanced, and clearing of steeper slopes is ongoing. However, the major factor affecting recent population trends for *P. tonganus* and *P. samoensis* has been mortality, largely from hunting, rather than habitat limitation. Craig et al. (1994) has summarized the results of roost counts and diurnal surveys, which began in 1987 on Tutuila after local legislation was enacted to halt commercial export of bats to the Marianas (see following). These authors showed that bat counts dropped precipitously following a cyclone in 1990. While some bats died or starved, increased opportunistic hunting was the primary cause of postcyclone mortality. Hunters reported double the typical annual harvest in the year following that storm. A second severe cyclone less than 2 years later compounded effects on bat populations.

Craig et al. (1994) estimated an 80%–90% population decline for both species over 5 years and tentatively concluded that 200–400 *P. samoensis* and 1,500–2,500 *P. tonganus* were on the island in late 1992. Diurnal roost counts of *P. tonganus* on Tutuila have shown a steady increase from 1,700 in 1991 to 5,700 in 1996. Radiotracking indicates that diurnal flight observations underestimate *P. samoensis* numbers, which may be closer to 1,000 (A. Brooke, personal communication). Perhaps partly as a consequence of continued illegal hunting, roosts for both species are concentrated on steep forested slopes in undeveloped areas. A substantial area of forest is contained within the recently established National Park of American Samoa, but this may not be adequate to maintain viable populations of bats and other wildlife if forest clearance in other areas continues at the current rate.

A study conducted in Western and American Samoa found differences between the two *Pteropus* species in response to a postcyclone reduction in food availability (Pierson et al. 1996). The less abundant, regionally endemic *P. samoensis* remained in the forest, initially feeding largely on leaves, while the more widespread *P. tonganus* foraged extensively on residual or fallen fruit in inhabited areas and experienced higher mortality from human hunters and predation by domestic animals. This study found that forest reserves on Savai'i, Western Samoa, appeared to protect small numbers of *P. samoensis* because their limited diurnal foraging movements remained within the reserve where hunting was banned. *Pteropus tonganus* frequently flew outside the reserve to feed and was likely subject to heavier hunting pressure.

The remnant *P. mariannus mariannus* population on Guam, usually found roosting as a single colony on a U.S. military base (a de facto reserve), has been monitored since 1981 (Wiles et al. 1995). These authors reported seasonal fluctuations in population size, with 200–400 animals present from June to September and 400–750 from November to February. Wiles and Glass (1990) reported observations of offshore bat flights among the Marianas and used changes in colony size to infer interisland group movements, including several (60 km over the ocean) between Guam and Rota in the Commonwealth of the Northern Marianas (CNMI). Wiles et al. (1995) suggested that the seasonal fluctuations in bat numbers on Guam reflect annual movement between the two islands. Predation on nonvolant juvenile bats by the introduced snake *Boiga irregularis* apparently prevents local recruitment (Wiles 1987b), so that the persistence of the Guam population could depend on movement from Rota. Historically, bats were heavily hunted on Guam, but poaching has not been a threat to the resident bat colony in recent years (G. Wiles, personal communication). Guam is an unusually affluent, regional transportation hub with rapidly expanding tourist facilities and related infrastructure. In 1993, the U.S. government created a wildlife refuge from relatively intact forest on military lands. While this has met with some local support, there has also been considerable resistance, based partly on pre–World War II private land claims to these areas (Wiles 1994). In addition to development pressures, remaining forest areas are subject to ongoing degradation by introduced ungulates (Wiles et al. 1995).

Rota harbors the only substantial population of *P. mariannus* in the southern islands of the CNMI (Stinson et al. 1992). Estimates of bat numbers on Rota from 1986 until a major cyclone in January 1988 were 2,000–2,500. Subsequent counts dropped to approximately 1,000, owing to postcyclone hunting and possibly emigration. The population has remained at this lower level through 1995, probably because of chronic illegal hunting (Worthington and Taisacan 1996). Wiles et al. (1989) surveyed the less-developed islands north of Saipan and estimated that 7,450 bats were present in 1983. Illegal market hunting has since increased. Also, a survey of Anatahan identified rapidly growing feral goat populations as a serious new threat to forest habitat (Marshall et al. 1995). The CNMI has a regulatory framework for achieving sustainable bat harvest, but public and institutional resistance to restrictions on bat hunting and trade are strong. Habitat conservation areas have been outlined, but enforcement of local wildlife laws remains minimal. Wiles and Glass (1990) suggested, as a way to rationalize the varied taxonomic and regulatory status of populations on different islands, that all *P. mariannus* in the southern Marianas should be managed as a single unit.

Recent reports of *Emballonura semicaudata,* the only microchiropteran bat in much of Micronesia and Polynesia, suggest that a patchy, rangewide decline is in progress. At the eastern limit of its current range in Polynesia, Grant et al. (1994) reported a rapid reduction of *E. semicaudata* to only a few individuals in American Samoa. They noted that similar declines, from hundreds or perhaps thousands of bats to isolated individuals, seem to have occurred contemporaneously for colonies on Upolu, Western Samoa. These declines lag by decades similar changes for populations of *E. semicaudata* and cave swiftlets (*Collocalia vanikorensis*) in Guam and adjacent islands in the CNMI (Lemke 1986; Steadman 1992).

In Samoa, however, swiftlet populations that share caves with *E. semicaudata* have remained stable or increased. Grant et al. (1994) discussed the role of a severe cyclone in the reduction of *E. semicaudata* populations in American Samoa, but did not link the long-term declines to the explanations offered for the Marianas—military destruction of caves, extensive pesticide use for vector control, and guano mining. Recent surveys in the Marianas have generally confirmed a pattern reported by Lemke (1986) suggesting that *E. semicaudata* is extinct on Guam and Rota and absent on all other islands, except Aguiguan (no human inhabitants) and perhaps Saipan (Wiles et al. 1995; Worthington and Taisacan 1996). Island-by-island patterns of persistence for bat and swiftlet populations in the Marianas are also not congruent (D. Worthington, personal communication).

Flannery (1995) related a long-term resident's report of similar declines of *E. semicaudata* on Viti Levu, Fiji, from abundance in the 1950s to virtual absence in the 1990s. The informant attributed declines to roost disturbance and the burning of forests near caves. Flannery (1995) also reported a small colony of *E. semicaudata* residing in an inaccessible cave roost on Taveuni, Fiji, presumably in the 1990s. This species may have also declined on the isolated island of Rotuma, Fiji. Clunie (1985) reported the bats as present in multiple thousands in caves and observed foragers "in far larger numbers than is usual in Fiji." In 1993–1994, Cox (personal communication) visited Rotuman caves several times and noted a marked reduction in numbers, consistent with reports offered by local residents. In surveys of more than 50 caves on 'Eua, Tonga, in 1988–1989, only a single colony of 25 *E. semicaudata* was observed (Koopman and Steadman 1995).

In several other areas of Micronesia, *E. semicaudata* has been recently present. Wiles and Conry (1990) reported a cave colony of 200 in Palau. Observations at multiple sites in Palau in 1991 showed the species to be widespread and common, with foraging movements by several thousand bats observed (G. Wiles, personal communication). In 1989, tens of foraging *E. semicaudata* and swiftlets were also seen at dusk over the streets on Moen Island, Chuuk, but, as observed earlier by Bruner and Pratt (1979), this species is considerably less common on Pohnpei (Rainey, unpublished observations).

A common cause for these seemingly parallel declines is not evident. Both the southern Marianas (with few bats) and Palau (with many) share a history of intense military activity during World War II and former use of DDT for insect control (Baker 1946), but the Marianas have few bats and Palau has many. Unlike the Marianas, Palau has numerous small outlying karstic islands that likely offer more extensive bat refuges. Chronic roost disturbance is a well-documented cause for declines in cave-dwelling bats, especially at sites close to concentrations of humans and domestic or feral animals. Colonies of *E. semicaudata* persist, however, close to habitations on densely populated, heavily altered Moen, Chuuk, and are declining on much larger, less densely populated islands such as Viti Levu, Fiji, and Upolu, Western Samoa. Introduced pathogens, which are apparently responsible for declines of native Hawaiian birds in relatively intact habitat (Atkinson et al. 1995), offer another possible mechanism. Resources for wildlife conservation often focus on charismatic, endemic "flagship" species, but the regional decline of this small insectivorous bat deserves careful scrutiny.

Systematics and Molecular Genetics

In the 1980s, use of protein electrophoresis in surveys of Australian Microchiroptera revealed or confirmed numerous well-differentiated sibling species and sometimes considerable geographic differentiation within species (Adams et al. 1982; Baverstock et al. 1987). More recent population- and species-level surveys of Pacific pteropodids, using several molecular techniques, offered contrasting results with both evolutionary and management implications. For example, protein electrophoresis of *Cynopterus* populations in Indonesia revealed relatively small genetic distances among species and little evidence of geographic differentiation within islands, suggesting the ocean is the primary barrier to gene exchange (Schmitt et al. 1995). Peterson and Heaney (1993) compared *Haplonycteris fischeri*, a Philippine endemic restricted to primary forest, and populations of the wide-ranging *Cynopterus brachyotis* in the Philippines and noted higher levels of genetic differentiation among populations of *H. fischeri*. In contrast, Kitchener et al. (1993) used electrophoretic analysis to examine variation, on several Indonesian islands, in *Aethalops alecto*, a small pteropodid restricted to montane forest. They concluded that populations were neither well differentiated nor genetically isolated.

An extensive study of the highly migratory Australian flying-fox, *Pteropus scapulatus*, using both allozymes and random amplified polymorphic DNA, showed low levels of population structure and indicated that this species, now independently managed by several jurisdictions, was effectively panmictic (Sinclair et al. 1996). Webb and Tidemann (1996) reported similar results for *Pteropus alecto* and *Pteropus poliocephalus* and suggested they should be managed as migratory species. Their analysis of putative hybrids from areas of sympatry revealed only one or two fixed alleles in 23 loci differentiating *P. alecto* from *P. poliocephalus* or *Pteropus conspicillatus*, which suggests relatively recent divergence among these species (Webb and Tidemann 1995).

The low levels of intraspecific differentiation in *Pteropus* and *Cynopterus* suggest that populations of *Pteropus* (or other large- to moderate-sized unspecialized pteropodids) on small oceanic islands can generally be viewed as single management units. The inference that *P. mariannus* regularly crosses as much as 60 km of ocean (Wiles and Glass 1990; Wiles et al. 1995) highlights the need to use molecular approaches to identify appropriate management units in species that lack obvious morphological discontinuities and whose ranges encompass multiple islands separated by significant water gaps.

While the description of new species and subspecies of smaller pteropodid and microchiropteran bats from the more diverse communities of the western Pacific has continued to the present (Kitchener et al. 1994), higher-level megachiropteran systematics has remained essentially that of Andersen (1912), which is inexplicit by modern standards. Mickleburgh et al. (1992) presented a conservation ranking scheme for geographic areas that incorporated weights for both "taxonomic distinctiveness" and species richness using a morphologically based cladogram of megachiropteran genera (by J. E. Hill; but see also Heaney 1991). More recently, Colgan and Flannery (1995), Kirsch et al. (1995), and Springer et al. (1995) presented independent molecular data sets on megachiropteran systematics. While these inevitably differ, they are consistent with some earlier morphological and biochemical studies (Hood 1989; Haiduk 1983) in one major finding: the morphologically specialized nectar-feeding genera, historically grouped in the subfamily Macroglossinae, are not a clade. With new tools and renewed interest in this topic, we can expect a

gradual consensus on phylogenetic relationships that will provide new criteria for conservation ranking schemes.

Threats to Bat Populations

Among declining vertebrate species, one can make a heuristic division between those that are lost with their habitat and those that vanish long before their habitat. The latter are usually animals that are either large (e.g., sirenians) or are selectively hunted for their perceived high value to humans (e.g., musk deer, birds with unusual plumes). Although bats have relatively small body mass and might be expected to closely track habitat loss, low reproductive output and, for some species, aggregated roosting habits at traditional sites make them vulnerable to selective hunting. Loss or degradation of forest habitat is clearly an important threat to the long-term persistence of bat populations on islands. The relative importance of habitat loss, exploitation, and other factors in determining population trends is however quite varied among islands and generally is poorly known. A few of the more discrete threats to island bat populations are discussed next.

International Trade

Partly as a consequence of the large body size of some Megachiroptera (and the limited array of animals available on isolated oceanic islands), human consumption of bats is a significant factor affecting bat populations on Indo-Pacific islands and in adjacent areas of Asia. Attitudes toward bats as food vary across religions, cultures, and geography (Kirch and Yen 1982; Fujita and Tuttle 1991). In Micronesia, for example, residents of both the Marianas and Yap have a long tradition of consuming bats. In the Marianas, they are highly favored (Sheeline 1991), while in Yap they are "not esteemed" (Falanruw and Manmaw 1992). To the east in Chuuk and Pohnpei, bats are viewed as somewhat repellent and are generally not part of the local diet (Rainey, unpublished observations).

The major focus for international trade in bats (primarily *Pteropus* spp.) has been Guam and the adjacent CNMI. Wiles (1992) summarized trade history and tabulated legal imports, showing that subsequent to local depletion bats were initially acquired from nearby islands. In the 1980s the radius of trade expanded, reaching as far as Papua New Guinea and the Philippines. The mean number of bats recorded as imported to Guam in 1981–1989 was roughly 13,000 annually (Wiles 1992). The mean number of bats imported annually into the CNMI in 1986–1989 was 3,300 (Stinson et al. 1992). In some instances, small islands such as Yap and American Samoa noted rapid declines in *Pteropus* populations and imposed local restrictions on export hunting (Craig and Syron 1992; Falanruw and Manmaw 1992).

In 1989, seven central and west Pacific small island *Pteropus* species were added to Appendix I of the Convention on Trade in Endangered Species of Wild Flora and Fauna (CITES), while the remaining species of *Pteropus* and all species of the allied genus *Acerodon* were placed on Appendix II (Braütigam and Elmqvist 1990). This listing obliged countries who were parties to the treaty (including the United States) to cease international trade in Appendix I species and monitor trade in Appendix II taxa. After enforcement of CITES provisions on Guam in 1990, legal imports from 1990–1993 came from Palau, which was still under U.S. jurisdiction (annual mean: for Guam, 7,688 bats; for CNMI, 5,755) (Wiles et al. 1997). This practice continued until late 1994 when Palauan independence ended legal trade in *Pteropus mariannus pelewensis*. Also in 1994, *Acerodon jubatus* and *Acerodon lucifer* (the latter probably extinct) were transferred to Appendix I, based on a petition from the Philippines, which emphasized ongoing illegal trade into Guam and CNMI.

Subsequent to closure of the legal trade from Palau, documented imports have essentially ceased and prices within CNMI have risen to more than US$50 per bat (Worthington and Taisacan 1996). It is generally presumed that, as a response to market forces, illegal hunting has increased in the northern Marianas, and that some international smuggling also occurs (Wiles 1994; Worthington and Taisacan 1996). No legal barrier exists to commercial imports of Appendix II or unlisted species from a number of countries. There are, however, practical barriers, such as market reluctance to accept unfamiliar species, complications regarding trade arrangements, and the risk of flight delays spoiling highly perishable cargo. In the past, local preference for large, strongly scented *Pteropus* with few parasites led to preferential market hunting and pricing. For example, buyers visiting Samoa preferred the less common *P. samoensis* to *P. tonganus*. Smaller *Pteropus* from Palau and the Federated States of Micronesia (FSM) were initially of less interest. Acceptance of smaller cave-dwelling pteropodids with obvious tails (e.g., *Rousettus*) was also poor (G. Phocas, personal communication; Sheeline 1991; D. Worthington, personal communication).

Reemerging proposals to import *Rousettus* from mainland Asia or *Pteropus* from Australia (G. Wiles, D. Worthington, personal communication) increase the risk that people who prepare frozen bats would be exposed to lethal pathogens, such as lyssaviruses, which are present in at least some species of Australian *Pteropus* (Fraser et al. 1996; Young et al. 1996). Although these risks may not alter the behavior of individual consumers, public health authorities

will be obliged to reinterpret import regulations for vectors of communicable diseases to allow international trade to expand along these lines. The only recurring international trade outside of the Marianas for bats listed on CITES involves a few hundred *Pteropus* exported annually from Vanuatu to New Caledonia (J. Caldwell, personal communication).

Introduced Predators

Although introduced predators have had a serious impact on island biotas (Atkinson 1989), they have not generally been implicated as the primary agents in the long-term declines or extinction of island bats. A notable exception is the fate of two species endemic to New Zealand, *Mystacina robusta* and *Mystacina tuberculata*. The last known population of *M. robusta* went extinct during an irruption of *Rattus rattus* in the mid-1960s (Daniel 1990). Also, a key role can be inferred for rats and other introduced predators in the extensive range contraction of both species throughout New Zealand (Daniel and William 1984; Daniel 1990). The tendency for *Mystacina* species to roost and forage close to the ground may account for their differential vulnerability (Daniel and William 1984).

Even though rats (especially *R. rattus*) and feral cats commonly climb trees and rock surfaces, bats roosting in tree canopies, in bole cavities, and on the ceilings of caves have managed to coexist with these predators in most settings. Their ability, lacking in birds, to move threatened nonvolant young has likely helped. On Christmas Island in the eastern Indian Ocean, Tidemann et al. (1994) found that the canopy-roosting endemic *Pteropus melanotus natalis* was vulnerable to feral cat predation when individuals foraged near the ground in fruiting shrubs. Limited observations of a second endemic, *Pipistrellus murrayi* (included in *Pipistrellus tenuis* by Koopman 1993), showed that it forms small groups in trees rather than caves (Tidemann 1985). Despite these introduced predators and considerable hunting of flying-foxes by humans, both bat species were present in substantial numbers in the 1980s (Tidemann 1985; Tidemann et al. 1994). However, as discussed for Samoa, single tropical cyclones can sharply increase the short-term vulnerability of canopy frugivores and nectarivores to introduced terrestrial predators.

The best-documented example of an introduced non-human predator reducing island bat populations is *Boiga irregularis*, an arboreal Australasian snake on Guam. Probably arriving in post–World War II military shipments from the Admiralty Islands, this snake slowly increased to high density, eliminating most of the island's avifauna and apparently preventing local recruitment in *Pteropus mariannus*

(Savidge 1987; Wiles 1987b; Rodda et al. 1992; Wiles et al. 1995). Extensive investigations, including control and containment methods, suggested that this snake will persist on Guam indefinitely (McCoid 1991), leaving long-term prospects for the local bat population in doubt. Guam is a regional center for military transport, and individuals of *B. irregularis* have been recovered on Saipan, Kwajalein, Oahu, and Diego Garcia (McCoid 1991). Because it is likely that dispersal will continue and that vigilance on islands receiving shipments may not always be sufficient, research on emergency eradication methods is important (McCoid 1991). Investigations of reproduction in Australian *B. irregularis* indicated that sperm storage may enhance its success as an invader (Whittier and Limpus 1996). Greene's (1989) observations on the diet and habits of other *Boiga* spp. suggested that they too pose serious invasion risks. Differences that permit coexistence between *Boiga* and *Pteropus* in portions of their natural range are unknown, but it is likely that snake densities are lower.

A nocturnal, commensal snake from southeast Asia, *Lycodon aulicus,* which was first detected on Christmas Island in 1987, now appears established (Fritts 1993). Its maximum reported size (84 cm total length [TL]; Fritts 1993) and limited gape (H. Greene, personal communication) make predation on adult *P. m. natalis* highly improbable. Its broad dietary habits and arboreality suggest that it may, however, be a major threat to the survival of the much smaller *Pipistrellus murrayi.* A scenario, similar to that observed in Guam, of high snake densities causing the extinction of indigenous vertebrates of lower fecundity, including bats, is likely unless this snake can be quickly eliminated.

Another possible instance of an introduced predator having an impact on an island bat population is the decline of *Coleura seychellensis.* Racey and Nicoll (1984) outlined a history of reports, indicating that the species was "very common" in 1868 and reduced to 16 individuals in their recent surveys. Suggested causes of decline are loss of forest habitat, human disturbance of roosts, and occupancy of roost caves by barn owls *(Tyto alba).* Barn owls were introduced to the Seychelles in 1949 (Cheke and Dahl 1981), and have been shown elsewhere to be significant predators on insectivorous bats (Speakman 1991).

Mass Mortality from Disease

Anecdotes of introduced diseases decimating island vertebrates are not uncommon (Simberloff 1995), but only recently has there been recognition that mass mortality from pandemic disease is a rare but widespread phenomenon among Pacific island flying-fox populations in the post-European-contact era (Flannery 1989, 1995; Pierson and Rainey 1992) (Table 23.3). Perhaps because such events might go undetected (or not occur) in less-colonial species, reports all concerned *Pteropus,* which form large aggregations at traditional roosts. Descriptions are dramatic, with incapacitated bats falling from the sky and carcasses or bones accumulating in piles at roost sites (Coultas 1931; Degener 1949; Flannery 1989). Few or no bats of the affected species were seen after epidemics, and, consistent with the population biology of *Pteropus,* recovery is reported as slow (Stair 1887; Flannery 1989, 1995). The two oldest reports link these events to simultaneous epidemics in humans or domestic animals (Stair 1887; Coultas 1931).

As Flannery (1989) pointed out, high mortality favors the hypothesis of a human-introduced pathogen to which bats had no prior exposure. Again by analogy to Hawaii, because few bats that come into contact with humans are later released, transfer of pathogens to forest-dwelling bats ten-

Table 23.3
Mass Mortalities of *Pteropus* from Disease on Indo-Pacific Islands

Locality	Species	Date	Comments	Source
Samoa	*P. tonganus* and/or *P. samoensis*	1839	Prior epidemic among residents; gradual bat population recovery; species not identified	Stair 1887
Kosrae (FSM)	*P. mariannus ualanus*	1926–27	Measles and dysentery epidemic in residents; thousands of bats died; few survivors	Coultas 1931
Vanua Levu, Fiji	*P. tonganus*	Before 1949	Hundreds dead under roosts	Degener 1949
New Caledonia	*P. ornatus*	Early 1960s	Formerly common; no subsequent population recovery under hunting pressure	Flannery 1995
Manus, Papua New Guinea	*P. neohibernicus hilli*	1985	Masses dead in roost areas; slow recovery; sympatric *P. admiralitatum* unaffected; unaffected *P. neohibernicus* on adjacent islands	Flannery 1989
Bougainville and Buka, Solomon Island	*P. rayneri grandis*	1987	Unaffected populations on adjacent islands	Flannery 1989

tatively suggests an arthropod or other animal vector. The lack of evidence for recurrent mass mortalities from disease in better-studied temperate zone bat populations suggests that these events are not part of the prehuman evolutionary history of island bats. However, birds transport pathogens and arthropod vectors with broad host ranges to islands (Olsen et al. 1993). Interisland transportation by humans continues to improve, increasing the prospect of dispersing microbes, arthropod vectors, and alternate hosts. Any new bat epidemic that is detected deserves careful study, given the risks of stochastic mortality to the long-term persistence of small populations.

Global Climate Change

Current models of global warming suggest that oceanically buffered climates in tropical regions are expected to show small increases in temperature (Lighthill et al. 1994), the consequences of which are unknown for island vertebrates. Two other projected changes, rising sea levels and altered cyclone regimes, would however almost certainly have major impacts on island biotas. In the absence of local tectonic uplift, sea level rise threatens current terrestrial communities on low-relief atolls, which account for many of the world's remote oceanic islands. On such islands, relatively large areas of land would be lost with only a small rise in sea level (17–26 cm by 2030 and increasing thereafter). The decreasing lens of fresh groundwater and increased effects from salt spray would also affect vertebrate consumers because decreasing plant diversity will reduce food resources (Roy and Connel 1991).

Roy and Connel (1991) discussed prospects for humans who inhabit nations composed largely of atolls (e.g., Kiribati or Maldives), concluding that the current pattern of economic emigration will likely be transformed during the next several decades into an exodus of environmental refugees. For countries (e.g., FSM, Cook Islands) that include both scattered low atolls and uplifted carbonate or emergent volcanic islands (typically with greater population and infrastructure), this migration may be partly internal but will increase human density and demand for natural resources on already crowded islands.

A few bat species and populations endemic to low-relief atolls would be directly threatened by rising sea levels (e.g., *Pteropus howensis* on Ontong Java; *P. mariannus ulthiensis* on Ulithi; the *P. insularis* population on Namonuito Atoll, FSM (Rainey and Pierson 1992); *P. phaeocephalus* in the Mortlocks, FSM; *P. giganteus ariel* in the Maldives). Several other species (e.g., *P. tonganus* in Fiji and Tonga, *P. insularis* on Chuuk, *P. mariannus* subspp. in Yap and Palau) occur both on high- and low-lying islands and thus are less acutely threatened. How-

ever, at a scale of a few kilometers, lost or altered vegetation will eliminate foraging habitat and refuges from human hunting. At a scale of many kilometers, the distribution of even small numbers of bats and patches of foraging habitat over several islands offers escape in space and time from the recurring but relatively narrow paths of tropical cyclones. The risk of extinction increases as the total geographic range of a species declines toward an area potentially swept by cyclones in a high-frequency year.

Mangrove forests are recognized as an important coastal habitat for bats, especially *Pteropus*. They provide relatively protected sites for aggregated roosting (e.g., *P. vampyrus* on Timor [Goodwin 1979], *P. molossinus* on Pohnpei, and *P. mariannus ualanus* on Kosrae [Rainey 1990]), with some common trees, notably *Sonneratia*, seasonally offering an important nectar resource (Start and Marshall 1976). Inputs of sediment from major rivers may allow mangrove forests on continental coastlines to keep pace with projected sea level rise (Jelgersma et al. 1993). On small islands where terrestrial inputs to coastal sediments are much lower, sea level rise, in the absence of tectonic uplift, will exceed soil accretion in mangrove forests, making forest death a likely scenario (Ellison and Stoddart 1991).

An altered tropical cyclone regime could be a major consequence of global warming. These cyclic disturbances play a key role in shaping forest structure and species composition, and, for low-lying islands, they can alter coastal geomorphology (Whitmore 1974; Shaw 1983; Tanner et al. 1991; Stoddart and Walsh 1992; Elmqvist et al. 1994; Foster and Boose 1995). Marked reductions in bat populations, followed by deferred reproduction, accompany severe cyclones (Cheke and Dahl 1981; Wiles 1987a; Pierson and Rainey 1992; Stinson et al. 1992; Craig et al. 1994; Gannon and Willig 1994; Grant et al. 1994; Pierson et al. 1996). Increased cyclone frequency or intensity would bode ill for the long-term persistence of low-fecundity vertebrate populations on small islands.

Assuming higher surface temperatures for tropical seas, as predicted by global climate models, several studies have projected increases in the maximum intensity, mean intensity, frequency, and latitudinal range of tropical cyclones and modeled their biological or economic consequences (Emanuel 1987; Gable and Aubrey 1990; O'Brien et al. 1992). This is a controversial topic. A recent review suggested that the latitudinal range of cyclones will not increase and that current climate models cannot evaluate what will happen to frequency or average intensity of cyclones (Lighthill et al. 1994). Some investigators, however, have argued strongly that maximum possible intensity will increase (Emanuel 1995). Reanalysis of relatively detailed cyclone data for the North Atlantic, Caribbean, and Gulf of

Mexico shows a significant decline in maximum intensity during the past 50 years (Landsea et al. 1996).

Conclusions

The zooarchaeological record of remote Indo-Pacific islands, which is drawn largely from limited samples in the less-diverse faunas of Polynesia, shows that bats of the genus *Pteropus* generally survived extensive prehistorical defaunation. Their survival relative to ground-nesting seabirds or numerous extinct flightless endemic land birds is not surprising. However, survival of bats relative to smaller, canopy-dwelling, nectarivorous or frugivorous birds likely reflects widespread cultural preferences for hunting birds, combined with the high vulnerability of avian eggs and young to predators. Extinctions and extirpations of island bats, from the prehistoric to the present, include many that are too poorly documented to interpret. Several losses and declines, however, underline the vulnerability of bats to overhunting (even with limited technologies), especially those species that roost colonially in caves or tree cavities.

Endemic taxa that rely on lowland or lower montane primary forest have undergone substantial population reductions with forest clearance. Yet, in areas where bat hunting is not intense, several species have persisted, albeit at low numbers, on islands with high human densities. Several such species, now confined to dwindling forest fragments, are being aided by conservation intervention. Especially on atolls, traditional agricultural practices on islands, which emphasize tree crops, may increase the abundance and predictability of fruit resources for bats and thus enhance the carrying capacity for those *Pteropus* that have catholic diets. Frugivorous bats that roost in secondary forest can often coexist successfully with man so long as their real or perceived impact on fruit crops does not trigger extermination efforts. Recent status assessments suggest that pteropodid populations are highly vulnerable to overhunting, and ready availability of guns is a significant risk factor. Cultural perspectives on bats can play a key role in determining prospects for survival.

During the 1970s and 1980s, commercial harvest of bats to supply the traditional luxury food markets in Guam and the Northern Marianas had a pervasive influence on the *Pteropus* populations on a number of islands in Polynesia and Micronesia. A partial exception to this pattern was that of bat populations in Palau, which declined during one period of commercial hunting but recovered and appeared to remain relatively common during a second, perhaps less-intense period (Wiles et al. 1997). While demand for bats is still high in the relatively affluent Marianas, little legal international trade now takes place. Limited evidence suggests that demand is partially met with bats taken illegally within CNMI under a regime of minimal enforcement. Also, forest degradation by expanding populations of introduced herbivores threatens bat habitat on some islands.

Accumulating data on molecular variation in bats from Australia and the Pacific Islands suggest that, for larger habitat generalists, practical management units based on current water gaps among islands will approximate units that would be delineated by genetic surveys. However, some montane forest endemics have shown more within-population genetic variation.

Although similar bats in Australasia and elsewhere coexist with a wide range of predators, the introduced arboreal snake *Boiga irregularis* apparently prevents local recruitment in the *Pteropus* population on Guam. Guam is a transportation hub, and transport of snakes to other islands has already occurred. Introductions of predators pose a severe threat to the near-term survival of endemic island vertebrates, including bats, even where habitat is still relatively intact. A second threat of uncertain importance, but which can also disperse through intact habitat, is disease. Outbreaks have been documented to cause severe reductions of *Pteropus* population at several localities.

While projected anthropogenically induced climate change is rapid relative to the Pleistocene record of natural climate change, its effects, except for inundation of low atolls, on remote island bat populations will likely be undetectable against the greater magnitude and pace of change induced by human population growth, resource consumption, and habitat alteration.

Acknowledgments

I thank A. Brooke, E. Pierson, D. Steadman, E. Towle, G. Wiles, and D. Worthington for discussions of island biology, literature suggestions, or unpublished reports. I also thank K. Koopman, American Museum of Natural History, and J. E. Hill and P. D. Jennings, British Museum (Natural History), for discussions and access to collections; and John Caldwell, World Conservation Monitoring Center, for CITES statistics. A. Brooke, T. Kunz, E. Pierson, M. Power, P. Racey, G. Wiles, and D. Worthington reviewed draft manuscripts. P. Racey, A. Hutson, and S. Mickleburgh deserve special thanks for the difficult task of preparing the Old World Fruit Bat Action Plan, which has stimulated interest and activity in island bat conservation biology.

Literature Cited

Adams, M., P. R. Baverstock, C. R. Tidemann, and D. P. Woodside. 1982. Large genetic differences between sibling species of bats, *Eptesicus,* from Australia. Heredity 48:435–438.

Adler, G. H. 1992. Endemism in birds of tropical Pacific islands. Evolutionary Ecology 6:296–306.

Andersen, K. 1912. Catalogue of the Chiroptera in the Collections of the British Museum. Vol. 1. Megachiroptera. British Museum (Natural History), London.

Atkinson, C. T., K. L. Woods, R. J. Dusek, L. S. Sileo, and W. M. Iko. 1995. Wildlife disease and conservation in Hawaii: Pathogenicity of avian malaria (Plasmodium relictum) in experimentally infected I'iwi (Vestiaria coccinea). Parasitology 111:S59–S69.

Atkinson, I. A. E. 1989. Introduced animals and extinctions. In Conservation for the Twenty-First Century, D. Western and M. Pearl, eds., pp. 54–69. Oxford University Press, New York.

Austad, S. N. 1993. Retarded senescence in an insular population of Virginia opossums Didelphis virginiana. Journal of Zoology (London) 229:695–708.

Austad, S. N., and K. E. Fischer. 1991. Mammalian aging, metabolism, and ecology: Evidence from bats and marsupials. Journal of Gerontology 46:B47–B53.

Baker, R. H. 1946. Some effects of the war on the wildlife of Micronesia. Transactions of the North American Wildlife Conference 11:205–213.

Baverstock, P. R., M. Adams, T. Reardon, and C. H. S. Watts. 1987. Electrophoretic resolution of species boundaries in Australian Microchiroptera. III. The Nycticeiini: Scotorepens and Scoteanax (Chiroptera: Vespertilionidae). Australian Journal of Biological Sciences 40:417–434.

Bräutigam, A., and T. Elmqvist. 1990. Conserving Pacific island flying foxes. Oryx 24:81–89.

Bruner, P. L., and H. D. Pratt. 1979. Notes on the status and natural history of Micronesian bats. Elepaio 40:1–4.

Cheke, A. S., and J. F. Dahl. 1981. The status of bats on Western Indian Ocean Islands with special reference to Pteropus. Mammalia 45:205–238.

Clout, M. N., and J. L. Craig. 1995. The conservation of critically endangered flightless birds in New Zealand. Ibis 137:S181–S190.

Clunie, F. 1985. Notes on the bats and birds of Rotuma. Domodomo 3:153–160.

Colgan, D. J., and T. F. Flannery. 1995. A phylogeny of Indo-West Pacific Megachiroptera based on ribosomal DNA. Systematic Biology 44:209–220.

Coultas, W. F. 1931. Whitney South Seas Expedition journals, Vol. W. Journal and letters of William F. Coultas, Vol. II, November 1930 to December 1931 (unpublished). American Museum of Natural History, New York.

Cox, P. A., T. Elmqvist, E. D. Pierson, and W. E. Rainey. 1991. Flying foxes as strong interactors in South Pacific Island ecosystems: A conservation hypothesis. Conservation Biology 5:448–454.

Craig, P., and W. Syron. 1992. Fruit bats in American Samoa: Their status and future. U.S. Fish and Wildlife Service Biological Report 90(23): 145–149.

Craig, P., P. Trail, and T. E. Morrell. 1994. The decline of fruit bats in American Samoa due to hurricanes and overhunting. Biological Conservation 69:261–266.

Daniel, M. J. 1990. Greater short-tailed bat. In The Handbook of New Zealand Mammals, C. M. King, ed., pp. 131–135. Oxford University Press, Auckland, New Zealand.

Daniel, M. J., and G. R. William. 1984. A survey of the distribution, seasonal activity, and roost sites of New Zealand bats. New Zealand Journal of Ecology 7:9–25.

Degener, O. 1949. Naturalist's South Pacific Expedition: Fiji. Paradise of the Pacific, Honolulu.

Ellison, J. C., and D. R. Stoddart. 1991. Mangrove ecosystem collapse during predicted sea-level rise: Holocene analogues and implications. Journal of Coastal Research 7:151–166.

Elmqvist, T., W. E. Rainey, E. D. Pierson, and P. A. Cox. 1994. Effects of tropical cyclones Ofa and Val on the structure of a Samoan lowland rain forest. Biotropica 26:384–391.

Emanuel, K. A. 1987. The dependence of hurricane intensity on climate. Nature (London) 326:483–485.

Emanuel, K. A. 1995. Global climate change and tropical cyclones. 1. Comments. Bulletin of the American Meteorological Society 76:2241–2243.

Falanruw, M. V. C. 1988. On the status, reproductive biology, and management of fruit bats of Yap. Micronesica 21:39–51.

Falanruw, M. C., and C. J. Manmaw. 1992. Protection of flying foxes on Yap Islands. U.S. Fish and Wildlife Service Biological Report 90(23): 150–154.

Flannery, T. F. 1989. Flying foxes in Melanesia: Populations at risk. Bats 7(4): 5–7.

Flannery, T. F. 1991. A new species of Pteralopex (Chiroptera: Pteropodidae) from montane Guadalcanal, Solomon Islands. Records of the Australian Museum 43:123–130.

Flannery, T. F. 1995. Mammals of the Southwest Pacific and Moluccan Islands. Cornell University Press, Ithaca, N.Y.

Flannery, T. F., P. V. Kirch, J. Specht, and M. Spriggs. 1988. Holocene mammal faunas from archaeological sites in island Melanesia. Archaeology in Oceania 23:89–94.

Foster, D. R., and E. R. Boose. 1995. Hurricane disturbance regimes in temperate and tropical forest ecosystems. In Wind and Trees, M. P. Coutts and J. Grace, eds., pp. 305–339. Cambridge University Press, New York.

Fraser, G. C., P. T. Hooper, R. A. Lunt, A. R. Gould, L. J. Gleeson, A. D. Hyatt, G. M. Russell, and J. Attenbelt. 1996. Encephalitis caused by a lyssavirus in fruit bats in Australia. Emerging Infectious Diseases 2:239–240.

Fritts, T. H. 1993. The common wolf snake, Lycodon aulicus capucinus, a recent colonist of Christmas Island in the Indian Ocean. Wildlife Research 20:261–266.

Fujita, M. S., and M. D. Tuttle. 1991. Flying foxes (Chiroptera: Pteropodidae): Threatened animals of key ecological and economic importance. Conservation Biology 5:455–463.

Gable, F. J., and D. G. Aubrey. 1990. Potential impacts of contemporary changing climate on Caribbean coastlines. Ocean and Shoreline Management 13:35–67.

Gannon, M. R., and M. R. Willig. 1994. The effects of Hurricane Hugo on bats of the Luquillo experimental forest of Puerto Rico. Biotropica 26:320–331.

Goodwin, R. E. 1979. The bats of Timor: Systematics and ecology.

Bulletin of the American Museum of Natural History 163:75–122.

Grant, G. S. 1994. Observations of *Pteropus tonganus* on Niue, South Pacific Ocean. Bat Research News 35:64–65.

Grant, G. S., S. A. Banack, and P. Trail. 1994. Decline of the sheath-tailed bat *Emballonura semicaudata* (Chiroptera: Emballonuridae) on American Samoa. Micronesica 27:133–137.

Greene, H. W. 1989. Ecological, evolutionary, and conservation implications of feeding biology in Old World cat snakes, genus *Boiga* (Colubridae). Proceedings of the California Academy of Sciences 46:193–207.

Haiduk, M. W. 1983. Evolution in the family Pteropodidae (Chiroptera: Megachiroptera) as indicated by chromosomal and immunoelectrophoretic analyses. Ph.D. dissertation, Texas Tech University, Lubbock.

Hayssen, V., and T. H. Kunz. 1996. Allometry of litter mass in bats: Maternal size, wing morphology, and phylogeny. Journal of Mammalogy 77:476–490.

Heaney, L. R. 1991. An analysis of patterns of distribution and species richness among Philippine fruit bats (Pteropodidae). Bulletin of the American Museum of Natural History 206:145–167.

Heaney, L. R., and P. D. Heideman. 1987. Philippine fruit bats: Endangered and extinct. Bats 5(1): 3–5.

Hill, J. E. 1958. Some observations on the fauna of the Maldive Islands. II. Mammals. Journal of the Bombay Natural History Society 55:3–10.

Hill, J. E., and W. N. Beckon. 1978. A new species of *Pteralopex* Thomas, 1888 (Chiroptera: Pteropodidae) from the Fiji Islands. Bulletin of the British Museum (Natural History), Zoology 34:65–82.

Holmes, D. J., and S. N. Austad. 1995. The evolution of avian senescence patterns: Implications for understanding primary aging processes. American Zoologist 35:307–317.

Holmes, M. 1994. A fruit bat survey of the Maldive Islands. Bat News 33:4–5.

Hood, C. S. 1989. Comparative morphology and evolution of the female reproductive tract in macroglossine bats (Mammalia, Chiroptera). Journal of Morphology 199:207–221.

James, H. F., T. W. Stafford, Jr., D. W. Steadman, S. L. Olson, P. S. Martin, A. J. T. Jull, and P. C. McCoy. 1987. Radiocarbon dates on bones of extinct birds from Hawaii. Proceedings of the National Academy of Sciences of the United States of America 84:2350–2354.

Jelgersma, S., M. Van der Zijp, and R. Brinkman. 1993. Sea level rise and the coastal lowlands in the developing world. Journal of Coastal Research 9:958–972.

Kirch, P. V., and D. E. Yen. 1982. Tikopia: The prehistory and ecology of a Polynesian outlier. Bernice P. Bishop Museum Bulletin 238:1–396.

Kirch, P. V., J. R. Flenley, D. W. Steadman, F. Lamont, and S. Dawson. 1992. Prehistoric human impacts on an island ecosystem, Mangaia, central Polynesia. Research and Exploration 8:166–179.

Kirsch, J. A. W., T. F. Flannery, M. S. Springer, and F. J. Lapointe. 1995. Phylogeny of the Pteropodidae (Mammalia: Chiroptera) based on DNA hybridisation, with evidence for bat monophyly. Australian Journal of Zoology 43:395–427.

Kitchener, D. J., S. Hisheh, L. H. Schmitt, and I. Maryanto. 1993. Morphological and genetic variation in *Aethalops alecto* (Chiroptera, Pteropodidae) from Jaba, Bali, and Lombok Is., Indonesia. Mammalia 57:255–272.

Kitchener, D. J., W. C. Packer, and I. Maryanto. 1994. Morphological variation in Maluku populations of *Syconycteris australis* (Peters, 1867) (Chiroptera: Pteropodidae). Records of the Western Australian Museum 16:485–498.

Koopman, K. F. 1993. Order Chiroptera. *In* Mammal Species of the World, D. E. Wilson and D. M. Reeder, eds., pp. 137–242. Smithsonian Institution Press, Washington, D.C.

Koopman, K. F., and D. W. Steadman. 1995. Extinction and biogeography of bats on 'Eua, Kingdom of Tonga. American Museum Novitates 3125:1–13.

Krzanowski, A. 1967. The magnitude of islands and the size of bats. Acta Zoologica Cracoviensia 12:282–345.

Kuroda, N. 1933. A revision of the genus *Pteropus* found in the islands of the Riu Kiu chain, Japan. Journal of Mammalogy 14:312–316.

Lack, D. 1976. Island Biology. University of California Press, Berkeley.

Landsea, C. W., N. Nicholls, W. M. Gray, and L. A. Avila. 1996. Downward trends in the frequency of intense Atlantic hurricanes during the past five decades. Geophysical Research Letters 23:1697–1700.

Lemke, T. O. 1986. Distribution and status of the sheath-tailed bat (*Emballonura semicaudata*) in the Mariana Islands. Journal of Mammalogy 67:743–746.

Lighthill, J., G. Holland, W. Gray, C. Landsea, G. Craig, J. Evans, Y. Kurihara, and C. Guard. 1994. Global climate change and tropical cyclones. Bulletin of the American Meteorological Society 75:2147–2157.

MacArthur, R. H., and E. O. Wilson 1967. The Theory of Island Biogeography. Princeton University Press, Princeton.

Marshall, A. P., D. J. Worthington, G. J. Wiles, C. C. Kessler, V. A. Camacho, E. M. Taisacan, and T. Rubenstein. 1995. A survey of the Marianas fruit bat (*Pteropus mariannus*) on Anatahan, Commonwealth of the Northern Marianas. Unpublished report, Division of Fish and Wildlife, Saipan, Commonwealth of the Northern Mariana Islands.

McCoid, M. J. 1991. Brown tree snake (*Boiga irregularis*) on Guam: A worst case scenario of an introduced predator. Micronesica (Supplement) 3:63–69.

McNab, B. K. 1994. Resource use and the survival of land and freshwater vertebrates on oceanic islands. American Naturalist 144:643–660.

Mickleburgh, S. P., A. M. Hutson, and P. A. Racey. 1992. Old World Fruit Bats: An Action Plan for Their Conservation. IUCN, World Conservation Union, Gland, Switzerland.

Nishinakagawa, H., M. Matsumoto, J. I. Otsuka, and S. Kawaguchi. 1994. Mammals from archaeological sites of the Jomon Period in Kagoshima prefecture. Journal of the Mammalogical Society of Japan 19:57–66.

O'Brien, S. T., B. P. Hayden, and H. H. Shugart. 1992. Global climatic change, hurricanes, and a tropical forest. Climatic Change 22:175–190.

Olsen, B., T. G. T. Jaenson, L. Noppa, J. Bunikis, and S. Bergstrom. 1993. A Lyme borreliosis cycle in seabirds and *Ixodes uriae* ticks. Nature (London) 362:340–342.

Peterson, A. T., and L. R. Heaney. 1993. Genetic differentiation in Philippine bats of the genera *Cynopterus* and *Haplonycteris*. Biological Journal of the Linnean Society 49:203–218.

Pierson, E. D., and W. E. Rainey. 1992. The biology of flying foxes of the genus *Pteropus*: A review. U.S. Fish and Wildlife Service Biological Report 90(23): 1–17.

Pierson, E. D., T. Elmqvist, W. E. Rainey, and P. A. Cox. 1996. Effects of tropical cyclonic storms on flying fox populations on the South Pacific islands of Samoa. Conservation Biology 10: 438–451.

Racey, P. A., and M. E. Nicoll. 1984. Mammals of the Seychelles. *In* Biogeography and Ecology of the Seychelles Islands, D. R. Stoddart, ed., pp. 607–626. Junk, The Hague, Netherlands.

Rainey, W. E. 1990. The flying fox trade: Becoming a rare commodity. Bats 8:6–9.

Rainey, W. E., and E. D. Pierson. 1992. Distribution of Pacific island flying foxes: Implications for conservation. U.S. Fish and Wildlife Service Biological Report 90(23): 111–121.

Rainey, W. E., E. D. Pierson, T. Elmqvist, and P. A. Cox. 1995. The role of pteropodids in oceanic island ecosystems of the Pacific. Symposia of the Zoological Society of London 67:47–62.

Reason, P. F., and W. J. Trewhella. 1994. The status of *Pteropus livingstonii* in the Comores. Oryx 28:107–114.

Reason, P. F., W. J. Trewhella, J. G. Davies, and S. Wray. 1994. Some observations on the Comoro rousette *Rousettus obliviosus* on Anjouan (Comoro Islands: Western Indian Ocean). Mammalia 58:397–403.

Rodda, G. H., T. H. Fritts, and P. J. Conry. 1992. Origin and population growth of the brown tree snake, *Boiga irregularis*, on Guam. Pacific Science 46:46–57.

Roy, P., and J. Connell. 1991. Climatic change and the future of Atoll states. Journal of Coastal Research 7:1057–1075.

Sanborn, C. C., and A. J. Nicholson. 1950. Bats from New Caledonia, the Solomon Islands, and New Hebrides. Fieldiana Zoology 31:313–338.

Savidge, J. A. 1987. Extinction of an island forest avifauna by an introduced snake. Ecology 68:600–668.

Schmitt, L. H., D. J. Kitchener, and R. A. How. 1995. A genetic perspective of mammalian variation and evolution in the Indonesian archipelago: Biogeographic correlates in the fruit bat genus *Cynopterus*. Evolution 49:399–412.

Shaw, W. B. 1983. Tropical cyclones: Determinants of pattern and structure in New Zealand's indigenous forests. Pacific Science 37:405–414.

Sheeline, L. 1991. Cultural Significance of Pacific Fruit Bats (*Pteropus* sp.) to the Chamorro People of Guam: Conservation Implications. World Wildlife Fund/Traffic USA, Washington, D.C.

Simberloff, D. 1995. Why do introduced species appear to devastate islands more than mainland areas? Pacific Science 49:87–97.

Sinclair, E. A., N. J. Webb, A. D. Marchant, and C. R. Tidemann. 1996. Genetic variation in the little red flying-fox *Pteropus scapulatus* (Chiroptera: Pteropodidae): Implications for management. Biological Conservation 76:45–50.

Smith, T. B., L. A. Freed, J. K. Lepson, and J. H. Carothers. 1995. Evolutionary consequences of extinctions in populations of a Hawaiian honeycreeper. Conservation Biology 9:107–113.

Speakman, J. R. 1991. The impact of predation by birds on bat populations in the British Isles. Mammal Review 21:123–142.

Springer, M. S., L. J. Hollar, and J. A. W. Kirsch. 1995. Phylogeny, molecules versus morphology, and rates of character evolution among fruit bats (Chiroptera: Megachiroptera). Australian Journal of Zoology 43:557–582.

Stair, J. B. 1887. Old Samoa or Flotsam and Jetsam from the Pacific Ocean. Southern Reprints, Papakura, New Zealand.

Start, A. N., and A. G. Marshall. 1976. Nectarivorous bats as pollinators of trees in west Malaysia. *In* Tropical Trees: Variation, Breeding, and Conservation, J. Burley and B. Styles, eds., pp. 141–150. Academic Press, London.

Steadman, D. W. 1986. Holocene vertebrate fossils from Isla Floreana, Galapagos. Smithsonian Contributions to Zoology 413: 1–103.

Steadman, D. W. 1991. Extinct and extirpated birds from Aitutaki and Atiu, southern Cook Islands. Pacific Science 45:325–347.

Steadman, D. W. 1992. Extinct and extirpated birds from Rota, Mariana Islands. Micronesica 25:71–84.

Steadman, D. W. 1995. Prehistoric extinctions of Pacific Island birds: Biodiversity meets zooarchaeology. Science 267:1123–1131.

Steadman, D. W., and B. Rolett. 1996. A chronostratigraphic analysis of landbird extinction on Tahuata, Marquesas Islands. Journal of Archaeological Science 23:81–94.

Stinson, D. W., P. O. Glass, and E. M. Taisacan. 1992. Declines and trade in fruit bats on Saipan, Tinian, Aguijan, and Rota. U.S. Fish and Wildlife Service Biological Report 90(23): 61–67.

Stoddart, D. R., and R. P. D. Walsh. 1992. Environmental variability and environmental extremes as factors in the island ecosystem. Atoll Research Bulletin 356:1–71.

Tanner, E. V. J., V. Kapos, and J. R. Healey. 1991. Hurricane effects on forest ecosystems in the Caribbean. Biotropica 23:513–521.

Thiollay, J. M. 1993. Habitat segregation and the insular syndrome in two congeneric raptors in New Caledonia: The white-bellied goshawk *Accipiter haplochrous* and the brown goshawk *Accipiter fasciatus*. Ibis 135:237–246.

Tidemann, C. R. 1985. A study of the status, habitat requirements, and management of the two species of bat on Christmas Island (Indian Ocean). Unpublished report, Australian National Parks and Wildlife Service, Canberra.

Tidemann, C. R., H. D. Yorkston, and A. J. Russack. 1994. The diet of cats, *Felis catus*, on Christmas Island, Indian Ocean. Wildlife Research 21:279–286.

Utzurrum, R. C. B. 1992. Conservation status of Philippine fruit bats (Pteropodidae). Silliman Journal 36:27–45.

Watling, D. 1995. Notes on the status of Kuhl's lorikeet *Vini kuhlii* in the Northern Line Islands, Kiribati. Bird Conservation International 5:481–489.

Webb, N. J., and C. R. Tidemann. 1995. Hybridisation between black *(Pteropus alecto)* and greyheaded *(P. poliocephalus)* flying-foxes (Megachiroptera: Pteropodidae). Australian Mammalogy 18:19–26.

Webb, N. J., and C. R. Tidemann. 1996. Mobility of Australian flying-foxes, *Pteropus* spp. (Megachiroptera): Evidence from genetic variation. Proceedings of the Royal Society of London B 263:497–502.

Weisler, M. I., and R. H. Gargett. 1993. Pacific island avian extinctions: The taphonomy of human predation. Archaeology in Oceania 28:85–93.

Whitmore, T. C. 1974. Change with time and the role of cyclones in tropical rain forest on Kolombangara, Solomon Islands. Paper 46, Commonwealth Forestry Institute, Oxford, U.K.

Whittier, J. M., and D. Limpus. 1996. Reproductive patterns of a biologically invasive species—the brown tree snake *(Boiga irregularis)* in eastern Australia. Journal of Zoology (London) 238: 591–597.

Wiles, G. J. 1987a. The status of fruit bats of Guam. Pacific Science 41:148–157.

Wiles, G. J. 1987b. Current research and future management of Marianas fruit bats (Chiroptera: Pteropodidae) on Guam. Australian Mammalogy 10:93–95.

Wiles, G. J. 1992. Recent trends in the fruit bat trade on Guam. U.S. Fish and Wildlife Service Biological Report 90(23): 53–60.

Wiles, G. J. 1994. The Pacific flying fox trade: A new dilemma. Bats 12(3): 15–18.

Wiles, G. J., and P. J. Conry. 1990. Terrestrial vertebrates of the Ngerukewid Islands Wildlife Preserve, Palau Islands. Micronesica 23:41–66.

Wiles, G. J., and M. S. Fujita. 1992. Food plants and economic importance of flying foxes on Pacific islands. U.S. Fish and Wildlife Service Biological Report 90(23): 24–35.

Wiles, G. J., and P. O. Glass. 1990. Interisland movements of fruit bats *(Pteropus mariannus)* in the Mariana Islands. Atoll Research Bulletin 343:1–6.

Wiles, G. J., J. Engbring, and D. Otobed. 1997. Abundance, biology, and human exploitation of bats in the Palau Islands. Journal of Zoology (London) 24:203–227.

Wiles, G. J., T. O. Lemke, and N. H. Payne. 1989. Population estimates of fruit bats *(Pteropus mariannus)* in the Mariana Islands. Conservation Biology 3:66–76.

Wiles, G. J., C. F. Aguon, G. W. Davis, and D. J. Grout. 1995. The status and distribution of endangered animals and plants in northern Guam. Micronesica 28:31–49.

Wilson, D. E., and J. Engbring. 1992. The flying foxes *Pteropus samoensis* and *Pteropus tonganus:* Status in Fiji and Samoa. U.S. Fish and Wildlife Service Biological Report 90(23): 74–123.

Wilson, D. E., and G. L. Graham, eds. 1992. Pacific Island Flying Fox Conference. U.S. Fish and Wildlife Service Biological Report 90(23): 1–176.

Woodell, S. R. J. 1979. The role of unspecialized pollinators in the reproductive success of Aldabran plants. Philosophical Transactions of the Royal Society B 286:99–108.

Worthington, D. J., and E. M. Taisacan. 1996. Fruit bat research: 1995 annual report. Division of Fish and Wildlife, Saipan, Commonwealth of the Northern Mariana Islands.

Wragg, G. M. 1995. The fossil birds of Henderson Island, Pitcairn Group. Natural turnover and human impact: A synopsis. Biological Journal of the Linnean Society 56:405–414.

Young, P. L., K. Halpin, P. W. Selleck, H. Field, J. L. Gravel, M. A. Kelly, and J. S. Mackenzie. 1996. Serologic evidence for the presence in *Pteropus* bats of a paramyxovirus related to equine morbillivirus. Emerging Infectious Diseases 2:239–240.

24
Geographic Patterns, Ecological Gradients, and the Maintenance of Tropical Fruit Bat Diversity
The Philippine Model

RUTH C. B. UTZURRUM

Like most tropical areas worldwide, Philippine rainforests have been lost to exploitation (resource extraction and conversion), monotypic reforestation, and replacement of native species with exotics. The archipelago's forests, estimated to cover 80% of its total land area in the 1800s, have been reduced to 12% with most of the decline occurring within the last six decades (Myers 1988; Kummer 1990). Given the archipelagic nature of the Philippines, anthropogenic degradation and fragmentation of the forest habitat could very well exacerbate effects of isolation on populations of plants and animals. By the late 1980s, 2 of the 26 species of Philippine fruit bats (Family Pteropodidae) were reported to be extinct (*Acerodon lucifer* and *Dobsonia chapmani;* Heaney and Heideman 1987). A recent rediscovery of a population of *A. lucifer* on Boracay Island (off Panay) awaits confirmation (E. E. Maro, personal communication; but see Heaney et al. [n.d.] questioning the species status of *Acerodon lucifer*). It is recognized that at least 5 of the 24 extant species (including *Acerodon jubatus, A. leucotis, Eonycteris robusta, Nyctimene rabori,* and *Pteropus leucopterus*) are seriously threatened by habitat destruction and hunting (Heaney and Utzurrum 1991; Mickleburgh et al. 1992; Utzurrum 1992).

In this chapter, I summarize results of elevational transect inventories conducted in recent years and use these data to assess the effects of habitat fragmentation on the maintenance of Philippine fruit bat diversity. Three questions are examined: (1) What are the patterns of species diversity in unfragmented (i.e., local gradients) and naturally fragmented areas (i.e., biogeographic) landscapes? (2) What impact will habitat fragmentation have on the maintenance of diversity? and (3) Do macro- and microgeographic patterns of community associations provide practical insights into conservation? The conservation implications of these questions are (1) to examine properties of populations and community assemblages that could define their response to effects of habitat fragmentation; (2) to identify measures that may mitigate effects of fragmentation given limited information; and (3) to assess research needs for the development of sound conservation strategies. Primary emphasis is given to the patterns of local gradient (rather than biogeographic patterns) because these bear the most relevance to the discussion of habitat fragmentation and its impact on the maintenance of species diversity.

Data Sources and Methods of Analysis

Study Sites

The Philippine Archipelago, lying between 4°40' N to 21°50' N latitude and 116°50' E to 136°35' E longitude, is tropical throughout its range. Habitat diversity is more marked along altitudinal than across latitudinal or longitudinal gradients, although variability in the local flora occurs in association with climatic subregions (Alcala 1976). Local gradients encompass three primary vegetation zones: lowland dipterocarp forest, montane forest, and mossy forest (dipterocarp, lower montane, and upper montane rain forest, in Whitmore 1984). The specific elevations at which these forest types occur vary among mountain sites, largely as a result of differences in maximum elevation and the amount and distribution of local rainfall (Whitmore 1984). Mountains that are at least 1,500 m in elevation support well-developed primary vegetational types over wide altitudinal ranges, and transition zones between major forest types exist as distinct bands (e.g., Mt. Guitinguiting, Sibuyan [Goodman and Ingle 1993]; Mt. Guisayawan, Negros [Heaney et al. 1989]; and Mt. Isarog, Luzon [Rickart et al. 1991]). Small mountains exhibit compression of vegetational zones, which thus occur at relatively lower elevations in what is known as the "Massenerhebung" effect (e.g., Mt. Pangasugan, Leyte [Heaney et al. 1989; Rickart et al. 1993]; and Mt. Konduko, Biliran [Rickart et al. 1993]) (Grubb and Whitmore 1966; Frahm and Gradstein 1991). Mt. Talinis, centered approximately at 9°16' N, 123°12' E on Negros Island, extends upward to 1850 m and exhibits well-defined lowland, montane, transitional montane-mossy, and mossy forests (see following section).

The degree of anthropogenic disturbance of natural habitats varies greatly among these sites. In most locations, primary lowland forest is absent below 500 m, although limited natural or replanted secondary forest may exist below this elevation. Mosaic patches of disturbance within primary forests resulting from small-scale timber extraction or shifting agriculture are, likewise, a common feature of the forests, even in areas designated as national parks.

Field Methods and Data Sources

The principal information on biogeographic patterns of distribution is that reported by Heaney (1986, 1991a) and Heaney and Rickart (1990). Additional details are derived from Heaney and Rabor (1982), Heaney et al. (1991), Utzurrum (1992), Goodman and Ingle (1993), and Vincguerra and Müller (1993). The summary presented here does not take into account the proposed change in the species status of *Acerodon lucifer* (Heaney et al. n.d.). Patterns of elevational gradients are summarized from a survey I conducted on Mt. Talinis (Negros Island) and from published studies on Biliran (Mt. Konduko: Rickart et al. 1993), Leyte (Mt. Pangasugan: Heaney et al. 1989; Rickart et al. 1993), and Negros (Mt. Guinsayawan: Heaney et al. 1981, 1989; Heideman and Heaney 1989) (Figure 24.1). Where available and relevant, unpublished data from recent studies are provided.

Data from the 1990 elevational transect study on Mt. Talinis, Negros Island, provided the focal point of analysis on elevational trends. Although previous studies on a neighboring mountain system, Mt. Guisayawan (Heaney et al. 1989; Heideman and Heaney 1989), were more extensive in scope and effort, a greater proportion of the netting was done at sites representing a single habitat type, and the full range of elevational sampling was not undertaken within the same year. Thus, results of the Mt. Guisayawan studies may be less comparable with data from more standardized sampling used in recent surveys conducted elsewhere in the Philippines.

Standardized surveys of the fruit bat fauna along elevational transects typically involved running a series of understory mist nets, during three to five nights, within a 50- to 100-m elevational band at sites corresponding to each major habitat type (Rickart 1993). On Mt. Talinis, I ran an elevational transect between May and July 1990 using six netting sites. The sites were (1) an area of low-intensity agriculture and secondary growth (500 m), characterized by stands of coconuts, coffee, and plots of cultivated vegetables and flowers, interspersed with patches of shrubs (*Melastoma* spp.), sawgrass (*Imperata* sp.), and scattered wild figs (*Ficus* spp.); (2) a naturally regenerated secondary forest (500 m) with remnant anthropogenic plants such as banana and abaca (*Musa* spp.), avocado (*Persea americana*), and jackfruit (*Artocarpus hetephyllus*); (3) primary lowland forest (750 m) punctuated by small (<0.5-ha) disturbed patches in various stages of regeneration; (4) primary montane forest (1,100 m) relatively free of and distant from disturbance; (5) transitional montane-mossy forest (1,250 m) with elements of previous anthropogenic disturbance (e.g., *Musa* spp. and *Bambusa* sp.); and (6) primary mossy forest (1,625 m) at the summit of the ridge system on which the five other sampling sites were located.

At each sampling location, ten nylon mist nets (12 m long × 2.6 m high) were run from 1800 to 0600 for five consecutive nights. External measurements (size measurements and body mass) were recorded for each fruit bat captured. All samples were identified to species following Heaney et al. (1987) and Ingle and Heaney (1992), and assessed for age and

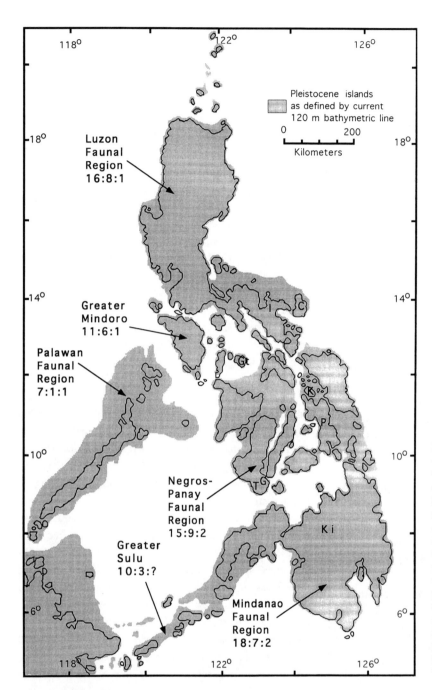

Figure 24.1. Extent of Pleistocene land connections (shaded areas, which correspond to the current 120-m bathymetric line) in comparison to the current topography (solid lines) of the Philippines. The first of the three numbers following each Pleistocene region name corresponds to the total number of fruit bat species present; the second number, the subset of this total that are endemic to the Philippines; and the third, the number of endemic species that are unique to the faunal region. Letters indicate the principal locations of study sites mentioned in the text: C, mountain on Catanduanes; G, Mt. Guisayawan, Negros; Gt, Mt. Guitinguiting, Sibuyan; I, Mt. Isarog, southeastern Luzon; Ki, Mt. Kitanglad, Mindanao; K, Mt. Konduko, Biliran Island; P, Mt. Pangasugan, Leyte; T, Mt. Talinis, Negros.

reproductive status by methods modified from Heideman (1987). Voucher specimens were deposited in the Field Museum of Natural History (Chicago), Philippine National Museum (Manila), Silliman University Museum of Natural History (Dumaguete City), and the teaching collection of the Department of Biology, Boston University.

Data Analysis

Analyses of data sets other than from Mt. Talinis included calculations of species richness (S, the total number of different species captured at each site), total abundance (expressed as numbers of bats captured per net-night), and relative abundance (number of bats of a given species captured per net-night). Similar treatments were applied to the Mt. Talinis data set. The Shannon–Wiener index of diversity (H') was also calculated for the Mt. Talinis (this chapter) and Mt. Pangasugan (Heaney et al. 1989; Rickart et al. 1993) elevational transect data. Within-mountain and between-mountain diversity indices were compared using a t-test (Magurran 1988). Large flying-foxes (*Acerodon* and *Pteropus* species) are customarily excluded from data analysis be-

Table 24.1

Distributions of the 26 Species of Philippine Fruit Bats (Pteropodidae) across Pleistocene Faunal Regions (sensu Heaney 1986) as Updated from Heaney (1991)

Distribution and endemism	Species[a]
Species widespread in Indo-Australia (6/26; 23%)	*Cynopterus brachyotis*
	Eonycteris spelaea
	Macroglossus minimus
	Pteropus hypomelanus
	Pteropus vampyrus
	Rousettus amplexicaudatus
Species shared with nearby archipelagos; of limited distribution in the Philippines (4/26; 15%)	*Dyacopterus spadiceus*
	Megaerops wetmorei
	Pteropus dasymallus[3]
	Pteropus speciosus
Endemic species widespread in oceanic Philippines (6/26; 23%)	*Acerodon jubatus*
	Eonycteris robusta
	Haplonycteris fischeri
	Harpyionycteris whiteheadi
	Ptenochirus jagori
	Pteropus pumilus
Endemic species on two or more Pleistocene islands (2/26; 8%)	*Pteropus leucopterus*[1,2]
	Nyctimene rabori[4,6]
Endemic species on only one Pleistocene island (8/26; 31%)	*Acerodon leucotis*
	Acerodon lucifer
	Alionycteris paucidentata
	Dobsonia chapmani
	Haplonycteris sp.[4]
	Otopteropus cartilagonodus
	Ptenochirus minor
	Pteropus sp.[5]

[a]Superscript numbers indicate sources of updates: (1) Heaney and Rabor 1982; (2) Heaney et al. 1991; (3) Utzurrum 1992; (4) Goodman and Ingle 1993; (5) Heaney 1993, and unpublished; and (6) Vinciguerra and Müller 1993.

cause they are not reliably sampled in understory net sets (Heaney et al. 1989; Heideman and Heaney 1989; Rickart et al. 1993). However, I included the smallest of the flying-foxes, *Pteropus pumilus*, in the analysis of the Mt. Talinis data. The nets for the Mt. Talinis transect were set on narrow ridges and were effective in capturing flying-foxes that commute over ridgetops.

Results

Pteropodids presently known from the Philippines range in size from 16 g (e.g., *Alionycteris paucidentata; Macroglossus minimus*) to more than 1 kg (e.g., *Acerodon jubatus* and *Pteropus vampyrus*). Of the 26 species present, 16 (62%) are restricted to the Philippines, including 6 species in the endemic genera: *Alionycteris* (1), *Haplonycteris* (2), *Otopteropus* (1), and *Ptenochirus* (2) (Heaney et al. 1987; Heaney 1991a; Utzurrum 1992).

Biogeographic Distribution Patterns

Biogeographic analysis of species richness among Philippine fruit bats shows distributions concordant with land masses formed during lowering of sea levels in the Pleistocene (Heaney 1991a). Six (23%) of the nonendemic species are widespread within the Philippines and in the Indo-Australian region; the other four (15%) are restricted within the Philippines and are shared with nearby islands (Table 24.1; see also Figure 24.1) (Heaney et al. 1987; Heaney 1991a). Among the endemic species, three basic patterns of geographic distribution emerge: (1) species that are widespread in oceanic Philippines (six); (2) species that are shared at least between two Pleistocene islands (two); and (3) species that are confined to only one Pleistocene island (eight) (Table 24.1; Figure 24.1) (Heaney and Rabor 1982; Heaney 1991a 1993; Utzurrum 1992; Goodman and Ingle 1993).

Data from well-inventoried present-day islands reveal a significant positive relationship between species numbers and island size (Heaney 1991a). However, this relationship does not hold true for endemic species. Luzon Island (108,171 km²) is the largest island to hold an endemic (i.e., *Otopteropus cartilagonodus*). The islands of Mindanao (99,078 km²; *Alionycteris paucidentata*), Negros (13,670 km²; *Dobsonia chapmani*), Panay (12,300 km²; *Acerodon lucifer*), Palawan (11,785 km²; *Acerodon leucotis*), and Mindoro (9,735 km²; undescribed *Pteropus* sp.) also have one endemic species each (Heaney 1991a, 1993; Heaney et al. n.d.). Sibuyan Island (463 km²; undescribed *Haplonycteris* sp.), an oceanic island with no historical connection to any of the Pleistocene islands, is the smallest Philippine island currently known to have an endemic bat species (Utzurrum 1992; Goodman and Ingle 1993).

The high degree of species overlap within and among Pleistocene islands suggests the importance of overwater colonization in shaping diversity and distribution (Heaney and Rickart 1990; Heaney 1991a). Measures of gene flow confirm these patterns (Peterson and Heaney 1993). Pleistocene land-bridge islands intermediate in size between Negros and Sibuyan that lack endemic species (e.g., Dinagat and Leyte) further support the importance of Pleistocene land connections, or conversely the lack thereof, in shaping speciation events within the archipelago. Species with disjunct distributions pose an interesting puzzle in our understanding of these faunal patterns and the mechanisms that shaped them. Pleistocene land connections among islands do not account for the disjunct distribution of *Nyctimene rabori* (in Cebu, Negros, and Sibuyan islands only; Heaney et al. 1987; Goodman and Ingle 1993, Vinciguerra and Müller 1993), or of *Pteropus leucopterus* (in Catanduanes, Dinagat, and northern Luzon; Heaney and Rabor 1982; Heaney et al. 1987, 1991). These cases suggest processes that have influenced extinction events in the past—or may simply reflect information gaps that need to be filled from more thorough inventories.

Elevational Gradients in Species Diversity and Abundance: General Trends

General patterns have emerged from elevational transect surveys: (1) species richness reaches its maximum in primary lowland forest and declines with elevation; (2) total abundance is highest in disturbed areas and decreases with elevation in forest habitats; and (3) species assemblages in disturbed areas tend to be characterized by the presence of geographically widespread species, whereas endemic species tend to be associated with forest habitats (Heaney et al. 1981, 1989; Heideman and Heaney 1989; Rickart et al.

1993). Endemic species are sometimes found in low numbers in moderately disturbed habitats provided that these are adjacent (within 1 km) to primary habitats (Heideman and Heaney 1989; Rickart et al. 1993).

Despite the abundance of fruit bats in urban orchards and agricultural areas far removed from forested sites, endemic species have not been recorded at these locations (Heideman 1987; Heaney et al. 1989; Heideman and Heaney 1989; Rickart 1993), with the exception of *Pteropus pumilus* (R. C. B. Utzurrum, unpublished records of captures in Dumaguete City [1986] and Siaton [1992] on Negros). In most cases, species at high elevations represent a subset of the lowland community, indicating a lack of high-elevation specialists. In 1992 and 1993, *Alionycteris paucidentata* was captured in considerable numbers on Mt. Kitanglad, Mindanao, but only at elevations above 1,500 m in montane and mossy forest (L. R. Heaney, personal communication).

Elevational Gradients on Mt. Talinis, Negros Island

Of the 15 species of fruit bats occurring in Negros Island, 9 were netted in this study (Table 24.2). The capture of *Cynopterus brachyotis* at 1,250 m is a new elevational record for the species; all other species have been recorded at elevations similar to or greater than in the present study (Heaney and Heideman, unpublished data). Missing from the samples were 3 species of flying-foxes (*Acerdon jubatus*, *Pteropus hypomelanus*, and *P. vampyrus*) not expected to be captured in understory nets, an uncommon endemic species, *Eonycteris robusta* (Utzurrum 1992), and the reportedly extinct *Dobsonia chapmani* (Heaney and Heideman 1987).

The total numbers of fruit bats captured were highest at the agricultural site (3.24 bats/net-night) and decreased with elevation in forest (from 1.9 bats/net-night in lowland forest to 0.22 bats/net-night in mossy forest; Table 24.2). This overall trend was true for all the nonendemic species as well as the endemic *Ptenochirus jagori* (Table 24.2). All other endemic species were uncommon or absent at the agricultural site. Instead, they were found in higher numbers in lower elevation forest (lowland or montane), although *Nyctimene rabori* was relatively uncommon even in forest habitat. These patterns of abundance and distribution are consistent with a previously observed ecological dichotomy between endemic and nonendemic species (Heaney et al. 1989). As predicted by Heaney et al. (1989), species richness was highest in lowland primary forest ($S = 9$) and lowest in mossy forest ($S = 3$) (Table 24.2). Unexpectedly high levels of species richness and abundance were recorded at the montane-mossy transitional zone.

Table 24.2

Summary of Captures of Fruit Bats (Pteropodidae) on Mt. Talinis, Negros Oriental, May–July 1990

	Site (and elevation)					
Species	A (500 m)	B (500 m)	C (725 m)	D (1,100 m)	E (1,250 m)	F (1,625 m)
Cynopterus brachyotis	0.74	0.68	0.54	0.08	0.10	0
Eonycteris spelaea	0.38	0.14	0.02	0	0.02	0
Haplonycteris fischeri	0.04	0.10	0.26	0.42	1.04	0.16
Harpyionycteris whiteheadi	0	0	0.04	0	0.10	0.04
Macroglossus minimus	0.94	0.20	0.54	0.10	0.36	0.02
Nyctimene rabori	0	0	0.06	0.06	0.02	0
Ptenochirus jagori	0.36	0.04	0.16	0	0	0
Pteropus pumilus	0.02	0	0.22	0	0.04	0
Rousettus amplexicaudatus	0.76	0.08	0.06	0	0	0
All nonendemics	2.82	1.10	1.16	0.18	0.48	0.02
All endemics (*)	0.42	0.14	0.74	0.48	1.20	0.20
All species	3.24	1.24	1.90	0.66	1.68	0.22
	(S = 7)	(S = 6)	(S = 9)	(S = 4)	(S = 7)	(S = 3)

Notes: Data are given as number of bats captured per net-night. Each site had a total netting effort of 50 net-nights. Site designations: A, mixed agriculture/secondary growth; B, secondary lowland forest; C, primary lowland forest; D, montane forest; E, transitional montane-mossy; F, mossy forest. An asterisk denotes endemic species.

Discussion

Biogeographic Information and the Design of a System of Protected Areas

Patterns of biogeographic distribution of species are relevant to fruit bat conservation at two levels. First, they provide a biological basis for the selection of important sites for protection. Second, island size and species diversity relationships revealed from biogeographic analysis provide estimates of areal requirements for the maintenance of species diversity based on estimated rates of colonization and extinction (Heaney 1986). Given the high degree of similarity in the composition of fruit bat assemblages between Pleistocene faunal regions and among islands of a region, a minimum of eight protected areas, to be located in the current islands of Luzon (one in the northern tip and a second in the southeastern peninsula), Mindanao, Negros, Panay, Palawan, Mindoro, and Sibuyan, may theoretically protect all 26 species (Table 24.3). Interestingly, the patterns of species richness and levels of endemism seen in fruit bats relative to Pleistocene faunal regions are concordant with those of other vertebrate groups, including nonvolant mammals (Heaney 1986, 1993; Heaney and Rickart 1990), birds (Dickerson 1928; Dickinson et al. 1991), and amphibians and reptiles (Brown and Alcala 1970; Hague et

al. 1986). These zoogeographic patterns also overlap well with phytogeographic patterns of diversity and endemism (D. A. Madulid, personal communication; Heaney 1993). Thus, a parks system modeled on the biogeography of these more speciose vertebrates and plants would subsume protection of the fruit bats.

In the recent Integrated Protected Areas Systems (IPAS) initiative toward the redevelopment of the Philippine parks system (IUCN 1991), recommended priority areas based on mammalian and floristic diversity patterns have led to the inclusion of sites in Luzon, Negros, and Mindanao (Heaney 1993). This plan will incorporate habitats for 85% (22 of 26) of all fruit bats, including 75% (12 of 16) of the endemic species (see Table 24.3). It should be emphasized that areal size requirements based on estimated rates of colonization and extinction should be treated as conservative guidelines, given that these and other evolutionary processes have occurred in a context of habitats (i.e., continuous expanse of forests) largely different from present-day conditions (i.e., discontinuous patches of forests). Thus, whenever possible, the largest continuous area of suitable habitat available at each priority site should be chosen.

There are perceived political barriers against the designation of small islands for protection even when strong biological reasons exist (see Utzurrum 1991 for discussion).

Table 24.3

Theoretical Percentages of Species That Would Be Protected by a Designated Park in the Philippine Islands

Island (and area)	All species (total, 26)		Endemic species (total, 16)	
	No.	Cumulative %	No.	Cumulative %
Mindanao (99,078 km²)	17	65%	8	50%
Luzon (108,171 km²)	+3	77%	+2	62%
Negros (13,670 km²)	+2	85%	+2	75%
Panay (12,300 km²)	+1	88%	+1	81%
Mindoro (9,735 km²)	+1	92%	+1	88%
Palawan (11,785 km²)	+1	96%	+1	94%
Sibuyan (463 km²)	+1	100%	+1	100%

Notes: Mindanao is ranked first because it has the most species. Luzon and Negros follow, based on the number of species that these islands will add to the theoretical protected pool. The last four islands are ranked on the basis of conservation priority of the additional species unique to each theoretical reserve (see Heaney 1993; Heaney and Utzurrum 1991; Utzurrum 1992). Four of six forest parks designated by the Integrated Protected Areas System (IPAS) will be located in Mindanao (Mt. Kitanglad), Luzon (Palanan Wilderness Area and Subic Bay), and Negros (Mt. Kanlaon).

Such may be the case for Sibuyan Island. At least five undescribed species of mammals were recently (i.e., since 1990) discovered on this island (one fruit bat, *Haplonycteris* sp., and four murids, *Apomys* (2), *Chrotomys* (1), and *Tarsomys* (1); Goodman and Ingle 1993). Its degree of isolation and relative state of "underdevelopment," however, elements inherently favoring species persistence, could work against its selection for protection under the national parks scheme. Ongoing political machinations exist that exploit the rich timber resources on Mt. Guitinguiting (N. Ingle, personal communication), but the island's relative isolation from trade and communication precludes the national visibility that could highlight a need to include it in a short list of priority parks. Instead, protection of areas such as Sibuyan, and other sites supporting single island endemics, may depend on the development of a network of secondary regionally managed conservation parks that complements the national parks system (Utzurrum 1991).

There are many potentially species-rich and biogeographically interesting areas that remain relatively unknown in the Philippines. Recent efforts to systematically inventory local mammalian fauna have been concentrated mostly on a latitudinal band extending from southern Luzon (in the north) to northern Mindanao (in the south) (see Figure 24.1). Although Palawan has been very attractive to researchers of birds (Dickinson et al. 1991) and, in part, of mammals (Heaney 1986), its bat fauna is poorly studied (Heaney 1991a). The northernmost regions of Luzon (especially the northeastern border) still support extensive forests, yet most of the recent surveys in the area focused primarily on birds (Mallari and Jensen 1993). These research trends reflect in part the opportunistic nature of research work in the Philippines, in terms of funding, political stability, and the expertise available.

Habitat Affinity and Its Implications for Conservation

It is widely recognized that forests are essential for the conservation of biodiversity in tropical regions. More importantly, the design and management of protected areas should incorporate provisions for different affinities among species for gradients in vegetational structure and composition and abiotic conditions throughout the local range of a forest habitat. Thus, a basic understanding of local patterns of species distribution and species–habitat associations is of utmost importance.

Lowland forests are essential for the maintenance of maximum local and, therefore, overall diversity of Philippine fruit bats (see *Results*). This requirement necessitates the inclusion of lowland forest into parks or reserves. Forests at low elevations may also be critical for the persistence of higher-elevation forests. These upper-elevation habitats, in turn, may be integral to the maintenance of lowland forest diversity. Montane forest habitats on Philippine mountains have experienced episodes of expansion and retraction associated with climate changes in the Pleistocene (Heaney 1991b). Although the relevance of these historical changes in vegetational cover to speciation events is more apparent for rodents (Heaney and Rickart 1990), they may have influenced speciation events or patterns of habitat specialization in fruit bats as well. The distributional pattern of *Alionycteris paucidentata* on Mt. Kitanglad indicates that other upper-elevation habitat specialists among the fruit bats are likely to occur in association with broadly distributed and well-developed montane-mossy forests. As on Mt. Kitanglad, these conditions may occur in mountains exceeding 1,500 m in elevation. However, few Philippine mountains bearing elevations close to 2,000 m or greater have been surveyed. Furthermore, we have a very limited understanding of the nature and dynamics of microgeographic preferences among and within fruit bat species. While surveys indicate lowland forests as the local centers of species richness, they do not necessarily reveal to what extent adjoining tracts of forests contribute to the maintenance of this diversity. For example, preliminary analysis of within-species differences in the elevational distribution of *Otopteropus cartilagonodus* on Mt. Isarog and Zambales revealed elevational segregation between males and females of the species, a mechanism that may reduce intersexual competition (Ruedas et al. 1994).

The importance of montane and mossy forest habitats for the maintenance of biodiversity is even more crucial for nonvolant mammals. Philippine murid rodents exhibit peaks in species richness and endemism in montane and mossy forests (Heaney and Rickart 1990; Rickart et al. 1991). In many of the mountains surveyed, patterns of avian diversity mirror that of fruit bats, with lowland forest as the locus of maximum species richness. However, endemism itself may be centered in the upper-elevation forests, as was documented in the northern Sierra Madre (Mallari and Jensen 1993). Together, these findings indicate the need for a full elevational complement of forest habitats in any designated conservation site if maximum species diversity is to be protected.

Impacts of Forest Fragmentation on Fruit Bat Diversity

Conservation of biodiversity must address whether and how contemporary degradation and fragmentation of habitats within islands will influence ecological processes and, thus, local patterns of species assemblage and distribution. If strong preference for forest habitats inhibits fruit bat movements over fragmented landscapes, despite their inherent vagility, then mechanisms of species maintenance and geographic structuring may be affected. Indeed, genetic variation in *Cynopterus brachyotis* (a nonendemic) and *Haplonycteris fischeri* (an endemic) suggests that the reduced gene flow seen in the latter species relates, in part, to its greater affinity for specific habitats (Peterson and Heaney 1993). The occurrence of an endemic species of *Haplonycteris* on Sibuyan Island demonstrates evolutionary stability in small isolated populations (Peterson and Heaney 1993). Thus, it is difficult to predict whether increased fragmentation of once continuous populations (within islands) in the recent past will shift evolutionary processes toward the negative trajectories associated with increasingly smaller population sizes (see Lande 1988).

Elevational gradients in fruit bat diversity and abundance on Negros and Leyte islands illustrate how habitat fragmentation or degradation affects local community structure. Negros and Leyte were parts of two different Pleistocene islands but exhibit a moderately high degree of similarity in their fruit bat fauna (Heaney 1991a). Both islands support 13 extant fruit bat species, with 12 species shared by both areas (Heaney et al. 1989). Whereas Negros Island has *Nyctimene rabori*, Leyte has *Ptenochirus minor*, both of which are endemic and share an affinity for forest habitats (Heaney et al.

1989; Rickart 1993). Thus, they may be considered ecological equivalents for the purpose of comparing general patterns between the two islands. On Mt. Talinis (Negros), primary forest was absent from below the 500-m elevation but extended over a greater distance to a summit of 1,850 m. Disturbed patches within the primary forest were not uncommon. The primary forest in Mt. Pangasugan (Leyte) was compressed over a narrower elevational range because the summit was lower (1,150 m). Disturbances within forests were minimal and confined within the lower 300-m elevation; lowland forest graded into agricultural areas below 200 m. Mt. Talinis has sharper topographical features consisting of deep, steep gulleys bisecting sharp narrow ridges. Thus, the interfaces between low- and high-elevation zones, and between forest and agricultural or secondary growth areas, were greater there than on Mt. Pangasugan.

On Negros, endemic species, such as *Haplonycteris fischeri* and *Ptenochirus jagori,* were absent at elevations below 500 m, with the exception of *Pteropus pumilus* (Figure 24.2). This lack coincides with the absence of forest habitat below this elevation. Conversely, the persistence of endemic species near sea level on Leyte coincides with the lower extent of forest on Mt. Pangasugan.

Both areas exhibited maximum species richness in primary lowland forest (Table 24.4). The transition from lowland to montane forest was the upper limit of a significant shift in species richness. Diversity indices of sites above lowland forest do not differ significantly, although values show a general downward trend with increasing elevation (Magurran 1988: t-test on H' values and variances of H', $p > 0.05$) (Table 24.4). Thus, lowland forest zones may define the upper elevational boundary of maximum fruit bat diversity. Species diversity levels (H') do not differ statistically (Magurran 1988: t-test on H' and variance of H', $p > 0.05$) between pairs of equivalent habitats even when these zones occurred at different elevations on each mountain (Table 24.4). Attenuation in species numbers occurred at the transition from montane to mossy forest on Mt. Pangasugan, but not on Mt. Talinis. On Mt. Talinis, species richness was higher in the montane-mossy transition forest than expected. I attribute this to the atypical occurrence of the nonendemic species *Cynopterus brachyotis* and *Eonycteris spelaea* at the site (see Table 24.2).

On Mt. Guinsayawan (northeast of Talinis), neither of these species occurred above upper montane forest (Heaney et al. 1989), and unpublished data from Mt. Isarog (in Luzon; L. R. Heaney, personal communication) revealed the same pattern of elevational distribution for these species. I interpret this unexpectedly high level of species richness at the transition zone of montane and mossy forests on Mt. Talinis as an upward range extension of species typical of lowland

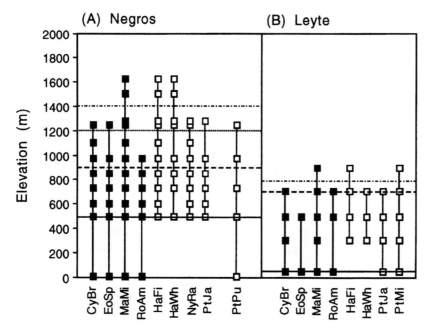

Figure 24.2. Comparative elevational distribution of fruit bats on (**A**) Negros Island (a composite from Mt. Guinsayawan and Mt. Talinis studies) and (**B**) Leyte (Mt. Pangasugan). Species are identified by the first two letters of the genus and species names (see Table 24.1 for list). The horizontal lines denote lower elevational limits of the various types of forests: solid line, primary forest; dashed line, montane forest (approximate); dotted line, transition into mossy forest; dashed-and-dotted line, well-developed mossy forest. Primary forest on Negros begins at an elevation of 500 m, whereas on Leyte it is still present at 50 m. Maximum elevations at the study sites were 1,800 m (Negros) and 1,150 m (Leyte).

or disturbed habitat in response to habitat disturbance at this site and its increased proximity to cultivated fields on adjoining slopes. This response should be differentiated from increased diversity that may occur at zones where communities form ecotones (Ricklefs 1979). In this particular case, the latter phenomenon does not truly apply because the fruit bat species found in the adjoining montane and mossy sites were not distinct from each other.

I draw two points of relevance from the preceding com-

parisons for the maintenance of species diversity. First, changes in the quality and quantity of forest habitats alter the nature of species assemblages within forest types by affecting both the relative numbers among species and the types of species present. As a corollary, these results suggest that (1) light to moderate levels of habitat disturbance, where the primary forest structure and composition and climatic conditions are retained, result in higher species richness than would be expected in undisturbed forest of

Table 24.4

Total Abundance of Fruit Bats and Measures of Their Species Richness, Diversity, and Evenness along Two Mountains

Site	Elevation (m)	Abundance (bats/net-night)	No. of bat species	Diversity $(H')^a$	Evenness
Mt. Pangasugan, Leyte Island[b]					
Agriculture and disturbed lowland forest	50	6.08	6	1.690*	0.943
Lowland forest, 2 sites:					
(1) Primary forest, disturbed	300	0.60	6	1.277*	0.712
(2) Primary forest	500	0.82	8	1.829*	0.880
Primary montane forest	700	0.97	7	1.065	0.547
Primary mossy forest	950	0.45	3	0.730	0.660
Mt. Talinis, Negros Island					
Agriculture and lowland secondary growth	500	3.24	7	1.610*	0.831
Lowland forest, 2 sites:					
(1) Secondary forest	500	1.24	6	1.360*	0.759
(2) Primary forest, disturbed	750	1.90	9	1.792*	0.816
Primary montane forest	1,100	0.66	4	1.047	0.755
Transitional montane-mossy forest	1,250	1.68	7	1.157	0.595
Primary mossy forest	1,625	0.22	3	0.760	0.691

[a]An H' value (the Shannon–Weiner index) marked by an asterisk (*) differs significantly from the H' value of the next higher elevation on the same mountain.

[b]Mt. Pangasugan data are from Heaney et al. (1989) and Rickart et al. (1993). Note that, unlike Mt. Talinis, Mt. Pangasugan had no transitional montane-mossy forest.

similar type and elevation; (2) the changes in the community structure relating to moderate habitat disturbance result primarily from local range extensions of nonendemic species that are typically associated with disturbed habitats and are rare or absent in primary forest (especially above lowland forest); (3) large-scale habitat disturbances resulting in degradation of the principal forest structure, the alteration of associated climate conditions, and (or) marked fragmentation of formerly continuous tracts of forests may result in lower levels of species richness than would be expected; and (4) the decrease in species richness in heavily disturbed or fragmented forest habitats is associated primarily with the disappearance of endemic species or with their increasing rarity.

Second, differences in the topographical features of mountain forests influence the responses of both plant and fruit bat community structure. Steepness, ruggedness, and irregularity in topography determine the degree of interface between habitat bands as well as the depth and expanse of a given band. Disturbance of similar scales may have different effects on two landscapes of dissimilar topographical features. Hence, efforts should be made toward the analysis of landscape features as they may influence local biotic communities.

In general, I predict that the upper montane zone will constitute the upper elevational limit of maximum fruit bat diversity. As forests at lower elevations disappear and the lower edges of forest progressively shrink upward in elevation, we may see shifts in the region of highest diversity from lower to higher forest regions. However, this "elevational retreat" may reach its limit when conditions (food, roosting, climatic) of the environment necessary for supporting viable populations in themselves become limiting, as may be the case in mossy forests. While this prediction results from studies associated with anthropogenic destruction of forests, habitat disturbances resulting from natural catastrophes (e.g., hurricanes) may generate similar results. One difference between these two types of disturbances is this: Although natural catastrophes do not typically generate sustained destructive stresses, anthropogenic processes typically do so.

Future Conservation Research Needs

Further inventories of continuous habitat gradients in the tropics are expected to demonstrate the close association between forest habitat and endemic species. In this context, comparisons of elevational distributions against a backdrop of changing habitat are useful as preliminary indices of the nature of diversity patterns and the ecological processes that may affect local community structure. Arguably, a more thorough analysis of relationships between habitat fragmentation and changes in species diversity requires quantitative measurements of habitat disturbance, such as relative areal coverage, spatial geometry, and extent of edge habitats, and the impact on local fruit bat assemblages.

Additionally, there is a need to examine the actual processes and mechanisms that underpin community structuring and correlations among species diversity, population structure, and habitat quality. These include: (1) identification and quantification of those elements of the habitat of direct importance to fruit bats, specifically food resources and roosts; (2) determination of critical ranges of environmental conditions (temperature and humidity) that are physiologically compatible with the persistence of species at given habitats; and (3) overlaying spatial analyses on analyses at temporal scales to assess how annual or seasonal dynamics of fruit bat activities (e.g., reproduction) will influence their responses to habitat changes on a spatial scale.

Conclusions

The potential effects of habitat fragmentation on the maintenance of biodiversity are varied (see Lande 1988 and Terborgh 1992 for recent reviews). Species extinctions are possible end results of these effects. It is widely believed that such extinctions result from demographic stochasticity rather than accumulations of deleterious genetic changes (Lande 1988). In the final analysis, both demographic and genetic stochasticity often become relevant only after populations have been decimated to inviable numbers. Hence, the pivotal issue for the conservation of biodiversity is the prevention of declines in population sizes. The studies summarized in this chapter strongly indicate that changes in fruit bat communities can occur within the brief time scale in which forest habitats are being destroyed. In this regard, it is apparent that the prevention of further forest destruction is most crucial for the long-term maintenance of fruit bat diversity.

Acknowledgments

Financial support for the 1990 field research on Mt. Talinis was provided by the American Society of Mammalogists (grants-in-aid for research), Bat Conservation International, Inc., Chicago Zoological Society, Marshall Field and Ellen Thorne Smith Funds of the Field Museum of Natural History, the Lubee Foundation, Inc., and Boston University. I thank F. Catalbas, M. Furacan, E. Maro, and L. Tag-at for their valuable assistance in the field. Many other studies on Philippine fruit bats cited, in which I have had the opportunity to collaborate, were supported by the U.S. National

Science Foundation (BSR-8514223), the John D. and Catherine T. MacArthur Foundation, and the Field Museum of Natural History (with L. R. Heaney as P.I.). L. R. Heaney, T. H. Kunz, E. A. Rickart, and J. O. Seamon provided valuable comments on this manuscript. I am especially thankful to L. R. Heaney, P. D. Heideman, and E. A. Rickart for providing the intellectual atmosphere and research opportunities that have shaped my interest in Philippine bat research.

Literature Cited

Alcala, A. C. 1976. Philippine Land Vertebrates: Field Biology. New Day Publishers, Quezon City.

Brown, W. C., and A. C. Alcala. 1970. The zoogeography of the herpetofauna of the Philippine islands, a fringing archipelago. Proceedings of the California Academy of Science 38:105–130.

Dickerson, R. E. 1928. Distribution of life in the Philippines. Bureau of Science Monograph (Manila) 21:1–322.

Dickinson, E. C., R. S. Kennedy, and K. C. Parkes. 1991. The birds of the Philippines: An annotated checklist. British Ornithologists Union Checklist 12:1–507.

Frahm, J.-P., and S. R. Gradstein. 1991. An altitudinal zonation of tropical forests using bryophytes. Journal of Biogeography 18:669–678.

Goodman, S. M., and N. R. Ingle. 1993. Sibuyan Island in the Philippines: Threatened and in need of conservation. Oryx 27:174–180.

Grubb, P. J., and T. Whitmore. 1966. A comparison of montane and lowland rain forest in Ecuador. II. The climate and its effects on the distribution and physiognomy of the forest. Journal of Ecology 54:303–333.

Hague, P., J. Terborgh, P. Winter, and J. Parkinson. 1986. Conservation priorities in the Philippine archipelago. Forktail 2:83–91.

Heaney, L. R. 1986. Biogeography of mammals in SE Asia: Estimates of rates of colonization, extinction, and speciation. In Island Biogeography of Mammals, L. R. Heaney and B. D. Patterson, eds., pp. 127–165. Academic Press, London.

Heaney, L. R. 1991a. An analysis of patterns of distribution and species richness among Philippine fruit bats (Pteropodidae). Bulletin of the American Museum of Natural History 206:145–167.

Heaney, L. R. 1991b. A synopsis of climatic and vegetational change in Southeast Asia. Climatic Change 19:53–61.

Heaney, L. R. 1993. Biodiversity patterns and conservation of mammals in the Philippines. Asia Life Sciences 2:261–270.

Heaney, L. R., and P. D. Heideman. 1987. Philippine fruit bats: Endangered and extinct. Bats 5(1): 3–5.

Heaney, L. R., and D. S. Rabor. 1982. Mammals of Dinagat and Siargao islands, Philippines. Occasional Papers of the Museum of Zoology of the University of Michigan 699:1–30.

Heaney, L. R., and E. A. Rickart. 1990. Correlations of clades and clines: Geographic, elevational, and phylogenetic distribution patterns among Philippine mammals. In Vertebrates in the Tropics, G. Peters and R. Hutterer, eds., pp. 321–332. Museum Alexander Koenig, Bonn.

Heaney, L. R., and R. C. B. Utzurrum. 1991. A review of the conservation status of Philippine land mammals. Association of Systematic Biologists of the Philippines Communications 3:1–13.

Heaney, L. R., P. C. Gonzales, and A. C. Alcala. 1987. An annotated checklist of the taxonomic and conservation status of land mammals in the Philippines. Silliman Journal 34:32–66.

Heaney, L. R., P. D. Heideman, and K. M. Mudar. 1981. Ecological notes on mammals in the Lake Balinsasayao region, Negros Oriental, Philippines. Silliman Journal 28:122–131.

Heaney, L. R., P. C. Gonzales, R. C. B. Utzurrum, and E. A. Rickart. 1991. The mammals of Catanduanes Island, Philippines. Proceedings of the Biological Society of Washington 104:399–415.

Heaney, L. R., P. D. Heideman, E. A. Rickart, R. B. Utzurrum, and J. S. H. Klompen. 1989. Elevational zonation of mammals in the central Philippines. Journal of Tropical Ecology 5:259–280.

Heaney, L. R., D. S. Balete, L. Dolar, A. C. Alcala, A. Dans, P. C. Gonzales, N. Ingle, M. Lepiten, W. Oliver, E. A. Rickart, B. R. Tabaranza, Jr., and R. C. B. Utzurrum. 1998. A synopsis of the mammalian fauna of the Philippine islands. Fieldiana Zoology (new series) 88:1–61.

Heideman, P. D. 1987. The reproductive ecology of a community of Philippine fruit bats (Pteropodidae, Megachiroptera). Ph.D. dissertation, University of Michigan, Ann Arbor.

Heideman, P. D., and L. R. Heaney. 1989. Population biology and estimates of abundance of fruit bats (Pteropodidae) in Philippine submontane forest. Journal of Zoology (London) 218:565–586.

Ingle, N. R., and L. R. Heaney. 1992. A key to the bats of the Philippine islands. Fieldiana Zoology (new series) 69:1–44.

IUCN (International Union for Conservation of Nature and Natural Resources). 1991. Protected Areas of the World: A Review of National Systems. Vol. 1: Indomalaya, Oceania, Australia, and Antarctica. Island Press, Washington, D.C.

Kummer, D. M. 1990. Deforestation in the post-war Philippines. Ph.D. dissertation, Boston University, Boston.

Lande, R. 1988. Genetics and demography in biological conservation. Science 241:1455–1460.

Magurran, A. E. 1988. Ecological Diversity and Its Measurement. Princeton University Press, Princeton.

Mallari, N. A. D., and A. Jensen. 1993. Biological diversity in northern Sierra Madre, Philippines: Implication for conservation and management. Asia Life Sciences 2:101–102.

Mickleburgh, S. P., P. A. Racey, and A. M. Hutson, eds. 1992. Old World Fruit Bats: An Action Plan for the Family Pteropodidae. IUCN Press, Gland, Switzerland.

Myers, N. 1988. Environmental degradation and some economic consequences in the Philippines. Environmental Conservation 15:205–214.

Peterson, A. T., and L. R. Heaney. 1993. Genetic differentiation in Philippine bats of the genera Cynopterus and Haplonycteris. Biological Journal of the Linnean Society 49:203–218.

Rickart, E. R. 1993. Diversity patterns of mammals along elevational and diversity gradients in the Philippines: Implications for conservation. Asia Life Sciences 2:251–260.

Rickart, E. A., L. R. Heaney, and R. C. B. Utzurrum. 1991. Distribution and ecology of small mammals along an elevational transect in southeastern Luzon, Philippines. Journal of Mammalogy 72:458–469.

Rickart, E. A., L. R. Heaney, P. D. Heideman, and R. C. B. Utzurrum. 1993. The distribution and ecology of mammals on Leyte, Biliran, and Maripipi islands, Philippines. Fieldiana Zoology (new series) 72:1–62.

Ricklefs, R. E. 1979. Ecology, 2nd Ed. Chiron Press, New York.

Ruedas, L. A., J. R. Demboski, and R. V. Sison. 1994. Morphological and ecological variation in *Otopteropus cartilagonodus* Kock, 1969 (Mammalia: Chiroptera: Pteropodidae) from Luzon, Philippines. Proceedings of the Biological Society of Washington 107:1–16.

Terborgh, J. 1992. Maintenance of diversity in tropical forests. Biotropica 24:283–292.

Utzurrum, R. C. B. 1991. Philippine island biogeographic patterns: Practical applications for resource conservation and management. Association of Systematic Biologists of the Philippines Communications 3:19–32.

Utzurrum, R. C. B. 1992. Conservation status of Philippine fruit bats (Pteropodidae). Silliman Journal 36:27–46.

Vinciguerra, L. B., and R. A. Müller. 1993. Neue Erkenntnisse über die Verbreitung des Röhrennasen-Flughunds *Nyctimene rabori* Heaney & Peterson, 1984 auf den Philippinen (Mammalia: Pteropodidae). Jahrbuch des Naturhistorisches Museum der Stadt Bern 11:125–129.

Whitmore, T. C. 1984. Tropical Rain Forests of the Far East, 2nd Ed. Oxford University Press, Oxford.

Taxonomic Index

Acerodon, 334
    A. jubatus, 154, 334, 342, 345
    A. leucotis, 342, 345, 346
    A. lucifer, 329, 330, 334, 342, 343, 345, 346
Aethalops alecto, 333
Alionycteris paucidentata, 345, 346, 348
Ametrida, 142, 150
    A. centurio, 154, 290
Anoura
    A. caudifer, 154, 287, 290
    A. geoffroyi, 154, 290
Anthops, 27, 72–88
    A. ornatus, 28, 31, 32, 33, 35, 39
    Southern fossil record, 74–75, 77, 82
Antrozoinae, 3, 5, 6, 8, 9, 10, 11, 12
Antrozous
    A. dubiaquercus, 298, 302, 306
    A. pallidus: conservation, North America, 312, 313, 316; distribution, North America, 310, 311; muscle biology, 128, 135; teeth, 149, 154
Antrozous (Bauerus). See *Bauerus*
Apomys, 348
Archaeonycteris, 11, 13
Artibeus, 1, 59–69, 134, 147, 148
    A. amplus, 60
    A. anderseni, 290
    A. aztecus, 298
    A. cinereus, 290

A. concolor, 68, 69, 290. See also *Koopmania concolor*
A. fimbriatus, 60, 61, 63, 64, 66, 67, 290
A. fraterculus, 60, 61, 63, 64, 66, 67, 68
A. fulginosus, 68
A. glaucus, 290
A. hirsutus, 60, 62, 63, 64, 66, 68; conservation, Middle America, 298, 302, 304
A. inopinatus, 60, 62, 63, 64, 66, 68; conservation, Middle America, 302, 306; habitat, 300; in Middle America, 298
A. intermedius, 60, 62, 63, 64, 66, 67
A. jamaicensis, 60, 62, 63, 64, 66, 67, 68; distribution, Brazil, 290; food search and location, 189; loss of habitat, 287; muscle biology, 128, 131, 135, 136; teeth, 149, 154; wing membrane, 122
A. lituratus, 44, 45, 46, 54, 60, 61, 62, 63, 64, 66, 67; cladograms, 47–49; conservation, Brazil, 286, 287, 288; distribution, Brazil, 290; loss of habitat, 287, 288; muscle biology, 132, 135, 136; teeth, 154
A. obscurus, 60, 62, 63, 64, 66, 67, 68, 290
A. phaeotis, 154
A. planirostris, 60, 62, 63, 64, 65, 66, 68, 291
A. toltecus, 154, 298
Asellia, 27, 35
    A. patrizii, 28, 31, 32, 33, 39
    A. tridens, 28, 31, 32, 33, 37, 39; echoloca-

tion call frequency, 170; Southern fossil record, 74, 75, 77, 78, 82
    Southern fossil record, 72–88
Aselliscus, 27, 35–36
    A. stoliczkanus, 28, 31, 32, 33, 37, 39
    A. tricuspidatus, 28, 31, 32, 33, 36, 39; Southern fossil record, 74, 75, 77, 78, 82, 83
    Southern fossil record, 72–88
Aspidontus, 165
Australian flying-fox. See *Pteropus scapulatus*
Australonycteris clarkae, 83

Balantiopteryx io, 298, 302, 306
Bauerus, 297, 298
bentwing bat, large. See *Miniopterus schreibersi*
big brown bat. See *Eptesicus fuscus*
big-eared bat, Townsend's. See *Corynorhinus townsendii*
Brachipposideros
    B. collongensis, 79
    B. khengkao, 80
    B. nooraleebus, 35, 74, 75, 77, 78, 82, 84
    Southern fossil record, 72–88
Brachyphylla, 145
    B. cavernarum, 154
    B. nana, 154
brown long-eared bat. See *Plecotus auritas*
bulldog bats. See *Myotis (Leuconoe); Myotis (Pizonyx)*
    greater. See *Noctilio leporinus*

California leaf-nosed bat. See *Macrotus califor-*
  nicus
Cardioderma
  C. cor, 30, 154
  teeth, 145
Carollia, 1, 43–57
  C. brevicauda, 43, 44, 45, 46, 50, 51, 52–53,
    55–56; cladograms, 47–49, 50, 52; distri-
    bution, Brazil, 290
  C. castanea, 43, 44, 45, 46, 50, 51, 52, 53, 55;
    cladograms, 47–49, 50, 52; distribution,
    Brazil, 290
  C. perspicillata, 43, 44, 45, 46, 50, 51, 52, 53,
    56, 154; cladograms, 47–49, 50, 52;
    cochlear adaptations, 233, 240; distribu-
    tion, Brazil, 290; loss of habitat, 287,
    288; muscle biology, 136
  C. subrufa, 43, 46, 50, 51, 52, 53, 56;
    cladograms, 47–49, 50, 52; in Middle
    America, 298
  muscle biology, 134
  seed dispersal by, 299
Carolliinae, phylogeny, 43
Centronycteris maximiliani, 289
Centurio
  C. senex, 154
  skull and teeth, 141, 147, 153
Chaerephon
  C. bivitatta, 268
  C. chapini, 265–266, 268
  C. jobensis, 327–328
  C. nigeriae, 268
  C. pumila, 266, 267, 268
Chalinolobus
  C. dwyeri, 273, 276
  C. gouldii, 277
  C. morio (cave form), 273
  C. picatus, 273
Cheiromeles torquatus, 154
Chiroderma, 61
  C. doriae, 287, 291
  C. trinitatum, 291
  C. villosum, 62, 154, 291
  teeth, 150
Chiroptera
  conservation: African woodlands, 261–269;
    Australia, 271–279; Brazil, 282–292;
    Europe, 249–257; Indo-Pacific Islands,
    326–337; Middle America, 295–306;
    Philippines, 342–352
  hindlimb morphology and blood feeding,
    157–165
  muscle biology, 127–137
  in North America, 309–320
  skull and teeth, 140–154
Choeroniscus
  C. godmani, 154
  C. intermedius, 154, 290
  C. minor, 290
  teeth, 146
Choeronycteris
  C. mexicana, 154, 300, 302, 310, 311
  teeth, 146
Chrotomys, 348
Chrotopterus
  C. auritus, 301; distribution, Brazil, 290; loss

of habitat, 287; omnivory, 164; teeth,
    154; wings, 105
  teeth, 151–152
Cloeotis, 27
  C. percivali, 28, 31, 32, 33, 37, 39; habitat
    associations, 268; Southern fossil record,
    74, 75, 77, 78, 82
  Southern fossil record, 72–88
Coelops, 27, 35, 72–88
  C. frithi, 28, 31, 32, 33, 37, 39; Southern fos-
    sil record, 74, 75, 77, 78, 82
  C. robinsoni, 28, 31, 32, 33, 39
Coleura seychellensis, 264, 335
Cormura brevirostris, 289
Corynorhinus
  C. rafinesquii, 310, 311, 315
  C. townsendii: conservation, North Amer-
    ica, 312, 313, 314, 316–317, 319; distribu-
    tion, North America, 310, 311; ecosys-
    tem role of, 317; flight behavior, 317,
    318
Craseonycteridae, 4, 5, 6, 8, 9, 11, 12
Cynopterus, 152, 154, 332
  C. brachyotis: conservation, Philippines,
    349, 350; distribution, Philippines, 345,
    346, 347, 350; systematics and molecular
    genetics, 332–333
  conspicillatus, 333
Cyttarops alecto, 289

Daubenton's bat. See *Myotis daubentonii*
Dermanura, 59–69
  D. anderseni, 60, 62, 63, 64, 66
  D. azteca, 60, 62, 63, 64, 66, 67, 69
  D. cinerea, 60, 62, 63, 64, 65, 66
  D. concolor, 60
  D. glauca, 60, 62, 63, 64, 66, 69
  D. gnoma, 60, 62, 63, 64, 66, 69
  D. phaeotis, 60, 62, 63, 64, 66, 69
  D. tolteca, 60, 62, 63, 64, 66, 69
  D. watsoni, 60, 62, 63, 64, 66, 69
Dermoptera, 10, 158
Desmodontinae, 159, 202
Desmodus
  D. rotundus: conservation, Brazil, 286, 287;
    distribution, Brazil, 291; food search and
    location, 202, 301; hindlimb morphol-
    ogy and blood feeding, 159, 160–161,
    162, 164, 165; muscle biology, 132, 133,
    134, 135–136, 137; roost, 301; teeth, 154;
    wing, 127
  muscle biology, 136
  teeth, 145, 153
Diaemus
  D. youngi: distribution, Brazil, 291; food
    search and location, 202, 301; hindlimb
    morphology and blood feeding, 159,
    160–161, 162, 164, 165; muscle biology,
    135, 136
  muscle biology, 134
Diclidurus, 191
  D. albus, 289
  D. ingens, 289
  D. isabellus, 289
  D. scutatus, 289
Didelphis virginiana, 327

Diphylla
  D. ecaudata: distribution, Brazil, 291; distri-
    bution, North America, 310, 311; food
    search and location, 202, 301; hindlimb
    morphology and blood feeding, 159,
    160–161, 162, 164, 165; teeth, 154
  teeth, 153
disk-winged bats. See *Thyroptera*
Dobsonia, 152
  D. chapmani, 329, 330, 342, 345, 346
  D. moluccensis, 154, 273, 278
dog-faced bat. See *Molossops planirostris*
Dyacopterus spadiceus, 345

eastern horseshoe bat. See *Rhinolophus mega-*
  phyllus
Ectophylla, 142, 150
  E. alba, 154, 300, 302
Eidolon helvum, 154, 264
elegant myotis. See *Myotis elegans*
Emballonura semicaudata, 327, 328, 332
Emballonuridae, 4, 5, 6, 8, 9, 11, 12, 80, 81, 83
  conservation, Middle America, 297, 298
  distribution, Brazil, 289
  diversity, Brazil, 284
  food search and location, 191, 199
  habitat loss, 275
  roosts, 276
Emballonuroidea, 5, 11
Enchisthenes, 60–68
  E. hartii, 60, 62
Eonycteris
  E. major, 154
  E. robusta, 342, 345
  E. spelaea, 154, 345, 346, 347, 349, 350
epauletted bat. See *Sturnira*
Epomophorus, 268
  E. wahlbergi, 263
Epomops buettikoferi, 154
Eptesicus
  E. brasiliensis, 291
  E. capensis, 263
  E. diminutus, 287, 291
  E. furinalis, 287, 291
  E. fuscus: cochlear adaptations, 231, 232,
    233, 236, 237, 238; conservation, North
    America, 314, 315; cortical computa-
    tional strategies, 205–218, 238; distribu-
    tion, North America, 310, 311; ecosys-
    tem role of, 318; food search and
    location, 192, 197–198; muscle biology,
    128, 129, 135; sensorimotor integration
    in sonar, 220–229; skeletal properties,
    119; wing membrane, 122
  E. nilssoni, 104, 251, 252
  E. serotinus, 154, 250, 252, 255–256
  food search and location, 192
  habitat associations, 268
  taxonomy, 264
  wings, 105
Erophylla, 150
  E. sezekorni, 154
Euderma maculatum, 310, 311, 312, 317, 318,
  319
Eumops
  E. auripendulus, 292

E. bonariensis, 292

E. glaucinus, 292, 310, 311, 313

E. hansai, 292

E. perotis: conservation, North America, 313, 318–319; distribution, Brazil, 292; distribution, North America, 310, 311; flight behavior, 318; hindlimb morphology and blood feeding, 158; musculoskeletal system, 127; teeth, 154

E. underwoodi, 154, 310, 311

evening bats. See *Eptesicus*; *Lasiurus*; *Nyctalus*; *Pipistrellus*; Vespertilionidae

false vampire bats. See *Chrotopterus auritus*; *Vampyrum spectrum*
　Indian. See *Megaderma lyra*

Falsistrellus mckenziei, 273

fig-eating bats. See *Arbiteus*

Findley's black myotis. See *Myotis findleyi*

fish-eating bat. See *Myotis vivesi*

flat-headed myotis. See *Myotis planiceps*

flying-foxes. See *Pteropus*
　Australian. See *Pteropus scapulatus*
　gray-headed. See *Pteropus poliocephalus*

free-tailed bats. *See* Molossidae
　Mexican. See *Tadarida brasiliensis*

frog-eating fringe-lipped bat. See *Trachops cirrhosus*

fruit-eating bats
　hairy. See *Artibeus hirsutus*
　highland. See *Artibeus aztecus*

Furipteridae, 4, 5, 6, 8, 9, 10, 11, 12
　conservation, Middle America, 297, 298
　distribution, Brazil, 291

Furipterus horrens, 291

Genoways' little yellow bat. See *Rhogeessa genowaysi*

ghost-faced bats. See *Macroderma gigas*; Megadermatidae

Glossophaga
　G. commissarisi, 290
　G. leachii, 298
　G. longirostris, 154, 290
　G. morenoi, 298
　G. soricina, 44, 45, 46, 47–49, 54, 154, 290

Glossophaginae, 52, 202, 287

Gould's wattled bat. See *Chalinolobus gouldii*

gray bat. See *Myotis grisescens*

gray-headed flying-fox. See *Pteropus poliocephalus*

greater bulldog bat. See *Noctilio leporinus*

greater horseshoe bat. See *Rhinolophus ferrumequinum*

Guatemalan myotis. See *Myotis cobanensis*

hairy fruit-eating bat. See *Artibeus hirsutus*

hairy-legged vampire bat. See *Diphylla ecaudata*

hairy-tailed bat. See *Lasiurus castaneus*

Haplonycteris, 345, 346, 348
　H. fischeri, 332, 333, 345, 347, 349
　H. sp., 345

Harpyionycteris whiteheadi, 154, 345, 347, 350

Hassianycteris, 11, 13

Hawaiian hoary bat. See *Lasiurus cinereus semotus*

highland fruit-eating bat. See *Artibeus aztecus*

Hipposideridae, 4, 5, 6, 8, 9, 10, 11, 12
　food search and location, 193, 198
　phylogeny, 27–40, 72–88
　roosts, 276
　Southern origin for the, 72–88
　wings, 105

Hipposideros, 1, 27, 35–38
　cochlear adaptations, 232, 234
　echolocation call frequency, 169–182
　H. abae: phylogeny, 28, 31, 32, 33, 39; Southern fossil record, 85
　H. armiger: phylogeny, 28, 31, 32, 33, 35, 37, 39; Southern fossil record, 74, 75, 77, 79, 82, 85
　H. ater: echolocation call frequency, 173; phylogeny, 28, 31, 32, 33, 37, 39; Southern fossil record, 74, 75, 77, 82, 84
　H. beatus: phylogeny, 28, 31, 32, 33, 39; Southern fossil record, 84
　H. bernardsigei, 73, 74, 75, 77, 79, 82, 83, 84
　H. bicolor, 72; echolocation call frequency, 173; phylogeny, 28, 31, 32, 33, 35, 36, 37, 39; Southern fossil record, 79, 83, 84
　H. breviceps, 84
　H. caffer: habitat associations, 268; insects in diet, 263; phylogeny, 28, 31, 32, 33, 37, 39; Southern fossil record, 74, 75, 77, 79, 82, 84
　H. calcaratus: phylogeny, 28, 31, 32, 33, 39; Southern fossil record, 74, 75, 77, 78, 81, 82, 83, 84
　H. camerunensis, 28, 31, 32, 33, 39, 84
　H. cervinus: conservation problems and resolution, 273; echolocation call frequency, 171, 172, 173, 174, 175–176, 178; phylogeny, 28, 31, 32, 33, 37, 39; Southern fossil record, 74, 75, 77, 82, 84
　H. cervinus labuanensis, 175–176
　H. cineraceus: echolocation call frequency, 173; phylogeny, 28, 31, 32, 33, 37, 39; Southern fossil record, 84
　H. commersoni: phylogeny, 28, 31, 32, 33, 34, 35, 36, 37, 39; Southern fossil record, 74, 75, 77, 79, 82, 84, 85
　H. commersoni commersoni, 154
　H. commersoni gigas, 154
　H. coronatus, 84
　H. corynophyllus: phylogeny, 28, 31, 32, 33, 39; Southern fossil record, 74, 75, 77, 79, 82, 83, 84
　H. coxi, 84
　H. crumeniferus, 84
　H. curtus, 28, 31, 32, 33, 39, 84
　H. cyclops: phylogeny, 28, 31, 32, 33, 39; Southern fossil record, 72, 74, 75, 77, 78, 79, 82, 84
　H. demissus, 83, 85
　H. diadema: conservation problems and resolution, 273; echolocation call frequency, 173; phylogeny, 28, 31, 32, 33, 35, 37, 39; Southern fossil record, 72, 74, 75, 77, 78, 79, 82, 83, 85

H. dinops, 28, 31, 32, 33, 34, 39, 85

H. doriae, 84

H. dyacorum: echolocation call frequency, 173; phylogeny, 28, 31, 32, 33, 39; Southern fossil record, 84

H. edwardshilli, 79, 83, 85

H. felix, 80

H. fuliginosus, 28, 31, 32, 33, 39, 84

H. fulvus: phylogeny, 28, 31, 32, 33, 36, 37, 39; Southern fossil record, 74, 75, 77, 79, 82, 84

H. galeritus: echolocation call frequency, 171, 173, 176, 178; phylogeny, 28, 31, 32, 33, 39; Southern fossil record, 79, 84

H. galeritus galeritus, 176, 178

H. galeritus insolens, 176

H. halophyllus, 28, 31, 32, 33, 39, 84

H. hypophyllus, 37

H. inexpectatus, 28, 31, 32, 33, 34, 39, 85

H. inornatus form, 273

H. jonesi, 28, 31, 32, 33, 39, 84

H. lamottei, 84

H. lankadiva: phylogeny, 28, 31, 32, 33, 34, 37, 39; Southern fossil record, 74, 75, 77, 79, 82, 85; teeth, 154

H. larvatus: echolocation call frequency, 173; phylogeny, 28, 31, 32, 33, 37, 39; Southern fossil record, 74, 75, 77, 79, 82, 85

H. lekaguli, 28, 31, 32, 33, 37, 39, 85

H. lylei: phylogeny, 28, 31, 32, 33, 39; Southern fossil record, 74, 75, 77, 78, 81

H. macrobullatus, 28, 31, 32, 33, 39, 84

H. maggietaylorae: phylogeny, 28, 31, 32, 33, 39; Southern fossil record, 74, 75, 77, 81, 82, 83, 84

H. marisae, 28, 31, 32, 33, 39, 84

H. megalotis: phylogeny, 28, 31, 32, 33, 34, 35, 39; Southern fossil record, 73, 77, 79

H. muscinus: phylogeny, 28, 31, 32, 33, 36, 39; Southern fossil record, 74, 75, 77, 79, 81, 82, 83, 84

H. naquam, 84

H. nooraleebus, 73. See also *Brachipposideros nooraleebus*

H. obscurus, 28, 31, 32, 33, 40, 84

H. papua: phylogeny, 28, 31, 32, 33, 40; Southern fossil record, 74, 75, 77, 78, 82, 83, 84

H. phymaeus, 84

H. pomona: echolocation call frequency, 173; phylogeny, 28, 31, 32, 33, 40; Southern fossil record, 84

H. pratti: phylogeny, 28, 31, 32, 33, 35, 37, 40; Southern fossil record, 72, 74, 75, 77, 81, 82, 85; teeth, 154

H. pygmaeus, 28, 31, 32, 33, 40

H. ridleyi: echolocation call frequency, 171, 173, 175; phylogeny, 28, 31, 32, 33, 40; Southern fossil record, 84

H. ruber: phylogeny, 28, 31, 32, 33, 40; Southern fossil record, 74, 75, 77, 79, 82, 84; teeth, 154

H. sabanus, 28, 31, 32, 33, 40, 84

H. schistaceus, 28, 85

Hipposideros (continued)

H. semoni: conservation problems and resolution, 273; phylogeny, 28, 31, 32, 33, 36, 40; Southern fossil record, 74, 75, 77, 78, 79, 82, 84, 85

H. species A, echolocation call frequency, 173

H. species B, echolocation call frequency, 173

H. speoris: cochlear adaptations, 240; muscle biology, 131; phylogeny, 28, 31, 32, 33, 37, 40; Southern fossil record, 74, 75, 77, 78, 79, 82, 85; wings, 105

H. stenotis: conservation problems and resolution, 273; phylogeny, 28, 31, 32, 33, 40; Southern fossil record, 80, 83, 84, 85

H. terasensis, 28, 31, 32, 33, 37, 40

H. turpis: phylogeny, 28, 31, 32, 33, 37, 40; Southern fossil record, 74, 75, 77, 82, 85

H. wollastoni: phylogeny, 28, 31, 32, 33, 40; Southern fossil record, 74, 75, 77, 79, 82, 83, 84

Southern fossil record, 72–88

Hipposideros (Brachipposideros), 72, 79

Hipposideros (Pseudorhinolophus), 35, 72

Histiotus

H. alienus, 291

H. montanus, 291

H. velatus, 287, 291

hoary bat. See *Lasiurus cinereus*

hog-nosed bat. See *Choeronycteris mexicana*

Honduran fruit-eating bat. See *Artibeus inopinatus*

Hornovits tsuwape, 13

horseshoe bats. See *Rhinolophus*

eastern. See *Rhinolophus megaphyllus*

greater. See *Rhinolophus ferrumequinum*

lesser. See *Rhinolophus hipposideros*

Hyaenodon, 151

Hyaenodontinae, 151

Hylonycteris, 297

H. underwoodi, 154, 298, 300, 302

Ia io, 154

Icaronycteris, 11, 13

I. index, 13

Idionycteris phyllotis, 310, 311, 319

Indian false vampire bat. See *Megaderma lyra*

Kerivoula, 268

K. papuensis, 273

K. picta, 73

Kerivoulinae, 3, 5, 6, 8, 10, 11, 12

Koopmania, 59–69, 63, 64, 65, 66

K. concolor, 62, 67, 68. See also *Arbiteus concolor*

Labroides, 165

Laephotis botswannae, 268

large bentwing bat. See *Miniopterus schreibersi*

large-eared pied bat. See *Chalinolobus dwyeri*

Lasionycteris, 105

L. noctivagans, 301, 310, 315, 318

Lasiurus, 45

food search and location, 191

L. blossevillii, 310, 311, 313

L. borealis, 318; distribution, Brazil, 291; distribution, North America, 310, 311; extirpation, Indo-Pacific islands, 328; loss of habitat, 287; taxonomy, 313; teeth, 154

L. castaneus, 300, 302

L. cinereus: distribution, Brazil, 291; distribution, North America, 310, 311; flight behavior, 318; teeth, 154; wing membrane, 122

L. cinereus semotus, 312

L. ebenus, 287, 291

L. ega, 291, 310, 311, 313

L. egregius, 287, 291, 302

L. intermedius, 310, 311

L. seminolus, 310, 311

L. xanthinus, 310, 311, 313

Lavia frons, 73, 263

leaf-chinned bats. See Mormoopidae

leaf-nosed bats. See *Arbiteus; Dermanura; Koopmania*

New World. See Phyllostomidae

Old World. See Hipposideridae; Rhinolophidae

Leptonycteris

conservation, North America, 312

L. curasoae: conservation, North America, 314, 315; distribution, North America, 310, 311; flight behavior, 318; food search and location, 193, 194; pollination by, 299; teeth, 154

L. nivalis, 300, 302, 304, 310, 311, 314

lesser horseshoe bat. See *Rhinolophus hipposideros*

Lichonycteris obscura, 154, 287

Limnocyoninae, 151

Linchonycteris obscura, 290

Lionycteris spurrelli, 154, 290

Lonchophylla

L. bokermanni, 287, 290

L. dekeyseri, 290

L. handleyi, 154

L. morax, 290

L. thomasi, 154, 290

Lonchophyllinae, 202

Lonchorhina, 142, 150

L. aurita, 154, 290

long-eared bat, brown. See *Plecotus auritus*

long-eared myotis. See *Myotis evotis*

long-nosed bats. See *Leptonycteris*

Lord Howe Island bat. See *Nyctophilus howensis*

Machaeroidinae, 151

Macroderma, 83

M. gigas: conservation, 273, 279; habitat loss, 276–277; teeth, 149, 150, 151, 154; wings, 104

teeth, 151, 152

Macroglossinae, 333

Macroglossus

M. minimus, 278, 345, 347, 350

M. sobrinus, 154

Macrophyllum macrophyllum, 290

Macrotus, 161

M. californicus: conservation, North America, 313; distribution, 300, 310, 311; hind-

limb morphology and blood feeding, 158; musculoskeletal system, 127; teeth, 154

omnivory, 164

Megachiroptera

hindlimb morphology and blood feeding, 158

skull and teeth, 140–154

wings, 99

Megaderma, 83

M. lyra, 30; cochlear adaptations, 231, 233, 236, 237, 238; teeth, 154

M. spasma, 30, 73

Megadermatidae, 4, 5, 6, 8, 9, 11, 12

food search and location, 193

in Hipposideridae phylogeny, 30, 81

roosts, 276

Megaerops wetmorei, 345

Megaloglossus woermanni, 154

Melonycteris melanops, 154

Mesophylla macconnelli, 291

Mexican free-tailed bat. See *Tadarida brasiliensis*

Mexican long-nosed bat. See *Leptonycteris nivalis*

Microchiroptera

food search and location, 183–195

hindlimb morphology and blood feeding, 158

skull and teeth, 140–154

Southern origin for, 72–88

Micronycteris

M. behnii, 290

M. brachyotis, 290

M. hirsuta, 290

M. megalotis, 154, 287, 290

M. minuta, 290

M. nicefori, 290

M. pusilla, 290

M. schmidtorum, 290

M. sylvestris, 290

omnivory, 164

Miller's myotis. See *Myotis milleri*

Mimetillus moloneyi, 104

Mimon

M. bennettii, 154, 290

M. crenulatum, 290

miniature yellow bat. See *Rhogeessa mira*

Miniopterinae, 3, 5, 6, 8, 10

Miniopterus

migration, 264

M. schreibersi, 131, 263, 276, 277

Miophyllorhina, 73, 81

Molossidae, 4, 5, 6, 8, 9, 11, 12, 80

conservation, Middle America, 298

distribution, 292, 310

diversity, Brazil, 284

food search and location, 191, 194, 199, 200

habitat loss, 275

muscle biology, 136

roosts, 276

teeth, 140

Molossops

M. abrasus, 292

M. ater, 292

M. greenhalli, 292

M. matogrossensis, 292
M. molossus, 292
M. neglectus, 292
M. planirostris, 292, 301
M. temminckii, 292
Molossus
    M. ater, 135, 286
    M. molossus, 154
    M. molossus aztecus, 297
Monophyllus
    M. plethodon, 154
    M. redmani, 149, 152, 154
Mops
    M. condylura, 268
    M. midas, 263, 268
Mormoopidae, 4, 5, 6, 8, 9, 11, 12
    conservation, Middle America, 298
    distribution, 289, 310, 311
    food search and location, 192, 193, 199
    muscle biology, 136, 137
Mormoops, 142, 150
    M. blainvilli, 233
    M. megalophylla, 135, 154, 310, 311
Mormopterus setirostra form, 273
Mosia nigrescens, 73
mouse-eared bats. See *Myotis*
mouse-tailed bats. See Rhinopomatidae
Murina
    M. florium, 273
    M. sp. from Iron Range, 273
Murininae, 3, 5, 6, 8, 9, 11, 12
Musonycteris, 141, 146, 152, 297
    M. harrisoni, 154, 298, 300, 302, 304
mustached bat, New World. See *Pteronotus parnellii*
Myotinae, 5, 6, 8, 9, 10, 11, 12
Myotis
    conservation ecology in Europe, 253
    food search and location, 194
    habitat, 250, 251
    hindlimb morphology and blood feeding, 158
    M. albescens, 291
    M. auriculus, 310, 311
    M. austroriparius, 310, 311, 312, 313, 315, 316
    M. brandti, 251, 256
    M. californicus, 310, 311
    M. ciliolabrum, 310, 311
    M. cobanensis, 298, 302
    M. daubentonii, 251, 256
    M. elegans, 298, 302, 306
    M. emarginatus, 253
    M. evotis: conservation, North America, 313, 316; distribution, North America, 310, 311; ecosystem role of, 317; flight behavior, 317, 318
    M. findleyi, 298, 302, 306
    M. fortidens, 298
    M. grisescens: conservation, North America, 312, 313, 314, 316; distribution, North America, 310, 311; flight behavior, 318
    M. keenii, 310, 311, 313, 319
    M. leibii, 310, 311
    M. levis, 291

M. lucifugus: conservation, North America, 312, 314, 315, 316; cortical computational strategies, 205; distribution, North America, 310, 311; ecosystem role of, 317, 318; muscle biology, 129, 131, 133–134, 135, 136; skeletal properties, 119; taxonomy, 313; wing membrane, 122
M. macrodactylus, 129
M. milleri, 298, 300, 302, 306
M. myotis, 154, 189, 194
M. mystacinus, 256
M. nattereri, 251
M. nigricans, 287, 288, 291
M. nigricans carteri, 297
M. occultus, 310, 311, 313
M. peninsularis, 298, 302, 306
M. planiceps, 298, 302, 306
M. riparius, 291
M. ruber, 287, 291
M. septentrionalis, 310, 311, 313
M. simus, 291
M. sp. from Kimberley, 273
M. sodalis: conservation, North America, 312, 313, 314, 316; distribution, North America, 310, 311
M. thysanodes, 310, 311, 313
M. velifer, 127, 154, 158, 310, 311, 316
M. vivesi, 298, 300, 301, 302, 306
M. volans, 310, 311, 317
M. yumanensis: conservation, North America, 312, 315, 316; distribution, North America, 310, 311; ecosystem role of, 318; flight behavior, 318
Myotis (Leuconoe), 201
Myotis (Pizonyx), 201
    M. vivesi, 159
Mystacina robusta, 334
Mystacinidae, 4, 5, 6, 8, 9, 11, 12, 13, 83
Myzopodidae, 4, 5, 6, 8, 9, 11, 12, 13

Natalidae, 4, 5, 6, 8, 10
    conservation, Middle America, 297, 298
    distribution, Brazil, 291
    wings, 105
Natalus stramineus, 291
Necromantis, 80
    N. adichaster, 73
New World leaf-nosed bats. See Phyllostomidae
New World mustached bat. See *Pteronotus parnellii*
Noctilio, 161
    N. albiventris, 289
    N. leporinus: diet, 301; distribution, Brazil, 289; food search and location, 189, 193, 195, 201, 202; hindlimb morphology and blood feeding, 159; teeth, 154
Noctilionidae, 4, 5, 6, 8, 9, 11, 12
    conservation, Middle America, 297, 298
    distribution, Brazil, 289
    food search and location, 199, 201
    wings, 105
Noctilionoidea, 5, 7, 11, 83
noctule bat. See *Nyctalus noctula*
Notonycteris, 164

N. magdalenensis, 51
Notopteris macdonaldi, 154, 327, 328
Nyctalus, 191
    N. lasiopterus, 154
    N. leisleri, 253
    N. noctula: food search and location, 189, 192; foraging behavior, 252, 253; habitat, 250, 251, 252; roosts, 253; use of bat boxes, 255; wings, 104
Nycteridae, 4, 5, 6, 8, 9, 11, 12
    food search and location. See Megadermatidae
    in Hipposideridae phylogeny, 30, 81, 83
    wings, 105
Nycteris, 145
    N. grandis, 30, 154, 263, 265, 267
    N. hispida, 30
    N. thebaica, 30, 263, 264, 268
Nycticeinops schlieffeni, 268
Nycticeius
    N. humeralis, 310, 311
    N. schlieffeni, 263
Nyctimene, 279
    cf. *N. cephalotes*, 278
    N. cephalotes form, 273
    N. draconilla, 154
    N. major, 154
    N. rabori, 342, 345, 346, 347, 349, 350
    N. robinsoni, 278
    N. sanctacrucis, 329
    N. vizcaccia, 273, 278
Nyctinomops, 310
    N. aurispinosus, 292
    N. femorosaccus, 311
    N. laticaudatus, 292
    N. macrotis, 292, 310, 311
Nyctophilinae, 3, 5, 6, 7, 9, 11, 12
Nyctophilus, 105
    N. daedelus form, 273
    N. howensis, 272, 277
    N. timoriensis (central form), 273
    N. timoriensis (eastern form), 273
    N. timoriensis (western form), 273

Old World leaf-nosed bats. See Hipposideridae; Rhinolophidae
Otomops martiensseni, 104, 154
Otonycteris hemprichi, 154
Otopteropus cartilagonodus, 345, 346

Palaeochiropteryx, 11, 13
Palaeophyllophora, 72–88
    P. quercyi, 73, 74, 75, 77, 82
Paracoelops, 27, 72–88
Paranyctimene raptor, 154
peninsular bat. See *Myotis peninsularis*
Peropteryx, 191
    P. kappleri, 154, 289
    P. leucoptera, 289
    P. macrotis, 289
Phylloderma stenops, 154, 290
Phyllonycteris, 150
    P. poeyi, 154
Phyllostomidae, 4, 5, 6, 8, 9, 11, 12
    conservation, 286, 297, 298, 302
    distribution, 290–291, 310, 311

Phyllostomidae (continued)
diversity, Brazil, 284
food search and location, 187, 193, 201–202
hindlimb morphology and blood feeding, 158, 159, 161, 164–165
muscle biology, 131, 136, 137
phylogeny, 37; of Artibeus, Dermanura, and Koopmania, 59–69; of Carollia, 43–54
skull and teeth, 141, 143, 145
wings, 105
Phyllostominae, 43, 44, 164
Phyllostomus
P. discolor, 154, 290
P. elongatus, 154, 290
P. hastatus: distribution, Brazil, 290; food search and location, 191; loss of habitat, 287; muscle biology, 131; omnivory, 164; teeth, 154; wings, 104
Pipistrellus, 105, 189, 192, 200, 264, 268
P. abramus, 129
P. capensis, 268
P. hesperus, 310, 311
P. kuhlii, 189, 252
P. murrayi, 334, 335
P. nathusii, 252, 256
P. pipistrellus, 256, 264; food search and location, 194, 198; foraging behavior, 252; habitat, 250, 251, 252; roosts, 253; wings, 104
P. quadridens, 189
P. subflavus, 310, 311
P. tasmaniensis, 73
P. tenuis, 334
P. zuluensis, 263
Platyrrhinus
P. brachycephalus, 291
P. helleri, 291
P. infuscus, 291
P. lineatus, 287, 288, 291
P. recifinus, 287, 291
Plecotus, 253
P. auritus, 98, 104, 127, 250, 251, 254–255
P. austriacus, 256
P. mexicanus, 298
Promops nasutus, 292
Pseudorhinolophus schlosseri, 79
Ptenochirus
P. jagori, 345, 347, 349, 350
P. minor, 345, 346, 349
Pteralopex, 329–330
P. acrodonta, 330
P. atrata, 330
P. pulchra, 330
Pterodon, 151
Pteronotus, 161
P. gymnonotus, 289
P. parnellii, 193, 289; cochlear adaptations, 231–242; cortical computational strategies, 205, 206, 207; muscle biology, 131, 135, 136; teeth, 154; wing membrane, 122
P. personatus, 289
P. quadridens, 232, 233, 236, 237, 238
Pteropodidae, 6, 8, 80, 81, 105, 276
Pteropus
conservation, 272, 277, 278, 333–337

food search and location, 202
morphological evolution, 327
muscle type, 94
P. admiralitatum, 329
P. alecto, 278
P. brunneus, 273, 278
P. caniceps, 329
P. conspicillatus, 272, 273, 278, 333
P. dasymallus, 328, 329, 330, 345
P. dasymallus formosus, 329, 330
P. giganteus ariel, 330, 336
P. howensis, 336
P. hypomelanus, 329, 345
P. insularis, 336
P. leucopterus, 342, 345, 346
P. livingstonii, 330–331
P. macrotis, 273, 278
P. mariannus, 328, 331–332, 333, 334–335, 336
P. mariannus mariannus, 330, 331
P. mariannus natalis, 335
P. mariannus pelewensis, 329, 330, 334
P. mariannus ualanus, 335, 336
P. mariannus ulthiensis, 336
P. melanotus natalis, 334
P. molossinus, 336
P. neohibernicus hilli, 335
P. niger, 328, 329
P. ornatus, 329, 335
P. phaeocephalus, 336
P. pilosus, 329, 330
P. poliocephalus: conservation, 273, 278; habitat loss, 275; predation by humans, 275–276; skeleton, 110–116, 119; systematics and molecular genetics, 333; teeth, 154
P. pumilus, 345; conservation, Philippines, 349; distribution, Philippines, 345, 346, 347, 350
P. rayneri grandis, 335
predation by humans, 275–276
P. rodricensis, 328
P. samoensis, 327, 328, 331, 334, 335
P. scapulatus, 73, 154, 276, 278, 333
P. seychellensis, 264
P. seychellensis comorensis, 331
P. sp., 345
P. sp. nov., 273, 278
P. speciosus, 345
P. subniger, 328, 329
P. tokudae, 329, 330
P. tonganus, 336; conservation, Indo-Pacific islands, 330, 331, 334, 335; exterpation, Indo-Pacific islands, 327, 328
P. vampyrus, 154, 336, 345
P. vetulus, 328, 329
wings, 103
Pygoderma, 145, 153, 291
P. bilabiatum, 154, 287

Rhinolophidae, 4, 5, 6, 8, 9, 10, 11, 12, 105
food search and location, 193, 198, 199
in Hipposideridae phylogeny, 28, 30, 31, 34, 36–37, 38, 75, 77, 80, 84
roosts, 276
Rhinolophina, 80

Rhinolophoidea, 5, 7, 11, 81, 83, 241
Rhinolophus, 35, 36, 37, 72–88
cf. R. borneensis, 173
cf. R. pusillus, 173
cochlear adaptations, 232, 234, 236
echolocation call frequency, 169–182
food search and location, 193
habitat associations, 268
R. acuminatus, 37, 173
R. affinis, 74, 75, 77, 173
R. blasii, 154
R. borneensis, 173
R. celebensis, 29, 30
R. clivosus, 29, 30
R. coelophyllus, 30
R. creaghi, 171, 172, 173, 174, 177, 178
R. ferrumequinum, 189, 253–254
R. hildebrandti, 267
R. hipposideros, 29, 30, 170, 178, 254
R. lepidus, 173
R. luctus, 30, 36, 154
R. macros form, 273
R. macrotis, 173
R. malayanus, 29, 30, 173
R. megaphyllus, 74, 75, 77, 83, 276
R. philippinensis, 173, 175, 273
R. priscus, 80
R. robinsoni, 173
R. rouxi: cochlear adaptations, 231–242; cortical computational strategies, 205, 206, 207, 238; echolocation call frequency, 170, 178
R. rufus, 154
R. sedulus, 173
R. shameli, 173
R. simulator, 74, 75, 77
R. sinicus, 29, 30
R. species A, echolocation call frequency, 173
R. species B, echolocation call frequency, 173
R. stheno, 173
R. thomasi, 173, 178
R. trifoliatus, 173
Rhinonycterina, 80
Rhinonycteris, 27
R. aurantius, 27, 31, 32, 33, 35, 40; Southern fossil record, 73, 74, 75, 77, 78, 82, 83
R. tedfordi, 73, 74, 75, 77, 78, 82
Southern fossil record, 72–88
Rhinophylla, 43
cladogram, 52
R. alethina, 52
R. fischerae, 44, 45, 46, 53, 55; cladograms, 47–49; distribution, Brazil, 290
R. pumilio, 44, 45, 46, 54, 55; cladograms, 47–49; distribution, Brazil, 290
Rhinopoma
R. hardwickei, 73
R. microphyllum, 30
Rhinopomatidae, 4, 5, 6, 8, 9, 10, 11, 12
food search and location, 191
in Hipposideridae regression analysis, 30
Rhogeessa
R. aeneus, 298
R. alleni, 302

R. genowaysi, 298, 302, 306
R. gracilis, 298, 302
R. mira, 298, 302, 304
R. parvula, 298
R. tumida, 291
Rhynchonycteris naso, 289
Riversleigha williamsi, 73, 74, 75, 77, 78, 82
round-eared bat. See *Tonatia evotis*
Rousettus, 334
R. aegyptiacus, 100–101, 127, 268
R. amplexicaudatus, 73, 345, 347, 350
R. angolensis, 154
R. obliviosus, 331

sabertooth blenny. See *Aspidontus*
Saccolaimus
S. mixtus, 273
S. peli, 154
S. saccolaimus, 273
Saccopteryx
S. bilineata, 289
S. canescens, 289
S. gymnura, 287, 289
S. leptura, 289
Sanborn's long-nosed bat. See *Leptonycteris curasoae*
Sauromys petrophilus, 263
Scandentia, 10
Scleronycteris ega, 290
Scotonycteris zenkeri, 154
Scotophilus, 264, 266–267, 267–268
S. borbonicus, 263, 267, 329
S. dingani, 267
S. gigas, 267
S. leucogaster, 263, 267
S. nigrita, 267
S. nigrita gigas, 154
S. viridis, 263, 267
serotine bat. See *Eptesicus serotinus*
sheath-tailed bats. See *Diclidurus*; Emballonuridae; *Peropteryx*; *Taphozous*
short-tailed fruit bat. See *Carollia*
silver-haired bat. See *Lasionycteris noctivagans*
Sinopa, 151
slit-faced bats. *See* Nycteridae
Soricidae, 136
Sphaeronycteris toxophyllum, 154, 291
spotted bat. See *Euderma maculatum*
stenodermatines, 52, 145, 147, 153, 287
Sturnira, 299
S. bidens, 291
S. lilium, 154, 287, 288, 291
S. mordax, 298, 302

S. tildae, 291
Syconycteris australis, 154, 278
Syndesmotis, 72, 79, 80
S. megalotis, 35, 72, 74, 75, 77, 82

Tadarida, 194
T. aegyptiaca, 268
T. brasiliensis: cochlear adaptations, 231, 232, 233, 235, 236, 237, 238, 240, 242; conservation, 286, 312, 319; distribution, 292, 310, 311; ecosystem role of, 317, 318; muscle biology, 128, 129, 135, 136; skeletal properties, 119, 120; teeth, 154; as tourist attraction, 299; wing membrane, 122
T. condylura, 263
T. pumila, 263
T. teniotis, 104
Taphozous, 105, 191
T. australis, 273
T. georgianus, 73
T. kapalgensis, 273
T. nudiventris, 154
T. troughtoni, 273, 276
Tarsomys, 348
Thomas' sac-winged bat. See *Balantiopteryx io*
Thyroptera, 103
T. discifera, 291
T. tricolor, 291, 300–301
Thyropteridae, 4, 5, 6, 8, 11, 12
conservation, Middle America, 297, 298
distribution, Brazil, 291
Tomopeas ravus, 1, 3–4
Tomopeatinae, 3, 5, 6, 8, 9, 11, 12
Tonatia
T. bidens, 290
T. brasiliense, 290
T. carrikeri, 290
T. evotis, 298, 302, 306
T. schulzi, 290
T. silvicola, 154, 290
Townsend's big-eared bat. See *Corynorhinus townsendii*
Trachops, 145
T. cirrhosus, 154, 164, 201, 290
Triaenops, 27, 72–88
T. furculus, 28, 31, 32, 33, 40
T. persicus, 28, 31, 32, 33, 37, 40; Southern fossil record, 74, 75, 77, 78, 80, 82
Troughton's sheathtail bat. See *Taphozous troughtoni*
trumpet-nosed bat. See *Musonycteris harrisoni*
tube-nosed bats. See *Nyctimene*
Tylonycteris, 105

Underwood's long-tongued bat. See *Hylonycteris underwoodi*
Uroderma, 60, 61, 63, 64, 65, 66, 67, 150
U. bilobatum, 62, 122, 154, 291
U. magnirostrum, 291

vampire bats. See *Desmodus rotundus*; *Diaemus youngi*; *Diphylla ecaudata*
Vampyressa
V. bidens, 291
V. brocki, 291
V. nymphaea, 302
V. pusilla, 287, 291
Vampyrodes caraccioli, 291
Vampyrops helleri, 135
Vampyrum, 151, 152
V. spectrum, 154, 164, 287, 290, 301
Van Gelder's bat. See *Antrozous dubiaquercus*
Vaylatsia, 37, 72, 73, 81
cf. *prisca*, 79
Vespadelus douglasorum, 273
Vespertilio, 253
V. murinus, 104
Vespertilionidae, 4, 5, 11, 12, 13, 34
conservation, 249, 250, 298, 302
distribution, 291, 310, 311
diversity, Brazil, 284
food search and location, 194, 199
muscle biology, 136, 137
phylogeny, 37
roosts, 276
vespertilionids. See *Eptesicus*; *Lasionycteris*; *Nyctophilus*; *Pipistrellus*; *Tylonycteris*; Vespertilionidae; Vespertilioninae
Vespertilioninae, 3, 5, 6, 7, 8, 9, 10, 11, 12, 13
Vespertilionoidea, 5, 11, 80, 83, 131

Western mastiff bat. See *Eumops perotis*
white bat. See *Ectophylla alba*
white-winged vampire bat. See *Diaemus youngi*
wrinkle-faced bat. See *Centurio*

Xenorhinos, 72–88
X. halli, 73, 74, 75, 77, 78, 82

Yangochiroptera, 5, 11, 12, 13
yellow bat
Genoways' little. See *Rhogeessa genowaysi*
miniature. See *Rhogeessa mira*
Yinochiroptera, 5, 7, 11, 12, 13
Yuma myotis. See *Myotis yumanensis*

Subject Index

African bats
    conservation strategies, 266–268
    human population impact, 261
    threats to regional populations, 268–269
African woodlands, 264
artificial roosts
    bat boxes, 252, 255, 257
    bridges, 266
    buildings, 266
audition. *See* hearing
auditory cortex
    cortical potentials, 213
    cortical range (computational) maps,
      207–208
    delay-tuned neurons, 209–210
    extracellular potentials, 211–216
    range acuity, 209
    wavelet transforms, 213, 216–217
Australian bats
    conservation needs, 272–277
    conservation planning, 278–279
    fossil record, 72–73

bat boxes, 252, 255, 257
bat populations, factors affecting
    building restoration, 257
    deforestation, 274–275
    diseases and parasites, 263
    eutrophication and channelization,
      317

habitat fragmentation, 274, 287, 342
human attitudes, 333
human disturbance, 315
human population growth, 261
human predation and consumption, 274,
    275, 333
intensive agriculture, 254, 287
international trade, 333–334
introduced predators, 334–335
mining, 276–277
pesticides, 277, 332
pollution, 256–257, 277, 317
predation, 316
roost disturbance, 276–277, 287
biodiversity, hidden, 256, 264
biogeography
    Africa, 13, 261–269
    Australia, 34
    Brazil, 282–292
    Ethiopia, 13, 34, 13, 261–269
    Europe, 13
    Indo-Australia, 13
    Middle America, 295–306
    Neotropics, 13, 282–292, 295–306
    New World, 13, 295–306, 309–313
    North America, 13, 309–313
    Old World, 13, 261–269, 295–306
    Orient, 34
    Palaearctic, 13, 27, 34
    Philippines, 345–348

blood-feeding. *See* sanguinivory
Brazilian bats
    biomes, 283–284
    conservation history, 286
    conservation status, 288–292
    distribution and abundance, 286, 289–
      292
    endemism, 285, 385
    species richness, 284–285
    threats to populations, 287–288
Bremer support analysis, 2, 8–9, 11, 46, 59, 61,
    67–68

carnivory, morphological adaptations for,
    143–152, 150
caves
    as roosts, 268, 313
    gated, 314
characters, descriptions of, 13–20, 30
chromosome morphology, 36–37, 44
cladistic analysis, 5–6, 13, 31–32, 35–36, 43–52
climate change, global, 336–337
cochlea, 171–177
    acoustic fovea, 235
    broadband tuning, 232–234
    frequency maps, 234–235
    narrowband tuning, 232–233
    structural and functional maps, 235
    tuning curves, 233–234
community assemblages, 347

conservation issues
Africa, 261–265, 268
Australia, 272–277
Brazil, 287–288
Indo-Pacific Islands, 327–330, 333–337
Middle America, 297–306
North America, 312–318
Philippines, 349–351
United Kingdom, 253
conservation priorities
community inventories, 330
ecological and conservation research, 272, 275
habitat protection, 264, 316–317
population monitoring, 330–332
recovery plans, 264
roost protection, 263, 276
species assessment, 330–332
taxonomic problems, 256, 264, 266, 272
constant frequency (CF) echolocation signals
Doppler-shift compensation, 170, 186–187, 193, 232–234
narrow-space CF bats, 193

dentition, 142–146
dietary specializations
carnivory, 140–141, 151–152, 184, 190
frugivory, 140–141, 152, 184, 190
insectivory, 140–141, 149–151, 184, 190, 262
nectarivory, 140–141, 152, 184, 190
omnivory, 190
piscivory, 184
sanguinivory, 141, 190
DNA
applications to conservation, 332–333
applications to systematics, 45–52, 332–333
DNA-DNA hybridization, 7, 10
mitochondrial DNA (mtDNA), 44–46, 60–61
nucleotide sequence data, 13
ribosomal DNA (rDNA), 61–62
Doppler-shift compensation, 170, 186–187, 193, 232–234

echolocation
acoustic fovea, 186
acoustic glint, 186
acoustic image, 210
attack rate, 200–201
Doppler-shift compensation, 170, 186–187, 193, 232–234
See also hearing
echolocation, call structure
horseshoe bats, 171–178
interspecific variation, 174–175
intrapopulation variation, 175–178
echolocation, computational strategies
cortical range map, 208
range perception, 206–207, 209
target representation, 206
temporal representation, 207, 209
echolocation, perceptual tasks
approach sequence, 197–198
classification, 184
detection, 183–184

localization, 184–185
pursuit sequence, 212
terminal phase, 198, 199
echolocation, signal types
constant frequency (CF), 185–187
frequency-modulated (FM), 185–187
quasi-constant frequency (QCF), 185
ecological services
cultural, 299
ecotourism, 299
nutrient transfer, 317–318
pollination, 299
predation on insects, 298, 317–318
seed dispersal, 299
ecomorphology, 135–137, 143–154
endangered species, 287
endemism, 13, 285, 297–298, 300, 337, 347–349
extinctions, 277, 297, 300, 327–330

feeding guilds, 194–195, 201–202
flight
aerodynamics of, 101–102
agility, 104
ecomorphology, 135–137
energetic consequences of, 123–124
evolution of, 136, 137
hovering, 102, 136
maneuverability, 104–105
speed, 105–106
flight musculature
arrangements, 93–95
electrical activity, 128
fiber types, 93–94, 129–132
function, 129–132
histochemistry, 129–132
mechanical activity, 128–129
myosin isoforms, 133–135
foraging habitats
background-cluttered space, 188–189, 191–192
highly cluttered space, 188–189, 192
uncluttered space, 188–191
foraging tactics
approach behavior, 184
ecological constraints, 184
perceptual problems, 183–184
search behavior, 184
search modes, 190
fossil record, 13, 27, 35–36, 51, 72–73, 81
frequency-modulated (FM) echolocation calls
broadband, 185–187
narrow-space, 193–194
frugivory, morphological adaptations for, 143–152

gleaning bats
approach sequence, 201
food acquisition, 201–202
perceptual tasks, 201
global climate change, 336–337

habitat associations
Africa, 267–268
Australia, 271–277
Brazil, 284–286
Middle America, 295–297

North America, 309–317
Philippines, 347–351
United Kingdom, 251–256
hearing
hearing systems, 183
peripheral auditory system, 231–242
sensorimotor feedback, 221–229
spatial perception, 220–221
higher-level relationships, 1–13
hindlimb morphology
myological variation, 161
osteological variation, 161
passive digital lock, 157–159
human impacts on bats
agriculture, 254, 287
attitudes, 333
building restoration, 257
deforestation, 274–275
disturbance, 276–277, 287, 315
eutrophication and channelization, 317
global climate change, 336–337
habitat fragmentation, 274, 287, 342
international trade, 333–334
introduced diseases, 335–336
introduced predators, 334–335
pollution, 256–257, 277, 317
population growth, 261
predation and consumption, 274, 275, 333

Indo-Pacific Island bats
evolutionary and ecological context, 326
extinctions and extirpations, 327–330
global climate change, effects of, 336–337
management strategies, 332–333
population monitoring, 330–332
status assessment, 330–332
threats to populations, 333–335
insectivory
following lesions, 223
inferior colliculus, role in, 225–227
laboratory studies, 222
morphological adaptations for, 143–152
interfamilial relationships, 10–11
international trade, 333–334

karyology, 36–37, 44

landscape ecology
anthropogenic factors, 252
corridor concept, 251
linear-landscape elements, 251
locomotion
flight, 101–107, 123–124, 135–137, 161
quadrupedal, 161, 164–165

Middle American bats
areas of special concern, 303–306
conservation value, 298–300
distributions, 302–303
endemism, 297
regional conservation strategies, 303
species richness, 297, 298, 303, 304–305
vulnerability to extinction, 300–301
migration, 264
mitochondrial DNA (mtDNA), 54–56

molecular systematics, 43–52, 59–69
  conservation applications of, 332–333
  cytochrome *b* gene, 62–66
monophyly, 1, 3, 5, 8–10, 46, 49–50, 67, 69

nectarivory, morphological adaptations for, 143–152
New World leaf-nosed bats (Phyllostomidae), 59–69
North American bats
  diversity and distribution, 309, 310–311
  ecosystem services, 317–318
  foraging habitats, 316–317, 318
  nutrient hotspots, 317–318
  roosting habitats, 314–316
  species richness, 309, 310, 311
  status assessment, 318–319

Old World leaf-nosed bats (Hipposideridae)
  characters and character states, 39–40, 86–87
  cladograms, 31–33, 35–36, 75, 77, 82
  cochlear morphology, 171–179
  echolocation call structure, 169–179
  fossil record, 72–73, 75, 80–84
  karyotypes, 37
  origin of, 34–36, 81–83
  phylogenetic relationships, 73–84
  radiations, 80–84
  skull morphology, 171–179
organ of Corti
  basilar membrane, 235–238
  cochlear tuning, 236–238
  hair cells (stereovilli), 240–241
  tectorial membrane, 239–240
  See also cochlea

palate, 142–147, 150, 152–153
parsimony analysis, 10, 33, 46, 47–50, 64–66, 74–75, 76
passive digital lock, 157–159
Philippine bats
  biogeographic patterns, 345–348
  conservation needs, 351
  elevational gradients, 347, 349
  fruit bat diversity, 349
  habitat affinities, 349–351
  protected areas, design of, 347–348
  species richness, 344–345
phylogenetic analysis, 2, 4–8, 27, 61–62
  bootstrap analysis, 8–9, 11, 61, 68, 74
  Bremer support analysis, 9, 11, 61, 65–66, 68
  cytochrome *b* data as basis, 61
  Hennig86 analysis, 76, 77, 78–79
  majority consensus cladogram, 30–33
  maximum-likelihood method, 30, 32, 36
  mitochondrial DNA (mtDNA), 44–46
  Phylogenetic Analysis Using Parsimony (PAUP), 8, 46–47, 61–62, 73–78
  ribosomal DNA (rDNA), 61–62

total evidence analysis, 7
phylogenetic hypotheses, 4–7, 8, 10–12, 31–33, 47–50, 64–66, 74–75, 82
phylogenetic relationships
  Artibeus, 59–69
  Carollia, 43–52
  Dermanura, 59–69
  Hipposideridae, 35–38, 79–84
  Koopmania, 59–69
pollution, 256–257
protovampires, 164–165

roost fidelity, 252
roosting habitats, artificial
  bat boxes, 314
  bridges, 266, 314
  buildings, 266, 314
  dams, 314
  mines, 268, 313
roosting habitats, natural
  animal burrows, 268
  caves, 268, 313
  crevices, 300, 314
  furled leaves, 300
  tree hollows, 265, 268, 313–316
roost types
  summer (maternity), 313
  fall (transient), 313
  night, 313
  winter (hibernacula), 313

sanguinivory
  arboreal-feeding hypothesis, 164
  cladistic analysis, 43–42
  dentition, 141
  dentition hypothesis, 163–164
  dietary specialization, 141, 190
  ectoparasite-feeding hypothesis, 162–163
  mitochondrial DNA (mtDNA) analysis, 43–52
  vampire bats, 159
  wound-feeding hypothesis, 163
short-tailed fruit bats (*Carollia*)
  consensus cladograms, 46–50
  mitochondrial DNA (mtDNA) sequence data, 54–56
  molecular systematics, 43–52
  parsimony analysis, 46–47
  phylogenetic relationships, 49
skull
  cochlea, 171–177
  function, 145–147
  morphology, 141–142
species distributions
  local population declines, 313
  North American bats, 309, 310, 311
  patchy distributions, 312
  range contractions, 313
  seasonal distributions, 312
  taxonomic issues, 313

species richness, 284–285, 297, 298, 303, 304–305, 309, 310, 311, 345–351
street lamps
  foraging sites, 252–253
  implications for conservation, 252–253
superior colliculus
  echo delay, 224
  function model, 227, 229
  microstimulation, 225–226
  motor map, 225–227
  neurophysiology, 223–227

teeth, 142–154
tooth function
  canines, 147–148
  molariform teeth, 148–150

United Kingdom, autecological studies
  brown long-eared bat, 254–255
  Daubenton's bat, 256
  greater horseshoe bat, 253–254
  lesser horseshoe bat, 254
  serotine bat, 255–256
United Kingdom, surveys
  acoustic monitoring, 250
  church surveys, 253
  National Bat Habitat Survey, 249–251
  national roost surveys, 253

vampire bats
  blood-feeding habits, 159
  hindlimb morphology, 160–162
  phylogeny, 159–160

wing adaptations, 98–105, 109–124
wing-beat frequency, 94–97
wing bones
  dactylopatagium, 113–114
  plagiopatagium, 112
wing design, 93–96
  aerodynamics, 101–102
  aspect ratio, 105–106, 135
  boundary layer, 99
  camber, 99
  flight muscles, 93–95
  skeletal-muscular system, 96–99
  torsion resistance, 116
wing loading, 103, 104–105, 106, 135
wing membranes
  aerodynamic function, 123
  general, 97–102, 109–110, 112
  mechanical properties, 120–123
wing motion, 111, 114–115
wing shape, 103–104
wing skeleton, 97–99
  bone loading, 110–114
  cortical thickness, 117–118
  flexibility, 118
  material properties, 119–120
  strain and stress profiles, 110–111, 113–115
  torsional loading, 112